# ICL Contribution to Landslide Disaster Risk Reduction

**Series Editor**

Kyoji Sassa, The International Consortium on Landslides, ICL, Kyoto, Japan

The ICL Contribution to Landslide Disaster Risk Reduction book-series publishes integrated research on all aspects of landslides. The volumes present summaries on the progress of landslide sciences, disaster mitigation and risk preparation. The contributions include landslide dynamics, mechanisms and processes; volcanic, urban, marine and reservoir landslides; related tsunamis and seiches; hazard assessment and mapping; modeling, monitoring, GIS techniques; remedial or preventive measures; early warning and evacuation and a global landslide database.

More information about this series at http://www.springer.com/series/16332

Vít Vilímek • Fawu Wang • Alexander Strom •
Kyoji Sassa • Peter T. Bobrowsky •
Kaoru Takara
Editors

# Understanding and Reducing Landslide Disaster Risk

## Volume 5 Catastrophic Landslides and Frontiers of Landslide Science

Springer

*Editors*
Vít Vilímek
Department of Physical
Geography and Geoecology
Faculty of Science
Charles University
Prague, Czech Republic

Alexander Strom
Geodynamics Research Center LLC
Moscow, Russia

Peter T. Bobrowsky
Geological Survey of Canada
Sidney, BC, Canada

Fawu Wang
College of Civil Engineering
Tongji University
Shanghai, China

Kyoji Sassa
International Consortium on Landslides
Kyoto, Japan

Kaoru Takara
Graduate School of Advanced Integrated
Studies in Human Survivability (Shishu-kan)
Kyoto University
Kyoto, Japan

*Associate Editors*
H. B. Havenith
UR Geology, University of Liège
Liège, Belgium

Gonghui Wang
Disaster Prevention Research Institute
Kyoto University
Kyoto, Japan

Ping Lu
College of Surveying and Geo-Informatics
Tongji University
Shanghai, China

Baator Has
Asia Air Survey
Tokyo, Japan

Stefano Luigi Gariano
CNR IRPI (Research Institute for
Geo-Hydrological
Protection of the Italian National
Research Council)
Perugia, Italy

ISSN 2662-1894 ISSN 2662-1908 (electronic)
ICL Contribution to Landslide Disaster Risk Reduction
ISBN 978-3-030-60318-2 ISBN 978-3-030-60319-9 (eBook)
https://doi.org/10.1007/978-3-030-60319-9



Cover illustration: The 2018 Baige landslide in the upper reaches of the Jinsha River, China. Photo was made by A. Strom on April 7, 2019

This Springer imprint is published by the registered company Springer Nature Switzerland AG
The registered company address is: Gewerbestrasse 11, 6330 Cham, Switzerland

ICL and Springer created a new book series "ICL Contribution to Landslide Disaster Risk Reduction" in 2019 which is registered as ISSN 2662-1894 (print version) and ISSN 2662-1908 (electronic version). The first books in this series are six volume of books "Understanding and Reducing Landslide Disaster Risk" containing the recent progress of landslide science and technologies from 2017 to 2020.

# Organizational Structure of the Fifth World Landslide Forum

## Organizers

International Consortium on Landslides (ICL)
Global Promotion Committee of International Programme on Landslides (IPL-GPC), including: United Nations Educational, Scientific and Cultural Organization (UNESCO), World Meteorological Organization (WMO), Food and Agriculture Organization (FAO), United Nations Office for Disaster Risk Reduction (UNDRR), United Nations University (UNU), International Science Council (ISC), World Federation of Engineering Organizations (WFEO), International Union of Geological Sciences (IUGS), International Union of Geodesy and Geophysics (IUGG)
Kyoto University (KU), Japan Landslide Society (JLS), Japanese Geotechnical Society (JGS), Japan Society for Natural Disaster Science (JSNDS) and Japan Association for Slope Disaster Management (JASDiM)

## Co-sponsors

Cabinet Office (Disaster Management Bureau) of Japan; Ministry of Foreign Affairs of Japan (MOFA); Ministry of Education, Culture, Sports, Science and Technology-Japan (MEXT); Ministry of Land Infrastructure, Transport and Tourism (MLIT); Ministry of Agriculture, Forestry and Fisheries (MAFF); Science Council of Japan (SCJ); Japan International Cooperation Agency (JICA); Japan Society of Civil Engineers (JSCE); Japanese Society of Irrigation, Drainage and Rural Engineering (JSIDRE); Japan Society of Erosion Control Engineering; Japan Society of Engineering Geology.

## Supporting Organizations with Finance

Tokyo Geographical Society
International Union of Geological Sciences (IUGS)
Association for Disaster Prevention Research, Kyoto, Japan

## Organizing Committee

**Honorary Chairpersons**
Audrey Azoulay, Director-General of UNESCO*
Mami Mizutori, Special Representative of the United Nations Secretary-General for Disaster Risk Reduction*
Petteri Taalas, Secretary-General of WMO*
Qu Dongyu, Director-General of FAO*
David Malone, Under-Sectretary General of the Unitred Nations and Rector of UNU
Daya Reddy, President of ISC
Gong Ke, President of WFEO
Qiuming Cheng, President of IUGS
Kathryn Whaler, President of IUGG
Qunli Han, Executive Director of Integrated Research on Disaster Risk (IRDR)
Walter Ammann, President and CEO of Global Risk Forum GRF Davos, Switzerland
Juichi Yamagiwa, President of Kyoto University, Japan
Angelo Borrelli, Head of the National Civil Protection Department, Italian Presidency of the Council of Ministers, Italy
Darko But, Director General of the Administration for Civil Protection and Disaster Relief of the Republic of Slovenia, Slovenia
Akifumi Nakao, Director, International Cooperation Division, Disaster Management Bureau, Cabinet Office, Japan
Kazuyuki Imai, Director General of Sabo Department, Ministry of Land Infrastructure, Transport and Tourism, Japan*
Chungsik Yoo, President of the International Geosynthetics Society
Rafig Azzam, President of the International Association for Engineering Geology and the Environment
(*to be confirmed)

**Chairpersons**
Kyoji Sassa, Professor Emeritus, Kyoto University; Secretary General of ICL
Peter T. Bobrowsky, Geological Survey of Canada; President of ICL
Kaoru Takara, Kyoto University, Japan; Executive Director of ICL

**Members**
Željko Arbanas (University of Rijeka, Croatia)
Snježana Mihalić Arbanas (University of Zagreb, Croatia)
Nicola Casagli (University of Firenze, Italy)
Fausto Guzzetti (Department of Civil Protection, Italy)
Matjaž Mikoš (University of Ljubljana, Slovenia)
Paola Reichenbach (Research Institute for Geo-Hydrological Protection, National Research Council, Italy)
Shinji Sassa (Port and Airport Research Institute, Japan)
Alexander Strom (Geodynamics Research Center LLC, Russia)
Binod Tiwari (California State University, Fullerton, USA)
Veronica Tofani (University of Firenze, Italy)
Vít Vilímek (Charles University in Prague, Czech Republic)
Fawu Wang (Tongji University, China)

**Chairpersons of Local Organizing Committee**
Kaoru Takara (Kyoto University)
Daisuke Higaki (Japan Landslide Society)
Ikuo Towhata (Japanese Geotechnical Society)

**Secretary Generals**

Ryosuke Uzuoka (Disaster Prevention Research Institute, Kyoto University)
Kazuo Konagai (International Consortium on Landslides)
Khang Dang (International Consortium on Landslides)

**International Scientific Committee**

Beena Ajmera, North Dakota State University, USA
Snježana Mihalić Arbanas, University of Zagreb, Croatia
Željko Arbanas, Faculty of Civil Engineering, University of Rijeka, Croatia
Amin Askarinejad, Technische Universiteit Delft, Delft, The Netherlands
Peter T. Bobrowsky, Geological Survey of Canada, Sidney, Canada
Michele Calvello, University of Salerno, Italy
Giovanna Capparelli, Universita degli Studi della Calabria, Rende, Italy
Nicola Casagli, University of Florence, Italy
Yifei Cui, Tsinghua University, Beijing, China
Sabatino Cuomo, University of Salerno, Fisciano, Italy
Khang Dang, International Consortium on Landslides, Kyoto, Japan
Elias Garcia-Urquia, National Autonomous University of Honduras, Tegucigalpa, Honduras
Stefano Luigi Gariano, Research Institute for Geo-Hydrological Protection, CNR, Perugia, Italy
Daniele Giordan, Research Institute for Geo-Hydrological Protection, CNR, Italy
Fausto Guzzetti, Department of Civil Protection, Italy
Baator Has, Asia Air Survey, Tokyo, Japan
Hans-Balder Havenith, Universite de Liege, Liege, Belgium
D. P. Kanungo, Central Building Research Institute (CBRI), Roorkee, Uttarakhand, India
Oded Katz, Geological Survey of Israel, Jerusalem, Israel
Kazuo Konagai, International Consortium on Landslides, Kyoto, Japan
Doan Huy Loi, International Consortium on Landslides, Kyoto, Japan
Ping Lu, Tongji University, Shanghai, China
Olga Mavrouli, University of Twente, Enschede, The Netherlands
Matjaž Mikoš, Faculty of Civil and Geodetic Engineering, University of Ljubljana, Slovenia
Alessandro C. Mondini, Research Institute for Geo-Hydrological Protection, CNR, Italy
Veronica Pazzi, Department of Earth Science, University of Florence, Florence, Italy
Dario Peduto, Department of Civil Engineering, University of Salerno, Fisciano, Italy
Paola Reichenbach, Research Institute for Geo-Hydrological Protection, CNR, Italy
Paola Salvati, Research Institute for Geo-Hydrological Protection, CNR, Italy
Katsuo Sasahara, Kochi University, Japan
Kyoji Sassa, International Consortium on Landslides, Kyoto, Japan
Shinji Sassa, Port and Airport Research Institute, Japan
Andrea Segalini, University of Parma, Italy
Hendy Setiawan, Universitas Gadjah Mada, Yogyakarta, Indonesia
Alexander Strom, Geodynamics Research Center LLC, Moscow, Russia
Kaoru Takara, Kyoto University, Japan
Faraz Tehrani, Deltares, Delft, The Netherlands
Binod Tiwari, California State University, Fullerton, California, USA
Veronica Tofani, University of Florence, Italy
Ryosuke Uzuoka, Kyoto University, Kyoto, Japan
Vít Vilímek, Faculty of Science, Charles University, Prague, Czech Republic
Fawu Wang, College of Civil Engineering, Tongji University, Shanghai, China
Gonghui Wang, Kyoto University, Kyoto, Japan
Mike Winter, Winter Associates Limited, Kirknewton, UK
Hiromitsu Yamagishi, Hokkaido Research Center of Geology (HRCG), Sapporo, Japan

**Local Organizing Committee**

Shinro Abe, Okuyama Boring Co., Ltd.
Kiminori Araiba, Fire and Disaster Management College
Shiho Asano, Forestry and Forest Products Research Institute
Has Baator, Asia Air Survey Co., Ltd.
Hiromu Daimaru, Forestry and Forest Products Research Institute
Khang Dang, International Consortium on Landslides
Mitusya Enokida, Japan Conservation Engineers & Co., Ltd.
Kumiko Fujita, International Consortium on Landslides
Kazunori Hayashi, Okuyama Boring Co., Ltd.
Daisuke Higaki, The Japan Landslide Society
Kiyoharu Hirota, Kokusai Kogyo Co., Ltd.
Kazuo Konagai, International Consortium on Landslides
Taketoshi Marui, MARUI & Co., Ltd.
Satoshi Nishikawa, Nagoya University
Keisuke Oozone, OYO Corporation
Katsuo Sasahara, Kochi University
Kyoji Sassa, International Consortium on Landslides
Shinji Sassa, Port and Airport Research Institute
Go Sato, Teikyo Heisei University
Nobuyuki Shibasaki, Nippon Koei Co., Ltd.
Nobuo Sugiura, Japan Association for Slope Disaster Management
Kaoru Takara, Kyoto University
Keisuke Takimoto, GODAI KAIHATSU Corporation
Yoko Tomita, Public Works Research Institute
Ikuo Towhata, The Japanese Geotechnical Society
Kenichi Tsukahara, Kyushu University
Ryosuke Tsunaki, Sabo & Landslide Technical Center
Taro Uchida, Saitama University
Mie Ueda, International Consortium on Landslides
Ryosuke Uzuoka, Kyoto University
Fawu Wang, Tongji University
Hiroshi Yagi, Yamagata University
Hiromitsu Yamagishi, Shin Engineering Consultants Co., Ltd.
Maki Yano, OSASI Technos Inc.

# Foreword by Mami Mizutori

More landslides can be expected as climate change exacerbates rainfall intensity. The long-term trend of the last 40 years has seen the number of major recorded extreme weather events almost double, notably floods, storms, landslides, and wildfires.

Landslides are a serious geological hazard. Among the host of natural triggers are intense rainfall, flooding, earthquakes or volcanic eruption, and coastal erosion caused by storms that are all too often tied to the El Niño phenomenon. Human triggers including deforestation, irrigation or pipe leakage, and mine tailings, or stream and ocean current alteration can also spark landslides. Landslides can also generate tsunamis, as Indonesia experienced in 2018.

Globally, landslides cause significant economic loss and many deaths and injuries each year. Therefore, it is important to understand the science of landslides: why they occur, what factors trigger them, the geology associated with them, and where they are likely to happen.

Landslides with high death tolls are often a result of failures in risk governance, poverty reduction, environmental protection, land use and the implementation of building codes. Understanding the interrelationships between earth surface processes, ecological systems, and human activity is the key to reducing landslide risk.

The Sendai Framework for Disaster Risk Reduction, the global plan to reduce disaster losses adopted in 2015, emphasizes the importance of tackling these risk drivers through improved governance and a better understanding of disaster risk.

One important vehicle for doing that is the Sendai Landslide Partnerships 2015–2025 for global promotion of understanding and reduction of landslide risk facilitated by the International Consortium on Landslides (ICL) and signed by the leaders of 22 global stakeholders, including the UN Office for Disaster Risk Reduction (UNDRR), during the Third UN World Conference on Disaster Risk Reduction in Sendai, Japan.

The Sendai Landslide Partnerships—featured on the Sendai Framework Voluntary Commitments online platform—helps to provide practical solutions and tools, education, and capacity building, to reduce landslide risks.

The work done by the Sendai Partnerships can be of value to many stakeholders including civil protection, planning, development and transportation authorities, utility managers, agricultural and forest agencies, and the scientific community.

UNDRR fully supports the work of the Sendai Landslide Partnerships and ICL and looks forward to an action-oriented outcome from the 5th World Landslide Forum to be held in November 2020 in Kyoto, Japan. Successful efforts to reduce disaster losses are a major contribution to achieving the overall 2030 Agenda for Sustainable Development.

Mami Mizutori
United Nations Special Representative of the Secretary-General for Disaster Risk Reduction

# Foreword by the Assistant Director-General for the Natural Sciences Sector of UNESCO for the Book of the 5th *World Landslide Forum*

As the world slowly recovers from the COVID-19 global pandemic, and looking back at the way this crisis developed, it becomes evident that as a global community we were not prepared for an event of this scale. Although not commonly perceived as such, biological hazards such as epidemics are included in the Sendai Framework for Disaster Risk Reduction 2015–2030. In that sense, the preparedness approach for a pandemic is very similar to that of a geophysical natural hazard such as landslides.

Although natural hazards are naturally occurring phenomena, the likelihood of their occurrence and of associated disasters is rising. Climate change, urban pressure, under-development and poverty and lack of preparedness are increasingly transforming these natural hazards into life-threatening disasters with severe economic impacts. Therefore, Disaster Risk Reduction (DRR) is gaining momentum on the agenda of the UN system of Organizations including UNESCO. While the Sendai Framework for Disaster Risk Reduction 2015–2030 is the roadmap for DRR, other global agendas including the Sustainable Development Goals, the Paris Climate Agreement and the New Urban Agenda have targets which cannot be attained without DRR.

In shaping its contribution to those global agendas, UNESCO is fully committed in supporting its Member States in risk management, between its different mandates and disciplines and with relevant partners. The International Consortium on Landslides (ICL) is UNESCO's key partner in the field of landslide science. The Organization's support to the Consortium is unwavering. Since ICL was established in 2002, the two organizations have a long history of cooperation and partnership and UNESCO has been associated with almost all of ICL activities. I am very glad that ICL and UNESCO are mutually benefitting from their collaboration.

The 5th World Landslide Forum (WLF5) is expected to represent a milestone in the history of landslide science particularly for scientists and practitioners. One of the major outcomes of WLF5 will be the Kyoto 2020 Commitment for global promotion of understanding and reducing landslide disaster risk (KLC2020). This commitment is expected to strengthen and expand the activities of the Sendai Landslide Partnership 2015–2025. With UNESCO already engaged as a partner, the adoption of this international commitment will raise global awareness on landslide risk and mobilize wider partnerships that draw together stakeholders from all levels of society, across different regions, sectors and disciplines.

It is my great pleasure to congratulate the organizers for holding this event and assure you that UNESCO is fully committed in contributing to its success. As part of that contribution, our Organization is proud to host a session on landslides and hazard assessment at UNESCO-designated sites such as natural World Heritage sites, biosphere reserves and UNESCO Global Geoparks. This session aims to assess landslide impacts on our shared cultural and natural heritage, providing the best opportunity to generate public awareness and capacity development for landslide disaster reduction.

I am confident that WLF5 will contribute to further advance the knowledge of both scientists and practitioners regarding landslide disaster risk reduction. This book paves the way for the science, knowledge and know-how which will feature in the deliberations of the Forum. UNESCO commends all of the contributors to this publication. I look forward to an enhanced collaboration between UNESCO and ICL in future activities and undertakings.

Shamila Nair-Bedouelle
Assistant Director-General for Natural Sciences
UNESCO

# Preface I

## Understanding and Reducing Landslide Disaster Risk

### *Book Series: ICL Contribution to Landslide Disaster Risk*

The International Consortium on Landslides (ICL) was established in pursuance of the 2002 Kyoto Declaration "Establishment of an International Consortium on Landslides," with its Statutes adopted in January 2002. The Statutes define the General Assembly of ICL as follows: in order to report and disseminate the activities and achievements of the Consortium, a General Assembly shall be convened every 3 years by inviting Members of the International Consortium on Landslides, individual members within those organizations, and all levels of cooperating organizations and individual researchers, engineers, and administrators. The General Assembly developed gradually prior to, during and after its first meeting in 2005. In the light of the 2006 Tokyo Action Plan, the Assembly was further facilitated at, and following the First World Landslide Forum held in November 2008. On the occasion of each of its triennial forums, ICL publishes the latest progress of landslide science and technology for the benefit of the whole landslide community including scientists, engineers, and practitioners in an understandable form. Full color photos of landslides and full color maps are readily appreciated by those from different disciplines. We have published full color books on landslides at each forum. In 2019, ICL created a new book series "ICL Contribution to Landslide Disaster Risk Reduction" ISSN 2662-1894 (print version) and ISSN 2662-1908 (electronic version). Six volumes of full color books *Understanding and Reducing Landslide Disaster Risk* will be published in 2020 as the first group of books of this series.

## The Letter of Intent 2005 and the First General Assembly 2005

The United Nations World Conference on Disaster Reduction (WCDR) was held in Kobe, Japan, 18–22 January 2005. At this Conference, ICL organized session 3.8 "New international Initiatives for Research and Risk Mitigation of Floods (IFI) and Landslides (IPL)" on 19 January 2005 and adopted a "Letter of Intent" aimed at providing a platform for a holistic approach in research and learning on 'Integrated Earth System Risk Analysis and Sustainable Disaster Management'. This Letter was agreed upon and signed, during the first semester of 2005, by heads of seven global stakeholders including the United Nations Educational, Scientific and Cultural Organization (UNESCO), the World Meteorological Organization (WMO), the Food and Agriculture Organization of the United Nations (FAO), the United Nations International Strategy for Disaster Risk Reduction (UNISDR-currently UNDRR), the United Nations University (UNU), the International Council for Science (ICSU-Currently ISC), and the World Federation of Engineering Organizations (WFEO).

The first General Assembly of ICL was held at the Keck Center of the National Academy of Sciences in Washington D.C., USA, on 12–14 October 2005. It was organized after the aforementioned 2005 World Conference on Disaster Reduction (WCDR). ICL published the first full color book reporting on Consortium activities for the initial 3 years, 2002–2005 titled "Landslides-Risk analysis and sustainable disaster management". In the preface of this book, the Letter of Intent for Integrated Earth System Risk Analysis and Sustainable Disaster Management was introduced. Results of the initial projects of the International Programme on Landslides (IPL) including IPL C101-1 Landslide investigation in Machu Picchu World Heritage, Cusco, Peru and previous agreements and MoU between UNESCO, ICL and the Disaster Prevention Research Institute of Kyoto University including UNESCO/KU/ICL UNITWIN Cooperation programme were published as well in this book.

## The 2006 Tokyo Action Plan and the First World Landslide Forum 2008

Based on the Letter of Intent, the 2006 Tokyo Round-Table Discussion—"Strengthening Research and Learning on Earth System Risk Analysis and Sustainable Disaster Management within UN-ISDR as Regards Landslides"—towards a dynamic global network of the International Programme on Landslides (IPL) was held at the United Nations University, Tokyo, on 18–20 January 2006. The 2006 Tokyo Action Plan—Strengthening research and learning on landslides and related earth system disasters for global risk preparedness—was adopted. The ICL exchanged Memoranda of Understanding (MoUs) concerning strengthening cooperation in research and learning on earth system risk analysis and sustainable disaster management within the framework of the United Nations International Strategy for Disaster Reduction regarding the implementation of the 2006 Tokyo action plan on landslides with UNESCO, WMO, FAO, UNISDR (UNDRR), UNU, ICSU (ISC) and WFEO, respectively in 2006. A set of these MoUs established the International Programme on Landslides (IPL) as a programme of the ICL, the Global Promotion Committee of IPL to manage the IPL, and the triennial World Landslide Forum (WLF), as well as the concept of the World Centres of Excellence on Landslide Risk Reduction (WCoE).

The First World Landslide Forum (WLF1) was held at the Headquarters of the United Nations University, Tokyo, Japan, on 18–21 November 2008. 430 persons from 49 countries/regions/UN entities were in attendance. Both Hans van Ginkel, Under Secretary-General of the United Nations/Rector of UNU who served as chairperson of the Independent Panel of Experts to endorse WCoEs, and Salvano Briceno, Director of UNISDR who served as chairperson of the Global Promotion Committee of IPL, participated in this Forum. The success of WLF1 paved the way to the successful second and third World Landslide Forum held in Italy and China respectively.

## The Second World Landslide Forum 2011 and the Third World Landslide Forum 2014

The Second World Landslide Forum (WLF2)—Putting Science into Practice—was held at the Headquarters of the Food and Agriculture Organization of the United Nations (FAO) on 3–9 October 2011. It was jointly organized by the IPL Global Promotion Committee (ICL, UNESCO, WMO, FAO, UNDRR, UNU, ISC, WFEO) and two ICL members from Italy: the Italian Institute for Environmental Protection and Research (ISPRA) and the Earth Science Department of the University of Florence with support from the Government of Italy and many Italian landslide-related organizations. It attracted 864 participants from 63 countries.

The Third World Landslide Forum (WLF3) was held at the China National Convention Center, Beijing, China, on 2–6 June 2014. A high-level panel discussion on an initiative to create a safer geoenvironment towards the UN Third World Conference on Disaster Risk Reduction (WCDRR) in 2015 and forward was moderated by Hans van Ginkel, Chair of Independent Panel of Experts for World Centers of Excellence (WCoE). In a special address to this high-level panel discussion, Irina Bokova, Director-General of UNESCO, underlined that countries should be united to work against natural disasters and expressed commitment that UNESCO would like to further deepen cooperation with ICL. Ms. Bokova awarded certificates to 15 World Centres of Excellence.

## The Sendai Landslide Partnerships 2015 and the Fourth World Landslide Forum 2017

The UN Third World Conference on Disaster Risk Reduction (WCDRR) was held in Sendai, Japan, on 14–18 March 2015. ICL organized the Working Session "Underlying Risk Factors" together with UNESCO, the Japanese Ministry of Land, Infrastructure, Transport and Tourism (MLIT) and other competent organizations. The session adopted ISDR-ICL Sendai Partnerships 2015–2025 (later changed to Sendai Landslide Partnerships) for global promotion of understanding and reducing landslide disaster risk as a Voluntary Commitment to the World Conference on Disaster Risk Reduction, Sendai, Japan, 2015 (later changed to Sendai Framework for Disaster Risk Reduction). After the session on 16 March 2015, the Partnerships was signed by Margareta Wahlström, Special Representative of the UN Secretary-General for Disaster Risk Reduction, Chief of UNISDR (UNDDR), and other representatives from 15 intergovernmental, international, and national organizations. Following the Sendai Landslide Partnerships, the Fourth World Landslide Forum was held in Ljubljana, Slovenia from 29 May to 2 June in 2017. On that occasion, five volumes of full color books were published to disseminate the advances of landslide science and technology. The high-level panel discussion on 30 May and the follow-up round table discussion on 31 May adopted the 2017 Ljubljana Declaration on Landslide Risk Reduction. The Declaration approved the outline of the concept of "Kyoto 2020 Commitment for global promotion of understanding and reducing landslide disaster risk" to be adopted at the Fifth World Landslide Forum in Japan, 2020.

## The Fifth World Landslide Forum 2020 and the Kyoto Landslide Commitment 2020

The Fifth World Landslide Forum was planned to be organized on 2–6 November 2020 at the National Kyoto International Conference Center (KICC) and the preparations for this event were successfully ongoing until the COVID-19 pandemic occurred over the world in early 2020. The ICL decided to postpone the actual Forum to 2–6 November 2021 at KICC in Kyoto, Japan. Nevertheless, the publication of six volumes of full color books *Understanding and Reducing Landslide Disaster Risk* including reports on the advances in landslide science and technology from 2017 to 2020 is on schedule. We expect that this book will be useful to the global landslide community.

The Kyoto Landslide Commitment 2020 will be established during the 2020 ICL-IPL Online Conference on 2–6 November 2020 on schedule. Joint signatories of Kyoto Landslide Commitment 2020 are expected to attend a dedicated session of the aforementioned Online Conference, scheduled on 5 November 2020 which will also include and feature the Declaration of the launching of KLC2020. Landslides: Journal of the International Consortium on Landslides is the common platform for KLC2020. All partners may contribute and publish news and reports of their activities such as research, investigation, disaster reduction administration in the category of News/Kyoto Commitment. Online access or/and hard copy of the Journal will be sent to KLC2020 partners to apprise them of the updated information from other partners. As of 21 May 2020, 63 United Nations, International and national organizations have already signed the KLC2020.

## Call for Partners of KLC2020

Those who are willing to join KLC2020 and share their achievements related to understanding and reducing landslide disaster risk in their intrinsic missions with other partners are invited to inform the ICL Secretariat, the host of KLC2020 secretariat (secretariat@iclhq.org). The ICL secretariat will send the invitation to the aforementioned meeting of the joint signatories and the declaration of the launching of the KLC2020 on 5 November 2020.

## Eligible Organizations to be Partners of the KLC2020

1. ICL member organizations (full members, associate members and supporters)
2. ICL supporting organization from UN, international or national organizations and programmes
3. Government ministries and offices in countries having more than 2 ICL on-going members
4. International associations /societies that contribute to the organization of WLF5 in 2021 and WLF6 in 2023
5. Other organizations having some aspects of activities related to understanding and reducing landslide disaster risk as their intrinsic missions.

Kyoji Sassa
Chair of WLF5/
Secretary-General of ICL
Kyoto, Japan

Peter T. Bobrowsky
President of ICL
Sidney, Canada

Kaoru Takara
Executive Director of ICL
Kyoto, Japan

## Appendix: World Landslide Forum Books

| WLF | Place/participants | Title | Editors | Publisher/pages |
|---|---|---|---|---|
| WLF0 (1st General Assembly) 2005 | Washington D.C., USA 59 from 17 countries/UNs | Landslides-Risk Analysis and Sustainable Disaster Management | Kyoji Sassa, Hiroshi Fukuoka, Fawu Wang, Goghui Wang | Springer/377 pages ISBN: 978-3-540-2864-6 |
| WLF1 2008 | Tokyo, Japan 430 from 49 countries/regions/UNs | Landslides-Disaster Risk Reduction | Kyoji Sassa, Paolo Canuti | Springer/649 pages ISBN: 978-3-540-69966-8 |
| WLF2 2011 | Rome, Italy 864 from 63 countries | Landslide Science and Practice Vol. 1 Landslide inventory and Sustainability and Hazard Zoning | Claudia Margottini, Paolo Canuti, Kyoji Sassa | Springer/607 pages ISBN: 978-3-642-31324-0 |
| | | Vol. 2 Early Warning, Instrumentation and Monitoring | | Springer/685 pages ISBN: 978-3-642-31444-5 |
| | | Vol. 3 Spatial Analysis and Modelling | | Springer/440 pages ISBN: 978-3-642-31309-7 |
| | | Vol. 4 Global Environmental Change | | Springer/431 pages ISBN: 978-3-642-31336-3 |
| | | Vol. 5 Complex Environment | | Springer/354 pages ISBN: 978-3-642-31426-1 |
| | | Vol. 6 Risk Assessment, Management and Mitigation | | Springer/789 pages ISBN: 978-3-642-31318-9 |
| | | Vol. 7 Social and Economic Impact and Policies | | Springer/333 pages ISBN: 978-3-642-31312-7 |
| WLF3 2014 | Beijing, China 531 from 45 countries/regions/UNs | Landslide Science for a Safer Geoenvironment Vol. 1 The International Programme on Landslides (IPL) | Kyoji Sassa, Paolo Canuti, Yueping Yin | Springer/493 pages ISBN: 978-3-319-04998-4 |
| | | Vol. 2 Methods of Landslide Studies | | Springer/851 pages ISBN: 978-3-319-05049-2 |
| | | Vol. 3 Targeted Landslides | | Springer/717 pages ISBN: 978-3-319-04995-3 |
| WLF4 2017 | Ljubljana, Slovenia 588 from 59 countries/regions/UNs | Advancing Culture of Living with Landslides Vol. 1 ISDR-ICL Sendai Partnerships 2015-2025 | Kyoji Sassa, Matjaž Mikoš, Yueping Yin | Springer/585 pages ISBN: 978-319-53500-5 |

(continued)

| WLF | Place/participants | Title | Editors | Publisher/pages |
|---|---|---|---|---|
| | | Vol. 2 Advances in Landslide Science | Matjaž Mikoš, Binod Tiwari, Yueping Yin, Kyoji Sassa | Springer/1197 pages ISBN: 978-319-53497-8 |
| | | Vol. 3 Advances in Landslide Technology | Matjaž Mikoš, Željko Arbanas, Yueping Yin, Kyoji Sassa | Springer/621 pages ISBN: 978-3-319-53486-2 |
| | | Vol. 4 Diversity of Landslide Forms | Matjaž Mikoš, Nicola Casagli, Yueping Yin, Kyoji Sassa | Springer/707 pages ISBN: 978-3-319-53484-8 |
| | | Vol. 5 Landslides in Different Environments | Matjaž Mikoš, Vít Vilímek, Yueping Yin, Kyoji Sassa | Springer/557 pages ISBN: 978-3-319-53482-4 |
| WLF5 | 2020 (publication) 2021 (Forum) | Understanding and Reducing Landslide Disaster Risk Vol. 1 Sendai Landslide Partnerships and Kyoto Landslide Commitment | Kyoji Sassa, Matjaž Mikoš, Shinji Sassa, Peter T. Bobrowsky, Kaoru Takara, Khang Dang | Springer In Process |
| | | Vol. 2 From mapping to hazard and risk zonation | Fausto Guzzetti, Snježana Mihalić Arbanas, Paola Reichenbach, Kyoji Sassa, Peter T. Bobrowsky, Kaoru Takara | |
| | | Vol. 3 Monitoring and early Warning | Nicola Casagli, Veronica Tofani, Kyoji Sassa, Peter T. Bobrowsky, Kaoru Takara | |
| | | Vol. 4 Testing, modelling and risk assessment | Binod Tiwari, Kyoji Sassa, Peter T. Bobrowsky, Kaoru Takara | |
| | | Vol. 5 Catastrophic landslides and Frontier of Landslide Science | Vit Vilimek, Fawu Wang, Alexander Strom, Kyoji Sassa, Peter T. Bobrowsky, Kaoru Takara | |
| | | Vol. 6 Specific topics in landslide science and applications | Željko Arbanas, Peter T. Bobrowsky, Kazuo Konagai, Kyoji Sassa, Kaoru Takara | |

# Preface II

## Catastrophic Landslides and Frontiers of Landslide Science

Landslides belong to the most catastrophic of natural phenomena. Their direct and also indirect effects, such as the formation of dammed lakes and outburst floods, have the potential to claim thousands of lives. In order to understand the causes and consequences of landslides it is necessary to study several specific themes, which are discussed in this book and may help in understanding and reducing landslide disaster risk. The papers included in this volume describe various aspects of the causes (e.g. climate change), direct triggers (earthquakes and rainstorms), and the primary and secondary effects of landslides for a better understanding of the process chain: prerequisite—impulse—process—response. These themes are closely interrelated with other aspects of landslide studies discussed in other volumes of the series, "Understanding and Reducing Landslide Disaster Risk".

Several of the papers presented herein discuss the roles of both global and local long-term processes such as the ongoing climate change, which predetermine the formation of landslides in various parts of the World. Considering the variety of environmental conditions around the globe in which climate, soil and rock weathering lead to significant differences in landslide susceptibility, the authors describe case studies from Europe, Asia, North America, the Pacific region, and even the marine environment (the Norwegian–Greenland Sea). Phenomena such as creep and deep-seated gravitational slope deformations, which often precede catastrophic slope failures, are discussed as well.

Among the direct triggers of large catastrophic landslides, earthquakes and extreme rainfall seem to be the most important factors. Detailed studies of seismically and rainstorm induced landslides help to understand the role of such triggering factors. Almost one third of the papers included in this volume discuss various aspects of the effects of earthquakes on slope stability and on landslide-related phenomena such as liquefaction, which attracted the attention of researchers after the 2018 Palu-Donggala earthquake in Indonesia. Several papers describe the spatial distribution of multiple landslides triggered by the 2012 Hejing earthquake in China the 2015 Gorkha earthquake in Nepal and the 2018 Hokkaido earthquake in Japan. The delayed effects of earthquakes on landslide processes are described by an example from the area affected by the 2008 Wenchuan earthquake in China.

Some of the papers discuss the possible effects of seismic strong motion on the formation of historic and prehistoric landslides in the Greater Caucasus in Russia, in the Swiss and Italian Alps, and in the Carpathians in Romania.

Climate change has long been a hot topic. Its effect on landslides becomes clearer after long-term observation. Recently, more attention has been paid to environmental change in terms of its effect on landslides, and especially in areas undergoing development. This book describes the efforts being made in Hong Kong and cold regions of China, and highlights the importance of considering climate and geoenvironmental changes in landslide disaster reduction.

Landslides are closely associated with other hazards in the sense of causes and response. That is why we need to study the prerequisites, triggers, and regional distribution of catastrophic landslides, as well as their classification. Another important issue discussed in the book is the secondary effects of slope failures such as the formation of natural dams and outburst floods, as well as the effects of slope failure on the banks of existing lakes, which can sometimes be even more catastrophic than the direct effects of slope failure. Such phenomena are analysed through examples from Argentina, Nepal, the Far East of Russia and Iran. In addition, more general aspects of landslide dam hazard assessment are discussed. One paper discusses the possible association of natural processes such as glacial retreat, landslides and volcanic eruptions through an example from British Columbia, Canada.

Innovations have been created throughout the world worldwide in order to help understand the causes and consequences of landslides. The use of Unmanned Aerial Vehicles (UAVs) has developed rapidly in several earth sciences applications, including landslide characterisation and monitoring. UAVs provide strong support for hazard and risk management activities, especially with the introduction of and advances in the miniaturization of traditional and new generation sensors. Through several case studies on landslide investigations, one paper in the book provides an overview of several sensors and techniques using UAVs for landslide detection, characterization and monitoring. Another example is a web-based disaster and risk reduction system (ARAS), which is used to evaluate landslide susceptibility and conduct hazard mapping in the Middle Black Sea region of Turkey and to minimize the undesired consequences of landslides.

We hope this volume will provide readers with new, interesting and useful information that will facilitate further progress in understanding and reducing landslide disaster risks both locally and globally.

Prague, Czech Republic — Vít Vilímek
Shanghai, China — Fawu Wang
Moscow, Russia — Alexander Strom

# Contents

# Part I

# Catastrophic Landslides with Different Triggers

2020 Kyoto Japan

# Rock Avalanches: Basic Characteristics and Classification Criteria

Alexander Strom

**Abstract**

Rock avalanches represent the specific type of flow-like landslides—dry granular flows—that pose major threat to population in mountainous regions and in the adjacent plains. Being extremely mobile, they can affect areas up to dozens of square kilometers, extending sometimes for more than 10 km from the feet of the collapsing slopes. The internal structure of their deposits is characterized by intensive fragmentation of inner parts overlain by much coarser carapace. Such internal structure is typical of the vast majority of large-scale rock slope failures, both long runout and forming compact blockages in narrow river valleys. Therefore, all of them should be classified as rock avalanches, rather than as rock slides. Three additional classification criteria closely related to rock avalanche mobility and allowing more strict definition of a particular rock avalanche are discussed, i.e. the confinement conditions, debris distribution along the rock avalanche path, and directivity of debris motion. Besides providing information on debris motion mechanism(s), these characteristics predetermine the assessment of the exposure of elements at risk that might be affected by rock avalanche. It is demonstrated that transformation from the block slide to granular flow depends somehow on the morphology of the transition-deposition zone and on the mechanical properties of the basal surface, but is independent from the type and mechanical properties of the host rocks.

**Keywords**

Rock avalanche • Rock slide • Classification • Fragmentation • Internal structure

## Introduction

Classification of any natural phenomena is an important step of its study. It fully relates to landslides or, in broader sense, to slope processes of various types. Their clear and logical definitions help researchers, besides better understanding of landslides' nature, to "talk one language" using the same terms when describing similar features or phenomena. This paper is focused on the definition and classification of rock avalanches—one of the most dangerous type of landslides, leaving aside other landslide types listed in (Hungr et al. 2014). Features that can and should be classified as rock avalanches are often described in the literature either as "rock avalanche" or as "rock slide" ("rockslide"), thus it seems to be important to propose more strict definition of these terms and their usage.

One more problem to which this article is addressed is that none of the landslide classifications commonly used worldwide (Varnes 1954, 1978; Hutchinson 1968, 1988; Cruden and Varnes 1996), including the latest one (Hungr et al. 2014), differentiate landslide types characterizing initiation of slope failure and those characterizing further motion of a landslide. However, their kinematics and motion mechanism at these stages, on which classification proposed in (Hungr et al. 2014) is based, could change drastically. Such transformation fully relates to rock avalanches.

Hungr and his co-authors (2014) divide all "landslides" (slope processes), at the first level, in six groups: fall, topple, slide, spread, flow and slope deformation (see column 1 of their final Table No 5). It should be noticed that the last type —"slope deformation"—does not provide any strict mechanical meaning, unlike the five other types.

Such characteristics as fall, slide, flow, and spread can be applied to many landslides starting from their motion initiation and up to final stabilization, when they reach slope foot and stop. Topples often convert into slides (Nichol et al. 2002), but may remain as topples for a very long time—up to centuries and even millennia. However, when we deal

A. Strom (✉)
Geodynamics Research Center LCC., 3rd Novomikhailovsky passage, 9, Moscow, 125008, Russia
e-mail: strom.alexandr@yandex.ru


V. Vilímek et al. (eds.), *Understanding and Reducing Landslide Disaster Risk*,
ICL Contribution to Landslide Disaster Risk Reduction,
https://doi.org/10.1007/978-3-030-60319-9_1

with flow slides in rock ("rock avalanches" according to Hungr et al. 2014), the situation changes significantly. It seems that none of them really initiates as a dry granular flow—large-scale rock slope failure always starts as a slide of any type, and, to become real rock avalanche, its mechanics should change from slide to flow. Since "type of movement" is the primary classification criterion (Hungr et al. 2014), such motion type evolution must be somehow considered in the rock avalanche classification system.

Processes preceding rapid motion of a rock mass downslope, such as primary, secondary and tertiary creep (see, e.g. Qin et al. 2006), are left aside in the following analysis to avoid excessive complexity.

The special classification of flow slides based on characteristics of the affected material was proposed in (Hungr et al. 2001). They strictly differentiate "debris avalanches" defined as "*a very rapid to extremely rapid shallow flow of partially or fully saturated debris on a steep slope, without confinement in an established channel*" from "rock avalanches" that originate in bedrock. It should be noticed, however, that the term "debris avalanche" is also used commonly by researchers studying large-scale sectoral collapses on volcanoes (Schuster and Crandell 1984; Capra 2011) whose size and motion mechanism(s) are closer to those of rock avalanches rather than of debris avalanches within the meaning of these terms after (Hungr et al. 2001).

Study of numerous non-volcanic rock avalanches, from the Central Asia region mainly (Strom and Abdrakhmatov 2018), though not exclusively, demonstrate significant variability of their deposits' shape: some moved strictly ahead producing narrow tongues of fragmented debris, some form wider fan-shape bodies, while other turn up to right angle affecting areas that otherwise could be considered as safe, etc. Such variability, on the one hand, provides information on debris motion mechanism(s), and, on the other hand, predetermines the exposure of elements at risk that might be affected.

All the above-mentioned problems are interrelated: type of rock avalanche depend on its motion mechanism and, vise-versa, the latter can be derived from the shape of rock avalanche deposits and their internal structure. All these could and should be reflected somehow in the rock avalanches classification.

## General Characteristic Features of Rock Avalanches

Hungr and his co-authors (2014) defined rock avalanche as an "*extremely rapid, massive, flow-like motion of fragmented rock from a large rock slide or rock fall*". I have to argue that failure of more than one million cubic meters of rock (the commonly accepted lower limit of such phenomenon, though descriptions of rock avalanches with smaller volume can be found in the literature) quite rarely occur as a fall and that vast majority of rock avalanches originate just as rock slides. Role of volume in differentiation of the phenomena starting as rock falls and as rock slides was mentioned in Hungr et al. (2001).

It is important that this definition does not imply long runout. It is just noticed that rock avalanches are "*extremely rapid*". High speed of motion due to significant momentum gained during the initial descend can result, however, in quite variable effects, depending, first, on the confinement conditions.

It can be illustrated by large slope failures of the approximately same size (about 1.5–2 km$^3$ in volume) that occurred in unconfined and in frontally confined conditions. The first one—the Koman rock avalanche (Kurdiukov 1950, 1964; Strom 2014; Robinson et al. 2015; Reznichenko and Davies 2015; Reznichenko et al. 2017) moved across wide Alai intermountain depression (39.54°N, 72.69°E) and had total runout of about 34 km that is impossible for slowly moving landslide (Fig. 1). This feature would be classified as rock avalanche univocally.

However, other rock slope failures of the same size that blocked narrow and deep valleys by rather compact natural dams have been called "rockslides" in most of publications.

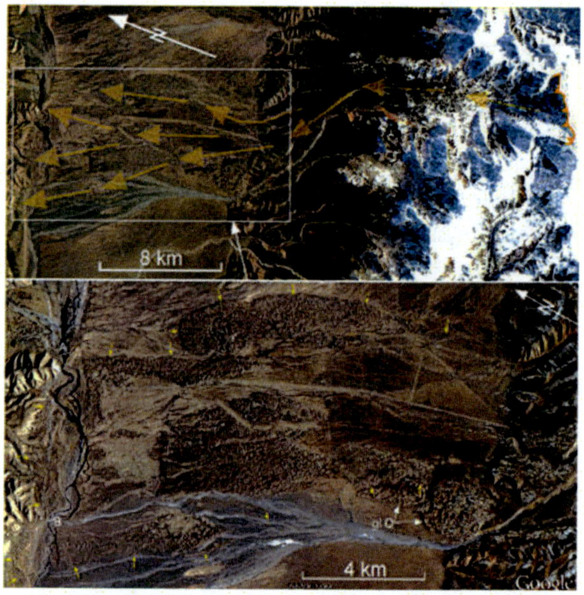


**Fig. 1** The Koman rock avalanche. Above—general view. Orange line—assumed headscarp; transparent arrows show direction of debris motion. Below—closer space view of the deposits, whose original lateral limits are marked by yellow arrows; glQ—pre-slide glacial deposits; A and B sites where the internal structure of the deposits can be observed in details. Google Earth images (after Strom and Abdrakhmatov 2018, with permission of Elsevier)

Such term was used for the breached Late Pleistocene blockage in the Kokomeren River valley, Central Tien Shan (Figs. 2, 3, 41.93°N, 74.22°E) (Strom 1994; Abdrakhmatov and Strom 2006; Hartvich et al. 2008; Strom and Abdrakhmatov 2018). Same term was used by Strom and Abdrakhmatov (2018) for the Big Dragon landslide dam in Eastern Tien Shan (42.6°N, 82.4°E, China that had crossed deep valley and had runout of ca. 4 km only, but with 370 m high runup (Fig. 4). In the latter case term "rock avalanche" was applied to describe just part of its body that had moved 4.3 km downstream. But was it correct?

No, it was incorrect! Much more correct is classifying both of them as rock avalanches in frontally confined conditions. Both compact landslides (the term "landslide" is used as a general term related to any type of slope failures) moved at a very high speed—in the Kokomeren case it is not so obvious, but in the Big Dragon case the extreme velocity of ca. 300 km/h can be derived from the distinct 370 m high runup. Besides, rock avalanche definition proposed in Hungr et al. (2014) points out "*massive, flow-like motion of fragmented rock*" as a characteristic feature of rock avalanche that fully corresponds to what can be observed in the completely dissected body of the Kokomeren landslide. Various lithologies that can be distinguished in its headscarp area due to their different color (Fig. 2) remain in the same mutual positions in the deposits, despite being intensively fragmented (Fig. 3).

Similar characteristic features of the deposits' interiors were observed at numerous other Central Asian landslides that originated on high rocky slopes (Abdrakhmatov and Strom 2006; Strom and Abdrakhmatov 2018), in the Alps (Dufresne et al. 2018), in the Caucasus (Strom 2004, 2006), in the Karakoram and Himalaya (Hewitt 2002, 2006; Weidinger et al. 2014), in New Zealand (McSaveney and Davies 2006). Similar features were found in Tibet in the narrow and deep valleys of the Jinsha River (Fig. 5) and of its tributaries where collapsing rock masses could not move far away due to the well pronounced confinement.

Since direct assessment of the prehistoric landslides motion velocity is not always possible, the critically important characteristic features allowing strict distinguishing between "rock slides" and "rock avalanches" are those indicating "*massive, flow-like motion of fragmented rock*" that can be derived from the study of the internal structure and grain-size composition of the deposits.

**Fig. 3** Right-bank remnant of the Kokomeren rock avalanche deposits. Grain size composition of the blocky carapace and of the internal comminuted part of the deposits can be seen in the insets. Well-expressed varicolored badland is composed of the heavily comminuted granite (pink pyramids) and metasediments (gray and whitish pyramids) (modified after Strom and Abdrakhmatov 2018, with permission of Elsevier)

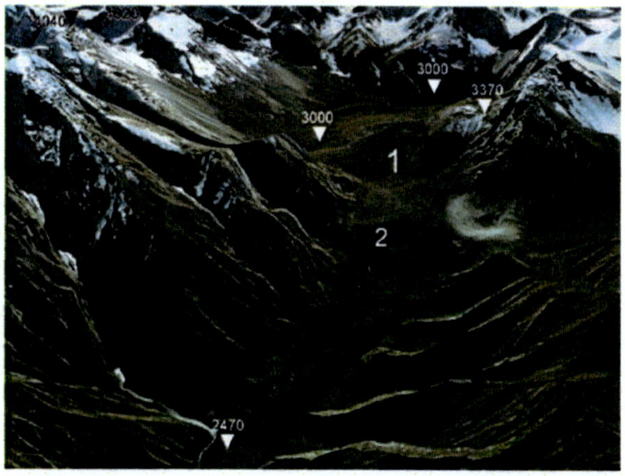


**Fig. 4** The Big Dragon Lake rock avalanche dam with 370 m run-up of its frontal part accompanied by the 4.3 km long deflected secondary rock avalanche that originated from the secondary scar 1 and moved up to 2470 m a.s.l. with an intermediate secondary scar 2. 3D Google Earth view (after Strom and Abdrakhmatov 2018, with permission of Elsevier)

**Fig. 2** Overview of the Kokomeren rockslide. White arrow at the lower right part of the photo marks active fault that, presumably, could trigger this slope failure. Base of the deposits is at ~1800 m a.s.l. S—much smaller "satellite" rockslide (after Strom and Abdrakhmatov 2018, with permission of Elsevier)

**Fig. 5** Intensively fragmented debris of the compact rock avalanche that collapsed from the left-bank slope of the Jinsha River valley, Sichuan, China (29.36°N, 99.067°E) with unmixed 'layers' of different lithologies

We should also take in mind that intensive fragmentation producing particles up to the nanometric size (Adushkin et al. 2006; Reznichenko et al. 2012) combined with retention of the host massif structure could not occur within massif composed of hard rock and moving downslope slowly, but is, somehow, the derivate of rapid motion typical of rock avalanches (Davies et al. 2020).

Such combination appears in moving rock mass subjected to combined action of the fluctuating lithostatic pressure and shearing (Dubovskoi et al. 2008; Strom and Pernik 2013). Thus, it can be concluded that most of large-scale catastrophic rock slope failures all over the World should be classified as rock avalanches.

Classification of rock slope failures in narrow valleys that form intact natural dams, whose internal structure cannot be observed directly, either as a "rock slide" (block slide) or as a "rock avalanche" (flow slide) is more difficult. However, even in such cases sound conclusions can be based on combination of landforms typical of either type. It can be exemplified by the 1911 Usoi landslide in Pamir (38.27°N, 72.6°E) about 2.2 $km^3$ in volume that forms $\sim$ 600 m high natural dam of the Sarez Lake. As it is clearly seen on high resolution space images (Fig. 6), central part of the Usoi blockage is formed by a gigantic block more than 800 $\times$ 600 m in plan view with well-preserved NNW-striking bedding and intensive fracturing stretching in the ENE direction. Presence of such block could indicate that this wedge slide retained as the block slide up to motion halt.

However, outside this central block, the deposit's surface is covered by conical molards (Fig. 7) – landforms typical of rock avalanches (Cassie et al. 1988), rather than of block slides. These observations, along with the assumed analogy with numerous deeply incised landslide dams of the comparable size such as the Köfels in the Alps (Erismann 1979; Sørensen and Bauer 2003), the Djashilkul and the Kokomeren in Central Asia (Abdrakhmatov and Strom 2006; Strom and Abdrakhmatov 2018), allows assumption that the Usoi landslide moved as a massive flow of the fragmented rocks carrying the tremendous "intact" block on top of it as an iceberg and, thus, also should be classified as rock avalanche rather than as rock slide.

Absence of mixing of different lithological units that was mentioned above already (see Figs. 3 and 5) can be observed not only in the outcrops incised in rock avalanche deposits, but also in plan views. One of the most impressive examples is the breached Kudara-Pasor blockage in Pamir, north of the Sarez Lake (38.39°N, 72.58°E) (Fig. 8).

This tremendous, 3 $\times$ 3.7 km in size, and up to 250 m thick body $\sim$ 1.5 $km^3$ in volume is composed of the concentric belts of various lithologies hundreds meters wide well visible in the general view. The frontal belt originated from the Quaternary deposits displaced from the valley bottom; the following yellow and dark brown belts—from the bedrock; the proximal belt, likely, from glacial deposits from an old glacial trough. Moreover, even separate layers in the intensively crushed dark central belt can be traced easily (see inset in Fig. 8).

Such absence of mixing indicates that style of rock avalanches motion should be characterized not as a flow of any type that might be either turbulent or laminar, but just as a laminar flow that makes them drastically different from other highly mobile type of the flow-like landslides—debris flows. The latter most often represent turbulent flow producing more or less homogenous and completely mixed deposits (Fig. 9).

This difference should be considered mandatorily for the numerical modeling of rock avalanches, otherwise their simulation will be incorrect and will convert just in fitting of coefficients in the motion equation that might be completely wrong.

## Transition from the Initial Sliding to Flow and Multi-stage Classification of Rock Slope Failures

As it was mentioned already in the Introduction, rock avalanches initiate as large rock slides and convert into flows during their motion. At the same time rock slope failure of sliding types may not convert into granular flow and detached rock mass can move as a block slide, retaining its integrity

**Fig. 6** Central "intact" block on top of the Usoi landslide dam. Left—the KFA-3000 space image made in 80th; right—high resolution modern space image of the central part of its body marked by white rectangle: white arrow indicates the monitoring station; yellow arrow—the approximate position of the photograph shown in Fig. 7

**Fig. 7** Numerous conical molards on top of the Usoi blockage (after Strom and Abdrakhmatov 2018, with permission of Elsevier)

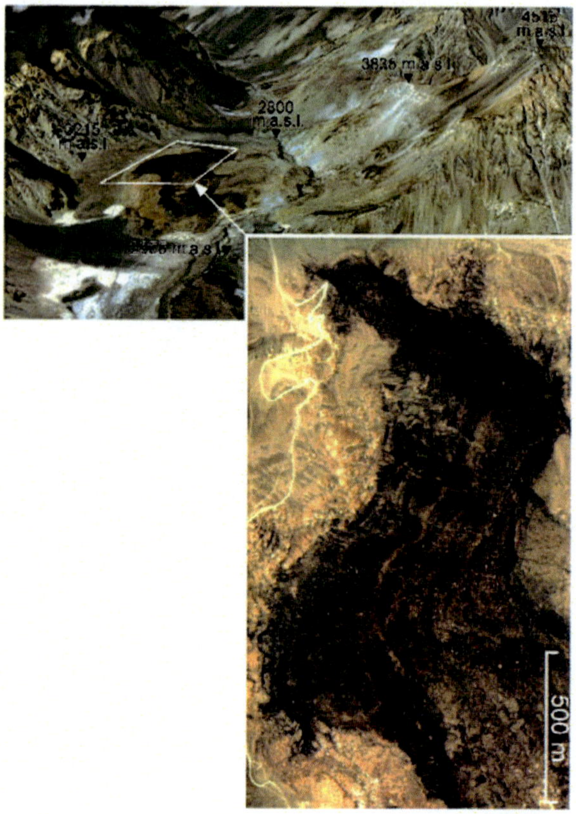


**Fig. 8** Unmixed lithological units of the Kudara-Pasor rock avalanche deposits and retention of the individual layers in its fragmented central part visible in the inset

**Fig. 9** Homogenous internal structure of each of two generations of debris flow deposits (DF-1 and DF-2) overlying Pleistocene alluvial deposits (aQ) on the right bank of the Kokomeren River (41.917°N, 74.223°E)

completely or partially up to the halt of motion. Presence or absence of motion style transformation can be considered for the rock slide and rock avalanche classification.

Rock slides that move as slides up to the end are classified according to the system, presented in Table 5 of Hungr et al. (2014). However, retention of the integrity of moving rocky block is rather rare. Just few such examples were found among ca. 1000 case studies identified in Central Asia Region (Strom and Abdrakhmatov 2018).

One of such block slides can be seen in Fig. 10 ('050', numbers corresponds to the Central Asian rock slide and rock avalanche database presented in Strom and Abdrakhmatov 2018). Its body, about 11 million $m^3$ in volume, can be classified as a rotational or wedge rock slide and differs drastically from the nearby rock avalanches ('042', '042a') that originated as wedge or as irregular rock slides. Both slope failures occurred on the slopes composed of Paleozoic granite.

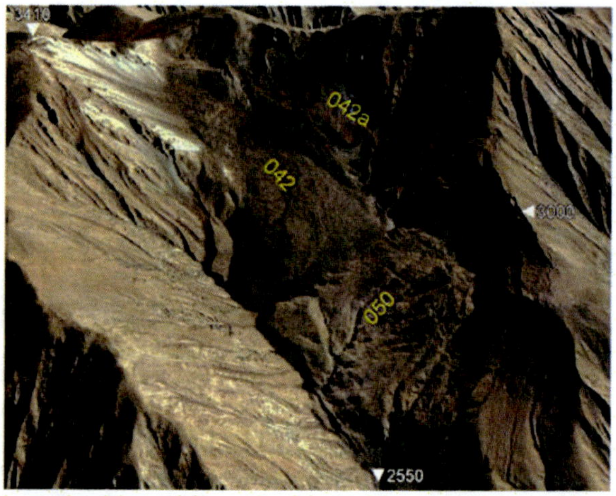


**Fig. 10** Rotational block slide (050) and rock avalanches (042, 042a) in granite in the Sarmin-Ula Range (Eastern Tien Shan, 42.49°N, 85.737°E) (after Strom and Abdrakhmatov 2018, with permission of Elsevier)

It should be pointed out that this feature is the only one real block slide of 19 large-scale rock slope failures that form a cluster within an area of 40 × 20 km only (Strom and Abdrakhmatov 2018). All other rock slope failures within this cluster transformed into flow slides with rather long runout and should be classified as rock avalanches two of which occurred just nearby (see Fig. 10).

Very large block slides, ca. 250-300 million $m^3$ in volume, presumably of the translational or the compound types that did not collapse completely and rest within their headscarps, are shown in Fig. 11: A—in the Muzart River basin (42.042°N, 80.765°E), B—in the Duzakhdara River valley, (37.675°N, 72.398°E). The second one can fail catastrophically in future.

One more example of the compound rock slide is the Beshkiol landslide, about 10 $km^3$ in volume in the Central Tien Shan (41.41°N, 74.48°E)—the largest one in the entire Central Asia region (Strom 1998; Strom and Abdrakhmatov 2018) (Fig. 12). Though its internal structure and state of the material have not been studied yet, such conclusion can be made based on its morphology visible on space images. The upper, steep part of its sliding surface crosscuts the Paleozoic carbonate rocks, while its main, nearly horizontal part inherits flat bedding in the Neogene terrigenous rocks.

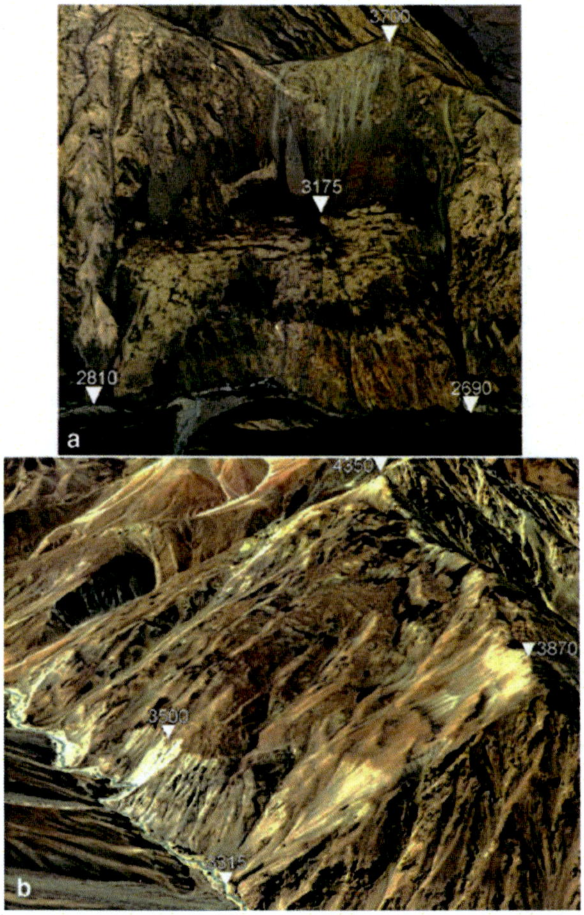


**Fig. 11** Block slides that did not collapse completely. **A**—in the Muzart River basin; **B**—in the Duzakhdara River valley, Central Pamir. 3D Google Earth views, modified from Strom and Abdrakhmatov (2018) with permission from Elsevier

**Fig. 12** The compound Beshkiol rock slide. Fragment of the KFA-3000 satellite image

Sometimes transformation of the rock slide into rock avalanche is just partial—the proximal part of the moving body remains as a block slide, while its frontal part converts into rock avalanche with the impressive evidence of the dry granular flows. It can be exemplified by the gigantic Waikaremoana bedrock landslide in New Zealand that had slid 2 km downslope along the gently dipping bedding plane as a giant block about 1.2 $km^3$ in volume (planar rock slide) while its frontal part had disintegrated forming the rock avalanche (Davies and McSaveney 2011).

One more impressive example of such a partial transformation is the Ornok landslide in the Central Tien Shan, on the left-bank slope of the Kokomeren River valley at 41.716°N, 74.223°E (Strom and Abdrakhmatov 2018) (Fig. 13). Slope failure occurred on the slope composed of the Paleozoic metasediments (PZ in the cross-section, Fig. 13B) thrusted over Neogene conglomerates (N, ibid). Distinct fault zone dividing these units is marked in Fig. 13A by red arrows. It can be seen well on the slopes of the Tabylgaty River (Tab, ibid).

The proximal part of this feature is a classical rotational block slide that displaced both Paleozoic and Neogene rocks marked as PZ-RS and N-RS in the cross-section and as 1 & 2 in the cross-section (Fig. 13B) and on the panoramic view in Fig. 13C. The scree accumulated at the slope base and

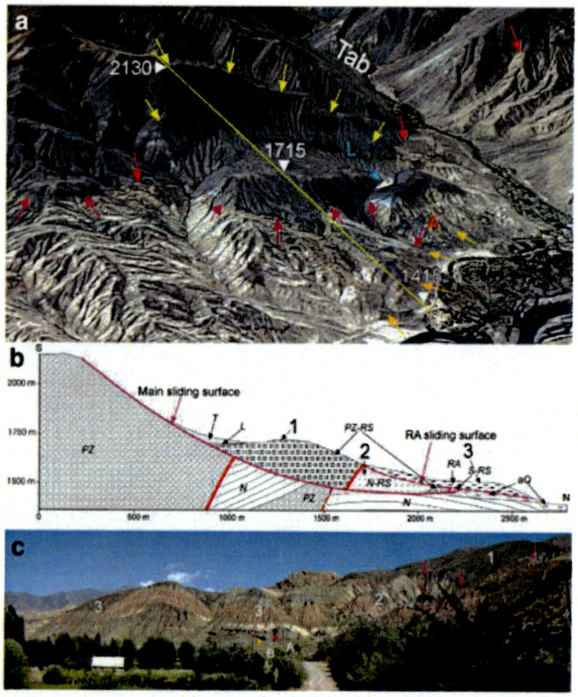


**Fig. 13** The rotational Ornok rock slide accompanied by rock avalanche. A—3D Google Earth view; B—the schematic cross-section along the yellow line; C—the distant view on the landslide body. Further explanation in the text Modified from Strom and Abdrakhmatov (2018) with permission from Elsevier

displaced by block slide is marked as S-RS. The headscarp is marked in Fig. 13A by yellow arrows and the displaced section of the fault zone by purple arrows. One more fault displaced by landslide (see Fig. 13B) was found at the outcrop marked by 'A' in Fig. 13C.

The frontal part of this feature is at least ~ 1 km long rock avalanche (3 in Figs. 13B and C), partially eroded (its present-day front is marked in Fig. 13A by orange arrows) with impressive pseudo-stratified body composed of the "layers" of the fragmented but unmixed Paleozoic and Neogene rocks interbedded with "layers" of scree. L and T in Fig. 13B—lake sediments and overlaying talus in the proximal depression of the rotated block. Rock avalanche (RA) overlaid the alluvium of the Kokomeren River terraces (aQ) that can be observed at the outcrop 'B' in Fig. 13C.

Considering drastic difference between motion mechanisms typical of rock avalanches at their initiation (sliding) and during subsequent motion (flow) the addendum to the landslide classification system (Hungr et al. 2014) can be proposed allowing more strict classification of rock slope failures that either experienced transformation from block to flow slides or not (Table 1). According to the addendum, type name of the landslide that originate on the slope composed of bedrock should consists of three words (except rock falls): first one indicates type of the initial movement (planar, wedge, rotational, irregular, compound); second (rock)—that it originated in bedrock, third—type of the resultant motion (either slide or flow, and, thus, avalanche). For those that initiate as flexural topples but finally collapsed, it might be difficult to define if such motion mechanism really preceded the sliding stage, so they could be classified in the same way.

## Additional Classification Criteria

Further classification of rock avalanches discussed hereafter is based on morphological characteristics of their deposits and on their relationships with the relief of the transitional and depositional zones. These classification criteria include: (1) the confinement conditions, (2) debris distribution along the rock avalanche path, and (3) debris motion directivity.

Besides providing information on debris motion mechanism(s), these characteristics predetermine the assessment of exposure of elements at risk that might be affected by rock avalanche.

## Confinement and Rock Avalanche Deposits' Plan View Shape

Confinement is the most obvious and easily assessed characteristics affecting rock avalanche motion. Its effect on the geometrical parameters of rock avalanches was investigated and proved statistically (Shaller 1991; Nicoletti and Sorriso-Valvo 1991; Strom et al. 2019). Three main groups —the unconfined, the frontally and the laterally confined rock avalanches can be selected according to this criterion (Fig. 14).

### Unconfined Rock Avalanches

Unconfined rock avalanches move over planar surfaces (wide river valleys, bottoms of neotectonic depressions, sometimes, flat glaciers) and, thus, their motion is governed by the internal processes and by the basal friction only. Rock avalanches of this type can be divided, further, into three subtypes based on the shape of their bodies: the mono-directional (see Fig. 14C), the fan-shaped (Fig. 15), and the isometric (pancake-shaped) (Fig. 16).

Those of the mono-directional subtype (see Fig. 14C) move strictly forward, according to the direction of the momentum gained during the initial collapse, and form tongues of debris which maximal width is nearly the same as the headscarp base width or just slightly larger. Since momentum gained during the initial descent is a vector value (unlike the kinetic energy), rock avalanches of this subtype demonstrate its most strict conservation.

**Table 1** Multistage classification of large-scale rock slope failures

| Failure initiation Type subtype of movement | | Intermediate type of movement | Final type of movement | Landslide type, its number according to Table 5 of Hungr et al. (2014) if applicable and case studies exemplifying the particular type (reference) |
|---|---|---|---|---|
| Fall | | | Fall | Rock fall (1) |
| Topple | Block | Fall | Fall | Rock fall (1) |
| | Flexural | Slide along the newly formed surface | Slide | Rotational rock slide (6); La Clapiére[1] |
| | | | Flow | Rotational rock avalanche; Akaishi Mountains[2] |
| Slide (along the pre-existing surface) | Planar | | Slide | Planar rock slide (7); Waikaremoana (its main part)[3] |
| | | | Flow | Planar rock avalanche; Flims[4], Avalanche Lake[5], Ak-Kiol Lower[6] |
| | Wedge | | Slide | Wedge rock slide (8) Ak-Kiol[6] |
| | | | Flow | Wedge rock avalanche; Usoi[6], Aksai-Kiolsu[6] |
| Slide (along the newly formed surface) | Rotational | | Slide | Rotational rock slide (6); Liard Plateau[1] |
| | | | Flow | Rotational rock avalanche; Ornok (its frontal part)[6] |
| | Compound | | Slide | Compound rock slide (9); Beshkiol[6] |
| | | | Flow | Compound rock avalanche; Mingteke[6], Aidyn-Ula[6] |
| | Irregular | | Slide | Irregular rock slide (10)* |
| | | | Flow | Irregular rock avalanche; Khait[6], Seit[6] |

1—Hungr et al. 2014; 2—Chigira and Kiho 1994; 3—Davies and McSaveney 2011; 4—Pollet et al. 2005; 5—Evans et al. 1994; 6—Strom and Abdrakhmatov 2018

*All large irregular rock slides that I could find had converted into rock avalanches

Other unconfined rock avalanches, in contrast, spread not only straight forward, but also sidewise forming either isometric or fan-shaped bodies. Such sidewise motion cannot be supported by the momentum gained during the initial descend, since this vector is directed ahead. Thus there should be some other source of energy that cause rock avalanche debris spreading in the transverse direction.

Good example of the more or less isometric "pancake-shaped" body is the Atdjailau planar rock avalanche in the Inylchek River valley, Tien Shan (42.155°N, 79.457°E) (Fig. 15).

Fan-shaped cases can be exemplified by relatively small wedge or irregular rock avalanche in Chinese Tien Shan at 42.28°N, 87.31°E, about 18 $Mm^3$ only, or by the giant Yimake wedge or compound rock avalanche 1.4 $km^3$ in volume (Yuan et al. 2013; Strom and Abdrakhmatov 2018) at the eastern foot of the Chinese Pamir, at 39.2°N, 76.15°E (Fig. 16).

Significant sidewise spreading of debris seems to be rather common for rock avalanches that moved either on glaciers like the 1964 Serman Glacier rock avalanche (McSaveney 1978) or rock avalanches triggered by the 2002 Denaly Fault earthquake (Jibson et al. 2006), or on well saturated flood plains such as the 2006 Leyte rock avalanche in Philippines (Sassa et al. 2010). Same could be assumed for the prehistoric Atdjailau rock avalanche shown in Fig. 15 that moved over the flat flood plain of the wide glacial valley with very shallow water table that could cause some liquefaction of the surficial alluvium due to dynamic loading of moving rock avalanche.

Despite that some of the past fan-shaped rock avalanches moved over the bottoms of the depressions covered by merged alluvial fans as those shown in Fig. 16, it can be hypothesized that presence or absence of the sidewise debris spreading might depend on shear strength of the base over which rock avalanche moves (Strom 2006; Strom and Abdrakhmatov 2018).

Most of rock avalanches, when they just escape from the source zone, move as a rather thick body undergoing gradual thinning during its further motion. It can be derived from the numerous laterally confined rock avalanches accompanied by the distinct trimlines much above the resultant surface of the deposits (Fig. 17) (this phenomenon cannot be observed at other confinement conditions). Such thinning, being governed by gravity, produce debris spreading in all directions, both forward and sidewise.

However, the 'additional' forces produced by such thinning should be not very high in comparison with forces

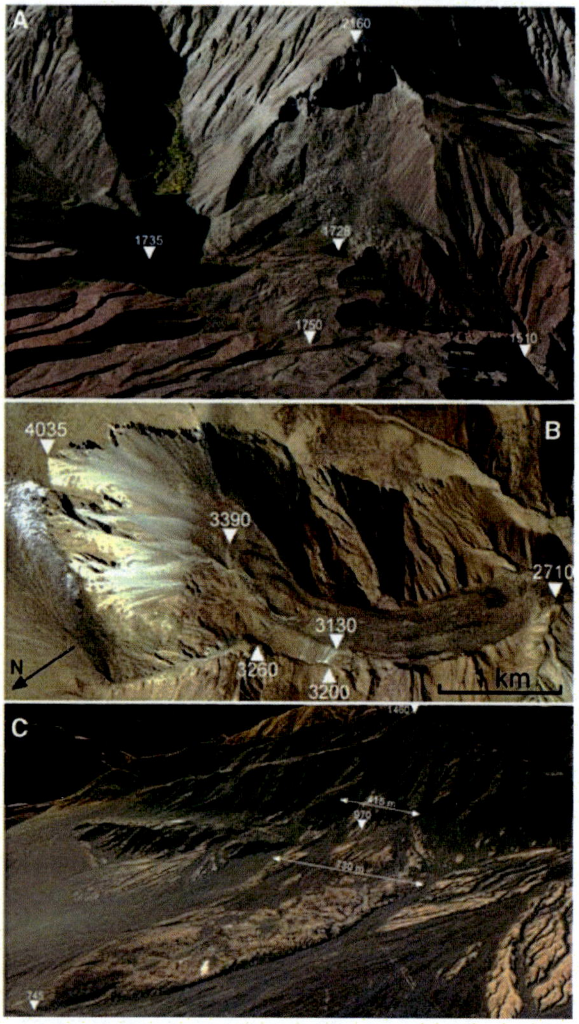


**Fig. 14** Rock slope failures with different types of confinement: A—frontally confined Aik-Kiol rockslide dam, Central Tien Shan (41.726° N, 73.96°E), B—laterally confined rock avalanche in the Sarmin-Ula Mountains, Eastern Tien Shan (42.467°N, 85.588°E), C—unconfined rock avalanche in the Eastern Tien Shan (42.699°N, 87.955°E). Modified from Strom and Abdrakhmatov (2018) with permission from Elsevier

inherited from the directed motion under rock avalanche own momentum.

If the material of the surface over which rock avalanche moves is strong enough, as it can be expected for dry gravelly alluvial fans, the additional force produced by debris thinning and acting in the transverse direction, might be too small to overcome it and thinning would be compensated by rock avalanche body elongation in the same direction as it moves under its own momentum only. It might increase the runout.

If, however, rock avalanche moves over glacier ice, or snow-covered plain (the latter can explain formation of fan-shape bodies of rock avalanches that moved over depression bottoms composed of alluvial fans shown in Fig. 16), or flood plain with shallow ground water layer that might liquefy and eject under vibration produced by rapidly moving rock avalanche, the basal friction can be very low. In such case the flattening debris can overcome this friction force and spread sidewise as well as forward.

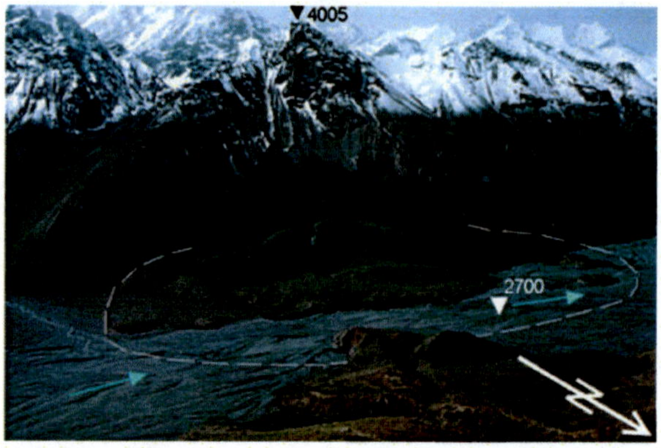


**Fig. 15** Remnants of the strongly eroded isometric body of the Atdjailau rock avalanche

The unconfined mono-directional rock avalanches can have longer runout, but, at the same time, can affect smaller area than fan-shaped or isometric rock avalanches of the same volume and height drop. It should be considered assessing the elements at risk exposure. However, this intuitive assumption has not been yet confirmed statistically due to lack of unconfined case studies that would satisfy equality of possible combinations of the input parameters—i.e. volume and height drop.

## Laterally Confined Rock Avalanches

Laterally confined rock avalanches usually form narrow "streams" of debris bounded by valley walls unless they enter wider valleys where either retain same width (see Fig. 17) or spread sidewise forming fan-shaped distal blades as it occurred at the 7.4-km-long 1949 irregular Khait rock avalanche in the Southern Tien Shan (39.19°N, 70.88°E) that formed a 2.6-km-wide distal fan at the flat bottom of the Yarhych River valley (Fig. 18). Same could be observed at the above mentioned Yimake rock avalanche (see Fig. 16-left) that had moved first 5 km along the narrow valley before spreading as an unconfined fan-shaped body.

Both direction of motion and the resultant shape of the laterally confined rock avalanches are determined by the topography of the valleys along which they have to move. It might be straight, or sinuous (see Fig. 17), and even can turn almost at a right angle as the 6.48 km long Karakystiak rock avalanche did (Fig. 19).

This laterally confined wedge rock avalanche in Southern Kazakhstan (42.58°N, 73.13°E) originated at the

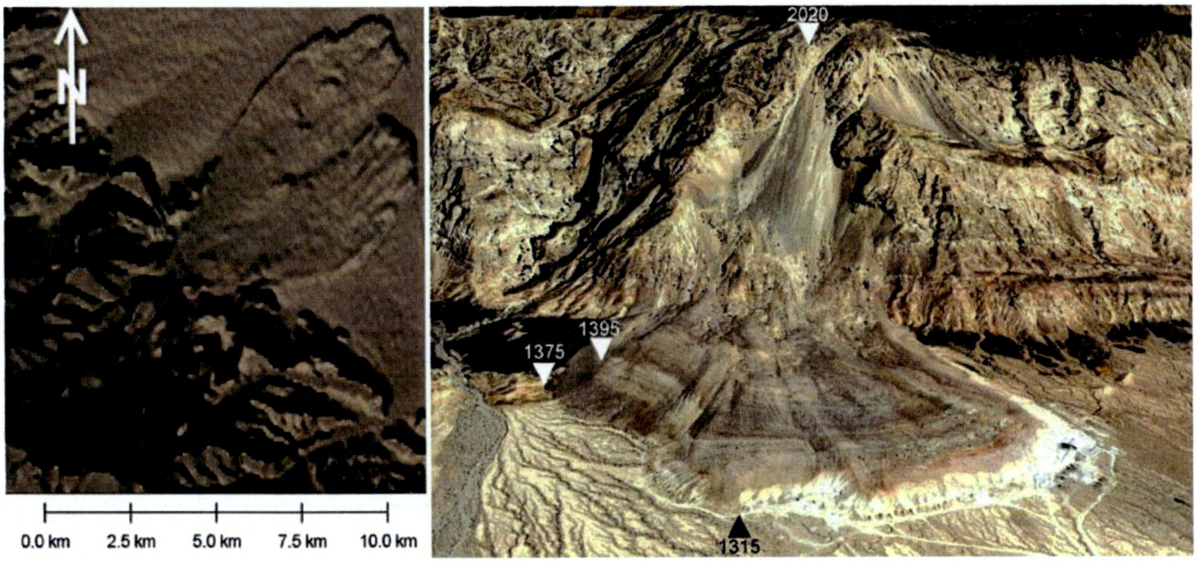


**Fig. 16** Examples of the fan-shaped rock avalanches. Left—the Yimake rock avalanche at the eastern foot of the Chinese Pamir; right—rock avalanche in Chinese Tien Shan with the succession of the host massif layers well-preserved in the deposits (compare with Fig. 8). Modified from (Strom and Abdrakhmatov 2018) with permission from Elsevier

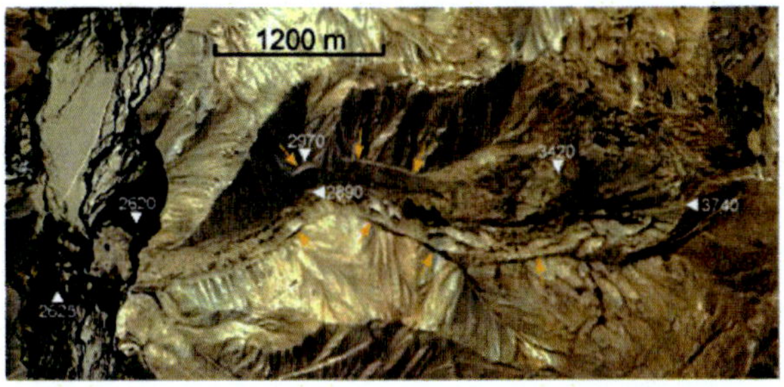


**Fig. 17** The 4.3 km long laterally confined irregular rock avalanche in Afghanistan at 36.59°N, 71.39°E that left trimlines up to 80 m above the deposits' surface (marked by orange arrows). Google Earth image. Modified from Strom and Abdrakhmatov (2018) with permission from Elsevier

**Fig. 18** The 1949 Khait laterally confined irregular rock avalanche with distal fan-shaped blade (between elevation marks 1590 and 1545 m a.s.l. Google Earth image. Modified from Strom and Abdrakhmatov (2018) with permission from Elsevier

funnel-shape headscarp up to 1.95 km wide (between elevation marks 3500 and 3470 m a.s.l.) and moved down-valley towards the Busuleankiol River leaving distinct trimlines up to 200 m above the surface of the deposits (marked at 2780 and 2930 m. a.s.l.). After reaching the Busuleankiol River it turned right for about 90° and moved downstream for about 3 km towards its halt at 2230 m. a.s.l.

So strict determination of the laterally confined rock avalanches motion by the morphology of their travel path makes further classification of this type unnecessary.

Laterally confined rock avalanches, at least those from Central Asia region, demonstrate lack of correlation between maximal height drop ($H_{max}$) and runout (L), while for the unconfined and frontally confined rock avalanches such correlation is well pronounced with high $R^2$ values (Fig. 20). It, in turn, results in poor correlation between runout and product of landslide volume and maximal height drop (V $H_{max}$)—the parameter somehow proportional to the potential energy released during rock avalanche motion that appeared to have best ratios with parameters characterizing rock avalanche mobility (Strom and Abdrakhmatov 2018; Strom et al. 2019). $H_{max}$ here corresponds to $H_1$ in Fig. 21.

Surprisingly, but correlation between V × $H_{max}$ and area affected by rock avalanche, both total and that of rock

**Fig. 19** The 6.48 km long laterally confined Karakystiak rock avalanche (After Strom and Abdrakhmatov 2018 with permission from Elsevier)

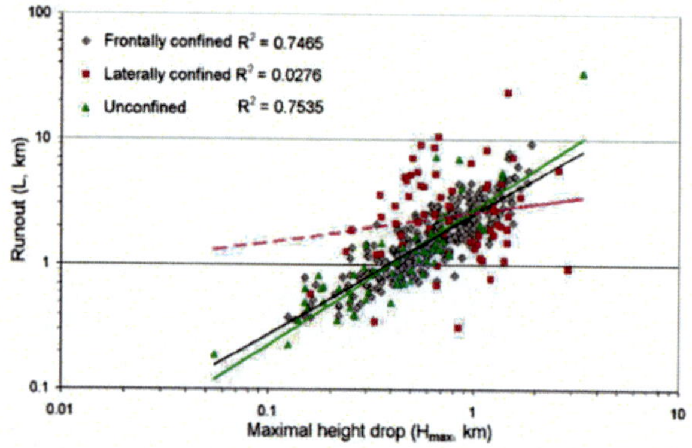


**Fig. 20** Relationships and corresponding $R^2$ values of log-log regressions between runout (L) and maximal height drop ($H_{max}$) for rock avalanches (rockslides) with different confinement conditions. After Strom and Abdrakhmatov (2018) with permission from Elsevier

avalanche deposits only, is quite high, regardless of the confinement type (Strom and Abdrakhmatov 2018; Strom et al. 2019).

## Frontally Confined Rock Avalanches

Frontally confined rock avalanches differ from those of other two types described above, first of all because they move not only downslope and stop due to basal and internal friction, but also upslope, against gravity force (Fig. 20). While friction might depend on such parameters as debris volume, its thickness or velocity, energy loss associated with the upslope material ascent depends neither of these factors nor of the mechanical properties of debris.

Besides we cannot exclude direct impact of the rock avalanche front with steep confining valley slope when large part of the kinetic energy transforms into elastic or non-elastic deformation of the slope massif and, mainly, of rock avalanche body. Processes governing motion of frontally confined rock avalanches follow different physical principles than those of rock avalanches in other confining

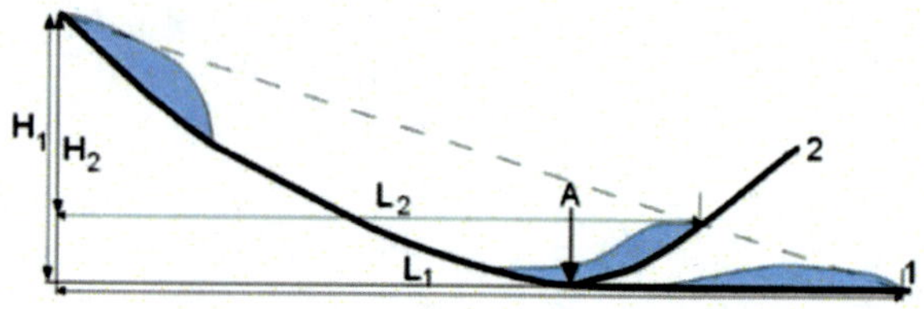


**Fig. 21** Scheme of the Primary rock avalanche (see next section) without (1) and with (2) frontal confinement. H/L ratio is the same for both cases. The kinetic energy and momentum of moving debris when it reached point 'A' should be the same, but physical regulations governing its motion beyond this point should differ significantly (After Strom and Abdrakhmatov 2018 with permission from Elsevier)

conditions and we can hardly estimate how far debris of any frontally confined rock avalanche could move in the absence of such confinement.

Additional peculiarity of rock avalanches in frontally confined conditions is that they almost always form landslide dams, and further typification of such rock avalanches is important for dam stability assessment.

Shape of the frontally confined rock avalanches depends on several factors: the relationships between failure volume and valley width, valley profile (V-shape or U-shape), straightness or sinuosity of the opposite (confining) valley wall. They predetermine how significant the confinement could be. However, more studies should be performed to estimate quantitative effect of these factors.

The morphology of the confining valley slope predetermine the formation of a landslide dam either of the "compact" or of the "widened" subtypes. Compact dams form if the confining slope has a concave shape or if rock avalanche enters a tributary mouth. Such lowering acts as a trap where rock avalanche consumes most of its momentum that prevents the transverse (up- and/or downstream) debris spreading (Strom 2010; Strom and Abdrakhmatov 2018).

Compact subtype with upstream and downstream slope angles close to angle of repose of rock avalanche debris can be exemplified by the breached Aksu landslide dam in the same-name river west from Bishkek City at 42.54°N, 74.0°E (Fig. 22). Here the gigantic wedge failure more than 1 $km^3$ in volume had transformed into rock avalanche that entered deep lowering at the left-bank slope of the valley.

One more example is the partially breached Chong-Tash dam in the Central Tien Shan (42.171°N, 74.616°E). Here most of momentum gained during ca. 450 m descend of about 11 million $m^3$ of the Ordovician metamorphosed conglomerates was consumed by an abnormal, ca. 120 m high runup along the smooth bend of the opposite valley slope (Fig. 23).

If, however, rapidly moving debris collides with a strait or, moreover, convex (in plan view) opposite slope of the valley, it results in momentum transfer and after such

**Fig. 22** The Aksu wedge rock avalanche (Northern Tien Shan) that formed compact dam about 400 m high. 3D Google Earth view from the headscarp side. Distance between elevation marks 1785 and 1570 m a.s.l. is 2.5 km that is almost the same as the headscarp base width. Modified from Strom and Abdrakhmatov (2018) with permission from Elsevier

collision rock avalanche body spreads in the transverse direction up- and/or downstream the valley (Strom 2010; Strom and Abdrakhmatov 2018) forming widened dam body. Such phenomenon can be demonstrated by example of the 1911 Usoi dam (see Figs. 6, 24 and 25).

Here the approximate axial line of rock avalanche motion was directed towards the very steep left-bank slope of the Murgab River valley immediately downstream of the Shadau River mouth that played a role of the 'trap'. The momentum of the true left (upstream) part of moving debris that entered in this trap was consumed due to which the upstream slope of the blockage remained very steep (see Fig. 24).

The axial and the downstream parts of this rock avalanche, in contrast, collided directly with a very steep slope that resulted in significant transverse spreading and formation of the impressive secondary scarps (see Fig. 25). These scarps are visible well not only now days, but could be recognized also on the detailed topographic map of the blockage made several years after dam's formation

**Fig. 23** The Chong-Tash wedge rock avalanche with abnormally high runup that formed partially breached compact dam. 3D Google Earth view. Red arrows mark fracture above the headscarp crown (After Strom and Abdrakhmatov 2018 with permission from Elsevier)

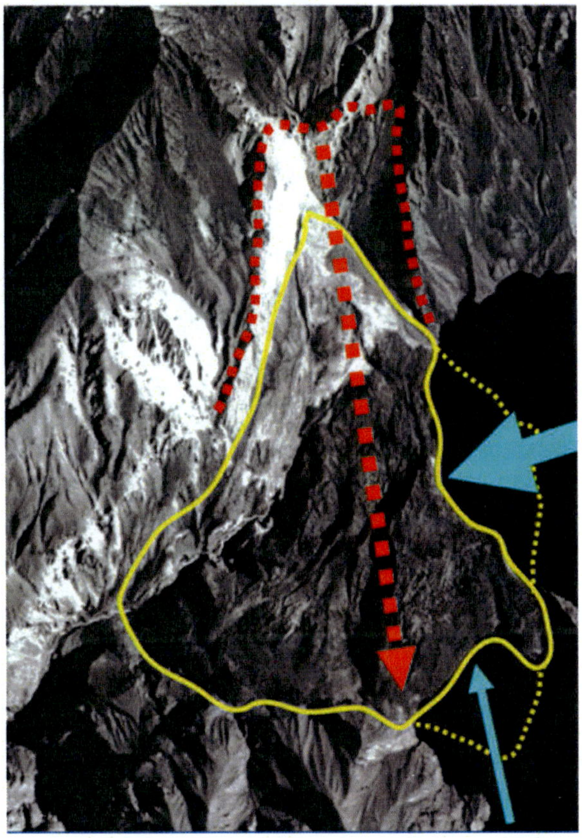

**Fig. 24** KFA-1000 space image of the Usoi wedge rock avalanche dam. Dotted red line—headscarp, thick blue arrow—the Murgab River valley filled by the Sarez Lake, thin blue arrow—the Shadau River valley filled by the same-name lake, bold yellow line—visible boundary of the dam body, dotted yellow line—assumed underwater boundary, dotted red arrow—approximate axis of rock avalanche motion

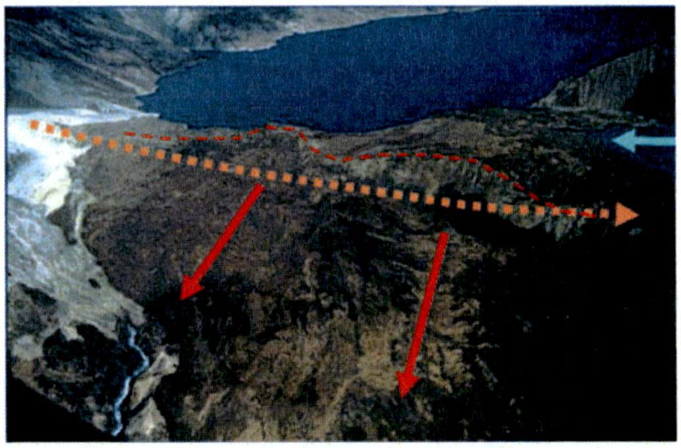

**Fig. 25** The helicopter view on the Usoi dam. The Sarez Lake at the background; blue arrow—the Shadau River valley filled by the same-name lake; dotted orange arrow—approximate axis of rock avalanche motion; thin red dashed line—secondary headscarp; red arrows—direction of the secondary debris spreading

(Preobrajensky 1920; Kolesnikov 1929), thus such downstream widening of the dam occurred, most likely, just during blockage formation. This case study can be treated as an intermediate type between the primary rock avalanche with frontal confinement and the secondary rock avalanche (Strom 2006, 2010; Strom and Abdrakhmatov 2018)—another classification criterion that will be described in the next section.

Landside dam widening can take place without creation of the pronounced landforms indicating transverse debris spreading. In such cases this phenomenon can be deduced if the upstream and/or downstream slope angles are much less than the standard angle of repose of rock avalanche debris as it was observed, for example at the upstream slope of the Djashilkul landslide dam in Northern Tien Shan (42.78°N, 76.362°E) (Strom and Abdrakhmatov 2018, p. 154).

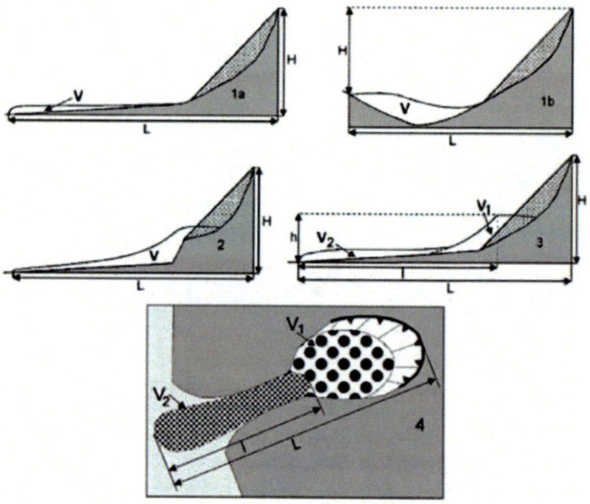


**Fig. 26** Rock avalanche types based on the along-way debris distribution. 1—primary rock avalanche: 1a—in unconfined or laterally confined conditions, 1b—in frontally confined conditions; 2—jumping rock avalanche; 3, 4—secondary rock avalanches of the "classical" and the "bottleneck" subtypes correspondingly. H—height drop (vertical distance between the headscarp crown and the deposits tip); h—that of the secondary rock avalanche; L—runout; l—secondary rock avalanche runout; V—entire volume, V1—volume of the compact part of the secondary rock avalanche, V2—that of its avalanche-like part (After Strom and Abdrakhmatov 2018 with permission from Elsevier)

## Debris Distribution Along the Rock Avalanche Path

This classification criterion can be used to characterize rock avalanche regardless of the confinement conditions and of the morphology of the depositional area. Three main types can be selected—primary, jumping, and secondary rock avalanches (Fig. 26) (Strom 1996, 2006, 2010; Strom and Abdrakhmatov 2018). It was proposed (ibid) that such an "along way" debris distribution depends on the morphology of the slope base where the initial accelerated downslope motion gives place to the motion under gained momentum.

The additional indicator of either jumping or secondary types is the shape of the slope of their compact parts just above the more mobile "avalanche-like" parts.

### Primary Rock Avalanches

The Primary type (see Fig. 26-1a, b) can be identified regardless of the confinement conditions and is characterized by debris accumulation at the distal part of the runout mainly. It can be exemplified by the unconfined rock avalanches shown in Figs. 14C and 16.

One of the best examples of the Primary type with lateral confinement is the Seit rock avalanche in the Central Tien Shan at 42.112 N, 74.15 E (Fig. 27). Practically all collapsed material accumulated within the last kilometer of its more than 3 km long runout and only small patches of debris remain on the slopes at the first half of its travel distance. At the end of its travel path this laterally confined rock avalanche had been terminated by sharp valley bend.

Debris of rock avalanches of the primary type in frontally confined conditions also accumulates in the frontal part, but such case studies are usually characterized by the significant runup on the opposite slope (see Figs. 8, 23). Same type of motion can be also ascribed to the main part of the Big Dragon natural dam with the extreme, 370 m high runup (see Fig. 4), and to the Usoi blockage (see Fig. 25) though entirely these features represent rock avalanches of the secondary type described hereafter.

From the hazard viewpoint it is important that primary rock avalanches in frontally confined conditions form natural dams with prominent lowering at their proximal parts often composed of relatively loose material (e.g. the Usoi dam), less resistant to overflow. It makes such dams most prone to catastrophic breach as compared to blockages with other geometry where the incision would start through more stable blocky carapace or even though the bedrock of the confining valley slope (Korchevskiy et al. 2011; Strom and Abdrakhmatov 2018).

### Jumping Rock Avalanches

Main characteristic feature of the Jumping type (see Fig. 26-2) is their sliding surface that daylights much above the foot of a steep slope so that rockslide body really jumps like from the springboard. It results in formation of a compact body with convex slope above the more mobile part, which thickness decreases gradually towards its terminus.

Mechanism resulting in such morphology can be explained in the following way. When the front of such rock avalanche have collided already with the valley bottom, its proximal part that originates from top of the ridge and, thus,

**Fig. 27** The Seit primary irregular rock avalanche in laterally confined conditions with partial frontal confinement (After Strom and Abdrakhmatov 2018) with permission from Elsevier)

**Fig. 28** The deflected jumping Northern Karakungey wedge (or irregular) rock avalanche in Paleozoic granite. Dashed arrow shows the direction of the initial slope failure. Part of its debris that had moved in the transverse direction downstream is shown in the inset (After Strom and Abdrakhmatov 2018 with permission from Elsevier)

have larger unit potential energy, is still sliding down, due to large linear dimensions of the entire rock mass involved in slope failure. This part of debris also falls down and accumulates on top of the debris portion that is already at place, compressing it and affecting it dynamically. The latter fluidizes and extrudes from under gradually accumulating tail part, spreading as a dry viscous granular flow (Strom 2006; Strom and Abdrakhmatov 2018).

Jumping type of rock avalanches can be exemplified by the 1882 Elm rock avalanche that jumped from the quarry bottom and whose deposits' thickness was maximal at the foot of the slope and gradually decreased towards the distal edge (Heim 1882; Hsü 1975). Several rock avalanches of this type were described in Central Tien Shan, such as the Northern Karakungey one at 41.87°N, 74.24°E (Fig. 28).

Their further classification depends on the constancy or inconstancy of the direction of its mobile part motion against the direction of the initial slope failure. In the Elm case it was the same, while in the Northern Karakungey case mobile part of debris turned almost 80° on the right (see Fig. 28). Such directivity as the additional classification criterion will be discussed hereafter.

## Secondary Rock Avalanches

Rock avalanches of the secondary type (see Fig. 26-3, 4) are characterized by distinct bipartite morphology with a proximal compact part, accompanied by long-runout avalanche-like part. Among case studies described above, the Big Dragon (see Fig. 4), the Usoi (see Figs. 6, 24, 25) rock avalanches and the unnamed rock avalanches shown in Figs. 14B and 17 can be ascribed to the secondary type. The Ornok rotational rock avalanche (see Fig. 13) can be ascribed to such type too, presumably.

Secondary rock avalanches can be subdivided further into "Classical" (Fig. 26-3), and "Bottleneck" (Fig. 26-4) subtypes. The "Classical" subtype is characterized by a concave shape of the frontal or sidewise slopes of their compact proximal parts—the so-called "secondary scar" that marks the boundary between the compact and highly mobile avalanche-like portions of debris (Fig. 29; see also site marked by '1' in Fig. 4).

The "Bottleneck" subtype is also characterized by bipartite mass distribution, either with (see site marked by '2' in Fig. 4), or without pronounced secondary scar. Case study without secondary scar can be exemplified by the Snake-Head rock avalanche in Northern Tien Shan at 42.35° N, 74.34°E (Fig. 30).

Such effect appears when rapidly moving mass of debris enters sharp valley constriction. The boundary between the proximal compact and highly mobile distal parts is linked with such landform.

The abovementioned Snake-Head case study is one of the most impressive examples of the secondary rock avalanches with the "Bottleneck" effect. Despite rather small amount of the collapsed rocks—5–7 million $m^3$ only—it has an extremely low fahrböschung of 0.14 (for the entire failure from the topmost part of the headscarp crown till the deposits tip) and 0.11 (if measured for the avalanche-like part only). Such values of the H/L ratio are typical of rock avalanches with volumes exceeding several $km^3$ (Hsü 1975; Strom and Abdrakhmatov 2018; Strom et al. 2019), thus, indicating an extreme mobility of this particular feature.

It was hypothesized that formation of secondary rock avalanches of both subtypes is caused by the collision of rapidly moving rock mass with some obstacle (opposite slope, valley bottom or valley constriction). Such collision results in the momentum transfer from the rapidly decelerating portion of debris that accumulates at the proximal part of the deposition zone to its portion retaining possibility of further motion (Strom 2006, 2010; Strom and Abdrakhmatov 2018). Same phenomenon is responsible for formation of the "widened" rock avalanche dams exemplified above by the Usoi case study (see Figs. 24, 25). Much more

**Fig. 29** Upper part of the Chongsu secondary wedge rock avalanche about 7 million m$^3$ in volume in the Central Tien Shan (41.99°N, 74.02°E). PS—the primary scar from which the entire rock mass collapsed; SS—the secondary scar on the frontal slope of the compact body; RA-S—surface of the secondary rock avalanche; black arrows mark the trimlines left on the valley slopes (After Strom and Abdrakhmatov 2018 with permission from Elsevier)

pronounced widening of the frontally confined rock avalanche dam can be seen, for example, at the Karasu Lake wedge rock avalanche in the Central Tien Shan at 41.57°N, 73.22° 200-250 million m$^3$ in volume (Fig. 31). In this case the collision with an opposite slope resulted in formation of a voluminous secondary rock avalanche that involved ~10% of the collapsed rock mass and moved downstream for 1.4 km.

While in the Chongsu (see Fig. 29) and the Snake-Head (see Fig. 30) cases portion of debris that retained possibility of further motion moved in the same direction as the initial slope failure, secondary rock avalanches of the Big Dragon (see Fig. 4), the Usoi (see Figs. 6, 24, 25) and the Karasu-Lake (see Fig. 31) case studies changed their direction drastically. This phenomenon can be used as an additional classification criterion and will be discussed in the next section.

## Debris Motion Directivity

Describing the jumping and the secondary types of rock avalanches I mentioned that in some cases their mobile parts moved in the same direction as the initial slope failure, while in other cases they turned, sometimes almost at a right angle.

Such directivity of motion can be used as one more classification criterion allowing selection of two additional types of rock avalanches—unidirectional and deflected. Usually this criterion can be applied to the frontally confined features: unidirectional are shown in Figs. 29, 30, deflected—in Figs. 4, 6, 24, 25, 28, 31.

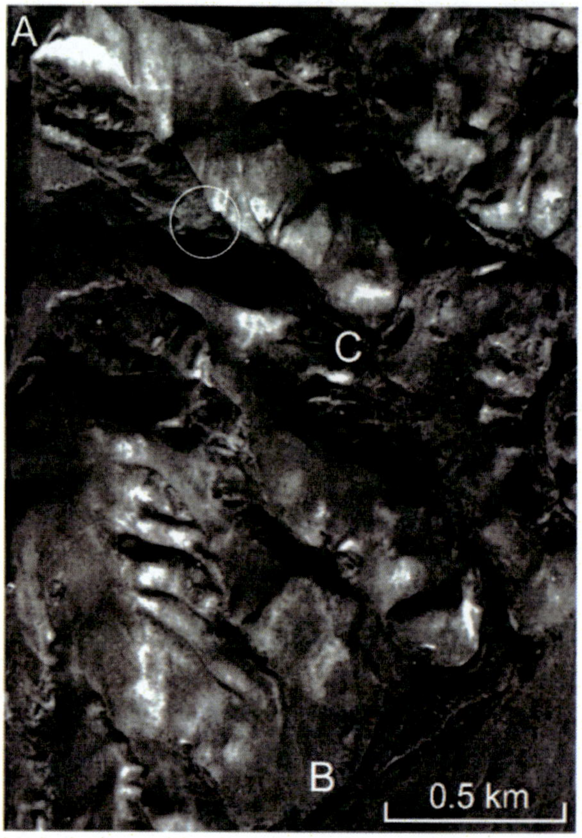


**Fig. 30** The Snake-Head rock avalanche of the Bottleneck subtype. The initial massive rock slope failure at **A** was accompanied by a rock avalanche that traveled along the narrow gorge for about 3 km up to **B**; **C** central part of the transitional zone where debris of landslides from the valley slopes constriction is marked by a circle. Fragment of KFA-3000 space image (After Strom and Abdrakhmatov 2018 with permission from Elsevier)

However, sometimes the deflection can be observed at the laterally confined rock avalanches too. Most of them can be classified as unidirectional (see Figs. 14B, 17, 27, and 30). But in several cases, the collapsed rock mass collides with an obstacle but do not stop and turns and moves almost entirely (as a primary type) downstream a river valley or dry gulley, as it can be seen in Figs. 18 and 19, or in Figs. 32 and 33.

Such directivity plays an important role in the assessment of rock avalanche hazard and of the associated risks. Numerous studies provide rather reliable and statistically representative assessment of the area that can be affected by rock avalanche and of the distance of its front from the foot of the collapsing slope (Sheidegger 1973; Hsü 1975; Davies 1982; Li 1983; Shaller 1991; Nicoletti and Sorriso-Valvo 1991; Kobayashi 1993, 1997; Corominas 1996, Kilburn and Sørensen 1998; Legros 2002, 2006; Strom et al. 2019).

**Fig. 31** The Karasu Lake wedge rock avalanche accompanied by the deflected secondary rock avalanche that involved ~10% of the collapsed rock mass. Its distinct secondary scar is marked by yellow triangles. T—the pronounced right-side trimline. 3D Google Earth view (After Strom and Abdrakhmatov 2018 with permission from Elsevier)

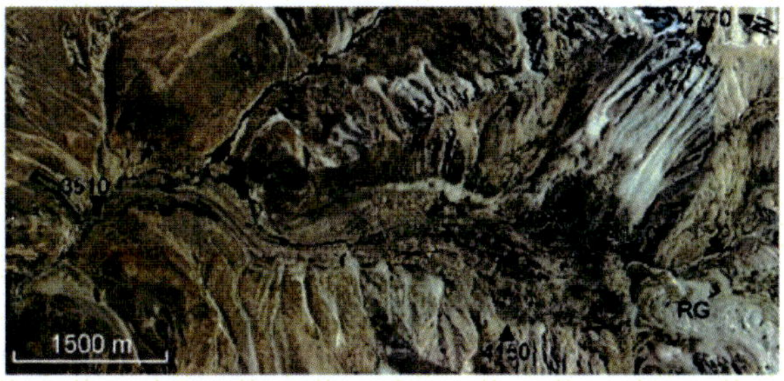


**Fig. 32** The deflected laterally confined Urmochdara rock avalanche in South-Western Pamir (37.423°N, 71.782°E). RG—rock glacier. Google Earth image. Modified from Strom and Abdrakhmatov 2018 with permission from Elsevier

**Fig. 33** The 5.7 km long deflected laterally confined Upper Kashkasu rock avalanche in Central Tien Shan (41.867°N, 74.15°E). Its debris raised for about 100 m on the opposite slope before it turned right. The headscarp is visible in the inset. The elevation mark 3125 m a.s.l. indicated the headscarp crown of a smaller rockslide whose body merges the main rock avalanche. 3D Google Earth view. Modified from Strom and Abdrakhmatov 2018 with permission from Elsevier

But at several sites described above, if we would work there prior to slope failure, it would be difficult to anticipate that parts of the valleys that are not only far away from the unstable slopes, but are located, as one can say, "round the corner", could be affected (see e.g. point with elevation mark 2470 m a.s.l. in Fig. 4, or that with elevation mark 2230 m. a.s.l. in Fig. 19). Nevertheless rock avalanche debris had reached these areas that otherwise, most likely, would be considered as being safe.

## Do Rock Types Influence Rock Avalanche Mobility?

Following the same approach as in the update of Varnes classification of landslide types (Hungr et al. 2014), type of rock involved in the large-scale slope failure has not been used herein as rock avalanche classification criterion. Unlike for landslides in soil for which type of soil is indicated, just "rock" was mentioned as the material affected by slope failure for landslides that originate in bedrock (see Table 5 in Hungr et al. 2014)

Indeed, study of numerous examples of such features all over the world demonstrate that the "*extremely rapid, massive, flow-like motion of fragmented rock*" can occur if large amount of almost any type of rock collapses catastrophically. Rock avalanches mentioned above originated on slopes composed of granite (see Figs. 10, 14B, 19, 27, 28, 29, 33), gneiss (see Fig. 17), limestone (see Figs. 15, 31), terrigenous rocks metamorphosed at various level (see Figs. 5, 16, 30). Some of the collapsed slopes are composed of several types of rocks, both metamorphic, igneous, sedimentary, with quite variable mechanical properties (see Figs. 1, 2, 3, 6, 8, 13).

In the outcrops of rock avalanche deposits we can see that, regardless of the host rock(s) strength, their outer parts are composed of coarse carapace (see upper left inset in Figs. 3, 6, 34), while their internal parts—of heavily shattered material (Figs. 3, 5, 35, 36).

These examples demonstrate that classifying rock slope failures by rock types we do not provide any additional information related to rock avalanche motion mechanism(s). Rock type along with rock massif structure could be informative in the context of type of the initial rock slope failure. Planar rock slides are common in sedimentary or metamorphic rocks with well pronounced bedding or schistosity and monoclinal structure that can be exemplified by Flims (Pollet et al. 2005; von Poschinger et al. et al. 2006) or Seimareh (Shoaei and Ghayoumian 2000) case studies, rock slides on slope composed of granite are, mainly, irregular (see Fig. 27) or, sometimes, wedge, etc.

In contrast, the mechanical properties of the substrate over which rock avalanche moves, plays an important role in debris motion, as it was described above for unconfined case studies.

## Discussion and Conclusions

As it was demonstrated above, vast majority of large-scale bedrock landslides, both with long runout and forming compact blockages in narrow river valleys should be classified as rock avalanches, rather than as rock slides.

Generally, the transformation from the block slide to flow (or absence of such transformation) depends neither on the type of the initial slope failure (planar, rotational, wedge, irregular, compound) nor on type of rocks involved in slope failure (sedimentary, igneous, metamorphic, etc.), nor on their mechanical properties (hard rock such as PR-PZ gneiss or granite or relatively weak rocks such as Neogene conglomerate).

**Fig. 36** Strong red gneiss fragmented up to powder at the upstream part of the Mini-Köfels rockslide (Central Tien Shan 41.906°N, 74.28° E). Rock avalanche deposits rest on the alluvium of the ~10 m high strath terrace of the Kokomeren River

**Fig. 34** Blocks of weak Neogene conglomerates on top of the Lower Ak-Kiol planar rock avalanche about 30 million $m^3$ in volume in Central Tien Shan (41.705°N, 74.284°E)

**Fig. 35** The homogenous mass of crushed Neogene conglomerates (pebbles in sandy-gravelly matrix) of the inner part of the same Lower Ak-Kiol rock avalanche

Such independence on composition and mechanical properties of the host rock seems to be the effect of the dynamic fragmentation (McSaveney and Davies 2006; Davies et al. 2020) that is the so powerful process that it transforms even hard rock blocks whose shape and size are predetermined by the structural irregularities of the host massive such as bedding, fracturing, foliation, etc., into mixture of the pebble- and gravel-size angular fragments with powder matrix, which mechanical properties are governed just by grain-size composition. It is important that fragmentation runs in a very short time—just during rock slide—rock avalanche motion that lasts tens of seconds—minutes only.

The multistage classification of rock slope failures that describes both initial type (mechanism) of movement, according to (Hungr et al. 2014) and its final type, i.e. sliding or flow presented in Table 1, can be complemented with classifications based on three geomorphic characteristics described above, i.e. (1) the confinement conditions, (2) debris distribution along the rock avalanche path, and (3) debris motion directivity (Fig. 37).

These types and subtypes are interrelated and use of any of them to describe a particular rock avalanche should be determined by the main goal of its study.

For example, the Big Dragon (see Fig. 4) and the Karasu Lake (see Fig. 31) case studies can be defined either as the secondary deflected rock avalanches of the classical subtype (if we are interested in assessment of the affected area dimensions), or as a primary widened rock avalanche dam in frontally confined conditions (if the main hazard depends on landslide dams' stability). Besides the Big Dragon can be

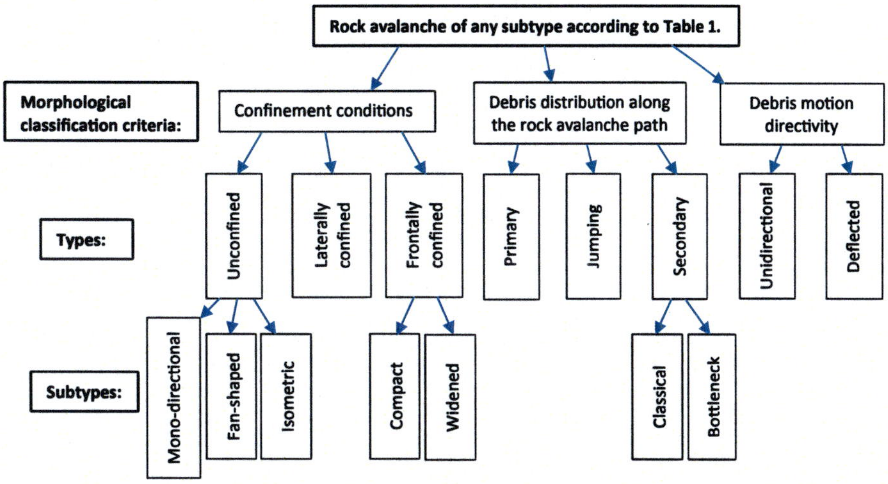


**Fig. 37** Multilevel classification of rock avalanches

characterized as the planar rock avalanche, and the Karasu Lake—as the wedge rock avalanche.

Rather often the geomorphic conditions change along the travel path so that rock avalanche type or subtype changes. It can be exemplified by the 1949 Khait rock avalanche (see Fig. 18) that could be classified initially as the primary deflected laterally confined irregular rock avalanche, while later, when it entered the Yarhych River valley, as unconfined fan-shaped unidirectional one.

Thorough analysis of the resultant morphology of rock avalanche deposits allows better understanding of factors leading to their either shape. It might be, for example, the morphology of the slope base where the initial accelerating downslope motion converts into motion under its own momentum gained during the initial descent (Strom 2006, 2010). This factor, likely, predetermines formation of the primary, jumping or secondary rock avalanches.

The mechanical properties of the basal surface over which the unconfined rock avalanche moved (dry alluvial fans, saturated flood plain, ice, surface covered by snow, etc.), probably along with thickness of moving debris, predetermine formation of the mono-directional, fan-shaped or isometric bodies.

Shape of the opposite valley slope with which frontally confined rock avalanche collides, along with the collision angle, governs the formation of either primary or secondary rock avalanches, as well as of either compact or widened natural dams, and also determine if rock avalanche will be of the deflected or unidirectional type.

Quantitative assessment of the role of these and other factors can be done if very large rock slide/rock avalanche databases (larger than those available at present) will be analyzed. Use of such databases will allow statistically representative samples characterized by combination of several factors. For example, if we would like to study role of mechanical properties of the substrate on the geometrical characteristics of rock avalanche deposits we will need dozens of the mono-directional, fan-shaped or isometric unconfined cases that moved over surfaces for which different properties have been measured or, at least, hypothesized.

To my knowledge, up to now only the effect of confinement on such characteristics as runout and affected area was analyzed quantitatively (Shaller 1991; Nicoletti and Sorriso-Valvo 1991; Strom and Abdrakhmatov 2018; Strom et al. 2019).

One of the largest databases that was compiled recently for the Central Asia region includes ca. 1000 cases, about 600 of which have been characterized quantitively (Strom and Abdrakhmatov 2018). Most of them (more than 300 cases) are the frontally confined features. Further progress in the study of the quantitative relationships between parameters characterizing rock slope failure (volume, height drop, headscarp base width, slope angle) and the resultant rock avalanche parameters (runout, affected area, etc.) considering all factors described above, will be possible if much more case studies will be used. It requires compilation of the worldwide database of large-scale rock slope failures.

**Acknowledgements** I want to thank Prof. Hengxing Lan who persuaded me to present more detailed classification of rock avalanches. I' also grateful to Drs. Maurice McSaveney and Salvatore Martino for valuable comments.

## References


Abdrakhmatov K, Strom A (2006) Dissected rockslide and rock avalanche deposits; Tien Shan Kyrgyzstan. In: Evans SG, Scarascia Mugnozza G, Strom A, Hermanns RL (eds) Landslides from massive rock slope failure. NATO science series: IV: earth and environmental sciences, vol 49. Springer, New York, pp 551–572

Adushkin VV, Pernik LM, Popel' SI, Strom AL, Shishaeva AS (2006) Formation of nano- and micro-range particles in rocky slopes

collapse. Nano- and micro-scale particles in geophysical processes. Institute of Geospheres Dynamics, RAS, Moscow (in Russian)
Capra L (2011) Volcanic natural dams associated with sector collapses: textural and sedimentological constraints on their stability. In: Evans SG, Hermanns R, Strom AL, Scarascia Mugnozza G (eds) Natural and artificial rockslide dams. Lecture notes in earth sciences, vol 133. Springer, Heidelberg, pp 279–294
Cassie JW, Van Gassen W, Cruden DM (1988) Laboratory analogue of the formation of molards, cones on rock-avalanche debris. Geology 16:735–738
Chigira M, Kiho K (1994) Deep-seated rockslide-avalanches preceded by mass rock creep of sedimentary rocks in the Akaishi Mountains, central Japan. Eng Geol 38:221–230
Corominas J (1996) The angle of reach as a mobility index for small and large landslides. Can Geotech J 33:260–271
Cruden DM, Varnes DJ (1996) Landslide types and processes. In: Turner AK, Schuster RL (eds) Landslides: investigation and mitigation, U.S. Transportation Research Board, Special Report, No. 247, pp 36–75
Davies TR (1982) Spreading of rock avalanche debris by mechanical fluidization. Rock Mech 15:9–24
Davies TR, McSaveney MJ (2011) Rock-avalanche size and runout—implications for landslide dams. In: Evans SG, Hermanns R, Strom AL, Scarascia Mugnozza G (eds) Natural and artificial rockslide dams. Lecture notes in earth sciences, vol 133. Springer, Heidelberg, pp 441–462
Davies TRH, Reznichenko NV, McSaveney MJ (2020) Energy budget for a rock avalanche: fate of fracture surface energy. Landslides 17:3–13
Dubovskoi AN, Pernik LM, Strom AL (2008) Experimental simulation of rockslide fragmentation. J Min Sci 44(2):123–130
Dufresne A, Ostermann M, Preusser F (2018) River-damming, Late-Quaternary rockslides and rock avalanches in the Ötz Valley region (Tyrol, Austria). Geomorphology 310:153–167
Erismann TH (1979) Mechanisms of large landslides. Rock Mech 12:15–46
Evans SG, Hungr O, Enegren EG (1994) The Avalanche Lake rock avalanche, Mackenzie Mountains, Northwest Territories, Canada: description, dating and dynamics. Can Geotech J 31:749–768
Hartvich F, Mugnai F, Proietti C, Smolkova V, Strom A (2008) A reconstruction of a former rockslide-dammed lake: the case of the Kokomeren River valley (Tien Shan, Kyrgyzstan). Poster presentation at the EGU conference, Vien
Heim A (1882) Der Bergsturz von Elm. Deutsch. Geol. Gesell. Zeitschr. 34:74–115
Hewitt K (2002) Styles of rock avalanche depositional complex in very rugged terrain, Karakoram Himalaya, Pakistan. In: Evans SG (ed) Catastrophic landslides: effects, occurrence and mechanisms. Rev Eng Geol 15:345–378
Hewitt K (2006) Rock avalanches with complex run out and emplacement, Karakoram Himalaya, Inner Asia. In: Evans SG, Scarascia Mugnozza G, Strom A, Hermanns RL (eds) Landslides from massive rock slope failure. NATO science series: IV: earth and environmental sciences, vol 49. Springer, pp 521–550
Hsü KJ (1975) Catastrophic debris streams (Sturzstroms) generated by rock falls. Geol Soc Am Bull 86:129–140
Hungr O, Evans SG, Bovis MJ, Hutchinson JN (2001) A review of the classification of landslides of the flow type. Environ Eng Geosci 7 (3):221–238
Hungr O, Leroueil S, Picarelli L (2014) Varnes classification of landslide types, an update. Landslides 11:167–194
Hutchinson JN (1968) Mass movement. In: Fairbridge RW (ed) Encyclopedia of geomorphology. Reinhold Publishers, New York, pp 688–695
Hutchinson JN (1988) General report: morphological and geotechnical parameters of landslides in relation to geology and hydrogeology. In: Proceedings of 5th international symposium on landslides, Lausanne, vol 1, pp 3–35
Jibson RW, Harp EL, Schulz W, Keefer DK (2006) Large rock avalanches triggered by the M 7.9 Denali Fault, Alaska, earthquake of 3 November 2002. Eng Geol 83:144–160
Kilburn CRJ, Sørensen S-A (1998) Runout length of sturzstroms: the control of initial conditions and of fragment dynamics. J Geophys Res 103B:17877–17884
Kobayashi Y (1993) A hypothesis for reduced resistance in large landslides. In: Safety and environmental issues in rock engineering. Proceedings of the ISRM international symposium Lisboa, 21–24 June 1993. I. Balkema Rotterdam, pp 335–339
Kobayashi Y (1997) Long runout landslides riding on basal guided wave. In: Marinos K, Tsiambaos S (eds) Engineering geology and the environment. Balkema, Rotterdam, pp 761–766
Kolesnikov VS (1929) Brief description of the trip to the Sarez Lake in 1925. In: Proceedings of the Central Asian geographical society, vol XIX, Tashkent (in Russian)
Korchevskiy VF, Kolichko AV, Strom AL, Pernik LM, Abdrakhmatov KE (2011) Utilisation of data derived from largescale experiments and study of natural blockages for blast-fill dam design. In: Evans SG, Hermanns R, Strom AL, Scarascia Mugnozza G (eds) Natural and Artificial Rockslide Dams. Lecture notes in earth sciences, vol 133. Springer, Heidelberg, pp 617–637
Kurdiukov KV (1950) Ancient rockfalls in the Alay Range valleys. Prob Geogr 21:191–204 (in Russian)
Kurdiukov KV (1964) Newest tectonic movements and traces of the extreme seismic shocks on the northern slope of the Zaalai Range. In: Activated zone of Earth crust, newest tectonic movements and seismicity. Moscow, pp 153–159 (in Russian)
Legros F (2002) The mobility of long-runout landslides. Eng Geol 63:301–331
Legros F (2006) Landslide mobility and the role of water. In: Evans SG, Scarascia Mugnozza G, Strom A, Hermanns RL (eds) Landslides from massive rock slope failure. NATO science series: IV: earth and environmental sciences, vol 49. Springer, New York, pp 233–242
Li T (1983) A mathematical model for predicting the extent of a major rockfall. Z Geomorphol 27:473–482
McSaveney MJ (1978) Sherman glacier rock avalanche. In: Voight B (ed) Rockslides and avalanches 1, natural phenomena. Elsevier, Amsterdam, pp 197–258
McSaveney MJ, Davies TRH (2006) Rapid rock-mass flow with dynamic fragmentation: inferences from the morphology and internal structure of rockslides and rock avalanches. In: Evans SG, Scarascia Mugnozza G, Strom A, Hermanns RL (eds) Landslides from massive rock slope failure. NATO science series: IV: earth and environmental sciences, vol 49. Springer, Heidelberg, pp 285–304
Nicoletti PG, Sorriso-Valvo M (1991) Geomorphic controls of the shape and mobility of rock avalanches. Geol Soc Am Bull 103:1365–1373
Nichol S, Hungr O, Evans SG (2002) Large-scale brittle and ductile toppling of rock slopes. Can Geotech J 39:773–780
Pollet N, Cojean R, Couture R, Schneider J-L, Strom AL, Voirin C, Wassmer P (2005) A slab-on-slab model for the Flims rockslide (Swiss Alps). Can Geotech J 42:587–600
Preobrajensky IA (1920) The Usoi landslide. Geol Comm Papers Appl Geol 14:21 p. (in Russian)
Qin SQ, Jiao JJ, Li ZG (2006) Nonlinear evolutionary mechanisms of instability of plane-shear slope: catastrophe, bifurcation, chaos and physical prediction. Rock Mech Rock Eng 39(1):59–76

Reznichenko N, Davies T (2015) The gigantic Komansu rock avalanche deposit in the glaciated Alai Valley, Northern Pamir of Central Asia. In: Lollino G et al (eds) Engineering geology for society and territory, vol 2. Springer International Publishing, Switzerland, pp 895–898

Reznichenko NV, Davies TRH, Shulmeister J, Larsen SH (2012) A new technique for identifying rock-avalanche-sourced sediment in moraines and some palaeoclimatic implications. Geology 40 (4):319–322

Reznichenko NV, Andrews GR, Geater RE, Strom A (2017) Origin of large hummock deposits "chukuryi" in Alai Valley, Northern Pamir: geomorphological and sedimentological investigation. Geomorphology 285:347–362

Robinson TR, Davies TRH, Reznichenko N, De Pascale GP (2015) The extremely long runout Komansu rock avalanche in the Trans Alay range, Pamir Mountains. Southern Kyrgyzstan. Landslides 12:523–535

Sassa K, Nagai O, Solidum R, Yamazaki Y, Ohta H (2010) An integrated model simulating the initiation and motion of earthquake and rain induced rapid landslides and its application to the 2006 Leyte landslide. Landslides 7:219–236

Schuster RL, Crandell DR (1984) Catastrophic debris avalanches from volcanoes. In: IV international symposium on landslides proceedings, Toronto, vol 1, pp 567–572

Shaller PJ (1991) Analysis and implications of large Martian and terrestrial landslides, Ph.D. thesis. California Institute of Technology, Pasadena, CA, USA

Sheidegger AE (1973) On the prediction of the reach and velocity of catastrophic landslides. Rock Mech 5:231–236

Shoaei Z, Ghayoumian J (2000) Seimareh landslide, Western Iran, one of the World's largest complex landslides. Landslide News 13:23–27

Strom AL (1994) Mechanism of stratification and abnormal crushing of rockslide deposits. In: Proceedings 7th international IAEG congress 3. Balkema, Rotterdam, pp 1287–1295

Strom AL (1996) Some morphological types of long-runout rockslides: effect of the relief on their mechanism and on the rockslide deposits distribution. In: Senneset K (ed) Landslides. Proceedings of the seventh international symposium on landslides, Balkema, Trondheim, Norway, Rotterdam, pp 1977–1982

Strom AL (1998) Giant ancient rockslides and rock avalanches in the Tien Shan Mountains, Kyrgyzstan. Landslide News 11:20–23

Strom AL (2004) Rock avalanches of the Ardon River valley at the southern foot of the Rocky Range, Northern Caucasus, North Osetia. Landslides 1:237–241

Strom AL (2006) Morphology and internal structure of rockslides and rock avalanches: grounds and constraints for their modelling. In: Evans SG, Scarascia Mugnozza G, Strom A, Hermanns RL (eds) Landslides from massive rock slope failure. NATO science series: IV: earth and environmental sciences, vol 49, pp 305–328

Strom AL (2010) Evidence of momentum transfer during large-scale rockslides' motion. In: Williams AL, Pinches GM, Chin CY, McMorran TG, Massei CI (eds) Geologically active. Proceedings of the 11th IAEG congress, Auckland, New Zealand, 5–10 September 2010. Tailor & Francis Group, London, pp 73–86

Strom A (2014) Catastrophic Slope Processes in Glaciated Zones of Mountainous Regions. In: Shan W et al (eds) Landslides in cold regions in the context of climate change, environmental science and engineering. Springer International Publishing, Switzerland, pp 3–10

Strom A, Pernik L (2013) Modeling of debris crushing during rock avalanche motion. Geophys Res Abst 15, EGU 2013 - 1373

Strom A, Abdrakhmatov K (2018) Rockslides and rock avalanches of Central Asia: distribution, morphology, and internal structure. Elsevier, Netherlands, 449p. ISBN 978-0-12-803204-6

Strom A, Li L, Lan H (2019) Rock avalanche mobility: optimal characterization and the effects of confinement. Landslides 16:1437–1452

Varnes DJ (1954) Landslide types and processes. In: Eckel EB (ed) Landslides and engineering practice, Special Report 176. Transportation Research board, National Academy of Sciences, Washington, DC, pp 11–33

Varnes DJ (1978) Slope movement types and processes. In: Schuster RL, Krizek RJ (ed) Landslides—analysis and control. National Academy of Sciences Transportation Research Board Special Report 176, pp 12–33

von Poschinger A, Wassmer P, Maisch M (2006) The flims rockslide: history of interpretation and new insights. In: Evans SG, Scarascia Mugnozza G, Strom A, Hermanns RL (eds) Landslides from massive rock slope failure. NATO science series: IV: earth and environmental sciences, vol 49. Springer, Heidelberg, pp 329–356

Weidinger JT, Korup O, Munack H, Altenberger U, Dunning S, Tippelt G et al (2014) Giant rockslide from the inside. Earth Planet Sci Lett 389:62–73

Yuan Z, Chen J, Owen LA, Hedrick KA, Caffee MW, Li W, et al (2013) Nature and timing of large landslides within an active orogen, eastern Pamir, China. Geomorphology 182:49–65

WLF 2020 Kyoto Japan

# Study on the Phenomena of Liquefaction Induced Massive Landslides in 28 September 2018 Palu-Donggala Earthquake

Paulus P. Rahardjo

**Abstract**

Liquefaction Induced Mass Landslide in Palu Donggala Earthquake on 28 September 2018 was one of the rare and biggest event on these types of landslides in the world. The phenomena was the most complete occurence since all mechanism of liquefaction and liquefaction induced landslides were represented. On these particular event, the author has conducted study on the liquefaction mechanism based on field observation including running drone in several areas, conducting soil investigations, work on analysis and collect as much as data from local people. This paper discusses Liquefaction Mechanism in these unique and spectacular sites (mainly in four areas: Balaroa, Petobo, Jono Oge and Sibalaya) because the earthquake seems to trigger liquefaction by multidirectional vibration and of particular interest is because the earthquake are near faults with shallow earthquake focus of about 10 km below the city. The extraordinary distance of liquefaction flow and lateral spreading is one of unique phenomena which is believed to be caused by the existence of initial artesian pressure and significant vertical acceleration causing the soil loosing contact stress. Layers of sands and clays or silts might have caused significant force to the liquefied sands and flow laterally. Instead of surface phenomena, the main objectives of this paper is also to discuss the results of CPTu tests conducted for analysis and fact findings on liquefaction phenomena. Each data of CPTu yields liquefaction potential index which is used to characterize the severity of ground damage and discuss mitigation and risk reduction in the future.

**Keywords**

Liquefaction • Lateral spreading • Liquefaction induced landslides • Liquefaction potential index

## Introduction

Palu-Donggala Earthquake of 28 September 2018 has caused severe damages and high fatalities as mentioned in many reports after the earthquake (Rahardjo 2019; Geotechnical Engineering Center, Unpar 2019 etc.). This earthquake has resulted in the most complete phenomena including ground cracks, tsunamis, liquefaction, buildings and infrastructure collapses and of particular rare event is the tremendous liquefaction induced flow slides, that might be the biggest of its kind in the world where damages are massive in scale. This liquefaction induced flow slides has obtained many attention from Indonesia and overseas.

In this paper, the author focuses research and observation on the phenomena at some severely affected areas in Balaroa, Petobo, Lolu Village, Jono Oge, and Sibalaya where the damages are the most severe among other regions. The last three sites are easily seen from the main road of Palu-Palolo roads by public transports. The impact of the earthquake is so devastating and almost impossible to survive from the moving materials, flowing all over the area and might has caused more than 3000 fatalities. It is obvious that the community of Palu and Sigi Regency have never thought about such disastrous event and not prepared.

## Tectonic of Sulawesi and 28 September 2018 Palu Donggala Earthquake

Tectonic of Sulawesi is very complex since it is the location where three big tectonic plates colliding to each other i.d. Indo-Australian plate moving north, Pacific Plate which is

P. P. Rahardjo (✉)
Department of Civil Engineering, Universitas Katolik Parahyangan, Bandung, 40141, Indonesia
e-mail: rahardjo.paulus@gmail.com


V. Vilímek et al. (eds.), *Understanding and Reducing Landslide Disaster Risk*, ICL Contribution to Landslide Disaster Risk Reduction,
https://doi.org/10.1007/978-3-030-60319-9_2

moving west and Eurasia plate moving east and meet in Sulawesi. In this area subduction and collision is still very active (Thein et al. 2014).

The shape of Sulawesi island is a result of complex history of rotating collisions of Continental block, island arcs and Sunda block which is the south eastern end of Eurasian plate. Nowadays, Sulawesi island is obliquely cut by the main fault as left-lateral Central Sulawesi Fault System = CSFS). This fault system consists of two fault zone, Palu-Koro (NNW-PKF) and Matano (WNW-MF) (Fig. 1). Within the last 5 millions years the slip rate may vary, however the slip rate of PKF is presently on the order of 38 ±8 mm/yr base on geodetic survey (GPS) (Walpersdorf et al. 1998).

Tectonic earthquake at Palu and Donggala Regency on Friday 28 September 2018 occured 18.02 Mid Indonesia time (WITA) to have long duration of 3–7 min with Mw = 7.4 Epicenter 0.18°LS dan 119.85°BT at distance of 26 km north of Donggala at Sulawesi Tengah, and depth 11 km. Position of the main shock is shown on Fig. 2.

## Measured Acceleration and Attenuation Function

Acceleration is an important parameter to be used for liquefaction potential analysis. Hence it is relevant to derive an attenuation function based on local data so that liquefaction potential can be calculated at specific locations. Figure 3 shows acceleration wave form developed by Jica and BMKG for Palu-Donggala Earthquake zoom up arround main shock. Measured data at Balaroa results in resultant acceleration 400 gal which is the resultant of maximum acceleration in north south direction 281 gal and east and west 203 gal causing a horizontal acceleration of 333 gal and vertical direction 335 gal. This high magnitude of vertical acceleration is due to the fact that Balaroa is located near fault area. Correlation of E-W, N-S acceleration with vertical acceleration to have average acceleration 49–51% from horizontal acceleration.

BMKG published data on measured acceleration from a number of stations in the direction of North (N), East (E) and vertical (z). Resultants of these acceleration were plotted to obtain attenuation function and compared to old simple equations of Esteva and Donovan. It is shown on the plot that Donovan equation was still in close proximity to the field data (Fig. 4).

## Liquefaction Phenomena in Palu and Sigi Regency

In general liquefaction occur during earthquakes and has caused failures of the ground causing severe damages of the buildings and infrastructures on the surface. The soil phase changes from solid into liquid due to substantial drops of its shear strength. Saturated loose to medium sands are among the most vulnerable to liquefaction. Liquefaction in Palu earthquake is the main cause of the infrastructure and housing damages. However, in Palu, liquefaction is not new, some earthquakes in the past had caused liquefaction, but no people take care on these, maybe because not much damages and number of fatality was low or none. Palu people has introduced the word "nalodo" long time ago before the last earthquake to explain how solid material sands turn the liquid materials.

**Sand Boils** and **Lateral Spreading** are found in many places in Palu and Sigi Regency. In Lateral Spreading the liquefied sand layer does not penetrate the non liquefied layer or only partially comes up the surface and then this liquefied layer flow laterally to bring blocks of soil on top like flowing soil cakes. Figure 5 illustrates the lateral spreading phenomena where these soil blocks are still on top of the liquefied layer.

**Flow Liquefaction** or Liquefaction Induced Flow Failures refer to situation where initially the soil has initial shear stress such as in slope. Flow liquefaction could be triggered even in very gentle slope of 2—3% gradient. The soils turn into liquid and viscous material and flow laterally. Houses, cars and other materials sink and swallowed in the sandy mud. The distance of travel of the liquefaction flow in Palu and Sigi Regency can be very far from 300 to 1000 m away depending on the initial density of the sands, gradient of topography and the excess pore pressures. This extra ordinary phenomena is believed to be due to the additional forces of artesian pressure and the high vertical acceleration (Rahardjo 2019). There are also facts that due to breakage of the Gumbasa irrigation channel, substantial water flowed to west direction moving the liquefied soils further. The height of the water was indicated on the stains or dirts shown on houses as high as 2 m. Figure 6 shows the effect of liquefaction flow in Petobo area where houses were carried away as far as 700 m.

There are 4 areas known to have serious impact by liquefaction in Palu and Sigi Regency, i.e. Balaroa, Petobo, Jono Oge and Sibalaya (Fig. 7). From the geology of Palu

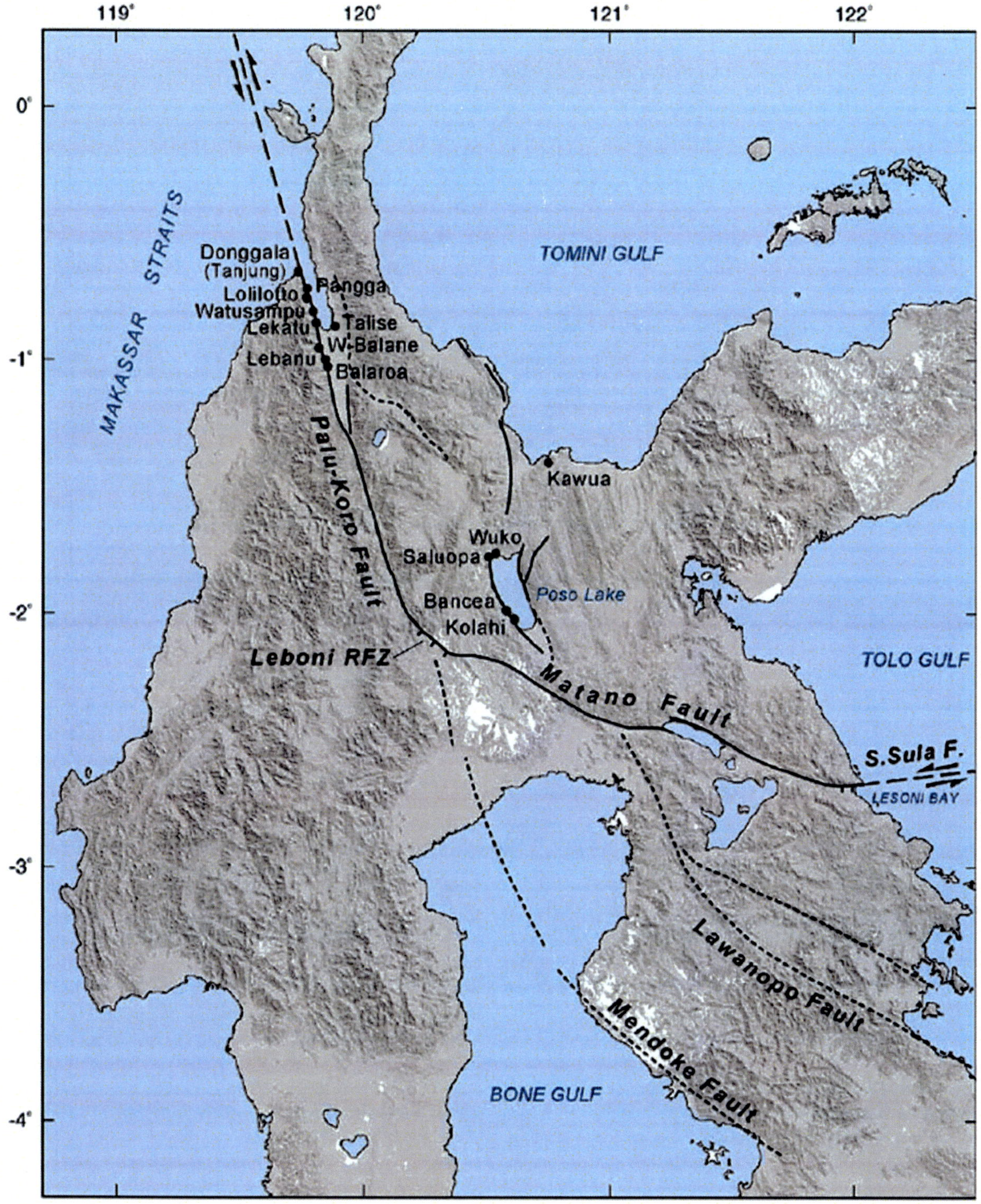


**Fig. 1** Two fault system, Palu koro and Matano Fault (Walpersdorf et al. 1998)

**Fig. 2** Main shocks and after shock at Palu earthquake 28 September, 2018 (BMKG 2018)

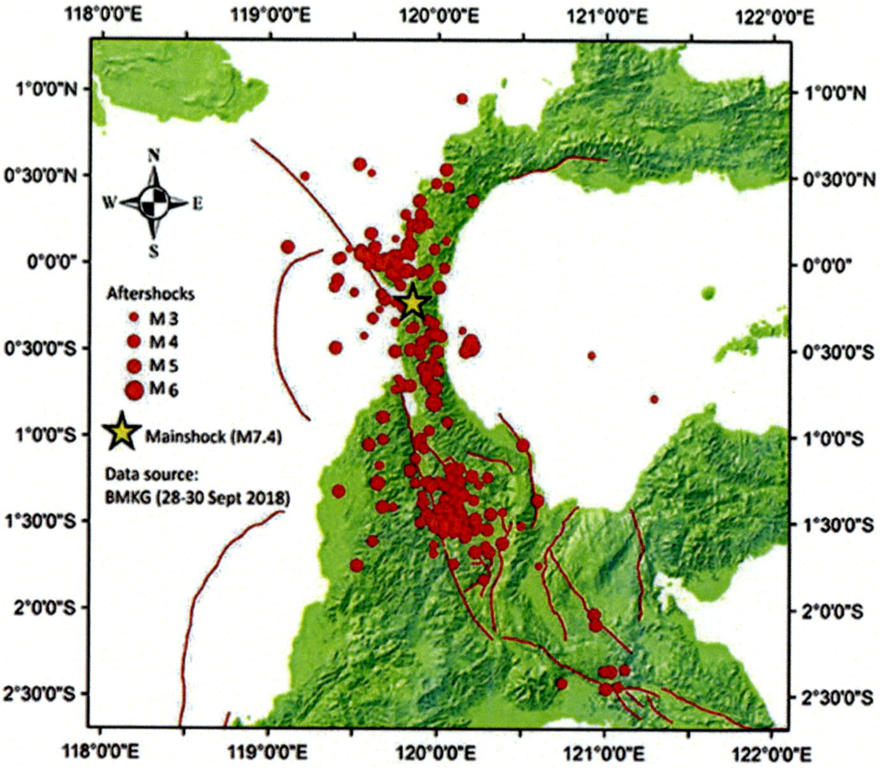


and Sigi, majority of these areas are dominated by sandy soils. Liquefaction mainly occured in saturated sandy soils or silty sands.

The sands in Palu is quaternary deposits (molasses deposit) which are located among faults segments. When the lower layer sediments moved then the upper layer lost its strength and stiffness and then liquefy. The requirements for liquefaction to occur are the material type (sandy soils and saturated), its stiffness or density, mostly poorlu graded, the magniture of earthquake (Mw $\geq$ 5.0). Usually ground water table is high. In Palu particularly those area on the east side of Palu, the elevation of water table is somewhat higher due to Gumbasa Irrigation Channel on the east side of Petobo (Fig. 8).

## Liquefaction Phenomena in Balaroa

Palu–Koro Fault pass through Balaroa hence it is very likely that Balaroa was very badly damaged by the earthquake Fig. 8 shows the position of Balaroa, Petobo and Jono Oge. In Balaroa area, as many as 1747 buildings collapsed or squeezed at the end of the flow. Some buildings have been moved by 200 m or more.

Visually the upper layer of soil in Balaroa that the upper part sediments recently deposited as sand and boulders. The underlying layer is loose to coarse sands in saturated conditions. Many spring water were found in this area and the type of failure was clearly liquefaction induced slope failures around the crown and become flow liquefaction. Figure 9 shows the crown of the Balaroa landslides.

Figure 10 shows boulders in the mid of sandy tuff, and the upper part was dry. Liquefaction was seen at depth 5.0–7.0 m where sand materials wet and water pounding can be seen. Coarse sands were detected in Balaroa liquefied layer (Fig. 10). It was reported that a few seconds to a few minutes after the quake, people witnessed water spurting vertically and then suddenly the ground dropped and flow and drew houses, roads, cars and human beings in up and down mechanism. Even many buildings including the big mosque moved laterally as shown on Fig. 11 (Figs. 12 and 13).

## Liquefaction Phenomena in Petobo

Petobo is located south of Palu Airport. This area has the most dramatic destruction affected by liquefaction. The area with severe destruction is about 0.9 km wide and 2.1 km long as shown on Fig. 14 (https://geologi.co.id). Based on the geology of this area, consisting of recent alluvial deposits (Qa) with high water table (average –3.0 m). On the east side boundary is Gumbasa irrigation channel. From the local

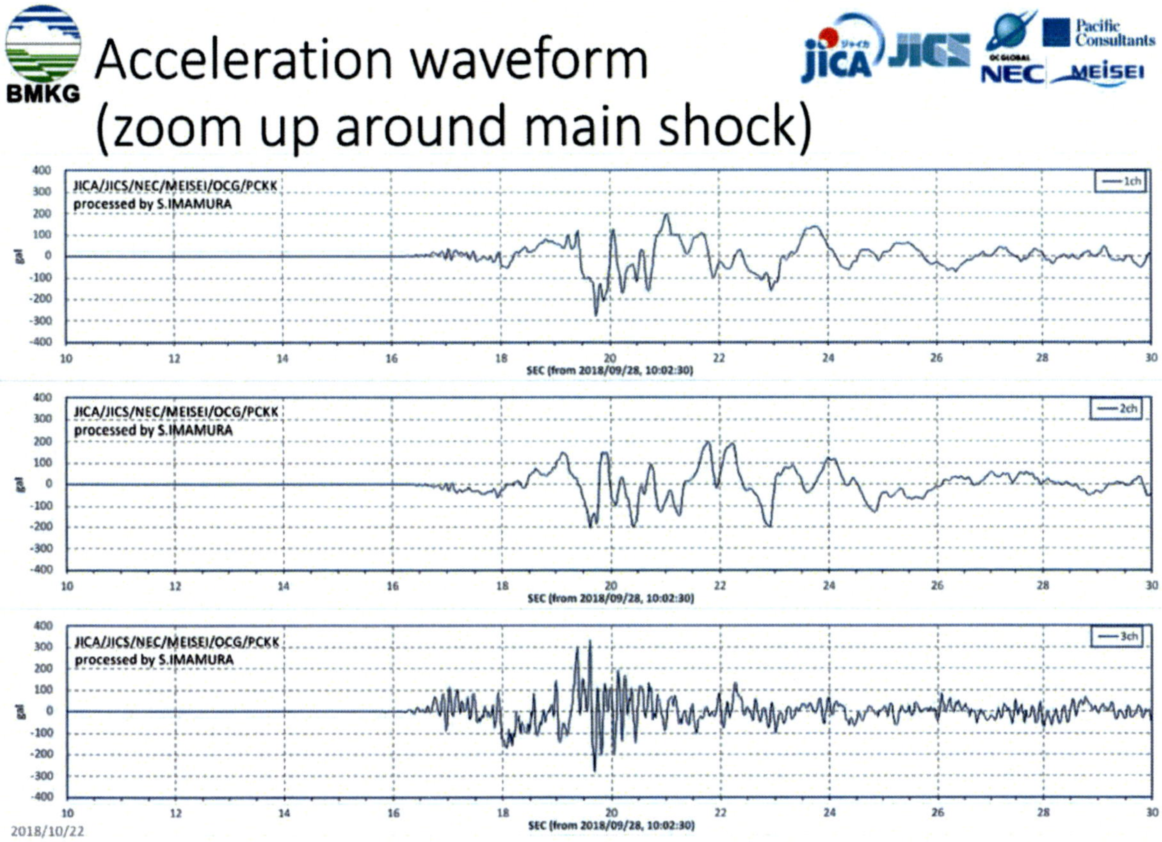


**Fig. 3** Acceleration wave form of Palu Donggala Earthquake (BMKG-JICA)

people information, initially water table was far below ground surface, but the existence of paddy field and the irrigation channel for more many years has increase the level of water table. It could be due to water pounding on the rice fields and seepage from the irrigation channel to west direction (Fig. 15).

From the comparison of before and after earthquake, it can be seen how many houses were damaged by liquefaction flow slides. In Petobo, 744 houses were destructed or burried into liquefied sandy mud. The destruction in Petobo is not the same as in Balaroa, where the houses were mostly still on ground, while in Petobo several houses were either halve buried in the mud and halve sink into the mud, due to liquefied layer on the surface.

At the boundary of liquefied and non liquefied area is Gumbasa Irrigation Chanel, the channels were severely damaged and water flow out of the channel flow together toward the west area. Interestingly, the eastside of the Gumbasa Irrigation Channel has very little or no damage (Fig. 16). Several discussion point out that the liquefied area is largely paddy field that was saturated for many years while the east part is safe because water table in this area is deeper (Fig. 17).

During survey, the team of Universitas Katolik Parahyangan also conducted several CPTu in location where liquefaction was severe and in unliquefied area to confirm between what is visually exposed condition and the results of CPTu. Cone Penetration Tests were also conducted at the irrigation channel area on December 2018 as shown on Fig. 18. The CPT revealed the existence of sandy soils about 10 m from ground surface and then followed by thin silty clay layers. Based on visual damage and CPT data, it can be predicted that the thickness of liquefied layer at Petobo is about 11.0 m. However when the CPTu test were conducted, there were no water table detected. The low friction ratio of the CPT data is consistent with the loose condition of the soils.

Based on the assessment of Liquefaction Potential in this area, the LPI is higher then 15 and classified as highly susceptible to liquefaction. The actual condition is shown on Fig. 19 which is consistent.

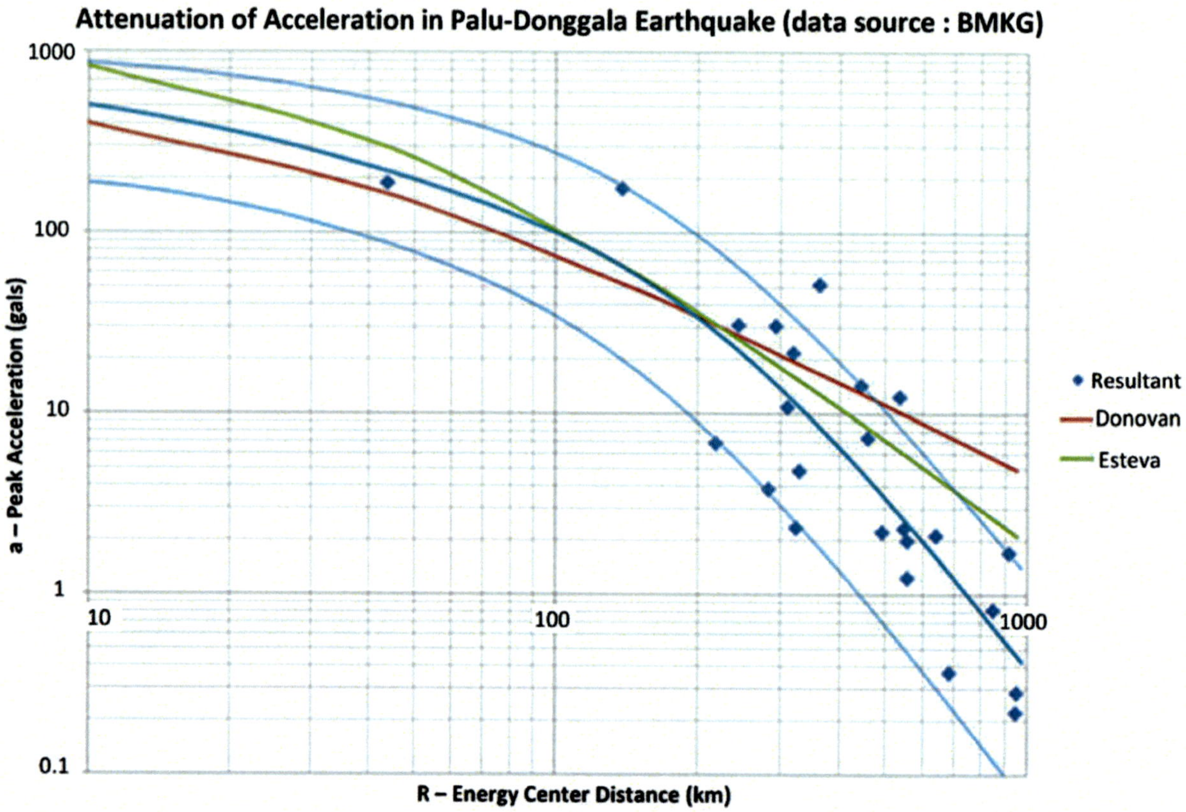


**Fig. 4** Ploted acceleration data as attenuation function in Palu Earthquake (Geotechnical Engineering Center, Universitas Katolik Parahyangan, 2019)

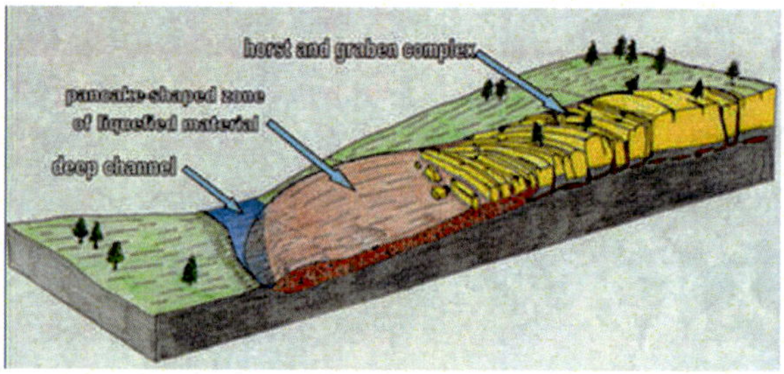


**Fig. 5** Illustration of lateral spreading

## Liquefaction Phenomena in Jono Oge

Liquefaction Phenomena at Jono Oge is very similar that in Petobo except that the liquefaction flow failures also moved liquefied soils away due to mix with water from the river. Hence the distance travelled by the liquefied sands is also much longer. Figure 20 shows how liquefaction mixed water results in severe damage and illustrates the area affected by liquefaction at Jono Oge, before and after the earthquake. Also in this area, the east boundary of liquefied and non liquefied sands is Gumbasa Irrigation Channel. The main road Palu-Palolo was washed away 700 m long, liquefaction flow as far as 400–1200 m because of the average slope and viscosity of the liquefied materials due to mixed with water from the river and the irrigation channel (Fig. 21).

Initial water table just before the earthquake is high and during investigation, the team found water table is much lower. The breakage of the irrigation channel was due to liquefaction, and subsequently the water caused further flow of all liquefied material like flood. The observation through a house near the missing road show that water flow could be 2 m high together with the debris (Fig. 22).

## Liquefaction Phenomena at Lolu Village

Lolu Village is located between Petobo and Jono Oge, and belong to Jono Oge district. At this particular location, the liquefaction Phenomena is not directly related to Gumbasa Irrigation but may be the elevation of the water table subject to the irrigation of the paddy field. There has been phenomena of lateral spreading in this area. The destruction depend on the mechanism of liquefaction.

**Fig. 6** Liquefaction flow failures in Palu, 2018

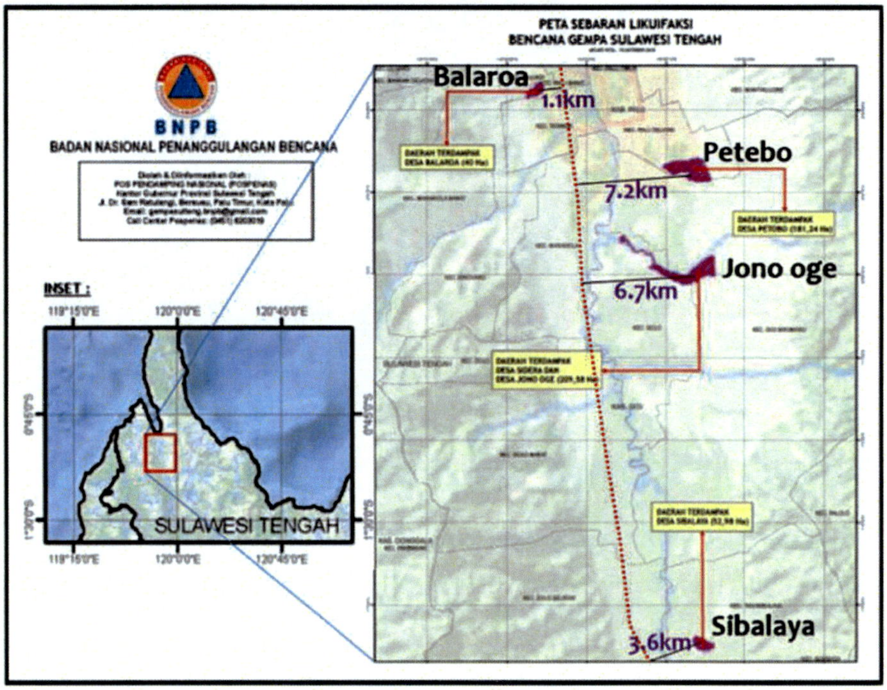


**Fig. 7** Locations of four major areas of lateral spreading and liquefaction flow (BNPB 2018)

Sand boils at Lolu village were found in many places. Sand boils did not always cause ground failures, but surely an indication of liquefied sand below ground surface, the mechanism may be interacted by the possibility of lateral spreading. Water of high pressure come up to the surface through ground cracks and distributed in many areas. Figure 23 shows example of the sand boils at Lolu village. Figure 24 shows sandblows at Lolu village on free ground.

Ishihara (2019) suggested mechanism lateral spreading is due to liquefied sand underneath and flow up due to sufficient pressure by the generation of the excess pore pressure and the existence of aquifer pressure or artesian pressure (Fig. 25).

At this site, lateral spread has been detected very clearly where blocks of soils flow on top of liquefied layer as far as 20–100 m away. Houses and infrastructure were displaced

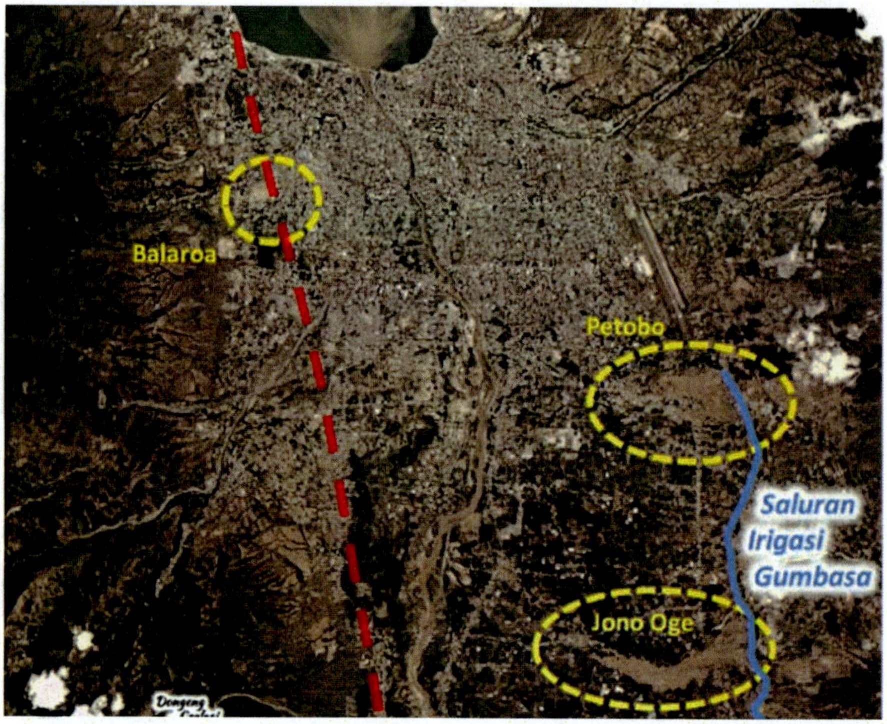


**Fig. 8** Balaroa located near fault, Petobo and Jono Oge on the east side near Gumbasa Irrigation Channel

**Fig. 9** Crown of landslides in Balaroa

laterally as a whole blocks. Figure 26 shows phenomena of lateral spreading at Lolu village.

## Gas Station and Nearby Housing

Issue on the lateral spreading of gas station and BTN housings is very well known. This location is easily seen from the main road. The road has shifted 23 m to west and the gas station moved 35 m. Other houses north of gas station even displaced 63 m approximately. Figure 27 shows original location and the new location of each building. The distance of old and new positions were measured using google map (Figs. 28 and 29).

It is of interest that the buildings at this site displaced at different distance. The north parts moved farther and this makes shifting of the buildings in torsion. Subsequently the damage is more severe (Fig. 30).

## Perumnas BTN Housing

Area where lateral spreading has displaced new housing complex called Perumnas BTN about 100 m. The displacement reached this far but all the shifted buildings were carried on ground "together" by the "flowing liquefied soil layer". So half block of the housing were away from their original positions.

It could be imagined that the upper block of soils did not liquefy but acting together as broken blocks and flow as a whole toward west. Several sand blows was observed. Figures 31 and 32 show the area of Lolu Village where the

**Fig. 10** Crown of the Balaroa landslides consisting of Boulders in the mid of tuffaceous sands

**Fig. 11** Liquefied sand materials in Balaroa

Perumnas BTN has been away from their original location (Fig. 33).

CPTu was conducted at BTN Perumnas housing location and the result is shown on Fig. 34. The data illustrate that the site was dominated by loose sandy layers interbedded with clay layer. Some researchers have discussed the possibility on the effect of "water film" where the soils liquefy down below but did not penetrate the upper layers. Hence the excess pore water pressure push the liquefied layer to the west.

The damage of the buildings are severe, the floors was broken and also the roofs but overall buildings still in the same pattern as previously and the local road although distorted up and down and broken but still on its place, and it can be concluded that the block has been displaced in more or less rigid manner.

## Liquefaction Phenomena in Sibalaya

Sibalaya is located in the south area of Sigi Regency that also experienced massive liquefaction. The area belongs to Tanambulawa district where phenomena of flow liquefaction has destructed Gumbasa Irrigation Channel. Liquefaction flow slides were also the cause to displace the village including paddy field, socker field, elementary school, roads and several houses 350 m away. Water table was recorded at –2 m below ground surface but after the quake, the water table dropped to −18 m which according to local people was the initial ground water table before irrigation of this area started. Out of 4 CPTu and two drillings, it can be concluded that the area is dominated by fine sands and silty sand, some times with interbedded clay. This area was liquefied in total of 51 hectares with liquefaction flow from southeast to northwest direction 1 km length.

Figure 35 shows the affected area and location of soil investigation where it is indicated as red, meaning that the area is highly susceptible to liquefaction with Liquefaction Potential Index (LPI) higher than 15 as predicted using Cone Penetration Test (CPTu). One point on the east of irrigation chanel has low LPI (less than 5), which is similar like those in Petobo and Jono Oge due to low water table (Figs. 36, 37 and 38).

**Fig. 12** Houses and the mosque were carried by flow liquefaction about 200 m away. Antara Foto/Irwansyah Putra/Reuters PALU/Published on October 02, 2018

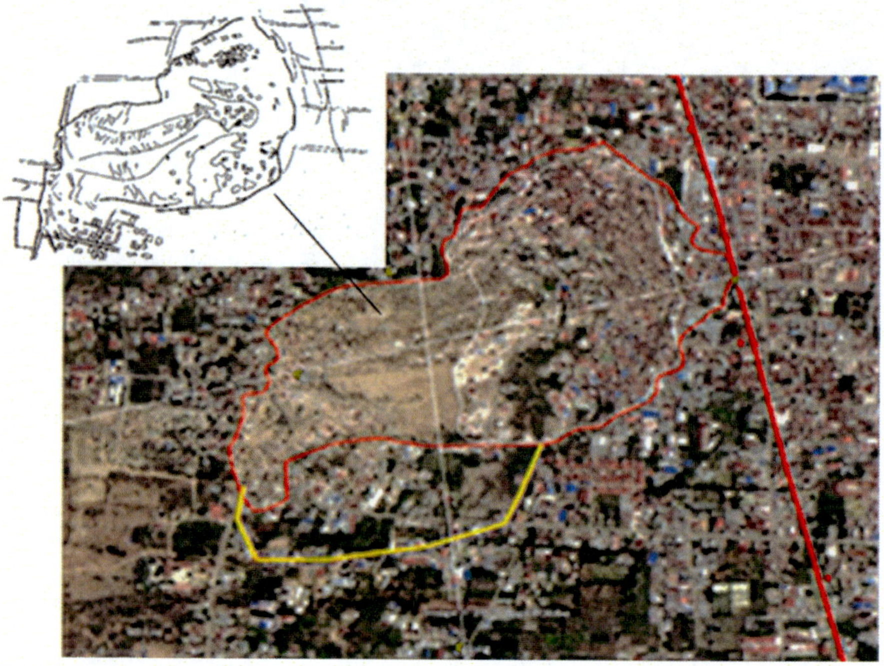

**Fig. 13** Damages of houses in Balaroa (Ishihara 2019)

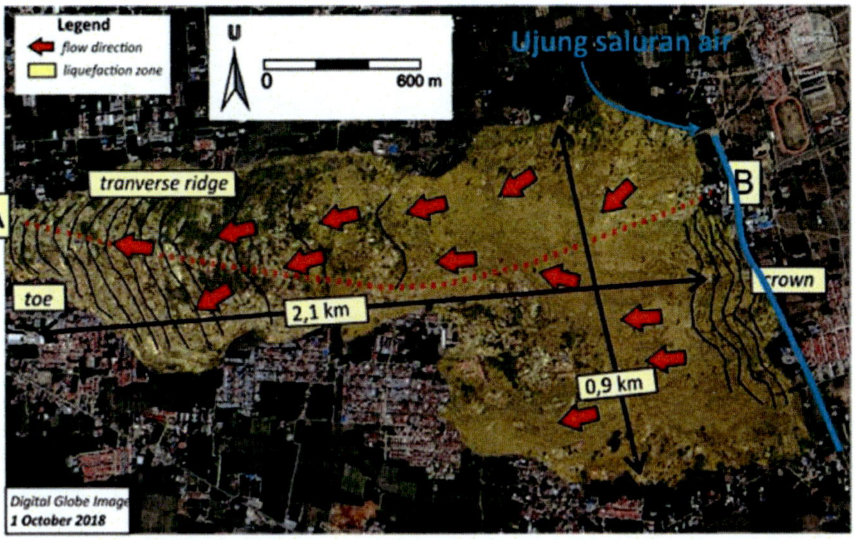


**Fig. 14** Liquefied area in Petobo and the position of Gumbasa Irrigation Channel (https://geologi.co.id)

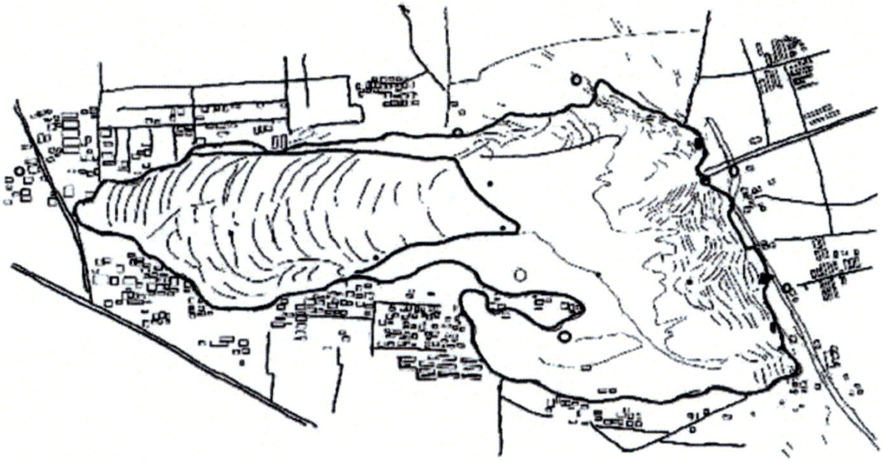

**Fig. 15** Mechanism of liquefaction, lateral spread and liquefaction flow slides in Petobo (Ishihara 2019)

The contribution of Gumbasa Irrigation is very significant to the saturation of the soils. It takes water from Gumbasa River and bring water to Petobo to irrigate 13,000 ha of paddy field.

## Results on Liquefaction Potential Analysis in Palu-Donggala and Sigi Area Using CPTu

For liquefaction potential assessment, 4 criteria suggested by Krammer (1996) has been conducted based on:

1. Historical Criteria
2. Geological Criteria
3. Compositional Criteria
4. State Criteria.

For Palu, Donggala dan Sigi Area, these criteria is investigated and the results are to be implemented to develop microzoning of the liquefaction potential. The soil investigation consisted of 6 bore holes and 36 CPTu. The use of CPTu for liquefaction assessment were suggested by some researchers (Shibata and Teparaksa 1987; Rahardjo 1989).

1. Historical Criteria
   Important Information from liquefaction behavior of sand can be investigated in the field after earthquake where liquefaction occured in the past it will be repeated assuming that water table does not change (Youd 1984). Study on the liquefaction cases can be utilised to identify specific area where liquefaction may occur in the future. Youd (1991) described a number of example where historical data from liquefaction were recorded and used

**Fig. 16** Boundary of the liquefied and non liquefied area is Gumbasa Irrigation Channel (Geotechnical Engineering Center, Unpar—2019)

**Fig. 17** Paddy Field flow as liquefied material and moved the houses toward west

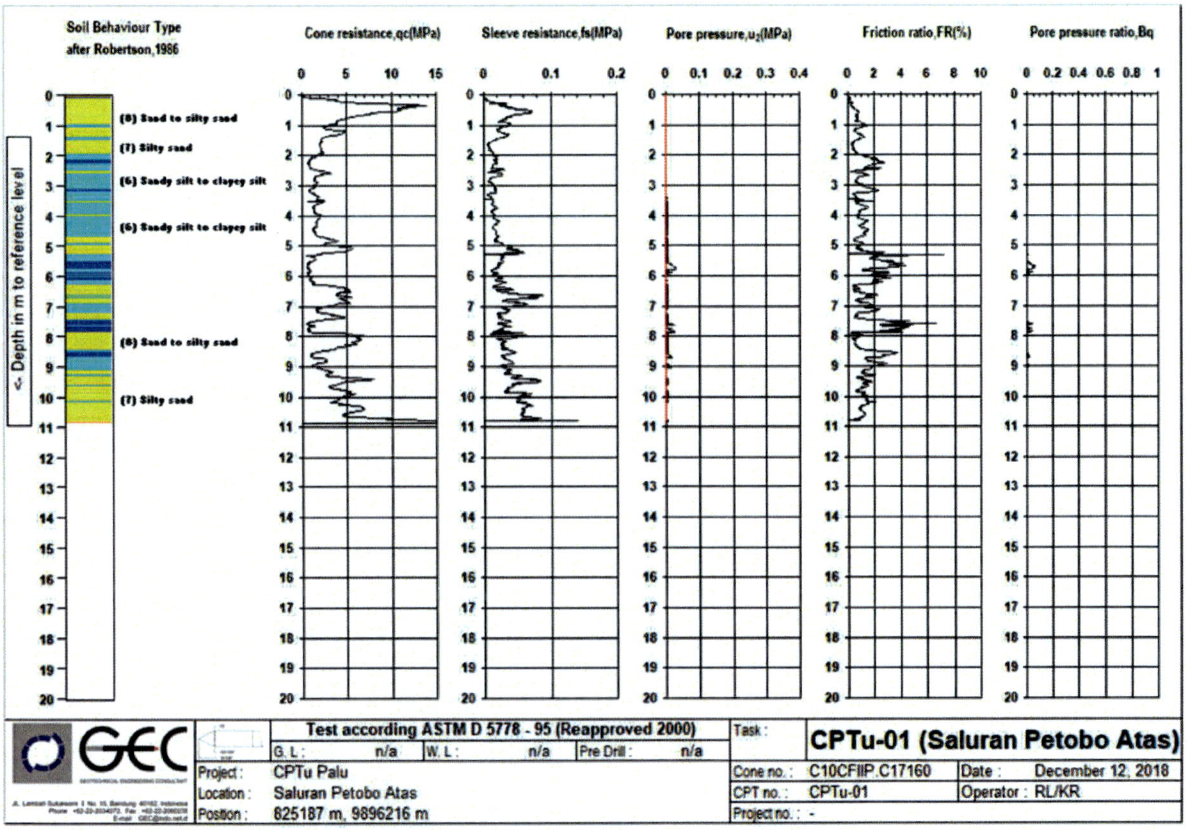


**Fig. 18** CPT data at Petobo was conducted from the base of Gumbasa Irrigation Channel

**Fig. 19** Area where CPTu was taken in Petobo, at location of the damage of Gumbasa Irrigation Channel

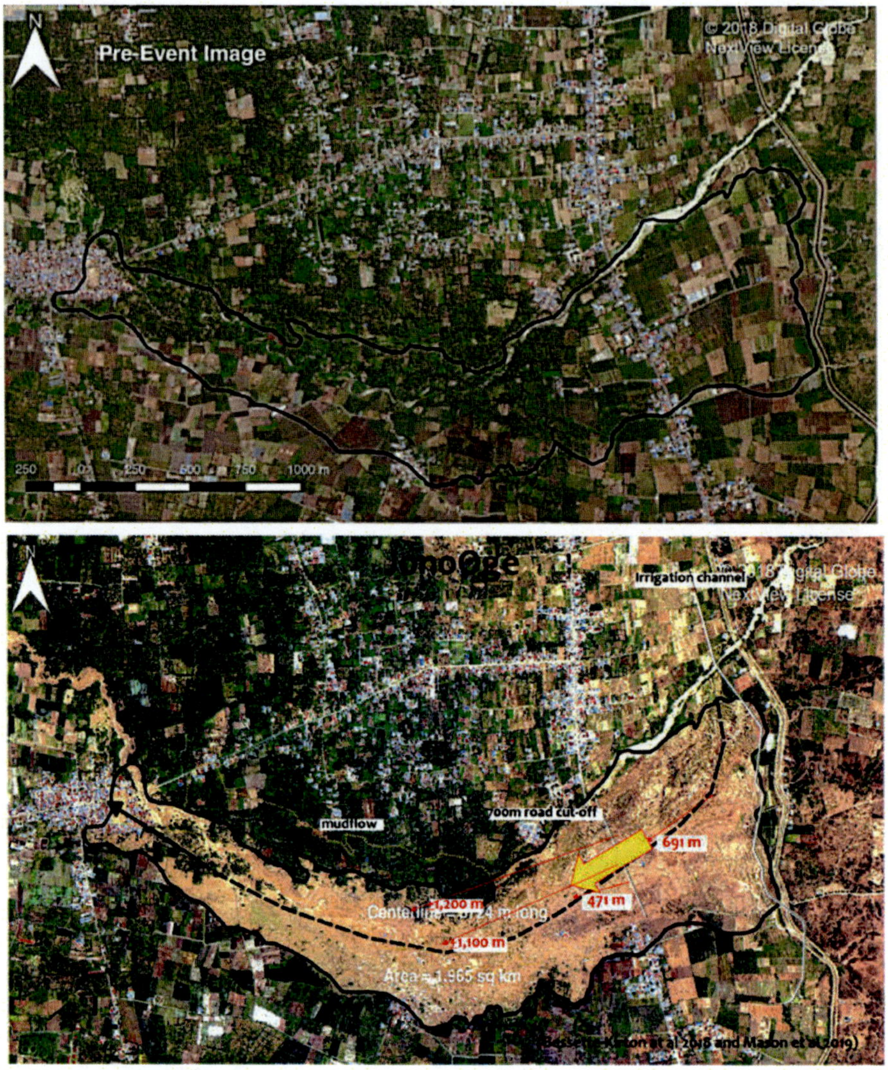


**Fig. 20** Area affected by liquefaction at Jono Oge before and after the earthquake

**Fig. 21** Mechanism of liquefaction flow at Jono Oge (Ishihara 2019)

**Fig. 22** A house at Jono Oge shows the height of the mud and water during liquefaction flow

**Fig. 24** Sand boils at Lolu Village near housing complex

**Fig. 23** Sand boils at Lolu Village

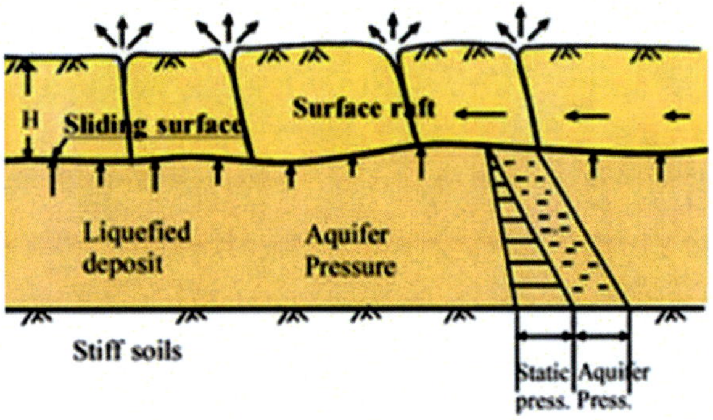


**Fig. 25** Mechanism of lateral spreading mixed with sand boils (Ishihara 2019)

to develop map for liquefaction phrone area.

Ambraseys and Menu (1988) used data from all over the world from shallow earthquakes and recommended boundary where liquefaction might or might not occur. For Palu area, data may be collected from USGS for earthquakes from 1900 to present. The earthquake data is selected to have magnitude M > 5.0 and the distance should be 500 km away. The reason is, for M = 9, at distance of 500 km, liquefaction is still possible. The results of earthquake data are presented in Fig. 39 and then plotted on Ambrasseys graph to find wether liquefaction is possible (Fig. 40). From this figure one can conclude that liquefaction may happen in Palu area, and at least five 5 earthquakes have caused liquefaction events.

2. Geological Criteria

   Alluvial soils are subject to easily liquefy. Based on geological map, almost all Palu valley are youg alluvium susceptible to liquefaction. All area in the investigation, Balaroa, Petobo, Jono Oge, Lolu Village and Sibalaya are located in alluvial area.

3. Compositional Criteria

   The third criteria is basically assessed based on the grainsize distribution. Recommended boundary of grainsize has been proposed by Tsuchida (1970) based

**Fig. 26** Lateral spreading at Lolu Village

**Fig. 27** Offset of buildings and infrastructure at Lolu Village by use of google map (Rahardjo 2019)

on data in Niigata, Alaska and other areas. For study in Palu, the soils sampels were collected from liquefied area as shown on the following figure. The conclusion is that the grainsize of area in Palu fall into category of easily liquefiable (Fig. 41).

4. State Criteria

   The most important criteria that we still need to consider is the state criteria. The vulnerability of soil against liquefaction depend on the tendency of the soil to generate excess pore pressure during seismic loading. Loose saturated sands will generate higher excess pore pressure. The density of sands is therefore the key parameter and can be measured indirectly by standard penetration test (SPT) or Cone Penetration Test (CPT).

   Liquefaction assessment is by recommendation of approach initiated by Seed and Idriss (1967), and comparing *cyclic stress ratio* (CSR) based on seismic loading and *cyclic resistance ratio* (CRR) of the soil which can

**Fig. 28** Gas Station and buildings at Lolu Village that were displaced due to lateral spreading

**Fig. 29** Condition after earthquake, gas station was displaced 35 m to west (rear view)

be approximated by cone resistance of CPT ($q_c$), SPT *blow count* (N-SPT), or shear wave velocity, ($V_s$).

*Cyclic stress ratio* (CSR) generated by earthquake at depth z, can be expressed from soil profile as average cyclic shear stress indicated by 65% of *cyclic shear stress ratio*:

**Fig. 30** Distorsion outside and inside the building that was displaced as far as 63 m due to lateral spreading

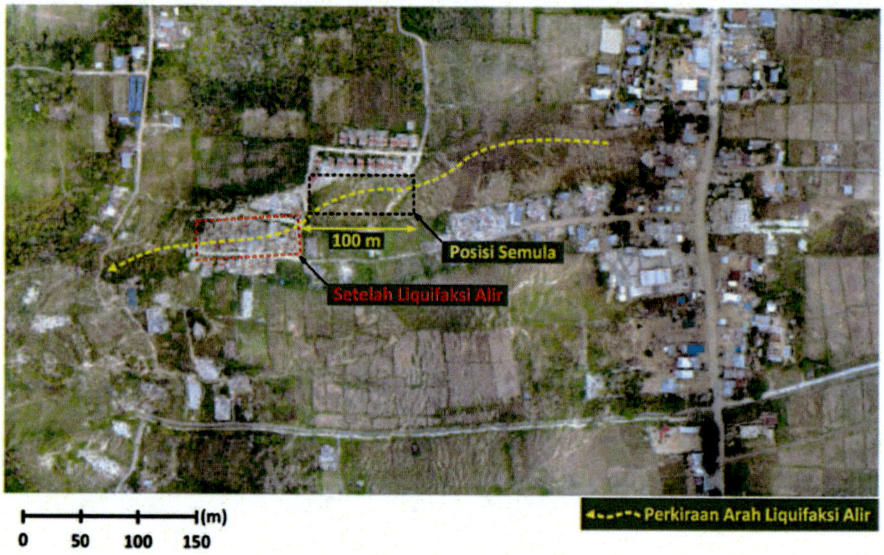


**Fig. 31** Situation map of Lolu village showing Gas Station and BTN Housing (Rahardjo 2019)

**Fig. 32** Moving blocks of BTN housing

**Fig. 33** Bird eye view of BTN Perumnas housing after the earthquake

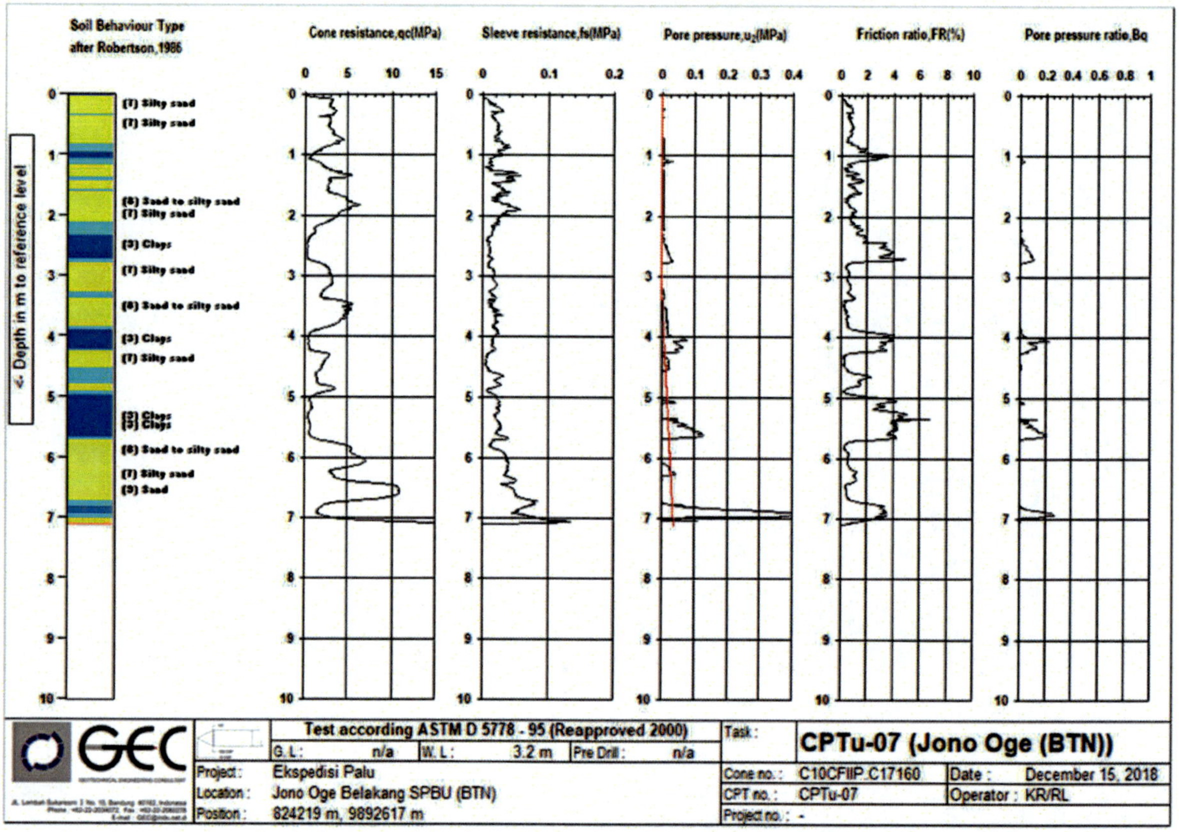


**Fig. 34** Results of CPTu at BTN Perumnas housing

**Fig. 35** Area in Sibalaya affected by liquefaction flow failures and results of CPTu based LPI

**Fig. 36** Damages of the Gumbasa Irrigation Channel in Sibalaya due to liquefaction

**Fig. 37** Failures of Irrigation Channel has caused flood in the mid of the liquefied soils and washed out the road, the buildings and paddy fields

**Fig. 38** The extend of liquefaction flow failures on the villages

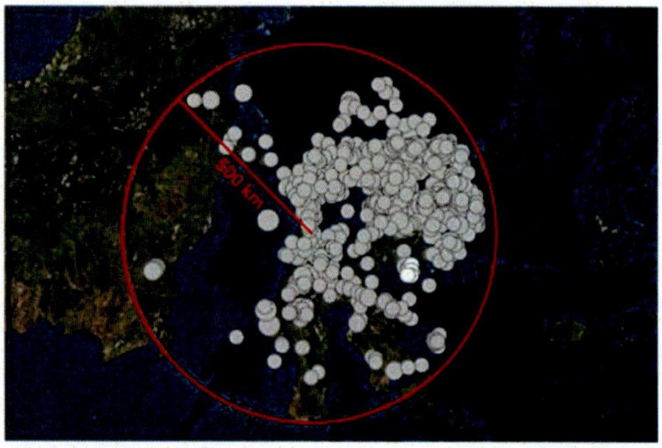


**Fig. 39** Earthquake data from 1900–2019 with M > 5 within 500 km from Palu (data from USGS)

$$CSR_{M,\sigma'_v} = 0.65 \frac{\tau_{max}}{\sigma'_v} \frac{a_{max}}{g} r_d \quad (1)$$

where $\sigma'_v$ = *vertical effective stress* at depth z, $a_{max}/g$ = *maximum horizontal acceleration (as a fraction of gravity)* at ground surface and $r_d$ = reduction factor due to soil deformability. The liquefaction resistance can be

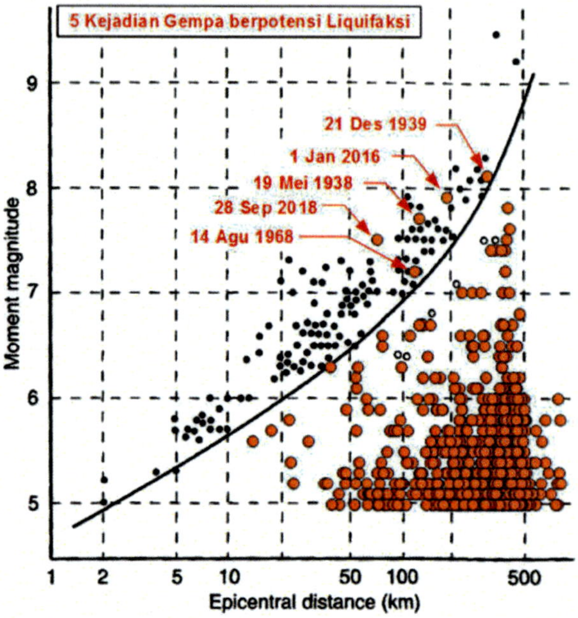


**Fig. 40** Historical earthquakes in Palu plotted in Ambrasseys graph

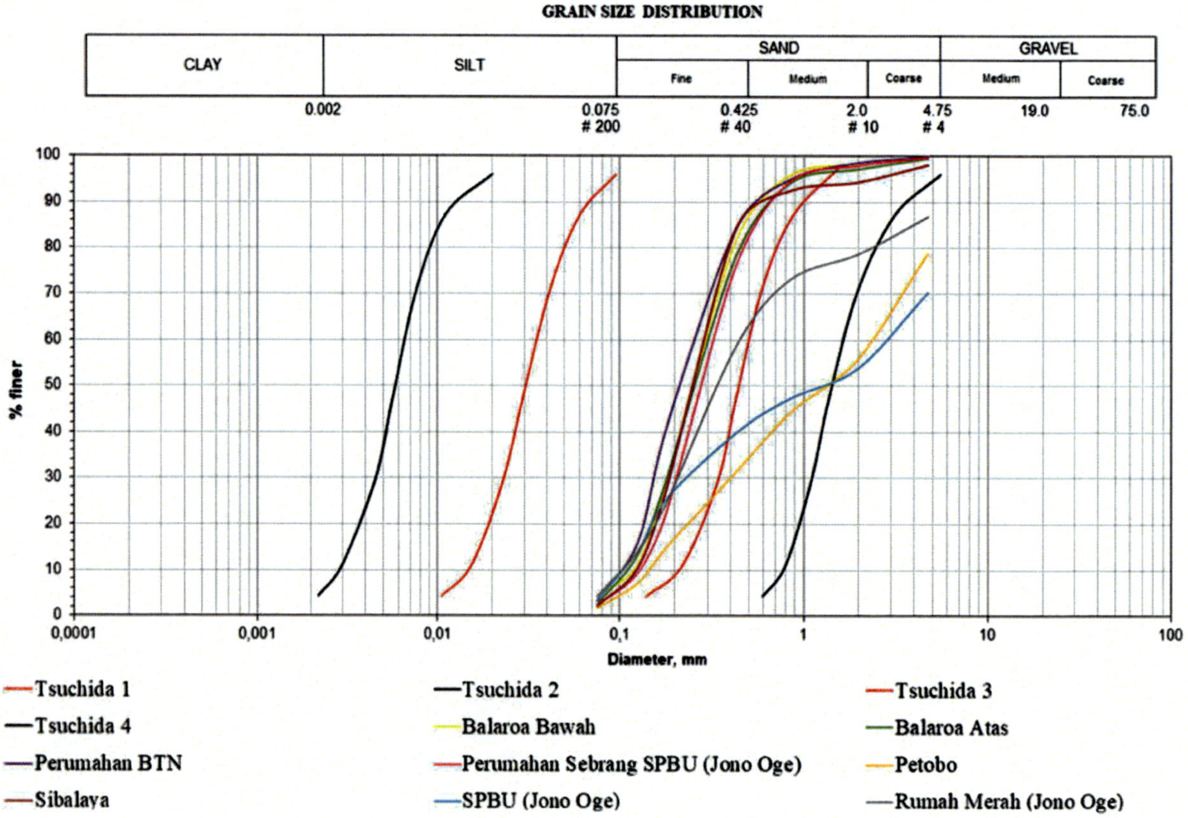


**Fig. 41** Selected grain sizes of sand samples in Palu, plotted in Tsuchida charts (1970)

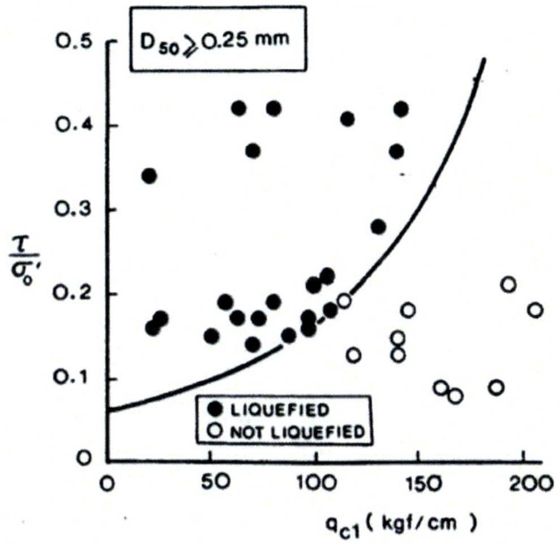


**Fig. 42** Method to obtain CRR using CPT (Shibata and Teparaksa 1988)

represented by Cyclic Resistance Ratio (CRR) as shown in Fig. 42 suggested by Shibata and Teparaksa (1988).

In the end, CSR is compared to CRR as shown on Fig. 42. All areas in Palu, Donggala and Sigi Regency, CPT have been conducted and the results were plotted on Palu map to show the liquefaction potential index (LPI). The liquefaction potential can be identified by the value of LPI as no liquefaction (LPI = 0), low potential (0 < LPI < 5), moderate (5 < LPI < 15) and high (LPI > 15). Settlement and Liquefaction Severity Index are also calculated and it shows there is consistency of the results and the field condition, where liquefaction potential index is high, the damage is also severe.

The results of analysis were then tabulated such as seen on Table 1 where at each location of CPTu, the LPI, LSI and settlement are shown (Fig. 43).

**Table 1** Results of liquefaction potential analysis in Palu-Donggala earthquakes

| Pengujian | Magnitudo Gempa (Mw) | Jarak Episenter (km) | Akselarasi (g) | LPI | Settlement (cm) | LSI |
|---|---|---|---|---|---|---|
| CPTu-01 | 6.37 | 12.181 | 0.245 | 15.60 | 30.63 | 5.40 |
| CPTu-02 | 6.37 | 12.579 | 0.243 | 6.44 | 11.28 | 5.10 |
| CPTu-03 | 6.37 | 11.351 | 0.251 | 1.25 | 2.21 | 6.17 |
| CPTu-04 | 6.37 | 11.337 | 0.251 | 20.67 | 29.69 | 6.18 |
| CPTu-05 | 6.37 | 8.649 | 0.268 | 1.03 | 1.34 | 10.23 |
| CPTu-06 | 6.37 | 8.648 | 0.268 | 15.04 | 16.42 | 10.23 |
| CPTu-07 | 6.37 | 8.816 | 0.267 | 15.20 | 17.71 | 9.87 |
| CPTu-08 | 6.37 | 19.474 | 0.201 | 1.15 | 2.02 | 2.26 |
| CPTu-09 | 6.37 | 18.926 | 0.204 | 5.70 | 8.38 | 2.38 |
| CPTu-10 | 6.37 | 13.535 | 0.237 | 8.69 | 17.37 | 4.45 |
| CPTu-11 | 6.37 | 17.781 | 0.211 | 11.58 | 13.63 | 2.68 |
| CPTu-12 | 6.37 | 20.129 | 0.198 | 0.00 | 0.00 | 2.13 |
| CPTu-13 | 6.37 | 6.909 | 0.22 | 0.00 | 0.00 | 7.05 |
| CPTu-14 | 6.37 | 6.487 | 0.222 | 0.00 | 0.00 | 7.93 |
| CPTu-15 | 6.37 | 19.853 | 0.199 | 20.93 | 34.81 | 2.18 |
| CPTu-16 | 6.37 | 17.984 | 0.209 | 2.39 | 4.21 | 2.62 |

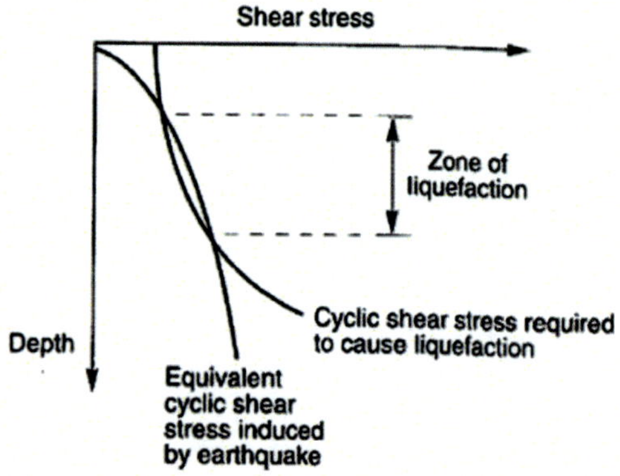


**Fig. 43** Assessment of liquefaction potential by simplified method (Seed and Idriss 1967)

## Conclusion Summary

(1) Liquefaction phenomena in Palu, Donggala and Sigi is a specific and spectacular where all liquefaction mechanism is represented such as sand boils, lateral spreading and flow liquefaction.

(2) The Gumbasa Irrigation has been one of significant contribution of liquefaction due to creating high water table, this plan may need to be reviewed for future or land use may be limited

(3) Mapping of liquefaction potential in the city is vital to direct the people to grow awareness as well as looking for safe locations and new policy on liquefaction hazard. Government must regulate and control. In the future for building permit, at least CPT is the minimum requirement and at least liquefaction potential can be done.

(4) A number of CPTu dan drillings were conducted and used for analysis, some data show high potentials of liquefaction and some other proved to be not susceptible to liquefaction. The results are consistent with the real field condition

(5) The extraordinary distance of liquefaction flow and lateral spreading is one of unique phenomena which is believed to be caused by the existence of initial artesian pressure and significant vertical acceleration causing the soil loosing contact stress.

## References


Ambraseys NN, Menu JM (1988) Earthquake induced ground displacement. Earthq Eng Struct Dyn 16(17):985–1006

BMKG (2018) Ulasan Guncangan Tanah Akibat Gempa Bumi Donggala, 28 September 2018. Bidzzang Seismologi Teknik

Ishihara (2019) Flow and lateral spreads of liquified ground following earthquakes. In: Lecture international conference on landslides and slope stability

Kramer SL (1996) Geotechnical earthquake engineering. Prentice Hall, New Jersey

Pusat Studi Geoteknik Unpar (2019) Menyelisik Untaian Bencana Palu Sigi Donggala (Lessons learned from Palu Sigi Donggala Earthquake 28 September 2018, in bahasa Indonesia, report of Geotechnical Research Institute, Universitas Katolik Parahyangan

Rahardjo PP (1989) Evaluation of liquefaction potential of silty sands based on cone penetration test. Dissertation submitted to Virginia Tech University, USA, as partial fulfilment for Ph.D. degree

Rahardjo PP (2019) Liquefaction phenomena in Palu Donggala earthquake of 28 September 2018. In: Proceedings, seminar on rehabilitation and mitigation for palu earthquake, Bandung 25 January 2019

Seed HB, ldriss (1967) Analysis of soil liquefaction: niigata earthquake. J Soil Mech Found Eng ASCE 93(SM-3):83–108

Shibata T, Taparaksa W (1987) Evaluation of cpt-based liquefaction assessment method using cyclic triaxial test. In: Proceedings of the 9th South East Asian geotechnical conference Bangkok

Shibata T, Taparaksa W (1988) Evaluation liquefaction potentials of soils using cone penetration tests. Soil Found 28(2):49–60

Thein P, Pramumijoyo S, Brotopuspito K, Kiyono J, Wilopo W, Setianto A (2014) Microtremors HVSR correlation with sub surface geology and ground shear strain at Palu City, Central Sulawesi Province, Indonesia. Int J Innov Sci Math

Tsuchida (1970) Prediction and counter measure against the liquefaction in sand deposits. In: Seminar on port and harbour and research institute, pp 3.1–3.33

Walpersdorf A, Rangin C, Vigny C (1998) GPS compared to long-term geologic motion of the north arm of Sulawesi. Earth Planet Sci Lett 159:47–55

Youd TL (1984) Recurrence of liquefaction at the same site. In: Proceedings of the 8th world conference on earthquake engineering, vol 3. Prentice Hall, Inc. Englewood Cliffs, NJ

# The Krasnogorsk Landslide (Northern Caucasus): Its Evolution and Modern Activity

Igor K. Fomenko, Oleg V. Zerkal, Alexander Strom, Daria Shubina, and Ludmila Musaeva

**Abstract**

This article describes large landslide on the left-bank slope of the Kuban River Valley, Northern Caucasus, Russia. Its evolution along with the conditions of its triggering and activation are described. Landslide that took place in 2016 was just is a partial reactivation of a much larger ancient landslide. While modern landslide was triggered by the climatic factors, slope stability assessment allows assumption that formation of the ancient landslide could be induced by seismic activity.

**Keywords**

Slope stability • Landslide hazard assessment • Seismically induced landslides • Northern Caucasus

## Introduction

The Krasnogorsk landslide is located 40 km to the south of the Cherkessk City on the left side of the Kuban River valley, where it crosses the elevated foothill plains of the Northern Caucasus. Since at least 2007 active slope deformations affected an area of about 6 hectares. Landslide was activated, most likely, in 2016 when it had moved rapidly and its "tongue" part reached the river bank (Fig. 1). Possible further activation of the landslide displacements that might affect larger area creates potential danger of the partial river blocking.

The main tasks of the work carried out were to identify the main factors of landslide activity and assess the role of seismicity in the development of the slope deformations. Complex nature of the structure of the slope affected by landslide was revealed and it was shown that the modern landslide is just the activated part of a much larger ancient landslide (Fig. 2).

## Meteorological and Climatic Conditions and Their Role in the Landslide Activity

The climate of the region is formed under the influence of circulation processes in the southern zone of middle latitudes.

A characteristic feature of the region's climate is a relatively warm winter, with frequent change in cooling with retaining snow cover and thaws with complete melting of snow. The average annual air temperature is ca. +9.2 °C. Negative monthly average temperatures are observed from December to February. The coldest month is January and the hottest one is July. The temperature of the coldest day was −23 °C. In the warm season, the air temperature is 24.8–28.8 °C. The annual distribution of precipitation is uneven. The warmest period of the year (from April to October) accounts the largest amount of precipitation of 456 mm, about 79% of the annual amount, the cold period (November to March)—124.2 mm, about 21% (Fig. 3). The monthly average rainfall up to 94.8 mm was observed in June.

The average annual rainfall in the study territory is 580 mm (at the Cherkessk weather station, working since

I. K. Fomenko · D. Shubina (✉)
Ordzhonikidze Russian State Geological Prospecting University, Moscow, 117997, Russia
e-mail: ddshubina@gmail.com

I. K. Fomenko
e-mail: ifolga@gmail.com

O. V. Zerkal
Moscow State University, Moscow, 119991, Russia
e-mail: igzov@mail.ru

A. Strom
Geodynamics Research Center LLC, Moscow, 125008, Russia
e-mail: strom.alexandr@yandex.ru

L. Musaeva
JSC Mosoblhydroproject, Moscow District, Dedovsk, 143532, Russia
e-mail: lutikhonova@yandex.ru


V. Vilímek et al. (eds.), *Understanding and Reducing Landslide Disaster Risk*, ICL Contribution to Landslide Disaster Risk Reduction,
https://doi.org/10.1007/978-3-030-60319-9_3

**Fig. 1** Common view of the Krasnogorsk landslide. Road section affected by landslide is 120 m long

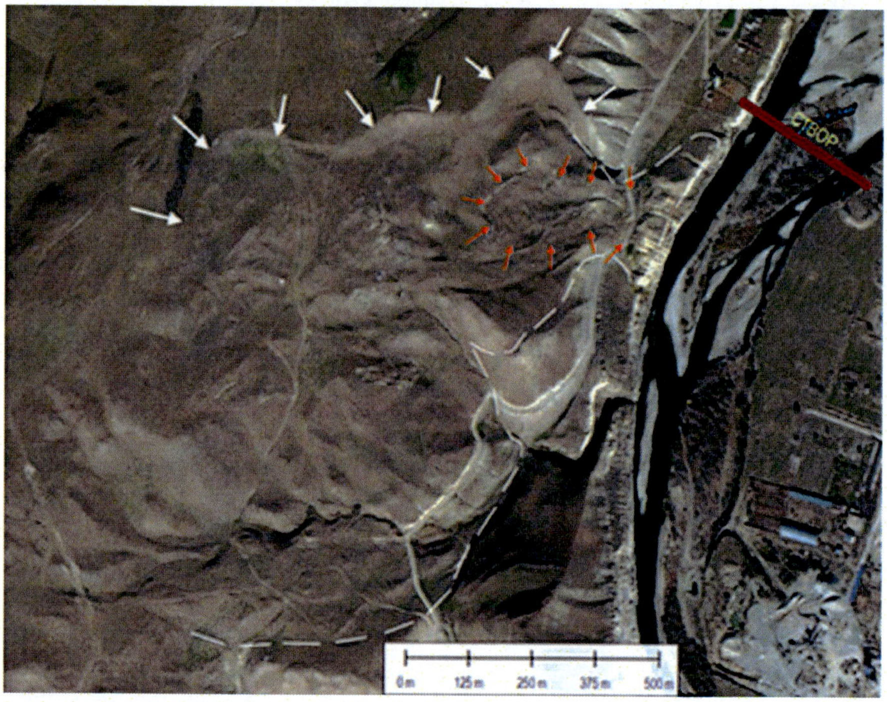


**Fig. 2** Borders of the modern (red arrows) and ancient (white arrows and dashed line) landslide bodies at the Krasnogorsk site. Bold brown line – the proposed dam site

1891). In recent decades there has been a significant increase in climate anomalies, including intensive rainstorms. For example, on June 20, 2002 the daily rainfall reached 99.4 mm (17% of the annual norm). Large yearly amount of precipitation was recorded in 2004 (Fig. 4). Shortly after, the first signs of slope destabilization within the Krasnogorsk site were recorded (Fig. 5).

The 2016, when landside activated, was rather rainy too. Another important factor of slope destabilization could be the seismicity, whose effect on the slope stability will be discussed hereafter.

## Geomorphological and Geological Conditions

The Kuban River is the largest one in the Northern Caucasus. Krasnogorsk landslide is located on the left side of the river valley in the transition zone from the Caucasus

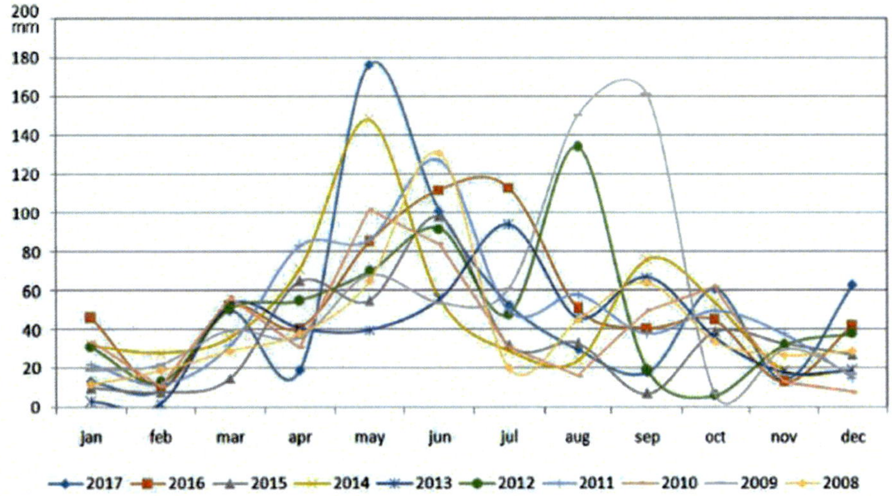


**Fig. 3** Monthly distribution of precipitation during the period of 2008–2017 according to the data from the Cherkessk weather station

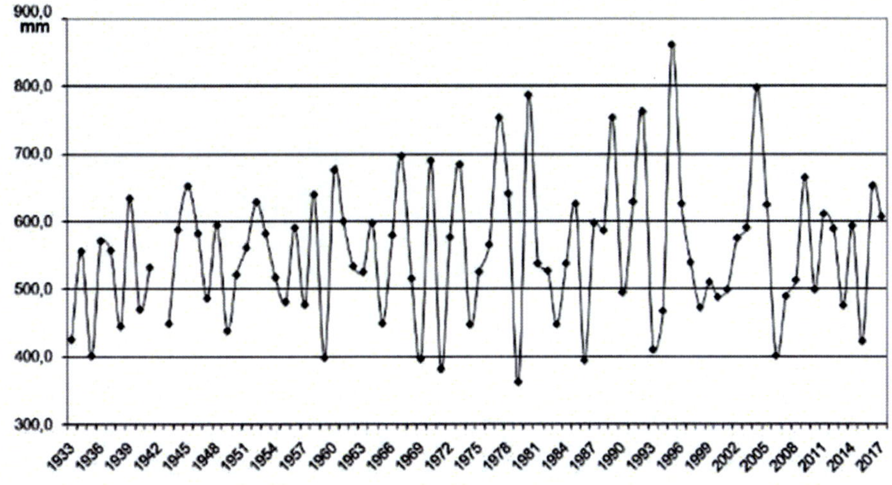


**Fig. 4** Average annual precipitation during the period of 1933–2017 according to the data from the Cherkessk weather station

Mountains to the foothill plains. The relief of the territory cut by the river valley can be classified as a cuesta.

The heights of the watersheds adjacent to the valley vary from 850–950 to 1100–1450 m. Local slopes are up to 200 m high and up to 30° steep. Just at the study section the left-bank slope of the Kuban River valley dips gradually towards the river that flows in the 400–500 m wide box-shaped canyon with almost vertical walls up to 40–50 m high composed of the bedrock. The opposite – right side of the valley is occupied by hundreds of meters wide strath terrace. The flow rate of the Kuban River ranges from 2 to 6 m/s.

The ancient landslide had affected the 200–250 m high slope over a length of up to 2.5 km. Landslide sliding surface nowadays daylights much above the flood plain and is not affected by floods.

The area is composed of the gently dipping, nearly horizontal Jurassic sediments (Sergeev 1978; Milanovsky and Koronovsky 1987; Shempelev et al. 2003). Lower Cretaceous sediments outcrop outside the landslide area (Fig. 5).

The Jurassic sedimentary rocks (of the Pliensbach, Aalenian and Bajocian stages) are represented by rhythmically interbedding gravelstone, sandstone, siltstone and mudstone with thin coal seams. Some layers are enriched by coaly detritus. Siltstone, often with a cleavage structure, amounts from 10 to 50% of the cross-section. They have significant porosity, high water-retaining capacity and soften with soaking.

The Aalenic strata overly the Pliensbach unit nonconformably. These sediments are slightly coarser than those of the Pliensbach stage. Sandstone and gravelstone of this age has calciferous highly ferruginized cement of the pore-filling to matrix-supported types.

The Bajocian strata up to 15 m thick compose the upper part of geological section. They consist of weak easily weathering dark-grey and black mudstones. Weathered rock

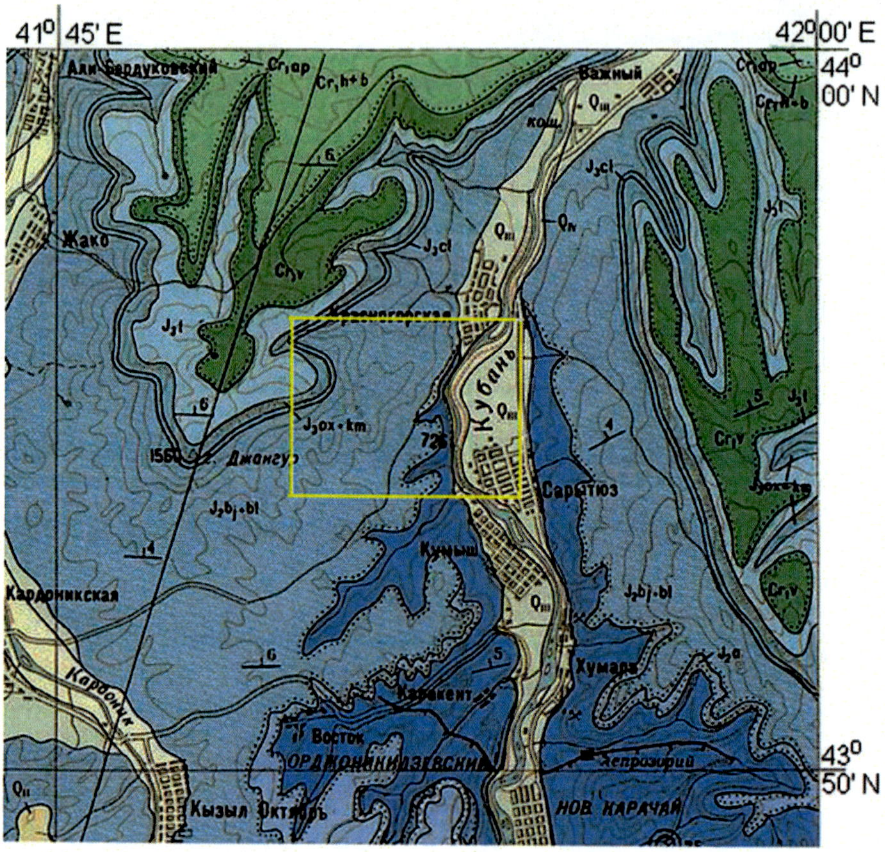


**Fig. 5** Fragment of the 1:200 000 State Geological map (sheet K-37-VI). Jurassic sedimentary rocks are in blue, Cretaceous rocks—in green, Quaternary deposits—in light yellow. Yellow rectangle marks landslide area shown in Fig. 2

at the top of the mudstone layer is composed of silty loams with inclusions (up to 20%) of siltstone debris.

Layers dip gently towards north. Few local steeply dipping faults of different orientation that were not shown on the State geological map (see Fig. 5) have been discovered at the study area during engineering-geological site investigations that have been carried out for the hydraulic scheme located just downstream from the landslide site (see Fig. 2). Besides, numerous fractures that form hierarchic block structure of the Jurassic sediments have been detected. Fractures' density increases along the fault zones.

The bedrock is overlain by the several meters thick talus and outwash. They are represented by brown loams with inclusions (up to 5–10%) of rubble and blocks.

The Jurassic sediments are water-saturated. Groundwater recharged by the infiltration through weathering cracks, was discovered by drilling at a depth of 2.4–35 m. It has sulphate-bicarbonate composition and extra pressure increasing with depth.

The study territory belongs to the seismically prone zone. The anticipated intensity of strong motion for unchanged soil conditions on the daylight surface was revealed in the following intervals (in the MSK-64 scale points):

- from 7.2 to 8.3 points with a 10% probability of the exceedance within 50 years;
- from 7.4 to 8.6 points with a 5% probability of the exceedance within 50 years;
- from 8.0 to 9.1 points with a 1% probability of the exceedance within 50 years.

## The Krasnogorsk Landslide Evolution

A detailed study of the general structure of the slope at the Krasnogorsk site showed that the modern landslide represents the small fragment of the much larger ancient landslide up to 1 km long and up to 450 m wide that was activated in the last decade (see Fig. 2). According to the drilling and geophysical data thickness of the ancient landslide body exceeds 30 m. The total volume of rocks involved in it is about 12 million cubic meters.

The modern active Krasnogorsk landslide is up to 400 m long (in the direction of displacement) and about 15–18 m thick. Its width is up to 200 m, tapering in the middle part to 120 m, and the total volume is about 1 million cubic meters.

The lower part of the landslide body is divided by a block of undisturbed rocks up to 30 m wide into two tongues 90 m and 70 m wide.

In the upper part, the landslide body is composed of chaotic blocks and rubble of Jurassic sediments (sandstones and siltstones) with silty filling of the interblock space. In the lower part landslide deposits are represented by intermittent Bayocian clay formations. Block inclusions are occasional.

According to the remote sensing data available, the first evidence of its activation has been recorded in 2007 when cracks appeared at its crown and at the lowermost part between the road and the river bank (Fig. 6). It was moving gradually with a slow rate and several years later, in 2013, the deforming zone became more evident, especially at the upper and lower parts of the slope (Fig. 7). It shows that modern landslide started developing as a planar block slide (Hungr et al. 2014). Later on, its motion style became more complex.

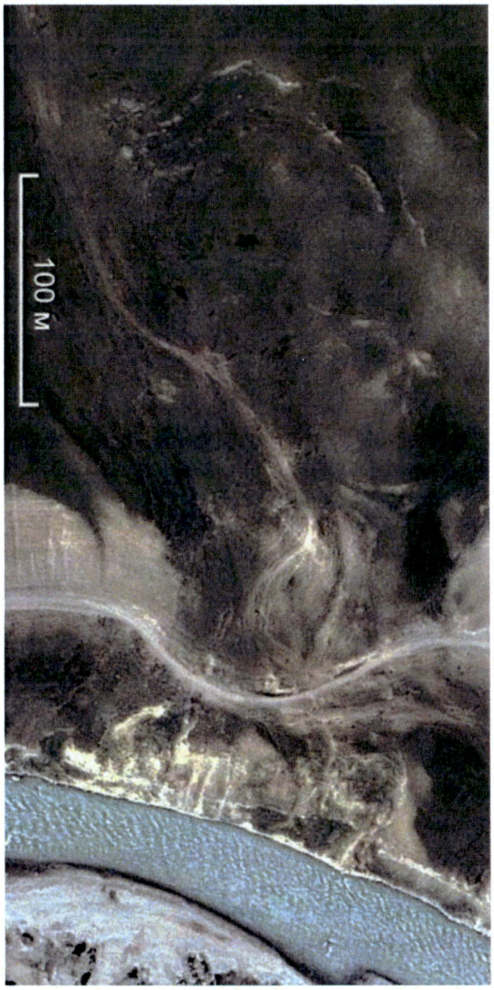


**Fig. 6** Space image of the Krasnogorsk landslide made on July 13, 2007

**Fig. 7** Space image of the Krasnogorsk landslide made on February 2, 2013

In Summer, 2016 (the exact date is unknown) it activated, destroying the road (that was rebuilt later) and its tongue fell from the 45–50 m high cliff on the floodplain (Fig. 8). We should notice that during these years the Kuban River repeatedly changes its course within the box-like canyon irrespective of landslide.

Further activation of the modern landslide could cause entry of the significant volume of debris (about 0.3 million $m^3$) into the Kuban River. If larger parts of the ancient landslide will be activated, the volume of the material that can be removed into the stream will significant increase that could cause:

- partial blocking of the stream with its reformation (formation of additional branches) and its shift towards the opposite (right) bank of the valley, which will cause the

**Fig. 8** Space image of the Krasnogorsk landslide made on September 10, 2016, after its last activation

sudden activation of erosion processes and the retreat of the bank;
- intensive siltation of the reservoir of the Krasnogorsk HPP that is under construction downstream with negative effects on its operation and energy production.

Apparently, the similar situation occurred in the considered section of the Kuban river valley in the past, at the end of the Late Pleistocene, being caused by the movement of an ancient landslide. Small remnants of its deposits are still visible on the ca. 40 m high left-bank terrace aside the tongue of the modern landslide.

# The Slope Stability Assessment

## Methods and Data

Peculiarities of the geological conditions and the structure of the slope can significantly affect the result of assessing the rockslides stability (Zerkal and Fomenko 2016). In this regard, when compiling the geomechanical model of the slope, special attention was paid to the geological structure of the area were the Krasnogorsk landslide has been formed.

The slope stability was quantified by the Morgenstern-Price method (Morgenstern and Price 1965).

The seismic impact was taken into account by the Newmark method according to the "coupled analysis of flexible blocks slip" scheme (Newmark 1965). Two synthetic accelerograms were used as input data for modelling of the seismic effect – with a maximum peak acceleration of 0.33 and 0.44 g correspondingly (Fig. 9).

## Results of the Slope Stability Modelling

Calculations performed by the Morgenstern-Price method showed that the lower and the upper zones can be identified

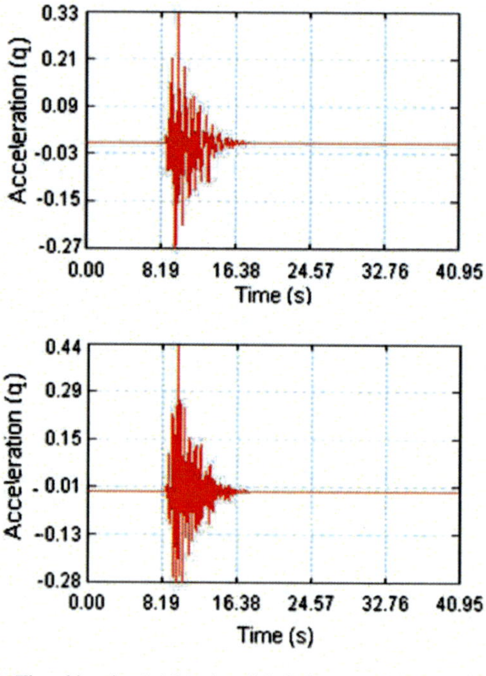


**Fig. 9** Time histories 1 (above) and 2 (below) used for calculations

at the Krasnogorsk landslide site. Considering rock properties characteristics derived during site investigations, the lower landslide zone, corresponding to the modern landslide, is unstable even without seismic impact. However, despite the low strength of the material, the upper part of the slope remains stable.

The slope stability quantitative assessment taking into account the seismic effect showed that slope loses its stability both in the lower and upper parts even with a horizontal acceleration about 0.003 g only.

The calculations showed that under seismic impact with PGA 0.33 g, the magnitude of seismically induced landslide displacements for the upper part of the slope can reach 28.58 cm (Fig. 10), and 17.59 cm for the lower part of the slope (Fig. 11). At the same time, under seismic action with PGA of 0.44 g and higher frequency, the calculated values

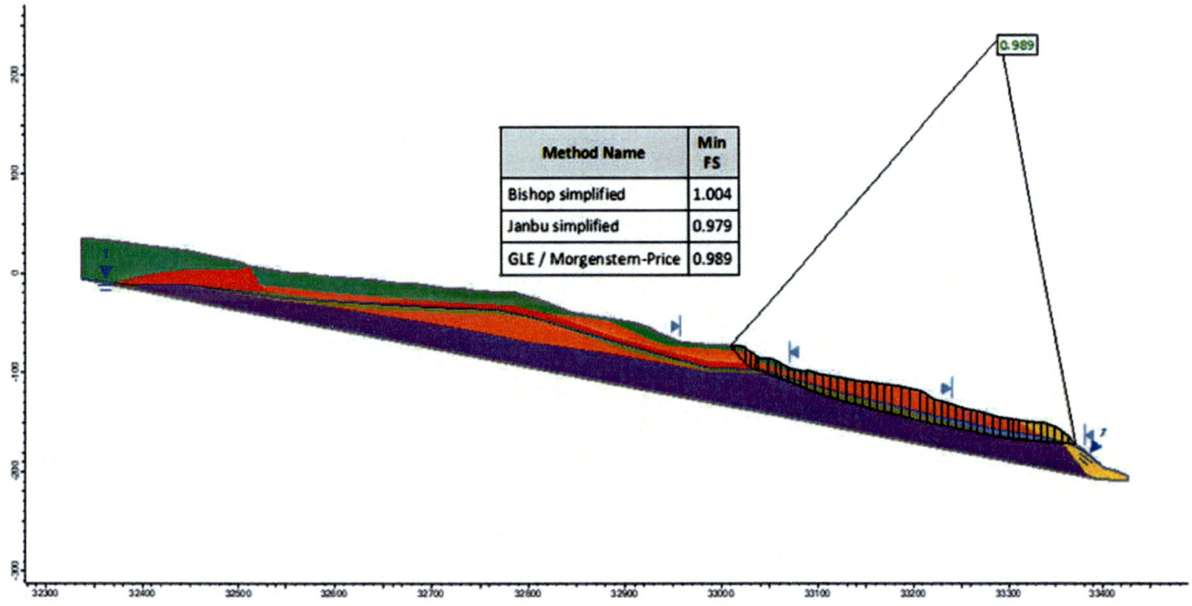


**Fig. 11** Stability assessment results obtained by the Newmark method for the lower part of the slope (with a maximum peak acceleration of 0.33 g)

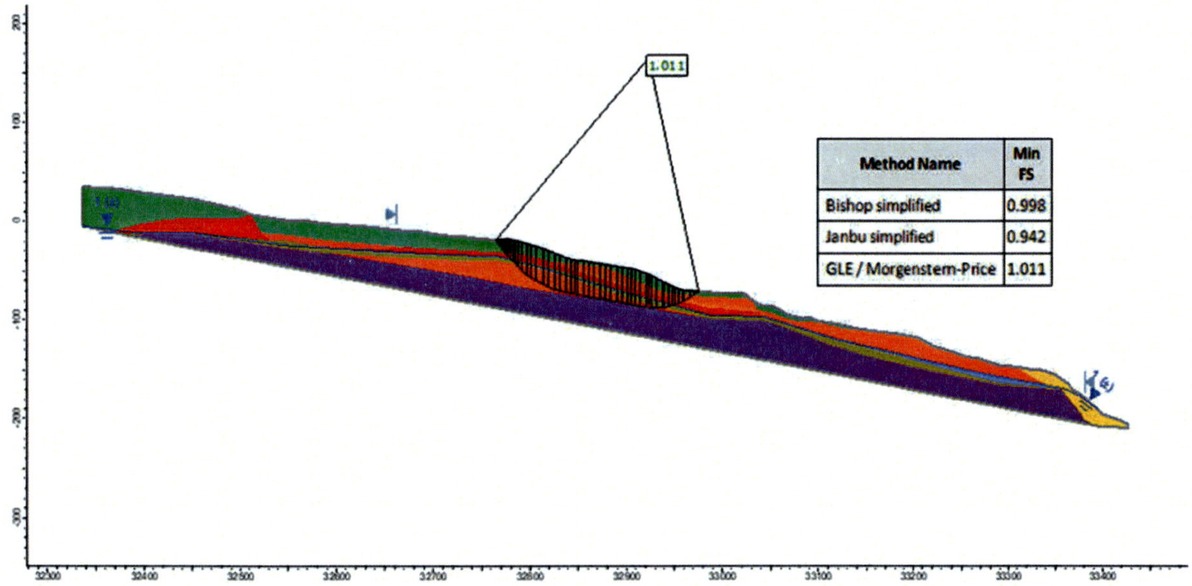


**Fig. 10** Stability assessment results obtained by the Newmark method for the upper part of the slope (with a maximum peak acceleration of 0.33 g)

of seismically induced landslide displacements were lower and amounted to 16.82 cm and 8.98 cm for the upper and lower parts of the slope.

These results are not trivial, since, despite larger PGA value of the time history 2 the resultant displacements are lower, than those based on time history 1. Thus, the latter one should be used to estimate slope stability. It demonstrates that the PGA value itself is not the deterministic one for slope stability calculations and that duration and spectral characteristics of strong motion are quite important too.

It casts doubt on use of the pseudostatic analysis (recommended by the Eurocode) for slope stability assessment (see also Biondi et al. 2007). Three values should be used, at a minimum, for comparison of the pseudostatic calculations with the calculations performed by use of the Newmark method: 1—peak horizontal acceleration ($k_{h,max}$); 2—acceleration at which the displacement starts but overall slope remains stable ($k_{h,eq}$); 3—calculated threshold acceleration when slope becomes unstable ($k_{h,c}$).

According to such approach number of peaks on the acceleration time history with accelerations exceeding (in modulus) the $k_{h,eq}$ value (N) multiplied by $k_h/k_{h,eq}$ ratio is more important than the PGA value itself.

It results in accumulation of displacements that might exceed displacement produced by single PGA pulse (if the latter alone does not cause slope failure).

## Conclusions

The present landslide site study showed that the modern landslide has been formed within the ancient landslide body due to its partial activation.

Study of factors supporting slope failure at the Krasnogorsk site showed that the modern landslide is associated with climate factors. Modern deformations develop within much larger ancient landslide whose main part is still stable.

The quantitative slope stability assessment shows that overall slope destabilization at the Krasnogorsk site could be triggered by the seismic strong motion only. It allowed conclusion that the ancient Krasnogorsk landslide could be considered as seismically triggered.

Another interesting research's result is lower values of seismically induced slope deformations obtained by the Newmark method for higher values of horizontal peak ground acceleration. Results of calculations performed show that the peak acceleration value alone is not the determinative for assessing stability of slopes affected by seismic strong motion.

## References

Biondi G, Cascone E, Rampello S (2007) Performance-based pseudo-static analysis of slopes. In: Proceedings of 4th international conference on earthquake geotechnical engineering, Greece

Hungr O, Leroueil S, Picarelli L (2014) Varnes classification of landslide types, an update. Landslides 11:167–194

Milanovsky EE, Koronovsky NV (ed) (1987) Geology and minerals of the Greater Caucasus - the science. Moscow, 268p. (in Russian)

Morgenstern NR, Price VE (1965) The analysis of the stability of general slip surfaces. Geotechnique 15(1):79–93

Newmark N (1965) Effects of earthquakes on dams and embankments. Geotechnique 15(2):139–160

Sergeev EM (ed) (1978) Engineering geology of the USSR, vol. 8. Caucasus, Crimea, Carpathians. Publishing House of Moscow University, Moscow, 366p. (in Russian)

Shempelev AG, Shvets AI, Zolotov EE, Feldman IS (2003) Geological and geophysical section along the Kuban profile. Tectonics and geodynamics of the continental lithosphere: materials XXXVI tectonic meeting. T. 2. GEOS. Moscow. S. 301–305 (in Russian)

Zerkal OV, Fomenko IK (2016) Landslides in rocks and their stability assessment. Engineering Geology, 4:4–21 (in Russian)

Check for updates

2020 Kyoto Japan

# Earthquake-Triggered Landslides and Slope-Seismic Waves Interaction Inferring Induced Displacements

Salvatore Martino, Celine Bourdeau, Josè Delgado, and Luca Lenti

**Abstract**

Earthquake-induced landslide mass mobility is a topic of particular relevance for the analysis of earthquake-induced ground effects scenarios. The landslide masses already existing on the slopes interact with the seismic waves that propagate from the bedrock, giving rise to effects of amplification of the seismic motion at specific frequencies connected to their geometry and their dynamic properties. The quantification of the earthquake-induced displacements expected in landslide masses through numerical models under dynamic conditions highlights how, especially for medium-low energy levels of the seismic input, the displacements thus obtained are generally higher than those computed by conventional approaches (e.g. Newmark method applied to the hypothesis of rigid or deformable block and related semiempirical relations). A series of case studies has also proved that the geometry of significantly dislodged landslide masses (i.e. segmented into kinematically distinct portions, namely "blocks") due to their geomorphological evolution over time, significantly controls the seismic-induced displacements obtained by numerical models. In particular, the results highlight that the maximum displacements computed through the numerical models do not correspond to seismic inputs whose characteristic periods coincide to those of the resonance or of the length of the landslide mass but are more directly connected to the smaller dimensions of the individual blocks in which the landslide mass is segmented.

**Keywords**

Earthquake-induced landslides • Earthquake-induced displacements • Numerical modeling

## Introduction

The evaluation of displacements associated with landslides triggered by earthquakes is one of the main objectives to provide scenarios of earthquake-induced effects even on large areas, which can be used for an appropriate planning and management of the territory. This requirement is particularly relevant in areas of intense urbanization, where civil constructions and lifelines (such as transport routes, aqueducts, gas pipelines, etc.) can be significantly damaged by the reactivation of landslides associated with earthquakes.

In this specific case, the damage induced by these landslides is combined to the effect of the shaking causing a significant increase in the risk associated with the seismic event. Already at the beginning of the 2000s, Bird and Bommer (2004), based on world-wide statistics, found that the damage caused by landslides triggered by earthquakes can be mostly associated with infrastructural elements widely distributed throughout the territory, while seismic shaking in itself is responsible for major damage to individual buildings. Starting from the second half of the last century, numerous approaches have been proposed to evaluate displacements induced by earthquake-triggered landslides. Ambraseys and Srbulov (1995) highlighted the differentiation between coseismic displacement (also called

S. Martino (✉)
University of Rome "Sapienza" – Earth Science Department and Research Centre for Geological Risk (CERI), P.le a. Moro 5, 00185 Rome, Italy
e-mail: salvatore.martino@uniroma1.it

C. Bourdeau · L. Lenti
Institut Français des Sciences et Technologies des Transports, de l'Aménagement et des Réseaux (IFSTTAR), GERS-SRO, Univ Gustave Eiffel, 77447 Marne-la-vallée, France
e-mail: celine.bourdeau@ifsttar.fr

L. Lenti
e-mail: luca.lenti@ifsttar.fr

J. Delgado
Dpt. Ciencias de la Tierra y del Medio Ambiente, Universidad de Alicante, Ap. Correos 99, 03080 Alicante, Spain
e-mail: jose.delgado@ua.es


V. Vilímek et al. (eds.), *Understanding and Reducing Landslide Disaster Risk*,
ICL Contribution to Landslide Disaster Risk Reduction,
https://doi.org/10.1007/978-3-030-60319-9_4

primary displacement) and post seismic displacements (secondary and tertiary), the latter due mostly to the dissipation of pore water overpressures generated in the soil during the seismic shaking (primary consolidation) followed by post seismic inertial settlements (i.e. due to creep process). The mechanical motivations of this temporal distribution of earthquake-induced displacements justify that the secondary and tertiary earthquake-induced displacements are higher than coseismic ones (even of an order of magnitude).

The coseismic displacement due to sliding is traditionally calculated using approaches that assume an effective seismic action above a critical acceleration threshold, kpedex (Newmark 1965). Such displacements can be conditioned by wave resonance phenomena within landslide mass if a viscoplastic rheology is assumed for landslide mass instead of the classic rigid-body behavior (Rathje and Bray 2000). A recently published review concerning the study of landslides triggered by earthquakes (Fan et al. 2019) essentially refers to the possibility of using these approaches to obtain earthquake-induced landslide based on probabilistic scenarios on a large scale also by the use of equations that correlate the seismic action (i.e., magnitude) to the critical acceleration to obtain the expected cumulated coseismic displacement value. However, the approaches based on the quantification of the coseismic displacement so far cited are not able to consider the post seismic effect of the displacement of the mass of landslide that far exceeds that of the coseismic one and to take into account the effect related to the much more complex interaction of the mass of landslide to the propagation of seismic waves on the slope (Delgado et al. 2011; Lenti and Martino 2012). In this regard, an approach has been recently proposed, defined as Characteristic Period Based (CPB), aiming at quantifying the earthquake-induced displacement expected in a landslide mass performing a deterministic analysis that accounts for possible interactions with seismic inputs characterized by different physical properties, such as Arias intensity and the frequency content (Lenti and Martino 2003). The CPB approach allows a deterministic solution of the earthquake-induced landslide mobility, which outputs a distribution of horizontal displacements vs. the ratios between characteristic periods relating to the thickness ($T_s$) and to the length ($T_l$) of the landslide mass compared to the characteristic period of the input earthquake ($T_m$) obtained in agreement with Rathje and Bray (2000). This approach, already tested on individual case studies (Martino et al. 2016; Bourdeau et al. 2017) is reproposed here with the aim of comparing the expected earthquake-induced displacements with those obtained with conventional analytical and semiempirical approaches as well as highlighting that, for more complex landslide geometries, the maximum expected displacements do not correspond to the basic parameters foreseen by the CPB approach.

## Case studies

Among the four landslide cases considered here (Fig. 1), two have undergone historical reactivations following earthquakes. Three cases are located in Andalusia (Spain), and are represented by the landslides of Diezma, Güevéjar and Albuñuelas; Güevéjar was reactivated by the Lisbon, 1755, and Andalucia, 1884, earthquakes (epicentral distances of >500 km and 50 km, respectively), while Albuñuelas was reactivated by the Andalucia earthquake (epicentral distance of 45 km). The fourth case is located in Turkey at the locality of Büyükçekmece, about 20 km west of Istanbul and about 15 km north of the North Anatolian Fault Zone which passes into the Sea of Marmara. The materials involved vary from shale in the case of the landslide of Diezma (Delgado et al. 2015) and of Büyükçekmece (Bourdeau et al. 2017) to sandy silts in the case of Güevéjar (Martino et al. 2016) to clayey silts in the case of Albuñuelas (Rodrıguez-Fernandez and Sanz de Galdeano 2006).

The volumes of landslides are of the order of one million cubic meters for Diezma and of dozens million cubic meters for the remaining ones. The landslide length (Fig. 2) varies from about 500 m for Diezma to about 700 m for Albuñuelas, about 1 km for Güevéjar and up to almost 2 km for Büyükçekmece. The maximum thickness of the landslide is about 20 m for Diezma, about 50 m for Güevéjar and Büyükçekmece and about 150 m for Albuñuelas. Given the shape of the landslide mass, only at Albuñuelas the rotational component of the movement is not negligible, which also caused the tilt of the houses in the historic center of the Albunuelas village (Fernandez et al. 2009).

## Results

The analysis of the earthquake-induced displacements for the four considered landslides was carried out by numerical modeling with the finite differences code FLAC 7.0 (Itasca 2011). Parameter values are derived from literature (Fernandez et al. 2009; Martino et al. 2016; Bourdeau et al. 2017). The procedure as well as the assumed boundary conditions followed the approach proposed by Lenti and Martino (2013) and involved the use of 15 seismic inputs

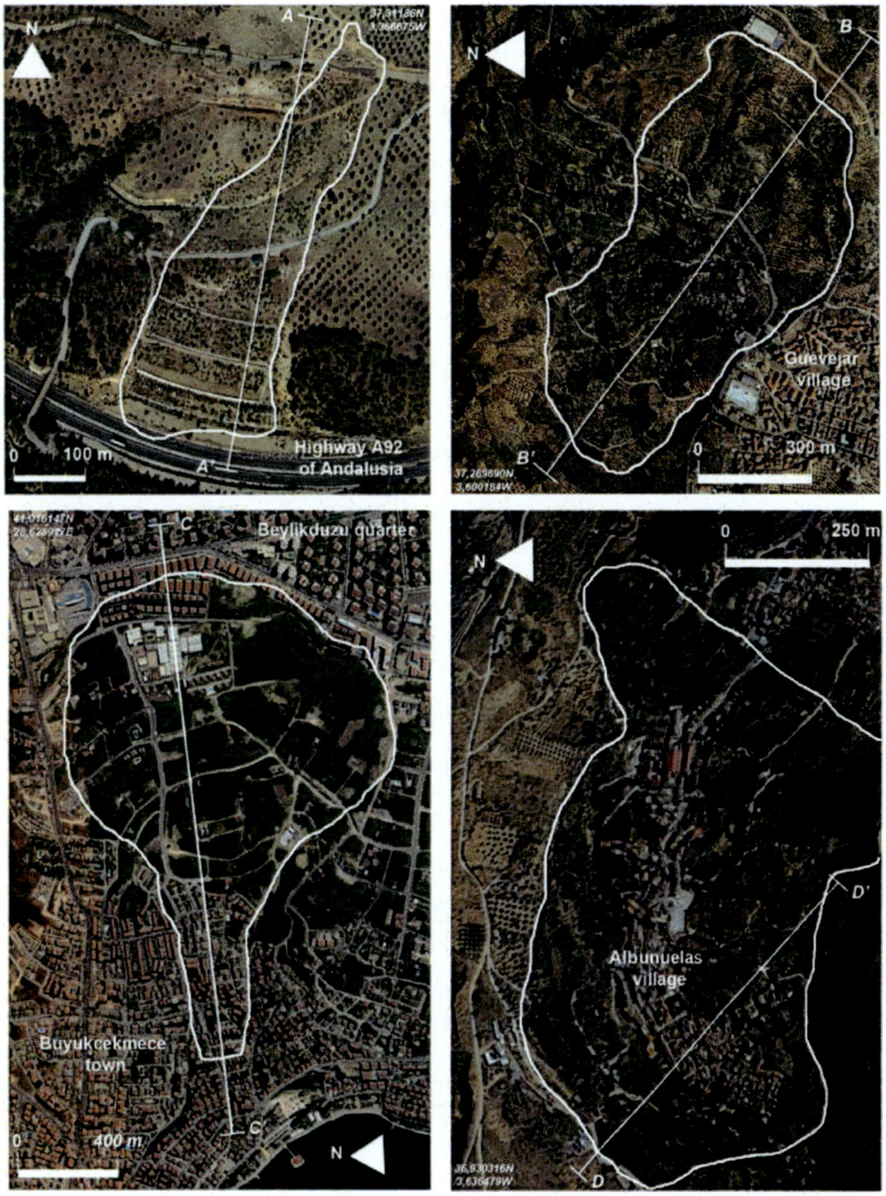


**Fig. 1** Satellite views of the here considered landslide case-studies (white contour) with traces of profiles reported in Fig. 2. Top left: Diezma (Spain); Top right: Güevéjar (Spain); Bottom left: Buyukcekmece (Turkey); Bottom right: Albuñuelas (Spain)

(selected from the global K-NET and COSMOS databases) that were applied at the base of the numerical domain as vertical upward-propagating SV stress waves. The selected inputs were characterized by Arias Intensity values of the order of 0.1 m/s and PGA values ranging from 0.04 to 0.35 g. The results obtained are shown in Fig. 3 which represent the distribution of the maximum horizontal displacement values (x-disp) obtained in at least 5% of the entire landslide mass as a function of the ratios between characteristic periods.

This choice gives the possibility to adopt the more conservative hypothesis on the extent of the displacements calculated by emphasizing the trend of landslide mass behaviour. The distribution of x-disp values vs. characteristic ratios reported in Fig. 3 reveals that the maximum x-disp is strictly related to the value of the two characteristic ratios $T_s/T_m$ (where $T_s$ was computed considering an average thickness of the landslide mass) and $T_l/T_m$. In general, it is possible to observe the presence of an absolute x-disp maximum and several relative maxima. The absolute

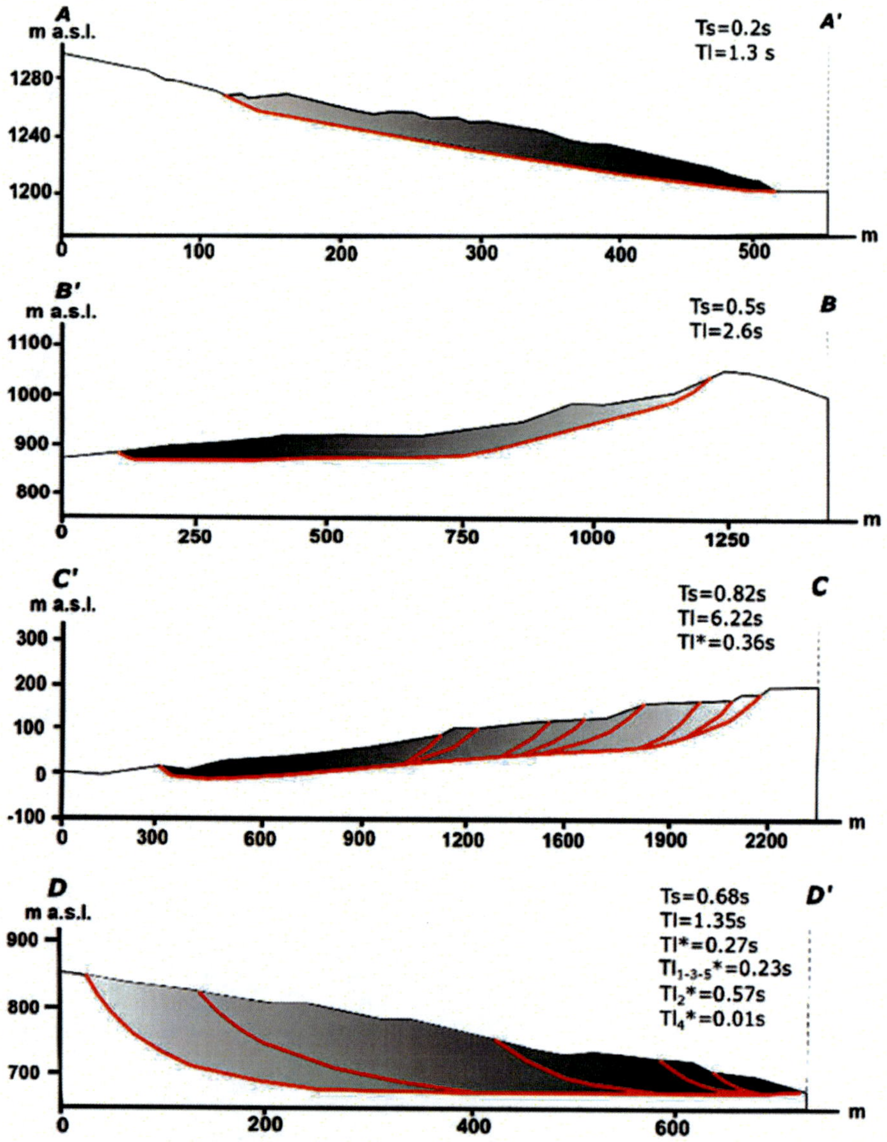


**Fig. 2** Landslide sections along traces of Fig. 1 with characteristic periods ($T_s$, $T_l$ and $T_l$*) reported (reference landslide blocks are numbered from the top to the bottom of the slope)

maximum generally coincides with lower values of both the characteristic ratios; in the case of Diezma and Güevéjar the peak of maximum expected displacement coincides with a characteristic ratio $T_l/T_m$ close to the theoretical value of 0.5 while in the cases of Büyükçekmece and Albuñuelas it is shifted to higher values of the same ratio which leads to consider a prevalent role of the block length (i.e. $T_l^*/T_m$ ratios). On the other hand, respect to the results reported by Lenti and Martino (2013), at the here considered Arias Intensity magnitude of the selected inputs, it is not possible to deduce from the distribution of maximum x-disp the effect connected to the resonance of the landslide mass thickness which, instead, should be related to a theoretical value of the characteristic $T_s/T_m$ ratio equal to 1.

## Discussion

Table 1 summarizes the results of the calculation of the coseismic displacements for the four landslides considered here taking into account different reports proposed in the literature by various authors. For all landslides they are considered PGA with a return time (Tr) of 475 years while for the landslides of Güevéjar and Albuñuelas only, for which a historical reactivation is known, PGA values refer to the last triggering event (Table 2).

In all the considered scenarios, displacement values are of the order of a few millimeters or they are null since the PGA is below the critical pseudostatic acceleration threshold. If

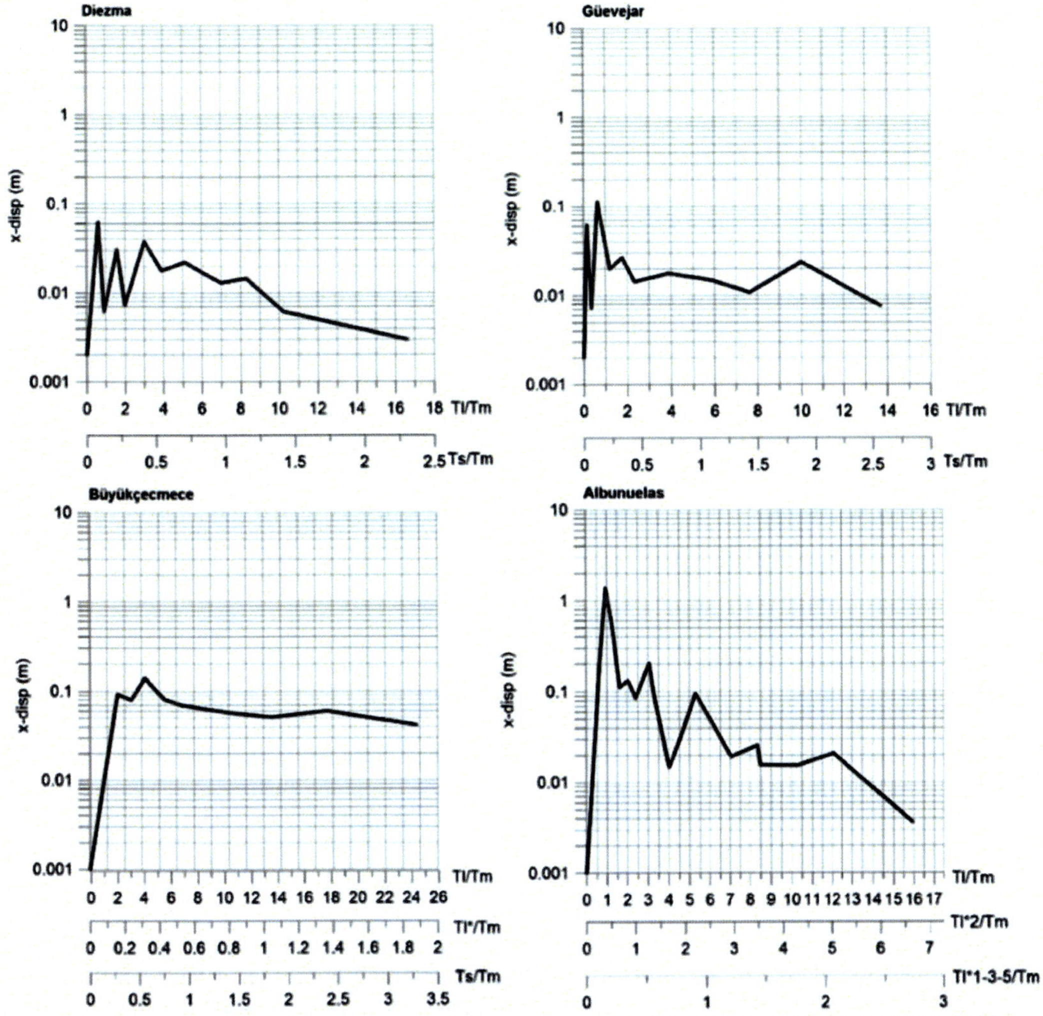


| Location | Date (yyyymmdd) | Epicentral distance (km) | Mw | PGA (m/s²) | Arias (m/s) | $T_m$ (s) |
|---|---|---|---|---|---|---|
| Tohoku (Japan) | 20110311 | 9700 | 8.9 | 4.4E-01 | 1.9E-01 | 16.50 |
| Alpi Apunane (Italy) | 20130621 | 300 | 5.2 | 4.0E-01 | 1.3E-01 | 7.37 |
| Serbia | 20101103 | 640 | 5.4 | 5.5E-01 | 1.7E-01 | 4.02 |
| Montenegro | 19790415 | 21 | 7.0 | 2.1E+00 | 7.3E-01 | 0.73 |
| Duzce (Turkey) | 19991112 | 11 | 7.2 | 7.3E-01 | 6.2E-02 | 0.16 |
| Friuli (Italy) | 19760911 | 9 | 5.6 | 2.3E+00 | 2.7E-01 | 0.64 |
| Izmit (Turkey) | 19990817 | 9 | 7.6 | 1.6E+00 | 7.1E-01 | 0.56 |
| Umbria Marche (Italy) | 19970926 | 3 | 5.7 | 3.4E+00 | 7.0E-01 | 0.50 |
| Umbria Marche (Italy) | 19980403 | 21 | 5.2 | 2.6E-01 | 1.2E-02 | 0.19 |
| South Iceland | 20000621 | 11 | 6.5 | 4.4E+00 | 1.2E+00 | 0.38 |
| Friuli (Italy) | 19760911 | 15 | 5.6 | 2.3E+00 | 3.7E-01 | 0.34 |
| Friuli (Italy) | 19760511 | 4 | 5.0 | 3.0E+00 | 3.7E-01 | 0.30 |
| Umbria (Italy) | 19840429 | 27 | 5.7 | 1.8E+00 | 1.4E-01 | 0.16 |
| Umbria Marche (Italy) | 19971016 | 1 | 4.4 | 3.4E+00 | 2.0E-01 | 0.08 |
| Kozani (Greece) | 19950513 | 17 | 6.6 | 7.7E-01 | 6.5E-02 | 0.20 |

**Fig. 3** Horizontal displacements versus characteristic periods referred to the here considered landslide case studies as they are defined in Fig. 2. Multiple x-axes are referred to different characteristic period ratios referred to: average thickness ($T_s/T_m$), total length ($T_l/T_m$) and block length ($T_l*/T_m$), each block is indicated with an arabic number

compared with the displacement values deriving from the dynamic analysis, the latter are two to three orders of magnitude higher (therefore from a few tens of centimeters to one meter) limited to the lowest values of ratios between the characteristic periods $T_l/T_m$ (generally less than 4). In case of higher values of that ratio, obtained coseismic displacements are of the same order of magnitude to those computed using the analytical relationships reported in literature. In the cases of the landslides of Diezma and Albuñuelas this is particularly evident when the ratio $T_l/T_m$

**Table 1** Coseismic displacements computed by SLAMMER (https://www.usgs.gov/software/slammer) for the landslides of Diezma (DZ), Güevejar (GV), Büyükçekmece (BU), Albuñuelas (AL) in case of earthquakes with Tr of 475 yrs (Arias Intensity in the order of $10^{-1}$ m/s), according to different literature relationships based on kpedex and PGA: Hsieh and Lee (2011) (R1); Saygili and Rathje (2008) (R2); Jibson (2007) (R3)

| Landslide | kpedex (g) | PGA(g) Tr (475yrs) | $R_1$ (cm) | $R_2$ (cm) | $R_3$ (cm) |
|---|---|---|---|---|---|
| DZ | 0.077 | 0.105 | 0.4 | 0.2 | 0.1 |
| GV | 0.180 | 0.150 | 0.0 | 0.0 | 0.0 |
| BU | 0.500 | 0.350 | 0.0 | 0.0 | 0.0 |
| AL | 0.500 | 0.110 | 0.0 | 0.0 | 0.0 |

**Table 2** Coseismic displacements computed by SLAMMER for the landslides of Güevejar (GV) and Albuñuelas (AL) for the last historical triggering earthquakes (Arias Intensity in the order of $10^{-2}$), according to different literature relationships: Hsieh and Lee (2011) (R1); Saygili and Rathje (2008) (R2); Jibson (2007) (R3)

| Landslide | kpedex (g) | PGA(g) Tr (475yrs) | $R_1$ (cm) | $R_2$ (cm) | $R_3$ (cm) |
|---|---|---|---|---|---|
| GV | 0.180 | 0.047 | 0.0 | 0.0 | 0.0 |
| AL | 0.500 | 0.043 | 0.0 | 0.0 | 0.0 |

exceeds 10, i.e. the period proper to the landslide mass is of one order of magnitude lower than that characteristic of the earthquake.

For both the Albuñuelas and Güevéjar landslides, historically reactivated by earthquakes with epicentral distances greater than 45 km, the maximum displacements obtained for low $T_l/T_m$ ratios justify such reactivation since high to very high $T_m$ should have forced the landslide masses as the lowest frequencies of the seismic waves reach greater distances from the epicenter.

The non-exact correspondence between the ratio $T_l/T_m$ equal to 0.5 and the maximum displacement obtained by the numerical modeling, allows to hypothesize that the earthquake-induced mobility of the landslide masses analyzed here is influenced by their internal structure, and in particular by the presence of portions (blocks) of the landslide mass having their own kinematic freedom and therefore predisposed to interact with the seismic input based on their own characteristic periods.

This could be the case for the Albuñuelas and Büyükçekmece landslides. Figure 3 shows for these two landslides a multiple x-axis in the mobility schedules; for the Albuñuelas landslide each axis is calculated for a period $T_l$ representative of one of the dimensions of the blocks present in the landslide ($T_l$* numbered from 1 to 5) while for both landslides an axis is reported relative to the average size of the blocks ($T_l$*). This comparison shows that for both landslides the maximum expected mobility is associated better with a $T_l^*/T_m$ ratio referred to the size of the blocks rather than the entire mass of landslide. Research perspectives include to study: (i) the role of heterogeneity within the landslide mass; (ii) more advanced solutions to compute an integral displacement for each landslide block, also considering the physics of seismic wave propagation inside a landslide mass; (iii) specific laboratory experiments to account for the role of pore water pressures (Setiawan et al. 2017).

## Conclusions

Earthquake-induced landslide mass displacements appear to be influenced by the geometric properties and the structure of the landslide mass, such as its subdivision into portions (blocks) with different kinematic freedom. In the four landslide cases considered here, the maximum displacements induced by earthquakes that can be calculated through a numerical analysis under dynamic conditions are greater up to three orders of magnitude if compared to those calculated by Newmark methods. In particular, in the case of the two landslides of Albuñuelas and Güevéjar, historically reactivated by earthquakes with epicentral distance greater than 40 km, the most significant mobility at low $T_l/T_m$ ratios at least partly justifies the attitude to this reactivation.

**Acknowledgements** The Authors wish to thank Ph.D. Danilo D'Angiò, Dr. Mara Mita, Dr. Alessandra Noviello, Dr. Manuela Palmas, Dr. Antonella Sacco and Dr. Fabiola Tammaro, for their support in the numerical modeling. This work was partially founded by the Spanish Minestry of Economy and EU FEDER funds (project EPILATES, CGL2015-65602-R).

## References

Ambraseys N, Srbulov M (1995) Earthquake induced displacements of slopes. Soil Dyn Earthq Eng 14:59–71

Bird JF, Bommer JJ (2004) Earthquake losses due to ground failure. Eng Geol 75:147–179

Bourdeau C, Lenti L, Martino S, Ozel O, Yalcinkaya E, Bigarrè P, Coccia S (2017) Comprehensive analysis of the local seismic response in the complex Büyükçekmece landslide area (Turkey) by engineering-geological and numerical modelling. Eng Geol 218:90–106

Delgado J, Garrido J, López Casado C, Martino S, Peláez JA (2011) On the far field occurrence of seismically induced landslides. Eng Geol 123:204–213

Delgado J, Garrido J, Lenti L, Lopez-Casado C, Martino S, Sierra FJ (2015) Unconventional pseudostatic stability analysis of the Diezma landslide (Granada, Spain) based on a high-resolution engineering-geological model. Eng Geol 184:81–95

Fan X et al (2019). Earthquake-induced chains of geologic hazards: patterns, mechanisms, and impacts. Rev Geophys. 1–83 https://doi.org/10.1029/2018RG000626

Fernandez P, Irigaray C, Jimenez J, El Hamdouni R, Crosetto M, Monserrat O, Chacon J (2009) First delimitation of areas affected by ground deformations in the Guadalfeo River Valley and Granada metropolitan area (Spain) using the DInSAR technique. Eng Geol 105:84–101

Hsieh SY, Lee CT (2011) Empirical estimation of the Newmark displacement from the Arias intensity and critical acceleration. Eng Geol 122(1):34–42

Hungr O, Leroueil S, Picarelli L (2014) The Varnes classification of landslide types, an update. Landslides 11:167–194

ITASCA (2011). FLAC 7.0: User manual, License number 213-039-0127-18973 (Sapienza University)

Jibson RW (2007) Regression models for estimating coseismic landslide displacement. Eng Geol 91:209–218

Lenti L, Martino S (2012) The interaction of seismic waves with step-like slopes and its influence on landslide movements. Eng Geol 126:19–36

Martino S, Lenti L, Delgado J, Garrido J, Lopez-Casado C (2016) Application of a characteristic periods-based (CPB) approach to estimate earthquake-induced displacements of landslides through dynamic numerical modelling. Geophys J Int 206:85–102

Newmark NM (1965) Effects of earthquakes on dams and embankments. Geotechnique 15(2):139–159

Rathje EM, Bray JD (2000) Nonlinear coupled seismic sliding analysis of earth structures. J Geotech. Geoenviron Eng ASCE 126 (11):1002–1014

Rodríguez-Fernández J, Sanz de Galdeano C (2006) Late orogenic intramontane basin development: the Granada basin, Betics (Southern Spain). Basin Res 18:85–102

Saygili G, Rathje EM (2008) Empirical predictive models for earthquake-induced sliding displacements of slopes. J Geotech Geoenviron Eng 134:790–803

Setiawan H, Sassa K, Takara K, Miyagi T, Fukuoka H (2017) Initial pore pressure ratio in the earthquake triggered large-scale landslide near Aratozawa Dam in Miyagi Prefecture, Japan. Poc Earth Plan Sci 16:61–70

2020 Kyoto Japan

# Slope Deformation caused Jure Landslide 2014 Along Sun Koshi in Lesser Nepal Himalaya and Effect of Gorkha Earthquake 2015

H. Yagi, G. Sato, H. P. Sato, D. Higaki, V. Dangol, and S. C. Amatya

**Abstract**

Although there is no commonly accepted methodology for evaluation of deep-seated landslide susceptibility, the progress on remote sensing technology such as aerial photos, satellite images, precise DEM and InSAR data have made it easier to detect ground movement at time interval and cost-effective. Combining these techniques and ground truth, the authors attempt to understand the nature of Jure landslide in Nepal that occurred due to heavy rain in August 2014. They have also studied the effect of Nepal Gorkha earthquake on the Jure landslide. Gravitational deformation, so-called "rock creep" has proceeded for long time forming thick weathering layer at the Jure landslide site. Small and shallow deformations have occurred in and around the site since a decade or before. However, significant enlargement of the landslide due to the earthquake in 2015 was not detected by InSAR because the quake occurred at the end of dry season.

**Keywords**

Gravitational creep • Deep weathering • Optical satellite image • Nepal gorkha earthquake 2015 • InSAR

H. Yagi (✉)
Yamagata University, Kojirakawa, Yamagata 990-8560, Japan
e-mail: yagi@e.yamagata-u.ac.jp

G. Sato
Teikyo Heisei University, Nakano, Tokyo, 170-8445, Japan
e-mail: gsato@thu.ac.jp

H. P. Sato
College of Humanities & Sciences, Nihon University, Sakura-josui, Tokyo, 156-8550, Japan
e-mail: satou.hiroshi37@nihon-u.ac.jp

D. Higaki
Hirosaki University (Formerly), Hirosaki, Japan
e-mail: a9024@n-koei.co.jp

D. Higaki
Nippon Koei Co., Ltd., Chiyodaku, Tokyo, 102-0083, Japan

V. Dangol
Tri-Chandra Campus, Tribvan University (formerly), Kathmandu, Nepal
e-mail: vdangol@yahoo.com

S. C. Amatya
Department of Water Induced Disaster Management (Formerly), Lalitpur, Nepal
e-mail: amatyasc@gmail.com

## Introduction

There is no generally accepted methodology for susceptibility evaluation of slope deformation such as deep-seated landslide. Gravitational rock creep is a long-term phenomenon. One of the typical morphology created by gravitational rock creep is high-relief mountain with gentle ridge top surrounded by steep valley slopes (Yagi and Hayashi 2017). Micro topography due to gravitational deformations, e.g. uphill-facing scarplets and linear depressions in ridge-top might be predictive signs of sudden collapse of mountain. They are detectable by aerial photo interpretation. However, the time scale of such micro topography sometimes ranges as long as a few ten ka (Yagi and Hayashi 2017). However, aerial photos are not taken each year. Hairline cracks or steps of a few ten centimeters on slope are direct signs of the slope movement, but such small displacement is also very difficult to detect on aerial photos even in the scale of 1/20,000.

Optical satellite images are taken periodically and their resolution has been steadily improved. And some of them are freely available at a website such as Google Earth Pro, which shows satellite images of disaster-affected areas in time series of several to ten years. Synthetic Aperture Radar interferometry images analysis (InSAR) data is recently introduced to detect slope deformation (Sato et al. 2013; Sato and Une 2016). InSAR image can detect slight displacement on the slope occurring in time scale of a few


V. Vilímek et al. (eds.), *Understanding and Reducing Landslide Disaster Risk*, ICL Contribution to Landslide Disaster Risk Reduction,
https://doi.org/10.1007/978-3-030-60319-9_5

months or much shorter period depending on the tracking interval of the satellite.

Combining the above-mentioned techniques of detecting slope movement in different time scales, the authors have tried to uncover a clue as to what predictive signs of deep-seated landslides have preceded on mountain slopes. The Jure landslide in Nepal Lesser Himalaya is chosen as a sample study site (Fig. 1).

## Outline of Jure Landslide in 2014

A large-scale landslide occurred at Jure village on the right bank of the Sun Koshi river, central Nepal on 2nd August 2014 (Fig. 1). It occurred during the mid-monsoon season, it rained 10 to 30 mm per day for a few days. There was heavy precipitation titaling 140 mm for five and six days before the event (Fig. 2).

The landslide area externally from the crown to the toe is as follows: width is 900 m, slope length is 1,500 m and max depth is 200 m. Elevation of the crown top is 1,550 m a.s.l (Fig. 2). The volume of the slide, however, is approximately only six million cubic meters, considering from a volume of the detritus burying the valley bottom of the Sun Koshi.

The sliding occurred in two phases at 1:00 AM and 2:36 AM (Nepal Government Ministry of Irrigation 2014). The latter one was larger and much destructive than the former one, causing a local quake in Richter scale of 3.3 (Nepal Government Ministry of Irrigation 2014).

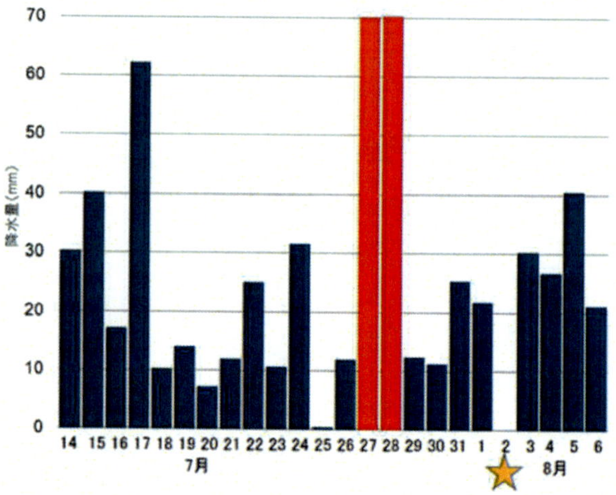


**Fig. 2** Precipitation in Jure area around the event

The landslide mass flushed down onto the valley bottom of 780 m a.s.l. and completely closed the Sun Koshi River channel and formed a landslide dam, the width, length and depth of which are 350 m, 67 5 m and 50 m, respectively. Overflow from the temporary lake initiated eleven and half hours later after the damming. Breaching and outburst of the natural dam occurred on 11th Sep, causing a debris flow. Sudden water level down to 20 m occurred. It damaged Sun Koshi barrage for a hydropower station located in two km downstream. Hydraulic bore caused by the outburst was observed fifty kilometers in downstream from the landslide

**Fig. 1** Jure landslide and its location

site and the water level of Sun Koshi River rose about two meters there (Pettley 2014).

The authors carried out field investigation around the landslide. 3D interpretation of aerial photos taken in the 1990s covering in and around the landslide was carried out to recognize topographic feature of the landslide site and the outline of topography along Sun Koshi and Bhote Koshi valleys. Time series changes of surface rupture in and around the landslide site were also carried out using optical satellite images of Google Earth after 2000 until May 2016 after the Nepal-Gorkha Earthquake 2015. Detected surface ruptures were digitized as polygons on Google Maps and turned into shape files. They were projected onto a detailed map of 10 m contour interval created from the AW3D (Advanced World 3D Topographic data) DEM before the landslide, which is based on the satellite observations from ALOS "Daichi" with a grid size of 5 m.

Nepal Gorkha Earthquake 2015 of Mw. 7.8 struck this area and its largest aftershock with a magnitude of 7.3 Mw hit it again, that triggered many landslides. The authors carried out an inventory mapping of the earthquake induced landslides by interpretation of satellite images of Google Earth. The result is also projected on a topographic map created by ASTER dataset. Slight displacement on the slope due to the earthquake was also detected by InSAR image analysis, using ALOS-2/PALASAR-2 data.

## Topography and Geology in and Around the Jure Landslide

Sun Koshi is one of the four main river systems in Nepal. Bhote Koshi, one of the uppermost stream of Sun Koshi, has incised the Great Himalayan range deeply, originated from the northern side of the Great Himalaya (Fig. 3). Relative height of Sun Koshi valley at the Jure site is ca 1,400 m and its mean slope angle is 30-50 degree. Profile of valley flank sometimes shows a shoulder shape called a break of slope or a nick point, forming an inner valley (Fig. 4). That crossing the Jure landslide site before the event was most distinct, showing stepped terrain, composed of steep and gentle slopes with three distinct breaks of slopes (Figs. 3 and 4). They are namely Lower Break, Middle Break and Higher Break in ascending order, respectively. They are locating at ca. 900–920 m a.s.l., 1,050–1,100 m a.s.l. and 1,400 m a.s.l. at the Jure site, respectively. The Lower Break separates a lower steep slope and a lower gentle slope. The middle Break separates a middle steep slope and a middle gentle slope. The upper Break also separates an upper steep slope and an upper gentle slope. The lower and middle gentle slopes sometimes include flat parts, consequently they are fluvial terrace origin or landslide origin. The lower gentle slope is an originally fluvial terrace, continuing from the upper course to the lower course. And it was covered by scree (colluvium of alluvial

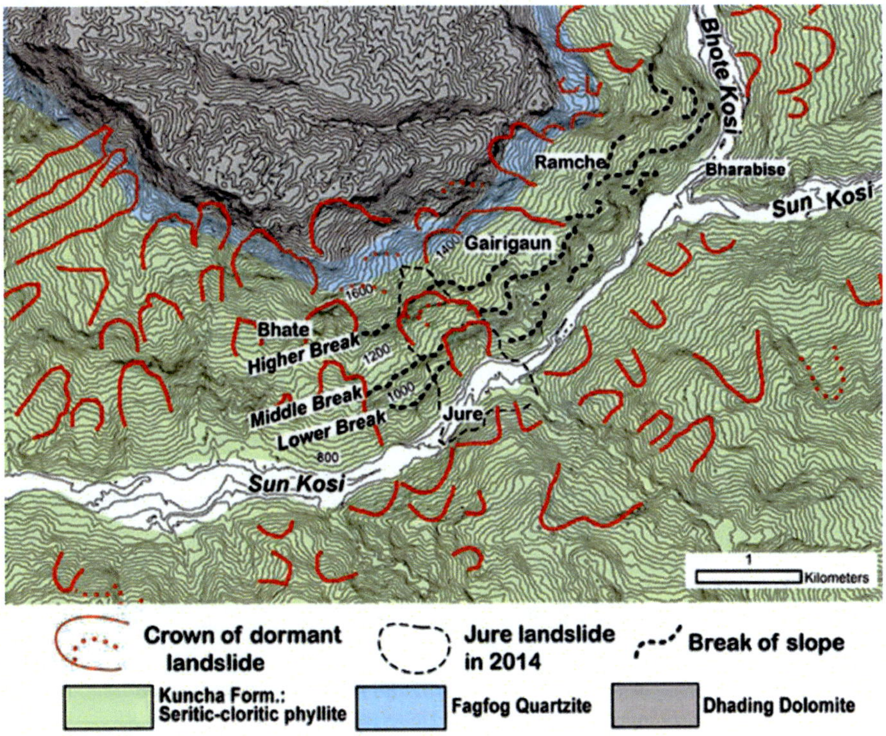


**Fig. 3** Geomorphological and geological map along Sun Koshi in and around Jure site (contour interval:20 m)

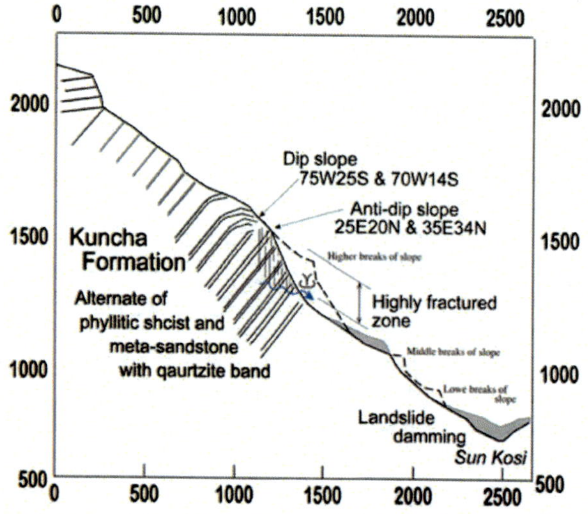


Fig. 4 Topographic and geological profile at Jure site

Fig. 5 Before and after the landslide at Jure

cones) from the middle steep slope. Jure landslide event on 2nd August 2014 occurred on the upper steep slope and the upper gentle slope as a source area of the landslide.

This area is widely underlain by Lesser Himalayan meta-sediments (Figs. 3 and 4; DMG 2005). Kuncha formation with alternating phyllitic schist and meta-sandstone with quartzite band is distributed below the upper nick point. It predominantly dips to the north with a high angle. Kuncha Formation is overlain by Fagfog Quartzite, which on the other hand is overlain by Dhading dolomite with a thrust in between. Dhading dolomite forms the upper-most steep slope just like a big wall fringing the upper most part of the Sun Koshi valley. A gentle denudated surface develops behind the upper nick point as an erosion surface on the top.

## Landslide Topography and Geological Deformation Around the Source Area of Jure Landslide

Aerial photo interpretation clarified that dormant deep-seated coherent landslides marked by a horse-shoe shaped crown are distributed at Ramche, Garigaun and Bhate villages along a right bank of the valley flank (Fig. 3). They develop at the higher position of the valley flank deforming the upper gentle slope at the Jure site. So the upper gentle slope was only widely remained around a source area of the Jure landslide 2014 until it occurred, though the upper steep slope developed as landslide scarp at its base.

The Jure landslide looks very big, wholly collapsing below the Higher Break to a river floor. However, slopes around the Higher Break only collapsed in this event. Two pictures of the landslide site taken before and after the event indicate that the landslide occurred only on the upper gentle slope and the upper steep slope (Fig. 5). The lower part of the valley slope was shallowly eroded by the passage of detritus derived from the landslide. Because the picture taken after the event shows that small spur and vegetation cover just below the Lower Break are well remained at the landslide site (Fig. 5).

The newly formed main cliff of the landslide in 2014 gave us insight into a geological background of the landslide. A remaining part of the upper gentle slope, source area of the Jure landslide, is composed of highly fractured and weathered rock of light-reddish color (Fig. 6). Thickness of the highly fractured and weathered zone is ca 150 m (Figs. 4 and 6). It overlies less fractured bluish rock. A boundary of those strata is very straight and clear. The boundary is located at ca 1,350 m a.s.l. Saturated ground water through highly fractured zone springs out along the boundary (Figs. 4 and 6).

Schist of the Kuncha formation distributed just along the landslide crown dips to the south (Loc.1 in Fig. 6), though its trend is north (Loc.2 in Fig. 6). That implies that a series of gravitational deformation such as creep and topple of bed-rock had gradually occurred in the upper part of the source area before the event.

The widely remained the upper gentle slope fringed by the upper steep slope around the Jure landslide site implies that it has been overburdened, considering its specific profile. And the upper gentle slope is composed of highly fractured and weathered rock as thick as 150 m. As a result, gravitational deformation such as rock creep and subsequent topple had continued for a long period. It is the geomorphological and geological background of the Jure landslide in 2014.

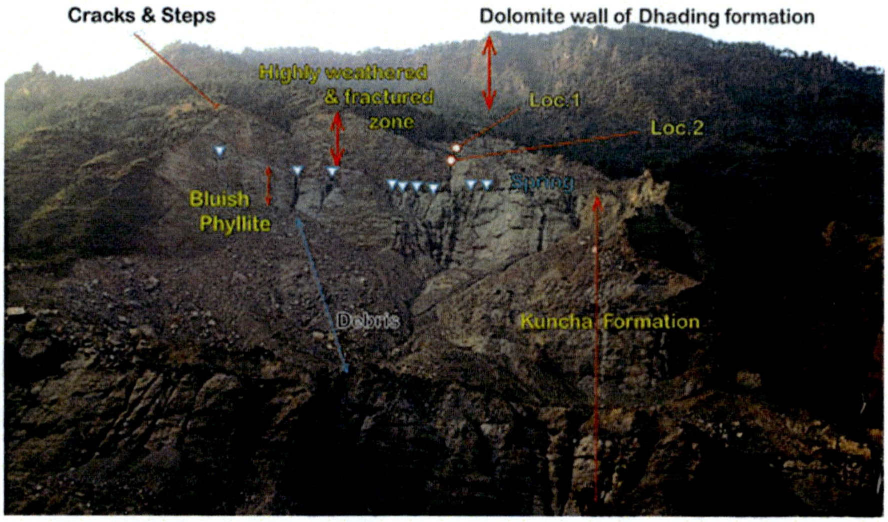


**Fig. 6** Geological section of the Jure landslide

## Chronological Change of Slope Failures Before the Event of the Jure Landslide in 2014

Time-series change of slope rupture in and around the source area of the Jure landslide 2014 was analyzed, using satellite images covering the Jure site to understand whether any predictive signs were able to detect before the event the area of the Jure landslide 2014 was detected from a satellite image taken on 10th August 2014.

We interpreted satellite images of Google Earth dated on 8th March, 2004, 18th February, 2009, 25th October, 2012, and 13th November, 2013, respectively. From those images, we detected some scars without vegetation cover due to slope failures as shallow surface rupture. We projected the location of these scars on the detail map with a contour interval of 10 m (Fig. 7). Consequently we can show their changes of time series.

In the satellite image taken on 8th March 2004 (Fig. 7 A-1), the slope failures of which width were less than 40 m were distributed inside the area of the Jure landslide 2014 and they were concentrated along a contour line of 1000 m a.s.l. Furthermore, a slope failure was also observed in a western side of its source area at ca 1,400 m a.s.l.

In the satellite image take on 18th February 2009 (Fig. 7A-2), the number of slope failures along the contour line of 1000 m had increased. Furthermore, a relatively wider slope failure occurred on a lower part of Fig. 7 Time series occurrence of slope failures around Jure the middle steep slope, located at 1260 m a.s.l. Another slope failure occurred western slope out of the area at ca 1,500 m a.s.l.

As per the image taken on 25th October 2012 (Fig. 7 A-3), the number of slope failures along the contour line of 1000 m had decreased; however, the slope failure occurring at the 1260 m point had retrogressively developed up to 1,300 m a.s.l. and became as wide as 190 m.

As per the image taken on 13th February 2013 (Fig. 7 A-4), the number of slope failures along the 1,000 m contour line had increased again and the area of slope failure at 1,300 m a.s.l. had expanded, compared to those observed in the previous images.

In the satellite image taken on 8th Nov. 2014 (Fig. 7B-1), retrogressive expansion of the main cliff is also recognized after the event in August.

## Aftermath of Nepal Gorkha Earthquake 2014 to the Slope in and Around the Jure Landslide Site

Mega earthquake with a magnitude of Mw 7.8 occurred in the central Himalayan area in April, 2015 and caused many landslides in the Central Himalayan foothill area including the Jure site (Tsuo et al. 2017; Yagi and Hayashi 2017, Yagi et al., 2018). Sun Koshi and Bhote Koshi watersheds were also strongly affected by its big aftershock of Mw 7.3 occurred on 12th May. The number of the landslides is more than 1,200 due to the earthquakes only in the northern upstream area from the Jure landslide site (Tsuo et al. 2017). However, most of them are shallow landslides of which scales are 0.05 to 0.3 ha in the source area (Tsuo et al. 2017). Any conspicuous expansion caused by deep-seated landslide did not occur at the earthquake, though retrogression of the crown of the Jure landslide occurred as slope failures or rock-fall due to the earthquake tremor (Fig. 7B-2). However, its scale is as small as a few ten meters and is not comparable to the Jure landslide in 2014, though Chadichaur town located on a valley bottom just downstream from the Jure landslide suffered from severe

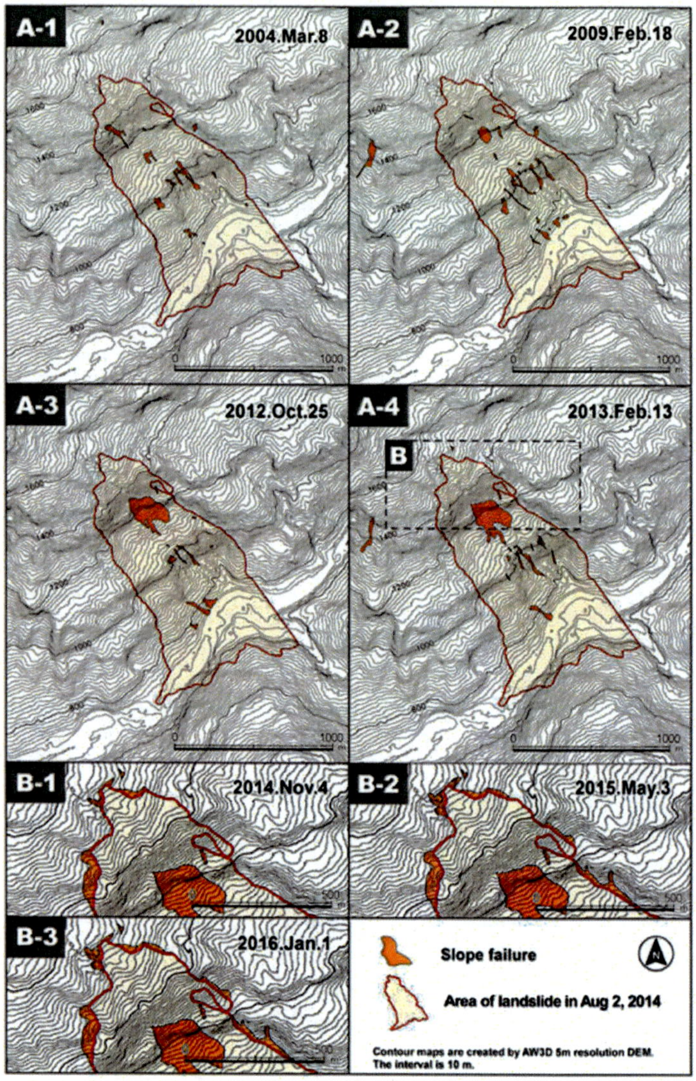


**Fig. 7** Time-series evolution of Jure landslide after 2004

damage due to the two earthquakes. Another satellite image taken after the monsoon season in 2015 also shows any expansion of the crown (Fig. 7B-3). It is very interesting that any dormant deep-seated coherent landslide was not re-activated by the earthquake in this watershed.

Many shallow landslide, slope failures were detected due to the earthquakes, however, slight deformations of ground surface without scarplets or scars were difficult to be interpreted by the optical images. In order to detect such slight deformations, time-series InSAR (Synthetic Aperture Radar interferometry) images observed before and after the earthquake were utilized in this study.

InSAR images were produced from PALSAR-2 (Phased Array type L-band SAR-2) data observed in different two dates. InSAR image shows surface deformation as phase change of interference pattern (fringe), e.g., color bar at the bottom of Fig. 8 means that if fringe shows color change of blue- > red- > yellow- > blue, land surface would move away from a satellite; vice versa, close to the satellite. In these cases, the amount of deformation is ca. 12 cm along line of sight from the satellite to the land surface. Local landslide surface deformation using InSAR image has already been reported (e.g., Sato and Une 2016). PALSAR-2 antenna is on the satellite of ALOS-2 (the Advanced Land Observing Satellite-2) that JAXA (Japan Aerospace Exploration Agency) launched in 2014. PALSAR-2 data used in this study were listed in Table 1. ALOS-2 (the Advanced Land Observing Satellite-2)/PALSAR-2 (Phased Array type L-band SAR) data were provided by courtesy of JAXA (Japan Aerospace Exploration Agency).

To eliminate fringe from topographic effect in wide area and to extract local fringe that reflects only the surface deformation, a 30-m-resolution SRTM-DEM (Shuttle Radar Topography Mission-Digital Elevation Model) was used in

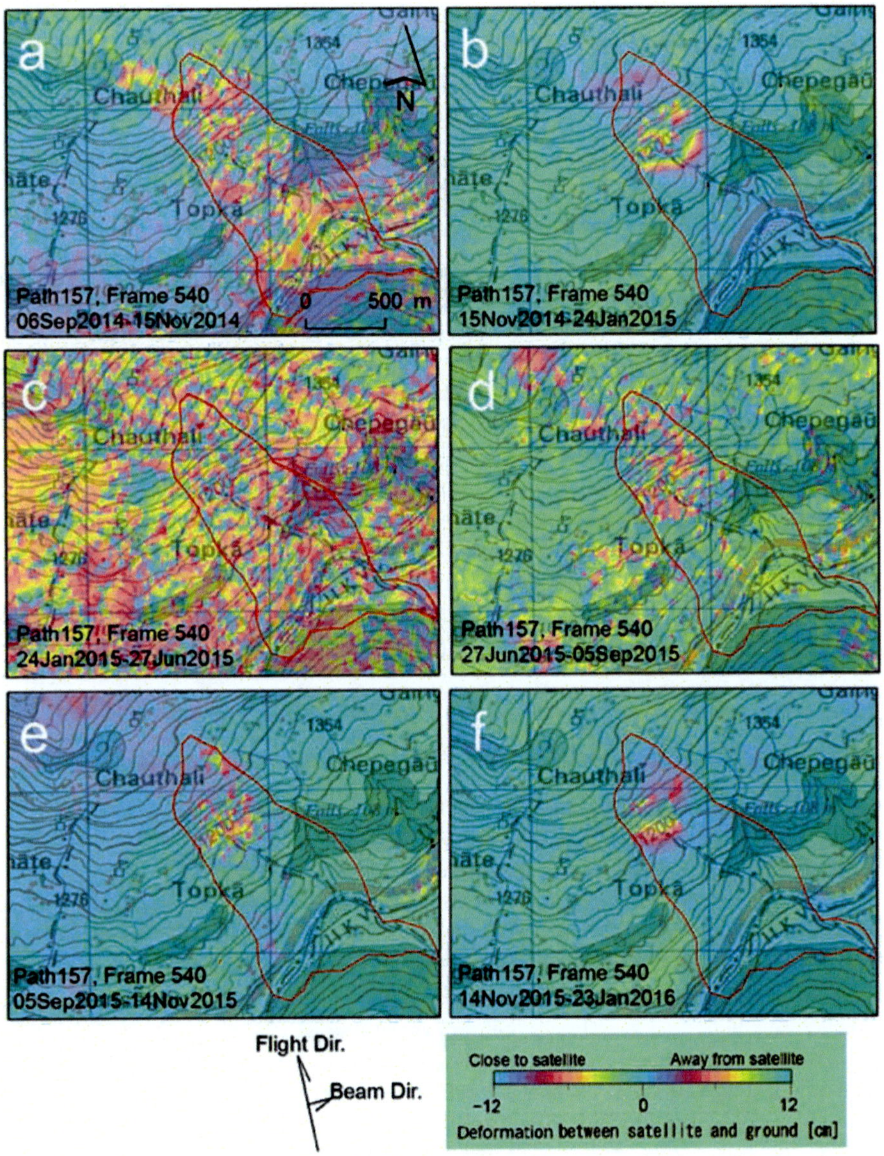


**Fig. 8** Time-series InSAR images after Jure landslide

preprocessing. If large elevation change (e.g., elevation reduction at 100 m and more) was caused by landslide and if pre-landslide DEM is used in the preprocessing, fringe from topographic effect would not be fully eliminated and the remained fringe would show apparent local surface deformation. Therefore, it is required to consider this remained effect in interpreting the InSAR image. Figure 8 shows time-series InSAR images observed after the Jure landslide. The resolution of the InSAR image is most of the same resolution of the SRTM-DEM, 30 m. Figure 8a is the InSAR image just after the landslide, in the late monsoon (wet) season, and it shows no clear fringe. However, in the figure incoherent (like sprayed sand) patterns appears both in the Jure landslide area and on the lower steep slope along Bhote Koshi, it infers that debris in the area moved randomly and unstable surface mass remains there, i.e., no surface deformation due to deep-seated landslide is identified for whole area of Jure landslide. Incoherent patterns are also identified in the whole image of Fig. 8c that was observed before and after the 2015 Gorkha earthquake. Incoherent patterns extended for whole the study area, therefore, it is difficult to evaluate surface deformation only due to reactivation of the Jure landslide. Figure 8d, e also shows incoherent patterns in mid monsoon to early post monsoon period. But the area which shows incoherent patterns was limited in the upper part of the Jure landslide, it infers debris was unstable and randomly and slightly moved in the part in mid to late monsoon seasons. Figure 8b shows clear fringes

**Table 1** PALSAR-2 data used in this study

| Path | Frame | Master/Slave | Orbit | BPERP (m) | Study area |
|---|---|---|---|---|---|
| 157 | 540 | Sep 6 2014/Nov 15 2014 | Asc, R | 151.7 | Sun Koshi-Bhote Koshi river watershed |
| 157 | 540 | Nov 15 2014/24 Jan 2015 | Asc, R | 27.8 | Sun Koshi-Bhote Koshi river watershed |
| 157 | 540 | Jan 24 2015/Jun 27 2015 | Asc, R | −127.3 | Sun Koshi-Bhote Koshi river watershed |
| 157 | 540 | Jun 27 2015/Sep 5 2015 | Asc, R | −143.9 | Sun Koshi-Bhote Koshi river watershed |
| 157 | 540 | Sep 5 2015/14 Nov 2015 | Asc, R | 28.9 | Sun Koshi-Bhote Koshi river watershed |
| 157 | 540 | Nov 14 2015/Jan 23 2016 | Asc, R | 141.9 | Sun Koshi-Bhote Koshi river watershed |

*Note* Length of baseline in perpendicular component; Des: descending orbit; Asc: ascending orbit; R: right-sided observation from the orbit

of blue -> red > yellow at the elevation of 1,150 m and 1,300 m. Figure 8f also shows the same fringe at the elevation of 1,200 m. These fringes, however, may be apparent ones produced by the DEM error mentioned above, because the DEM used in preprocessing of InSAR image production does not reflect the topographic change owing to the Jure land-slide. Therefore, the authors think that the upper part of the Jure landslide had been stable in the dry season.

## Concluding Remarks

Gravitational deformation and the subsequent fracturing and weathering for long period have proceeded at the Jure site located at the inner gorge showing distinct break of slope. There is much groundwater infiltrated deeply through the highly fractured zone. Consequently the authors consider that the source area of the Jure landslide in 2014 has transformed into a landside prone slope also from the hydrological aspect. And authors also think that it was triggered by heavy rain three days before the event in Monsoon season. However, much attention should be paid to surface ruptures or shallow slope failures occurred at the base of the source area and its marginal part for longer than 10 years. The concentrated distribution of scars along the contour line of 1000 m implies that abundant groundwater was supplied to the middle gentle slope from the upper slope and it caused shallow slope failure of small scale. Occurrence of slope failures at the base of the upper steep slope and its widening and subsequent retrogressive development directly predicted a big landslide, the event in 2014.

Highly fractured and weathered rock also heavily remained along the crown in October, 2014. The authors saw many cracks and up-hill facing scarplets along the main crown of the landslide then. They presumed that retrogressive collapse along the landslide would occur in the near future just after the event in 2014. However, no noticeable expansion occurred at the Nepal Gorkha earthquake in 2015 except shallow slope failures or rock falls. Other dormant coherent deep-seated landslides distributed along Bhote Koshi also did not show any signs of reactivation. InSAR analysis also showed coherent movement did not occur in and around the Jure Site. The authors attribute no expansion of the Jure site and no reactivation of the coherent deep-seated landslides by the quake to less groundwater in dry season, because the earthquakes occurred in the dry season with rare precipitation for six months since the end of monsoon. And a watershed behind the Jure site is not directly connected to watershed of glacial area, though the Sun Koshi – Bhote Koshi watershed is originated from the Himalayan glaciers. So it implies groundwater level was not enough to reactivate the deep-seated landslide.

## References

DMG (2005) Geological map of part of Sindhupalchok District (Barhabise area) 1/50,000. Department of Geology and Mines, Nepal Government

Ministry of Irrigation (2014) Report on Jure landslide, Mankha VDC, Sindhulpalchowk District. Nepal Government, 29pp

Pettley D (2014) Bhote Kosi Sun Koshi (corrected) landslide in Nepal – an updated, AGU Blogsphere. https://blogs.agu.org/landslideblog/2014/08/07/bhote-kosi-landslide-3/ (referred on 10th December 2017)

Sato HP,Okatani T, Nakano T, Koarai M, Iwahashi J, Une H (2013) Application and validation of differential SAR interferometry image to landslide detection through detailed field survey. J Japan Landslide Society 50(1):18–23 (in Japanese)

Sato HP, Une H (2016) Detection of the 2015 Gorkha earthquake-induced landslide surface deformation in Kathmandu using InSAR images from PALSAR-2 data. Earth, Planets Space 68:47. https://doi.org/10.1186/s40623-016-0425-1

Tsuo CY, Chigira M, Higaki D, Sato G, Yagi H, Sato HP, Wakai A, Dangol V, Amatya SC, Yatagai A (2017) Topographic and geologic controles on landslides induced by the 2015 Gorkha earthquake and its aftershoks: an example from the Trishuli Valley, central Nepal. Landslides. https://doi.org/10.1007/s10346-017-0913-9

Yagi H, Hayashi K (2017) Bell-shape index indicating top-heavy profile of high relief mountain and gravitational deformation. 4th Slope Tectonics, Kyoto 2017

Yagi H, Hayashi, K, Higaki D, Tsou CY, Sato G (2018) Dormant landslides distributed in upper course of Sun Kosi watershed and landslides induced by Nepal Gorkha Earthquake 2015. J Nepal Geol Soci 55(Special Issue):1–7

2020 Kyoto Japan

# Inventory of Landslides Triggered by the Hejing Ms6.6 Earthquake, China, on 30 June 2012

Chong Xu and Kai Li


## Abstract

Coseismic landslide inventorying is of great significance for understanding the occurrence, distribution, and risk assessment of earthquake-triggered landslides. Based on manual visual interpretation of high-resolution satellite images acquired before and after the event and field survey verification, this work prepared an inventory map of the landslides triggered by the 2012 Ms6.6 Heijing, Xinjiang, China earthquake. Results show that this earthquake triggered at least 453 landslides with a total area of 0.664 $km^2$ in the areas of VII and VIII seismic intensity. The area density of the landslides is 0.01% and the landslide point density is 0.07 $km^{-2}$. A north-east extending river valley west of the epicenter registered the highest landslide density. The reason of this is considered to be the strongest ground motion during the earthquake, the steep topographic relief, the large slope gradient, and accumulated loose materials at this site, which means the area is prone to landsliding. This study provides a typical case with weak landslide triggering capability which was associated with special local conditions.



## Keywords

Hejing earthquake • Coseismic landslides • Landslide inventory • GIS



C. Xu (✉)
National Institute of Natural Hazards, Ministry of Emergency Management of China, Beijing, 100085, China
e-mail: xc11111111@126.com

C. Xu · K. Li
Institute of Geology, China Earthquake Administration, Beijing, 100029, China

K. Li
School of Earth Sciences and Resources, China University of Geosciences, Beijing, 100083, China



V. Vilímek et al. (eds.), *Understanding and Reducing Landslide Disaster Risk*,
ICL Contribution to Landslide Disaster Risk Reduction,
https://doi.org/10.1007/978-3-030-60319-9_6

## Introduction

Earthquake-triggered landslides are an important secondary earthquake effect, which can not only cause severe casualties and economic losses but also seriously influence the future potential geological hazards in the affected areas (Keefer 1999; Xu et al. 2010, 2014a; Fan et al. 2019). For instance, landslides triggered by the 2008 Wenchuan, China, M7.9 earthquake killed about 30,000 people (Yin 2008; Xu et al. 2010), about 100,000 of the 234,000 people died in the 1920 Haiyuan earthquake directly by the coseismic landslides (Close and McCormick 1922; Xu et al. 2018a), and the 2018 Hokkaido earthquake-triggered landslides killed 36 people (Yamagishi and Yamazaki 2018). Therefore, earthquake-triggered landslides have received much attention in recent years. Accurate and detailed earthquake-triggered landslide inventorying is a vital foundation for study of the landslide cause, hazard assessment, and post-seismic reconstruction planning of affected areas (Harp et al. 2011; Guzzetti et al. 2012; Xu et al. 2014a; Xu 2015). In many previous major earthquakes, detailed earthquake-triggered landslide inventories have played an important role in related studies (Xu et al. 2014a, 2015; Martha et al. 2017; Xu et al. 2018b; Shao et al. 2019). However, there are still many important earthquake events without detailed and complete landslide records. Therefore, it is important to carry out remedial research for detailed earthquake-triggered landslide inventorying.

The Hejing Ms6.6 earthquake in Xinjiang, China, which occurred on June 30, 2012, is such an example. Its epicenter is 43.4°N and 84.8°E, with maximum intensity VIII and the focal depth 10 km. According to local government, 52 people were injured and 552,503 people were affected by the earthquake, with direct economic losses of nearly 2 billion yuan RMB (Chang et al. 2012; Ge et al. 2012). So far few studies on landslides caused by this earthquake have been made. For example, field surveys have been carried out soon after the quake and 93 coseismic landslides were located and described (Li et al. 2012). However, there is no detailed inventory of landslides triggered by the earthquake released. To fill this gap, based on the interpretation of high-resolution pre- and post-earthquake satellite images combined with field investigations, this study prepared a detailed coseismic landslide inventory in the intensity VII and VIII zones. Results show that this earthquake triggered at least 453 landslides with a total area of 0.664 km$^2$, mainly distributed in the intensity VIII zone. The high-density area of the landslide is located near the epicenter, especially on the slopes along the a northeast extending river valley west to the epicenter.

## The Hejing Earthquake

The Hejing Ms6.6 earthquake occurred at the junction of Hejing County and Xinyuan County, Xinjiang, China. The epicenter was located the intersection of the Narat fault, the southern margin of the South Awulale Mountain fault, and the Kashi River fault, a complex tectonic setting (Fig. 1a). Research suggested that the seismogenic structure is the Kashi River fault zone, characterized by dextral strike-slip and a relatively simple rupturing process (Ran et al. 2014; Wang et al. 2015). This fault zone is a large-scale thrust in the Tianshan Mountains, with a general trend of 290°. It is about 315 km long in China, dipping to the north with an inclination angle of 60°–80°. The fault is a boundary between the Yili Basin and the Borokonu Mountains (Yin et al. 2003). The zones of seismic intensity VII and VIII cover a total area of 6446 km$^2$, taken as the study area, where elevations range 1300–5200 m, higher in the north-east and lower in the southwest (Fig. 1b). According to the 1:500,000 Chinese Geological Map from China Geological Survey, the stratigraphy in the area is complicated (Fig. 1a).

## Data and Methods

### Data

The satellite images used in this study are shown in Fig. 2 and listed in Table 1, including the RapidEye images (Planet Team 2017) and Ikonos images. The pre- and post-quake RapidEye images cover most of the study area, except a very small area west to the study area. Although only post-quake Ikonos images with limited coverage were available, the area around the epicenter, the most severe area, was covered. In addition, more pre- and post-quake satellite images released by Google Earth platform are also a useful supplement. Based on the differences in the tone, shape, and texture characteristics of the pre- and post-quake images, coseismic landslides could be detected objectively and accurately.

### Methods

Interpretation of remote sensing images is one of the most important methods for landslide inventorying (Xu et al. 2014a; Santangelo et al. 2015; Gnyawali and Adhikari 2017; Tanyaş et al. 2017). The landslide identification in this study relied on the visual interpretation method by comparison of pre- and post-quake images, supplemented by selected field investigations. Because the images before and after the earthquake were close to the occurrence of the earthquake, so landslides occurred before and long after the earthquake can be excluded objectively. Each individual coseismic landslide was represented by a polygon. The source area, movement area, and accumulation area of landslides were not distinguished. Figure 3 is an example of the interpretation and delineation of coseismic landslides in an area of Gongnaisi Town. A red polygon represented a coseismic landslide. In ArcGIS and Google Earth, the boundaries of coseismic landslides were directly drawn, finally forming a file in the vector format, which was beneficial for subsequent landslide attribute information extraction, spatial analysis, and data dissemination.

Visual interpretation of high-resolution orthographic satellite images before and after earthquakes based on GIS platform is one of the most effective methods to carry out earthquake-triggered landslide inventorying. It has obvious advantages in the investigation of earthquake-triggered landslides in a large area. The spectral resolution, date, snow and ice coverage, and cloud coverage of satellite images have significant effects on the quality of the inventory map. We also conducted a rapid response survey and a scientific investigation to verify the visual interpretation inventory (Li et al. 2012).

## Results

The results show that at least 453 landslides were triggered by the Hejing earthquake in the study area (Fig. 4), with landslide number density (LND) 0.07 km$^{-2}$. These landslides are dominated by rock falls and shallow disrupted slides. The total area of the landslides is 0.664 km$^2$, which occupies 0.01% of the study area. The maximum individual size of the landslides is 19322.4 m$^2$. There are two landslides with an area of larger than 10000 m$^2$, 222 landslides between 1000 m$^2$ and 10000 m$^2$, 220 between 100 m$^2$ and 1000 m$^2$, and remaining 9 landslides are less than 100 m$^2$. The landslide point density map (Fig. 5) shows that the maximum value is 2.12 km$^{-2}$. The high-density area is located near the epicenter, especially on the slopes on both sides of a northeast trending river valley west to the epicenter. Probably because this area is close to the epicenter

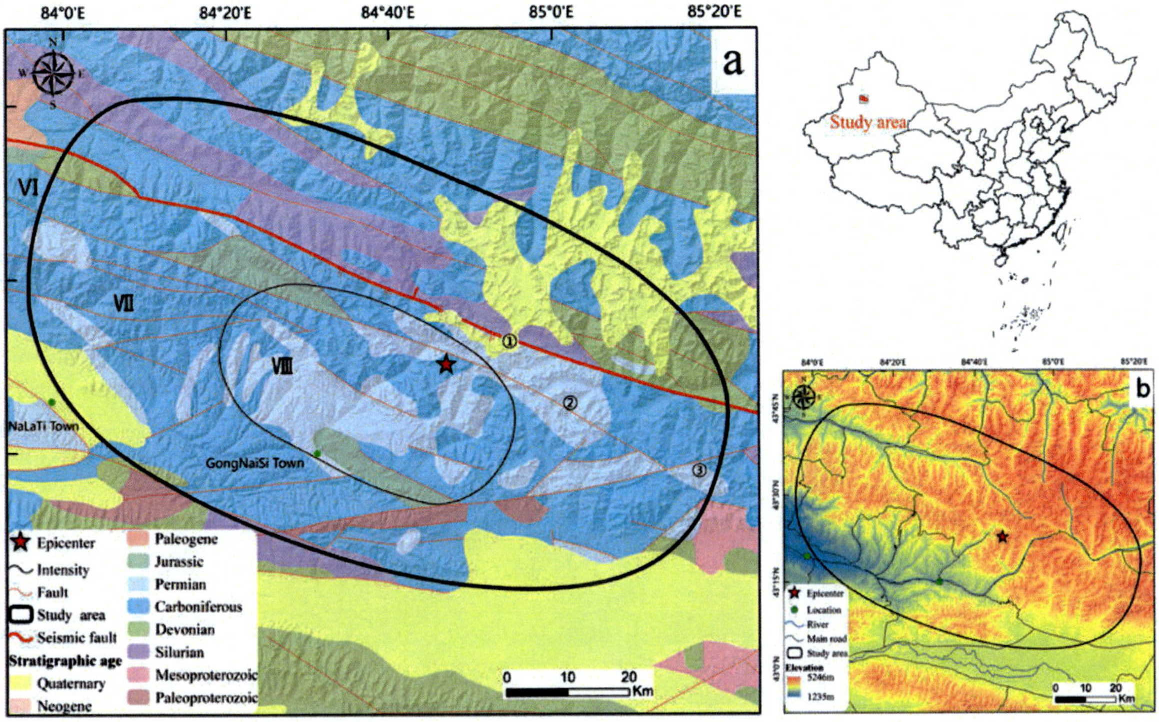


**Fig. 1** Maps showing geology and topography of the Hejing earthquake area. **a** Underlying strata and tectonic setting, 1-Kashi River fault; 2-South Awulale Mountains fault; 3-Narat fault. The fault and strata data are from China Geological Survey. **b** Shaded map showing topography of the study area

and experienced the strongest ground motion during the earthquake. Moreover, the steep topographic relief and large slope angles of the area as well as the loose material accumulated on both sides of the river resulted in high susceptibility to landsliding. Table 2 shows that the area of seismic intensity VIII registered more coseismic landslides and high landslide number density, which are 243 pieces and 0.21 $km^{-2}$, respectively, accounting for 53.6% of the total number. There are 210 landslides in the VII degree zone, with landslide density 0.04 $km^{-2}$.

## Discussion

The main purpose of this study was to provide a detailed and complete inventory of landslides triggered by the 2012 Heijing, China Ms 6.6 earthquake. In general, earthquake-triggered landslides are often characterized by a wide distribution area, large number, and high density, which make it difficult to obtain detailed and comprehensive information. Therefore, to obtain such data, remote sensing technology and images are required. In addition, the uncertainty of the occurrence of earthquakes and whether the seismic events have sufficient research value and receive attention by researchers are also limitations. Therefore, only decades of detailed inventories of landslides triggered by major earthquakes were released in recent decades globally (Harp et al. 2011; Xu 2015; Schmitt et al. 2017). The earliest inventory in GIS or vector format can be traced back to the 1994 Northridge earthquake, USA (Harp and Jibson 1995). This study provides a comprehensive and detailed inventory map of slope failures triggered an Ms6.6 earthquake, which adds a valuable case study to the global earthquake-triggered landslide database.

For regional landslide inventorying, the acquisition time, resolution, quality, and coverage of remote sensing images are the basis for inventorying maps (Xu 2015). The acquisition time of the pre- and post-quake remote sensing image should be as close as possible to the occurrence time of the earthquakes, so as to ensure that the pre-quake landslides and post-quake landslides are excluded. The satellite images used in this study included pre-earthquake images taken on June 13 and 22, 2012 and pre-earthquake images taken on August 6, 7, and 10, 2012. In addition, some factors that affecting coseismic landslides, such as human activities and rainfall, are not significant in this area, so it guarantees the time requirements of satellite images for interpretation of coseismic landslides. In terms of the effective coverage of

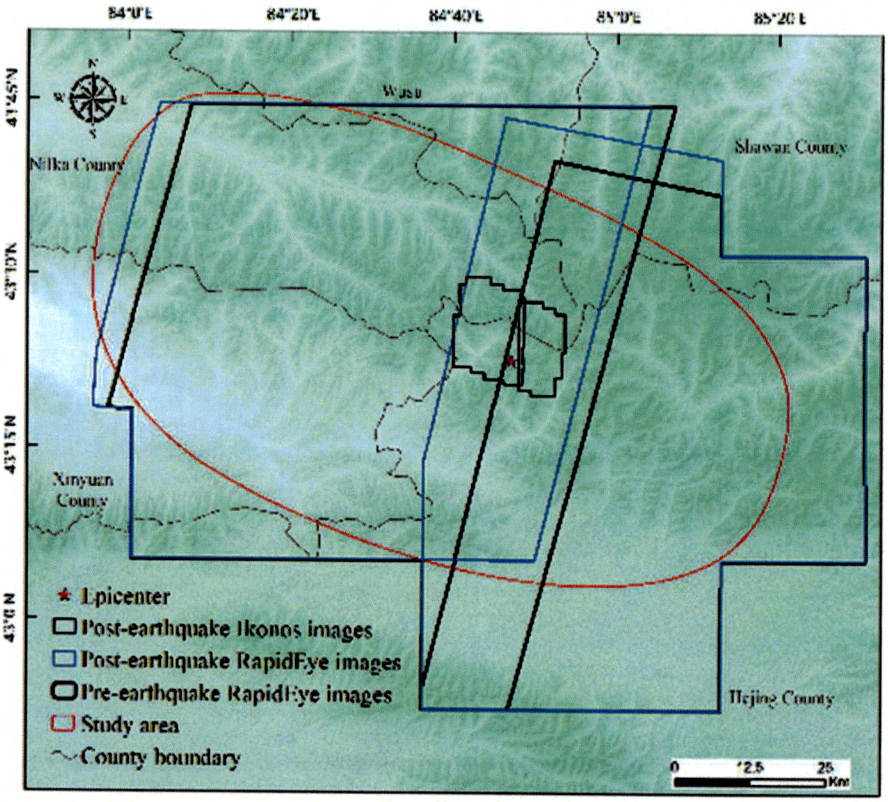


**Fig. 2** Coverage of satellite images for this study

**Table 1** Satellite images used in this work

| Date | Type | Resolution (m) |
|---|---|---|
| June 13, 2012 | RapidEye | 5 |
| June 22, 2012 | RapidEye | 5 |
| July 9, 2012 | RapidEye | 5 |
| August 6, 2012 | RapidEye | 5 |
| August 7, 2012 | Ikonos | 1 |
| August 10, 2012 | Ikonos | 1 |

satellite images, all the images used in this study were cloud-free, ensuring that the coverage of all the images was valid. Although a small part west of the study area was not covered by the RapidEye image, it was supplemented by pre- and post-earthquake images on the Google Earth platform. Although the periods between taking these images from Google Earth and the occurrence of the earthquake were relatively long, the area was far away from the epicenter and few landslides would occur. From the landslide distribution map (Fig. 4), the landslide density gradually decayed from the epicenter to the outer rings, and almost no co-seismic landslides occurred in the area without coverage of the RapidEye images. Therefore, there was little effect on the results and effective coverage of the satellite image was quite satisfactory. Besides, the resolutions of the two main images used in this study were 5 m and 1 m, respectively. The RapidEye image with a resolution of 5 m covered most of the study area. Although the 1-m resolution Ikonosimages only covered a small area, it was just the most affected area with the highest density of coseismic landslides, which met the requirement for image resolution in the meizoseismal area. In addition, the coverage area of the Ikonosimages was also covered by RapidEye images. The result from the interpretation of 5-m resolution RapidEye images was verified from the 1-m resolution Ikonosimages, which permitted to establish the interpretation signs and expert knowledge objectively for coseismic landslide interpretation on the 5-m resolution RapidEye images, helpful in obtaining an objective coseismic landslide inventory throughout the study area.

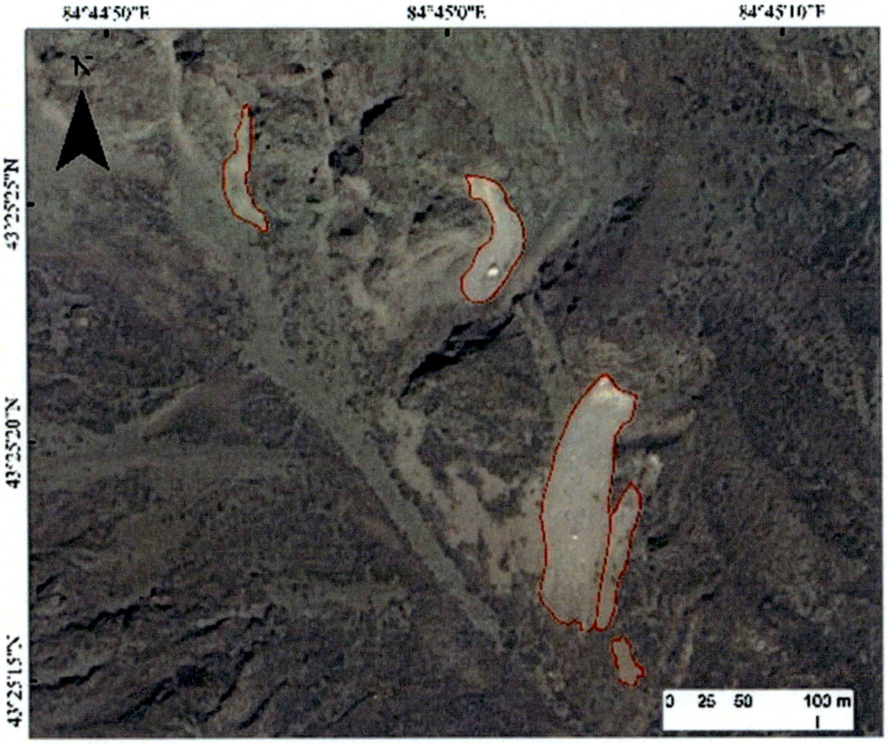


**Fig. 3** Interpretation of landslides in an area of Gongnaisi Town

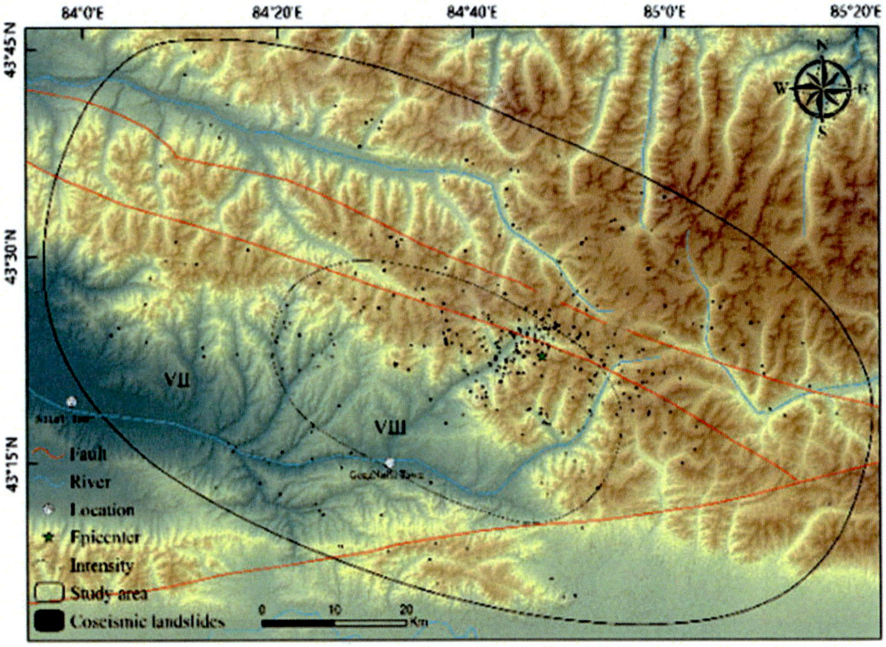


**Fig. 4** Map showing spatial distribution of landslides triggered by the 2012 Hejing earthquake

The completeness of the landslide inventory was also reflected in the objective interpretation of small- and moderate-scale landslides. Quite a few high-quality coseismic landslide inventories show that a landslide with an area about 100 m$^2$ can be detected (Harp and Jibson 1995; Liao and Lee 2000; Xu et al. 2014a, 2015). The results of this study show that although most of the landslides are of an area larger than 100 m$^2$, there still a few small-scale landslides that are exceptional. Each of these small-scale landslides covered at least 4 grids on 5 m-resolution images,

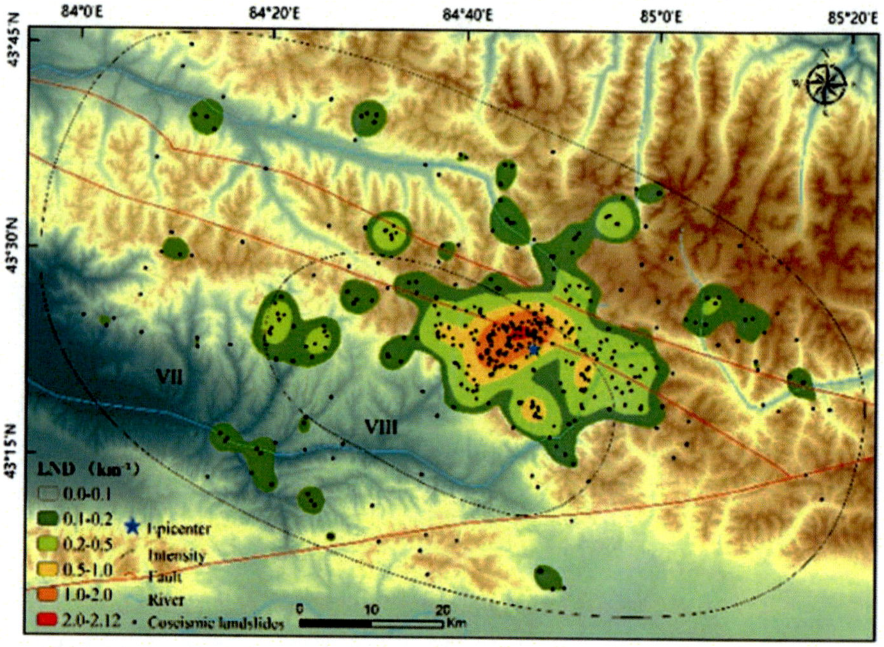


**Fig. 5** Landslide number density map related to the 2012 Hejing earthquake

**Table 2** Landslides in zones of different seismic intensity values

| Area (km$^2$) | Landslide area (km$^2$) | Landslide number | LND (km$^{-2}$) |
|---|---|---|---|
| VIII, 1175 | 0.21 | 243 | 0.21 |
| VII, 5271 | 0.04 | 210 | 0.04 |
| Total, 6446 | 0.664 | 453 | 0.07 |

which can be identified from the image objectively. Therefore, compared with other high-quality seismic landslide inventories, the inventory of this study for small- and moderate-scale landslides are also objective enough.

The magnitude of the Hejing earthquake is Ms6.6, while only 453 coseismic landslides induced were detected. No large-scale landslides were present in the inventory. Compared with several other earthquakes with similar magnitudes, such as the 2007 Chile Ms6.3 earthquake (Sepúlveda et al. 2010), the 2004 Niigata, Japan Ms6.8 earthquake (Chigira and Yagi 2006), the 2014 Ludian, China Ms6.5 earthquake (Xu et al. 2014b), the 2018 Tomakomai, Japan Mw6.6 earthquake (Shao et al. 2019), the severity of the landslides triggered by the 2012 Hejing earthquake is quite low. The analysis above shows that the quality of the resultant landslide inventory of this study is satisfactory. Why this event triggered relatively fewer coseismic landslides? The reason may be that the affected area has a low susceptibility to landsliding and the earthquake didn't generate surface ruptures, which have been shown to affect the severity and spatial patterns of coseismic landslides (Gorum et al. 2013; Xu 2014). On the other hand, there are relatively fewer studies on such earthquake events with weak capabilities to trigger landslides, leading to a lack of sufficient data for studying the mechanisms of significant differences in landslides triggered by earthquakes with similar moderate magnitudes. Therefore, this study provides a typical case study of inventory mapping of landslides triggered by a moderate earthquake with a weak capability of triggering landslides, including a high-quality coseismic landslide inventory.

## Conclusions

In this study, an inventory of landslides triggered by the 2012 Hejing, China, Ms6.6 earthquake was prepared. We delineated 453 coseismic landslides in the 6446 km$^2$ study area with landslide point density 0.07 km$^{-2}$. The total area of these landslides is 0.664 km$^2$, which accounts for 0.01% of the study area. The high-density area was located near the epicenter, especially on the slopes on both sides of a north-east extending river valley west of the epicenter. The reason may be that this area is close to the epicenter and thus experienced the strongest ground shaking during the earthquake. Moreover, the steep topographic relief and the accumulation of loose materials on the river slopes also contributed to high susceptibility to landsliding. We discussed the quality of the resultant landslide inventory from the perspectives of satellite image selection, interpretation criteria, comparison with landslides triggered by other earthquakes with similar magnitudes. It can be concluded that the inventory by this work is quite objective and complete, and the Hejing Ms6.6 earthquake has a weak capacity to trigger landslides, which was likely associated with special local conditions there.

**Acknowledgements** This study was supported by the National Key Research and Development Program of China (2018YFC1504703-3) and the National Natural Science Foundation of China (41661144037).

## References

Chang X, Tan M, Alimujiang Y, Li S, Tang L (2012) Disaster loss assessment and characteristic of seismic hazard of Xinyuan, Hejing earthquake with Ms6.6 on June 30th, 2012. Inland Earthq 26 (4):373–380

Chigira M, Yagi H (2006) Geological and geomorphological characteristics of landslides triggered by the 2004 Mid Niigta prefecture earthquake in Japan. Eng Geol 82(4):202–221

Close U, McCormick E (1922) Where the mountains walked. Natl Geogr Mag 41(5):445–464

Fan X, Scaringi G, Korup O, West AJ, van Westen CJ, Tanyas H, Hovius N, Hales TC, Jibson RW, Allstadt KE, Zhang L, Evans SG, Xu C, Li G, Pei X, Xu Q, Huang R (2019) Earthquake-induced chains of geologic hazards: patterns, mechanisms and impacts. Rev Geophys 57(2):421–503

Ge M, Tan M, Chang X, Alimujiang Y, Chen J, Luo J, Li S, Tang L (2012) Analysis on characteristics of seismic hazards and building construction in Xinyuan, Hejing earthquake with Ms6.6. Inland Earthq 26(4):360–364

Gnyawali KR, Adhikari BR (2017) Spatial relations of earthquake induced landslides triggered by 2015 Gorkha earthquake Mw = 7.8. In: Workshop on world landslide forum, pp 85–93

Gorum T, van Westen CJ, Korup O, van der Meijde M, Fan XM, van der Meer FD (2013) Complex rupture mechanism and topography control symmetry of mass-wasting pattern, 2010 Haiti earthquake. Geomorphology 184:127–138

Guzzetti F, Mondini AC, Cardinali M, Fiorucci F, Santangelo M, Chang K-T (2012) Landslide inventory maps: New tools for an old problem. Earth Sci Rev 112(1–2):42–66

Harp EL, Jibson RW (1995) Inventory of landslides triggered by the 1994 Northridge, California earthquake. US Geological Survey. http://pubs.usgs.gov/of/1995/ofr-95-0213/plate1.gif

Harp EL, Keefer DK, Sato HP, Yagi H (2011) Landslide inventories: The essential part of seismic landslide hazard analyses. Eng Geol 122(1–2):9–21

Keefer DK (1999) Earthquake-induced landslides and their effects on alluvial fans. J Sediment Res 69(1):84–104

Li S, Xu C, Chen J, Tan M, Wu G, Chang X, Alimujiang Y, Luo J (2012) Analysis on inducing factors of geological hazard caused by Xinyuan, Hejing earthquake with Ms6.6. Inland Earthq 26(4):365–372

Liao HW, Lee CT (2000) Landslides triggered by the Chi-Chi earthquake. In: Proceedings of the 21st Asian conference on remote sensing, Taipei, vol 1–2, pp 383–388

Martha TR, Roy P, Mazumdar R, Govindharaj KB, Kumar KV (2017) Spatial characteristics of landslides triggered by the 2015 Mw 7.8 (Gorkha) and Mw 7.3 (Dolakha) earthquakes in Nepal. Landslides 14(2):697–704

Planet Team (2017) Planet application program interface: in space for life on earth. planet company. San Francisco, CA, USA. https://api.planet.com

Ran H, Zhang Z, Zhao Q (2014) The focal-mechanism solution of the Xinyuan-Hejing Ms6.6 earthquake in Xinjiang on June 30, 2012. Earthq Res China 30(3):432–441

Santangelo M, Marchesini I, Bucci F, Cardinali M, Fiorucci F, Guzzetti F (2015) An approach to reduce mapping errors in the production of landslide inventory maps. Nat Hazards Earth Syst Sci 15(9):2111–2126

Schmitt RG, Tanyas H, Jessee MAN, Zhu J, Biegel KM, Allstadt KE, Jibson RW, Thompson EM, van Westen CJ, Sato HP, Wald DJ, Godt JW, Gorum T, Xu C, Rathje EM, Knudsen KL (2017) An open repository of earthquake-triggered ground-failure inventories. U.S. Geological Survey, No. 1064. https://pubs.er.usgs.gov/publication/ds1064

Sepúlveda SA, Serey A, Lara M, Pavez A, Rebolledo S (2010) Landslides induced by the April 2007 Aysén Fjord earthquake. Chilean Patagonia. Landslides 7(4):483–492

Shao X, Ma S, Xu C, Zhang P, Wen B, Tian Y, Zhou Q, Cui Y (2019) Planet image-based inventorying and machine learning-based susceptibility mapping for the landslides triggered by the 2018 Mw6.6 Tomakomai, Japan earthquake. Remote Sens 11(8):978

Tanyaş H, Westen CJ, Allstadt KE, NowickiJessee MA, Görüm T, Jibson RW, Godt JW, Sato HP, Schmitt RG, Marc O (2017) Presentation and analysis of a worldwide database of earthquake-induced landslide inventories. J Geophys Res: Earth Surf 122(10):1991–2015

Wang Q, Ji Z, Zhao C, Wang H, Nie X, Li Z (2015) Research on source rupture process of Ms6.6 earthquake of June 30, 2012, in the border area of Xinyuan and Hejing County, Xinjiang. Seismol Geol 37 (1):33–43

Xu C (2014) Do buried-rupture earthquakes trigger less landslides than surface-rupture earthquakes for reverse faults? Geomorphology 216:53–57

Xu C (2015) Preparation of earthquake-triggered landslide inventory maps using remote sensing and GIS technologies: principles and case studies. Geosci Front 6(6):825–836

Xu C, Dai F, Xu X (2010) Wenchuan earthquake induced landslides: an overview. Geol Rev 56(6):860–874

Xu C, Xu X, Yao X, Dai F (2014a) Three (nearly) complete inventories of landslides triggered by the May 12, 2008 Wenchuan Mw 7.9 earthquake of China and their spatial distribution statistical analysis. Landslides 11(3):441–461

Xu C, Xu X, Shen L, Dou S, Wu S, Tian Y, Li X (2014b) Inventory of landslides triggered by the 2014 Ms6.5 Ludian earthquake and its implications on several earthquake parameters. Seismol Geol 36 (4):1186–1203

Xu C, Xu X, Shyu JBH (2015) Database and spatial distribution of landslides triggered by the Lushan, China Mw 6.6 earthquake of 20 April 2013. Geomorphology 248:77–92

Xu C, Tian Y, Ma S, Xu X, Zhou B, Wu X, Zhuang J, Gao Y, Wu W, Huang X (2018a) Inventory and spatial distribution of landslides in IX-XI high intensity areas of 1920 Haiyuan (China) M8.5 earthquake. J Eng Geol 26(5):1188–1195

Xu C, Tian Y, Shen L, Xu X, Zhou B, Ma S, Huang X, Ma J, Chen X (2018b) Database of landslides triggered by 2015 Gorkha (Nepal) Mw7.8 earthquake. Seismol Geol 40(5):1115–1128

Yamagishi H, Yamazaki F (2018) Landslides by the 2018 Hokkaido Iburi-Tobu earthquake on September 6. Landslides 15(12):2521–2524

Yin Y (2008) Research on the geo-hazards triggered by Wenchuan earthquake, Sichuan. J Eng Geol 16(4):433–444

Yin G, Jiang J, Zhang Y (2003) Research on Kashi River fault in Yili, Xinjiang and its activity. Inland. Earthquake 17(2):109–116

2020 Kyoto Japan

# Pressure Head Dynamics on a Natural Slope in Eastern Iburi Struck by the 2018 Hokkaido Earthquake

Toshiya Aoki, Shin'ya Katsura, Takahiko Yoshino, Takashi Koi, Yasutaka Tanaka, and Takashi Yamada

**Abstract**

The 2018 Hokkaido Eastern Iburi Earthquake triggered numerous shallow landslides on slopes covered with thick pyroclastic fall deposits. These shallow landslides tended to occur on concave rather than convex slopes and their slip surfaces were very wet, indicating that water played an important role in landslide initiation. As a first step toward clarifying the role of water in these landslides, we used tensiometers to monitor pressure head dynamics on a natural hillside covered with thick pyroclastic deposits that remained in place throughout the earthquake. We found that on concave slopes, the lower part of the pyroclastic fall deposits throughout the weathered basement complex (sedimentary rock) were always wet. Notably, the interface between the pyroclastic fall deposits and weathered basement complex, which forms a potential slip surface for earthquake-induced landslides, was always at or near saturation. On convex slopes, the weathered basement complex was never saturated and showed greater pressure head fluctuation. We infer that the pyroclastic fall deposits over the basement complex tend to weather more easily and are more susceptible to intense ground motion on concave than on convex slopes and the landslide slip surface was saturated at the timing of the earthquake on concave slopes. We conclude that these factors contributed to the larger number of shallow landslides initiated on concave slopes.

**Keywords**

2018 hokkaido eastern iburi earthquake • Pyroclastic fall deposits • Pressure head • Shallow landslides

T. Aoki (✉) · T. Yoshino
Graduate School of Agriculture, Hokkaido University, Kita 9, Nishi 9, Kita-Ku, Sapporo, 0608589, Japan
e-mail: aokitoshiya6@eis.hokudai.ac.jp

T. Yoshino
e-mail: allizdog9160@eis.hokudai.ac.jp

S. Katsura · T. Yamada
Research Faculty of Agriculture, Hokkaido University, Sapporo, 0608589, Japan
e-mail: skatsura@for.agr.hokudai.ac.jp

T. Yamada
e-mail: tyamada@for.agr.hokudai.ac.jp

T. Koi · Y. Tanaka
Center for Natural Hazards Research, Hokkaido University, Sapporo, 0608589, Japan
e-mail: koi@cen.agr.hokudai.ac.jp

Y. Tanaka
e-mail: tanakay@cen.agr.hokudai.ac.jp

## Introduction

At 03:07:59.3 (JST) on September 6, 2018, the eastern Iburi region of Hokkaido, northern Japan was struck by an earthquake with a Japan Meteorological Agency (JMA) magnitude of 6.7 and maximum seismic intensity of 7 (JMA 7-stage seismic intensity scale) (Fig. 1). The 2018 Hokkaido Eastern Iburi Earthquake triggered numerous shallow landslides, especially in the towns of Atsuma and Abira. Figure 1c shows the distribution of landslides triggered by the earthquake. The total area and number of landslides were 43.4 km2 and 6116, respectively (Osanai et al. 2019). The basement complex (sedimentary rock) in this area is covered with thick pyroclastic fall deposits including volcanic ash, pumice, and scoria mainly derived from the Tarumae, Eniwa, Shikotsu, and Kuttara volcanoes, located 50–70 km southwest of this area (Fig. 1b) (Furukawa and Nakagawa 2010; Nakagawa et al. 2018); these deposits were the main components of the shallow landslides induced by strong seismic shocks during this event (Osanai et al. 2019).

Historically, shallow landslides caused by large-scale earthquakes in Japan tend to occur more frequently on convex slopes than concave slopes (e.g., Kawabe 1987; Saito et al. 1995; Nishida et al. 1996; Kinoshita et al. 2010);


V. Vilímek et al. (eds.), *Understanding and Reducing Landslide Disaster Risk*, ICL Contribution to Landslide Disaster Risk Reduction,
https://doi.org/10.1007/978-3-030-60319-9_7

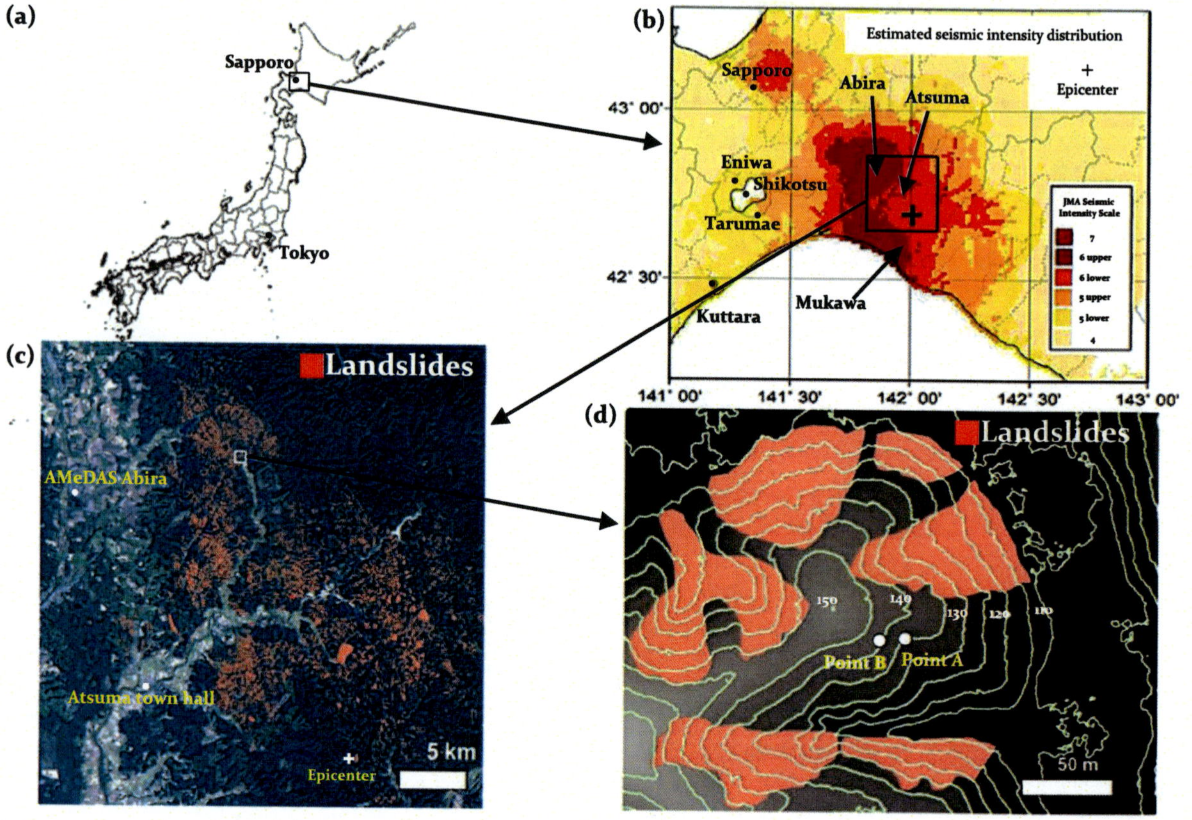


**Fig. 1** a Map of Japan (Geospatial Information Authority of Japan, modified by authors). b Map of central Hokkaido with estimated seismic intensity distribution (Japan Meteorological Agency, modified by the authors). c Satellite imagery of the earthquake-affected area (the Geospatial Infomation Authority of Japan, modified by the authors. This data source is as follows: Landsat8 images (GSI, TSIC, GEO Grid/AIST) and Landsat8 images (courtesy of the U.S. Geological Survey)) with landslide distribution indentified by Kita (2018). d Map of the monitored hillside

indeed, peak ground accelerations are significantly amplified at or near ridge crests (Meunier et al. 2008). However, Kasai and Yamada (2019) demonstrated that 65% of the landslides caused by the 2018 Hokkaido Eastern Iburi Earthquake occurred on concave slopes. Osanai et al. (2019) observed that shallow landslide slip surfaces tend to be very wet. Shortly (25–27 h) before the earthquake, Typhoon Jebi delivered 11 mm of rainfall to the surrounding area. Together, these findings suggest that water played an important role in the initiation of shallow landslides following the earthquake.

As a first step toward clarifying the role of water in the landslides that were triggered by the 2018 Hokkaido Eastern Iburi Earthquake, we conducted long-term monitoring of pressure head dynamics on a natural hillside covered by thick pyroclastic fall deposits that remained in place throughout the earthquake.

## Methods

### Study Site

Pressure heads were monitored on a forested hillside in Forest Compartment 111, which is managed by the Office of Forestry Management, Iburi General Subprefectural Bureau, Hokkaido Prefectural Government in northern Takaoka, Atsuma, Hokkaido, Japan. Shallow landslides were densely distributed in this area following the 2018 Hokkaido Eastern Iburi Earthquake (Fig. 1); however, landslides did not occur on this slope. The average slope angle is 22°; this slope angle class (20–25°) is characterized by a very high probability of landslides triggered by earthquake (Kasai and Yamada 2019). The basement complex in this area consists of sedimentary rock of the Neogene tertiary system called the Fureoi Formation (alternate layers of sandstone and mudstone) (Ozaki and Komatsubara 2014), and is covered with thick pyroclastic fall deposits, including Tarumae a (Ta-a: 1739 A.D.), c (Ta-c: 2.5 ka), d (Ta-d: 9.0 ka), the Eniwa a (En-a: 20 ka), Shikotsu 1 (Spfa-1: 40 ka), and Kuttara 1 (Kt-1: 43.5 ka). Tarumae b pyroclastic fall deposits (Ta-b: 1667 A.D.) are nearly absent in this area (Furukawa and Nakagawa 2010).

## Observation of Pressure Heads

The pressure head ($\psi$) was measured at two points (Fig. 1d). Point A is located at the upper part of a concave slope and Point B is located at the upper part of a convex slope.

The layer structure at Point A is shown in Fig. 2a. At this point, the layers consist of Ta-a, Ta-c, Ta-d, and the basement complex from top to bottom. Humus layers are found at the surface of Ta-a and sandwiched between Ta-a and Ta-c and between Ta-c and Ta-d. The middle to bottom of Ta-d and the basement complex are highly weathered. Pyroclastic fall deposit thickness is approximately 1.9 m.

The layer structure at Point B is shown in Fig. 2b. At this point, the layers consist of humus, Ta-a, Ta-c mixed with buried humus, Ta-d, and the basement complex from top to bottom. Pyroclastic fall deposit thickness is approximately 1.3 m. We installed six tensiometers at Point A and two at Point B. Tensiometer installation depths and the corresponding layers are listed in Table 1. The observation period was from July 10 to November 5, 2019 at Point A and from September 5 to November 5, 2019 at Point B. Because our field surveys of the surrounding shallow landslides and other field survey reports (Hirose et al. 2018; Chigira et al. 2019; Osanai et al. 2019; Wang et al. 2019) suggested that the landslide slip surface was located at the bottom of Ta-d or En-a, we infer that the slip surface would have been at the interface between the weathered Ta-d and the basement complex if the earthquake had triggered a shallow landslide on the monitored slope.

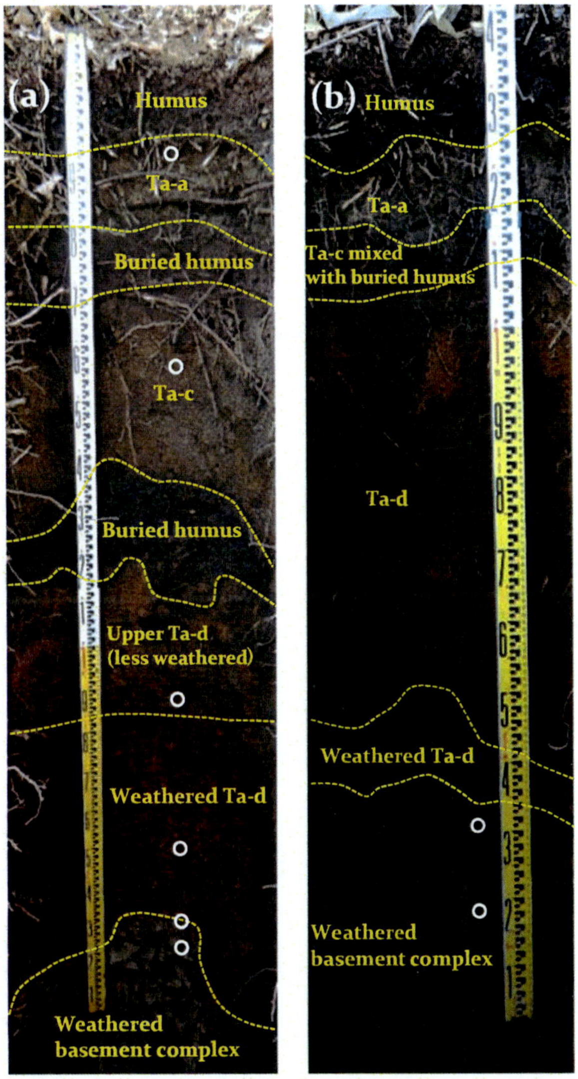


**Fig. 2** Layer structure at **a** Point A and **b** Point B. Open circles indicate tensiometer installation depths. Photographs were taken after the observation period, once the tensiometers had been removed

## Results and Discussion

### Pressure Head Dynamics in Each Layer

Figure 3a and 3b show the $\psi$ dynamics of each layer at Points A and B, respectively. Positive and negative $\psi$ values indicate saturated and unsaturated conditions, respectively. Rainfall data shown in Fig. 3a and 3b was obtained from the nearest weather station, which was the JMA Automated Meteorological Data Acquisition System (AMeDAS) Abira observation station, located 9.5 km west-southwest of the study site (Fig. 1c). Total rainfall during the observation period was 518.5 mm from July 10 to November 5 and 246.0 mm from September 5 to November 5, which was nearly equal to the average (512.7 and 219.6 mm, respectively).

At Point A (concave slope), Ta-a and Ta-c $\psi$ responded rapidly to rainfall, varying greatly throughout the observation period. Weak responses of $\psi$ to rainfall were recorded among the upper Ta-d, weathered Ta-d, the interface between Ta-d and weathered basement complex, and weathered basement complex. Notably, $\psi$ of the weathered Ta-d, the interface between Ta-d and weathered basement complex, and weathered basement complex showed damped rainfall responses, followed by very slow decrease after the

**Table 1** Tensiometer installation depths and corresponding layers

| Location | Depth (cm) | Layer |
|---|---|---|
| Point A | 10 | Ta-a |
| | 45 | Ta-c |
| | 114 | Upper Ta-d |
| | 152 | Weathered Ta-d |
| | 196 | Interface between Ta-d weathered basement complex |
| | 212 | Weathered basement complex |
| Point B | 129 | Weathered basement complex |
| | 142 | Weathered basement complex |

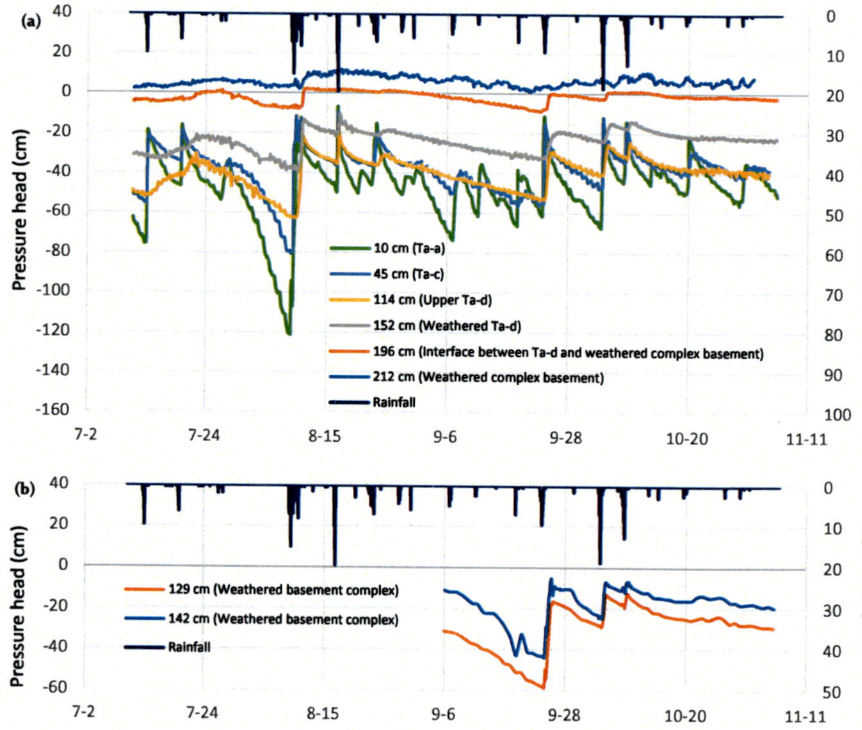


**Fig. 3** Pressure head dynamics at **a** Point A and **b** Point B

cease of rainfall. The interface between Ta-d and weathered basement complex was nearly saturated and became saturated in response to relatively large rainfall events (particularly at total rainfall > 50 mm). The weathered basement complex was always saturated.

Box-and-whisker plots of $\psi$ for each layer at Points A and B throughout the observation period are shown in Fig. 4. At Point A, the range of $\psi$ was greatest in Ta-a (–121.2 to –6.7 cm) and decreased as depth increased, resulting in the narrowest range in the weathered basement complex (0.9–12.9 cm). The mean and median of $\psi$ tended to increase with depth from Ta-a (–46.1 and –43.7 cm, respectively) to the weathered basement complex (6.3 and 6.0 cm, respectively). Thus, the deeper layer clearly remained wetter, with less variation in $\psi$.

At Point B (convex slope), the weathered basement complex remained unsaturated throughout the observation period and $\psi$ decreased to –40 to –60 cm during a dry period without large rainfall (Fig. 3b). The range of $\psi$ at depths of 129 and 142 cm was greater (–58.6 to –12.2 cm, and –43.9 to –5.1 cm, respectively) and the mean (–25.8 and –15.6 cm, respectively) and median (–21.5 and –11.8 cm, respectively) of $\psi$ were smaller than those of the weathered basement complex at Point A (Fig. 4). Thus, the weathered basement complex showed greater variation in $\psi$ and remained drier at Point B than at Point A.

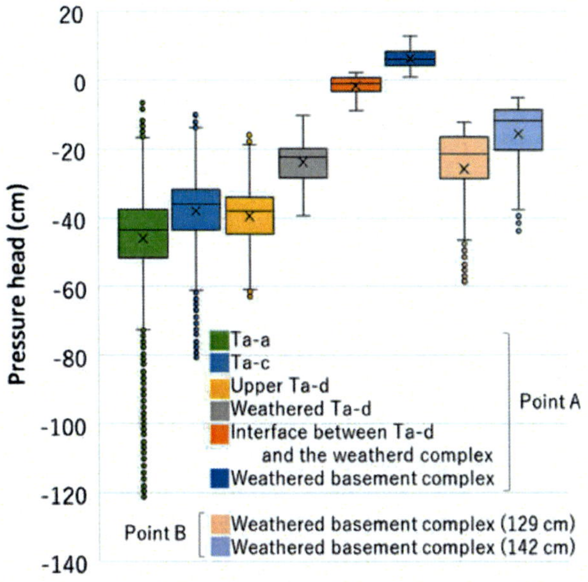


**Fig. 4** Box-and whisker plots of pressure heads observed in each layer at Points A and B. The interquartile range is represented by the box. Upper whiskers are maximum values $\leq 1.5$ times the interquartile range from the top of the box. Lower whiskers are minimum values $\geq 1.5$ times the interquartile range from the bottom of the box. Outliers are plotted as dots. Horizontal line and x indicate the median and mean, respectively

## The Role of Water in Earthquake-Triggered Landslides

On the concave slope (Point A), the interface between Ta-d and weathered basement complex was nearly saturated and became saturated in response to large rainfall events. The weathered basement complex remained saturated throughout the observation period. Since these conditions were observed during a period with average rainfall, we infer that these conditions have persisted for many years, more strongly promoting weathering of the pyroclastic fall deposits over the basement complex (i.e., Ta-d) on the concave slope. This conclusion is supported by the greater thickness of weathered Ta-d at Point A (60 cm) than at Point B (11 cm) (Fig. 2 and Table 2).

Table 2 shows the monthly total rainfall for June, July, and August 2018 (approximately 3 months before the earthquake) and 2019 at the Abira AMeDAS station. Monthly total rainfall was larger in 2018 than in 2019 during all months, and the 3-month total rainfall in 2018 was 1.92 times greater than that in 2019. The $\psi$ of the interface between Ta-d and the weathered basement complex (the potential landslide slip surface) at 03:00:00 (JST) on September 6, 2019 was approximately –2 cm. Considering the much greater rainfall in 2018, it is highly likely that the area surrounding the potential slip surface was saturated on slopes similar to those in our study site when the earthquake occurred.

Based on a field study, Li et al. (2020) reported that shallow landslides in the study area were initiated from liquefied failure of weak Ta-d and that Ta-d had lower shear resistance and anti-liquefaction strength than other layers, thus easily liquefying under intense ground motion. Increased weathering due to persistent saturated or nearly saturated conditions, together with saturated conditions near the potential slip surface when the earthquake struck the concave slope, as inferred from our observations, support the landslide mechanism suggested by Li et al. (2020) and explain why more landslides occurred on concave slopes than on convex slopes. However, it should be noted that $\psi$ at the similar depth at Points A and B (e.g., 114 cm at Point A and 129 cm at Point B) showed the similar $\psi$ dynamics (Fig. 3). This implies that wetness condition is determined by depth from the ground surface rather than the layer materials and that the area surrounding the potential slip surface can be always saturated or nearly saturated also on convex slopes where the pyroclastic fall deposits thickness is the same as on our concave slope (Point A).

The effect of the pyroclastic fall deposits thickness should be clarified in future work.

Based on our observation results, it is unlikely that the 11 mm of rainfall brought by Typhoon Jebi 25–27 h prior to the earthquake had a major effect on landslides triggered by the earthquake because similar rainfall amounts had little influence on $\psi$ of the weathered Ta-d through the weathered basement complex, including the potential slip surface; these layers were already sufficiently wet before the typhoon. However, more detailed investigation is needed because it is possible that (1) much more rainfall before the 2018 earthquake kept the layers wetter and the 11-mm rainfall merely expanded the depths of this water prior to the earthquake, (2) it rained more, since the catchment ratio of the rain gauge at the Abira station decreased due to strong, typhoon-induced winds (ca. 15 m/s), and (3) larger overall water content in the

**Table 2** Monthly rainfall from June to August in 2018 and 2019

| Monthly rainfall (mm) | | | | |
|---|---|---|---|---|
| Year | June | July | August | 3-month total |
| 2018 | 191.0 | 213.5 | 288.5 | 693.0 |
| 2019 | 72.5 | 79.0 | 209.0 | 360.5 |

layers, inferred from the observed rise in $\psi$ (especially in Ta-a and upper Ta-d) in response to rainfall made the layers heavier and more susceptible to sliding in response to strong seismic shocks.

## Conclusions

In this study, we conducted long-term monitoring of $\psi$ dynamics on a natural hillside covered with thick pyroclastic fall deposits where landslides did not occur after the 2018 Hokkaido Eastern Iburi Earthquake. The results indicate that the lower part of the pyroclastic fall deposits and the weathered basement complex remained wetter (nearly saturated or saturated) and had less $\psi$ fluctuation than the upper layers. These conditions likely promoted weathering of the pyroclastic fall deposits over the basement complex. These trends were stronger on the concave slope than on the convex slope, which may explain why coseismic landslides occurred more frequently on concave slopes than on convex slopes.

The results of this study represent a first step toward clarifying the important role of water in the initiation of landslides by earthquake. Further studies, including measurement of the hydraulic properties (e.g., hydraulic conductivity and water retention characteristics) of each layer, saturated/unsaturated seepage flow analysis, and slope stability analysis are needed to elucidate the role of water in this landslide mechanism.

**Acknowledgements** We are grateful to the Office of Forestry Management, Iburi General Subprefectural Bureau, Hokkaido Prefectural Government for their provision of the study site. We are also grateful to our colleagues for their support in the field. Airborne light detection and ranging (LiDAR) data were provided by the Construction Department, Hokkaido Regional Development Bureau, Ministry of Land, Infrastructure, Transport, and Tourism (MLIT), Japan. This study was partly supported by Japan Society for the Promotion of Science (JSPS) KAKENHI grants (nos. JP18H03819 and JP19H02393) and a grant from MLIT.

## References

Chigira M, Tajika J, Ishimaru S (2019) Landslides of Pyroclastic Fall Deposits Induced by the 2018 Eastern Iburi Earthquake with Special Reference to the weathering of Pyroclastics. DPRI Kyoto Univ, Kyoto, Japan, Annu 62(B):348–356. http://hdl.handle.net/2433/244968

Furukawa R, Nakagawa M (2010) Geological map of Tarumae Volcano 1:30,000. Geological map of volcanoes 15. Geological survey of Japan, AIST

Hirose W, Kawakami G, Kase Y, Ishimaru S, Koshimizu K, Koyasu H, Takahashi R (2018) Preliminary report of slope movements at Atsuma Town and its surrounding areas caused by the 2018 Hokkaido Eastern Iburi Earthquake. Report of the Geological survey of Hokkaido (90):33-44

Kasai M, Yamada T (2019) Topographic effects on frequency-size distribution of landslides triggered by the Hokkaido Eastern Iburi Earthquake in 2018. Earth Planets Space 71(89). https://doi.org/10.1186/s40623-019-1069-8

Kawabe H (1987) Landslides caused by Earthquake (2): landslide characteristics and estimation of landslide area rates. The Tokyo University Forests report, Tokyo, Japan, 77:91–142. http://hdl.handle.net/2261/23123

Kinoshita A, Yamaguchi S, Hirano Y, Fujinoki Y, Murata K, Yoshimatsu H (2010) Study of the geological and geomorphological characteristics of the collapse-type landslides triggered by earthquake. J Jpn Landslide Soc 47(1):34–41. https://doi.org/10.3313/jls.47.34

Kita K (2018) Inventory of landslides triggered by the 2018 Hokkaido Eastern Iburi earthquake. https://github.com/koukita/2018_09_06_atumatyou. Accessed 5 Mar 2019

Li R, Wang F, Zhang S (2020) Controlling role of Ta-d pumice on the coseismic landslides triggered by 2018 Hokkaido Eastern Iburi Earthquake. Landslides 17:1233–1250. https://doi.org/10.1007/s10346-020-01349-y

Meunier P, Hovius N, Haines JA (2008) Topographic site effects and the location of earthquake induced landslides. Earth Planet Sci Lett 275(3–4):221–232. https://doi.org/10.1016/j.epsl.2008.07.020

Nakagawa M, MIyasaka M, Miura D, Uesawa S (2018) Tephrostratigraphy in Ishikari Lowland, Southwestern Hokkaido: Eruption history of the Shikotsu-Toya volcanic field. J Jpn Geol Soc 124 (7):473–489. https://doi.org/10.5575/geosoc.2018.0038

Nishida K, Kobashi S, MIzuyama T (1996) Analysis of distribution of landslides caused by 1995 Hyougoken Nanbu earthquake referring to a database specialized for sediment related disasters. J Jpn Soc Eros Control Eng 49(1):19–24. https://doi.org/10.11475/-sabo1973.49.19

Osanai N, Yamada T, Hayashi S, Katsura S, Furuichi T, Yanai S, Murakami Y, Miyazaki T, Tanioka Y, Takiguchi S, Miyazaki M (2019) Characteristics of landslides caused by the 2018 Hokkaido Eastern Iburi Earthquake. Landslides 16(8):1517–1528. https://doi.org/10.1007/s10346-019-02106-7

Ozaki M, Komatsubara T (2014) 1:200,000 land geological map in the Ishikari depression and its surrounding area with explanatory note. http://www.gsj.jp/data/coastal-geology/GSJ_SGMCZ_S4_2014_03_b.pdf. Accessed 1 Mar 2020

Saito M, Araya T, Nakamura F (1995) Sediment production and storage processes associated with earthquake-induced landslides in Okushiri Island, 1993. J Jpn Soc Eros Control Eng 47(6):28–33. https://doi.org/10.11475/sabo1973.47.6_28

Wang G, Furuya G, Watanabe N, Doi I, Ma N (2019) On the features of landslides triggered by the 2018 Hokkaido Eastern Iburi Earthquake. DPRI Kyoto Univ, Kyoto, Japan, Annuals 62(A):48–56. http://hdl.handle.net/2433/-244887

2020 Kyoto Japan

# Investigation of 20 August 2019 Catastrophic Debris Flows Triggered by Extreme Rainstorms Near Epicentre of Wenchuan Earthquake

Dalei Peng, Limin Zhang, Hofai Wong, Ruilin Fan, and Shuai Zhang

**Abstract**

A strong earthquake could trigger a large number of co-seismic landslides and induce large amount of loose materials on steep slopes and in the gullies. Under strong rainfall conditions, these loose materials could induce devastating debris flows, which will endanger the resettled population and destroy the re-built infrastructures. From 19 to 20 August 2019, fourteen debris flows were triggered by extreme rainstorms near the epicentre of the Wenchuan Earthquake. Among the fourteen incidents, three of them produced debris flow dams, which changed the course of the Minjiang River and resulted in flooding at different parts of the reconstructed Miansi town. In addition, sixteen casualties, twenty two missing persons, and destruction of four main roads were reported. In this paper, one typical catchment, named as "Dengxi gully", near the epicentre of the Wenchuan earthquake (Sichuan Province, China) was chosen as a case study for remote sensing analysis, field investigation of landslide evolution and debris flow development before and after the catastrophic events. The debris flow in the study area was initiated in four stages: (a) generation of a large amount of loose materials from the Wenchuan Earthquake; (b) run-off erosion from co-seismic landslide material on hilly slopes and repeated mobilizations in steep channels over the years; (c) development of high intensity localised rainfall events; (d) wash out of accumulated materials in gully by the flood. The study of "8.20 debris flows" can provide a benchmark for analysis of long-term evolution of debris flows in order to identify potential continuing hazards in the earthquake-affected areas and make proper engineering decisions.

**Keywords**

Debris flows • Extreme rainstorms • Epicentre of earthquake • Wenchuan

## Introduction

Debris flow is considered to be a type of movement within the landslide classification which commonly induce serious disaster in mountain areas. Due to its great velocity with long run-out distance and high capacity, large and heavy objects such as tree and rock (Wasowski et al. 2011; Hungr et al. 2014; Zhang et al. 2014; Fan et al. 2018b) are transported. Debris flows can have very to extremely rapid flows of saturated debris in deep channels such as a gully or a ravine and they are major hazards worldwide causing considerable damages to structures and infrastructures (Fan et al. 2018a). Debris flow is usually initiated by the erosion and entrainment of hill slopes and channel materials by overland flow (Domènech et al. 2019; Bezak et al. 2020), or sometimes, triggered by the outbursts of reservoirs built in the channels (Coe et al. 1997). Many geomorphological papers view the role of debris flows like a very active agent of landscape evolution and sediment transfer in mountain areas (Fan et al. 2019a; Yunus et al. 2020).

In the evening of August 19 to the morning of August 20, heavy rainfall occurred in the Wenchuan County and

D. Peng · H. Wong
Faculty of Science and Technology, Technological and Higher Education Institute of Hong Kong, Tsing Yi, Hong Kong, China
e-mail: pdchbsz@vtc.edu.hk

H. Wong
e-mail: ceshfw@vtc.edu.hk

L. Zhang · R. Fan
Department of Civil and Environmental Engineering, The Hong Kong University of Science and Technology, Sai Kung, Hong Kong, China
e-mail: cezhangl@ust.hk

R. Fan
e-mail: rfanaa@connect.ust.hk

S. Zhang (✉)
MOE Key Laboratory of Soft Soils and Geoenvironmental Engineering, Zhejiang University, Zheda Road #38, Hangzhou, 310027, China
e-mail: zhangshuaiqj@zju.edu.cn


V. Vilímek et al. (eds.), *Understanding and Reducing Landslide Disaster Risk*, ICL Contribution to Landslide Disaster Risk Reduction,
https://doi.org/10.1007/978-3-030-60319-9_8

Wolong Special Administrative Region of Aba Prefecture in China. This rainstorm resulted in mountain torrents and debris flow disasters affecting eleven towns in the Wenchuan County at different levels. Among the eleven towns, those located at Shuimo, Sanjiang, Yingxiu, Mianfan and Gengda in the Wenchuan County have suffered from disastrous damages. For the Wenchuan County, the main disasters induced by the heavy rainfall were mainly mountain torrents and debris flows. There were fewer landslide disasters (Fig. 1). According to field investigations and news reports, fourteen debris flows were induced on 20 August 2019, in the Coutou gully, Banzi gully, Dengxigou, Chediguan gully, Xiazhuang gully, Qipan Gully, Xingfu gully, Longtan gully, Mushan gully, Shuimo town (Fig. 1). During this period of time, tens of thousands of tourists were stranded by the debris flows. It also caused sixteen casualties, twenty two missing persons and destruction of four main roads. The direct economic losses was estimated to be 3.6 billion yuan. Under the influence of debris flow deposits, the whole riverbed had been uplifted, which directly caused river water blocking and diversion, and brought many adverse effects to post-disaster reconstruction, road restoration, industrial and agricultural recovery of the study area. According to previous literatures (Fan et al. 2019b), the Dengxi gully did not occur large-scale debris flow with more material sources and strong activity areas after Wenchuan earthquake.

The aims of this paper are to: (1) investigate the typical debris flow near the epicentre of the Wenchuan earthquake on 20 August 2019, and (2) analyse the formation condition and disaster characteristics of the rainfall-induced debris flows.

## Regional Setting

The study area is the Dengxi ravine, with an area of 44 km$^2$, located in the town of Miansi. It is on the right bank of the Mingjiang River and 20 km away from the capital of Wenchuan County, Sichuan Province, China (Fig. 1a). It is 40 km away from the north of the Wenchuan earthquake epicentre (Fig. 1b). The relative elevation of the river valley is greater than 1100–1200 m, and the area in the vicinity of the river has typical alpine and canyon landscapes formed by strong erosion and down cutting (Fan et al. 2018a). The upslope elevation of the Dengxi gully is more than 4500 m asl, and the outlet of the valley is 1300 m asl.

The bedrock of the study area is primarily highly weathered Quaternary Jingningian Period intrusive granite, with intensely developed joints. The cover soil is thin and mainly composed of Quaternary glacial deposits, diluvium, colluvium and alluvium. Consequently, extensive collapses occurred during the Wenchuan earthquake and a tremendous amount of loose materials was retained on the steep hillslope and in the gullies, which developed massive co-seismic landslides (Shen et al. 2017).

The ravine is located within the Mianchi-Wenchuan zone of rainfall with an average annual rainfall of 719.7 mm between 1960 and 2015. The rainfall variability is large in different years. Most of the yearly rainfall occurs during the wet season from May to September. The groundwater is mainly magmatic rock fissure water with shallowly restored, where the supply is almost equal to the discharge (Zhang et al. 2012).

## Methods

A GIS analysis platform is adopted to interpret satellite images over the study area. Three sets of Satellite images with a resolution of 1 m taken on 27 December 2014, 10 m taken on 25 April 2019, and 1 m taken on 29 October 2019 were collected to map the locations of rainfall-induced landslides. These were also used to analyse the features of the debris flow source areas, the flow paths and flow deposits in the study region to provide data about the amount of loose material available for the initiation of debris flows. A digital elevation model (DEM) with a spatial resolution of 10 m was used to generate slope angles and elevations.

Field investigations were undertaken to confirm visual interpretation based on the remote sensing data. The average deposit thickness, erosion and entrainment depth of the ravine was measured using a laser range finder. Material sources of debris flows are investigated. A large number of high-resolution pictures of coarse deposition materials were taken for grid-by-number analysis and runout characteristics analysis.

The distribution of 24-h rainfall in the Wenchuan County is collected from Aba Prefecture Meteorological Bureau. Hourly rainfall records from the meteorological station near the town of Mozi and the Wenchuan County were used to determine the rainfall intensity/duration for each debris flow-producing rainstorm.

## Disaster Characteristics and Formation Condition

### Disaster Characteristics

Material sources findings and landslide inventory mapping are an essential first step in the analysis of debris flows and characteristics development in the study area. The evolution of loose materials in the present study can be quantified using satellite images (Fig. 2a–c) (Fan et al. 2019a). In the rainy season, hillslope deposits reactivate, where a considerable portion of these hillslope deposits either slid downhill

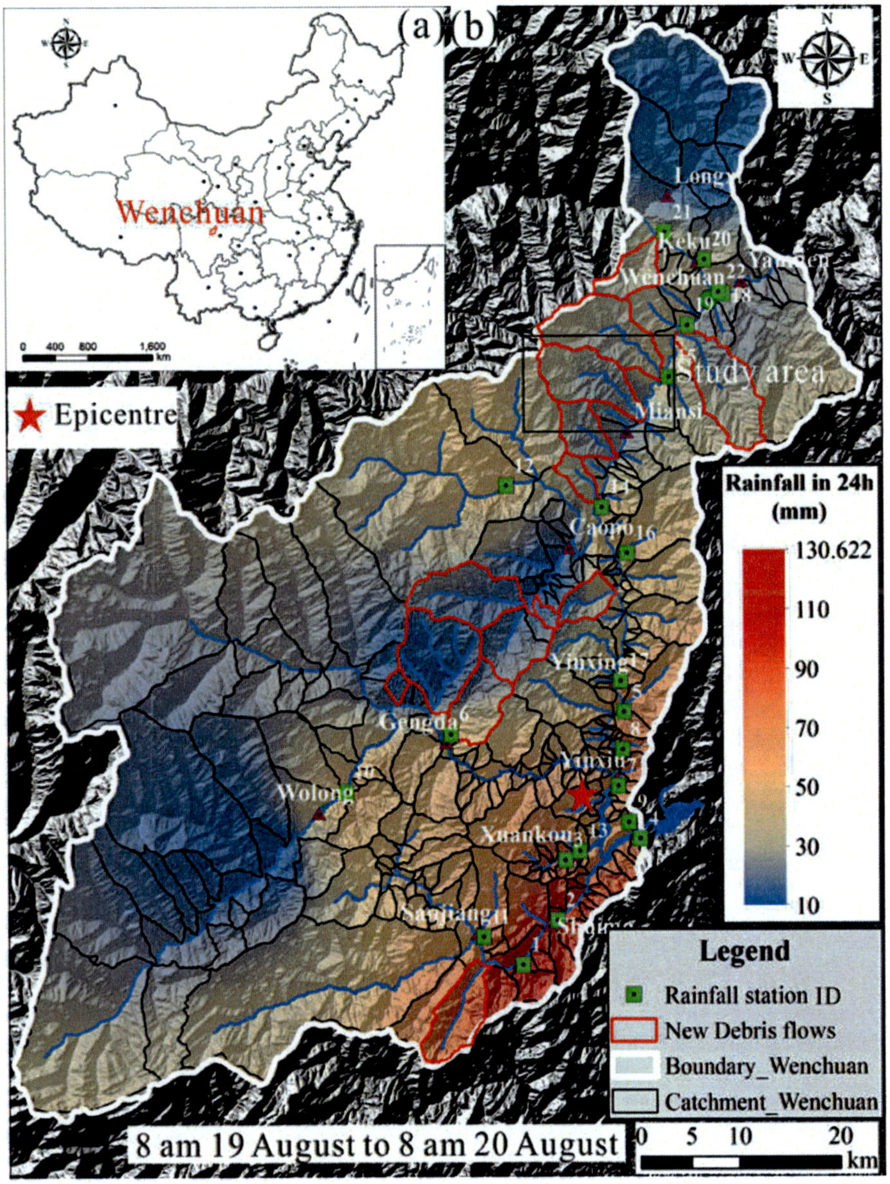


**Fig. 1** Location of study area. **a** Location of Wenchuan in China; **b** Distribution of debris flows and rainfall at 24 h from 8 am 19 August to 8 am 20 August occurred on 20 August 2019 in Wenchuan County and location of earthquake epicentre and rainfall stations

or washed into existing channels by surface runoff, and turn into channel deposits (Zhang and Zhang 2017). According to the material source, the area of the Dengxi gully is divided into the source area and flow-deposits area (Fig. 2d).

Interpretation of the remote sensing images and field observations in the Dengxi catchments of Miansi town showed: (1) according to remote sensing imaging taken on 27 December 2014, six years after the Wenchuan earthquake, the total volume of the 125 hillslope deposits shortly after the Wenchuan earthquake was $18.08 \times 10^6$ m$^3$ and the total volume of the channel deposits was $9.04 \times 10^6$ m$^3$ (Table 1); (2) according to field investigations and remote sensing imaging taken on 25 April 2019, before the "8.20" debris flows (Figs. 3 and 4), the number of fresh landslides was 24. The volume of fresh landslides and original landslides were $2.02 \times 10^6$ and $11.75 \times 10^6$ m$^3$, respectively. The total volume of the channel deposits was $12.76 \times 10^6$ m$^3$ (Table 2); (3) widespread shallow slope sliding were present on high elevation and steep slopes in areas with granitic bedrock. A total of 22 fresh landslides of various sizes were mapped with a total landslide area of 1.163 km$^2$ and a total volume of $2.59 \times 10^6$ m$^3$, which were much smaller than those occurred between 2008 and 2014 (Table 3). Moreover, an abundance of loose material accumulated in the channel after 20 August 2019 rainstorms (Fig. 2c).

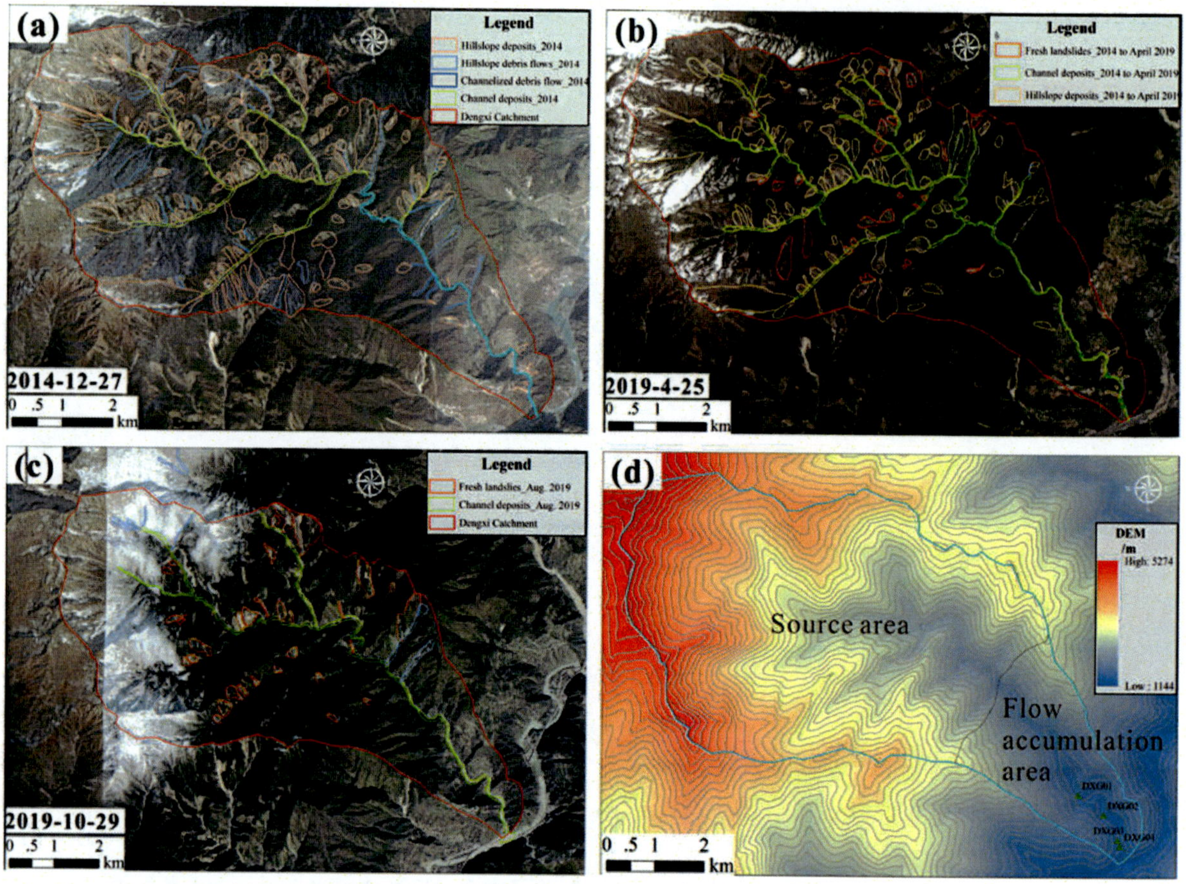


**Fig. 2** Remote sensing imaging from three different data shows the landslide and debris flow development in the Dengxi gully catchment. **a** Source materials after the Wenchuan Earthquake (taken on 12 December 2014); **b** Aerial photograph before the event (taken on 25 April 2019); **c** Aerial photograph after the event (taken on 27 September 2019); **d** The topographic map of the Dengxi debris flow and location of field test

**Table 1** Landslide materials identified through remote sensing imaging taken on 27 December 2014, six years after the Wenchuan earthquake

| Type | Number | Total area ($km^2$) | Volume ($10^6$ $m^3$) |
|---|---|---|---|
| Hillslope deposits | 125 | 8.610 | 18.08 |
| Channel deposits A | 8 | 0.862 | 8.61 |
| Channel deposits B | 1 | 0.216 | 0.43 |
| Total | – | 9.688 | 27.12 |

The co-seismic landslide materials were from run-off erosion from hillslopes and repeated mobilisations in steep channels between 2008 and 2019. The accumulated material in the gully was washed out by the flood. In rainfall conditions, the co-seismic landslide collapsed and deposits were washed away. From the initiation form, the debris flow is channel-activated. Through post-earthquake field experiment on sites, the mechanism of turning accumulations from floods to debris flows in channels at regions with strong earthquake occurrence are illustrated (see Fig. 3).

## Formation Conditions

At 8:30 a.m. on 20 August 2019, Wenchuan meteorological station issued the rainfall report for the past 24 h from 8:00 a.m. on August 19 to 8:00 a.m. on August 20, one station had heavy rainfall, 130.7 mm in Baishi Village, Shuimo town; 16 stations had heavy rainfall. From 0:00 a.m. to 7:00 a.m. on 20 August 2019, the maximum accumulated rainfall in Wenchuan county was 65 mm (Fig. 1). At 0:00 a.m. to 3:00 a.m. on 20 August 2019, the precipitation in Baishi

**Fig. 3** Field investigation photos. **a** The Dengxi gully debris flow (taken on 20 August 2019) (credit: Sichuan Highway Planning, Survey, Design and Research Institute Ltd); **b** Outlet of the Dengxi gully and the blocked Mingjiang River induced by debris flows (taken on 20 August 2019); **c** Fresh landslide (taken on 24 December 2019); **d** River bank erosion the downstream of a gully (taken on 24 December 2019); **e** Channel erosion in the upstream of a gully (taken on 22 December 2019); **f** Broken house 20 m above river course (taken on 24 December 2019)

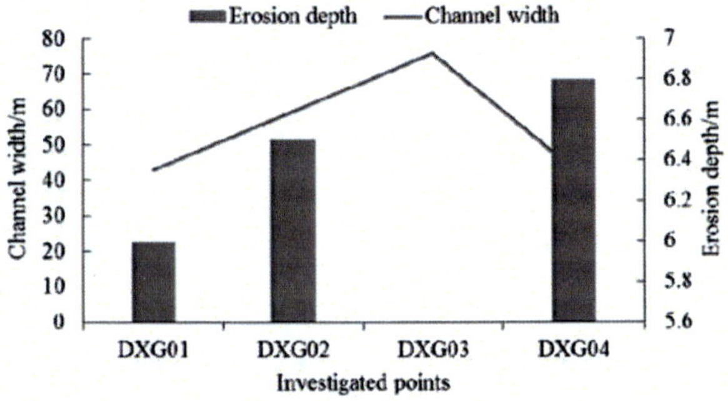


**Fig. 4** Erosion depth at different investigated sites (for location of points see Fig. 2d)

Village, Shuimo Town, Wenchuan County had reached 76.8 mm (Fig. 5). The nearest distance between the rainfall station and debris flow is 5 km, named Banzigou rainfall station and ID 15 with a rainfall value of 55.2 in 24 h.

According to the previous research reports (Zhou and Tang 2014; Fan et al. 2018a, 2019b), the ID threshold graph is associated with 85 debris flows events in the Wenchuan earthquake-stricken area (Fig. 6). It can be seen from Fig. 6 that the rainfall intensity inducing this event is far higher than the debris flow threshold ($I = 18.6D^{-0.87}$), which is located in the debris flow occurrence area of the model), nevertheless, the rainfall intensity is slightly smaller than the ID threshold

**Table 2** Landslide materials identified through field investigations and remote sensing imaging taken on 25 April 2019, before the "8.20" debris flows

| Type | Number | Total area ($km^2$) | Volume ($10^6$ $m^3$) |
|---|---|---|---|
| Fresh landslides | 24 | 0.869 | 2.02 |
| Original landslides | 98 | 5.043 | 11.75 |
| Channel deposits A | 19 | 1.198 | 8.39 |
| Channel deposits B | 1 | 0.324 | 4.37 |
| Debris flow runout material | * | * | * |
| Total | – | 7.434 | 26.53 |

**Table 3** Landslide materials identified through field investigations and remote sensing imaging taken on 29 October 2019, after the "8.20" debris flows

| Type | Number | Total area ($km^2$) | Volume ($10^6$ $m^3$) |
|---|---|---|---|
| Fresh landslides | 22 | 1.163 | 2.59 |
| Original landslides | 83 | 4.264 | 9.51 |
| Channel deposits A | 17 | 1.663 | 11.64 |
| Channel deposits B | 1 | 0.279 | 2.79 |
| Debris flow runout material | 1 | 0.058 | 0.87 |
| Total | – | 7.427 | 27.40 |

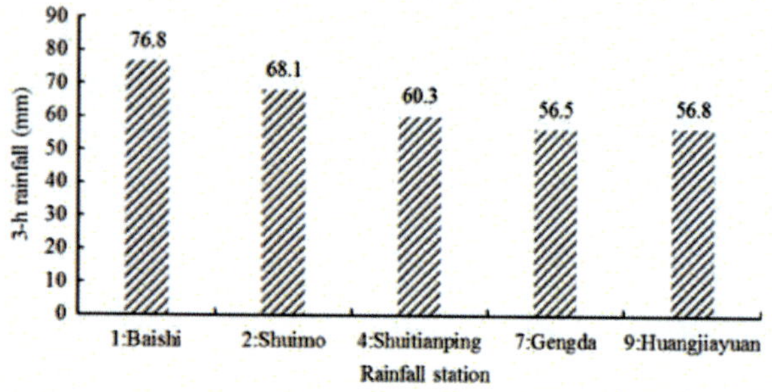


**Fig. 5** 3-hr rainfall before debris flow occurred in Wenchuan County meausured by five rain gauges

($I = 66.36D^{-0.79}$) proposed by Zhou and Tang (2014). In 2010, the cumulative rainfall of the "8.13" mass debris flow was 62.9 mm. The critical hourly rainfall intensity is 32.2 mm. The average hourly rainfall is 6.99 mm. The accumulated rainfall of this debris flow is 87.5 mm. The critical hourly rainfall intensity is 24.2 mm (SKLGP 2019).

According to local residents, there were many heavy rainfalls in the early stages of this event during the period from June to August. The early rainfalls raised the moisture content of the loose materials in the relevant areas and made it close to the saturated state. On this basis, heavy early rainfalls directly caused the occurrence of debris flow.

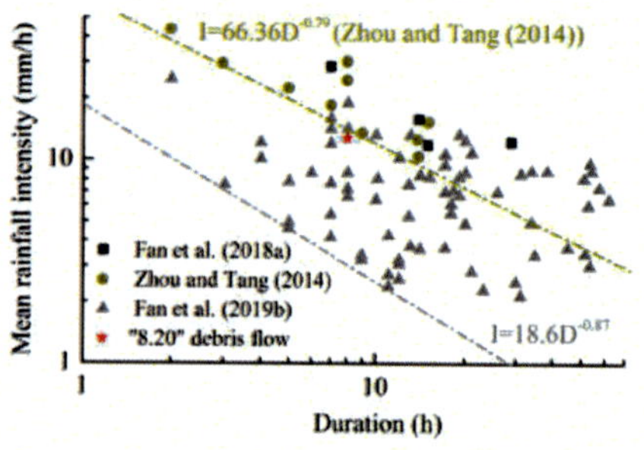


**Fig. 6** Compasion in rainfall intensities between the 8.20 debris flow and other debris flow events in the Wenchuan earthquake area

## Conclusion and Suggestion

The 8.20 debris flows showed the following main features:

(1) A total of fourteen catastrophic debris flows occurred in the same period from 19 to 20 August 2019 in the earthquake zone and destroyed local urban housing and infrastructure. According to typical Dengxi debris flow, a total of twenty two fresh landslides of various sizes were mapped with a total landslide area of 1.163 $km^2$ and a total volume of 2.59 × $10^6$ $m^3$, which were much smaller than those occurred between 2008 and 2014.

(2) Co-seismic landslide materials were run-off erosion from hillslopes and repeated mobilisations in steep channels during 2008–2019. From the initiation, most of the debris flows are channel-activated. Moreover, an abundance of loose material accumulated in the channel after 20 August 2019 rainstorms.

(3) The critical hourly rainfall intensity of "8.20" debris flows is 24.2 mm, which is much smaller than "8.13" mass debris in 2010 with a value of 32.2 mm. the rainfall intensity inducing this event is far higher than the debris flow threshold ($I = 18.6D^{-0.87}$).

(4) At present, rainfall stations are mostly located in urban areas or gullies, meaning it is impossible to truly reflect the rainfall in mountainous areas. In order to achieve timely and accurate meteorological warnings, it is necessary to increase the arrangement of rainfall stations in mountain areas. At the same time, it is necessary to carry out real-time meteorological early-warning of geological disasters.

**Acknowledgements** The work described in this paper was fully supported by a grant from the Research Grants Council of the Hong Kong Special Administration Region, China (No. UGC/FDS25/E11/17).

## References

Bezak N, Jež J, Sodnik J, Jemec Auflič M, Mikoš M (2020) An extreme may 2018 debris flood case study in northern slovenia: analysis, modelling, and mitigation. Landslides

Coe JA, Glancy PA, Whitney JW (1997) Volumetric analysis and hydrologic characterization of a modem debris flow near yucca mountain, nevada. Geomorphology 20(1):11–28

Domènech G, Fan XM, Scaringi G, van Asch TWJ, Xu Q, Huang RQ, Hales TC (2019) Modelling the role of material depletion, grain coarsening and revegetation in debris flow occurrences after the 2008 wenchuan earthquake. Eng Geol 250:34–44

Fan XM, Scaringi G, Domènech G, Yang F, Guo XJ, Dai LX, He CY, Xu Q, Huang RQ (2019) Two multi-temporal datasets that track the enhanced landsliding after the 2008 wenchuan earthquake. Earth Syst Sci Data 11(1):35–55

Fan XM, Zhan WW, Dong XJ, van Westen CJ, Xu Q, Dai LX, Yang Q, Huang RQ, Havenith H-B (2018) Analyzing successive landslide dam formation by different triggering mechanisms: the case of the tangjiawan landslide, sichuan, china. Eng Geol 243:128–144

Fan RL, Zhang LM, Wang HJ, Fan XM (2018) Evolution of debris flow activities in gaojiagou ravine during 2008–2016 after the wenchuan earthquake. Eng Geol 235:1–10

Fan XM, Scaringi G, Korup O, Joshua West A, van Westen CJ, Tanyas H, Hovius N, Hales TC, Jibson RW, Allstadt KE, Zhang LM, Evans SG, Xu C, Li G, Pei XJ, Xu Q, Huang RQ (2019) Earthquake-induced chains of geologic hazards: Patterns, mechanisms, and impacts. Rev Geophy

Hungr O, Leroueil S, Picarelli L (2014) The varnes classification of landslide types, an update. Landslides 11(2):167–194

SKLGP (2019) Preliminary investigation and analysis of the "8.20" heavy rainfall induced mountain flood and debris flow disaster in Sichuan Province. https://www.sklgp.cdut.edu.cn/info/1018/3892.htm

Shen P, Zhang LM, Chen HX, Gao L (2017) Role of vegetation restoration in mitigating hillslope erosion and debris flows. Eng Geol 216:122–133

Wasowski J, Keefer DK, Lee C-T (2011) Toward the next generation of research on earthquake-induced landslides: current issues and future challenges. Eng Geol 122(1–2):1–8

Yunus AP, Fan XM, Tang XL, Jie D, Xu Q, Huang RQ (2020) Decadal vegetation succession from modis reveals the spatio-temporal evolution of post-seismic landsliding after the 2008 wenchuan earthquake. Remote Sens Environ 236

Zhang S, Zhang LM (2017) Impact of the 2008 wenchuan earthquake in china on subsequent long-term debris flow activities in the epicentral area. Geomorphology 276:86–103

Zhang LM, Zhang S, Huang RQ (2014) Multi-hazard scenarios and consequences in beichuan, china: The first five years after the 2008 wenchuan earthquake. Eng Geol 180:4–20

Zhang S, Zhang LM, Peng M, Zhang LL, Zhao HF, Chen HX (2012) Assessment of risks of loose landslide deposits formed by the 2008 wenchuan earthquake. Natural Hazards Earth Syst Sci 12(5):1381–1392

Zhou W, Tang C (2014) Rainfall thresholds for debris flow initiation in the wenchuan earthquake-stricken area, southwestern china. Landslides 11(5):877–887

# Spatial Distribution of Lakes in the Central Andes (31°–36°), Argentina: Implications for Outburst Flood Hazard

2020 Kyoto Japan

Mariana Correas-Gonzalez, Stella Maris Moreiras, and Jan Klimeš

**Abstract**

This paper presents an inventory of lakes in the Central Andes region (31°–36° S) of Argentina by using high resolution satellite images. Lakes were classified according to the damming processes or associated landforms. Statistical analysis of main lakes attributes was carried out showing that larger lakes are associated with volcanic and karstic landforms, while smaller ones correspond to thermokarst supraglacial lakes. Lake distribution is forced by the regional topography being mainly located in the periglacial and glacial environment (above the 3730 m a.s.l.). Moreover, latitudinal distribution of different types of lakes differs. Larger water bodies in the Argentinean Central Andes are landslides dammed except to the 34°–35° S volcanic region. These findings set the first step in hazard analysis and highlight the need for carrying out further research to assess outburst flood hazard in the Arid Central Andes of Argentina.

**Keywords**

High mountain lakes • Natural dams • Outburst flood • Hazard • Arid Central Andes • Argentina

M. Correas-Gonzalez · S. M. Moreiras (✉)
Instituto Argentino de Nivología, Glaciología y Ciencias Ambientales (IANIGLA). CONICET, Avda. Ruiz Leal s/n. Parque General San Martín, 5500 Mendoza, Argentina
e-mail: moreiras@mendoza-conicet.gob.ar

M. Correas-Gonzalez
e-mail: mcorreas@mendoza-conicet.gob.ar

S. M. Moreiras
Departamento de Agrícola Aplicada, Facultad de Ciencias Agrarias, Universidad Nacional de Cuyo. Almirante, Brown 500, M5528AHB Chacras de Coria-Luján de Cuyo, Mendoza, Argentina

J. Klimeš
Institute of Rock Structure and Mechanics, The Czech Academy of Sciences, V Holesovickach 41 182 09, Prague 8, Czech Republic
e-mail: klimes@irsm.cas.cz

## Introduction

Lakes and natural dams are generally considered in a positive sense since they play an important ecological role for wildlife, provide freshwater for human consumption and crops irrigation, and they serve as nature tourist attractions. However, they can represent a potential danger if their dams are unstable possibly resulting in their catastrophic breach. Their stability depends on many factors such as lake volume, dam freeboard and material, geomorphological context, among others (Costa and Schuster 1988; Emmer et al. 2014; Emmer and Vilímek 2014).

Main lakes' inventories and classifications are based on dam forming processes and nature of its materials (Costa and Schuster 1988; Richardson and Reynolds 2000; Korup and Tweed 2007; Emmer et al. 2014, 2016; Iturrizaga 2014; Iribarren Anacona et al. 2015; Wilson et al. 2018). Lake type identification provides key information on its stability analysis. Glacial lakes and landslide-dammed lakes are more prone to instability turning clearwater floods into hyperconcentrated flows by erosion and sediment entrainment while moving down a valley, which highly increases peak discharge and volume that causes destruction downstream. For instance, Breien et al. (2008) reported an increase of eroded material by a factor of ten in the debris flow volume caused by a GLOF in Western Norway.

Catastrophic outburst floods have occurred in the Central Andes since the 18th century. One of the largest worldwide events was reported in 1914. An extraordinary Landslide Lake Outburst Flood (LLOF) occurred when the Carri-Lauquen lagoon dammed by a rock avalanche collapsed releasing $1.55 \times 10^9$ $m^3$ of water into the Barrancas River (36°30′S) (Groeber 1916; Hermanns et al. 2011; Colavitto et al. 2012; Ramos 2017). Besides, an unexpected GLOF (Glacial Lake Outburst Flood) took place in 1934, when an ice-dammed lake in the Plomo River basin (33° 07′ S) collapsed releasing $60 \times 10^6$ $m^3$ of water to the Mendoza River (King 1935; Harrison et al. 2015; Correas-Gonzalez


V. Vilímek et al. (eds.), *Understanding and Reducing Landslide Disaster Risk*,
ICL Contribution to Landslide Disaster Risk Reduction,
https://doi.org/10.1007/978-3-030-60319-9_9

et al. 2020). More recently, in 2005 a landslide-dammed lake collapsed in the Santa Cruz River (31° 40′ S) producing an outburst flood of 3.2 × 10$^7$ m$^3$ (D'odorico et al. 2009; Perucca and Angillieri 2009).

The high frequency of outburst floods in this region may indicate higher susceptibility for similar events in the future (Wilson et al. 2008; Correas-Gonzalez et al. 2020). Moreover, warming conditions related to global climate change scenario force glaciers retreatment probable increasing the glacial lake formation and GLOF occurrence in the Central Andes. Nonetheless, whether GLOF or LLOF events could be more plausible in this dry and tectonically active Andean region is not well understood. This study focuses on apprehending which is the origin and type of lakes in the Central Andes to understand probable dangerous outburst floods. In this sense, the inventory constitutes the first step in hazard analysis which has not been addressed in detailed in the study area.

## Study Area

The Central Andes span between the 31° to 36° S latitude, from the center of the San Juan province until the South of Mendoza province (Lliboutry 1998). At this latitude, the Andes range has a mean altitude of 3500 m a.s.l. with the highest elevations of the entire range between 31° and 32° with Aconcagua Mt. (6961 m a.s.l.) and Mercedario Mt. (6770 m a.s.l.); in contrast, the topography decreases towards the South. The region comprises nine main basins that drain the rivers to the lower lands, providing drinking water for human consumption, irrigation and hydroelectricity power generation as main cities are located at the foothills of the Andes (Fig. 1).

The study region is located in the South American Arid Diagonal since the Andes range acts as an effective topographic barrier for the air masses coming from the Pacific Ocean. Thus, the climate is continental and characterized by warm dry summers with some convective storms to the East and snow precipitations during the cold winter to the West. According to an altitudinal and latitudinal gradient, precipitations vary from 100 mm on the northeastern sector of San Juan to 900 mm on the southwestern part of Mendoza (Bruniard 1982).

These climatic and topographic conditions allow the existence of a very important glacial system that covers 1696 km$^2$ (IANIGLA-Inventario Nacional de Glaciares 2018) and includes uncovered glaciers, debris-covered glaciers and (in) active rock-glaciers. Uncovered glaciers are located at the highest altitude (above 4490 m a.s.l.) whereas active rock glaciers are located at a mean altitude of 3820 m a.s.l. in accordance with the location of the freezing level line determined at 3730 m a.s.l. by Drewes et al. (2018).

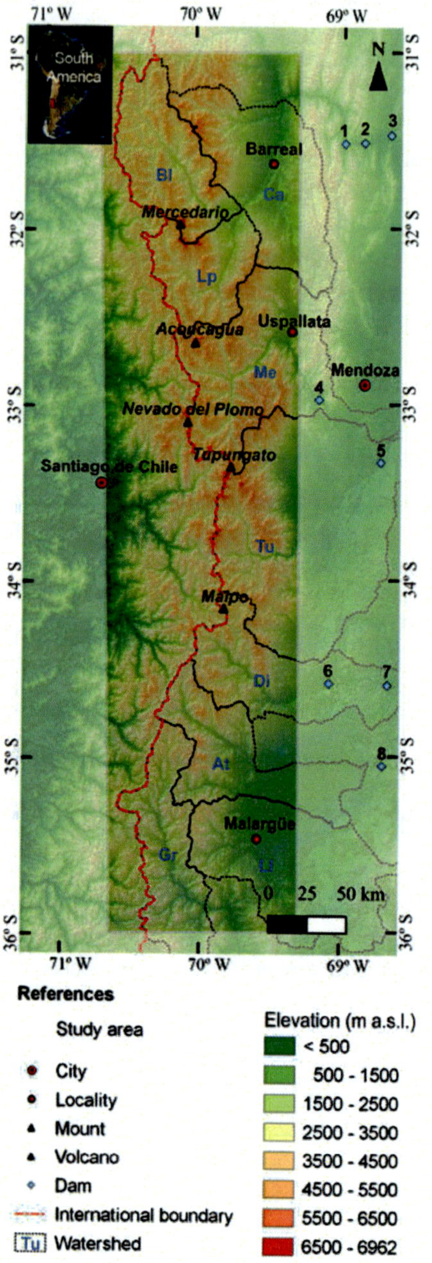


**Fig. 1** Location of the study area where main river basins, cities and dams are referred. Watershed names: Bl: Blanco; Ca: Calingasta; Lp: Los Patos; Me: Mendoza; Tu: Tunuyán; Di: Diamante; At: Atuel; Gr: Grande; Ll: Llancanelo. Dams 1: Los Caracoles, 2: Agua Negra, 3: Ullum, 4: Potrerillos, 5: El Carrizal, 6: Agua del Toro, 7: Los Reyunos and 8: El Nihuil

Streamflow variability of the Central Andes is highly dependent on winter snowpack accumulation and temperature increase during spring and summer (Lauro et al. 2019). On the other hand, during dry years, glacier melt water has been proved critical for maintaining river's minimum discharge (Masiokas et al. 2013).

This portion of the Andes is characterized by an important tectonic activity with historical records of MM > 7 earthquakes (Moreiras and Paez 2015) associated with the flat-slab segment of the Nazca plate subducting under the South America plate. This fact determines, among others, a very steep topographic gradient and the presence of narrow valleys particularly prone to the formation of landslide dammed lakes. However, climatic induced landslides have also been reported in the area (Moreiras 2006; Moreiras et al. 2018).

## Material and Methods

Identification of lakes was done manually by a single operator using high resolution satellite images available in Google Earth, Bing and ESRI provided by Qgis 3.4.15 (quick map services extension) covering a period of 15 years (2005–2020). Even though this is a time consuming approach, error sources are limited to one operator (Wilson et al. 2018) and it has been used in other landform inventories (e.g. Emmer et al. 2016; Vilímek et al. 2016).

Identified lakes were classified in seven major categories according to the lake dam-type and basin development (Table 1). Then, lakes were reclassified according to their surface and gathered into seven groups to map their distribution according to their size (Fig. 2).

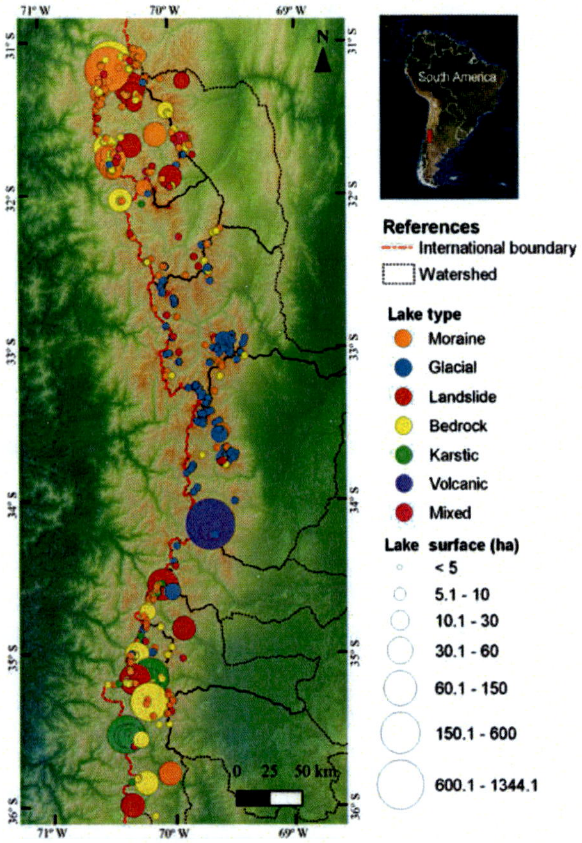


**Fig. 2** Spatial distribution and size of lakes in the Argentinean Central Andes

## Preliminary Results

### Lake Type Inventory

According to the inventory carried out, 658 lakes were identified along the Argentinean Central Andes (Table 2). These water bodies cover a total area of 3460 ha representing less than 0.1% of the study area (Fig. 2). The surface area of these lakes is very variable ranging from 1340 ha to some hundreds of square meters for the smaller ones. Only

**Table 1** Classification of lakes identified in the Central Andes region

| Lake type class | Sub class—Lake characteristics |
|---|---|
| **ML**. Lake associated to a moraine | ML—dammed by a frontal or lateral moraine<br>HML—over the hummocky moraine |
| **GL**. Lake associated to a glacier | RGL—dammed in the main or tributary valley by a rock glacier<br>DCL—dammed in the main or tributary valley by a debris covered glacier<br>DFL—dammed in the main or tributary valley by a debris free glacier<br>SGL—dammed by a surging glacier<br>STL—supraglacial lake (thermokarst) |
| **LL**. Lake associated to a landslide | LL—dammed by a landslide which blocked a main or tributary river<br>DL—dammed by a debris flow |
| **BL**. Lakes associated to glacial erosion | Bedrock dammed lake resulting from glacial erosion (also called tarn or corrie lake) |
| **KL**. Karstic lake | Water body related to karstic dissolution of rocks |
| **VL**. Volcanic lake | Lake from volcanic origin |
| **MXL**. Mixed type lake | Two or more combined elements hold the water into a lake. Ex. moraine and rockbar; moraine and protalus rampart |

**Table 2** Summary of the lake inventory

| Lake type | Lakes | | Surface | | | Mean alt (m a.s.l.) |
|---|---|---|---|---|---|---|
| | No | % | Total (ha) | % | Mean | |
| ML | 206 | 31.3 | 554.5 | 16.0 | 2.7 | 3694 |
| GL | 260 | 39.5 | 124.2 | 3.6 | 0.5 | 4242 |
| LL | 23 | 3.5 | 282.6 | 8.2 | 12.3 | 3267 |
| BL | 113 | 17.2 | 872.3 | 25.2 | 7.7 | 3767 |
| KL | 24 | 3.6 | 224.0 | 6.5 | 9.3 | 2865 |
| VL | 3 | 0.5 | 1375.2 | 39.7 | 458.4 | 3325 |
| MXL | 29 | 4.4 | 32.5 | 0.9 | 1.1 | 3754 |
| Total | 658 | 100 | 3465.3 | 100 | – | – |

30% of these lakes (N = 194) exceeds 1 ha of surface which contributes to 94.7% of the total lake area. Concerning the number of lakes according to their type, those water bodies associated with glaciers are the most important contributing into a 39.5% to the total identified lakes, while lakes dammed by moraine are the second group of importance with 31.3%. Then, follows bedrock dammed lakes (17.2%), mixed type lakes (4.4%), karstic lakes (3.6%), landslide dammed lakes (3.5%), and finally, volcanic lakes (0.5%).

Lakes' surfaces distribution according to lake type is very asymmetric as shown in the boxplot (Fig. 3). The largest one is the Diamante lagoon (33° S) of volcanic origin as it is dammed by basaltic lava of the Maipo volcano. This single lake represents 39.7% of the total area covered by water bodies. On the other hand, bedrock-dammed lakes represent 25.2% of the total area followed by moraine-dammed lakes (16%) and landslide-dammed lakes (8.2%). Finally, karstic lakes constitute 6.5% of the total area followed by glacial lakes (3.6%) and mixed type lakes (0.9%).

The spatial distribution according to lake type (colored circles) and lake size (proportional circles) is shown on a map (Fig. 2). In general, lakes are located in the upper basins and a spatial configuration can be identified with lakes of larger areas (> 10 ha) situated North of 32°30′S and South of 34° S, leaving a strip of lakes smaller than 10 ha in between.

There is a large group of lakes (n = 611) with areas smaller than 5 ha, among which 39% are supraglacial lakes associated with thermokarst located at an average altitude of 4249 m a.s.l. Even though at present these lakes are very small, due to ongoing deglaciation they might coalescence in the future, forming new lakes of greater dimension (Vilímek et al. 2016). This fact might be of relevance in the future evolution of the Andes landscape in the context of current climate change.

## Altitudinal Distribution

Lakes distribution is strongly forced by topography being located between 1924 and 5369 m a.s.l. (average elevation of 3878 m a.s.l.). Regardless of their type, half of the lakes are located above (below) 3951 m a.s.l. However, the altitudinal distribution of lakes decreases from North (31° S) to South (36° S) following the elevation of the Central Andes range. On the other hand, if we consider their types, glacial lakes (GL) are located on average 350 m higher than the other types of lakes. This makes sense as moraines remain below glaciers and landslides are more common along the valley downstream. Besides, the distribution of the lakes in accordance with the periglacial environment is striking: 64.2% of the lakes are located above the freezing line evidencing the main role of permafrost in lake generation (Fig. 4). Contrasting, karstic lakes (KL) are located at lower altitudes (1000 m below the general average).

## Latitudinal Distribution

Lake type classes' distribution contrast at different latitudes of the Argentinean Central Andes. Between 31°–32° S and 34°–35° S, lakes linked to moraines predominate, but in the intermediate latitudinal band glacial lakes do (Fig. 5). Great part of these glacial lakes (small ones) is related to thermokarst processes on covered rock glacier surface as was noted above. Bedrock-dammed lakes, commonly called tarn, are frequent at all latitudes especially in glacial valley headwater, while, the number of landslide dammed lakes increases between 31°–32° S. Moreover, greater water bodies are generated by damming landslides in the

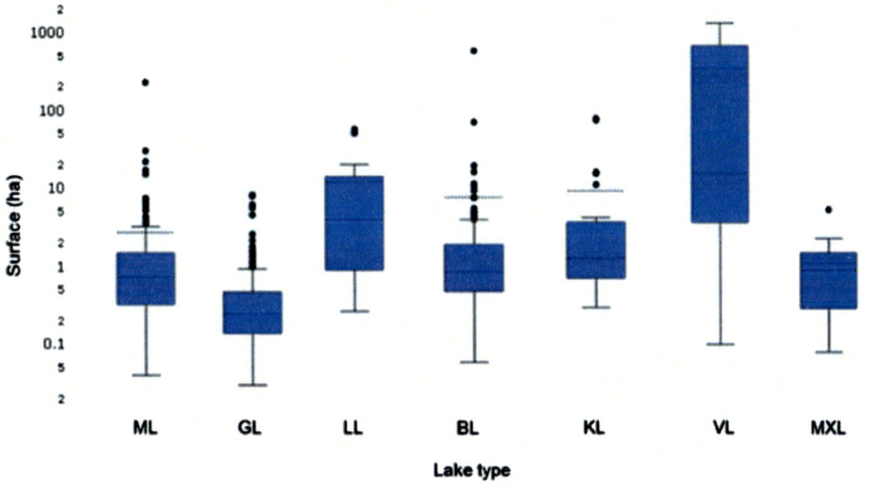


**Fig. 3** Boxplot of classified lakes versus lake surface in the Central Andes. Note that y-axis is in logarithmic scale (for explanation of the abbreviations see Table 1)

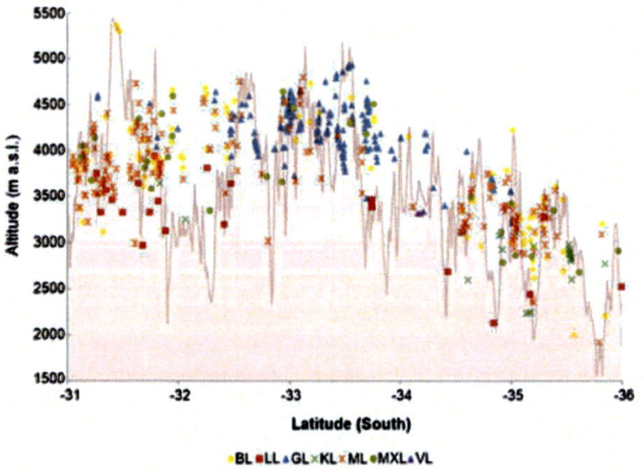


**Fig. 4** Topographic distribution of classified lakes in the Argentinean Central Andes (31°–36° S). The approximate topographic profile is red shaded

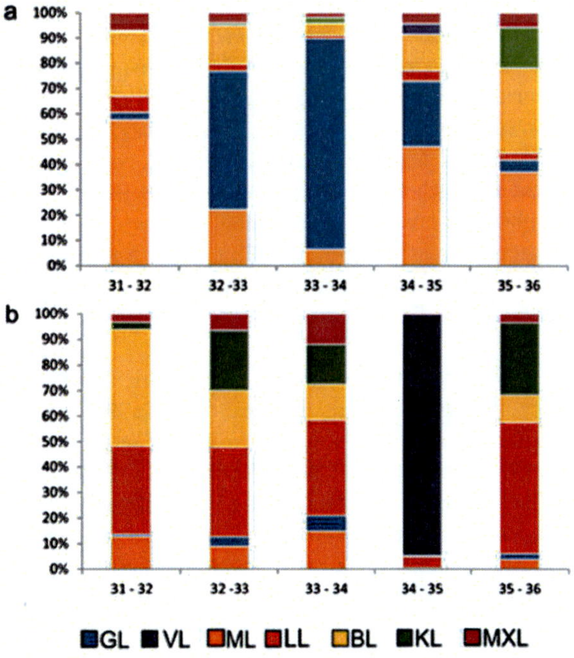


**Fig. 5** Lake type classes at different latitudes in the Argentinean Central Andes considering: **a** percentage of different lakes and **b** areas of different types of lakes

Argentinean Central Andes, except to the 34°–35° S where the very large volcanic Diamante lagoon plays the main role.

## Outburst Flood Hazard Considerations

Outburst flood hazard resulting from naturally dammed lakes breaching or overtopping is among the most important threats in mountainous regions around the world. For instance, glacial lake outburst floods have caused at least 393 deaths in the European Alps, 5745 deaths in South America and 6300 deaths in central Asia (Carrivick and Tweed 2016). According to Iribarren Anacona et al. (2015), in the Extra-tropical Andes, at least 100 floods have been produced from 31 lakes since the year 1750.

Specifically in the Central Andes region of Argentina, eight outburst floods have occurred in the last 250 years (Table 3). Six of these events were produced by the cyclic phenomenon of ice-damming of the Plomo River (33° 07′ S) by a surging glacier (King 1935; Harrison et al. 2015; Correas-Gonzalez et al. 2020). The most catastrophic event was completely unexpected and took place on 10 January 1934, when $6 \times 10^7$ $m^3$ where released from the unknown lake destroying the Cacheuta hydropower plant, several kilometres of the international road and causing 20 fatalities along its path. Six days later, a new flow of 1350 $m^3$ was produced, without damages (Correas-Gonzalez et al. 2020).

One previous event was induced for the year 1788 from historical records by Prieto (1986) and the other three events took place in February and March from 1985 (Bruce el al. 1987; Harrison et al. 2015; Correas-Gonzalez et al. 2020). More recently, in 2005 a landslide-dammed lake collapsed in the Santa Cruz River (31° 40′ S) producing an outburst flood of $32 \times 10^6$ $m^3$. This violent streamflow damaged the Aguas Negras hydropower plant after travelling 254 km and affected the supply of drinking water in San Juan capital city (D'odorico et al. 2009; Perucca and Angillieri 2009). Finally, one of the largest events known on Earth occurred in 1914 when the Carri-Lauquen lagoon, dammed by a rock avalanche, collapsed releasing $1.55 \times 10^9$ $m^3$ of water into the Barrancas River (36°30′S). This outburst flood reached a 93,000 $m^3$/s peak discharge and extended until the Atlantic Ocean some 1000 km away causing at least 300 fatalities (Groeber 1916; Hermanns et al. 2011; Colavitto et al. 2012; Ramos 2017).

The first steps for outburst flood hazard assessment, are the identification of their main characteristics such as type of lake, size, elevation and topographical setting as well as estimating the stored volume, dam freeboard, and its geometry (Aggarwal et al. 2017; Emmer et al. 2016; Wilson et al. 2018). After these fundamental data collection, different approaches have been proposed to evaluate the susceptibility of lakes to outburst flood, such as a combination of decision trees and scenarios (Emmer and Vilímek 2014); Analytic Hierarchy Processes (Aggarwal et al. 2017); Multi-Criteria Decision Analysis (Kougkoulos et al. 2018) or estimation of possible characteristics of the flood such as peak discharge, run out distance, flooding area, and probability by using empirical, analytical or numerical modelling (McKillop and Clague 2007; Klimeš et al. 2014; Somos-Valenzuela et al. 2016; Westoby et al. 2014).

The inventory of lakes presented in this work can be taken as the first step on outburst flood hazard assessment in the Argentinean Central Andes. Some basic geomorphological conclusions can be drawn from the nature of lakes, their extension and elevation, but this is not enough in terms of hazard assessment, which is beyond the scope of this study. Nevertheless, the existence of 28 lakes larger than 10 ha and historical records of outburst floods in the region suggest that the Central Andes might not be excluded from this type of hazard. Among the greater identified lakes, there are nine landslide-dammed lakes, seven bedrock-dammed lakes, five moraine-dammed lakes, five karstic lakes and two volcanic lakes. As lakes dammed by moraine or landslide material are more prone to collapse, we assume that they represent a potential hazard. On the other hand, the presence of surging glaciers in the region (Harrison et al. 2015; Falaschi et al. 2018) also implies the possibility of formation of new ice-dammed lakes in the future, which might cause new GLOFs.

Further research needs to be conducted to evaluate slope instability and susceptibility of mass movement impact on all kind of lakes to assess overtopping potential. Moreover, the potential enlargement of supraglacial lakes due to coalescence should be under consideration due to their location in the headwaters of Mendoza and Tunuyán basins.

Finally, lakes instability in the Central Andes shouldn't be ignored as Andean communities and main infrastructures key for economic development such as hydropower plants (Fig. 1) and critical routes (e.g. Bi-Oceanic Corridor MERCOSUR-Chile) are established along the main valleys. On the other hand, important cities such as San Juan (500,000 inhabitants) and Mendoza (1.2 million inhabitants), which are the economic centre of the region, are located at the foot of the Andes as well.

## Final Remarks

In the Arid Central Andes of Argentina, characterized by a regional dryness with a mean annual precipitation below 500 mm and active tectonic activity, 658 lakes were identified. This inventory certainly improves our knowledge about the spatial distribution and character of the lakes in the region. The last updated inventory developed by Wilson et al. (2018) considered 313 lakes classified in three types of lakes (moraine dammed-lakes, bedrock dammed lakes and glacial lakes) comprising a total surface of 650 ha. Our inventory expands the number of these three types of lakes up to 579 with a total area of 1,551 ha. Besides, new categories such as landslide-dammed lakes, karstic and volcanic lakes were taken into account.

A group of 28 lakes exceeds 10 ha representing 82.6% of the total surface of lakes. From this group, lakes dammed by loose material (LL and ML) are the most important in number and may pose a dangerous scenario in case of dam breach for human settlements and economic activities. Distribution of these lakes in the periglacial environment, where permafrost is being severely degraded due to the global warming, denotes a serious threat to dams′ stability in the future as ice support could decay in their loose material. Therefore, even though glacial lakes have called the attention all over the Andes due to the potential increasing GLOF hazard (Dussaillant et al. 2010; Iturrizaga 2014; Iribarren Anacona et al. 2015; Emmer et al. 2016; Wilson et al. 2018), dangerous scenario related to landslide-dammed lakes in the Arid Central Andes of Argentina remains unknown. The increased number of landslide-dammed lakes at 31–32° S matches very well with the landslide cluster associated with the intense neotectonic activity of this region (Junquera Torrado et al. 2019).

Therefore, landslide-dammed and moraine-dammed lakes stability and potential outburst flood hazard need urgent attention in this region. Understanding the current state and

**Table 3** Historical outburst in the study region (taken and modified from Correas-Gonzalez et al. 2020)

| Lake name /Type/ Location | Outburst flood characteristics | | | | | Source |
|---|---|---|---|---|---|---|
| | Date | Peak discharge (m3/s) | Volume ($m^3$) | Runout distance (km) | Consequences | |
| **Los Erizos / LL/** (31°40′ S, 70° 17′ W) | 12-11-2005 | 1,000 vs 7,960 | $32 \times 10^6$ | 254 | Destruction of one bridge and water intakes for irrigation and human consumption, halting of power generation | D'Odorico et al. (2009), Perucca and Angillieri (2009) |
| **Plomo / SGL /** (33°07′ S, 69° 59′ W) | 22-01-1788 | N/d | N/d | N/d | Halting of terrestrial communication between Argentina and Chile | Prieto (1986) |
| | 10-01-1934 | 3,000 | $60 \times 10^6$ | 210 | Halting of terrestrial communication, destruction of seven bridges, halting of power generation and 20 fatalities (damages estimated in 1.8 million U $D) | King (1935), Harrison et al. (2015), Correas-Gonzalez (2020) |
| | 16-01-1934 | 1,350 | N/d | 180 | None | |
| | 14-02-1985 | 321 | $54.1 \times 10^6$ | 180 | None | Bruce et al. (1987), Fernández et al. (1991), BDHI (2020) |
| | 22-02-1985 | 172.6 | $21 \times 10^6$ | – | None | |
| | 13-03-1985 | 175 | $20 \times 10^6$ | – | None | |
| **Carri-Lauquen / LL /** (36°30′ S, 70° 09′ W) | 29/12/1914 | 93,000 | $1.55 \times 10^9$ | 1,000 | Destruction of dozens of kilometers of railways, important loss of crops and cattle, destruction of many villages and 500 fatalities | Groeber (1916), Hermanns et al. (2011), Colavitto et al. (2012), Ramos (2017) |

possible evolution of lakes in the Andes is crucial for outburst flood hazard assessment as well as considering lakes as water reservoirs in changing environments marked by a decrease in snow precipitations.

Concerning to the potential risk of outburst to local communities in the Central Andes, major populated areas and main infrastructures are located downstream these lakes. Consequently, and contrasting with Patagonia where glacial lakes collapses are more frequent involving larger water volumes, but it is less populated (Dussaillant et al. 2010; Iribarren Anacona et al. 2015); in the Central Andes region outburst floods from lakes may represent a greater risk since population and infrastructure exposure is greater than in Patagonia.

**Acknowledgements** This research is part of the PhD thesis of M. Correas Gonzalez who is granted by CONICET. Funding research was get from the ANLAC Program of the University of Cuyo leaded by Moreiras.

## References

Aggarwal S, Rai SC, Thakur PK, Emmer A (2017) Inventory and recently increasing GLOF susceptibility of glacial lakes in Sikkim, Eastern Himalaya. Geomorphology, 295 (Supplement C): 39–54

BDHI (2020) Base de Datos Hidrológica Integrada. Secretaría de Infraestructura y Política Hídrica. https://bdhi.hidricosargentina.gob.ar/_. Last Accessed 09 April 2020

Breien H, De Blasio FV, Elverhøi A, Høeg K (2008) Erosion and morphology of a debris flow caused by a glacial lake outburst flood Western Norway. Landslides 5(3):271–280

Bruce RH, Cabrera GA, Leiva JC, Lenzano LE (1987) The 1985 surge and ice dam of Glaciar Grande del Nevado del Plomo Argentina. J Glaciol 33:131–132

Bruniard ED (1982) La diagonal árida argentina: un límite climático real. Revista Geográfica 95:5–20

Carrivick JL, Tweed FS (2016) A global assessment of the societal impacts of glacier outburst floods. Global Planet Change 144:1–16

Colavitto B, Orts DL, Folguera A (2012) El caso del Outburst Flood histórico de la laguna Derrumbe, Cholila, Chubut: Colapso de dique Morénico en la Cordillera Norpatagónica. Revista de la Asociación Geológica Argentina 69(3):457–465

Correas-Gonzalez M, Moreiras SM, Jomelli V, Arnaud-Fassetta G (2020) Ice-dammed lake outburst flood risk in the Plomo basin, Central Andes (33° S): Perspectives from historical events. Cuadernos de Investigación Geográfica 46(1):223–249

Costa JE, Schuster RL (1988) The formation and failure of natural dams. Geol Soc Am Bull 100(7):1054–1068

D'odorico PE, Pérez DJ, Sequeira N, Fauqué L (2009) El represamiento y aluvión del río Santa Cruz, Andes Principales (31° 40'S), provincia de San Juan. Revista de la Asociación Geológica Argentina 65(4):713–724

Drewes J, Moreiras S, Korup O (2018) Permafrost activity and atmospheric warming in the Argentinian Andes. Geomorphology 323:13–24

Dussaillant A, Benito G, Buytaert W, Carling P, Meier C, Espinoza F (2010) Repeated glacial-lake outburst floods in Patagonia: an increasing hazard? Nat Hazards 54(2):469–481

Emmer A, Klimeš J, Mergili M, Vilímek V, Cochachin A (2016) 882 lakes of the Cordillera Blanca: an inventory, classification, evolution and assessment of susceptibility to outburst floods. CATENA 147:269–279

Emmer A, Vilímek V (2014) New method for assessing the susceptibility of glacial lakes to outburst floods in the Cordillera Blanca Peru. Hydrol Earth Syst Sci 18(9):3461–3479

Emmer A, Vilímek V, Klimeš J, Cochachin A (2014) Glacier retreat, lakes development and associated natural hazards in Cordilera Blanca, Peru. In: Shan W, Guo Y, Wang F, Marui H, Strom A (eds) Landslides in cold regions in the context of climate change. Springer, Switzerland, pp 231–252. https://doi.org/10.1007/978-3-319-00867-7_17

Falaschi D, Bolch T, Lenzano MG, Tadono T, Lo Vecchio A, Lenzano L (2018) New evidence of glacier surges in the Central Andes of Argentina and Chile. Progress Phys Geogr: Earth Environ 42(6):792–825

Fernández PC, Fornero L, Maza J, Yañez H (1991) Simulation of flood waves from outburst of glacier-dammed lake. J Hydraulic Eng 117 (1):42–53

Groeber P (1916) Informe sobre las causas que han producido las crecientes del río Colorado (territorios del Neuquén y La Pampa) en 1914. Dirección General de Minas, Geología e Hidrología, Buenos Aires, Boletín 11, Serie B (Geología) 29 p

Harrison WD, Osipova GB, Nosenko GA, Espizua L, Kääb A, Fischer L, Huggel C, Craw Burns PA, Truffer M, Lai AW (2015) Chapter 13: Glacier surges. In: Shroder JF, Haeberli W, Whiteman C (Eds),Snow and ice-related hazards, risks and disasters. Elsevier, Netherlands, pp 437–485. ISBN: 978-0-12-394849-6

Hermanns RL, Folguera A, Penna I, Fauqué L, Niedermann S (2011) Landslide dams in the central andes of Argentina (northern Patagonia and the Argentine northwest). In: Evans SG, Hermanns RL, Strom A, Scarascia-Mugnozza G (Eds) Natural and artificial rockslide dams. Springer-Verlag Berlin Heidelberg. ISBN: 978-3-642-04764-0. 133:147–176

IANIGLA-Inventario Nacional de Glaciares. (2018) Resumen ejecutivo de los resultados del Inventario Nacional de Glaciares. IANIGLA-CONICET, Ministerio de Ambiente y Desarrollo Sustentable de la Nación. (27 p). https://www.glaciaresargentinos.gob.ar/?page_id=2571. Last Accessed 08 April 2020

Iribarren Anacona P, Mackintosh A, Norton KP (2015) Hazardous processes and events from glacier and permafrost areas: lessons from the Chilean and Argentinean Andes. Earth Surf Proc Land 40 (1):2–21

Iturrizaga L (2014) Glacial and glacially conditioned lake types in the Cordillera Blanca, Peru: a spatiotemporal conceptual approach. Prog Phys Geogr 38(5):602–636

Junquera TS, Moreiras SM, Sepúlveda S (2019) Distribution of landslides along the Andean active orogenic front (Argentinean Precordillera 31–33° S). Quatern Int 512:18–34

King WDVO (1935) El aluvión del río Mendoza de enero de 1934. La Ingeniería. Buenos Aires, Argentina, pp 309–313

Klimeš J, Benešová M, Vilímek V, Bouška P, Rapre AC (2014) The reconstruction of a glacial lake outburst flood using HEC-RAS and its significance for future hazard assessments: an example from Lake 513 in the Cordillera Blanca Peru. Nat Hazards 71(3):1617–1638

Korup O, Tweed F (2007) Ice, moraine, and landslide dams in mountainous terrain. Quatern Sci Rev 26(25–28):3406–3422

Kougkoulos I, Cook SJ, Jomelli V, Clarke L, Symeonakis E, Dortch JM, Edwards LA, Merad M (2018) Use of multi-criteria decision analysis to identify potentially dangerous glacial lakes. Sci Total Environ 621:1453–1466

Lauro C, Vich AIJ, Moreiras SM (2019) Streamflow variability and its relationship with climate indices in western rivers of Argentina. Hydrol Sci J 64(5):607–619

Lliboutry L (1998) Glaciers of South America. Glaciers of Chile and Argentina. In Satellite Image Atlas of Glaciers of the World: Vol. U. S. Geological Survey Professional Paper 1386. USGS. https://pubs.usgs.gov/pp/p1386i/chile-arg/intro.html

Masiokas MH, Villalba R, Luckman BH, Montaña E, Betman E, Christie D, Le Quesne C, Mauget S (2013). Recent and historic Andean snowpack and streamflow variations and vulnerability to water shortages in central-western Argentina. In: Climate vulnerability. Pielke, R. (Ed.) Elsevier, Netherlands. e-ISBN: 978-0-12-384704–1. 5: 213–227

McKillop RJ, Clague JJ (2007) A procedure for making objective preliminary assessments of outburst flood hazard from moraine-dammed lakes in southwestern British Columbia. Nat Hazards 41(1):131–157

Moreiras SM (2006) Frequency of debris flows and rockfall along the Mendoza river valley (Central Andes), Argentina: associated risk and future scenario. Quatern Int 158(1):110–121

Moreiras SM, Páez MS (2015) Historical damage and earthquake environmental effects related to shallow intraplate seismicity of central western Argentina. Geological Society, London, Special Publications 399(1):369–382

Moreiras SM, Vergara Dal Pont IV, Araneo D (2018) Were merely storm-landslides driven by the 2015–2016 Niño in the Mendoza River valley? Landslides 15:1–18

Perucca LP, Angillieri MYE (2009) Evolution of a debris-rock slide causing a natural dam: the flash flood of Río Santa Cruz, Province of San Juan—November 12, 2005. Nat Hazards 50(2):305–320

Prieto M. del R (1986) The glacier dam on the Rio Plomo: a cyclic phenomenon? Zeitschrift fur Gletscherkunde und Glazialgeologie 22(1): 73–78

Ramos VA (2017) El aluvión del Río Colorado de 1914: Primera Contribución geológica de Groeber desde su llegada a la Argentina. Revista de la Asociación Geológica Argentina 74(1):9–18

Richardson SD, Reynolds JM (2000) An overview of glacial hazards in the Himalayas. Quatern Int 65–66:31–47

Somos-Valenzuela MA, Chisolm RE, Rivas DS, McKinney DC (2016) Modeling a glacial lake outburst flood process chain: the case of Lake Palcacocha and Huaraz Peru. Hydrology Earth System Sci 20 (6):2519–2543

Vilímek V, Klimeš J, Červená L (2016) Glacier-related landforms and glacial lakes in Huascarán National Park Peru. J Maps 12(1):193–202

Westoby MJ, Glasser NF, Brasington J, Hambrey MJ, Quincey DJ, Reynolds JM (2014) Modelling outburst floods from moraine-dammed glacial lakes. Earth Sci Rev 134:137–159

Wilson R, Glasser NF, Reynolds JM, Harrison S, Iribarren Anacona P, Schaefer M, Shannon S (2018) Glacial lakes of the central and patagonian andes. Global Planet Change 162:275–291

2020 Kyoto Japan

# Rockfall/Rockslide Hazard, Lake Expansion and Dead-Ice Melting Assessment: Lake Imja, Nepal

Tomas Kroczek and Vit Vilimek

**Abstract**

The current global climate change is accelerating many natural processes that can lead to the rupture of dams of glacial lakes. One of these lakes is Lake Imja in the Khumbu area of Nepal. Three factors that influence the stability of the moraine dam were selected for analysis in this work—rockfalls/rockslides, growth of the lake area and melting of dead ice in the frontal moraine. The results of this study show that there is currently no risk of rockslide into the lake, which, however, may change due to the accelerating growth of the lake in the near future. The development of temperatures is also observed, where the increase in the last two decades is particularly striking. Crucial for the stability of the moraine dam is now the melting of dead ice at its core, as new thermokarst lakes are forming on the moraine surface, and a leak through the moraine in its southwestern part has also been discovered.

**Keywords**

GLOF • Rockfall • Glacial lake • Imja • Sagarmatha • Nepal

## Introduction

The Himalayan region is one of the most glaciated highland areas on Earth (Matthew 2013; Lala et al. 2018). In the context of climate change, local glaciers are melting and many new glacial lakes are emerging and increasing in size (Evans and Clague 1994; Watanabe et al. 2009; ICIMOD 2011; Rounce et al. 2016; Haritashya et al. 2018). Many of these lakes are dammed by moraines and can potentially be very dangerous if the dam is breached, which may occur due to various triggers including landslides, increasing lake area, dead ice melting and earthquakes, etc. (Emmer and Vilímek 2013; Rounce et al. 2016; Byers et al. 2019). Due to tectonic activity and rapid weathering, landslides are the most common factor causing glacial lake outburst floods (GLOFs) in Nepal (Falátková 2016).

One of the potentially dangerous lakes is Imja Lake, which is located in Sagarmatha National Park (Mount Everest) in the Khumbu area at an elevation of approximately 5,000 m a.s.l.. and is fed by melted water from the Imja and Lhotse Shar glaciers (Haritashya et al. 2018) on the eastern side of the lake. The north and south shores are made up of lateral moraines, and the west shoreline is marked by an ice-cored terminal moraine (Watanabe et al. 2009). The expansion of the lake began in the 1950s, when it was possible to see the first small supraglacial lakes (Watanabe et al. 2009; Thakuri et al. 2016; Haritashya et al. 2018). The expansion is one of the fastest in the region—in 2001 the length of the lake was 1647 m, while in 2006 it was 2017 m and today it is expanding only eastwards at the expense of the glaciers (Bajracharya and Mool 2009; Somos-Valenzuela et al. 2013, 2014). The side moraine is about 60 m above the lake surface and has steep slopes on the lake shore and gentle slopes on the other side (Hambrey et al. 2008). They are also lower and narrower in the direction of the terminal moraine and are highly eroded (ICIMOD 2011). The terminal moraine at the western end of the lake is approximately 50 m high, 536 m wide and 567 m long. It includes dead ice at a depth of about 15–20 m (Hambrey et al. 2008; ICIMOD 2011; Somos-Valenzuela et al. 2014). Runoff from the lake only flows through the moraine terminal (Watanabe et al. 2009). A buried ice allow the formation of stream lakes on the surface of the moraine (Hambrey et al. 2008). The depth of these lakes is a maximum of 15 m (5 m on average)

T. Kroczek (✉) · V. Vilimek
Department of Physical Geography and Geoecology, Faculty of Science, Charles University, Prague, Czech Republic
e-mail: tomas.kroczek@natur.cuni.cz

V. Vilimek
e-mail: vit.vilimek@natur.cuni.cz



V. Vilímek et al. (eds.), *Understanding and Reducing Landslide Disaster Risk*, ICL Contribution to Landslide Disaster Risk Reduction,
https://doi.org/10.1007/978-3-030-60319-9_10

(Lala et al. 2018). Further studies have been conducted on bathymetry with some floating ice issues (Somos-Valenzuela et al. 2014). The latest results show that the main lake is approximately 65 m deep (158 m maximum) with an area of 1.35 km2 and a volume of 88 × 106 m3 (Lala et al. 2018).

Glacial lakes can pose an important natural hazard—GLOF, which can be caused by many factors and is very difficult to measure as a complex system for best results. Many authors have performed assessments on Imja Lake, but have received different results. Today, leading scientists combine remote sensing with fieldwork to determine the level of danger. Due to the relative rapid growth of the lake, it is therefore important to regularly monitor sub-factors as various hazards may continue to develop (landslide impact area growths, pressure on frontal moraine increases, melting of dead ice, etc.).

## Methods

Three main factors for this evaluation were chosen: impact of rockfall or rockslide possibility; expansion of the lake; dead-ice in terminal moraine melting.

The first factor includes the possibility of an impulse wave being created due to slope movement into the lake, which can be created either by a rockslide or rockfall from the surrounding slopes or by the impact of ice while calving the glacier. Due to the gentle slope and small elevation difference between the lake surface and the surface of the glacier, however, calving of the glacier can be neglected and only the possibility of a rockfall with an impact on the lake will be evaluated.

The next factor is the stability and development of the terminal moraine, which is another crucial factor affecting the tendency of the lake to outburst. Not only its shape, but also the development of its morphology in relation to dead ice melting was monitored.

The third and last parameter is the expansion of the lake, whose speed is rapidly increasing in connection with climate change. This should be continuously monitored in order to evaluate the possibility of newly created natural hazards. In this work, the expansion of the lake is evaluated in the context of temperature development based on meteorological reanalysis.

Based on remote sensing and field survey, the southern slope of Imja Tse, located north of the lake, was selected for the evaluation of the rockfall hazard to the lake—see Fig. 1. There are both the disruption of lateral moraine and a lot of rocks from rockfalls were found under this slope during the field survey—see Fig. 2. The southfacing slope is also expected to be more susceptible to rockfalls due to solar insolation and therefore greater temperature fluctuation. The slope shape was measured using Google Earth in combination with field data using the Trupulse 200B laser rangefinder. The heel of the slope and the outer foot of the lateral moraine at the point of disruption were measured manually. SW RocFall (RocScience 2019) was used to analyze the rockfall trajectory. The area is still tectonically active, and it can be assumed that in addition to relatively small (non-threatening) debris, large rock blocks can also appear (Hyndman and Hyndman 2017).

The analysis was divided into three possible rockfall/rockslide scenarios—see Table 1. The density of the rocks for analysis was set at 2,650 kg/m$^3$ (mean value of density of paragneiss) and the friction was completely neglected to maintain the maximum possible impact.

The individual scenarios also contain a different number of potential segregation areas identified by remote sensing and field survey. In the first scenario, where the collapsed rocks are relatively small and could affect the stability of the side moraine rather than having a direct impact on the lake. Generally, it can be argued that fragments of this size can fall off from virtually anywhere on the slope.

Scenario 2 further evaluates a 270 m$^3$ rockfall, which could not only severely disrupt the side moraine, but also cause a shock wave as one of the main causes of dam breach or overtopping (Emmer and Vilímek 2014). For this scenario, three separation areas were defined within the assessed slope.

The last of the three scenarios evaluates a relatively extreme rockslide, which is a block of huge dimensions (100 × 50 × 10 m). Such dimensions are of a theoretical nature, however, the area is seismically active, so in the case of an earthquake, the likelihood of this phenomenon cannot be completely ruled out. For this scenario, only one separation area was defined, which is evident both from the

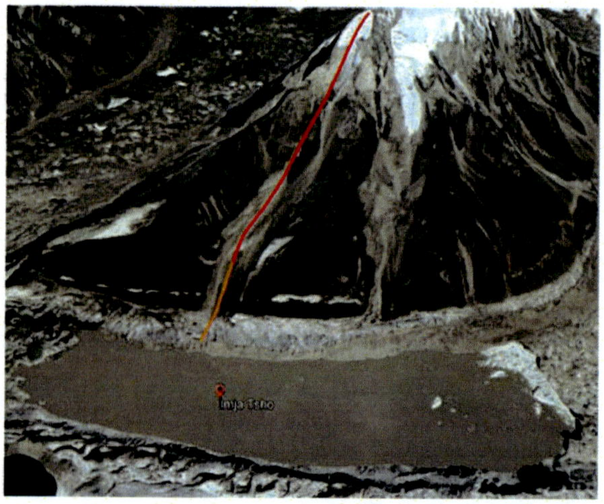

**Fig. 1** The slope of Imja Tse, where the rockfall analysis was performed

**Fig. 2** 1—The slope of Imja Tse, which was chosen as a potential slope for rockfalls which can affected the lake; 2—Disruption of the lateral moraine through which can rockfall or rockslide go over into the lake

**Table 1** Properties characteristics of individual scenarios of rockfall/rockslide

| | Block size | Number of potential segregation areas |
|---|---|---|
| Scenario 1 | 3 × 3 × 1 m | 4 |
| Scenario 2 | 30 × 30 × 3 m | 3 |
| Scenario 3 | 100 × 50 × 10 m | 1 |

remote sensing and when viewed from the foot of the slope. It is the part of the slope with the highest elevation. The directions of movement of the older falls from higher slope areas are on both sides of this separation area, and therefore the stability of this block may be disrupted.

The expansion of the lake from the 1950s to the present was also monitored. Due to the lack of availability of older images, as well as the frequent influence of cloud cover on their quality, the results of Thakuri et al. (2016) were used for 1962, 1975, 1992 and 2000. The remaining reporting years (2003, 2009, 2010, 2014, 2016 and 2017) were further quantified using Google Earth. Changes in the size of the runoff lakes on the surface of the terminal moraine were monitored on years 2009–2017. One cross section was measured using laser rangefinder—see Fig. 2. This was measured in order to compare the development of the moraine shape due to the melting of dead ice. A field survey of the moraine was also conducted to record possible melting of dead ice.

Due to the unavailability of continuous data from the nearest weather stations, weather analysis using the GHCN-CAMS dataset provided free of charge by the American NOAA research service was used to analyse the climate and temperature in this area. GHCN-CAMS reanalysis is a project introduced in 2008 (Fan and van den Dool 2008), which provides data on average monthly temperatures in the period 1948—to-date. This reanalysis was chosen for its relatively good resolution - the data is distributed at a resolution of 0.5° × 0.5° and is also considered suitable for use in monitoring snow melting, ice and its effect on runoff, etc. (Fan and van den Dool 2008).

## Results

### Rockfall/Rockslide Analysis

In assessing the rockfall, it was first necessary to define the potential separation areas of the individual fragments, which are usually the result of frost weathering, which is active in the area. The flow was analyzed on a 2,721 m long profile with an elevation difference of 1,110 m (Fig. 1). The restitution coefficient was maintained as 1 to maintain the maximum risk to the lake. The results of the three different rockfall scenarios will now be described in more detail.

#### Scenario 1

After entering the individual detachment zones and including the above-specified parameters for Scenario 1 into the RocFall model, in the vast majority of such rockfalls all the rock material reaches the foot of the slope. However, it does not continue past the side moraine and should not disturb it in any way. It can be noted that the span of the impact areas is about 100 m, but the distance to the edge of the moraine is approximately 150 m further at an elevation of 25 m.

#### Scenario 2

The analysis revealed that these fragments do not pose a major threat to the lake. The falling material will again end before the foot of the lateral moraine and about three-quarters of the thirty analyzed rockfalls will stop at the foot of the Imja Tse slope. As in Scenario 1, the impact range of these fragments is approximately 100 m. A slight difference can be seen in the distribution of the impact, but this does not make any change in terms of assessing the hazard to the lake.

### Scenario 3

Scenario 3 suggests that even such a great onrush should not affect the lake. Compared to the other scenarios, the distribution of impacts of possible rockfalls changes slightly, but the location and the range of impacts are still the same—see Fig. 3.

## Lake Expansion

One of the most important factors influencing the potential hazard from the lake is its expansion rate. Not only the area is increasing, but logically also the volume and in connection with it the hydrostatic pressure on the frontal moraine are increasing too and thus threatened by the moraine dam wall. The area of theoretical impact of slope movement into the lake is also increasing. Therefore, it is very important to monitor and quantify the ever-increasing expansion of the lake.

### Lake Area

Due to the ongoing climate change, however, different phases of the expansion rate of the lake take place, which Watanabe et al. (2009) divided into four stages. In extending the time series of lake area records until 2017 using data from Thakuri et al. (2016) and data obtained using Google Earth, five phases can be distinguished, where the last (fifth) phase is the fastest in terms of the increasing area—see Fig. 4.

While in phases 1 and 2 the lake mainly expanded towards the frontal moraine and lateral moraines and the expansion towards the glacier was negligible, in phases 3 and 4 the expansion occurred almost exclusively in a direction towards the glacier—about 48 m/year (Watanabe et al. 2009). With the extension of the monitored series to 2017, the expansion of the lake towards the glacier in 2009–2017 increased to 92 m/year and the difference in measurement between 2016 and 2017 is approximately 190 m.

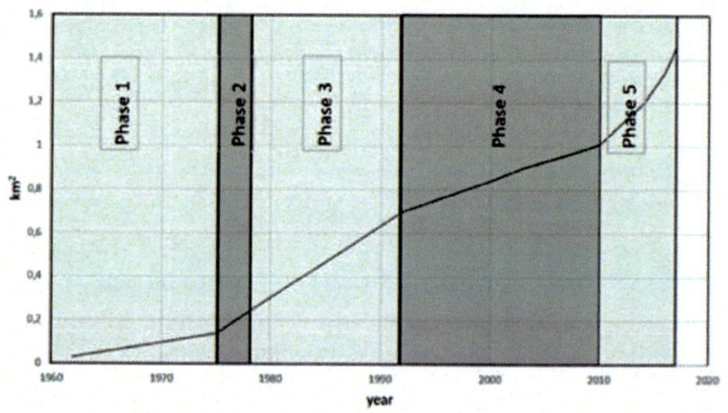


Fig. 4 Phases of Imja Lake expansion

### Temperature Change

To understand the expansion of the lake, it is necessary to evaluate its development in the context of climate change. For the purposes of this work, the average monthly temperatures were obtained by means of meteorological reanalysis in the years 1960–2017. Figure 5 shows a gradual increase in the average annual temperature, which has been striking especially since the turn of the millennium. The rising temperature also confirms a linear trend that is likely to increase in the coming period.

When evaluating the development of the annual course of average monthly temperatures—see Fig. 6—the temperature increase is evident especially in the summer months when the average monthly temperatures exceed 5 °C. At the same time, the period of positive temperatures has been extended to both spring and autumn months. There is also a relatively high increase in average temperatures in winter, spring and autumn, when, especially in spring and autumn months, we can assume positive daytime temperatures in conjunction with sunshine, which can cause continuous melting even during the year when the average monthly temperature is negative. The acceleration of the expansion of the lake and associated natural hazard are certainly a result of the overall

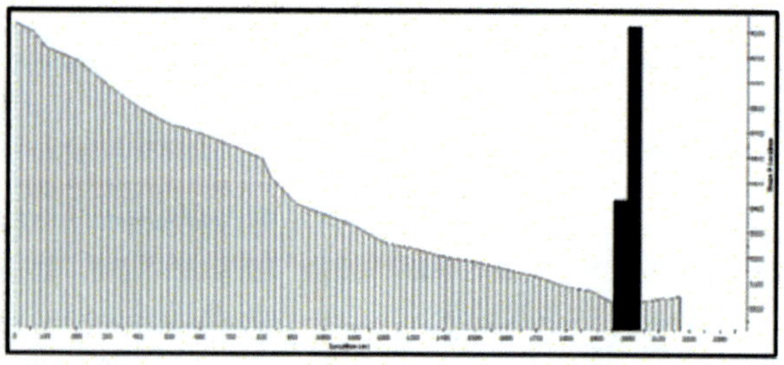

Fig. 3 Distribution of the frequency of impacts of rockfalls in Scenario 3 using RocFall

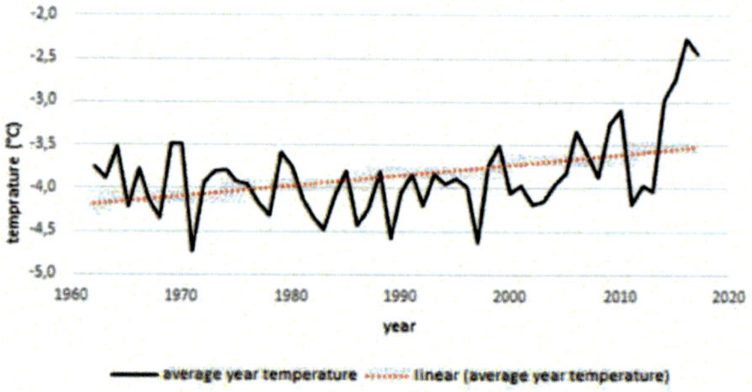


Fig. 5 Increasing year temperature averages

development of annual temperatures. The average monthly and annual average temperatures were correlated with the lake area in the monitored years using the Kendall τb correlation coefficient. Dependence was proved in the warm period of the year, when the strongest dependence is visible in the months of June, August and September. It can be argued that rising summer temperatures have a significant effect on the growth of the lake.

## Terminal Moraine

The stability of the frontal moraine is a very important factor. Not only is the hydrostatic pressure on the dam increasing due to the expansion of the main lake, but its stability is influenced by the melting of the dead ice contained in the moraine. The outflow from the lake, which flows through the frontal moraine, also has a significant influence on the shape of the moraine. For this reason, it is very important to continue to monitor the moraine in terms of changes in its morphology and subsidence.

### Moraine Morphology

The transverse profile of the moraine was measured during the field research (27.9029 N,86.9120E–27.8980 N, 86.9118E)—see Fig. 7. The transverse profile covers the entire width of the moraine and is located in close proximity to the shore of the main lake. The total width of the moraine was measured at 555 m, which, when evaluated according to Wang et al. (2008) ranks the lake as being potentially dangerous. The same methodology also evaluates the ratio of the width of the dam to its height and at the result of 0.03 the lake is again considered as being potentially dangerous. The profile shows a relatively marked segmentation of the moraine, when there are alternating round and sharp forms, which may be due to the presence and melting of buried ice, among other things. The average height of the moraine above the lake surface reaches approximately 20 m, which does not necessarily mean that the height of a possible flood wave must be higher in order to flow over the dam.

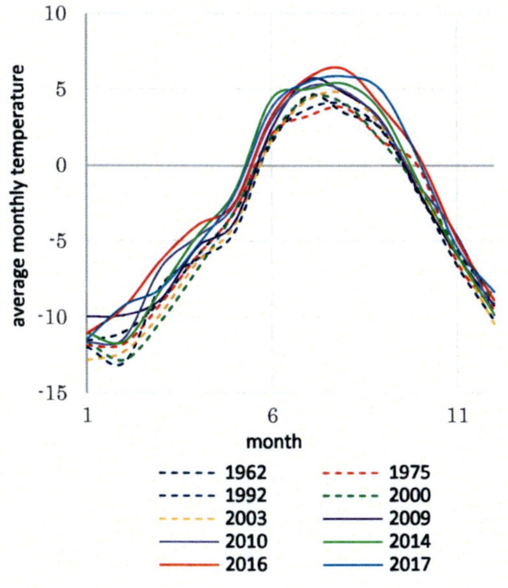


**Fig. 6** Annual development of temperatures using GHCN-CAMS reanalysis

### Flow and Thermokarst Ponds

In this work, development of lakes in the years 2009–2017 was recorded (see Fig. 8), thus in the years of the last phase of the fastest expansion of the main lake. As the temperature increased, these smaller lakes also increased outside the main lake. However, after the construction of a reinforced weir in 2016, when the water level was lowered, the total area of the lakes also decreased.

### Dead-Ice Melting

Dead-ice inside the terminal moraine is possibly the most important factor determining its stability. The melting rate of the buried ice is certainly related to the increase in the average annual temperatures in the area. Subsidence of the terminal moraine due to melting was noticeable especially in the years when the lake began to form. Watanabe et al. (1995, 2009) claim that the water level of the lake decreased by about 37 m between 1964 and 2006.

The lowering of the water level is certainly related to the sediment consolidation of the terminal moraine. Furthermore, Rounce et al. (2016) after further research showed that the lake level has stabilized over the last 15 years. However, the dead ice continues to melt and the associated formation of the thermokarst is confirmed by field observation on the surface of the moraine. Also during the field work, slopes with gravity graded material found in the melting, where backwasting (horizontal melting) predominates, were not found within the moraine. Therefore, it can be stated that melting of the ice is mainly by downwasting (vertical) and the main monitoring factors are therefore the deposition of the moraine and changes to its morphology. During the field research, secondary runoff from the lake was also found in the southwestern part of the frontal moraine. Nevertheless, it

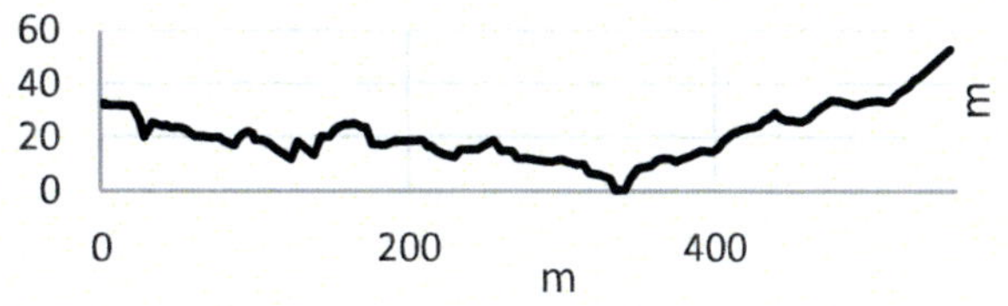


**Fig. 7** Transversal profile of terminal moraine

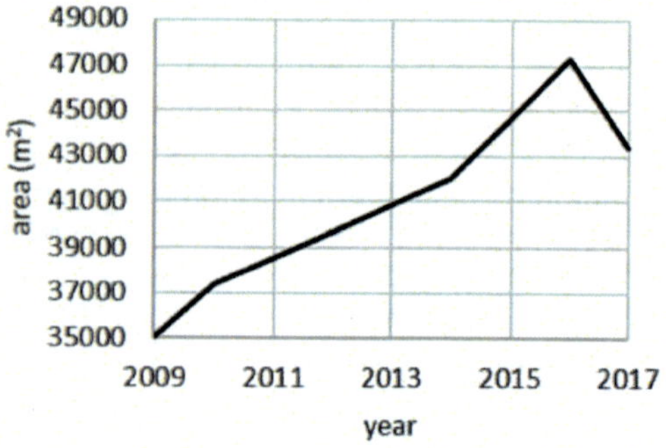


**Fig. 8** Flow and thermokarst ponds area development

is very difficult to distinguish whether this is negligible runoff created by melting of dead ice or a leak through the dam. However, if it were a leak, it could be one of the main factors affecting the stability of the dam and a natural hazard to Imja Lake in the future.

## Discussion

Assessment of the potential for GLOFs is a very complex issue and determining their probability is often difficult. Appropriate factors need to be selected for the potential hazard analysis, and these factors may differ to a varying degree for each lake evaluated. This problem is evident from the results of Rounce et al. (2016), who applied several methods to Lake Imja, with each method yielding different results. We considered the potential hazard of ice blocks falling from the contact glacier into the lake using the methodology of O'Connor et al. (2001), which has only two input parameters. However, the glacier cant and slope are neglected so the methodology evaluated the lake incorrectly as being endangered in this regard. A shock wave that would jeopardize the stability of the dam may be considered as improbable. A more complex method was proposed by Costa and Schuster (1988) and was used by Rounce et al. (2016) who determined extreme runoff and melting of dead ice in the terminal moraine as dangerous. However, the potential hazard to the lake by an earthquake was neglected. According to the field research performed after the earthquake in 2015, only small cracks up to tens of centi-meters were found on the surface of the moraine in the direction of the lake shore line, as well as new slope processes on the "lakeside" of the lateral moraine (Byers et al. 2017). It is important to consider an earthquake as a potential hazard, despite the fact that there has been no significant disruption since 2015. An earthquake is mentioned by Emmer and Vilímek (2014) as one of the possible causes of GLOFs. On the other hand, it is very difficult to secure the lake against earthquake hazards. The methodology proposed by Wang et al. (2008) brings different results. It is possible to monitor several dangerous parameters using this method (Rounce et al. 2016). However, it completely omits the possibility of slope movement into the lake, which is a very frequent trigger of GLOFs. For example, Falátková (2016) mentions an avalanche as the main cause of a GLOF in the Himalayas. Slope movement into the lake was evaluated by Bolch et al. (2011); however, this method neglects the melting of dead ice in the moraine, which is another cause of GLOFs in the Himalayas (Falátková 2016). The new method created by Rounce et al. (2016) monitors both potential dynamic and self-destructive disorders. When comparing the results of these authors with the results of the present work, one can find a consensus in the unlikely impact of slope movement on the lake. The lateral moraine of the lake acts as a barrier and protects the lake from rock falls and avalanches even in places where the debris of the falls accumulates and the side moraine is relatively disrupted under the slope.

Nevertheless, Rounce et al. (2016), do not exclude the possibility of an impact of slope movement in the coming decades, when the lake area will grow further. Lala et al. (2018) estimate the potential impact of slope movement into the lake around 2045 and consider the rock glaciers to the east of the lake shore as the main source areas of these movements—Fig. 9.

Therefore, Rounce et al. (2016) assess the lake as currently being prone to self-destructive failure, where, due to the width and length of the frontal moraine, increasing hydrostatic pressure may be neglected and the main potential threat is the melting of dead ice. The dynamics of ice melting in this work were recorded during field research, which found thermokarst lakes, which are a typical phenomenon of dead ice melting (Schomacker and Kjær 2008), as well as possible leakage through the damming moraine. Thermokarst lakes on the surface of the frontal moraine may be found at Lake Petrova lake in Kyrgyzstan, where Janský et al. (2009) observed even greater dynamics in the expansion of the lakes and recorded exposed ice along its shores. A similar development may be expected for Lake Imja.

The results of this work often correspond with the results of other authors. It is possible to agree with the statements of

**Fig. 9** Potential source zones of slopemass movements captured according to Lala et al. (2018)

Hambrey et al. (2008); Watanabe et al. (2009); Rounce et al. (2016); Lala et al. (2018) that the lake is currently not endangered by slope movements that could cause a shock wave. The expansion of the lake was observed, for example, by Somos-Valenzuela et al. (2014) or Watanabe et al. (2009), when these authors followed the development of the lake area, including outflow lakes. According to Rounce et al. (2016), however, it is important from the point of view of monitoring hydrostatic pressure and melting in the frontal moraine to monitor these ponds separately, which was done in this work.

The vast majority of authors consider the presence of dead ice in the moraine as a very important part of the hazard to Lake Imja and it is necessary to monitor the changes caused by the melting (Watanabe et al. 2009; Somos-Valenzuela et al. 2014; Rounce et al. 2016).

## Conclusions

Rockfall, development of lake area and dead ice melting were analysed. It was found that none of the three scenarios according to the size of the rockfalls pose a hazard to the lake through the lateral moraine disruption in the northern part of the lake. Nevertheless, it can also be said that the growth of the lake significantly accelerated in recent years and is likely to increase the probability of slope movements into the lake in the near future and another assessments of rockfall/rockslide, icefall, avalanche etc. should be done there—see Fig. 10 where potentially dangerous slopes are shown. The most important factor affecting the stability of the dam at present is the melting of dead ice in the damming moraine. The stability of this moraine is also influenced by the Lhotse Glacier located to the east of Imja Lake, and its rapid melting could also rapidly release accumulated water. Therefore, further research should focus on monitoring the continuous melting of the Lhotse Glacier and its effect on the stability of the damming moraine of Imja Lake.

To sum up:

- At present, there is no risk of the formation of an impulse wave after the impact of slope movement into the lake from the analysed slope, which was verified using the measured profile at the site of the lateral moraine disruption below the southern slope of Imja Tse and subsequent modelling of rockfalls with different scenarios.
- The expansion of the lake to the east is accelerating at the expense of the Imja glacier and Lhotse Shar glacier. Another expansion phase was determined.
- Meteorological reanalysis showed that not only the average annual temperature is increasing, but also the “scissors” in the course of the annual temperature cycle, when the summer temperatures significantly influence the expansion of the lake.
- The slow melting of dead ice in the frontal moraine is currently the most problematic parameter affecting the stability of the dam, creating new thermo-karst lakes on the surface of the moraine. During the field work, leakage in the damming moraine was also recorded.
- In the next few years, no GLOF is likely to occur, unless there is an earthquake, but it is necessary to continue in the hazard assessment.

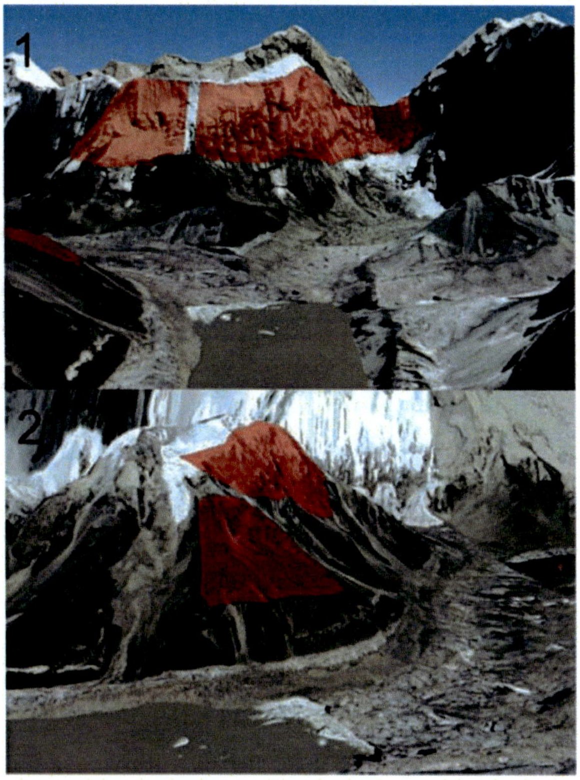


**Fig. 10** Potential source areas of slope mass movements which can affect the lake in next decades. 1—slopes in the eastern part of the lake, 2—slopes of Imja Tse (Island Peak)—using Google Earth

**Acknowledgements** Author would like to thank Grant Agency of Charles University (Project GAUK No. 1164320) for financial support.

## References

Bajracharya SR, Mool P (2009) Glaciers, glacial lakes and glacial lake outburst floods in the Mount Everest region Nepal. Ann Glaciol 50:81–86. https://doi.org/10.3189/172756410790595895

Bolch T, Peters J, Yegorov A et al (2011) Identification of potentially dangerous glacial lakes in the northern tian shan. Terrigenous Mass Movements Detect Model Early Warn Mitig Using Geoinf Technol 9783642254:369–398. https://doi.org/10.1007/978-3-642-25495-6_12

Byers AC, Byers EA, McKinney DC, Rounce DR (2017) A field-based study of impacts of the 2015 earthquake on potentially dangerous glacial lakes in Nepal. Himalaya 37:26–41

Byers AC, Rounce DR, Shugar DH et al (2019) A rockfall-induced glacial lake outburst flood, Upper Barun Valley Nepal. Landslides 16:533–549. https://doi.org/10.1007/s10346-018-1079-9

Costa JE, Schuster RL (1988) Formation and failure of natural dams. Bull Geol Soc Am 100:1054–1068. https://doi.org/10.1130/0016-7606(1988)100%3c1054:TFAFON%3e2.3.CO;2

Emmer A, Vilímek V (2014) New method for assessing the susceptibility of glacial lakes to outburst floods in the Cordillera Blanca, Peru. Hydrol Earth Syst Sci 18:3461–3479. https://doi.org/10.5194/hess-18-3461-2014

Emmer A, Vilímek V (2013) Review Article: Lake and breach hazard assessment for moraine-dammed lakes: an example from the Cordillera Blanca (Peru). Hazards Earth Syst Sci 13:1551–1565. https://doi.org/10.5194/nhess-13-1551-2013

Evans SG, Clague JJ (1994) Recent climatic change and catastrophic geomorphic proocesses in mountain environments. Geomorphology 10:107–128

Falátková K (2016) Temporal analysis of GLOFs in high-mountain regions of Asia and assessment of their causes. Acta Univ Carolinae, Geogr 51:145–154

Fan Y, van den Dool H (2008) A global monthly land surface air temperature analysis for 1948-present. J Geophys Res Atmos 113:1–18. https://doi.org/10.1029/2007JD008470

Hambrey MJ, Quincey DJ, Glasser NF et al (2008) Sedimentological, geomorphological and dynamic context of debris-mantled glaciers, Mount Everest (Sagarmatha) region Nepal. Quat Sci Rev 27:2361–2389. https://doi.org/10.1016/j.quascirev.2008.08.010

Haritashya UK, Kargel JS, Shugar DH et al (2018) Evolution and controls of large glacial lakes in the Nepal Himalaya. Remote Sens 10:1–31. https://doi.org/10.3390/rs10050798

Hyndman DW, Hyndman D (2017) Natural hazards & disasters. Cengage Learning

ICIMOD (2011) Glacial lakes and glacial lake outburst floods in Nepal. The World Bank

Janský B, Engel Z, Šobr M et al (2009) The evolution of Petrov lake and moraine dam rupture risk (Tien-Shan, Kyrgyzstan). Nat Hazards 50:83–96. https://doi.org/10.1007/s11069-008-9321-8

Lala JM, Rounce DR, McKinney DC (2018) Modeling the glacial lake outburst flood process chain in the Nepal Himalaya: Reassessing Imja Tsho's hazard. Hydrol Earth Syst Sci 22:3721–3737. https://doi.org/10.5194/hess-22-3721-2018

Matthew R (2013) Climate change and water security in the himalayan region. Asia Policy 16:39–44. https://doi.org/10.1353/asp.2013.0038

O'Connor JE, Hardison JH, Costa JE (2001) Debris flows from failures of neoglacial-age moraine dams in the Three Sisters and Mount Jefferson wilderness areas, Oregon

RocScience (2019) Rocscience Inc. 2001, RocFall Version 4.0—Statistical analysis of rockfalls. https://www.rocscience.com, Toronto, Ontario, Canada

Rounce DR, McKinney DC, Lala JM et al (2016) A new remote hazard and risk assessment framework for glacial lakes in the Nepal Himalaya. Hydrol Earth Syst Sci 20:3455–3475. https://doi.org/10.5194/hess-20-3455-2016

Schomacker A, Kjær KH (2008) Quantification of dead-ice melting in ice-cored moraines at the high-Arctic glacier Holmströmbreen, Svalbard. Boreas 37:211–225. https://doi.org/10.1111/j.1502-3885.2007.00014.x

Somos-Valenzuela M, Mckinney DC, Byers AC, et al (2013) CRWR Online Report 12-06 Bathymetric Survey of Imja Lake, Nepal in 2012 High Mountain Glacial Watershed Program Bathymetric Survey of Imja Lake, Nepal in 2012 Bathymetric Survey of Imja Lake, Nepal in 2012

Somos-Valenzuela MA, McKinney DC, Rounce DR, Byers AC (2014) Changes in Imja Tsho in the Mount Everest region of Nepal. Cryosphere 8:1661–1671. https://doi.org/10.5194/tc-8-1661-2014

Thakuri S, Salerno F, Bolch T et al (2016) Factors controlling the accelerated expansion of Imja Lake, Mount Everest region Nepal. Ann Glaciol 57:245–257. https://doi.org/10.3189/2016AoG71A063

Wang X, Liu S, Guo W, Xu J (2008) Assessment and Simulation of Glacier Lake Outburst Floods for Longbasaba and Pida Lakes, China. Mt Res Dev 28:310–317. https://doi.org/10.1659/mrd.0894

Watanabe T, Kameyama S, Sato T (1995) Imja glacier dead-ice melt rates and changes in a supra-glacial lake, 1989–1994, Khumbu Himal, Nepal: danger of lake drainage. Mt Res Dev 15:293. https://doi.org/10.2307/3673805

Watanabe T, Lamsal D, Ives JD (2009) Evaluating the growth characteristics of a glacial lake and its degree of danger of outburst flooding: Imja Glacier, Khumbu Himal Nepal. Nor Geogr Tidsskr 63:255–267. https://doi.org/10.1080/00291950903368367

# Formation of the 2018 Bureya Landslide, Far East of Russia

2020 Kyoto Japan

Oleg V. Zerkal, Aleksey N. Makhinov, Alexander Strom, Vladimir I. Kim, Michael E. Kharitonov, and Igor K. Fomenko

**Abstract**

The Bureya landslide was formed on December 11, 2018 in the Bureya River valley in the Far East of Russia. It affected metamorphic rocks of the Upper Proterozoic age. The peculiarity of this rock slope failure was that it occurred in winter when air temperature dropped from ca. −3 °C to −37 °C. Landslide had complex structure and was formed in several stages with variable displacement mechanism. The first stage of main displacement can be classified as wedge failure transformed into rock avalanche more than 700 m long (measured from the slope foot) that moved with velocity up to 25–26 m/s. Landslide collapsed into reservoir and formed the splash wave up to 60 m high that washed out the taiga forest on the opposite slope of the valley. During the second stage that followed the first one in few seconds, large block of rock (260 × 280 m) slid down from the eastern part of the headscarp and formed rock avalanche up to 860 m long. The mean velocity of its motion was ~23–25 m/s, while the maximal one in its front could reach ~60 m/s. During the last stage several smaller secondary slides occurred on the slopes of the main landslide body and within the main headscarp. The total volume of the affected rocks can be estimated as 25 million m$^3$, up to 12 million m$^3$ of which were displaced during the first stage, up to 11.8 million m$^3$—during the second stage and up to 1.2 million m$^3$—during the secondary landslides formation. The Bureya landslide formed the natural dam more than 70 m high and up to 550 m wide that split the reservoir into two parts, so that special measures had to be undertaken to restore normal water flow.

**Keywords**

Large-scale slope deformations • Rock avalanche • Multistage development • Landslide dam

## Introduction

The Bureya landslide was formed on December 11, 2018 in the middle reaches of the Bureya River in the Far East of Russia, filled at present by the Bureya reservoir (Figs. 1 and 2). This area is sparsely populated and, thus, no direct evidence of the slope deformations evolution prior to catastrophic failure are available. Our description of this extraordinary event is based on the analysis of the local geological and geomorphic information, remote sensing and field inspection data, seismological observations and on the numerical modelling of slope stability.

## Natural Conditions

### Geomorphology

Landslide occurred on the left-bank slope of the Bureya River valley that stretches here in east-west direction. Its headscarp is located almost opposite the mouth of the Sredniy (Middle) Sandar River—small right tributary of the

O. V. Zerkal (✉)
Moscow State University, Moscow, 119991, Russia
e-mail: igzov@mail.ru

A. N. Makhinov · V. I. Kim
Institute of Water and Ecological Problems, Far Eastern Branch of Russian Academy of Sciences, Khabarovsk, 680000, Russia
e-mail: ivep@ivep.as.khb.ru

V. I. Kim
e-mail: kim@ivep.as.khb.ru

A. Strom
Geodynamics Research Center LLC, Moscow, 125008, Russia
e-mail: strom.alexandr@yandex.ru

M. E. Kharitonov
Independent Researcher, Talakan, 676730, Russia
e-mail: kme8775@yandex.ru

I. K. Fomenko
Ordzhonikidze Russian State Geological Prospecting University, Moscow, 117997, Russia
e-mail: ifolga@gmail.com


V. Vilímek et al. (eds.), *Understanding and Reducing Landslide Disaster Risk*, ICL Contribution to Landslide Disaster Risk Reduction,
https://doi.org/10.1007/978-3-030-60319-9_11

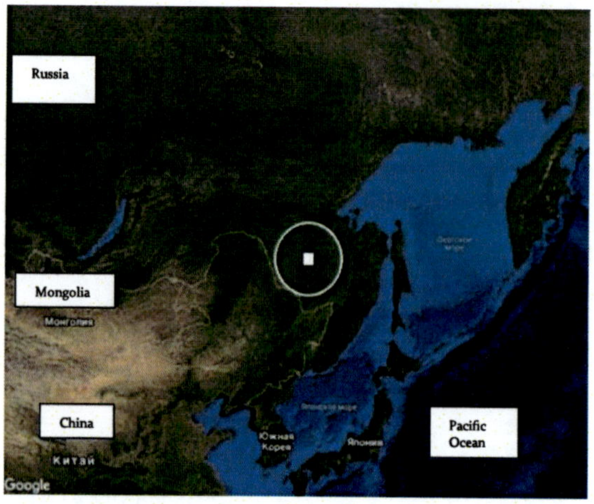


**Fig. 1** Site location

**Fig. 2** General view of the Bureya landslide from the helicopter. Photo by A.N. Makhinov, 25 December 2018

Bureya River. The latter crosses here the Bureya mountainous area—elatively low-mountainous terrain formed by the erosional-denudational processes.

The dome-like watersheds are at 600–640 m a.s.l., while river bottom is at ca. 190 m a.s.l. Valley is filled with reservoir, which water level on December 11, 2018 was 253 m a.s.l. Valley has a trapezoid-shape profile with steeper southern (left-bank) slope and more flat (10–12°) and elongated (more than 1 km long) northern (right-bank) slope. Southern slope up to ca. 310 m high dips 30° and further upslope flattens up to 22–24° and stretches up to local watershed.

## Climatic and Weather Conditions

Climate of the study region is characterized by negative mean annual air temperatures: −2.8 °C for the last 30-year long period, according to the nearest weather station "Chekunda" (~65 km north-east of the landslide site). Besides, quite significant annual (up to 70° and even more) and diurnal (up to 15° and more) temperature variations are typical.

Mean long-term annual precipitation at the same weather station is 669 mm, most of which falls during warm period. Significant increase of monthly precipitation was recorded in July 2018—up to 233 mm, while mean monthly precipitation for July is 146 mm only. Another specific weather condition that preceded landslide formation directly (from December 3 till 6–9, 2018) was sharp decrease of the air temperature from about −3 °C to −36° ÷ −37 °C (according to the Chekunda weather station). It resulted in formation of the stable ice cover in the reservoir up to 20 cm thick on December 8.

## Geological Conditions

Landslide site is located in the northern part of the Bureya tectonic block. It coincides, most likely, with of the junction of several faults or fracture zones, largest of which is the fault zone trending in north-eastern direction (Fig. 3). Orientation of the Bureya River valley also had been predetermined by the tectonics of the area, because rocks outcropping at it banks differ in age and composition.

Left bank of the river directly at the landslide side is composed of the biotite and two-mica schist with interbeds of meta-andesite stretching, generally, along the slope, and by amphibolite with marble lenses and gneiss of the Uril series (Erish 1964). They are intruded by numerous dikes of the granosyenite porphyry stretching across the slope. Metamorphic rock of the Uril series had been formed from the volcanogenic-sedimentary rocks of the Neoproterozoic age (Kotov et al. 2009). The opposite—right bank of the river is composed of the granites, presumably of the early and late Paleozoic age (Sorokin et al. 2010).

The study area belongs to the sporadic permafrost zone with temperatures varying −0.5 °C to −1 °C (Zerkal and Strom 2017). The permafrost thickness is 5–15 m, increasing at the watersheds up to 100 m (Kozlovsky 1988). Formation of large reservoir that has been filled first time up to its normal level (256 m a.s.l.) in 2009 affected the permafrost conditions significantly and, likely caused its thawing at the valley bottom and at the lower part of the north-facing slope where the landslide occurred.

Presence of the permafrost at this slope above the reservoir level can be derived from field observations. In January 2019, shortly after the landslide, when the first field inspection was carried out, large blocks of crushed bedrock (likely from the fault zones) and of loose Quaternary deposits, frozen and cemented with ice were observed in the frontal zone of rock avalanche deposits. Later, in June 2019, they thawed and formed the cone-shaped hummocks.

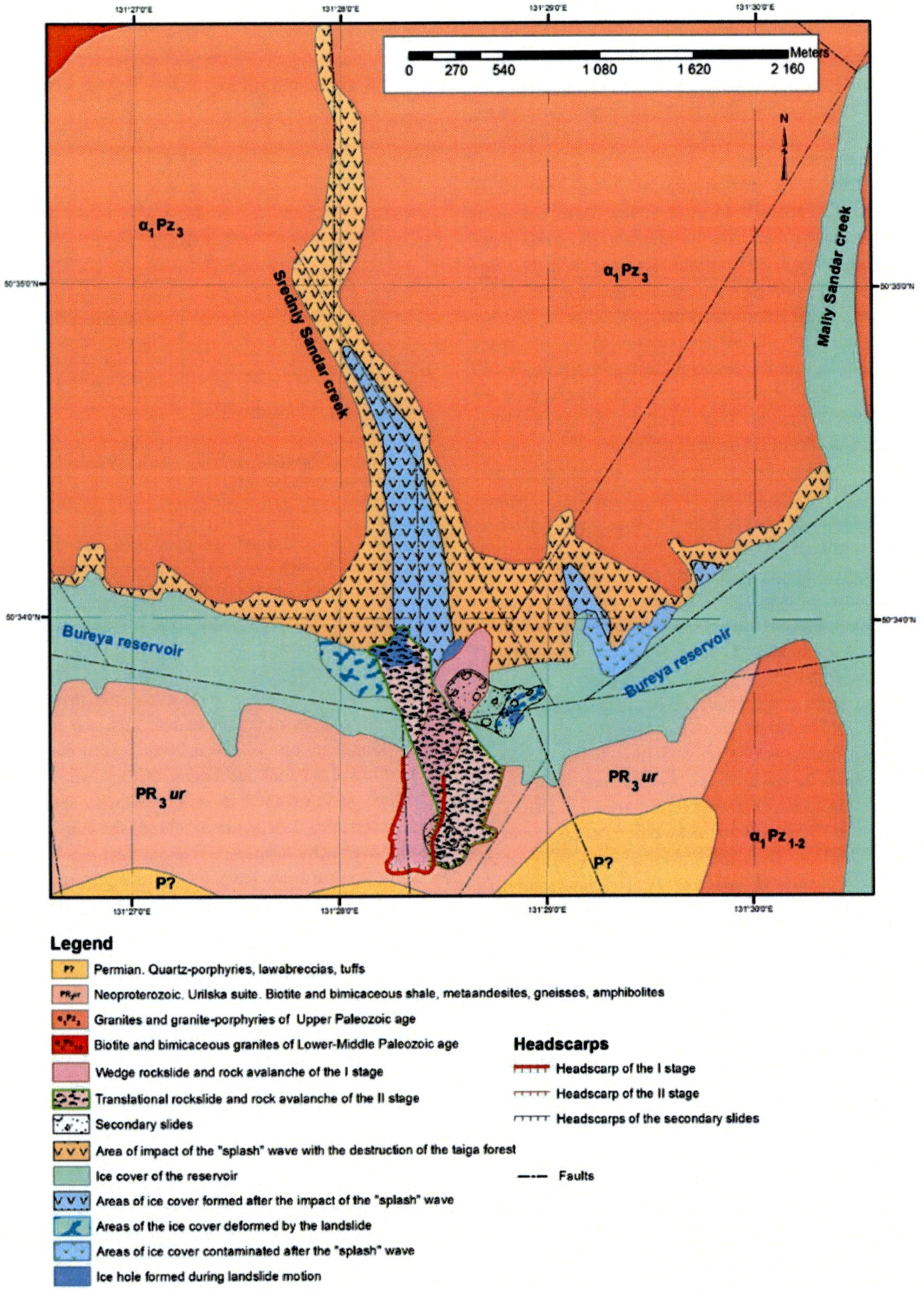


**Fig. 3** Map of the Bureya landslide and its impact zone (based on the results of satellite images interpretation, made by O.V. Zerkal). Geological structure of the region modified after (Muzylev 1959)

## Formation and Evolution of the Bureya Landslide

Several stages of the preparation and formation of the Bureya landslide can be identified: pre-slide slope deformations, main displacement that consists of two stages and final stage of slope deformations (Fig. 4).

**Initial stage**. Slope deformations at the left bank of the Bureya reservoir had started at least few days before its collapse. As it was found on the Sentinel-2B space image made on December 8, 2018, spring appeared near the slope base just above the frozen reservoir, about 100 m east of the western side of the future headscarp wall. Spring did not freeze despite very cold air temperature that could reach −36° ÷ −37 °C, and its discharge increased gradually till December 11, 2018 morning that could be observed on the successive space images. One more spring about 150 m higher could be identified on the last images.

Such release of ground water within the permafrost area indicates intensive fracturing on the slope that stipulates outflow of "warmer" water from the rock massif interior, below the permafrost zone.

**Main displacements**. According to the records made by local seismological network of the Bureya HPP, the intensive slope deformations had started at 15:49 (4:49 GMT), when the signal produced, most likely, by the surficial "source" was recorded.

Seismic records are complex, according to M.E. Kharitonov (Fig. 4). The entire record that lasted for about 4 min 25 s indicates three pulses of active deformations: "A"—that lasted for about 28–29 s, "B"—14–15 s long, and "C"—with low intensity shaking that continued for 3 min 37 s. Pulses "A" and "B" are separated by time interval of weaker shaking 9–10 s long. Shift from pulse "B" to "C" is gradual.

**First stage of main displacements**. The kinematic back analysis of slope stability demonstrates that large rock block at the north-facing slope could slide down as a wedge along the existing irregularities such as schistosity, existing tectonic fractures, dikes' contacts (Fig. 5) (Zerkal and Fomenko 2016). It was confirmed by data collected during field inspection at the site.

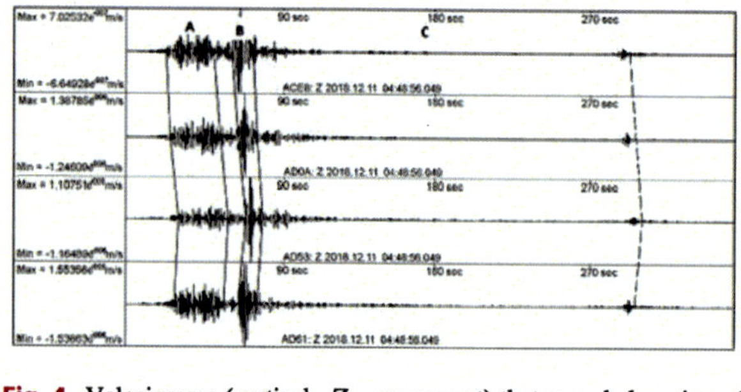


Fig. 4 Velosigrams (vertical - Z - component) that recorded motion of the Bureya rockslide. Pulses A, B, C are selected

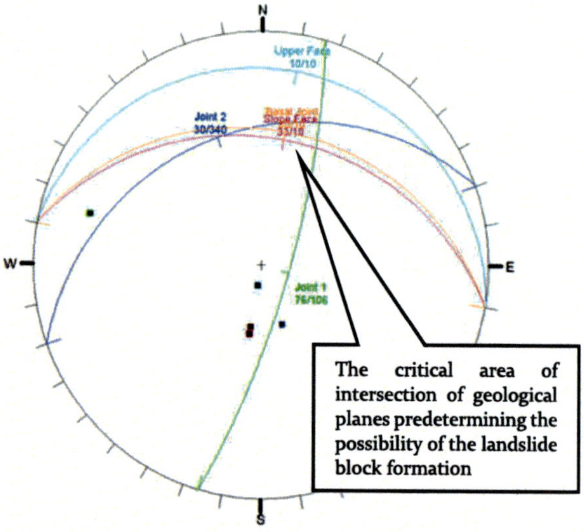


Fig. 5 Results of kinematic analysis of the north-facing slope stability conditions at the landslide site

Block that moved during first stage was located within the western part of the headscarp area that was finally formed on the slope. It was 700 m long in the downslope direction and up to 250 m wide. The initial wedge failure transformed into flow-like rock avalanche that moved in north–north-eastern direction and formed the body up to 740 m long and up to 270 m wide. According to field observations the initial succession of rock types typical of the source zone retained in rock avalanche deposits indicating absence of debris mixing during the emplacement.

Rock avalanche formed the natural dam that blocked the ~550 m wide Bureya River valley filled by the reservoir and split it into two parts. The calculated H/L ration is 0.19 and angle of reach (fahrböschung) is 11°. The entire volume of rocks that had moved during the first stage can be estimated as ~12 million $m^3$. These calculations were made considering that most of rock avalanche body is hidden under the Bureya reservoir which maximal depth here is up to 70 m. Taking in mind the runout measured from the slope foot (~740 m), and duration of the displacement that corresponds to the duration of pulse "A" on seismic record (28–29 s) we estimated ~25–26 m/s (~90 km/h) mean velocity of this rock avalanche motion.

Landslide caved into the reservoir and formed "splash wave" up to 60 m high that completely destroyed the taiga forest on the opposite right bank of the valley at a 550 m wide front. This wave entered the Sredniy Sandar River valley and moved upstream for 3.6 km where the forest cover was totally damaged on the banks leaving the bear land up to 300 m wide (Fig. 3).

Traces of the wave are visible up to 7 km upstream and up to 4 km downstream the Bureya valley, in line with north-north-eastern (slightly upstream) direction of landslide motion. Brocken ice and most of the trees washed out by the wave were later carried by the backward wave and deposited at the Bureya valley banks and over the ice cover where it remained safe. It can be assumed that two short low-amplitude peaks on seismic records between pulses "A" and "B" (see Fig. 4) reflect the splash wave and the backward wave.

**Second stage of main displacements**. Start of the second stage of landslide formation corresponds to the beginning of pulse "B" on seismic record. During this stage large block located east from the first stage headscarp collapsed (see Fig. 3). According to field observations the detachment zone formed at this stage coincides with western contact of the granosyenite porphyry dike that crosses the slope in the north-south direction. Set of similarly oriented dikes can be traced within the headscarp from the reservoir water level up to its upper edge.

Slope deformations of the second stage occurred, most likely, when the block bounded by granosyenite porphyry dike had lost its stability after wedge failure of the first stage. Shock produced by the backward wave that hit the base of the left-bank valley slope leaving ice blocks visible there could act as an additional trigger. All these resulted in detachment of the block 260 × 280 m in size and up to 15 m thick and its motion in the north–north-western direction for 350 m. This rock mass moved as a single block that can be revealed from the presence of trees that survived on its top, though being tilted towards the headscarp and forming the typical "drunken forest".

During the initial motion of this block it likely tilted towards the headscarp and was fragmented partially and some portion of debris was ejected forming rock avalanche 860 m long (from the slope foot) and up to 240–260 m wide. This body is composed of crushed debris and blocks of schist, metaandesite and breccia. Possibility of such mechanism of rockslide motion was demonstrated by the conceptual modelling (Zerkal et al. 2017).

Mean sliding velocity of the block that was displaced during second stage of main displacements and remained more or less intact was estimated as ~23–25 m/s (~84–90 km/h). At the same time velocity of its frontal part that moved far from the slope base could reach ca. 60 m/s (~216 km/h). Such estimate is based on the duration of the pulse "B" that was almost two times less than the duration of pulse "A". Such an abnormally high velocity supports the assumption of the additional forcing of moving debris (its ejection), and, probably, of its motion with lubrication, without direct contact with the base.

Total volume of rocks displaced at this stage was estimated as ca. 11.8 million $m^3$. H/L ratio is 0.16, and angle of reach (fahrböschung) is 9° only.

**Final stage of displacements**. Several relatively small failures occurred in the south-eastern part of the headscarp at the final stage of the Bureya landslide formation. Their total volume is estimated as ca. 1.2 million $m^3$ and such retrogressive sliding increased the headscarp size. Besides, several secondary landslides originated on the eastern slope of the dam formed during the first stage. Duration of these final motions can be estimated from last pulses on seismic records that lasted for 3 min 37 s (see Fig. 3).

## Discussion

The data collected show that the Bureya landslide evolved as a complex multi-stage rock avalanche. The following peculiarities deserve to be pointed out:

1. Large-scale rock slope failure was associated with rapid (within 2–3 days only) decrease of air temperature from −3 °C to −37 °C. It is evident that such fall in temperature resulted in fast freezing of the surficial part of the slope not less than 20 cm thick. This frozen 'layer' became an aquiclude that changed the draining conditions of the slope drastically.
2. Large volume (up to 25 million $m^3$ as a whole) and high speed of landslide motion (from ~25–26 m/s up to 60 m/s at different stages), while the slope height (~300 m) and it steepness (not more than 30°) were relatively low.
3. Landslide rushed into the reservoir about 500 m wide and up to ca. 70 m deep almost at a right angle (80°–90°) that formed the splash wave (tsunami) up to 60 m high. Such a high wave also indicates very high speed of the landslide (rock avalanche) motion.

So large and deep-seated rock slope failure could be triggered by rather specific combination of two factors—permafrost thaw at the lower part of the slope caused by reservoir impoundment and sharp—almost for 30° in few days temperature drop just before the collapse.

Appearance of the spring on the slope little higher than the reservoir level 3 days before the landslide that did not freeze despite very cold weather and despite assumed presence of permafrost dozens meters thick in the rock massif indicates that the ground water feeding the spring was warm enough. Thus, they could come out from the rock massif interior, from 'under the permafrost'. Rapid freezing of the surficial part of the slope that formed the aquiclude could result not only in rock mass watering, but also in deep-seated warming of the permafrost and weakening of the fractured zones at a depth comparable with the thickness of the landslide block.

Hence, change of lower boundary of the permafrost zone state and presence of ice interbeds along the fractures within the frozen rocks could be an important, though still unexplored factor that could affect overall stability of the slope. Their thawing could produce some 'lubricant' at the base of the displacing rock mass.

These peculiarities allow considering the 2018 Bureya landslide as a unique geological phenomenon that occurred in the permafrost zone of the Far East of Russia. Formation of such a large-scale rockslide associated with freezing rather than with thawing is a rare phenomenon that requires further study.

## Conclusions

Complex structure of large landslide formed on 11.12.2018 at the left-bank slope of the Bureya River valley at its east–west trending section was determined by its multi-stage formation and by variable mechanism of motion that changes at the successive stages of emplacement. Two main stages can be identified that is supported, first, by seismic records of the event provided by the local seismological network of the Bureya HPP and, second, by the fish tail shape of the deposits due to multidirectional debris motion.

Slope failure was preceded by fracturing that occurred when local air temperature dropped from −3 °C to −37 °C in few days.

Displacements at both stages started as rockslides that further transformed into rock avalanches that formed natural dam blocking the Bureya River valley. Mean velocities of rock avalanches motion are estimated as ca. 90 km/h. At the same time, considering much shorter duration of pulse "B" of the seismic record, we estimated much higher velocity of the second stage rock avalanche front—up to ca. 216 km/h. We hypothesized that so high velocity could be caused by ejection of some part of rock avalanche debris that accelerated.

Total volume of the displaced rock mass is estimated as ca. 25 million $m^3$; up to 12 million were displaced during the first stage, about 11.8 million—during the second stage, and about 1.2 million, during subsequent failures at the south-eastern part of the headscarp.

Landslide formed a blockage that split the reservoir of the Bureya HPP into two parts and special efforts had to be undertaken to restore the hydraulic interconnection of its upstream and downstream parts.

**Acknowledgements** Authors want to thank Alexander Kokovkin and Alexander Kudymov from Institute of Tectonics and Geophysics (Far Eastern Branch of Russian Academy of Sciences) who participated in the landslide field inspection.

## References

Erish LV (1964) Geological map of the USSR. Scale 1:200 000. Khingano-Bureinskaya series. Sheet M-52-XVIII. Explanatory note. Moscow (in Russian)

Kotov AB, Velikoslavinskii SD, Sorokin AA, Kotova LN, Sorokin AP, Larin AM, Kovach VP, ZagornayaN Yu, Kurguzova AV (2009) Age of the Amur group of the Bureya-Jiamusi Superterrane in the Central Asian fold belt: Sm-Nd isotope evidence. Dokl Earth Sci 428(5):637–640 ((in Russian))

Kozlovsky EA (ed) (1988) Geology of the BAM zone, vol 2. Hydrogeology and engineering geology. Nedra, Leningrad (in Russian)

Muzylev SA (ed) (1959) Geological map of the USSR. Scale 1:200 000. Khingano-Bureya series. M-52-XVIII. Gosgeotekhizdat Press, Moscow. (in Russian)

Sorokin AA, Kotov AB, Salnikova EB, Kudryashov NM, Anisimova IV, Yakovleva SZ, Fedoseenko AM (2010) Granitoids of the Tyrmo-Bureya complex in the northern Bureya-Jiamusi superterrane of the Central Asian fold belt: age and geodynamic setting. Russ Geol Geophy 51(5):717–728 (in Russian)

Zerkal OV, Fomenko IK (2016) Rockslides and their stability analysis. Enginernaya Geologiya (Eng Geol) 4:4–21 ((in Russian))

Zerkal OV, Frolova YuV, Strom AL (2017) The conceptual modeling of the style of rock massif destruction and of its influence on rockslides and rock avalanches formation. ISRM progressive rock failure, extended abstract of the conference, 5–9 June 2017. Ascona, Switzerland, pp 113–114

2020 Kyoto Japan

# Landslide Dam Hazards: Assessing Their Formation, Failure Modes, Longevity and Downstream Impacts

Regine Morgenstern, Chris Massey, Brenda Rosser, and Garth Archibald

**Abstract**

In the last few decades, >200 new natural (mainly landslide) dams have formed in New Zealand. Several of these dams, such as the: 1996/97 Mount Ruapehu tephra dam (formed by volcanic eruption); 2007 Young River landslide dam; largest ten landslide dams associated with the 2016 Kaikōura Earthquake; and 2019 Kaiwhata landslide dam, have been studied in detail to better understand their: (a) formation mechanisms; (b) material properties; (c) failure modes; and (d) downstream impacts. This paper outlines a method to assess the longevity and downstream impacts of landslide dams, post-formation, by adopting a combination of field techniques, forecast models and expert judgement based on the performance of past landslide dams, immediately post-event. To do this we use the following steps: (1) carry out initial breach and inundation modelling using existing information—done prior to visiting the site; (2) detailed, high-resolution topographic surveys of the dam and downstream area; (3) site-specific investigations to measure key parameters such as the volume and geometry of the dam and lake, and the particle size distribution of the dam materials; (4) dam breach modelling using empirical methods to identify dam failure scenarios; (5) numerical flood/debris inundation modelling to determine area of impact; and (6) overlaying dam failure and debris inundation scenario models on asset maps to identify people, buildings and other infrastructure that are potentially at risk. This study summarises the method for assessing the likelihood of dam failure and the potential downstream consequences, using the Hapuku River and Kaiwhata dams as case studies.

R. Morgenstern (✉) · C. Massey · B. Rosser · G. Archibald
GNS Science, 1 Fairway Drive, Avalon, Lower Hutt, 5010, New Zealand
e-mail: r.morgenstern@gns.cri.nz

**Keywords**

Landslide dam • Dam outburst flood • Kaikōura earthquake • Kaiwhata landslide • Rapid response

## Introduction

The rapid failure of landslide dams can pose a significant secondary hazard immediately following the formation of the initial landslide, whether it be earthquake- or non-earthquake induced. The impacts on life and critical infrastructure downstream from a dam can be devastating if it fails rapidly and the impounded water is released (e.g., Burmadinho dam disaster, Brazil, on 25 Jan 2019). A landslide dam can form if landslide debris blocks the watercourse in a valley, usually in response to: (a) a large earthquake - strong ground shaking, e.g., 2016 Kaikōura Earthquake; (b) a high intensity/prolonged rainfall event - increase in pore-water pressure, e.g., 2019 Kaiwhata landslide dam; (c) no obvious trigger - progressive failure, e.g., 2007 Young River landslide and dam; or (d) other process—such as volcanic eruption, e.g., 1996/97 Mount Ruapehu tephra dam. The longevity of the landslide dam in the landscape is determined by a complex interplay between: (1) the initial landslide source material, failure mechanism and debris runout distance; (2) the geometry of the valley and dam (shape, height and volume); (3) the characteristics of the water course, i.e., catchment area and stream power; and (4) impounded lake volume (e.g., Foster et al. 2011; Massey et al. 2010). These parameters can be highly variable and affect the dam's post-formation behaviour and persistence in the landscape, which could vary from a few hours to thousands of years. Dam failure modes can include: overtopping, headward/backward erosion, piping/throughflow and additional slope failure (e.g. slumping) (Foster et al. 2011). If the landslide dam gradually erodes over time, the lake level decreases slowly with minimal downstream flooding;


V. Vilímek et al. (eds.), *Understanding and Reducing Landslide Disaster Risk*, ICL Contribution to Landslide Disaster Risk Reduction,
https://doi.org/10.1007/978-3-030-60319-9_12

however, rapid failure of the dam can result in the sudden evacuation of impounded lake water, causing a debris flow/flood to travel long distances downstream. The likelihood of rapid dam failure increases during and immediately after rainfall events, especially if these occur within a year of dam formation and the catchment area is large in relation to the size of the dam and/or lake.

In New Zealand, due to the climate and tectonic setting, landslide dams triggered by rainfall and earthquake events are widely recognised, especially in regions with high relief, steep topography and/or soft Neogene sediments (Korup 2004). Korup (2004) complied 232 landslide dams into a national inventory (Fig. 1), of which, 85 (37%) are known to have failed. Data on the initial triggering mechanism of the landslide and its type, and dam failure modes are sparse, but where known, landslide dams appear to be mostly associated with shallow, earthquake-induced rock avalanches, and the majority are relatively short-lived (Korup 2004). In the last few decades, over 200 new natural (mainly landslide) dams have occurred and were documented during GeoNet rapid response studies following major landslide events, several of which have been studied in detail. These include the 2007 Young River landslide dam, landslide dams associated with the 2016 Kaikōura Earthquake and the 2019 Kaiwhata landslide dam. These studies have generated comprehensive datasets to better understand triggering mechanisms, failure modes, longevity and the downstream hazards posed by landslide dams.

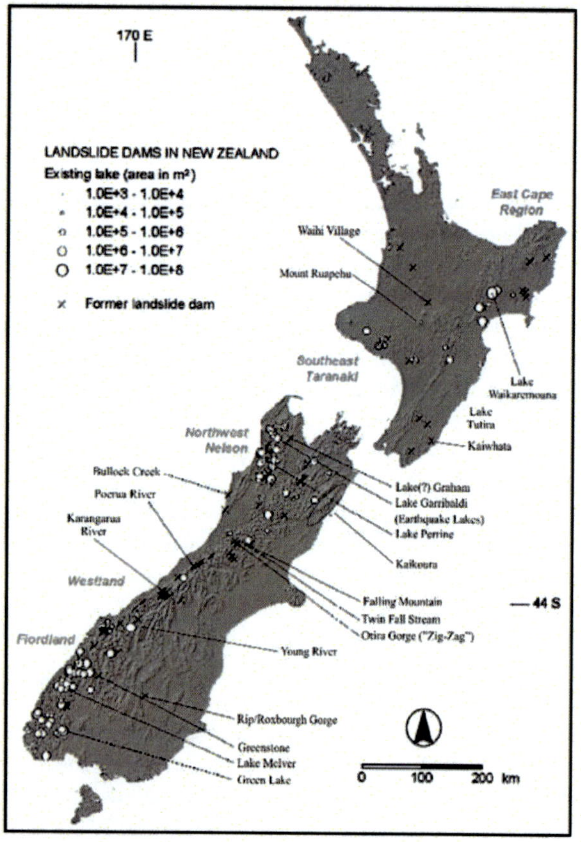


**Fig. 1** The location of landslide-dammed lakes and former landslide dams in New Zealand. Updated from Korup (2004)

## Methods

Responding to significant landslide and landslide dam-forming events in New Zealand is typically carried out by GNS Science under the GeoNet project (a national natural hazard monitoring programme). If a response is initiated, identifying and assessing the location, size, likelihood of catastrophic failure and potential downstream consequences become priorities to inform public safety risk (McSavaney et al. 2010). The method for assessing the likelihood of dam failure and the potential downstream consequences comprises a combination of field techniques, empirical forecast models and expert judgement based on the performance of past landslide dams. To do this we use the following steps:

### Step 1: initial breach and inundation modelling

During the initial stages of a GeoNet landslide response, the affected area is delineated and systematically searched. The method and time taken depends on the scale of the event and can be anything from quick and straightforward (if there is just one) to time-consuming and complex (if a whole region has been affected). This delineation is often an iterative process and can be done in various ways, such as by using remote sensing techniques (e.g. satellite imagery) or by visiting the site (e.g. ground-based or helicopter reconnaissance), and areas with highest risk to life are prioritised (Dellow et al. 2017). Once a dam has been identified and prior to visiting the site(s), initial breach and inundation models are generated using existing information. These help to plan the response and focus the site-specific assessments.

### Step 2: high-resolution topographic surveying

Detailed, site-specific surveying is undertaken to acquire high-resolution, 3D topographic data that can be used to estimate the volume and geometry of the dam, lake, and downstream flow-path, and to model potential dam-breach scenarios. Terrestrial Laser Scanner (TLS) and Unmanned Aerial Vehicle (UAV) surveys allow for rapid acquisition of topographic datasets to facilitate modelling during the immediate response phase. Although Light Detection and Ranging (LiDAR) data are preferred for scenario modelling, data acquisition can take months and is therefore not appropriate for the initial response phase. The topographic data, in combination with Real-Time Kinematic (RTK) point surveys, are then used to calculate key parameters, such as the lake level, dam height and cross-sectional profile, as well as estimate the volume of the landslide debris, dam and lake, which can then be used in the empirical assessment of dam failure mode and outflow discharge.

In addition to the site-specific TLS surveys using a Riegl VZ-2000i, a UAV survey of the entire site is also conducted using a Phantom 4 RTK in combination with AeroPoints and a Post-Processing Kinematic (PPK) workflow—a quick and accurate method of acquiring high-resolution imagery and topographic data of the site (Morgenstern et al. 2019). Although the UAV survey has a lower level of accuracy when compared with the TLS survey, the two survey methods are complimentary. The TLS survey captures highly accurate data of the dam for volume estimation and breach modelling, while the UAV covers a much larger area, allowing for the rapid and complete capture of ephemeral data that can be used for interpreting the wider area, from viewing the ground conditions above the source to understanding the floodplain geomorphology downstream from the dam. The UAV is also used to acquire photos and videos to gain a wider understanding of the entire site, or regions of the site that are unsafe to visit on foot.

Repeat topographic surveying and elevation differencing allows for change models of the site to be generated using data from different TLS, UAV, LiDAR and/or photogrammetry survey epochs, thus tracking dam evolution up to and after failure. These models can be used to pick out deformation and change in dam geometry over time, such as slumping, bulging and the development of outflow channels and seepage points, and areas of erosion and deposition. This allows us to determine how the dam is evolving, allowing us to better understand its behaviour, longevity and hazard.

#### Step 3: site-specific investigations

Field mapping and *in-situ* sampling and testing is conducted before and after the dam-breach event to define key parameters, such as the source material, failure mode, and location and flow rate of seepage points. Bulk sieve analysis is undertaken to understand the particle size distribution (PSD) of the debris forming the dam, and thus its likely inherent stability and longevity. This allows for the geotechnical characterisation of the dam material and for an engineering geological model of the landslide and landslide dam to be developed. During site visits, a pressure transducer is also sometimes installed to monitor the impounded lake level.

#### Steps 4 to 6: scenario breach and inundation modelling and risk analysis

Various scenarios (e.g. 'most likely', 'worst credible') are modelled using RApid Mass MovementS (RAMMS) software to forecast potential flood and debris inundation zones downstream from the dam. These are used to inform emergency services and affected members of the community on the potential impacts from a dam-breach event. From this, people and assets at risk can be identified and management plans, such as evacuation zones, put in place. This is initially done using the best estimates of input parameters such as dam volume, lake height and width, discharge rate and debris content. These forecast models are highly sensitive to the downstream channel geometry; therefore, it is important to use the most recent and highest-resolution topographic models available.

The initial breach models are often highly conservative, highlighting the importance of field data to verify input parameters and optimise model results. As more data are collected, the RAMMS simulations are refined and re-run. Post-breach empirical evidence (e.g. trash line surveys, lake level and rainfall data, eyewitness accounts) can be used to calibrate the models by back-analysing dam failure. This systematic process of data collection, field verification and model calibration are important to inform future hazards and risks posed by landslide dams.

## Results and Discussion

### Kaikōura Earthquake-Induced Landslide Dams

A shallow (14.1 km) magnitude ($M_w$) 7.8 earthquake occurred in Canterbury, New Zealand (epicentre near Waiau), at 12.03 a.m. on 14 November 2016 local time, resulting in the complex multi-fault rupture of >23 individual faults (Litchfield et al. 2018) (Fig. 2). The most obvious regional consequence of the strong ground shaking was widespread landsliding. Systematic mapping shows a distinct correlation between both the largest landslide volume (5 to 20 M $m^3$) and landslide density with the location of surface fault ruptures (Dellow et al. 2017; Massey et al. 2018b). Ground shaking was concentrated north of the epicentre, and c. 30,000 landslides and c. 200 significant landslide dams (Fig. 2) have been mapped over an area of c. 10,000 $km^2$ (Dellow et al. 2017; Massey et al. submitted). Detailed topographic surveys show that the largest landslide dam (Hapuku River; Fig. 3) had an initial volume of c. 23 (± 2) M $m^3$, where the debris travelled 2.7 km down-slope from its source and created a 100 m high dam. Ten other landslide dams with volumes ranging between 1 and 10 M $m^3$ also occurred.

The large number of landslides and landslide dams generated by the earthquake was due to both the strong and widespread ground shaking as well as the steep slopes and confined valleys of the area. The coseismic landslides are generally shallow, and 80% are classified as soil or rock avalanches (Massey et al. submitted). Field mapping and

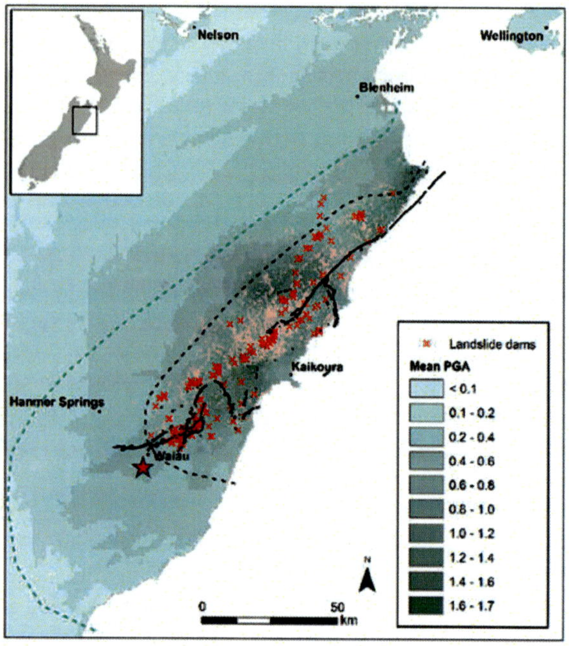


**Fig. 2** The location of significant landslides (pink polygons) and landslide dams (red crosses) generated during the 2016 Kaikōura Earthquake (epicentre shown by red star), overlain onto the mean Peak Ground Acceleration (PGA) experienced in the region. Areas with light to moderate (green dash) and severe (black dash) landslide damage are shown. Surface fault ruptures are shown by solid black lines

**Fig. 3** The Hapuku River landslide dam triggered by the 2016 Kaikōura Earthquake. The photo was taken on 16 Nov 2016 as the lake was filling, less than 48 h after the earthquake. Photo credit: Dougal Townsend

in situ sampling shows that two distinct types of landslide dams were generated, reflecting the geological and geotechnical properties of the source material: (1) first-time and reactivated rock and rock-block slides and slumps, mainly sourced from weak sedimentary rock (Neogene sandstone and siltstone); and (2) first-time rock and debris

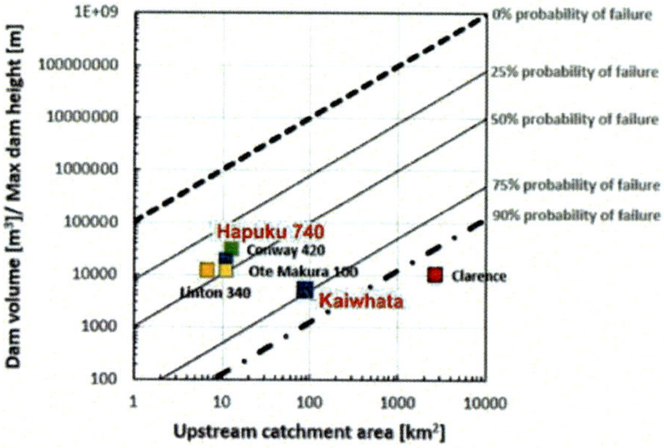


**Fig. 4** Landslide dam failure probability using the empirical-based Dimensionless Blockage Index (DBI)

avalanches, comprising course angular gravel, mainly sourced from stronger sedimentary rock (Late Jurassic to Early Cretaceous Torlesse greywacke and Late Cretaceous to Paleogene limestone). These landslides generated different dam morphologies with distinct material properties. The D50 (i.e., the 50th percentile of the PSD) of the greywacke dam samples was 5 mm to 10 mm, in contrast the D50 of the Neogene dams was <1 mm.

Although no landslide-related deaths were recorded, and the area affected by them was relatively remote and sparsely populated, the stability of the dams, and the potential hazard that a sudden catastrophic failure posed in generating a flood wave downstream, were a major concern in the months following the earthquake. To assess the stability of the dams, and thus the likelihood of dam failure, the empirical-based Dimensionless Blockage Index (DBI) (Massey et al. 2018a) was initially used (Fig. 4), which compares key dam parameters with upstream catchment area. The DBI indicated that the majority of these coseismic landslide dams were likely to be unstable (Massey et al. 2018a). The first dam (Clarence River) breached within 16 h of its formation, and most of the others failed within the first year. Several of the greywacke dam failures, including partial failure of the large landslide dam in the Hapuku River, were caused by the first post-earthquake high-intensity rainfall events associated with cyclones Debbie (23 Mar–7 Apr 2017) and Cook (14 Apr 2017) and ex-Tropical Cyclone Gita (20 Feb 2018). The breaches caused debris flows and floods to extend several kilometres downstream from the former dams (Massey et al. 2018b). Repeat surveying, field inspections and lake level monitoring show that most of the dams failed through a combination of overtopping and seepage through the dam, leading to partial headward erosion of the downstream face. For the dams that are still present in the landscape, such as the one in the Hapuku River, much of the remaining debris is still unstable and continues to erode (Massey et al. 2018b).

For the Kaikōura Earthquake landslide dams, the downstream debris inundation levels forecasted by the RAMMS models for the 'most likely' dam breach scenario, for the most part, matched those observed in the field post-failure, indicating that such forecast models are useful for assessing debris flood hazards from similar landslide dams (Massey et al. 2018a).

## Kaiwhata Landslide Dam

A significant landslide occurred on 1 June 2019 in the Kaiwhata River, Wairarapa, New Zealand, forming a landslide dam (Fig. 5A) with an extensive lake upstream (c. 1.1 M $m^3$ of water), and leading to the inundation and closure of the local Kaiwhata Road. In the following days, the rising lake level was a concern to both households upstream from the dam (inundation hazard) as well as downstream households and infrastructure (i.e. two bridges) should the dam fail catastrophically (dam outburst flood hazard). Although the Kaiwhata River has a low mean annual flow rate (1.55 $m^3$/sec), the relatively large upstream catchment area (85 $km^2$) combined with the dam geometry had a strong influence on its longevity in the landscape. An initial hazard assessment suggested the dam had a 75% probability of failure based on the DBI (Fig. 4), i.e., a volume of 110,000 $m^3$ and a dam height of 16 m. Initial empirical dam breach models (e.g., Costa and Schuster 1988) were used to derive potential failure discharges for different dam geometries and lake volumes assuming overtopping and rapid failure of the dam. These discharges were then used as inputs for inundation modelling using RAMMS software (refer to the method section for details).

Bulk sieve analysis of the dam material showed that it was reasonably erodible, consisting of mostly fine-grained material (soil and weathered rock) with scattered cobbles, boulders and woody debris. The D50 of the dam material was 3 mm (based on sieving 280 kg).

The initial failure of the landslide had occurred in 2017; field inspection and mapping found that the head of the landslide that reactivated in 2019 was in a Tertiary mudstone unit containing smectite clay, which when wet, is highly susceptible to landsliding. The body of the landslide occurred in Cretaceous mudstone with minor greensand and glauconitic sandstone, which are susceptible to slope instability by slumping, earth-flow and gully erosion. Analysis of the LiDAR data shows that there was active slumping on the slope immediately upstream of the landslide and an active earthflow present on the slope opposite. Furthermore, the region is undergoing active deformation relating to the compressional plate boundary regime and an inactive fault is mapped 150 m up-slope of the landslide head scarp, which has likely contributed to further weakening of this material.

**Fig. 5** The landslide dam that blocked the Kaiwhata River (**A**) before (6 June) and (**B**) after (14 June) failure

The combination of these weak materials, the geological setting and historic deforestation, coupled with the seasonal increase in rainfall prior to failure, are the likely cause for landslide activity at this location (Rosser et al. in prep).

Multiple field visits were conducted in the two weeks following dam formation to monitor evolution of the dam and calibrate the forecast models to provide updated information on the hazard posed by it. By 12 June, an outflow channel had formed around the toe of the landslide, and seepage and slumping were observed on the downstream face. On the afternoon of 13 June, lake level monitoring shows that the lake level increased rapidly after steady all-day rainfall, and at 9.15 p.m. the dam failed (Fig. 5B). This was due to a combination of overtopping, throughflow and slumping, and 1.1 M $m^3$ of water was released from the impounded lake in <1.5 h. Peak discharge was estimated at 900 $m^3$/sec, almost two orders of magnitude greater than the 100-year flood discharge of 15 $m^3$/sec. Trash line surveying (indicating the maximum extent and height of the flood water) and the presence of numerous terraces deposited immediately downstream of the dam show that failure occurred in multiple pulses. The flood wave inundated the entire floodplain with 1–2 m of water, up to a maximum

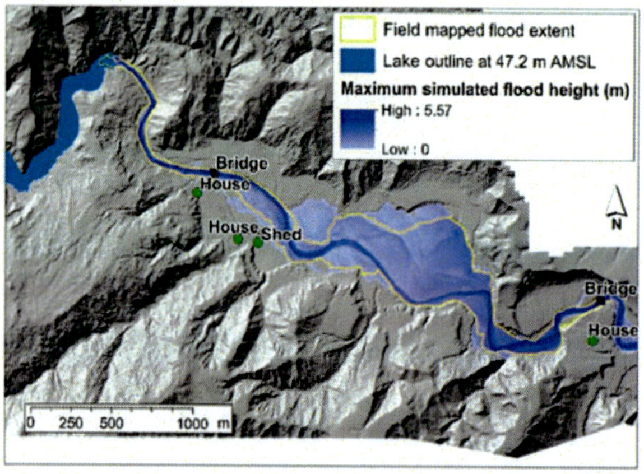


**Fig. 6** Comparison between forecast (blues) and field verified (yellow) inundation extent on the Kaiwhata River floodplain following failure of the landslide dam

depth of c. 6–7 m and width of 600 m, within at least the first 4 km downstream of the dam (Fig. 6), which was just below the freeboard on the bridge. These field observations matched well with the maximum simulated flood height shown in the updated forecast models. It took 1.5 h from dam failure to the emptying of the lake; no one was injured and no houses or infrastructure were damaged. There was very little sedimentation on the floodplain, but 115,000 $m^3$ of loose debris remains on the landslide and is vulnerable to erosion during future storm events.

## Summary and Conclusions

Most of the recent landslide dams in New Zealand have failed rapidly within the first year of their formation, usually because of overtopping caused by high intensity rainfall. For example, nearly all of the 200 significant dams formed from landslides triggered by the Kaikōura Earthquake were overtopped and partially removed during three main high-intensity rainfall events (Massey et al. 2018a). The Kaiwhata landslide dam was also overtopped and failed rapidly during a rain event. However, the 2007 Young River landslide dam and impounded lake are still present in the landscape although it has been overtopped multiple times since formation (Massey et al. 2013). These differences in longevity are thought to be a function of the PSD of the dam material, e.g., the D50 of several of the Kaikōura greywacke-sourced dams ranges from 5 to 10 mm and the D50 of the Kaiwhata dam is 3 mm, compared to a D50 of 1 m for the Young River dam.

Other failure modes have been observed, for example: the 1996/97 Mount Ruapehu tephra dam failed rapidly in 2007 due to backward erosion, releasing the impounded lake to form a lahar (Massey et al. 2010). Several of the landslide dams triggered by the 2016 Kaikōura Earthquake also started to fail post-formation via headward erosion, which would have likely led to failure of the dam if overtopping caused by significant rainfall, had not occurred.

Forecasting the post-formation behaviour of a landslide dam for hazard and risk assessment purposes is complex. The failure mode(s) and therefore longevity of a dam in the landscape is determined by a complex interplay between: (1) the initial landslide source material, landslide failure mechanism and runout distance; (2) the geometry of the valley and dam; (3) the characteristics of the water course, i.e., catchment area and stream power; and 4) the impounded lake volume.

To assess the likelihood and impacts of landslide dam failures, post-formation, we use a six step process: (1) initial breach and inundation models; (2) topographic surveys of dam and downstream area; (3) field mapping and sampling; (4) dam breach modelling using empirical methods; (5) numerical flood/debris inundation modelling; and (6) overlaying debris inundation scenario models on asset maps to identify people, buildings and other infrastructure potentially at risk.

**Acknowledgements** This work was funded by a Ministry of Business, Innovation and Employment (MBIE) Endeavour Fund and by the Earthquake Commission (EQC). The authors would like to thank Dougal Townsend for reviewing this abstract.

## References

Costa JE, Schuster RL (1988) The formation and failure of natural dams. Geol Soc Am Bull 100:1054–1068

Dellow GD, Massey CI, Cox SC, Archibald GC, Begg JG, Bruce ZR, Carey JM, Davidson J, Della-Pasqua FN, Glassey PJ, Hill MP, Jones KE, Lyndsell BM, Lukovic B, McColl S, Rattenbury MS, Read SAL, Rosser BJ, Singeisen C, Townsend DB, Villamor P, Villeneuve M, Godt J, Jibson R, Allstadt K, Rengers F, Wartman J, Rathje E, Sitar N, Adda AZ, Manousakis J, Little M (2017) Landslides caused by the $M_w$7.8 Kaikōura earthquake and the immediate response. Bull New Zealand Soc Earthq Eng. 50(2):106–116

Foster M, Fell R, Spannagle M (2011) Statistics of embankment dam failures and accidents. Can Geotech J 37:1000–1024

Korup O (2004) Geomorphometric characteristics of New Zealand landslide dams. Eng Geol 73(1/2):13–35

Litchfield NJ, Villamor P, Van Dissen RJ, Nicol A, Barnes PM, Barrell DJA, Pettinga JR, Langridge RM, Little TA, Mountjoy JJ, Ries WF, Rowland J, Fenton C, Stirling MW, Kearse J, Berryman KR, Cochran UA, Clark KJ, Hemphill-Haley M, Khajavi N, Jones KE, Archibald GC, Upton P, Asher C, Benson A, Cox SC, Gasston C, Hale D, Hall B, Hatem AE, Heron DW, Howarth JD, Kane TJ, Lamarche G, Lawson S, Lukovic B, McColl S, Madugo C, Manousakis J, Noble D, Pedley K, Sauer K, Stahl T, Strong D T, Townsend D B, Toy VG, Williams J, Woelz S, Zinke R (2018) Surface rupture of multiple crustal faults in the 2016 MW 7.8 Kaikōura, New Zealand, earthquake. Bull Seismolog Soc Am 108(3B): 1496–1520

Massey CI, Manville VR, Hancox GT, Keys HJ, Lawrence C, McSaveney MJ (2010) Out-burst flood (lahar) triggered by retrogressive landsliding, 18 March 2007 at Mt Ruapehu, New Zealand: a successful early warning. Landslides 7(3):303–315

Massey CI, McSaveney MJ, Davies T (2013) Evolution of an overflow channel across the Young River landslide dam, New Zealand. In: Margottini C, Canuti P, Sassa K (eds) Landslide science and practice. Vol 6, Risk assessment, management and mitigation, pp 43–49. Springer, Dordrecht. https://doi.org/10.1007/978-3-642-31319-6-7

Massey C I, Townsend D B, Dellow G D, Lukovic B, Rosser B J, Archibald, G C, Villeneuve M, Davidson J, Jones K E, Morgenstern R, Strong D, Lyndsell B, Tunnicliffe J, Carey J, McColl S (2018a) Kaikoura Earthquake Short-Term Project: landslide inventory and landslide dam assessments. GNS Science, Lower Hutt (NZ). GNS Science report 2018/19. 45p

Massey C I, Townsend D, Lukovic B, Jones K, Rhoades D, Morgenstern R, Rosser B, Ries W, Howarth J, Hamling I, Clark M, Petley D, Litchfield N, Olsen M (submitted) Landslide volume-area relationships for earthquakes are controlled by their type/failure mode and source material. J Geophys Res: Earth Surf

Massey C I, Townsend DB, Rathje E, Allstadt KE, Lukovic B, Kaneko Y, Bradley B, Wartman J, Jibson RW, Petley DM, Horspool NA, Hamling IJ, Carey JM, Cox SC, Davidson J, Dellow GD, Godt GW, Holden C, Jones KE, Kaiser AE, Little M, Lyndsell BM, McColl S, Morgenstern R, Rengers FK, Rhoades DA, Rosser BJ, Strong DT, Singeisen C, Villeneuve M (2018b) Landslides triggered by the 14 November 2016 $M_w$ 7.8 Kaikōura earthquake, New Zealand. Bull Seismologl Soc Am 108 (3B):1630–1648

McSaveney MJ, Massey CI, Dellow GD (2010) Landslide response and monitoring: the New Zealand GeoNet experience. In: Williams AL., Pinches GM., Chin CY., McMorran TJ., Massey CI (eds) Geologically active: delegate papers 11th congress of the international association for engineering geology and the environment, 5–10 September 2010, pp 443–450. Auckland, New Zealand

Morgenstern R, Archibald GC, Hill MP (2019) Phantom 4 RTK with a KlauPPK solution and AeroPoints: a guideline for workflow and best practices. GNS Science, Lower Hutt (NZ). GNS Science internal report 2019/06. 17p

Rosser BJ, Massey CI, Archibald GC, Morgenstern R (in prep) GeoNet response to Kaiwhata landslide dam. GNS Science report 2019/54

WLF
UNESCO · ICL · ISC · WFEO · FAO · UNDRR · WMO · UNU
2020 Kyoto Japan

# The Sedimentology and Internal Structure of Landslide Dams—Implications for Internal Erosion and Piping Failure: A Review

Chukwueloka A. U. Okeke

**Abstract**

Detailed information on the internal structure and sedimentological characteristics of landslide dams is important for natural dam hazard assessment studies. The materials composing landslide dams play a significant role in determining their overall longevity and stability. This, on the other hand, is linked to their source stratigraphy and the degree of fragmentation and pulverization undergone by the materials during their transport and emplacement history. Overall, the time of failure of landslide dams and the magnitude of the outburst flood depend on the internal characteristics of the impoundment, such as the grain size distribution, grain shape, degree of consolidation, presence of subsurface discontinuities, erodibility, and clay percentage. The evolution of internal erosion and piping in landslide dams has been ascribed to the relationship between the internal structure of the dam and the discharge rate through the impoundment. This paper presents a systematic review of the influence of the internal structure and sedimentological characteristics of landslide dams on the piping failure of natural river blockages. The paper highlights the processes of emplacement of landslide dams, the internal modification of the sediment during transport, the different facies assemblages and their role in the evolution of internal erosion and piping. Results obtained from the microtremor (MTM) chain array survey gave a better understanding of the evolutionary history of landslide dams. Furthermore, preliminary results obtained from laboratory experiments gave credence to the notion that the longevity and stability of landslide dams formed by translational mass movements are higher than those formed by rock avalanche processes.

**Keywords**

Landslide dam • Sedimentology • Internal structure • Seepage • Piping • Internal erosion

## Introduction

Landslide dams are formed by the partial or complete blockage of stream channels by displaced earth or rock materials. The probability of formation of landslide dams is determined by several factors including the volume of the displaced material, the type of landslide that formed the impoundment, the velocity of the displaced material, and the morphology of the stream channel. Schuster et al. (1998) identified four major factors that control the spatial distribution of landslide dams, based on the study of the landslide dams formed by the 1989 Loma Prieta earthquake in California (USA). These factors include seismic intensity (peak acceleration, duration of strong motion), topography and high slope gradient, lithology and weathering, and soil moisture and groundwater level. Landslide dams constitute a significant geomorphological hazard in geodynamically active regions of the world, which triggers many natural disasters that claim the lives of thousands of people every year (Schuster and Evans 2011; Wu et al. 2014).

Emergency countermeasures are often carried out on landslide dams immediately after their formation to stabilize the dam and avert danger posed by the release of catastrophic outburst floods. However, the implementation of effective countermeasures that could mitigate these disasters requires a good understanding of the sedimentology, internal structure and evolutionary history of these river blockages. For example, Shang et al. (2003) and Evans et al. (2011) gave a detailed account of the formation and failure of the Yigong River landslide dam in Tibet. The large-scale landslide ($3 \times 10^9$ $m^3$) dammed the Yigong River on April 9, 2000, and failed 62 days later (June 10), 7 days after the

C. A. U. Okeke (✉)
Department of Civil Engineering, College of Engineering, Covenant University, P.M.B 1023 Ota, Ogun State, Nigeria
e-mail: elo_destiny@yahoo.com


V. Vilímek et al. (eds.), *Understanding and Reducing Landslide Disaster Risk*, ICL Contribution to Landslide Disaster Risk Reduction,
https://doi.org/10.1007/978-3-030-60319-9_13

construction of an overflow channel on the dam crest. The outburst flood released by the dammed lake led to the displacement of about 50,000 people and caused a significant amount of damage to infrastructure. While many dam stability assessment methods have been developed to predict the longevity and stability of natural dams, the role played by their internal structure and sedimentological characteristics is still not fully understood.

Therefore, this paper presents a detailed review of the internal structure and material properties of landslide dams and their role on the internal erosion and piping failure of natural river blockages.

## Sedimentology of Landslide Dams

The study of the grain size distribution of materials composing landslide dams is not only crucial for evaluating their evolutionary history, but also for predicting their potential failure mechanisms (Casagli et al. 2003; Wassmer et al. 2004). This is because grain size distribution has a predominant effect on the longevity and the overall performance of landslide dams and controls their mechanical resistance to failure by either overtopping, piping or seepage (Wassmer et al. 2004). Landslide dams formed by rock avalanche processes are mainly composed of an unstructured fragmented mass of all grain sizes, excluding the uppermost layer, which is dominated by larger clasts and boulders (Davies and McSaveney 2011).

Field and laboratory investigation of the textural properties of several natural dam materials indicate that they are mostly poorly to very poorly sorted, matrix- or clast-supported, finely skewed, brecciated, stratified to massive sediments, and are usually armoured with angular boulders and sutured within a matrix of very fine materials (Capra 2011; Dufresne et al. 2016). The results of grain size distribution analysis carried out by Casagli et al. (2003) on 42 landslide dams in the Northern Apennines (Italy) showed that the diameter of the materials ranged from $1 \times 10^{-4}$ to $1 \times 10^{4}$ mm.

Many rock avalanche dams exhibit a dual system of internal structure that is usually made up of: (1) a highly crushed and matrix-supported basal layer, which is very erodible but has low hydraulic conductivity because of its low void ratio; and (2) an upper layer predominated mostly by a coarse blocky carapace of disjointed angular boulders with large void spaces which support internal erosion and piping (Davies and McSaveney 2011; Strom 2013). This dual system of internal structure signifies the low likelihood of piping failure than overtopping because the highly pulverized and widely graded materials of the basal layer limit the development of seepage in the dam.

## Facies Analysis

Several attempts have been made to classify the sedimentology of rock avalanche deposits using either the facies approach (Wassmer et al. 2004) or direct analysis (Casagli et al. 2003). A detailed review of several sedimentological data on landslide dams formed by rock avalanche processes revealed three unique features: crude inverse grading (Cruden and Hungr 1986); preservation of source stratigraphy (Strom 2006; Dufresne et al. 2016); and presence of highly fragmented but undisaggregated clasts (Davies and McSaveney 2011). A detailed sedimentological classification of some landslide dams is given below.

### Flims Sturzstrom (Swiss Alps)

According to Wassmer et al. (2004), the Holocene sturzstrom of Flims, Grisons, Switzerland, mobilized about 12 $km^3$ of Jurassic limestone, making it one of the largest mass movements on earth. The materials formed a 10 km 400 m blockage in the Upper Rhine River Valley. The authors characterized the materials composing the Flims sturzstrom into five sedimentary facies.

*The structured facies* comprises thick layers of rock predominated by small-scale oblique fractures that are differentiated by centimeter-thick fine-grained layers. This facies is located mostly in the internal and proximal parts of the deposit. The internal structure of this facies shows a remarkable similarity with the early stage of sturzstrom displacement. This characteristic has been attributed to the lithostatic pressure effects caused by the overburden material. *The intermediary facies* is a transitional zone marking the boundary between the structured facies and the granular facies. This facies dominates the margins and the uppermost layer of the deposits and hence is mainly composed of clasts embedded in a granular matrix. *The granular facies* is mainly composed of destructured breccia with clasts mixed in a sand matrix produced by friction between the rotating clasts. This facies is located mostly at the distal and lateral parts of the deposits. The effect of shear stress at different scales induced a rotational movement of the smallest clasts. *The brecciated facies* dominates the right side of the Rhine Valley where the avalanche ran into the foot of the slope. The breccia resulting from this collision is characterized by a low amount of matrix and a jigsaw-type deep fragmentation of the clasts. *The washed facies* comprises an accumulation of heterometric clasts without any fine matrix which dominates the surficial layer of the deposit. This facies can be attributed to the interaction between the overflowing lake water and the deposit which resulted in the winnowing of the material downstream from the dam crest.

## Volcanic Natural Dams

The sedimentary architecture of volcanic natural dams formed by volcanic-debris avalanche processes has been described by Glicken (1991) and Capra (2011) using two end-member facies: *block facies* and *mixed (matrix) facies*. The authors reported that both facies could be present, or one can dominate over the other. *The block facies* is mainly composed of debris-avalanche blocks (unconsolidated or poorly consolidated materials of the source rock), lacking a significant amount of matrix. The size of the debris blocks ranges from tens of centimeters to more than a hundred meters wide (Glicken 1991). *The mixed facies* comprises a variety of poorly graded, poorly sorted and unstratified aggregates representing all rock types derived from the source rock. Evidence of juvenile materials such as magmatic rocks, woody fragments, and clasts of different size ranges are usually embedded in the fine matrix.

## Rock Avalanche Deposits

Dunning and Armitage (2011) analyzed the sedimentological properties and the grain size distribution of five rock avalanche deposits in New Zealand and the Alpine region of Switzerland. The authors characterized the internal structure of the deposits into three distinct facies: *the carapace facies*, *the body facies*, and *the basal facies*. Dunning et al. (2005) asserted that the longevity and stability of rock avalanche dams are influenced by the thickness of the *Carapace facies*, the valley morphology and the filling rate of the reservoir.

*The carapace facies* is the coarsest part of a rock avalanche deposit constituting up to 30% of the total deposit thickness (Dunning and Armitage 2011). The facies is clast-supported and comprises large angular intact bedrock massifs detached from the parent rock during rapid runout. Material within this facies is mostly inversely graded and is separated by the underlying facies by a sharp boundary below which the material is intensely fragmented and matrix-supported. Permeability in this facies is very high, primarily due to the clast-supported fabric, and hence the large void spaces support filtration and restrain failure by overtopping. However, seepage erosion could occur at the boundary between the *carapace facies* and the *body facies* resulting in daylighting, dam-face saturation and subsequent breaching of the dam (Dunning et al. 2006; Davies and McSaveney 2011; Strom 2013).

*The body facies* represents the unit with the highest thickness within a rock avalanche deposit and is usually the most voluminous of all valley-confined deposits. Characteristic features of this facies include impact marks with radial fracture preserved on clasts, altered orientations and thicknesses, matrix-supported, highly fractured but relatively undisaggregated clasts, and preserved source stratigraphy. The grain size distribution and material properties of this facies vary with respect to its lithological attributes. This unique feature has been attributed to its preserved source stratigraphy.

*The basal facies* is the smallest unit by volume within a rock avalanche deposit and also constitutes materials altered by passing rock avalanche. The material forming this facies is derived from the erosion and entrainment of substrate materials (such as soft erodible valley fills and surface vegetation) into the base of the moving debris, and subsequent deposition of the modified mixtures of the substrate and rock avalanche material (Davies and McSaveney 2011).

## Internal Structure of Landslide Dams: Implications for Internal Erosion and Piping

The internal structure of landslide dams is one of the essential factors that control their longevity and stability (Swanson et al. 1986; Costa and Schuster 1988; Ermini and Casagli 2003; Stefanelli et al. 2016). The likelihood of initiation of internal erosion and piping has been linked to the type of mass movement that formed the dam. Landslide dams formed by translational slides are usually more stable than those formed by other processes such as rock avalanches, rock falls, debris flows and rotational slides (Costa and Schuster 1988; Ermini and Casagli 2003). This has been related to the ability of some landslide dam materials to preserve their source stratigraphy (Strom 2013). A typical example is the Usoi Dam (Fig. 1). The Usoi dam was formed in February 1911, by a strong earthquake that triggered a huge rockslide (with an estimated volume of 2.2 $km^3$) that dammed the Murghab River. The longevity and stability of the Usoi Dam have been related to the type of mass movement that formed the impoundment, coupled with the presence of clast-sized materials, including boulders with an average diameter of 2.5 m. The existence of numerous seepage outlets at the downstream face of the dam, almost at the same topographic level (approximately 3000 m a.s.l), gave further credence to the long-term stability of the dam (Strom 2013).

Landslide dams formed by blockslides have higher tendency to preserve their original stratigraphy because they are usually comprised of intact bedrock with a few fractures but generally do not comminute into smaller fragments. A better understanding of the fragmentation processes that occur during runout of blockslides is vital for assessing the potential for initiation of seepage and piping in the blockage. A conceptual model proposed by Davies and McSaveney (2011) has helped in the understanding of the mechanism of fragmentation of blockslides during runout.

**Fig. 1** Location of the Usoi natural dam on the Murghab River and the impounded lake (NASA earth observatory file image)

## Insights from Geophysical Investigations

Several studies on the internal structure of landslide dams have been conducted using passive geophysical techniques such as the multichannel analysis of surface waves (MASW) (Wang et al. 2013), and the microtremor (MTM) chain array survey (Wang et al. 2018). The application of the MTM chain array survey was motivated by the need to characterize the internal structure of landslide dams, assess their stability, and predict their potential failure mechanisms, including the time of failure and the flood hydrograph. The principles of the MTM chain array survey and its methodology have been described by Wang et al. (2018).

The MTM chain array survey was used to evaluate the internal structure of the Terano landslide dam in Niigata Prefecture, Japan. The landslide dam was formed in October 2004, by the $M_W$ 6.8 earthquake, which triggered a landslide that formed the dam. Extensive countermeasures have been completed on the dam site several years before the geophysical investigation was conducted in July 2014. The MTM survey was carried out on two survey lines: $CM - CM'$ and $TM - TM' - TM''$ (Fig. 2a). The phase velocity versus depth profiles along the two survey lines are shown in Figs. 2b, c. The results show that phase velocity generally increased with an increase in depth, indicating that the dynamic properties of the dam materials (e.g., density, Poisson's ratio and rigidity modulus) increased with an increase in depth. However, Wang et al. (2018) observed that the discordant trends of phase velocities within the upper and middle layers of the dam could be attributed to the impact force of the sliding mass on the valley floor, which led to the modification of the internal structure of the dam. Furthermore, the phase velocity versus depth profile of line $TM - TM' - TM''$ shows that the inclination angle of the sliding mass varies from 5 to 10°. Therefore, there is a very low likelihood of internal erosion and piping occurrence in the dam given the high phase velocity values (140 to 480 m/s) of the dam material and the effectiveness of the engineering countermeasures.

## Insights from Physical (Laboratory) Experiments

A series of laboratory-scale experiments were conducted to evaluate the longevity and stability of landslide dams formed by translational slope movements, where about 70% of the original stratigraphy of their source rock is usually preserved. Furthermore, additional experiments were conducted to investigate the failure mechanisms and the hydraulic behaviour of landslide dams formed by rock avalanche processes, with a view to determining the critical hydraulic gradients for failure ($i_f$) of landslide dams characterized by layered internal structure and dams formed by rock

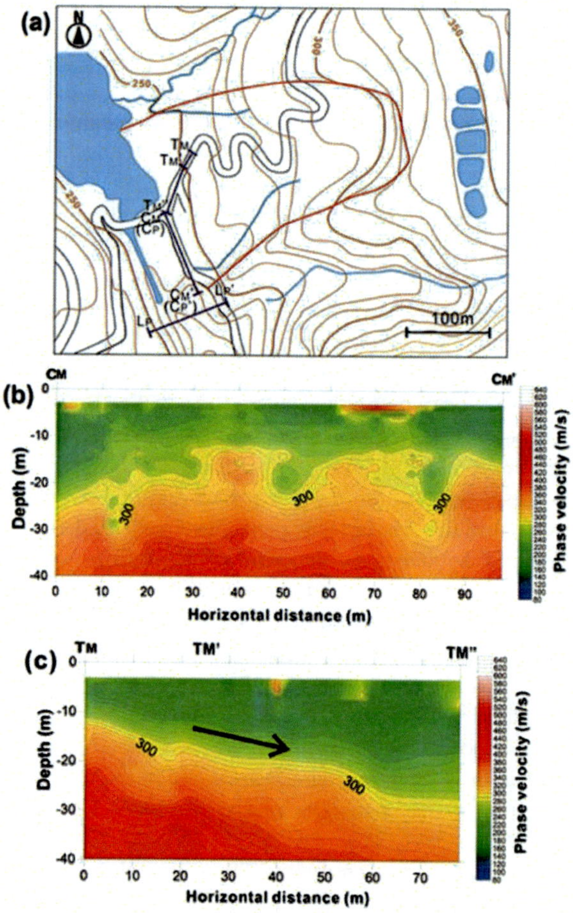


**Fig. 2** a Location map of the Terano landslide dam **b-c** Phase velocity versus depth profiles along survey lines $CM - CM'$ and $TM - TM' - TM''$

avalanche processes. To this end, a 0.25 m-thick dam model was built in a rectangular flume tank equipped with pore-pressure transducers and laser displacement sensors. A detailed description of the experimental setup and methodology employed in evaluating the stability of the dam models formed by rock avalanches has been done by Okeke and Wang (2016). The rock avalanche dam models were simulated with a mixture of silica sand no. 8 (SS8) and gravel in a ratio of 40:60. The SS8-gravel mix is essentially non-plastic with median grain size ($D_{50}$), void ratio, and hydraulic conductivity ($k$) of 0.28 mm, 0.71, and 3.8 × $10^{-5}$ m/s, respectively.

The layered landslide dam models were made up of 0.05 m-thick alternating layers of bentonite – SS8 mixtures (mix ratio, 30:70) and clay soil obtained from the Mihata landslide, in Izumo City, Japan. The physical properties of the materials used in building the layered dam models are given in Table 1. The dam models were built to achieve constant upstream and downstream slope angles of 40 and 55°, respectively. All the dam models were built using the moist tamping method and subsequently compacted to achieve a homogeneous mass. This was done by mixing the initially dry bentonite – SS8 mixture with a known volume of water to achieve a moisture content that was equal to the natural moisture content of the clay soil. During the experiment, steady-state seepage through the dam models was maintained by ensuring that the upstream water level was constant at about 0.05 m above the dam crest (Fig. 3a). The trends of pore-water pressures and hydraulic gradients through the dam models were used to determine the critical hydraulic gradients for failure of the dams.

The failure mechanism of the dam models was induced by seepage, irrespective of their internal structure (Figs. 3b, c). The results show that $i_f$ and the critical pore-water pressure ($P_{crit-1}$) were higher in the rock avalanche dam models (Figs. 4a, b) than in the layered dam models (Figs. 4c, d). However, the longevity of the layered dam models was higher ($T$ = 1600 s) than the rock avalanche dam models ($T$ = 920 s). These experimental results agree with the assertions made by Ermini and Casagli (2003), who

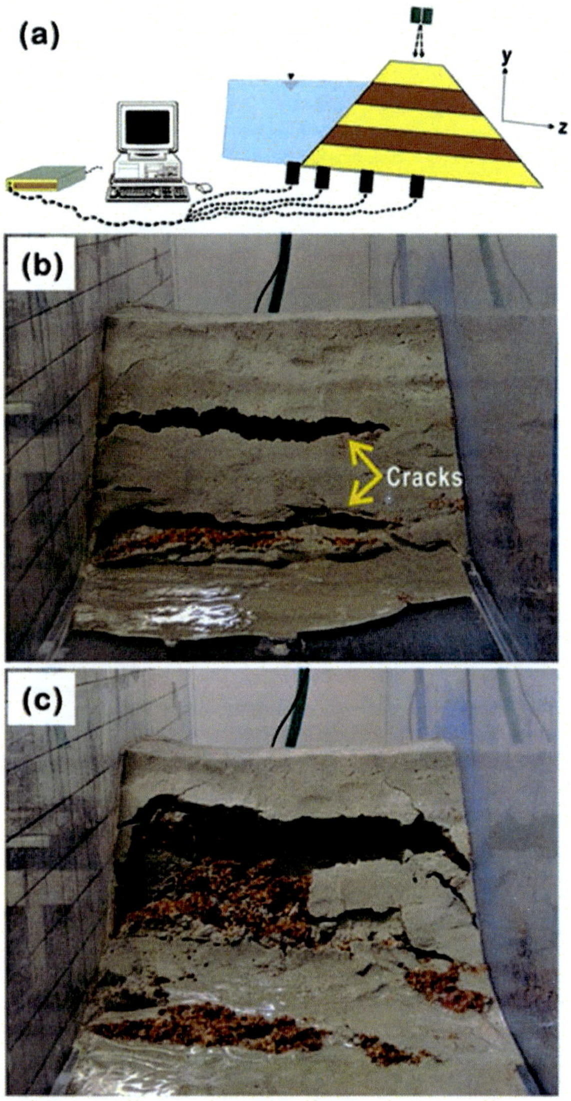


**Fig. 3** **a** Diagrammatic representation of the experimental setup **b** Seepage-induced cracking and undermining of the downstream slope, and **c** Downstream slope debuttressing and sloughing of the fluidized material

**Table 1** Physical properties of the layered dam materials

| Soil type/Physical parameter | Clay soil | Bentonite—SS8 mixture |
|---|---|---|
| $I_p$ (%) | 27 | NP |
| $D_{50}$ (mm) | 0.03 | 0.17 |
| $e_0$ | 1.42 | 0.89 |
| $\gamma_{dry}$ (Mg/m$^3$) | 1.14 | 1.21 |
| $k$ (m/s) | 6.5 × $10^{-7}$ | 4.8 × $10^{-6}$ |

$I_p$= plasticity index, $D_{50}$ = median grain size, $e_0$ = initial void ratio, $\gamma_{dry}$= dry density, $k$ = hydraulic conductivity, NP = non-plastic

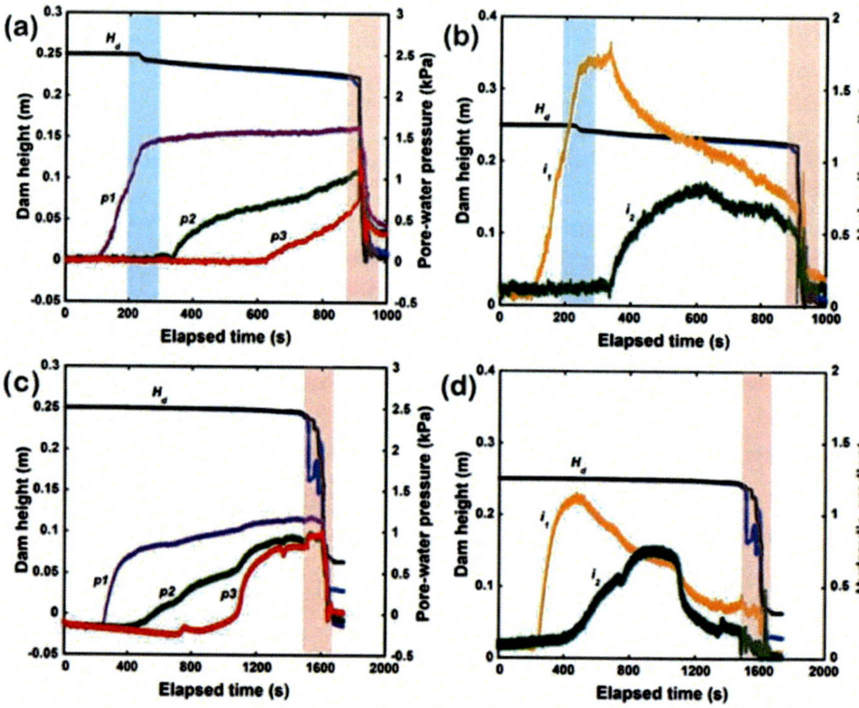


**Fig. 4** Experimental results showing trends of pore-water pressures and hydraulic gradients in dam models formed by: **a-b** rock avalanche processes, and **c-d** translational slope movement (layered dam) $H_d$ = dam height; $p1$, $p2$ and $p3$ = pore-water pressures; $i_1$ and $i_2$ = hydraulic gradients

observed that the longevity of dams formed by translational slides was higher than those formed by rock avalanche processes.

## Conclusions

The sedimentology and internal structure of landslide dams are two significant factors which control their longevity and stability. Extensive sedimentological analyses have helped in the classification of volcanic natural dams and rock avalanche dams into 2- and 3-member sedimentary facies. Consequently, the likelihood of dam failure by internal erosion and piping depends on the thickness of the uppermost layers – the *block facies* and the *carapace facies*.

The MTM chain array survey has proven to be an effective method of evaluating the internal structure of landslide dams, and for studying the evolutionary history of the landslide that formed the dam. This information is crucial for implementing long-term countermeasures and for predicting the failure mechanism of the dam and the potential outburst flood hydrograph.

Preliminary results obtained from the physical modelling of natural dams formed by translational mass movements (layered dams) and rock avalanche processes revealed that the layered dam models were more stable than the rock avalanche dam models. However, $i_f$ and $P_{crit-1}$ were lower in the layered dam models and higher in the rock avalanche dam models. These preliminary results agree with the notion that dams formed by competent bedrock movements last longer than those formed by rock avalanches and debris flows.

**Acknowledgements** This research benefited from JSPS KAKENHI Research Grant No. A-24241061 for landslide dam failure prediction.

## References

Capra L (2011) Volcanic natural dams associated with sector collapses: textural and sedimentological constraints on their stability. In: Evans SG, Hermanns RL, Strom A, Scarascia-Mugnozza G (eds) Natural and artificial rockslide dams. Springer, Berlin Heidelberg, pp 279–294

Casagli N, Ermini L, Rosati G (2003) Determining grain size distribution of the material composing landslide dams in the Northern Apennines: sampling and processing methods. Eng Geol 69(1):83–97

Costa JE, Schuster RL (1988) The formation and failure of natural dams. Geol Soc Am Bull 100:1054–1068

Cruden DM, Hungr O (1986) The debris of the Frank Slide and theories of rockslide–avalanche mobility. Can J Earth Sci 23(3):425–432

Davies TR, McSaveney MJ (2011) Rock-avalanche size and runout–implications for landslide dams. In: Evans SG, Hermanns RL, Strom A, Scarascia-Mugnozza G (eds) Natural and artificial rockslide dams. Springer, Berlin Heidelberg, pp 441–462

Dufresne A, Bösmeier A, Prager C (2016) Sedimentology of rock avalanche deposits–case study and review. Earth Sci Rev 163:234–259

Dunning SA, Armitage PJ (2011) The grain-size distribution of rock-avalanche deposits: implications for natural dam stability. In: Evans SG, Hermanns RL, Strom A, Scarascia-Mugnozza G (eds) Natural and artificial rockslide dams. Springer, Berlin Heidelberg, pp 479–498

Dunning SA, Rosser NJ, Petley DN, Massey CR (2006) Formation and failure of the Tsatichhu landslide dam. Bhutan. Landslides 3 (2):107–113

Dunning SA, Petley DN, Rosser NJ, Strom AL (2005) The morphology and sedimentology of valley confined rock-avalanche deposits and their effect on potential dam hazard. Landslide Risk Management. Taylor and Francis Group, London, pp 691–701

Ermini L, Casagli N (2003) Prediction of the behaviour of landslide dams using a geomorphological dimensionless index. Earth Surf Proc Land 28(1):31–47

Evans SG, Delaney KB, Hermanns RL, Strom A, Scarascia-Mugnozza G (2011) The formation and behaviour of natural and artificial rockslide dams; implications for engineering performance and hazard management. In: Evans SG, Hermanns RL, Strom A, Scarascia-Mugnozza G (eds) Natural and artificial rockslide dams. Springer, Berlin Heidelberg, pp 1–75

Glicken H (1991) Sedimentary architecture of large volcanic-debris avalanches. In: Smith GA, Fisher RV (eds.) Sedimentation in volcanic settings, SEPM, Tulsa, Special Publication 45:99–106

Okeke ACU, Wang F (2016) Critical hydraulic gradients for seepage-induced failure of landslide dams. Geoenviron Disasters 3(1):9

Schuster RL, Evans SG (2011) Engineering measures for the hazard reduction of landslide dams. In: Natural and artificial rockslide dams. Springer, Berlin, Heidelberg, pp 77–100

Schuster RL, Wieczorek GF, Hope DG (1998) Landslide dams in Santa Cruz County, California, resulting from the earthquake. US Geol Surv Prof Pap 1551:51–70

Shang Y, Yang Z, Li L, Liu DA, Liao Q, Wang Y (2003) A super-large landslide in Tibet in 2000: background, occurrence, disaster, and origin. Geomorphology 54(3):225–243

Stefanelli CT, Segoni S, Casagli N, Catani F (2016) Geomorphic indexing of landslide dams evolution. Eng Geol 208:1

Strom A (2006) Morphology and internal structure of rockslides and rock avalanches: grounds and constraints for their modelling. Landslides from massive rock slope failure. Springer, Netherlands, pp 305–326

Strom A (2013) Geological prerequisites for landslide dams' disaster assessment and mitigation in Central Asia. In: Wang FW, Miyajima M, Li T, Wei S, Fathani TF (eds) Progress of Geo-Disaster Mitigation Technology in Asia. Springer, Berlin Heidelberg, pp 17–53

Swanson FJ, Oyagi N, Tominaga M (1986) Landslide dams in Japan. In: Schuster RL (ed) Landslide dams: processes, risk and mitigation. Geotech Sp, vol 3. ASCE, New York, pp 131–145

Wang F, Okeke AC, Kogure T, Sakai T, Hayashi H (2018) Assessing the internal structure of landslide dams subject to possible piping erosion by means of microtremor chain array and self-potential surveys. Eng Geol 234:11–26

Wang G, Huang R, Kamai T, Zhang F (2013) The internal structure of a rockslide dam induced by the 2008 Wenchuan (Mw7. 9) earthquake. China. Engineering Geology 156:28–36

Wassmer P, Schneider JL, Pollet N, Schmitter-Voirin C (2004) Effects of the internal structure of a rock–avalanche dam on the drainage mechanism of its impoundment, Flims sturzstrom and Ilanz paleo-lake, Swiss Alps. Geomorphology 61(1):3–17

Wu CH, Chen SC, Feng ZY (2014) Formation, failure, and consequences of the Xiaolin landslide dam, triggered by extreme rainfall from Typhoon Morakot. Taiwan. Landslides. 11(3):357–367

2020 Kyoto Japan

# An Interdisciplinary Assessment of a Coal-Mining-Induced Catastrophic Landslide (Czech Republic)

Jan Burda and Vít Vilímek


## Abstract

The results are summarised of the correlation between long-term geotechnical monitoring, geological and geophysical surveys, finite element method modelling and their geomorphological interpretation in the area of North Bohemia affected by coal mining for more than 100 years. An underground mining-induced catastrophic landslide occurred in 1952 and had direct impacts on the local municipality, this still active and complex landslide has been well described since a detailed monitoring network was installed in the 1980s. Unique deep rock-mass monitoring is based on precise geodetical measurement in an exploration gallery from 1980. Several different types of deep-seated and near-surface landslide processes with different dynamics were identified during four decades of interdisciplinary monitoring, FEM modelling, ERT profiling and numerous other surveys. Landslide processes are influenced by anthropogenic activities in varying degrees, but climate factors have played the main triggering role in the recent past.



## Keywords

Complex landslide • Mining-induced landslide • Monitoring • ERT • FEM • Czech republic



J. Burda (✉)
Department of Geotechnics and Hydrogeology, Brown Coal Research Institute, Tř. Budovatelů 2830/3, Most, 434 01, Czech Republic
e-mail: burda@vuhu.cz

V. Vilímek
Department of Physical Geography and Geoecology, Charles University, Albertov 6, Prague, 128 43, Czech Republic
e-mail: vit.vilimek@natur.cuni.cz


## Introduction

In 1952, deep-seated human induced slope deformation damaged the village of Jezeří (also Eisenberg) situated in Most region (North Bohemia; Fig. 1). This catastrophic landslide occurred as a result of underground coal mining in the 1950s (Rybář 1997) and although the movement was relatively slow, ultimately led to the disappearance of the entire village. While there were 210 inhabitants in 56 houses in 1950, a few years later, in 1961, only the last 18 people were living in the village (Group Author 1978). Only the Jezeří chateau complex situated higher on the tectonically-fractured slope of the Krušné hory Mts. has been preserved to date. A system observing the impact of coal mining on the stability conditions of the entire locality was established in the 1980s, due to concerns about the stability of the mountain slope around the chateau and the former village of Jezeří (Marek 1983; Hurník 1982).

In the original primitive way, coal mining began on the outcrops in the early 17$^{th}$ century, but no evidence of this type of mining has been preserved to date. However, mining has been decisive since the mid-19$^{th}$ century. There were two underground mines in operation: Elly in Ervěnice and Grohmann in Dřínov. The Grohmann Mine (renamed Marshal Koněv in 1964) began operation in 1893. At first, the upper coal seam (approximately 2.5 m thick) was exclusively mined by the board-and-pillar method and panel working with the caving system method.

In 1945, the mining area was extended by the addition of new districts near the village of Jezeří. In 1942, the miners began using a new broadwall method with sublevel caving. In the early 1950s, coal mining completely switched to effective machine mining. All of these methods created extensive caving zones, subsequent irregular ground settlement and the initiation of deep-seated deformations. It first created depressions in the overlying sedimentary layers, which finally resulted in a catastrophic collapse during a landslide event that began in 1952 (Rybář 1997).


V. Vilímek et al. (eds.), *Understanding and Reducing Landslide Disaster Risk*,
ICL Contribution to Landslide Disaster Risk Reduction,
https://doi.org/10.1007/978-3-030-60319-9_14

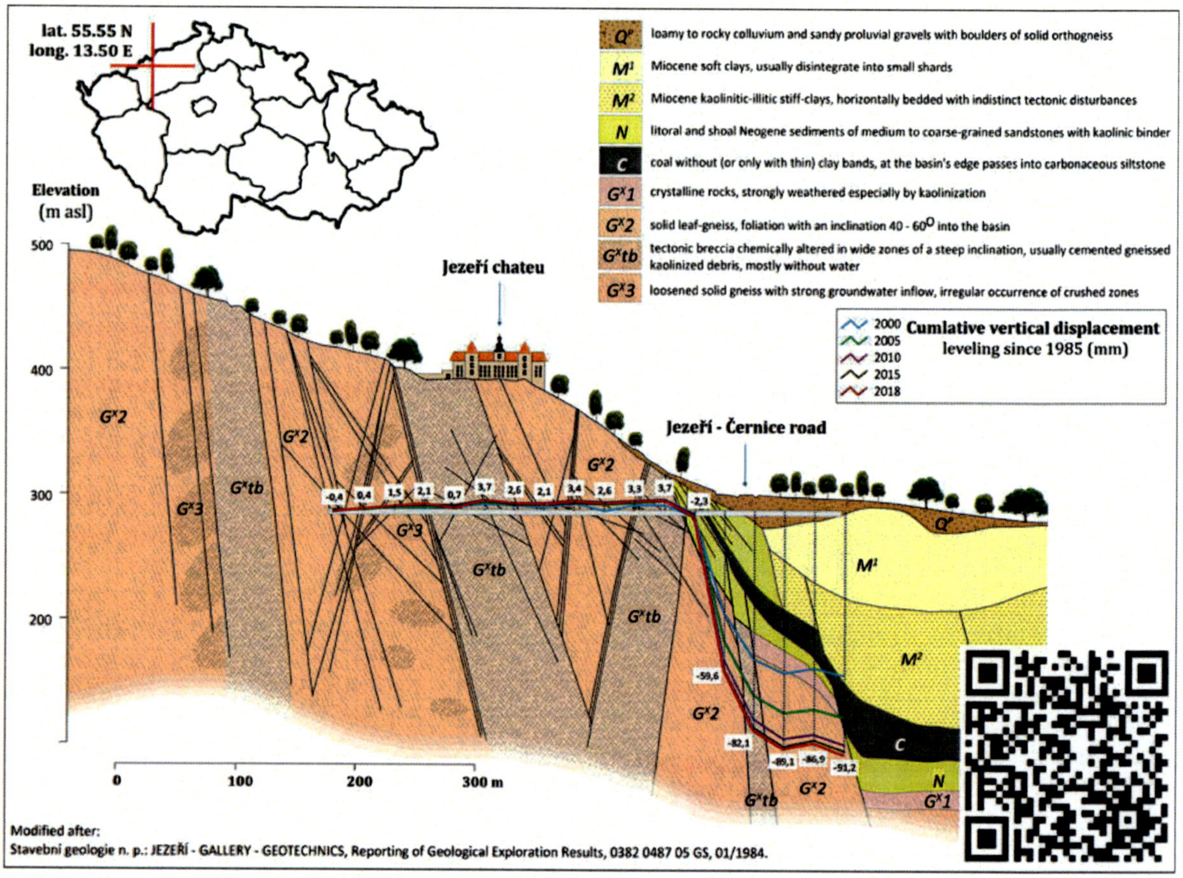


**Fig. 1** Geological cross-section along the exploration gallery; with projection of cumulative vertical displacements since 04/1985

Thereafter, mining continued in the form of open-pit exploitation, which further reduced the stability of the adjacent slopes and led to mass movement acceleration since the 1990s, which culminated in January 2011 and continues to date. The 1950s slope failure evolved into a large complex landslide (Dikau 2004) and represents one of the largest flow-like landslides in the Czech Republic (Burda et al. 2011).

The aim of this paper is to summarise the results of multidisciplinary research performed over the last few decades (e.g. Rybář and Dudek 1976; Rybář 1981, Hurník 1982; Marek 1983; Kalvoda et al. 1990, 1994; Rybář 1997; Chán et al. 2003; Rybář and Novotný 2005; Burda et al. 2011; Košťák et al. 2011; Burda et al. 2013; Tumajer et al. 2015; Burda et al. 2018a; Vanneschi et al. 2018, etc.); in order to demonstrate the need of a multidisciplinary approach to landslide investigation under difficult geological, structural as well as geomorphological conditions.

## Study Site

### General Conditions

The study area is located in North Bohemia (Czech Republic), where the ČSA open-pit mine is located at the edge of the Most Basin and the crystalline massif of the Krušné hory Mts.

With a depth of approximately 200 m, the ČSA open-pit mine (Fig. 2) was the deepest open-pit mine in the Czech Republic until 2009 and is formed by a high anthropogenic slope that passes smoothly (with an inclination of 10–15°) into the steep slopes of the Krušné hory Mts. behind the margin of the basin (Burda et al. 2011). A schematic geological cross-section of the area is provided in Fig. 1.

As a part of the Saxothuringian zone, the Krušné hory Mts. consist of crystalline complexes consolidated during

**Fig. 2** An aerial view of the study site and its close surroundings (photo: J Hodač 2012)

the Cadomian orogeny and the mantle of Lower Palaeozoic rocks weakly metamorphosed during Variscan metamorphism. The mountain range comprises orthogneisses and metagranites, which are adjacent to a series of crystalline shales, while the Most Basin comprises various Cenozoic sediments.

The boundary, the Krušné hory piedmont, has a graben structure (Váně 1985) with major boundary faults and a complicated geological structure (Marek 1983). Genetically, it belongs to the tectonic system of the Eger Graben (Domácí 1977).

## Most Basin

The Most Basin is located in the heart of the region and it is surrounded by the Krušné hory mountain range (to the north-west) and volcanic ranges of České Středohoří and Doupovské hory (to the east and south-east). The basin fill spans the time interval from the Oligocene to late Early Miocene.

The oldest Paleogene-age sediments belong to the Střezov Formation (Domácí 1977; Grygar and Mach 2013) and consist of stratigraphically heterogeneous clays, sandstones, and sands as well as Cretaceous quartzite, calcareous clays, marlites, and limestones. In addition, volcanic rocks, such as basalts, phonolites, and tuffs, have also been found in this complex.

The thick Miocene sedimentary complex of the Most Formation begins by gravel and sands of the Duchcov Member, passes to the Main Coal Seam—interrupted by irregularly distributed sandy-clayey bodies of the Žatec and Bílina Delta (Malkovský 1985)—of the Holešice Member. The overlying sedimentary complex comprises predominantly silty clays belonging to the Libkovice Member and attains a thickness of up to 175 m (the maximum thickness of the entire sedimentary fill was proved to be 231 m). Upper layers (20–60 m) of the Most Formation are disintegrated into shard fragments as a result of regelation processes during the Pleistocene (Marek 1983).

The Quaternary sediments predominately comprise coarse-grained gravels, sandy gravels, and clays with crystalline fragments. The thickness of these sediments reaches up to 40 m (with an average of approximately 10 m) along the mountain foothills, and the increasing thickness is associated with the alluvial fans or old fossil landslides. The maximum thickness of the Quaternary sediment deposits (70 m) was found near the study site below Mt. Jezeř (Burda et al. 2018b). These sediments commonly form aquifers, and their saturation depends on their thickness, grain size composition, and the clay content.

The gravels and alluvial fan sediments have good permeability—up to $10^{-4}$ $ms^{-1}$ (Burda et al. 2013). The groundwater flow system depends on the configuration of the Quaternary, and both the character of the basin sediments and the crystalline rocks. In general, the Quaternary basement inclines from the north to the south or south-east; hence, the groundwater flows in this direction (Burda et al. 2011). There are two inflow types in these aquifers, which pass into each other and complement each other. The first inflow type brings water from the crystalline slopes above the Most Basin outcrops. This type shows a direct dependence on atmospheric precipitation and causes inter-annual water table fluctuations of up to 1 m. The second type is represented by permanent groundwater inflow from a deep system of fractures in the crystalline massif. Kaolinised granite rocks, which are directly linked to the faults, create a nearly impermeable layer, and cause the formation of confined aquifers.

## Materials and Methods

### Geodetical Monitoring

The interpretation of levelling measurements is based on different monitoring periods. The special levelling circuit of the crystalline massif of the Krušné hory Mts. was established in 1983 (Kalvoda et al. 1990, 1994). However, regular measurements only continued until 1990 (spring period) when only part of the special polygon was included into another circuit (based on different stabilisation).

In 1985, these measurements were supplemented by precise levelling of 18 geodetic points in the 429 m long Jezeří exploration gallery drilled into the crystalline massif in 1980 (Marek 1983, Fig. 1). According to Pichler and Malá (2009), standard deviation per kilometre ($m_o$) is ±0.0002 m and deviation between backsight and foresight stations measuring (λ) does not exceed 0.0013 m in these measurements.

An automatic target recognition (ATR) monitoring system has been in place at the study site since 2005. In total, 50 geodetic points (with reflective prisms) placed on 3.5 m long standing pipes cemented to a depth of 2 m were implemented on the edge of Most basin to monitor the anthropogenic slopes of the nearby open-pit mine. The hourly measurements enable the precise calculation of daily accumulative and relative displacements.

### Geophysical Profiling

The inner structure of the landslide and its vicinity was studied using two-dimensional electrical resistivity tomography—ERT (Burda et al. 2011, 2013). Two-dimensional electrical resistivity tomography uses a high number of electrodes placed along a profile, and an automatic measuring unit that alternately switches the electrodes from current to potential (Loke and Barker 1996; Dahlin 2001).

Three profiles were measured using the Wenner–Schlumberger array in July 2010, April 2011, and October 2011 (Fig. 3). The profile (A–A′) was measured before landslide reactivation in July 2010 (Burda et al. 2011). The profile was 186 m long with an electrode span of 2 m. It crossed the most active part of the slope deformation from N–S (175°). The second profile (A-A″) was measured in the same course as the first profile in October 2011. The profile was 270 m long with an electrode span of 2 m. The third profile B–B′ was measured approximately parallel (160°) to the A-A′ and A-A″ profiles in April 2011. The profile was 590 m long with an electrode span of 5 m.

### Finite Element Modelling

FEM analysis was performed in order to simulate the evolution of shear strain values resulting from excavation of a 200 m thick layer of overburden soils and thus clarify and interpretation of previous monitoring results.

The analysis was performed using the 2D FEM analysis software RS$^2$. Laboratory specified material properties were used and a Mohr Coulomb failure criterion was adopted for the analysis, while the plastic response of the different materials was simulated using the values included in Table 1 (Vanneschi et al. 2018).

FEM deterministic analysis was performed using an enlarged profile as shown in Fig. 1, considering the presence of water in all of the geological strata.

### Analysis of Meteorological Factors

The influence of climatic factors on slope stability is generally accepted (Schuster and Wieczorek 2002), and has been confirmed in the study area by case studies (e.g. Rybář and Novotný 2005).

The data from the Jezeří meteorological station (Fig. 4), operated by Brown Coal Research Institute, situated near the landslide, were analysed to investigate the possible causes of reactivation. These were also compared to water tables

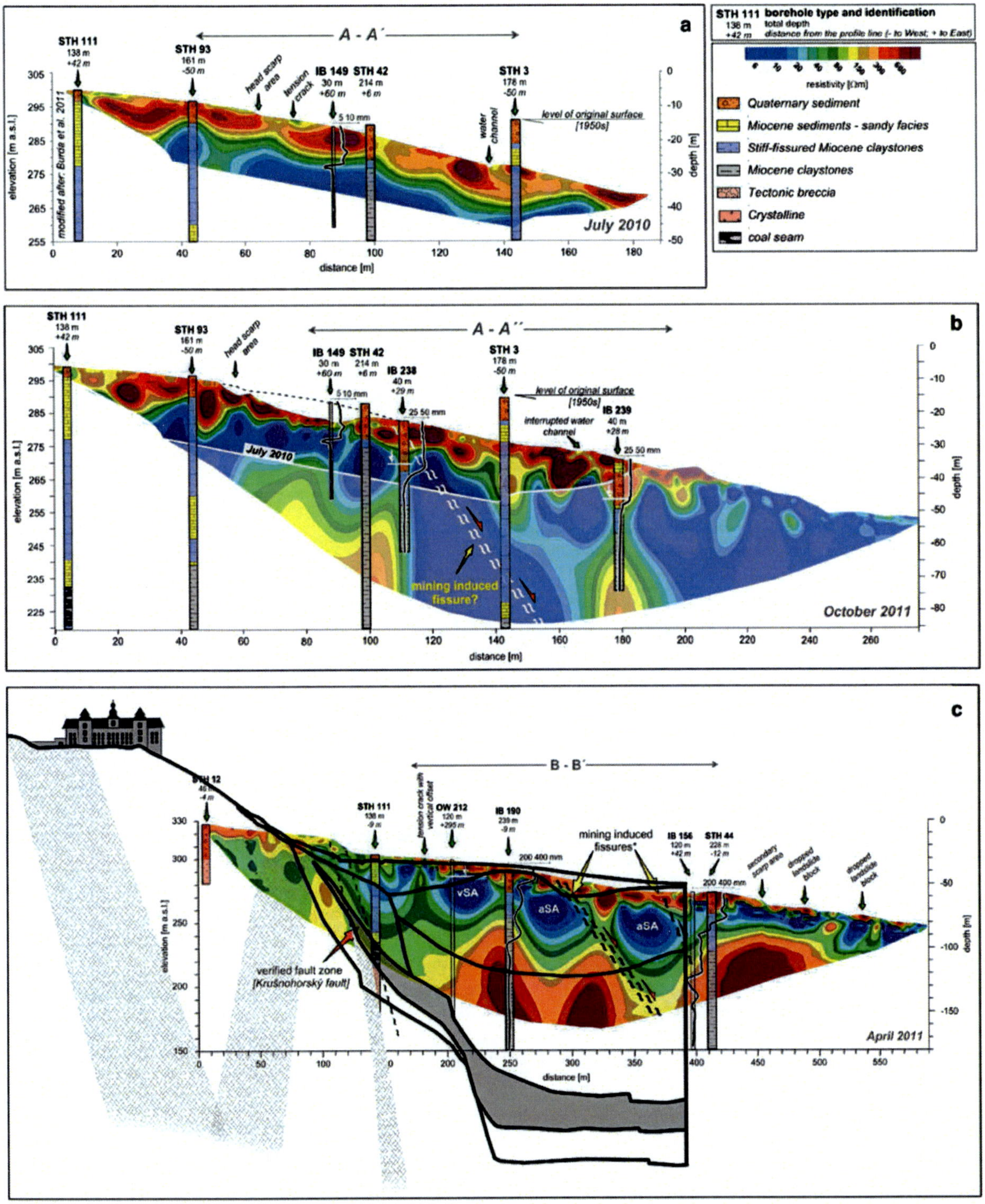

**Fig. 3** The ERT profiles taken across the landslide—locations depicted in Fig. 3 (**a** and **b**). The projection of ERT profile B-B′ into the geological cross-section from Fig. 1 (**c**). STH: structural test hole; IB: inclinometric borehole; OW: observation well; vSA: verified shallow aquifer; aSA: assumed shallow aquifer (modified after: Burda et al. 2013)

**Table 1** Statistical properties for each geological formation including Young's modulus and Poisson's ratio (Vanneschi et al. 2018)

| Material | Property | Mean | Rel. min. | Rel. max. |
|---|---|---|---|---|
| Rock basement | Cohesion(kPa) | 1000 | 500 | 500 |
| | Phi (°) | 42 | 5 | 5 |
| | Unit weight (kN/m$^3$) dry/sat. | 25 | 5 | 5 |
| | Young's modulus (MPa) | 100,000 | | |
| | Poisson's ratio | 0.2 | | |
| Underlying Neogene sediments | Cohesion(kPa) | 40 | 20 | 20 |
| | Phi (°) | 24 | 5 | 5 |
| | Unit weight (kN/m$^3$) dry/sat. | 18/20 | 5 | 5 |
| | Young's modulus (MPa) | 200 | | |
| | Poisson's ratio | 0.3 | | |
| Coal | Cohesion(kPa) | 50 | 25 | 25 |
| | Phi (°) | 25 | 5 | 5 |
| | Unit weight (kN/m$^3$) dry/sat. | 14/17 | 5 | 5 |
| | Young's modulus (MPa) | 1000 | | |
| | Poisson's ratio | 0.3 | | |
| Miocene claystones | Cohesion(kPa) | 100 | 50 | 50 |
| | Phi (°) | 18 | 5 | 5 |
| | Unit weight (kN/m$^3$) dry/sat. | 18/20 | 5 | 5 |
| | Young's modulus (MPa) | 5000 | | |
| | Poisson's ratio | 0.25 | | |
| Quaternary sediments | Cohesion(kPa) | 5 | 5 | 5 |
| | Phi (°) | 27 | 5 | 5 |
| | Unit weight (kN/m$^3$) dry/sat. | 16/20 | 5 | 5 |
| | Young's modulus (MPa) | 200 | | |
| | Poisson's ratio | 0.3 | | |

measured in boreholes OW 212, IB 238, IB 239 (Figs. 3 and 4) and soil water content in local soils (measured by two MicroLogs V3A).

## Results

### Description of the Site Morphology

There are two geomorphological types of slopes: (i) fault slope of the massif of the Krušné hory Mts. fluently passing to the anthropogenic slopes, (ii) anthropogenic slopes in the Most basin formed by mining activity (Fig. 2).

Regarding the fault slope, the landscape evolution has been predominantly influenced by neotectonics rather than by climatic oscillations. Considering the character and quantity of accumulations situated at the foot, it can be supposed that an intensive uplifting of the study area occurred in the Middle Pleistocene. Holocene up to the recent uplifting processes seem to be less intensive (Vilímek 1995).

Despite extensive relief transformation due to coal mining, relics of the 1950s landslide morphology (Špůrek 1974; Rybář 1987) are still visible and play a significant role in the most recent processes. By comparing a recent geomorphic map with old landslide inventories, an absolute match was found (Fig. 4). The active recently reactivated landslide can be described as a complex landslide (Dikau 2004) as it was characterised by a change in motion mechanics from sliding to flowing. The active part is more than 900 m long and, at most, 500 m wide with a total surface area of 188,000 m$^2$. The total volume is estimated to be between 2 and 4 mil. m$^3$ (Burda et al. 2011). The active headscarp is located at an elevation of approximately 295 m a.s.l. and has a typical amphitheatral shape with a height of up to 13 m. In the outcrop of the headscarp it is possible to see the soil, bouldery gneiss proluvium, weathered claystones, and interbedded overlying sands. Fragments, mainly from demolished building foundations, from the former village of Jezeří were found in the eastern part of the headscarp. The main scarp continues to the north-east in the form of a significant tension crack that runs for 200 m. This probably

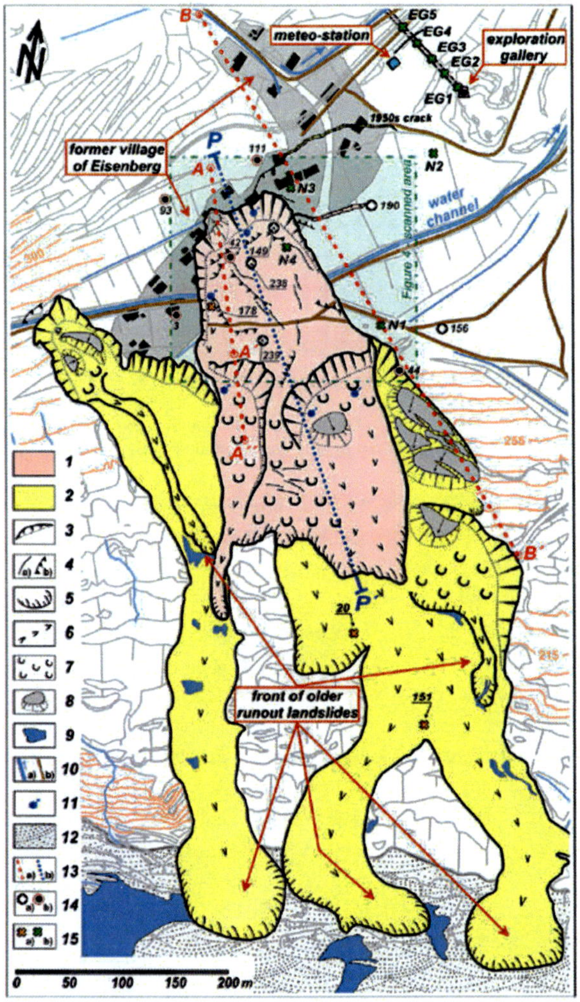


**Fig. 4** A geomorphological sketch map of the active complex landslide—(1) the most recent landslide, after January 2011; (2) older landslides within the landslide complex, from 1990s; (3) headscarps; (4) (a) tension cracks, (b) tension cracks with vertical offset; (5) accumulation toes; (6) earthflows; (7) landslide accumulation surfaces; (8) landslide blocks within the landslide complex; (9) shallow colluvial depression; (10) (a) brooks and channels, (b) road; (11) spring; (12) dump; (13) (a) ERT profile, (b) longitudinal profile across the landslide; (14) (a) inclinometric borehole (b) structural test hole; (15) observed geodetic points (a) ATR—reflective prism, (b) precise levelling—bench marks (Burda et al. 2013)

defines the unstable part of the reactivated landslide. On the western side, a smooth shear plane is exposed with an inclination of up to 30°. On the eastern side, a minor scarp with a height of 5–7 m defines the limit of the reactivated landslide body.

## The Internal Structure of the Landslide

The main landslide structures can be found on both profile A–A′ and profile A–A″ (Fig. 3). Profile A–A″ clearly shows that the landslide body predominately comprises coarse, gravelly slope deposits. The shear plane, however, passes through the sensitive claystones at depths of 15–20 m. This is a zone of sensitive clays, according to Pletichová (2006). The zone of very low resistivity between stations 230 and 280 reflects a water-saturated accumulation lobe formed by a mixture of Quaternary and weathered Miocene sediments. These sediments have been deposited on the eroded overburden benches of the open-pit mine. This accumulation has the characteristics of an earthflow and cannot be distinguished on the profile from the underlying conductive claystones.

Profile B–B′ (compared in Fig. 3 to the geological cross-section from Fig. 1) starts in clearly identifiable crystalline bedrock (paragneisses according to the geological map and borehole documentation). The bedrock loses some of its compactness towards station 100, as reflected by decreasing resistivity. The zone of very low resistivity between stations 100 and 150 is interpreted as the Krušnohorský Fault, which is also obvious from the geological cross-section at a similar distance. It is one of the most prominent and active faults along which the tectonic basin has descended. The fault is steeply inclined towards the basin at 75–80°. Its presence is corroborated by two of the borehole records, which penetrated the rock massif to depths of 90 m and 117 m, respectively (STH 111 and STH 6; not shown on Fig. 4). A high resistivity undulated surface layer is seen along the top of the profile at depths of between 10 and 15 m, although in places it reaches depths of 20–25 m (e.g. between stations 325 and 375) and it is virtually absent between stations 115 and 150. The thickness of 15 m at station 390 corresponds well to that found both in the nearby borehole STH 44 and in the geological cross-section. This is again made up of coarse gravel slope deposits, which also form the body of the nearby landslide. Underlying the high resistivity surface layer is a zone of very low resistivity and highly variable thickness (15–50 m). This zone most likely reflects soft wet clays, which are also observed below 15 m in boreholes IB 190 and STH 44. Similarly, this layer of soft sensitive clays is very well illustrated in the geological cross-section. Here, two incipient scarps were observed (at stations 185 and 275—see also in Fig. 3), as well as a verified shallow aquifer. The varying resistivities probably reflect the varying physical properties of the clay. The higher resistivities are associated with dryer and more compact claystones, possibly with a greater sand content. In some places, these dry claystones separate water saturated clays (e.g. station 330).

## Measurements on the Surface of the Crystalline Slope (1983–1990)

During the period of very precise levelling measurements (1983–1990) areas were identified with different degrees of deep loosening of the rock massif (Kalvoda et al. 1994). The slope stability in these areas is predisposed both by the existence of a fault slope and by the proximity of the open-pit mine. The movement trends of the geodetical points are climatic, tectonic as well as anthropogenic in origin. Kalvoda et al. (1994) concluded that the area of Jezeří chateu is the most hazardous especially because of the considerable probability of the origin and/or renewal of catastrophic slope movements. The Quaternary sediment cover on the crystalline slope close to Jezeří chateu revealed creep, even during the tendency to uplift the levelling circuit $Z_2$b 3 (Stolínová and Vilímek 2003).

## Evolution of the Complex Landslide Since the 1990s

The development of the landslide since 1990 can be investigated using the results of the precise levelling of points No. 1 to 5 (Fig. 4). The mean rate of movements at No. 1 and No. 2 was linear from 1993 to 2010 (6.4 mm/yr and 9.4 mm/yr, respectively). However, at No. 3 and No. 4, the mean rate of movement accelerated from 19.5 mm/yr to 31.5 mm/yr and from 21 mm/yr to 55.5 mm/yr, respectively after 2008 (Burda et al. 2013).

This acceleration culminated in December 2010–January 2011 when the slope collapsed along the 1950s tension crack (Burda et al. 2011). Detailed information about the development of this landslide reactivation was recorded by the automatic target recognition monitoring of reflective prism No. 178 (installed in June 2009, Fig. 4). This point showed quick, but linear deformation during the first year of monitoring as well as the first period of acceleration in spring 2010 and subsequent deceleration during summer 2010. The main movement occurred in January 2011, but the movement velocity started to increase during December 2010. Between 9 and 14 January 2011, the daily rate of movement increased steadily to 20 mm/day. On 15 January at 8:01 p. m., reflective prism No. 178 was measured for the last time, with a total cumulative displacement of 777 mm and relative displacement of 80 mm from the measurement one hour before (Burda et al. 2013).

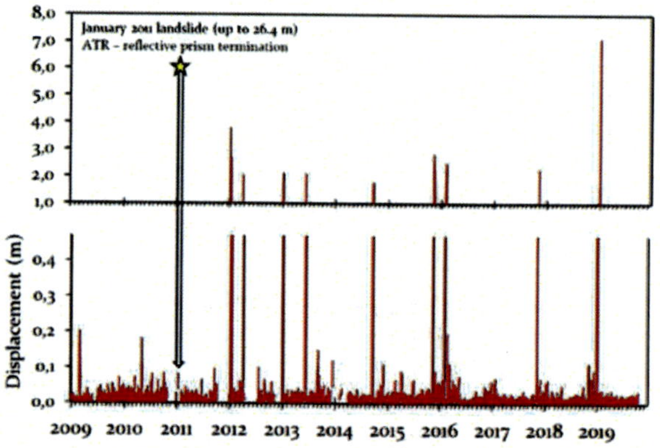


**Fig. 5** Evolution of the January 2011 landslide movement velocity since 2009 when reflective prisms were installed in the landslide area

During this event, the landslide body shifted horizontally, the upper part of rotated block, below the headscarp, moved by 12–14 m, while the middle part, including the road and artificial water channel, shifted by 23–25 m (Fig. 4). The area of the older headscarps in the lower part shifted up to 26.4 m. At elevations of 250–265 m a.s.l., the frontal part of the landslide body was thrust over the upper overburden bench, which means that the material lost its cohesion and was subsequently deposited upon the eroded overburden benches and older earthflows (Fig. 2).

After January 2011, several landslide phases (reactivations) were manifested by ATR monitoring (Fig. 5). These landslide phases usually accelerated during October or November (2011, 2014, 2015, 2016, 2017 and 2018) and culminated during November (only in 2015 and 2018) and more likely in January (2012, 2013, 2019) or February (2016). The maximum velocity during the main movement activity reached up to 1.014–3.778 m/day, respectively 7.096 m/day during the 16 January 2019 landslide reactivation, which was the largest displacement since January 2011.

## Measurements in the Exploration Gallery (Since 2000)

The results of levelling measurements since 2000 are shown in Fig. 6. Cumulative vertical shift was between +2.3 and −27.4 mm in the period 2000–2018, while an intensive vertical drop (from −19.6 to −27.4 mm; 1.03 to 1.44 mm/yr) is evident at points No. 1 to 5. The decrease of vertical subsidence rates is evident since the beginning of 2011, when the annual rates reached 0.37 to 0.46 mm/yr, while before 2011 it reached 1.48 to 2.17 mm/yr.

A projection of cumulative vertical displacement of all 18 geodetic points in the period 2000–2018 is shown in Fig. 1, where it is obvious that points No. 1 to 5 are in the basin

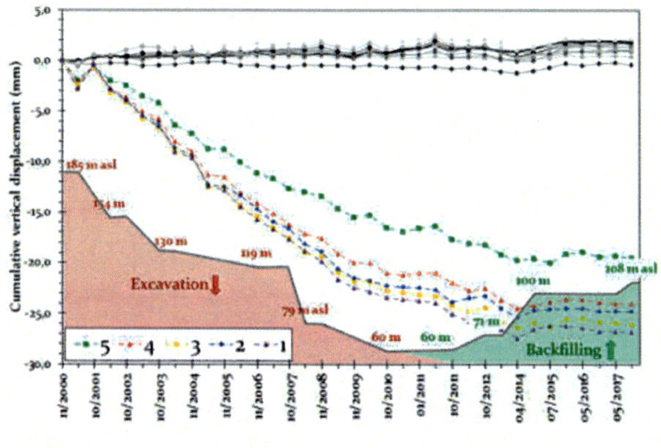


**Fig. 6** Levelling measurements in the exploration gallery compared to the excavation and backfilling processes

sedimentary fill, while points No. 6 to 18 are situated in the crystalline massif. Cumulative vertical displacement was between +0.8 to +2.3 mm in the period 2000–2018. The only exception is point No. 18, with a cumulative vertical shift of -0.3 mm (which is significantly below the measurement accuracy limit).

## Results of FEM Modelling

The FEM analyses show the development of shear strain values in deeper strata. The critical detected values of the FoS are equal to 1.09 and 1.35.

The critical FoS value in this case is 1.09 and replicates the observed shallow landslide affecting the Quaternary sediments (similar to the January 2011 landslide reactivation), indicating that the two different failure mechanisms (shallow landslide and deep-seated movements) have different FoS values of 1.35 and 1.09, respectively (Fig. 7).

To understand the reliability of these results, inclinometer data were used for a comparison. Two sets of inclinometer data were selected, 30 m deep IB149 situated in the landslide area (Fig. 7b) and 120 m deep IB156 situated near the landslide area (Fig. 7c). Even if the last readings date back to 2004, they still provide useful information. The IB 149 inclinometer data suggest the location of a shallow landslide associated with deformation at approximately 12 m below the surface. IB 156 suggests two distinct deformation horizons, one related to the shallow landslide and the other at a depth that may be indicative of deep-seated instability.

The data from the two different inclinometer locations were simulated and analysed during the numerical simulation in $RS^2$ (Vanneschi et al. 2018). The results of the simulation are shown in Fig. 7, the modelled deformation reflects the shallow and deeper movements observed in the inclinometer boreholes.

## Discussion and Interpretation of the Results

As mentioned above, there are two geomorphological types of slopes both characterised by different types of mass movements. Long-term monitoring suggests a minimal influence of extensive open-pit mining on the massif of the Krušné hory Mts. fault slope stability (or influence in a low order) but a significant influence on the anthropogenic tertiary soil-formed-slopes in the Most basin.

The crystalline fault slope is characterised by seasonal oscillation of levelling points depending on climatic conditioned volume changes of rocks (1-3 mm). A significant part of the levelling points measured on the surface in the 1980s has uplift; nevertheless, there were differences year to year. The uplift was also confirmed in the exploration gallery (1–3 mm since 2000), where a similar seasonal oscillation is evident as well. Subsidence associated with near-surface deformations was confirmed in the levelling points near the edge of the open-pit mine and in the areas with a higher degree of deep loosening of the rock massif (Kalvoda et al. 1994). Similar results were also confirmed by a dendroecological analysis of *Fagus sylvatica* species (Burda 2010). Growth disturbances were analysed in the period 1900–2008 and their increase is evident since the late 1940s with a culmination in 1959—1962 and 1972–1974 (Burda 2011). Most of the disturbed trees were located near the edge of the open-pit mine or close to the 1952 catastrophic landslide. Both studies (Kalvoda et al. 1994; Burda 2011) show the presence of creeping or sliding of Quaternary colluvial sediment.

A more significant effect of mining activities is evident from the results of basin sediment monitoring. These results manifested both near-surface deformations as well as deep-seated deformations of the entire sedimentary basin fill —both with different mass movement dynamics.

The highest movement dynamic (usually up to tens of meters per day during the main movement activity) is typical for a large reactivated landslide complex in the area of the former Eisenberg village (Fig. 4). This complex landslide is situated near the toe of the Krušné hory Mts., where geotechnical conditions are suitable for the occurrence of such a type of slope failure: (i) a relatively thick layer of Quaternary sediments, (ii) a thick layer of highly plastic soft clays, (iii) general geological conditions with unfavourable inclined strata, (iv) an undermined surface, (v) permanent groundwater inflow from both shallow Quaternary and deep crystalline aquifers, (vi) an anthropogenic slope long-term exposed for loosening and erosion.

The cumulative effect of these factors is adverse from a long-term perspective. It is almost impossible to artificially prevent groundwater inflow and thus water flows through the permeable Quaternary ($10^{-4}$ $ms^{-1}$) and immediately

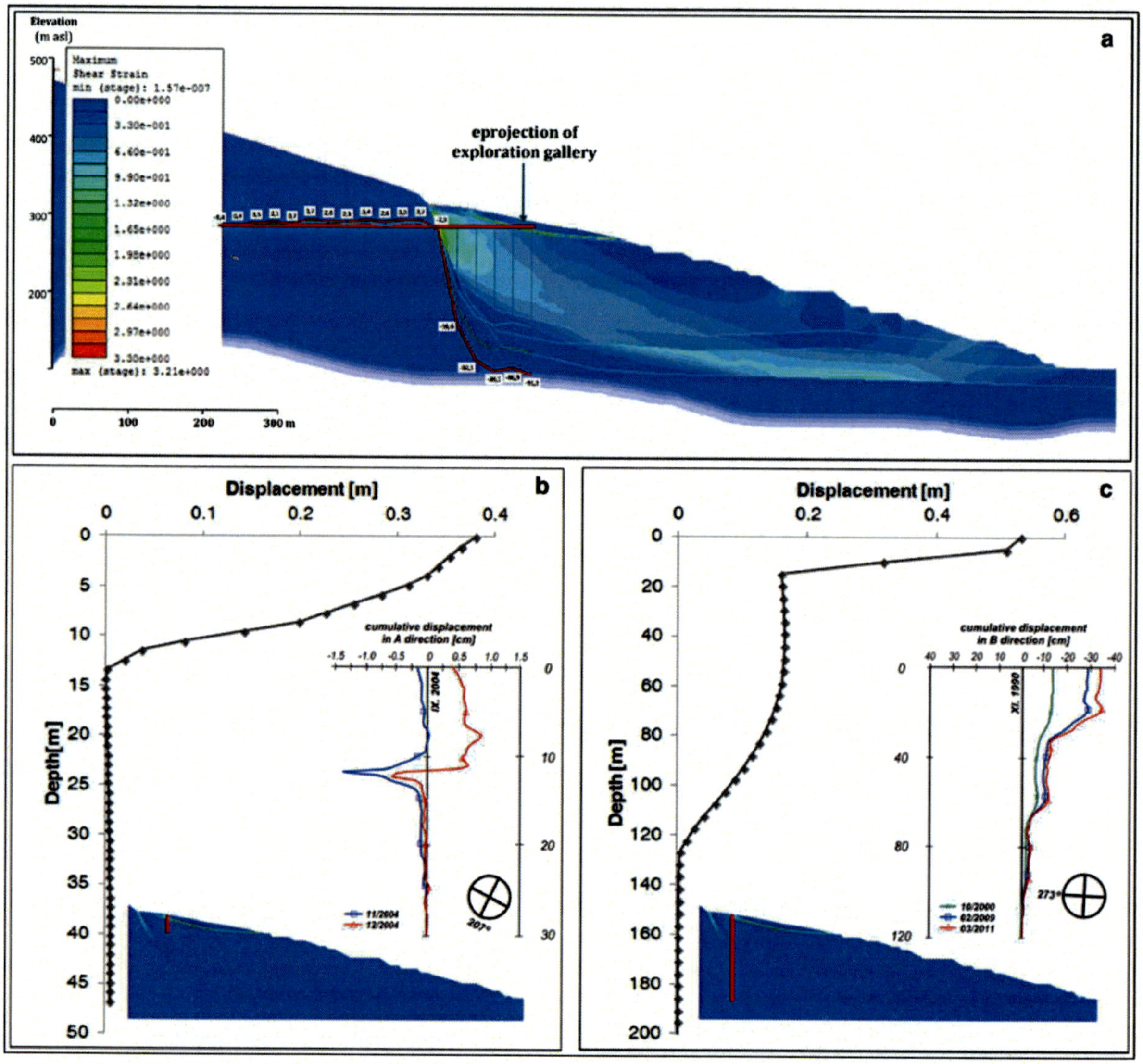


**Fig. 7** Shear strain development depicted in preliminary $RS^2$ numerical modelling with a projection of the exploration gallery measurements (*A*). Selected inclinometer boreholes—IB 149 (*B*), was placed within the landslide body; IB 156 (*C*) was placed near the January 2011 landslide. The measured cumulative displacement is compared to the results of the $RS^2$ inclinometer borehole analysis (*B* and *C*). Inset profiles show approximate inclinometer locations (modified after: Vanneschi et al. 2018 and Burda et al. 2013)

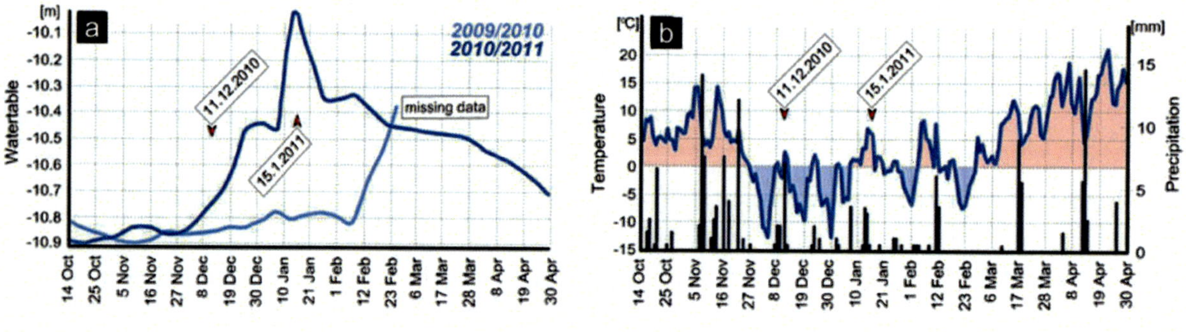


**Fig. 8** Water table increase in OW 212 (**a**) resulted from the increase in the average temperature during January (**b**)

infiltrates the underlying soft clays gravitationally. This local phenomenon was most probably potentiated by mining induced fissures occurred in the middle of the 20th century. This scenario is supported by the results of soil moisture monitoring (Fig. 9) as well as geophysical survey (Fig. 3).

While the values of soil water content (SWC) oscillate around 0.1 with occasionally rain-induced peaks (0.3 to 0.5) in Quaternary soils, soft clays have a significantly different regime. SWC values oscillate around 0.6 (which is also the peak value) for most of the time and very slowly decrease below 0.2 only at the end of the warmer and dryer period of the year (summer/autumn). These different regimes document the good permeability of the Quaternary sediments, which allow quick inflow of rainfall but also its subsequent quick saturation into soft clays. These soft sensitive clays are capable of receiving saturated water (a quick increase of SWC values up to 0.6), but its deeper infiltration into solid claystones as well as its horizontal outflow is not possible due to the impermeability of deeper clays. A layer of these water-saturated soft clays is clearly interpretable in geophysical profiles (Fig. 3) as low resistance zones with <10 Ωm between two high resistance zones (Quaternary and impermeable solid claystones both with >600 Ωm).

Permanent saturation and increasing shear strain of soft clays (Fig. 9) cause their long-term deformations, which can be observed both in the inclinometers (as shown in Fig. 7) and the levelling measurements (Fig. 6) months or years before the final climate triggered landslide initiation.

An explicit example of such a climate triggered landslide initiation/reactivation is a landslide event from January 2011 (Fig. 8). The winter of 2010–2011 was characterised by a greater than average snow cover in North Bohemia. In the study site and its surroundings, snow cover was the highest ever recorded (Burda et al. 2013). In addition, the progressive development of the winter snow cover was unusual when set against previous years. In 2010–2011, the peak snow cover at all stations was recorded in the second half of December, whereas in recent years it has normally occurred

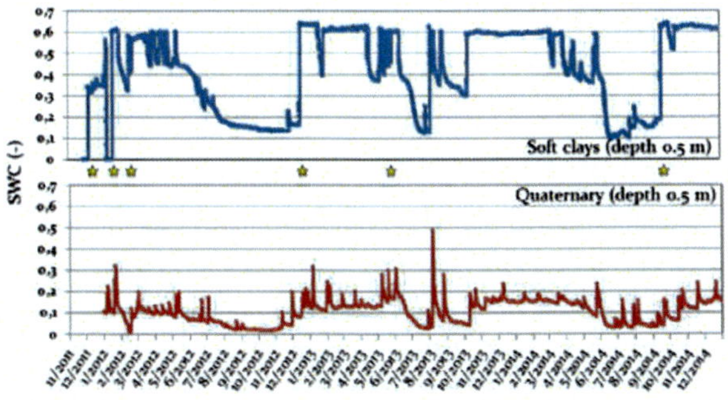


**Fig. 9** Soil water content development in different types of soils. The yellow star symbolises the main landslide accelerations

in the second half of January or in February. The rapid cooling in late November and early December was followed by considerable warming that began on 7 January. From 7 to 16 January, the temperature did not drop below 0 °C. The average daily temperature reached 7 °C on the 14th and 6°C on the 15th. The maximum temperature recorded during this period was 10.7 °C on the 16th. In Central Europe this type of December–January warming may occur on isolated days (Rybář 2001). The protracted nature of this warming event makes it unique, at least for the Jezeří meteorological station. It led to a rapid melting of the record snow cover with an estimated snow water equivalent of approximately 200 mm (Čekal et al. 2011) and an immediate rise in the water table (Fig. 8). The total precipitation measured at Jezeří was 7 mm (13–15 January). The records of other meteorological stations demonstrate that this rapid thaw occurred throughout the Krušné hory Mts., which was immediately followed by the landslide mobilisation (Fig. 5). However, the final movement acceleration occurred at the beginning of December 2010 when a separate warming event on 11 December led to partial melting of snow and subsequent rising of the water table. On 11 December, the average temperature was 2.8 °C and the daily maximum was 4.9 °C, while there was also 7 mm of rainfall (Fig. 8). The subsequent saturation of the material led to a rapid increasing of the water table on 13 December (i.e. with a two-day lag following the temperature peak).

It is obvious that mining activities led to stress-strain changes within the basin fill but the recent near-surface landslide activity was primarily triggered by climatic factors with a minimal delay after precipitation/snow-thaw peaks. Otherwise, the increase of shear strain in deeper strata (Fig. 7) was associated with a vertical shift of points No. 1 to 5 in the exploration gallery (-19.6 to -27.4 mm) and most probably occurred as a direct consequence of massif unloading during open-pit mining. The dynamic of vertical movements links very well to the evolution of the open-pit mine bottom, when the movements accelerate during excavation and stopped during subsequent backfilling. In addition to the FEM modelling, the existence of a nascent shear zone was confirmed by the inclinometers at the depths of 60–70 m (e. g IB 156). At these depths, we do not expect the direct influence of climate factors while the strata consists of solid claystones and is, therefore, impermeable. This was also confirmed by several previous studies (Rybář and Novotný 2005), which described that the stability of general slopes consisting of Tertiary claystones could be influenced with a two- or three-year delay by long-term climate trends. However, the character of the vertical movements measured in the front part of exploration gallery reflected mining activities with a minimal delay and thus we assume that these were key factors leading to the increase in shear strain (instead of climate factors, which are crucial in near-surface slope failures). Since the initiation of such a deep-seated slope failure did not occur when the bottom of the open-pit mine was at the lowest level (< 60 m a.s.l.) and FoS was critical (1.35), it may be expected that future backfilling will lead to long-term stabilisation of the general slope, which seems to be confirmed by measurements after 2014 (Fig. 6).

## Conclusions

The correlations between different interdisciplinary methods in the wider region of the exploration gallery near the former village of Eisneberg (Jezeří) show the following: (1) general geotechnical conditions are suitable for the occurrence of a landslide complex where the Quaternary sediment with very good permeability allows for the quick inflow of groundwater into sensitive soft clays; (2) shear strain is increasing in the strata of permanently saturated clays (which was confirmed by ERT profiling (zones with <10 Ωm) and SWC monitoring (quick increase and very gradual decrease of SWC values) and causes their long-term deformations observed by all of the measurements; (3) within these conditions landslide activity was immediately triggered by short-term climate peaks—rainfall or meltdown; (4) the expected influence of the open-pit mining is crucial in the sedimentary fill of the Most basin where a vertical shift of points No. 1 to 5 in the exploration gallery and deep deformations in IB 156 resulting from increasing shear strain in deeper strata due to massive anthropogenic unloading were confirmed.

These deep-seated deformations have never led to slope failure due to sliding and it can be expected that this will not happen in the future because the mining activity is almost over, and the process of backfilling has already started. On the other hand, movement of the near-surface landslide complex will not be affected by this backfilling process and, therefore, its future cyclic activity triggered by climate factors can be expected.

**Acknowledgments** The research was funded by the grant projects TAČR-TL03000506 and RFCS-SUMAD-847227. The authors wish to acknowledge the financial support provided by the Ministry of Education, Youth and Sports of the Czech Republic (Program Inter-Excellence, Inter-Vector, LG15007)

## References

Burda J (2010) Dendrogeomorfologická analýza dynamiky svahových pohybů v lokalitě Jezeří. Geografie 115(4):440–460

Burda J, Žižka L, Donhal J (2011) Monitoring of recent mass movement activity in anthropogenic slopes of the Krušné Hory Mountains (Czech Republic). Nat Hazards Earth Syst Sci 11:1463–1473

Burda J (2011) Spatio-temporal activity of mass movements in the Krušné Hory Mountains (Czech Republic): dendrogeomorphological case study. AUC Geographica 46(2):15–30

Burda J, Hartvich F, Valenta J, Smítka V, Rybář J (2013) Climate-induced landslide reactivation at the edge of the most Basin (Czech Republic) – progress towards better landslide prediction. Nat Haz Earth Syst Sci 13:361–374

Burda J, Svatý Z, Žižka L (2018a) Possibilities of uas in open-cast mine monitoring. Slope Stability Symposium 2018, 11–13 April 2018, Seville, Spain. pp 2037–2045

Burda J, Veselý M, Řehoř M, Vilímek V (2018b) Reconstruction of a large runout landslide in the Krušné hory Mts. (Czech Republic). Landslides 15(3):423–437

Čekal R, Ryglewicz M, Fryč T, Bořiková L, Ssuchá M, Přibyl J, Knotek R (2011) Zpráva o povodni v lednu 2011. ČHMÚ, Praha, p 73

Chán B, Mrlina J, Polák V (2003) Rozšíření a modernizace systému náklonoměrného monitorování rizikových svahů velkolomu ČSA v Mostě. Zpravodaj Hnědé uhlí 3:28–39

Dahlin T (2001) The development of DC resistivity imaging techniques. Comput Geosci 27:1019–1029

Domácí L (1977) Litostratigrafie třetihorních sedimentů v hnědouhelné severočeské pánvi. Acta Univ. Carol. Geol 1:75–80

Dikau R (2004) Mass Movement, in: Encyclopedia of Geomorphology. Goudie A (eds.), Routledge Ltd., London, ISBN 0–415–32737–7. pp 644–652

Group Author (1978) Retrospektivní lexikon obcí Československé socialistické republiky 1850–1970: počet obyvatelů a domů podle obcí a částí obcí podle správního členění k 1. lednu 1972 a abecední přehled obcí a částí obcí v letech 1850-1970. Part I., Vol 1, Vol. 2, Praha: Fed. stat. úřad: SEVT, 678 p, 506 p

Grygar T, Mach K (2013) Regional chemostratigraphic key horizons in the macrofossilbarren siliciclastic lower Miocene lacustrine sediments (Most Basin, Eger Graben, Czech Republic). Bull Geosci. 88 (3):557–571

Hurník S (1982) Endogenní geologické procesy a rozvoj velkolomů v severočeské hnědouhelné pánvi. Geol. Průzk. 5:129–131

Kalvoda J, Stemberk J, Vilímek V, Zeman A (1990) Analysis of levelling measurements of the Earth's surface movements on the geodynamic polygon Mikulovice - Jezeří in the Krušné hory Mts. Proc. 6th Int. IAEG Cong., 6–10 August 1990 Amsterdam. pp 1631–1637

Kalvoda J, Vilímek V, Zeman A (1994) Earth's surface movements in the hazardous area of Jezeří kastle. Krušné hory Mts. GeoJournal 32 (3):247–252

Košťák B, Mrlina J, Stemberk J, Chán B (2011) Tectonic movements monitored in the Bohemian Massif. J Geodyn 52:34–77

Loke MH, Barker RD (1996) Rapid least-squares inversion of apparent resistivity pseudosections by a quasi-Newton method. Geophys Prospect 44:131–152

Malkovský M. (1985) Geologie severočeské hnědouhelné pánve a jejího okolí. Geology of the North Bohemian Brown Coal Basin and its surroundings (in Czech). Academia, Ústřední ústav geologický, Praha, 424 p

Marek J (1983) Inženýrsko-geologický průzkumu stability zámku Jezeří v předpolí uhelného velkolomu. Geolog Průzk 25:234–236

Pichler E, Malá D (2009) Continuous evaluation of monitoring of the side slopes of the ČSA open-pit mine. Research report, Brown Coal Res. Inst., Most, 53 p

Pletichová M (2006) Hydrogeologická problematika v prostoru ochranného pilíře SKPJ. Zpravodaj Hnědé uhlí 3:12–16

Rybár J, Dudek J (1976) Vliv strukturně geologických poměrů na stabilitu svahů povrchových dolů u Kyjic. Sbor Geol Věd, HIG 13:29–49

Rybář J (1981) Inženýrskogeologické hodnocení stabilitách poměrů předpolí povrchových velkolomů při úpatí Krušných hor. Stabilitní řešení svahů a jejich zabezpeční, Sborník přednášek semináře, pp 76–93

Rybář J (1987) The engineering-geological zoning of the outcrop part of the North-Bohemia brown-coal basin at the footing of Krušné Hory Mts. Acta Montana 77:3–64

Rybář J (1997) Interpretation of data about technogenic activity at the toe of Krušné Hory Mts. affecting endogenous and exogenous processes in the rock environment. Acta Montana 106:9–24

Rybář J (2001) Bedeutung der klimatischen Verhältnisse für die Aktivität von Rutschungen (in German). Mitteilungen für Ingenieurgeologie und Geomechanik 4:89–104

Rybář J, Novotný J (2005) Vliv klimatogenních faktorů na stabilitu přirozených a antropogennih svahů. Zpravodaj Hnědé uhlí 3:13–28

Schuster RL, Wieczorek GF (2002) Landslide triggers and types in: Landslides–proceedings of the 1st European conference on landslides. In: Rybář J, Stemberk J, Wagner P (eds), 24–26 June 2002, Praha, Czech Rep., pp 59–79

Stolínová E, Vilímek V (2003) Geomorfologická interpretace geodetických měření v Krušných horách. Geomorfologický sborník 2, 22–23 April 2003, Nečichy, Czech Republic. pp 247–252

Špůrek M (1974) Sesuvné jevy u Dřínova na Mostecku. Věst Ustř Úst Geol 49:231–234

Tumajer J, Burda J, Treml V (2015) Dating of rockfall events using vessel lumen area in Betula pendula. IAWA J 36(3):286–299

Vanneschi C, Eyre M, Burda J, Žižka L, Francioni M, Coggan JS (2018) Investigation of landslide failure mechanisms adjacent to lignite mining operations in North Bohemia (Czech Rep.) through a limit equilibrium/finite element modelling approach. Geomorphology 320:142–153

Váně M (1985) Geologická stavba podkrušnohorského prolomu a jeho tektogeneze. Sbor. Geol. Věd. Geologie 40:147–181

Vilímek V (1995) Quaternary development of the Kateřinohorská Vault relief in the Krušné hory Mts. AUC Geographica, supplementum: 115–137

UNESCO · ICL · ISC · WFEO · FAO · UNDRR · WMO · UNU
WLF
2020 Kyoto Japan

# Could Glacial Retreat-Related Landslides Trigger Volcanic Eruptions? Insights from Mount Meager, British Columbia

Gioachino Roberti, Brent Ward, Benjamin van Wyk de Vries, Nicolas Le Corvec, Swetha Venugopal, Glyn Williams-Jones, John J. Clague, Pierre Friele, Giacomo Falorni, Geidy Baldeon, Luigi Perotti, Marco Giardino, and Brian Menounos

**Abstract**

Mount Meager, a glacier-clad volcanic complex in British Columbia, Canada, is known for its large landslides, as well as a major eruption about 2360 years ago. In 2010, after decades of glacier retreat, the south flank of Mount Meager collapsed, generating a huge (53 $Mm^3$) landslide. In 2016, fumaroles formed ice caves in one of the glaciers on the complex. This glacier is bordered by a large unstable slope presently moving about 3.5 cm per month. If this slope were to fail, a long-runout debris avalanche would reach the floor of the Lillooet River valley, with possible destructive effects on downstream infrastructure. The unloading of the volcanic edifice from an abrupt failure would also have unknown effects on the magmatic plumbing system. From geochemical, geophysical, and petrological data, we infer the presence of a magmatic chamber at 3–16 km depth. Based on numerical model simulations carried out to constrain the stress change, the failure would affect the stress field to depths of up to ~6 km, with changes in effective stress of up to ~4 MPa. The change in effective stress following such a landslide might destabilize the magmatic chamber and trigger an eruption. This result also suggests that a previously documented major flank collapse may have had a role in the 2360 cal yr BP eruption.

**Keywords**

Volcanic landslide • Stress changes • Eruption trigger • FEM • InSAR

## Introduction

Mount Meager is a glaciated volcanic complex in the southern Coast Mountains of British Columbia, with activity spanning the past ~2 Ma (Read 1979) (Fig. 1). The most recent eruption occurred 2360 years ago (Clague et al. 1995) and produced a tephra plume that tracked eastward into western Alberta (Hickson et al. 1999). Today, Mount Meager hosts an active hydrothermal system and is the site of frequent shallow earthquakes (Friele et al. 2008; Venugopal et al. 2017). It is known for its very large pre-historic and historic landslides (Friele et al. 2008), of which a 53 $Mm^3$ debris avalanche in 2010 is the largest landslide in Canadian history (Guthrie et al. 2012). Following the 2010 landslide, recent studies have focused on the ongoing instability on Mount Meager and its link to recent deglaciation and climate warming (Roberti et al. 2018a, b). These studies have identified many large unstable slopes, one of which, on the west flank of Plinth Peak (Fig. 1), is of particular concern as it moved 3 cm during a 24-day period during the summer of 2016 (determined by InSAR interferometry; see Roberti

G. Roberti (✉)
Minerva Intelligence, 850 W Hastings St #301, Vancouver, BC, Canada
e-mail: groberti@minervaintelligence.com

B. Ward · S. Venugopal · G. Williams-Jones · J. J. Clague
Earth Sciences Department, Simon Fraser University, Burnaby, BC, Canada

B. van Wyk de Vries · S. Venugopal
CNRS, IRD, OPGC, Laboratoire Magmas et Volcans, Université Clermont Auvergne, Aubiere Cedex, France

N. Le Corvec
111 rue Jules Ferry, 33200 Bordeaux, France

P. Friele
Cordilleran Geoscience, Squamish, BC, Canada

G. Falorni · G. Baldeon
TRE-ALTAMIRA Inc., Vancouver, BC, Canada

L. Perotti · M. Giardino
Earth Sciences Department, GeoSitLab - GIS and Geomatics Laboratory, University of Torino, Torino, Italy

B. Menounos
Geography Program and Natural Resources and Environmental Studies Institute, University of Northern British Columbia, Prince George, BC, Canada


V. Vilímek et al. (eds.), *Understanding and Reducing Landslide Disaster Risk*,
ICL Contribution to Landslide Disaster Risk Reduction,
https://doi.org/10.1007/978-3-030-60319-9_15

et al. 2018b). This slope is near an amphitheatre (A in Fig. 1) that hosts a retreating glacier and a fumarole field.

In this study, we present a numerical model to investigate the stress field evolution prior to and after a hypothetical collapse of the west flank of Plinth Peak. We discuss how this stress change could affect the hydrothermal and volcanic plumbing systems and lead to an eruption. We also document another large edifice collapse from Job Valley, which happened a few centuries before the 2360 cal yr BP eruption. This landslide originated from the above-mentioned amphitheatre near Plinth Peak that is now occupied by Job Glacier and the fumarole field. Discussion of decompression mechanisms after the eventual collapse of Plinth Peak and documentation of the collapse in Job Valley enable an examination of the possible role of the earlier edifice collapse in the 2360 cal yr BP eruption. That collapse might have decompressed the volcanic plumbing system and contributed to the eruption.

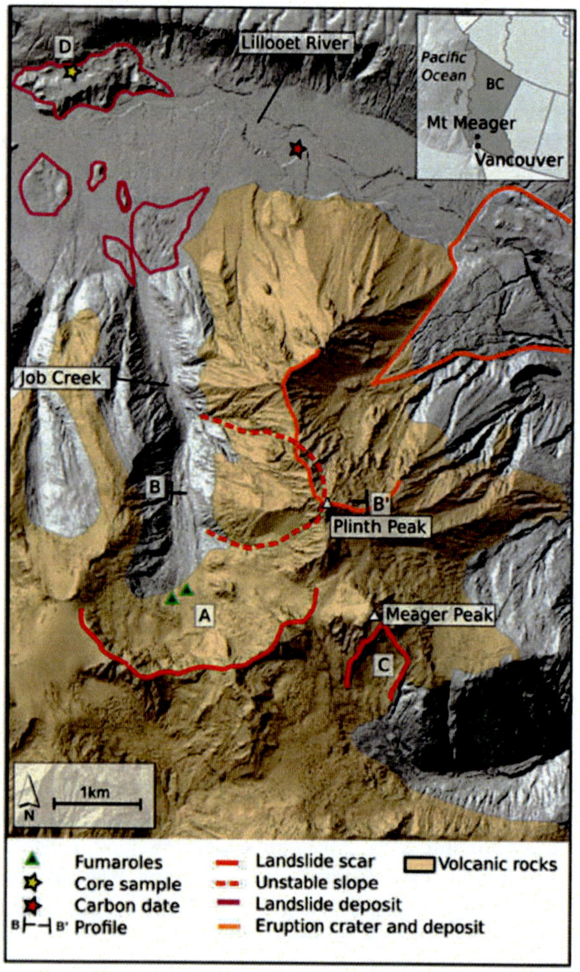


**Fig. 1** Job Valley, Plinth Peak, and the 2360 cal yr BP eruption crater. Scars, unstable slopes, landslides, eruption deposits, fumaroles, sample, and profile locations are indicated. **a** Job Valley amphitheatre. **b** Profile shown in Fig. 2. **c** 2010 landslide scar. **d** Job landslide deposit. Base map is a hillshaded rendering of Lidar data from 2015–2016

## Methods

We describe the geology and structure of the west flank of Plinth Peak and Job Valley from historical aerial photographs, Lidar data, literature review and field work. Field work at Plinth, Job Glacier, and the mouth of Job Valley was conducted in October 2017. We collected a core sample on the landslide deposit at the mouth of Job Valley to constrain the temporal relation between the landslide and the 2360 cal yr BP eruption. We used a 2D elastic finite element model (FEM) with COMSOL Multiphysics® to explore the effect of the collapse of Plinth Peak on the state of stress within the Mount Meager volcanic edifice. The FEM topography was built by superimposing E-W profiles of the west flank of Plinth (B-B' in Fig. 1) and of Job Valley 500 m to the south, assuming that the failure would remove the bulge of the west flank of Plinth Peak to reach the profile of Job Valley.

Two elastic domains were defined to represent both volcanic material (Plinth Peak) and the granitic basement rocks. We used Plinth Peak's 2360 cal yr BP eruption dacite to model the volcanic edifice and the Fall Creek Stock monzogranite to model the substratum rock (Campbell et al. 2016; Kelly Russell, personal communication 2018). Following previous work (Le Corvec et al. 2018), the FEM defined as a half-space, is loaded with gravitational forces. Except from its surface that is set free to move, the other boundaries are defined by a roller condition restricting displacements normal to the boundary. COMSOL allows to compute the differential stresses and the orientation of the stress field before and after the removal of the landslide material, subsequently allowing to compute the difference between the two cases.

## Results

### Plinth

A prominent knob forms the west flank of Plinth Peak on the east side of Job Valley about 2 km from the Job Glacier headwall. The knob has an area of $2.7 \times 10^6$ $m^2$ and rises 1000 m from Job Glacier to the ridge crest (Fig. 2). DInSAR analysis of Sentinel-1 satellite images show displacement of this part of the slope of $34 \pm 10$ mm over a 24-day period from July to August 2016 (Roberti et al. 2018a) and $30 \pm 10$ mm over a 12-day period in July 2017. The upper slope extends 400 m above the lower slope (Fig. 2) and has near-vertical cliffs and buttresses composed of horizontally layered rhyodacitic lava flows of the Plinth assemblage (p9f, Read 1979). SqueeSAR analysis of the 1992–2000 ERS data indicate LOS (line of sight) displacements up to 17 mm $yr^{-1}$. Analysis of the 2015–2018 Sentinel-1 data indicate LOS displacements of up to 10 mm $yr^{-1}$. 2D stress field modeling in COMSOL shows that removal of the bulging rock mass would affect the effective stress to a depth ~6 km by up to ~4 MPa (Fig. 2). Most of the decompression would occur at the toe of the slope, about 1 km downstream of the fumaroles.

## Job

A 2-km long, >100 m thick, hummocky landslide deposit is located at the Job Creek—Lillooet River confluence (D in Fig. 1). Considering the thickness and extent of the deposit, we tentatively relate the landslide to the formation of the large amphitheatre at the head of Job Valley (~$10^9$ $m^3$) (A in Fig. 1). This amphitheatre is the only feature large enough to be the source of the landslide deposit and furthermore has widespread hydrothermal alteration, consistent with the altered nature of the deposit. Given the size of this deposit, this landslide may be related to one of the distal volcanic debris flow and hyperconcentrated flow in the valley fill. In four cores 50 km downstream from Mount Meager there is a massive debris flow unit, about 4 m thick forming a 2 km valley-wide sheet, with an age bracketed between 2570 ± 40 14C yr BP and 2690 ± 50 14C yr BP, or between 2540–2970 cal yr old, of unknown proximal deposit correspondence. This event predates the 2360 cal yr BP eruption by 100–200 years. It is capped by a thin humic laminae, the incipient paleosol, carbon source for the younger bracketing age, and in turn is capped by a reversely graded, pumiceous hyperconcentrated flow unit related to the 2360 cal yr BP eruption. This debris flow unit is thus interpreted as a close precursor to the eruption.

# Discussion

## Plinth

In view of displacement rates, geometry, geological setting, and a low (1.084–1.077) factor of safety (FOS) calculated by (Hetherington 2014), the west flank of Plinth Peak can be considered to be at failure and near catastrophic collapse threshold. The fumaroles on Job Glacier may accelerate glacier melting and bring the slope closer to catastrophic failure. The base of the slope, weakened by glacier erosion and having been recently deglaciated, shows high deformation rates. It is likely to fail first, followed quickly by retrogressive failure involving the whole slope, similar to the evolution of the 2010 collapse (Roberti et al. 2018b). The failed rock mass would rapidly fragment, liquefy, and transform into a debris flow, as observed at other large landslides at Mount Meager (e.g. Roberti et al. 2017) and in other volcanic terrain (e. g. Scott et al. 2001) with a volume of $10^8$–$10^9$ $m^3$ (Class 9 debris flow; Jakob 2005). The hydropower infrastructure near the Keyhole Falls would be impacted, as it is within the reach of a $10^7$ $m^3$ debris flow. Even populated areas in Pemberton Meadows, which are within the reach of a $10^9$ $m^3$ debris flow, might be affected (see Friele and Clague 2004; Friele et al. 2008; Simpson et al. 2006; Hetherington 2014 for inundation zones of debris flows from Mount Meager).

Decompression following the collapse of Plinth Peak would influence the stress field to a depth of ~6 km. This sudden decompression would affect the near-surface hydrothermal system at Job Glacier, as well as the deeper magmatic plumbing system. The geothermal gradient in the

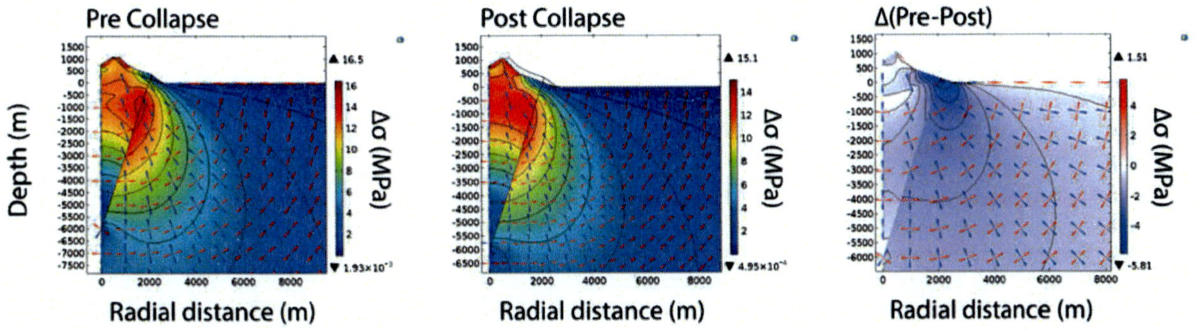


**Fig. 2** Differential stresses ($\Delta\sigma$) and stress field orientations before and after a hypothetical collapse of the knob west of Plinth Peak. The oblique line separates volcanic rocks to the left from basement rocks to the right. The change in effective stress after the collapse would be up to ~4 MPa and would reach a depth of ~6 km. The blue and red cones represent the direction of the maximum and minimum compressional stresses ($\sigma_1$ and $\sigma_3$), respectively

area is 80–100 °Ckm$^{-1}$, with recorded temperatures in geothermal explorative drill holes of up to 270 °C (Ghomshei et al. 2013). Shallow earthquakes (1–6.7 km depth; Friele et al. 2008) occur below Mount Meager. This seismic activity and the high geothermal gradient may indicate the presence of magma at shallow depths (3–10 km). The stress perturbation of the system by a large flank collapse from Plinth Peak might trigger a series of processes, including fracture and magma propagation, gas exolution, and magma differentiation and mixing that could lead to an eruption. Deeper magma (5–16 km, 1057–1142 °C, and 1.4–4.8 kbar; Venugopal et al. 2020) is also present below Mount Meager, as indicated by olivine melt inclusion analysis of basalts of the Mosaic assemblage (p10f, Read 1979). The decompression following the landslide would not affect these deeper magma reservoirs unless there were a telescoping of events downward through the system, and the localised decompression would not likely induce partial melting in the mantle.

### Job

A very large landslide such as this may have perturbed the stress field in the volcanic edifice, and may have triggered the ascent of deeper, less differentiated magma and its mixing with a shallow, more differentiated magma, triggering the 2360 cal yr BP eruption (Stasiuk et al. 1996). The interval between the landslide and the eruption could be related to the time that the magmatic system took to react to the decompression. The location of the vent, NE of the Job amphitheatre collapse scarp, might also have been conditioned by the large collapse. The stress field perturbation after the Job Valley collapse could have conditioned the location of the vent, lateral to the collapse scar, and triggered a sequence of events that led to the eruption as observed at other volcanoes (Delcamp et al. 2012; Zurek et al. 2015; Maccaferri et al. 2017).

## Conclusions

The high gravitational slope deformation rates and rapid glacier retreat near Plinth Peak are factors favouring large catastrophic slope failures. We applied a 2D finite element model to assess the state of stress after a catastrophic Plinth slope failure. Changes in effective stresses would reach a depth of up to ~6 km with up to ~4 MPa decompression. Such changes in confining pressure could trigger hydrothermal explosions from the shallow hydrothermal system at Job Glacier. Fracture propagation, magma migration, and gas exolution following decompression could also lead to a magmatic eruption. Such a landslide and eruption would impact infrastructure and communities near the volcano. The model of stress change and a new field description of a large rock avalanche source in upper Job Valley suggest that the past landslide may have had a role in the 2360 cal yr BP Mount Meager eruption. The large (~$10^9$ m$^3$) collapse that created the amphitheatre at the head of Job Valley preceded the eruption by, at most, a few centuries. Decompression of the magmatic system might have allowed new magma to rise and mix, leading to the 2360 cal yr BP eruption. The trigger for the collapse is still unknown but may be related to the hydrothermal activity and continued glacial erosion.

**Acknowledgements** Alex Wilson and Kelly Russell provided the data for Mount Meager rock properties used in the model. We thank Hazel Wong, Brad Woods, Erik Young, and Kai Friele for the help in the field. Financial support for the 2015 Lidar data acquisition was provided by the Natural Sciences Engineering Research Council of Canada through Discovery Grants to Clague and Menounos and a grant from the Canada Foundation for Innovation (Menounos). Natural Resources Canada (NRCan), through Melanie Kelman, provided funds for acquisition of the 2016 Lidar dataset. Fieldwork was supported by SFU Earth Science Department Chair grants to Ward. The InSAR analyses were provided by TRE-Altamira. Financial support for Roberti's PhD salary was partially provided by the "End of an Arc: The Remarkable Life and Death of a Volcanic Arc" project, a French-Canadian partnership.

## References

Campbell ME, Porritt L, Russell JK (2016) Forensic recovery of transient eruption parameters for the 2360 BP fall deposit, Mount Meager, British Columbia. J Volcanol Geotherm Res. 312:11–25. https://doi.org/10.1016/j.jvolgeores.2016.01.013

Clague JJ, Evans SG, Rampton VN, Woodsworth GJ (1995) Improved age estimates for the White River and Bridge River tephras, western Canada. Can J Earth Sci 32:1172–1179. https://doi.org/10.1139/e95-096

Delcamp A, Troll VR, van Wyk de Vries B, et al (2012) Dykes and structures of the NE rift of Tenerife, Canary Islands: A record of stabilisation and destabilisation of ocean island rift zones. Bull Volcanol 74:963–980. https://doi.org/10.1007/s00445-012-0577-1

Friele PA, Clague JJ (2004) Large holocene landslides from pylon peak. Southwestern British Columbia. Can J Earth Sci 182:165–182. https://doi.org/10.1139/E03-089

Friele PA, Jakob M, Clague JJ (2008) Hazard and risk from large landslides from Mount Meager volcano, British Columbia, Canada. Georisk Assess Manag Risk Eng Syst Geohazards 2:48–64. https://doi.org/10.1080/17499510801958711

Ghomshei M, Sadlier-Brown T, Hassani F (2013) Current status of the Pebble Creek (North Meager) Geothermal Project, 11–15 August 2013, Montreal, QC. Suthwestern British Columbia. 23rd World Min Congr, 10p

Guthrie RH, Friele P, Allstadt K et al (2012) The 6 August 2010 Mount Meager rock slide-debris flow Coast Mountains, British Columbia: Characteristics, Dynamics, and Implications for Hazard and Risk Assessment. Nat Hazards Earth Syst Sci 12:1277–1294. https://doi.org/10.5194/nhess-12-1277-2012

Hetherington R (2014) Slope stability analysis of mount meager, South-Western British Columbia. MSc Thesis, Michigan Technol Univ 68

Hickson CJ, Russell JK, Stasiuk MV (1999) Volcanology of the 2350 B.P. eruption of Mount Meager volcanic complex, British Columbia, Canada: Implications for hazards from eruptions in topographically complex terrain. Bull Volcanol 60:489–507. https://doi.org/10.1007/s004450050247

Jakob M (2005) A size classification for debris flows. Eng Geol 79:151–161. https://doi.org/10.1016/j.enggeo.2005.01.006

Le Corvec N, Muirhead JD, White JDL (2018) Shallow magma diversions during explosive diatreme-forming eruptions. Nat Commun 9:1–8. https://doi.org/10.1038/s41467-018-03865-x

Maccaferri F, Richter N, Walter TR (2017) The effect of giant lateral collapses on magma pathways and the location of volcanism. Nat Commun 8:1–11. https://doi.org/10.1038/s41467-017-01256-2

Read PB (1979) Geology of Meager Creek Geothermal Area, British Columbia. Geol Surv Canada Open File 603, map, 1 sheet, scale 120,000. https://doi.org/https://doi.org/10.4095/129507

Roberti G, Friele van P, Wyk de Vries B, et al 2017 Rheological evolution of the mount meager 2010 debris avalanche, southwestern british columbia Geosphere 13(2): 369–390. 0.1130/GES01389.1

Roberti G, Ward B, van Wyk de Vries B, et al (2018a) Landslides and glacier retreat at Mt. Meager volcano: Hazard and risk challenges. In: Geohazards7 Engineering Resiliency in a Changing Climate. June 3 - 6, 2018 Canmore, AB

Roberti G, Ward B, van Wyk de Vries B, et al (2018) Precursory slope distress prior to the 2010 Mount Meager landslide, British Columbia. Landslides 15:637–647. https://doi.org/10.1007/s10346-017-0901-0

Scott KM, Macias JL, Naranjo JA et al (2001) Catastrophic debris flows transformed from landslides in volcanic terrains: mobility, hazard assessment, and mitigation strategies. Professional Paper 1630:59. https://doi.org/10.3133/pp1630

Simpson KA, Stasiuk M, Shimamura K et al (2006) Evidence for catastrophic volcanic debris flows in Pemberton Valley, British Columbia. Can J Earth Sci 43:679–689. https://doi.org/10.1139/e06-026

Stasiuk M, Russell JK, Hickson C (1996) Distribution, nature, and origins of the 2400 BP eruption products of Mount Meager, British Columbia: linkages between magma chemistry and eruption. Geol Surv Canada Bullettin 486:27. https://doi.org/10.4095/208165

Venugopal S, Moune S, William-Jones G et al (2017) Gas emissions and magma source of the mount meager volcanic complex, Garibaldi Volcanic Belt, BC. In: IAVCEI scientific assembly. Portland, OR. 14–18 August 2017

Venugopal S, Moune S, Williams-Jones G et al (2020) Two distinct mantle sources beneath the Garibaldi Volcanic Belt: Insight from olivine-hosted melt inclusions Chem Geol 532 10.1016/j.chemgeo.2019.119346

Zurek J, Williams-Jones G, Trusdell F, Martin S (2015) The origin of Mauna Loa's Nīnole Hills: Evidence of rift zone reorganization. Geophys Res Lett 42:8358–8366. https://doi.org/10.1002/2015GL065863

# Rock Avalanches in the Upper Reaches of the Mzymta River, Russia

2020 Kyoto Japan

Andrey A. Ponomarev, Kai Kang, and Oleg V. Zerkal

**Abstract**

Large-scale slope instability is a very characteristic geological hazard in the entire Western Caucasus., Rock avalanches with volume up to 60 million $m^3$ and a displacement distance of more than 2–3 km are widely developed in this region, especially in the upper reaches of the Mzymta River basin. Such rock avalanches are dangerous for densely populated and developed areas, such as the Krasnaya Polyana village. This article describes a typical prehistoric rock avalanche in this region. The conditions of its formation and characteristic features of the structure are presented, based on interpretation of aerial photographs, field observations, digital geological models, as well as physical-mechanical testing of soil samples.

**Keywords**

Rock avalanche • Slope instability • Mzymta river • Western caucasus

A. A. Ponomarev (✉)
Geological Faculty, Moscow State University, Leninskie Gory, 1, Moscow 119991, Russia
e-mail: set092222@gmail.com

K. Kang
School of Environment and Civil Engineering, Jiangnan University, Wuxi, 214122, China
e-mail: kevinkang8@mail.ru

O. V. Zerkal
Geological Faculty, Moscow State University, Moscow, 119991, Russia
e-mail: igzov@mail.ru

## Introduction

Rock avalanches in the Krasnaya Polyana area are investigated. Krasnaya Pol yana is located in the district of Sochi, southern Russia, in the upper reaches of the Mzymta River, bordered from the North by the Psehako and the Achishkho Ridges, and by the Aibga Ridge from the South (Fig. 1). This region is of great importance for the tourism development and is one of the largest centers of skiing in Russia.

Rock avalanches (stürzstroms, RA) are large-scale (commonly greater than 1 million $m^3$ in volume) moving debris derived from the rock mass of the slopes, occurring (in the main phase of displacement) like a “dry flow” of rock debris and characterized by abnormal (compared to common landslides) speed and run-out distance (Hsu 1975).

The strong dissection of the terrain and the characteristic features of the geological structure of the territory under consideration causes widespread and various modern slope processes. However, the development of rock avalanches in the past has had the greatest impact on the current engineering and geological conditions of the Krasnaya Polyana area.

Such large-scale slope processes in the study area are characterized by specific complex conditions of formation and structural features.

The deposits of ancient rock avalanches are widespread in the upper part of the Mzymta River valley. Deposits of ancient rock avalanches create the gentle slopes of the valley and form flat areas on the territories of even more ancient negative landforms among the remnants of high terraces of the Mzymta River. These flat areas are widely used for construction.

This article discusses one typical rock avalanche in the Krasnaya Polyana area. The study is based on data obtained by field observations, remote sensing, photo-interpretation of high-resolution aerial photographs. The structure of the avalanche body is based on the results of extensive engineering and geological studies including boreholes,


V. Vilímek et al. (eds.), *Understanding and Reducing Landslide Disaster Risk*, ICL Contribution to Landslide Disaster Risk Reduction,
https://doi.org/10.1007/978-3-030-60319-9_16

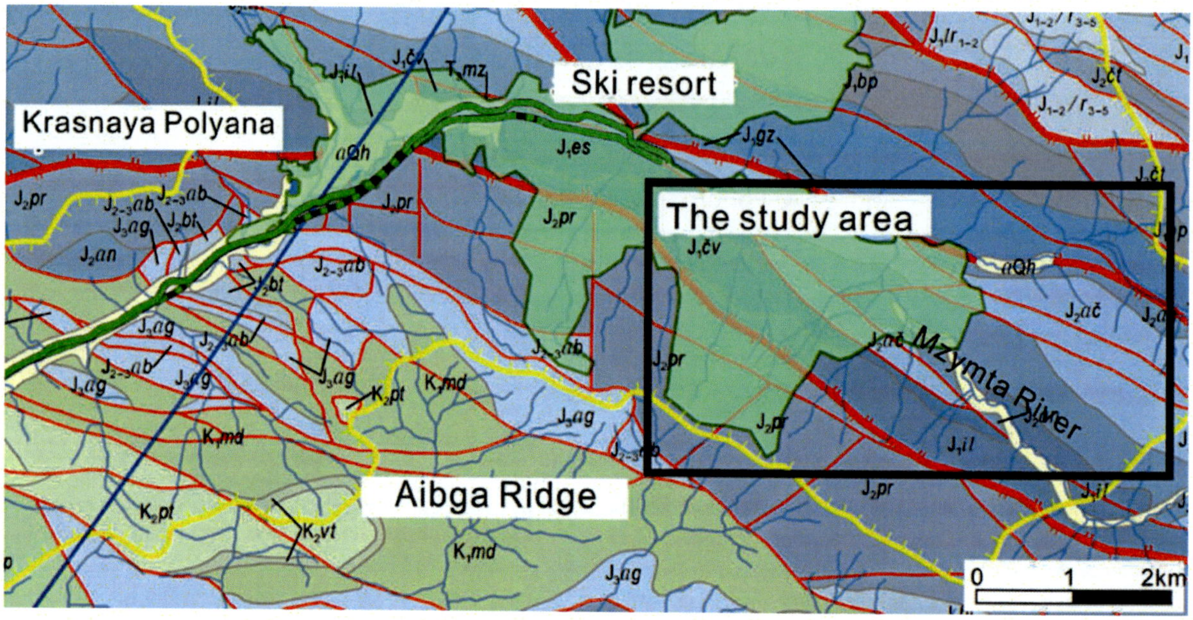


**Fig. 1** Geological map of the study area

geophysical work, as well as data from field and laboratory studies of the host rock and RA debris samples.

## Conditions of Rock Avalanches Formation

The studied territory is characterized by a predominance of the alpine-type relief with large elevation differences and steep slopes.

The elevation above sea level varies from 500–600 m (the Mzymta River valley) to 2,000–2,200 m (the peak of the Aibga Ridge). All rock avalanches in the study area occurred on the slopes of the Mzymta River valley. Modern valleys of large rivers of the southern slope of the Caucasus—the Mzymta, the Sochi, the Shah River, etc.) are heterogeneous and are characterized by numerous cranked bends. The Mzymta valley is characterized by the following structural feature: its right-bank slope length (from the ridge crest to the river) increases from 7 km in the upper reaches, up to 13–14 km in the downstream part, along with the general expansion of the valley, while the length of the left-bank slope length is almost constant and does not exceed 5 km. The short steep left slope is divided by numerous steep-sloping tributary valleys, many of which have hanging mouths. The spurs of the Aibga Ridge often end at the river sheer precipices. On the other hand, the right slope of the valley is characterized by the tributaries of the considerable length (Kochetov 1970). The Mzymta River valley has a wide accumulative bottom. The total width of the terrace area reaches 350–500 m; the height of the first terrace is 1.5–2.5 m above the river, while that of the second terrace is 5–7 m. The Mzymta River is undecutting its left bank. It is possible that the presence of this right-sided terrace and its morphological appearance are due to the neotectonic regime of the territory.

The geological deposits of the study area are characterized by the presence of the Middle Jurassic porphyrites and tuff-sandstones of porphyrite-unit ($J_2$pr) in the upper parts of the ridges with underlying mudstones of the Illarionov ($J_1$il), Achishkhinskaya ($J_2$ac) and Anchkhoy ($J_2$an) units. Porphyrites and tuff-sandstones of the porphyrite-unit are characterized by steep dip angle (65–70°) towards the free face. The mudstones of the underlying formations are characterized by gentle inclination in the western direction (Ponomarev et al. 2017).

The physical and mechanical properties of soil of the slopes near Krasnaya Polyana also vary significantly. While porphyrites and tuff-sandstones are characterized by high density (2.6–2.7 g/cm$^3$) and strength (saturated uniaxial compressive strength $R_{cw}$ = 60–70 MPa), the underlying mudstones are characterized by rather low physical and mechanical characteristics (density = 2.2–2.3 g/cm$^3$, $R_{cw}$ = 1.8–2.0 MPa) and can be destroyed very quickly by weathering. The soaking of mudstone samples in the laboratory takes several minutes (Ponomarev et al. 2017).

In the regional tectonic aspect, the research area is located in the Western part of the Central Caucasus and is composed of the Jurassic sedimentary and volcanic rocks folded and faulted in the Alpine tectonic epoch. In more detail, the area is distinguished by the Gagro-Java fold-block zone and the

Goitkh-Achipsin fold zone, divided by the Krasnopolyansky regional fault.

The study site is located within the Krasnaya Polyana structural unit, which is the part of the Gagro-Java folded-block zone. The section of the Krasnaya Polyana tectonic unit is composed of the sedimentary Aalenian deposits, of relatively small thickness (about 4 km), the Lower Jurassic volcanic-sedimenary rocks, and with widespread basic volcanic-clastic rocks. The tectonics of the Krasnaya Polyana structural unit is characterized by disharmonious folds crossed by three main fault systems: trending along the Caucasian Ridge, the sub-latitudinal and sub-meridional.

One of the important distinguishing features of the territory of the Krasnaya Polyana village is high seismicity. In accordance with the General Seismic Zoning Map of Russia, the territory of Sochi belongs to the 8-point seismicity zone on the MSK-64 scale for earthquake recurrence once every 500 years and to the 9-point seismicity for earthquake recurrence once every 1000 years and 5000 years.

The geological and geomorphological conditions described above are extremely favorable for the development of large-scale slope processes. The rock properties in the upper part of the Aibga Ridge region, contribute to the rock avalanche formation too.

The development of lateral erosion of the left side of the valley contributes to the formation of extensive negative landforms, which is undoubtedly a favorable factor for the formation of sufficiently long steep slopes. These circumstances together form a set of conditions for the development of rock avalanches within the study region.

The disintegration of landslide blocks during rock slope failure contributed to the formation of a significant number of crushed rock materials with a large amount of dispersed aggregates, which can be proved by an increase of dispersed materials in the direction of the rock avalanche. During its development, the rock avalanche initial volume increased by about 30% due to the rock crushing and entrainment.

Along with the further movement, blocks and coarse materials with dispersed aggregates of the avalanche went down to the Mzymta River, causing river damming and dammed lakes formation. Such lakes influenced the modern appearance of the river valley, as evidenced by traces of dammed lakes with the changed morphology of the valley.

## The Structure of Rock Avalanche

As an example, we describe the structure of the Pogranichnaya avalanche about 60 million $m^3$ in volume on the northern slope of the Aibga Ridge that originated at 2050 m a.s.l. and terminated at 700 m a.s.l in the channel of the Mzymta River. The runout is about 3.5 km (Fig. 2).

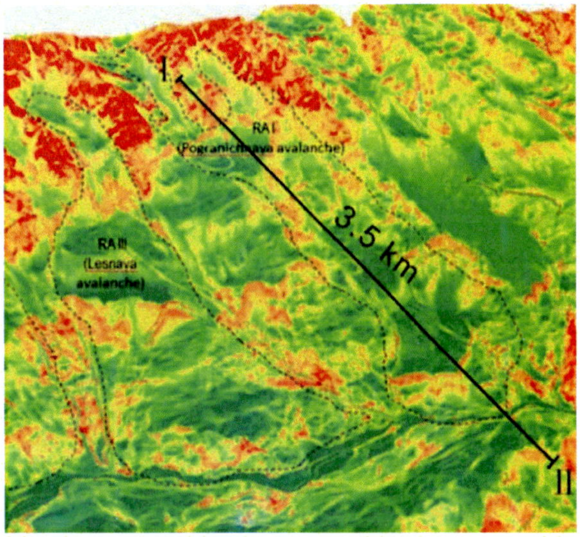


**Fig. 2** Pogranichnaya rock avalanche and position of the profile shown in Fig. 4. Colours indicate the slope angle of the surface: green—up to 15°, yellow—up to 25°, and red—more than 25°

Four zones can be identified with the entire length of the Pogranichnaya rock avalanche:

(1) Source zone (0–600 m from the top);
(2) Transition zone (600–1100 m from the top);
(3) Proximal zone (1100–2000 m from the top);
(4) Distal zone (2000–3500 m from the top).

The main influence on the structure of each zone is the distance of rock avalanche from the source zone and the relief on its way.

**The source zone** is located at altitudes of 1550–2000 m. The angles of inclination of the surface are quite large and can be up to 40°, however, before the shift to the transition zone there is a gentle (almost flat section) with minimal angles of inclination. These sites are regarded as a kind of "springboard" for moving rock avalanche deposits) (Strom and Abdrakhmatov 2018).

The Pogranichnaya rock avalanche source zone is covered by rubble of porphyrite, siltstone and tuff-sandstone with dark gray loam in the uppermost part (0–100 m from the top) and boulders of porphyrite and tuff-sandstone in the lower part of the zone. In the lower part of the zone one can see the proximal part of rock avalanche with a carapace composed of blocks of up to 2–3 m in size.

Despite the relatively small extent of the rock avalanche source zone, its size is about 600 m long, 500 m wide, and 200 m high. Moreover, the source zone consists of several rocky circuses.

In the source zone, disperse deposits are underlaid with the porphyrites and tuff sandstones of the porphyry unit ($J_2$pr).

**Fig. 3** Rubble of porphyrite, siltstone and tuff-sandstone with boulders in the distal part of rock avalanche

The thickness of the dispersed deposits in the source zone is small and amounts to 1–2 m in the upper part, while, in the lowest part, the thickness of the dispersed deposits with blocks of modern rockfalls can reach 10 m. Areas with significant thicknesses of modern block deposits are gently sloping surfaces and, apparently, contain a significant amount of secondary glacial deposits.

**The transition zone** is a section where the speed of the debris reached maximum values. The slope steepness of the transition zone is 25–35°, as shown in Fig. 4. In the transition zone of the Pogranichnaya rock avalanche, the thickness of the modern slope deposits does not exceed 1–2 m.

Within this zone, there is a boundary between the underlying deposits of porphyrites and tuff-sandstones ($J_2pr$) and low-strength mudstones ($J_1il$, $J_2ac$ and $J_2an$).

**The proximal zone** has the fuzziest distribution boundary, and for various places of the avalanche it has different sizes and lengths. However, despite the difference in size, this zone has a number of features.

From the geomorphological viewpoint, the proximal zone is in ancient flatted area where a significant amount of coarse material accumulated.

For the Pogranichnaya rock avalanche, the proximal zone is clearly expressed in the relief. Its deposits cover a gentle slope area at altitudes of 1250–1100 m above sea level and at a distance of 1000–2000 m from the source zone, flattening downslope. The steepness of the proximal zone is 20–25° in the upper part and gradually decreases to 5–10° in the lower part. This zone of rock avalanche ends with a clearly defined ledge more than 50 m high. In this area, important ski infrastructure facilities are located.

The deposits of this zone have a significant degree of heterogeneity. However, regarding this case, there is a tendency that coarse deposits decrease as they move away from the transition zone to the distal zone. Rock avalanche deposits in the proximal zone are crushed tuff-sandstone and porphyrites with loam (less than 20%) in the upper part of the section and boulders of tuff-sandstone, porphyrites and tuff in a loam (less than 10%) in the lower part. In coarse-grained avalanche deposits in the zone, a significant number of lumps of porphyrite and tuff-sandstone of various diameters are found everywhere. Some blocks have diameters exceeding 4–5 m. The presence of a high ledge between the proximal zone and the distal zone is associated with large internal friction angle of the deposits of the proximal zone due to a large number of large blocks. The presence of clay matrix in block deposits is associated with a later accumulation of slope deposits in gentle areas.

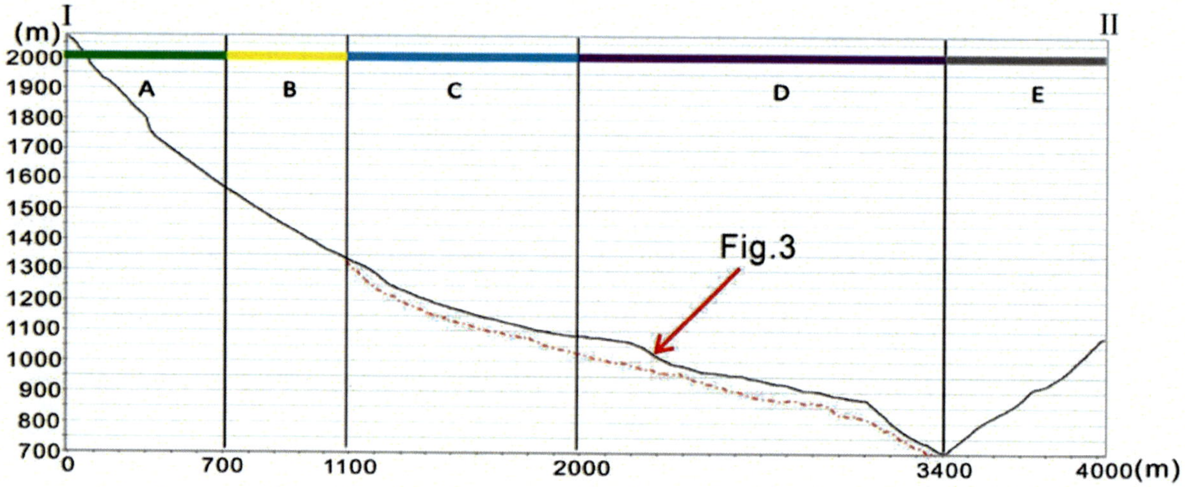


**Fig. 4** Profile of the Pogranichnaya avalanche: A—Source zone; B—Transition zone, C—Proximal zone, D—Distal zone, E— Landslide on the opposite side of the valley

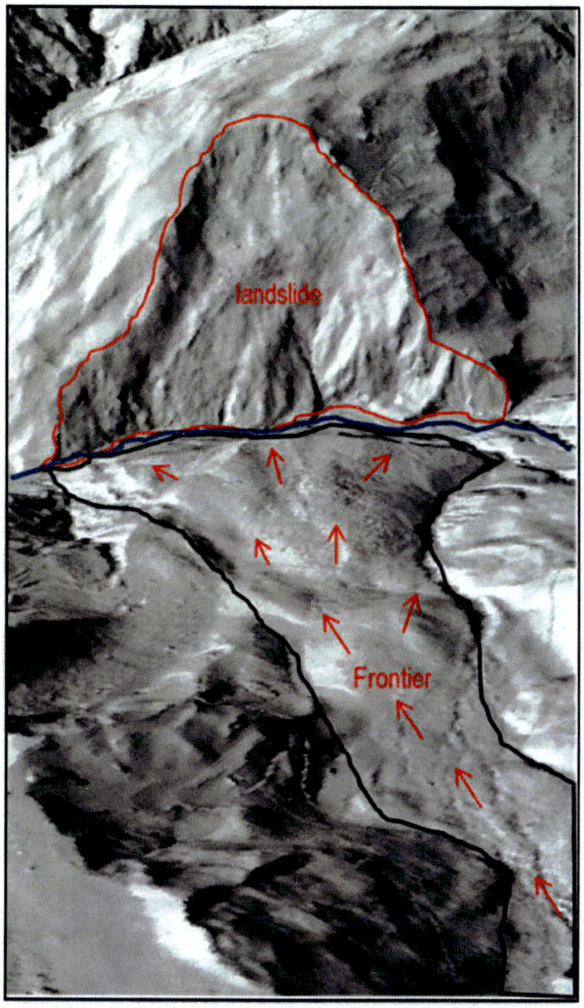


**Fig. 5** Landslide opposite the Pogranichnaya rock avalanche front

The thickness of the rock avalanche deposits in this zone is determined from drilling and geophysical studies. In the upper part of the zone, the thickness of deposits does not exceed 5–10 m, further increasing in thickness towards the flat section. Toward the end of the zone, the thickness of the deposits increases and amounts to 25–40 m.

**The distal zone** of the Pogranichnaya rock avalanche has a length of about a kilometer and begins at about 2 km from the source zone. Geomorphologically, this zone has a gentle surface with a length of 700 m, however, it has a steep slope area near the Mzymta River valley where it was eroded by the river. The steepness of the distal zone front is 35–45°.

The terminal part of the distal zone is located in the Mzymta River bed. In this section, the deposits had blocked the river, which subsequently led to a change of the channel direction. Within the part of the distal zone, modern landslide processes are most widely developed, which is due to the presence of dispersed deposits.

On the side of the Mzymta River valley, opposite the terminus of the Pogranichnaya rock avalanche, a landslide with volume of more than 7 million $m^3$ was revealed (Fig. 5). Formation of landslides at the opposite slopes of the valleys that had been blocked by rock avalanches is very common in the study region.

Considering this circumstance, the age of the Pogranichnaya rock avalanche can be determined as late Neopleistocene.

The distal zone of the Pogranichnaya rock avalanche is the most convenient for human activities due to the presence of flat areas, however, the modern active landslides are widespread here that require the prevention measures.

The deposits of the distal zone of the Pogranichnaya rock avalanche are similar to the deposits of the proximal zone with a tendency of the block dispersion increase (Fig. 3). The thickness of the rock avalanche deposits in the distal zone is more than 50 m, as shown in Fig. 4. Analysis and comparison of samples of coarse material from the distal zone and rockmass from the source zone show the complete identity of the composition and properties of the materials, which allows us to say that the blocks were brought into the valley by the rock avalanche.

## Conclusions

Rock avalanche deposits are widespread in the vicinity of the Krasnaya Polyana village. The deposits form ancient negative landforms over the vertical section the upper part of the left-side Mzymta River valley. In these areas, there are many buildings and structures of the largest ski resort in Russia.

The development of rock avalanche Pogranichnaya in the Krasnaya Polyana village is determined by specific tectonic, geomorphological and geological conditions of the region. The tectonic uplift and rapid incision contribute to the formation of steep slopes, which favorably affects the possibility of the development of rock avalanches. The specific geological structure with dense earth materials in the upper parts of the slope and less dense and easily weathering mudstones in the lower part also contributes to the involvement of a significant number of loose dispersed materials in the process of development of avalanche.

The study rock avalanche can be divided into four characteristic zones. Each zone has its own structural features and material properties. The development of ancient rock avalanches contributed to the activation of large-scale landslide processes on the opposite side of the Mzymta River valley, which significantly influenced the change in the shape of its channel. Analysis of position changes of the river bed allows us to determine the age of Pogranichnaya rock avalanches as Late Neopleistocene.

In general, despite only one rock avalanche was described in this study, its structure is quite typical for regions with similar geological condition.

Despite this rock avalanche was studied during site investigations for construction of the skiing facilities in the Krasnaya Polyana village, its mechanism as well as mechanism(s) of other rock avalanches is still not well understood. Undoubtedly, further study of this engineering-geological problem is necessary in the future.

## References

Hsü K (1975) Catastrophic debris streams (stürzstroms) generated by rockfalls. Geol Soc Am Bull 86:129–140

Kochetov NI (1970) The geomorphology of the valley of the Mzymta river (Western Caucasus) News of the University Geology and exploration. 6:21–27 (in Russian)

Ponomarev AA, Zerkal OV, Samarin EN (2017) Protection of the transport infrastructure from influence of landslides by suspension grouting. Procedia Engineering. 189:879–884

Strom A, Abdrakhmatov K (2018). Rockslides and rock avalanches of Central Asia: distribution, morphology, and internal structure. Elsevier. (ISBN: 978-0-12-803204-6). 449p

Check for updates

# Structural and Dynamic Numerical Models of Rockslides in the Carpathians and the Alps

UNESCO · ICL · ISC · WFEO · FAO · UNDRR · WMO · UNU
WLF
2020 Kyoto Japan

Emilie Lemaire, Anne-Sophie Mreyen, and Hans-Balder Havenith

**Abstract**

The stability of rock slopes is often guided by the structural geology of the rocks composing the slope. In this work, we analyse the influence of structural characteristics, and of their seismic response, on large and deep-seated rock slope failure development. The study is focused on the Tamins and Fernpass rockslides in the Alps and on the Balta and Eagles Lake rockslides in the southeastern Carpathians. These case studies are compared with catastrophic rock slope failures with ascertained or very likely seismic origin in the Tien Shan Mountains. The main goal is to identify features allowing to identify seismically induced deformation modes based on the source zone rock structures. We will present examples of classical anti-dip slope and along-strike rock structures that hint at a possible/partial seismic origin, but we will also consider a series of mixed structural types, which are more difficult to be interpreted. This morpho-structural study is supported by distinct element numerical modelling results showing that seismic shaking typically induces deeper seated deformation in initially 'stable' rock slopes. In addition, for failures partially triggered by dynamic shaking, these studies can help identify the contribution of the seismic factor to slope movements. The identification of the partial seismic origin on the basis of the dynamic response of rock structures can be particularly interesting for case histories in less seismically active mountain regions (in comparison with the Andes, Tien Shan, Pamirs), such as in the Alps and the Carpathian Mountains.

**Keywords**

Deep-seated failure • Bedding and joint orientation • Seismic deformation • Structural analysis • Discrete element modelling

## Introduction

Hazards related to massive rockslides affect most mountain regions of the World, especially those in tectonically active areas. However, the origin of the pre-historic failures is often unclear, as they are often the result of processes acting over longer terms (preparatory factors: e.g., slope oversteepening, weathering, glacier retreat and debuttressing) and of those acting over the short term (single or multiple triggers: e.g. single or multiple earthquakes, climatic events). In addition, several preconditioning factors influence the stability of rock slopes, such as the type of local lithology and the orientation of geological structures. Indeed, geological structures, such as bedding, ductile folds, discontinuities, brittle faults and fractures are known factors decreasing the stability of rock slopes according to their orientation and the (seismo-)tectonic deformation history (Stead and Wolter 2015). In this paper we will shortly review structural features that would allow distinguishing seismic trigger modes based on the source zone rock structures. Therefore, especially deep-seated and very large failures (with a volume of >$10^7$ $m^3$) will be analysed with regard to their structural characteristics. We will focus on the SE Carpathians and the Alps that have been exposed to numerous rockslide hazards, but for which the seismic contribution to rockslide hazards is not well defined (examples analysed include the 'Balta' and 'Eagles Lake' rockslides located in the SE Carpathians and the 'Tamins' and 'Fernpass' rockslides, respectively, in the Swiss and Austrian Alps—Fig. 1). The main goal of this work is to define the general influence of the rock structure,

E. Lemaire (✉) · A.-S. Mreyen · H.-B. Havenith
Department of Geology, Georisks and Environment, Liège University, Liège, Belgium
e-mail: emilie.lemaire@student.uliege.be


V. Vilímek et al. (eds.), *Understanding and Reducing Landslide Disaster Risk*,
ICL Contribution to Landslide Disaster Risk Reduction,
https://doi.org/10.1007/978-3-030-60319-9_17

**Fig. 1** Location of studied rockslides

and, more specifically, of its seismic response, on the stability of a slope.

## Seismotectonic Context of Studied Rockslides

### SE Carpathians

The SE bend of the Carpathian Mountains belt, Romania, shows possibly the highest sub-crustal earthquake activity in Central Europe. Even though the region presents a comparatively smooth relief, it is prone to slope instabilities due to intense precipitation and floods, but also seismic events (Balteanu et al. 2010). Especially the here studied Vrancea-Buzau region is known to be seismogenic and numerous intermediate-depth events were registered in the last three centuries. Magnitudes of these events vary between Mw 7.4–7.9. Historic reports of most M > 7 earthquakes (in 1802, 1838, 1940, 1977) also document geohazards such as groundwater rises, floods, liquefaction, cracks and landslides (see e.g. Georgescu and Pomonis 2012).

Here, we focus on the case studies of the Balta and Eagles Lake rockslides, both deep-seated failures located in the Flysch Carpathians, of unknown age, but marked by geomorphic and structural indices of potential co-seismic failures. Balta rockslide developed in a Paleogene schistose sandstone flysch marked by thick sandstone layers (general view shown in Fig. 3). The Eagles Lake rockslide was triggered from a crest made of Cretaceous sandstone flysch. These mass movements are marked by profound head scarps, and anti-dip slope bedding (40–50° for Balta, 20° for Eagles Lake). For both, a major part of the crushed mass accumulated in the middle part of the slope. For the Balta rockslide, the landslide toe spreads into the valley and most likely had dammed the river in previous times (Mreyen et al. 2017).

The pronounced morphology of the rockslide distinguishes itself from the vast number of superficial debris flows typical for the region. Views of this rockslide together with structural data are presented in the next section, in Figs. 3 and 4.

### The Alps

Earthquakes in the Alps take place at different depths and have the potential of reaching seismic magnitudes above 6 (e.g. Friuli 1976 and Basle 1356) and can therefore trigger landslides up to 50 km away from the epicentre (Lenhardt 2007).

There is a long record of landslide disasters in historical and recent times in the Alps, but there only a few documented cases of larger rockslides triggered by earthquakes. For instance, in 1348, six landslides were triggered from the Dobratsch Mountain by an earthquake in the Friuli region (northern Italy).

In the Eastern Swiss Alps (Grisons region), the ancient Tamins rockslide with a volume exceeding 1 km$^3$ has an unknown origin but presents some characteristics of a mass movement with at least partial seismic origin—such as the deep scarp morphology that was also highlighted above for the Balta rockslide in Romania. This rockslide collapsed from the Sennenstein Mountain and blocked the Rhine Valley. The Tamins rockslide is a very large mass movement, but it was far less studied than the neighbouring Flims slide. It is assumed that the Tamins event influenced the Films slide through the formation of the dam on Rhine River and the related impounding of a lake (Krietsch and Wolter 2016). Pollet et al. (2011) have dated the Flims event between 8000 and 9000 14C years BP and other dating measurements suggest that Tamins is about the same age as Flims (Poschinger et al. 2006).

In the western Austrian Alps, the Fernpass rock avalanche made of about 10$^9$ m$^3$ of rock material (Fig. 1) is also marked by a deeply incised scarp, cut into thick layers of dolomites, limestones and marls (Ostermann and Prager (2016). Rock materials of the massive slope failure split into two separate arms and were transported 12 km towards the North and 15.5 km towards the South. Prager (2010) mentioned that the particular shape of the scarp suggests that this along-strike wedge failure depends not only on the existing rheology and bedding conditions, but mostly due to convoluted intersections of fracture and uncemented fault systems. The AlpS Gmbh mentioned the possibility of a local earthquake in the area but the trigger process of the Fernpass rockslide is still debated. The age of the Fernpass event has

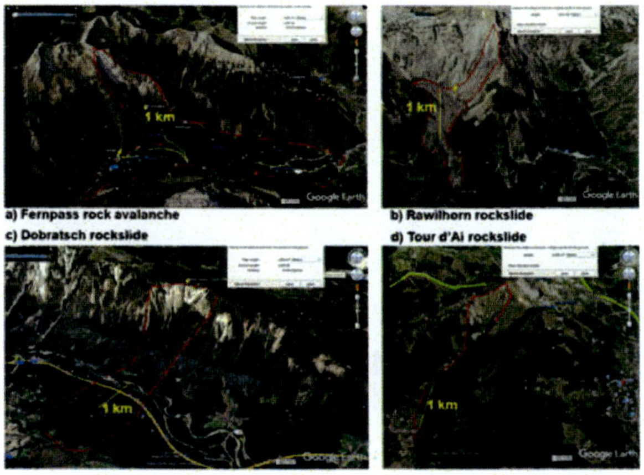


**Fig. 2** Google Earth® views of the **a** Fernpass rock avalanche, Austria; **b** Rawilhorn rockslide triggered by the 1946 Sierre earthquake (Switzerland); **c** Dobratsch rockslide triggered near Villach (Austria) in 1348; **d** Tour d'Ai rockslide triggered by the 1584 Aigle (Switzerland) earthquake (with uncertain outline)

been fixed at about 4.2 ka (Ostermann and Prager 2016). Google Earth® views of the Fernpass rock avalanche and the main rockslide that fell in 1348 from Dobratsch Mountain, both located in the Austrian Alps, are shown, respectively, in Figs. 2a, c. Figures 2b, d present two other proved seismically triggered mass movements, located in the Swiss Alps, the 1946 Rawilhorn and 1584 Tour d'Aï rockslides

## Geomodels of Studied Rockslides

### Inputs

3D geomodels were created for the Tamins and Balta rockslides with surface data and structural features. The Leapfrog Geological Modelling software, developed by Seequent, was used in order to display several types of structural data within a 3D model. For the development of these 3D models, surface data were extracted from the 11 m resolution TanDem-X digital elevation model (UTM 35 N) for the Balta site and from the Swiss ALTI3D digital elevation model for the Tamins site (Swisstopo). The aim of the model is to better outline the spatial relationship between the structural and morphological features of the site. The structural data were collected during field trips and were imported in the software using the Dip and Dip-azimuth data format. These structural data are displayed as disks oriented and inclined according the strike and dip and within a stereographic projection. A stereographic projection, also called stereonet, allows us to represent the 3-dimensional geometry of lines and planes and to display and assess complex structural data (Davis and Reynolds 1996).

## The Balta and Tamins Structural Geomodels

The Balta rockslide is a typical anti-dip slope case. The field measurements compiled into Google Earth® images (Fig. 3) and the stereographic projection (Fig. 4) are composed of three main structural features: the bedding planes and two families of joints. The bedding planes are mainly dipping S-SE, with a N48° E/40° S Fisher mean plane. Indeed, the figures below clearly highlight the anti-dip character of the bedding. The first family of joints (green), with the mean Fisher plane oriented N241° E/56° NW, favours instability, as dipping in the direction of the sliding. The second family of joints (blue), has a Fisher mean plane oriented N133° E/74° SW and displays patterns favouring detachment.

The structural data of the Tamins rockslide collected on the field and reported on Google Earth® images with strike and dip symbols (Fig. 5) and on a stereographic projection (Fig. 6) display bedding planes with a N171° E/46° W oriented Fisher mean plane (red dashed line), however, some planes are more oriented northwards or southwards. Indeed, the bedding planes measured on the eastern flank are dipping westwards, while near the top of the scarp the bedding is dipping southwards, and on the western flank it is dipping towards the north-west. Figures 5 and 6 outline the different orientations of the bedding planes. The Fisher mean planes of the bedding were calculated with a N179° E/57° W orientation for the eastern flank, N82° E/45° S orientation for the top of the scarp and N230° E/50° NW orientation for the western flank (see all results in Fig. 5). The bedding planes of the eastern and western flanks do not have a favourable dip orientation for sliding (towards the south), while the bedding planes observed on the top the scarp are dipping in the same direction as the main sliding movement. During the field work, it was noticed that the top of the western flank shows some destabilisation. Indeed, recent rock falls were observed and confirm the favourable trend of the upper part of the mountain for instabilities.

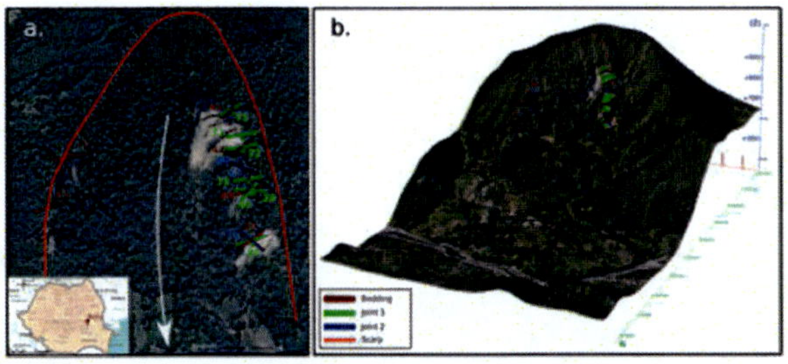


**Fig. 3** 3D geomodel of the Balta rockslide with a 3D stereonet **a** view towards the South, **b** view towards the SW and **c** view towards the W. The coloured planes are the Fisher mean planes and the disks mark the local strike and dip orientation

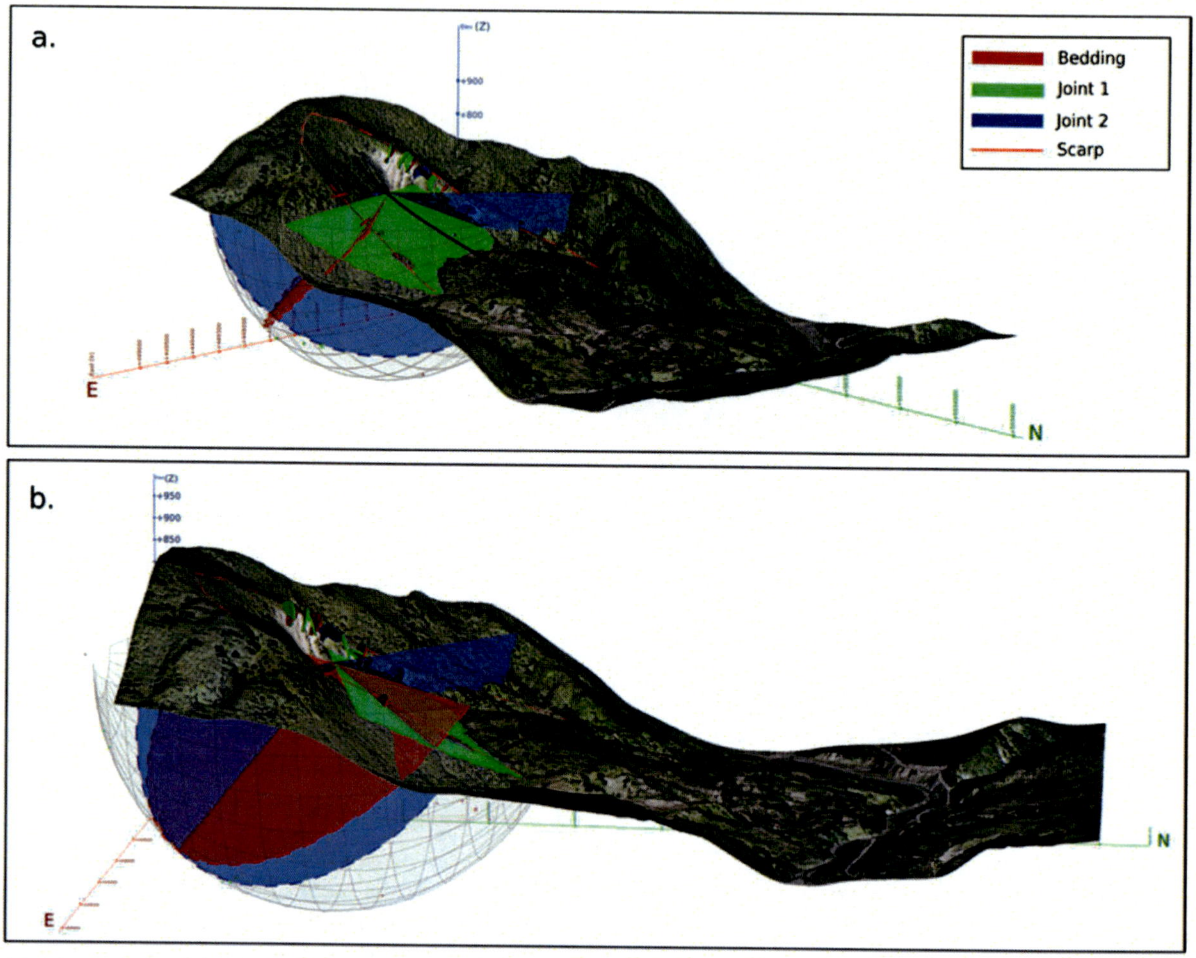


**Fig. 4** 3D geomodel of the Balta rockslide with a 3D stereonet **a** view towards the South, **b** view towards the SW and **c** view towards the W. The coloured planes are the Fisher mean planes and the disks mark the local strike and dip orientation

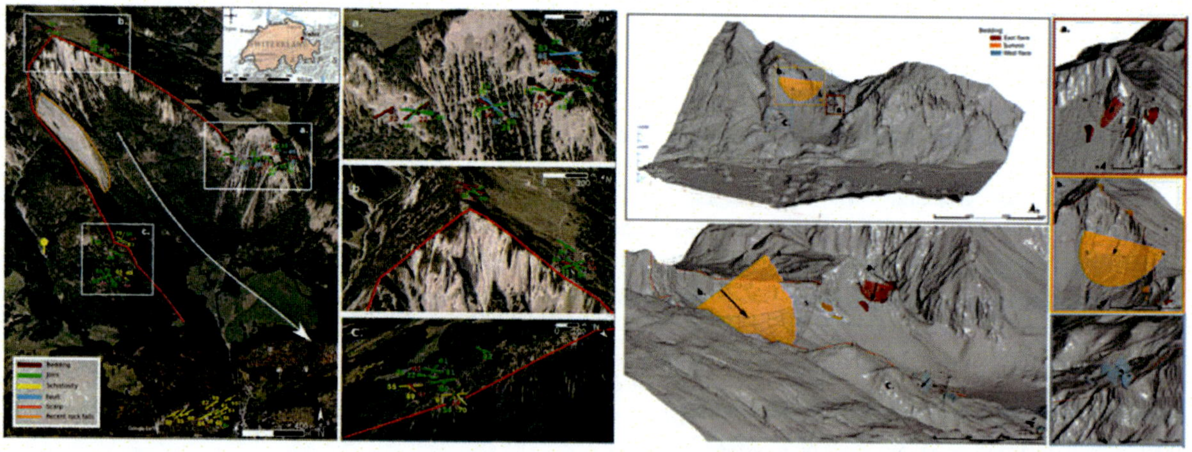

**Fig. 5** Left: Google Earth® views with the strike-dip symbols of the measurement made on the Tamins rockslide; the numbers near the strike-dip symbols are the value of the dip and the white arrow displays the direction of the slide. The red icon on the Swiss map show the location of the rockslide. Right: Five different views of the 3D model of the Tamins rockslide; the upper left panel gives a global view of the rockslide, the lower left panel if a zoom on the three stereonet representing **a** the bedding of the east flank, **b** the bedding of the summit and **c** the bedding of the west flank. The colour planes are the Fisher mean planes, the black arrows show the plunge of these planes and the red line outline the scarp of the rockslide.

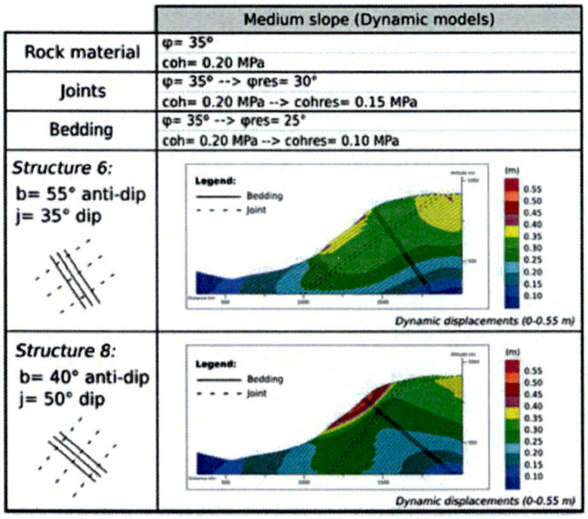


**Fig. 6** Dynamic modelling results (in terms of displacements) for the rock slope conditions closest to the Balta case, with anti-dip slope bedding structure

In the following section, some influence of bedding and joint orientation on static and dynamic deformation will be discussed on the basis of 2D numerical modelling results.

## 2D Numerical Modelling of Static and Dynamic Deformation

### Discrete Element Modelling

For the modelling we used the UDEC (Universal Distinct Element Code) 2D numerical code developed by Itasca® for discontinuous problems. The software allows to quantify the effects of loads and external stresses applied to jointed rock material by combining two numeric approaches: the distinct element method (DEM) and the finite-difference method (FDM). Here, we present the application of UDEC to estimate the initial slope stability conditions of the Balta rockslide and of two other variants of the initial slope (only for comparison, as those do not represent slope morphologies of the studied area – a smooth and a very steep slope). In addition to the observed 35–55° anti-dip slope bedding and associated orthogonal dip-slope joint orientation also four other structural models were tested. Those tests were completed both by applying static and dynamic loading (about 4 m/s$^2$ as peak acceleration), i.e. gravitational force and a synthetic Ricker wavelet, respectively.

For those models the geomechanical properties are summarized in Table 1 (considering that this is mainly a parametric study, average properties of sandstone rocks were used, applying lower values for the more gentle slope and larger shear resistance properties for the steeper slopes). For all models a volumetric weight ρ of 2400 kg/m$^3$ and a bulk modulus K of 10$^7$ kPa and a shear modulus G of 4 × 10$^6$ kPa were applied.

A possible slope surface prior to failure at the Balta site was created, based on analyses of the failed rock volume. The reconstructed slope morphology is designated as "medium slope" and supposedly represents the most realistic scenario. The two additional surface morphologies were created by decreasing ("gentle slope") and increasing ("steep slope") the slope angle, respectively, of the medium slope case. For those three slope cases, as indicated above, various rock structures have been designed. Structural parameters are summarized in Table 2.

By varying the slope angle and bedding-joint structure parameters as indicated in Table 2, fourteen static and corresponding dynamic modelled displacement sections were obtained. Here, we show (Fig. 6) only the displacement sections produced by dynamic modelling for the two cases closest to the Balta morphology and structure, with a medium slope and 40° to 55° anti-dip bedding structure. Those results indicate that some minor displacements occurred after simulated shaking, while the static conditions are totally stable. Actually, only the steep-slope case with 80° anti-dip slope bedding structure produced significant displacements (0.3 m) after gravitational loading. Indeed, the same bedding structure revealed to be the most unstable one, as 0.5 m displacements could be simulated for all slope morphologies, even the gentle one, after seismic inputs. Interestingly, the 80° dip slope bedding structure did not produce any significant (>0.3 m) displacement after shaking, for any slope morphology or structure. We associate this surprising

**Table 1** Geomechanical properties: friction angle $\varphi$ and cohesion for the bulk material; both are also used as peak values for the bedding and joint planes; corresponding residual values for the bedding are bfricr and bcohr and for the joints, jfricr and jcohr

| Slope | Φ<br>bfricr/jfricr (°) | Cohesion<br>bcohr/jcohr (kPa) |
|---|---|---|
| Gentle ~ 15° | 30<br>20/25 | 100<br>0/50 |
| Medium = Balta<br>~35° | 35<br>25/30 | 200<br>100/150 |
| Steep ~ 70° | 45<br>35/40 | 1000<br>300/500 |

**Table 2** Structural parameters, considering the '−' means 'anti-dip' (into the slope) and '+' means 'dip' (out of the slope)

| Slope | Bedding dip (°) variants | Joint dip (°) variants |
|---|---|---|
| Gentle ~ 15° | 10/−80 | −80/10 |
| Medium = Balta ~ 35° | 10/−10/80/−80/ 35/−40/50/−55 | −80/80/−10/10 −55/50/−40/35 |
| Steep ~ 70° | 10/−10/80/−80 | −80/80/−10/10 |

output to the toppling and detachment process that can affect the 80 anti-dip slope structure (also under static conditions, strongly accentuated by shaking). This process finally leads to the collapse of the whole upper part (above mid-slope position) of the steep and medium slope. For the latter the deformation can be very deep (more than 100 m deep) while only some 0.5 m deformation occurs near the surface over the convex curved part of the more gentle slope.

## Conclusions

In this paper we presented some structural geology information for several rockslide cases. Actually, some rock structures can have a strong influence on slope stability and may be more or less susceptible to seismically induced failures. The Balta rockslide in Romania, is the prime example of ancient massive slope failure, which origin is unknown but for which a seismic origin is very likely: that slope is made of an initially stable bedding structure, dipping into the slope with 35° to 50°, locally including very thick (>5 m) sandstone beds. A similar conclusion can be determined for the Eagles Lake rockslide. For the Tamins rockslide source zone, a mixed situation was found, as rocks outcropping in the lateral parts of the scarp are not really favouring massive slope failure, while the central zone is marked by a bedding dipping in the same direction as the sliding movement (towards the South). Now, as the Tamins rockslide is likely to the oldest one analysed here (even though we do not know the ages of the Romanian rockslides), so with an origin that is the closest to the glacier retreat and related debuttressing effects, a possible seismic influence cannot be easily proved for this mass movement. The much younger Fernpass rock avalanche, however, presents an untypical type of along-strike failure. The possible/partial seismic contribution to it cannot be excluded and is thus worth to be studied more in detail.

Static and dynamic numerical 2D models confirm that the medium slope case with anti-dip slope bedding structure (closest to the 'Balta model') is stable after gravitational loading and only minor deformation occurs for the applied seismic load (~0.4 g). The most unstable rock structure surprisingly appeared to be the one with an 80° anti-dip slope bedding orientation. This structure likely supports toppling and detachment, both in static and dynamic conditions. The possible/partial seismic origin of the aforementioned case of the Fernpass rock avalanche could, however, not be analysed here as the along-strike failure requires a back-analysis based on a 3D numerical modelling approach.

**Acknowledgments** We thank Anja Dufresne (RWTH Aachen, Germany) for the support in the structural geology mapping survey during June 2019 in the Alps. We also thank Mihai Micu (Romanian Academy Bucharest, Romania) for the support in the structural geology mapping survey during May 2019 in the Romanian Carpathian Mountains.

## References

Balteanu D, Chendes V, Sima M, Enciu P (2010) A country-wide spatial assessment of landslide susceptibility in Romania. Geomorphology 124:102–112

Davis GH, Reynolds S (1996) Structural geology of rocks and regions. 2nd edn, JohnWiley Sons, p 776

Georgescu ES, Pomonis A (2012) Building damage vs. territorial casualty patterns during the Vrancea (Romania) earthquakes of 1940 and 1977. 15th world conference on earthquake engineering, 24–28

Krietsch H, Wolter A (2016) Preliminary forensic investigation of the structural and geomorphological controls on the prehistoric Tamins rockslide in Grisons, Switzerland. In Landslides and Engineered Slopes. Experience, Theory and Practice; CRC Press, 2016, 1205–1210

Lenhardt W (2007) Earthquake-triggered landslides in Austria – Dobratsch revisited. Jahrb Geol Bundesanst 147:193–199

Mreyen AS, Micu M, Onaca A, Cerfontaine P, Havenith HB (2017) Integrated geological-geophysical models of unstable slopes in seismic areas. In Advancing Culture of Living with Landslides; Springer International Publishing, 269–279

Ostermann M, Prager C (2016) Field Trip 12 - Large-scale catastrophic rock slope failures in the Ötz-Inn-Loisach Valley region. Geo Alp 13:2257–276

Pollet N, Cojean R, Couture R, Schneider JL, Strom A, Voirin C, Wassmer P (2011) A slab-on-slab model for the Flims rockslide (Swiss Alps). Can Geotech J 42:587–600

Poschinger AV, Wassmer P, Maisch M (2006) The flims rockslide: history of interpretation and new insights. Landslides from massive rock slope failure. In: Evans SG, Mugnozza GS, Strom A, Hermanns RL (eds) Springer Netherlands: Dordrecht, pp 329–356

Prager C (2010) Geologie, Alter und Struktur des Fernpass Bergsturzes und tiefgründige Massenbewegungen seiner Umgebung (Tirol, Österreich). PhD thesis

Stead D, Wolter A (2015) A critical review of rock slope failure mechanisms: The importance of structural geology. J Struct Geol 74:1–23

Check for updates

2020 Kyoto Japan

# Quantitative Investigation of a Mass Rock Creep Deforming Slope Through A-Din SAR and Geomorphometry

Michele Delchiaro, Emanuele Mele, Marta Della Seta, Salvatore Martino, Paolo Mazzanti, and Carlo Esposito

**Abstract**

A Deep-Seated Gravitational Slope Deformation (DSGSD) affects the SE slope of the Siah-kuh anticline in its SE periclinal tip in the Ilam region (Zagros Mts., Iran), almost 30 km south of the Seymareh Landslide, which represents the largest landslide on Earth surface. The DSGDS is driven by a Mass Rock Creep (MRC) process and involves an area of about 8 km$^2$. The evolution of such a gravity-induced process is strictly related to the evolution of the of Dowairij River drainage system. River incision originated a stress release at the bottom of the slope which likely caused the initiation of the deformational process. The present study is part of a broader International Programme on Landslide project (Project IPL-237) focused on the role of time-dependent rock mass deformations and landscape evolution rates as predisposing factors for massive rock slope failures. In this regard, the preliminary results of an ongoing research are here presented focusing on the assessment of the present-day landscaping processes. Specifically, a geomorphological survey was carried out in this area firstly through the analysis and interpretation of remote data (Google Earth satellite optical images), which led to the first detection of possible gravity-induced landforms, such as evidence of bulging and lateral release within the deforming slope of the Siah-kuh fold-related ridge. To confirm and quantify the existence of ground displacement due to a MRC process, InSAR techniques were performed for the Siah-kuh slope and surrounding areas by processing 279 satellite Sentinel-1 (A and B) radar images of the ascending and descending orbit spanning from 06 October 2014 to 31 March 2019. Moreover, a quantitative morphometric evaluation was also performed through a morphometric index suitable for predicting the catchment-scale suspended sediment yield on the deformation area produced by the Dowairij River system. We derived the erosion rate of the drainage network responsible for the valley engraving which allows to estimate a starting time for MRC in the order of $10^1$ ka. The comparison between the valley erosion rate and the slope strain rate reveals a difference of almost one order of magnitude allowing to assume that the gravity induced process, identified from remote and field geomorphological survey, evolves faster and originates landforms which can be preserved by the drainage system of the Dowairij River.

**Keywords**

Mass Rock Creep • SAR interferometry • Tu index • Zagros Mts

M. Delchiaro (✉) · E. Mele · M. Della Seta · S. Martino · P. Mazzanti · C. Esposito
Department of Earth Sciences, Sapienza University of Rome, Piazzale Aldo Moro 5, 00185 Rome, Italy
e-mail: michele.delchiaro@uniroma1.it

E. Mele
e-mail: emanuele.mele94@gmail.com

M. Della Seta
e-mail: marta.dellaseta@uniroma1.it

S. Martino
e-mail: salvatore.martino@uniroma1.it

P. Mazzanti
e-mail: paolo.mazzanti@uniroma1.it

C. Esposito
e-mail: carlo.esposito@uniroma1.it

P. Mazzanti
Nhazca Co. Ltd, Spin-off Sapienza University of Rome, Via Vittorio Bachelet, 12, 00185 Rome, Italy

## Introduction

In the hillslope evolution of tectonically active landscapes, Mass Rock Creep (MRC) process (Chigira 1992) may become a primary factor for damaging rock masses so leading to slope failures that generate huge rock avalanches.


V. Vilímek et al. (eds.), *Understanding and Reducing Landslide Disaster Risk*, ICL Contribution to Landslide Disaster Risk Reduction,
https://doi.org/10.1007/978-3-030-60319-9_18

MRC acts on large time-space scale through a continuous and non-linear variation of the stress-strain state of entire portions of slopes, giving rise to processes known as Deep Seated Gravitational Slope Deformations (DSGSD).

More in particular, the MRC process starts at a time ($t_0$) when the rock mass has already cumulated elasto-plastic strains caused by the local stress variation. It follows a phase of Primary Creep (I), in which the strain rate decreases over time (Decelerating Creep) under constant stress conditions. In general, the process evolves into the Secondary Creep (II) stage, where the deformations increase with a steady trend (Steady State Creep). From the Secondary Creep stage the curve can evolve with different trends, depending on the deformation rate achieved: if it decreases, constant deformations are reached and the process is suspended; if the creep rate is kept constant, it remains secondary for a longer time; if the applied stresses are close to the peak ones, the Secondary Creep can evolve into Tertiary Creep (III), where the strain rate increases over time until the rock mass reaches failure conditions (Accelerating Creep) (Saito 1969 and Petley and Allison 1997), which can be anticipated in the case of transient external forcing (e.g. earthquakes). In this regard, MRC is significantly related to the timing of landscape evolution and tectonics (Bozzano et al. 2012; Della Seta et al. 2017). In mountain areas made of contrasting rheological layering, the evolution of the drainage network can de-stress and kinematically release rock masses, isolating portions of slope (carapaces) that start to deform by MRC process until the failure. Earthquakes are not to be neglected as a possible cause of acceleration of deformations already in progress although they are not the most widespread type of trigger (e.g., Malamud et al. 2004; Hovius et al. 2011; Li et al. 2014).

The object of this study is the Siah-kuh DSGSD, located at the tip of a fold in the Zagros Mountains (Iran), 30 km south of the Seymareh landslide (Delchiaro et al. 2019, 2020; and references therein), the largest landslide on Earth surface, detached from the flank of a more internal fold of the belt. Although the case study represents an excellent example in which tectonics and landscape evolution originated the predisposing conditions for MRC process development, the Siah-kuh slope deformation has not yet been described by the scientific community.

The study is framed in the International Programme of Landslides (Project IPL-237) whose topic focuses on the role of time-dependent rock mass deformations and landscape evolution rates as predisposing factors for massive rock slope failures. In this paper we present the preliminary results of the research project, consisting in the identification of the present-day landforms associated to the ongoing MRC deformation of the Siah-kuh slope. We also aim at quantifying the rates of ground displacements and to compare them with the ones of the drainage system erosion rate, to justify the preservation of gravity induced landforms. Such an analysis provides also some constraints for future stress-strain numerical modelling, aimed at quantifying the strain effects related to the ongoing creep process in the Siah-kuh slope as well as its possible evolution towards ultimate failure conditions.

To achieve these objectives, it has been necessary to:

(1) detect and analyze possible gravity-induced landforms through field and remote data;
(2) confirm and quantify the ground displacement rate through space-borne SAR Interferometry in order to verify the state of activity and better constrain the kinematic role of the detected landforms;
(3) derive the erosion rate produced by the Dowairij River system on the deforming area through geomorphometry.

## Geological Setting

The ongoing gravitational deformation of Siah-kuh covers an area of about 8 km$^2$ and is located along the periclinal tip of the homonymous fold near the SW boundary of the Zagros (Simply) Folded Belt.

The available geological map produced by National Iranian Oil Company (Setudehnia and Perry 1967) has been used as reference for the geological setting of the study area (Fig. 1) in addition to data collected during geological field survey.

The outcropping stratigraphy is that typical of a passive margin that evolves in a foreland basin, characterized by strength contrasts between stratigraphic units. Units exposed in the SE sector of the Siah-kuh fold (Fig. 1) range from the Upper Cretaceous Emam Hassan to the Upper Miocene Agha Jari Formations (James and Wynd 1965 and Alavi 2004). Specifically, the most ancient unit of the local stratigraphy is the considerably more calcareous Emam Hassan Member of the Gurpi Formation. The marly Pabdeh-Gurpi Formation overlies it. On top, the Asmari Formation creates the carbonatic carapace of the fold. The most recent units of the local stratigraphy include: the Gachsaran Formation, consisting of salt, anhydrite, colored marl and gypsum and the Agha Jari Formation, composed of marls and red sandstones. Such a rheological layering is mirrored in the geomorphology of the breached anticline in the NW part of the studied area, where less resistant units (Gurpi and Pabdeh-Gurpi Formations) and more resistant units (Emam Hassan and Asmari Formations) have been differentially eroded, respectively forming secondary valleys and ridges (Roberts and Evans 2013). The MRC process enucleated mainly in the Pabdeh-Gurpi marls and passively

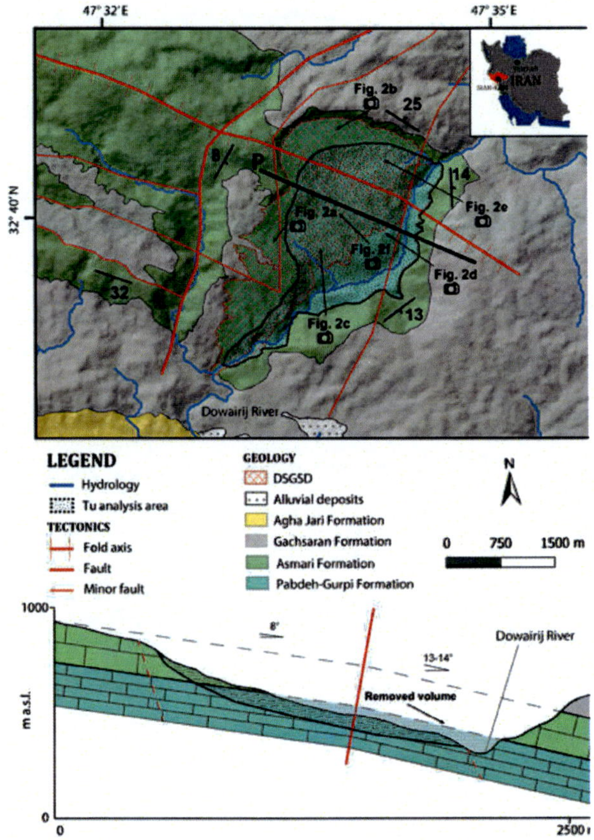


**Fig. 1** Geological map and cross section along trace P of the Siah-kuh deforming slope

involved the carapace of the Asmari limestone. The evolution of the gravitational instability is closely connected to the drainage evolution of Dowairij River, since its erosion produced the stress kinematic release at the base of the slope, likely starting the deformation process. For this reason, it could have begun as soon as the marly formation was uncovered by water erosion along the main river (see geological section in Fig. 1).

## Methodology and Results

### Field and Remote Surveying

A remote survey of the gravity induced landforms was carried out, based on the Google Earth satellite images (2018 Landsat Imagery) as well as a field cross-check was performed.

The results of remote surveying are shown in Figs. 2a–c while the ones from fieldwork are shown in Figs. 2d–f. As a main outcome, four zones characterized by different gravitational stress were identified: zone A (Fig. 2a) is referred to

**Fig. 2** Google Earth satellite images (**a–c**) and pictures taken during the field survey performed on August 2019 (**d–f**, see Fig. 1 for locations) which represent different typologies of gravity-induced landforms: **a** Tension cracks; **b** Scars; **c** Bulging; **d** Auxiliary landslide along the river gorge; **e** Scars and bulges; **f** Bulges

tension cracks in the Asmari Fm.; zone B (Fig. 2b, e) represents scars involving the contact between the Pabdeh-Gurpi and Asmari Fms.; finally, zone C figures compressional bulges developed mainly in the Pabdeh-Gurpi Fm (Fig. 2e, f). Moreover, as you can observe along the river gorge, an auxiliary landslide in the deformation system was detected (Fig. 2d) characterizing the zone D.

### Deformation Rate: Advanced SAR Interferometry

Satellite SAR Interferometry was applied in order to confirm above described evidences and, in particular, to quantify the ground displacement rate associated to the above described gravitational landforms. The analysis was performed using a multi-image interferometric approach that uses many images (STACK) characterized by the same acquisition geometry, to create interferograms related to a given area (Ferretti et al. 2000, 2001; Berardino et al. 2002; Hanssen 2005; Bert 2006). Among the various techniques the Persistent Scatterer Interferometry (PSI) (Ferretti et al. 2000, 2001) was used, based on the observation of time and space-coherent pixel, the so-called persistent scatterers, represented by anthropogenic and natural structures like buildings, antennas, exposed rocks, etc. The SAR scape software tool was used to process 147 Sentinel-1 SAR (Synthetic Aperture Radar) images for the ascending orbit in the time frame between 17 October 2014 and 31 March 2019, as well as 132 Sentinel-1 SAR images for the descending orbit in the time frame

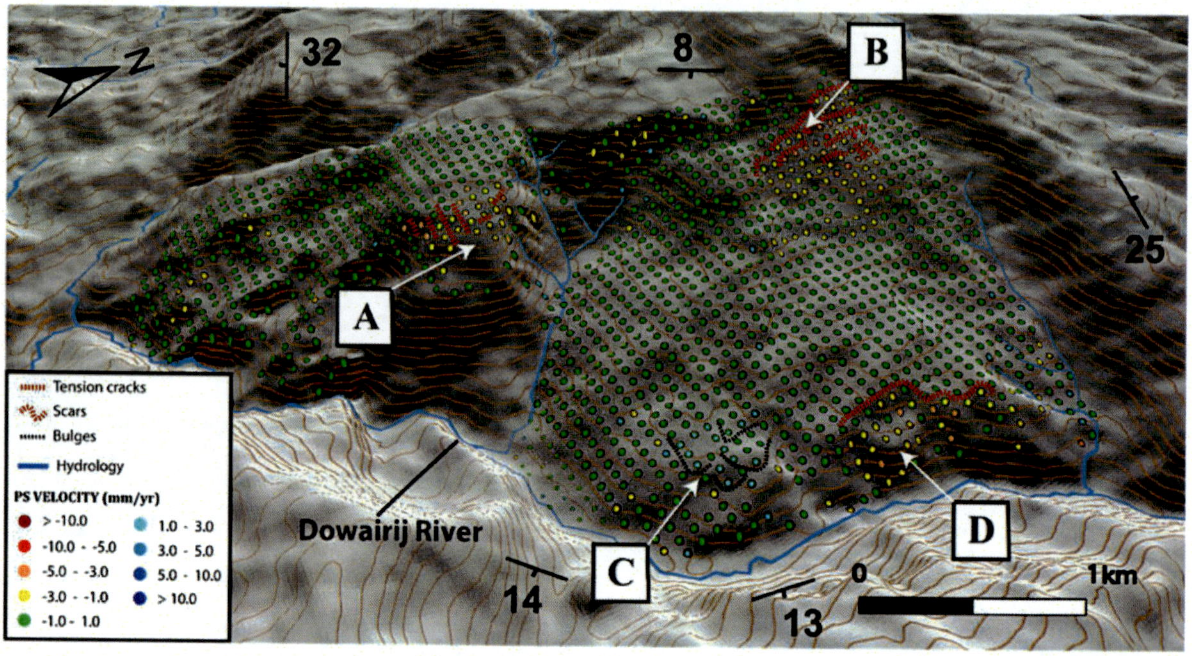


**Fig. 3** Surface velocity map of the vertical component displacement of the Siah-kuh slope obtained by using a 3D decomposition algorithm developed by Nhazca Co. Ltd

between 06 October 2014 and 26 March 2019. A coherence threshold of 0.75 has been used to obtain an improvement in the signal noise ratio but guaranteeing at the same time a good PS density. Then, the decomposition of vertical and horizontal displacements was performed by combining PS InSAR ascending and descending data by using a proprietary software provided by NHAZCA S.r.l. Figure 3 shows the vertical surface velocity map of the Siah-kuh slope and surrounding areas, thus allowing to identify different zones showing a downward displacement.

As it resulted by the performed analysis, the zones A and B are characterized by the highest negative displacement (respectively, about −4 and −6 mm $y^{-1}$) and are associated to tension cracks and scars observed by the previous phase of the investigation. Zone C is characterized by positive displacement rate (3 mm $y^{-1}$), likely caused by the bulging movement in the middle part of the deformation. Finally, zone D shows a negative displacement rate (about −2 mm $y^{-1}$), likely associated to the auxiliary landslide, whose main movement occurred before October 2014, since its displacement rate indicates a quiescent phase. The obtained rates are not significant for landslide failure forecasting purposes. In fact, several limits affect the ability of the interferometric technique to monitor deep seated gravitational slope deformations on the whole, especially during the pre-failure phase (Tertiary creep). The main limitation is related to the difficulty of transposing a decennial observation on an extremely longer process, especially if you want to catch the creep phase in which the process is located: in fact, there is a high risk of confusing short-term temporary change of the displacement trend with the transition between the different displacement trends that characterize the 3 different phases of the creep. Nonetheless, the good fit of the satellite interferometric evidences with the slope accessory landforms, which testify to an ongoing rock mass creep process, is encouraging.

## Erosion Rate: $T_u$ Index

In such a kind of gravitational instability the role of run-off and river erosion is twofold: on one hand, it can be a predisposing factor for gravity-induced deformational processed, while on the other hand, it can contrast the preservation of gravitational landforms. In order to quantify the erosive contribution offered by the drainage network, since it started cutting the Pabdeh-Gurpi Fm., a quantitative morphometric evaluation was performed, through the $T_u$ index (Della Seta et al. 2009; and references therein). The computation of such an index allowed to predict the catchment-scale suspended sediment yield on the deformation area produced by the drainage system, hierarchized according to the Strahler's method (Strahler 1957). The index is calculated using an empirical equation obtained empirically statistical correlations among measured catchment-scale suspended sediment yield and geomorphic

**Table 1** Removed volume of the Pabdeh-Gurpi Fm. and erosion rate in the deformation area due to water erosion as obtained through the $T_u$ index in order to calculate an estimate of the starting time of the deformation

| $\sum L$ (km) | a | Area ($km^2$) | $T_u$ (ton $km^{-2}y^{-1}$) |
|---|---|---|---|
| 74.10 | 1.72 | 9.18 | 6,887.44 |
| Ton by Area per Year | Erosion Rate ($mm\ yr^{-1}$) | Missing Pabdeh-Gurpi Volume ($m^3$) | MRC $t_0$ (yr) |
| $6.32 \cdot 10^4$ | 2.65 | $9.71 \cdot 10^3$ | $39.94 \cdot 10^3$ |

parameters of the drainage network calculated at the outlets of several Italian badlands areas developed on marly/clayey lithological units (similar to the Pabdeh-Gurpi Fm.). The experimental equation for calculating the $T_u$ index is as follows (1):

$$\log Tu = a + b \log D + c\Delta a \quad (1)$$

where a = 1.060, b = 2.797 and c = 0.140 are empirical coefficients. $D$ is the drainage density = $\sum L_i/A$ ($\sum L_i$ is the total length of the drainage segments in the drainage area; A is the drainage area); $\Delta a$ is the hierarchical anomaly index = $Ga/N1$ ($Ga$ is hierarchical anomaly number, i.e. the minimum number of Strahler's first order channels which would be necessary to make the drainage perfectly hierarchized; $N1$ is the number of Strahler's first order channels present in the drainage network). The reported equation has an $R^2$ of about 0.96 and the max error recorded between the measured $T_u$ and the calculated one is about 305 ton $km^{-2}$ $y^{-1}$, which correspond to a possible error in the order of 0.1 mm $y^{-1}$ considering a density value of 2.6 g $cm^{-3}$ and a drainage area of 9.18 $km^2$ same as the Siah-Kuh condition.

This index was already tested by Delchiaro et al. (2020) in a close (therefore climatically similar) area in the Zagros Mountains, in order to derive the erosion rate affecting the Seymareh landslide scar after the breaching of the landslide dam by the Seymareh River. In this case, the erosion rate derived from the analysis provided a starting time for the erosion which is coherent with the one obtained independently by OSL ages of a strath terrace cutting the landslide dam. Furthermore, the applicability of the morphometric method over a quite long time span had already been demonstrated in the study by Del Monte et al. (2014), in which the volume removed by water erosion processes over a time-span 300 ky long, in a badland area of Central Italy, was computed and compared to the one estimated through the $T_u$.

Table 1 shows a catchment-scale erosion rate of about 2.65 mm $y^{-1}$ obtained from the $T_u$ Index, after considering a density value for the Pabdeh-Gurpi Fm. of 2.6 g $cm^{-3}$. Such an erosion rate allowed also to calculate a time-span of ~ 40 ky needed by the drainage network, hierarchized according to the Strahler's method (Strahler 1957), to erode the missing volume of the Pabdeh-Gurpi Fm. from the bottom of Asmari Fm. to the present topography. Since the drainage network started eroding the Pabdeh-Gurpi Fm. ~40 ka, therefore this can be considered approximately as the starting time for the MRC process. In this calculation, the erosion rate was assumed to be spatially and temporally invariant and was compared to the calculated volume of the missing Pabdeh-Gurpi Fm. The latter was estimated in ArcGIS® using the *Trend* tool to reconstruct the top of the Pabdeh-Gurpi Fm. and the *Cut-Fill* tool to obtain the missing volume given the present topography. The above simplifications allow to infer just an order of magnitude of $10^1$ ka that can be associated to the time at which the MRC process could have been initiated.

## Discussions

Transient, tectonically active mountain landscapes are widely affected by landslides in response to gravity-induced disequilibrium on hillslopes. In this regard, forecasting the failure of a large hazardous landslide is a difficult task as it involves the evaluation of a large number of inter-related variables, among which the most important are: geometrical and geological complexity of the involved mass, nonlinearity of the time-displacements relationships, and superposition of seasonal effects (Crosta and Agliardi 2003; and reference therein).

The MRC process involving the Siah-kuh fold-slope can be regarded as a very suitable case study to output the relations between gravity-induced slope deformations, tectonics and landscape evolution as predisposing factors for large rockslide triggering.

Remote and field surveys allowed to identify four different zones characterized by different gravity-induced stress fields. The whereabouts of every ground displacement were well mirrored by the spatial location of the most rapidly moving radar targets. In detail, advanced SAR Interferometry was applied in order to confirm the above described evidence and, in particular, to quantify the ground displacement rate. It was measured a maximum present displacement rate of -6 mm $y^{-1}$ and a maximum total displacement of about 30 mm over 5 years. This value compared to the catchment-scale erosion rate resulting from

$T_u$ analysis of about 2.6 mm $y^{-1}$ justifies the preservation of gravity-induced landforms that would otherwise have been obliterated by the action of the surface run-off. An estimation of the order of magnitude of $10^1$ ka for a possible $t_0$ of the MRC was obtained assuming spatially and temporally invariant the provided erosion rate of about 2.65 mm $y^{-1}$. Over this time window, water erosion removed the volume of Pabdeh-Gurpi Fm. missing from the present topography to the bottom of the Asmari Fm. likely starting the Siah-kuh MRC slope deformation.

## Conclusions

In this study, preliminary results of an International Programme on Landslide project (Project IPL-237) aimed at quantifying the role of time-dependent rock mass deformations and landscape evolution rates in predisposing massive rock slope to failure, are presented. Specific goal of this research was to assess the present-day landscape evolution processes, quantifying their rates in order to identify the areas in which the prevalence of ground displacement over the drainage system erosion rate has allowed the preservation of gravity-induced landforms. In the more general perspective of understanding at which creep stage the Siah-kuh deforming slope is involved, a stress-strain numerical model of the slope in back- and forward-analysis could be a reliable tool to assess risk scenarios, after calibrating it on the presently collected evidences. In this regard, a time window for the considered MRC process was estimated; moreover, the comparison between slope strain rate and valley erosion rate justify the preservation of gravity-driven landforms respect to the drainage system erosion. As a future perspective, a more detailed assessment of past erosion rates from geomorphic markers, such as terraces sampled for OSL dating, as well as from landscape evolution modelling should be achieved to provide more quantitative constraints for the modelling of the MRC involving the Siah-kuh fold-slope.

## References


Alavi M (2004) Regional stratigraphy of the Zagros fold-thrust belt of Iran and its proforeland evolution. Am J Sci 304(1):1–20

Berardino P, Fornaro G, Lanari R, Sansosti E (2002) A new algorithm for surface deformation monitoring based on small baseline differential SAR interferograms. IEEE Trans Geosci Remote Sens 40(11):2375–2383

Bert MK (2006) Radar interferometry: persistent scatterers technique. Springer, The Netherlands

Bozzano F, Martino S, Montagna A, Prestininzi A (2012) Back analysis of a rock landslide to infer rheological parameters. Eng Geol 131–132:45–56

Chigira M (1992) Long-term gravitational deformation of rocks by mass rock creep. Eng Geol 32:157–184

Crosta G, Agliardi F (2003) Failure forecast for large rock slides by surface displacement measurements. Can Geotech J 40(1):176–191

Del Monte M, Della Seta M, Melelli L, Vergari F, Ciccacci S (2014) Intensità dell'erosione nel bacino idrografico del Rio Torbido (Bagnoregio, VT). Atti del Convegno Dialogo intorno al Paesaggio (Perugia, 19-22 Febbraio 2013). Cult Territori Linguaggi 4:107–116

Delchiaro M, Della Seta M, Martino S, Dehbozorgi M, Nozaem R (2019) Reconstruction of river valley evolution before and after the emplacement of the giant Seymareh rock avalanche (Zagros Mts., Iran). Earth Surf Dyn 7:929–947

Delchiaro M, Rouhi J, Della Seta M, Martino S, Nozaem R, Dehbozorgi M (2020) The giant Seymareh Landslide (Zagros Mts., Iran): a lesson for multi-temporal hazard scenario evaluation. In the Springer Book "Applied Geology" (in press)

Della Seta M, Del Monte M, Fredi P, Palmieri EL (2009) Space–time variability of denudation rates at the catchment and hillslope scales on the Tyrrhenian side of Central Italy. Geomorphology 107(3–4):161–177

Della Seta M, Esposito C, Marmoni GM, Martino S, Mugnozza GS, Troiani F (2017) Morpho-structural evolution of the valley-slope systems and related implications on slope-scale gravitational processes: New results from the Mt. Genzana case history (Central Apennines, Italy). Geomorphology 289:60–77

Ferretti A, Prati C, Rocca F (2000) Nonlinear subsidence rate estimation using permanent scatterers in differential SAR interferometry. IEEE Trans Geosci Remote Sens 38(5):2202–2212

Ferretti A, Prati C, Rocca F (2001) Permanent scatterers in SAR interferometry. IEEE Trans Geosci Remote Sens 39(1):8–20

Hanssen RF (2005) Satellite radar interferometry for deformation monitoring: a priori assessment of feasibility and accuracy. Int J Appl Earth Obs Geoinf 6(3–4):253–260

Hovius N, Meunier P, Lin CW, Chen H, Chen YG, Dadson S, Horng MJ, Lines M (2011) Prolonged seismically induced erosion and the mass balance of a large earthquake. Earth Planet Sci Lett 304(3–4):347–355

James GA, Wynd JG (1965) Stratigraphic nomenclature of Iranian oil consortium agreement area. AAPG bulletin 49(12):2182–2245

Li G, West AJ, Densmore AL, Jin Z, Parker RN, Hilton RG (2014) Seismic mountain building: Landslides associated with the 2008 Wenchuan earthquake in the context of a generalized model for earthquake volume balance. Geochem Geophys Geosyst 15(4):833–844

Malamud BD, Turcotte DL, Guzzetti F, Reichenbach P (2004) Landslide inventories and their statistical properties. Earth Surf Proc Land 29:687–711

Petley DN, Allison RJ (1997) The mechanics of deep-seated landslides. Earth Surf Proc Land 22:747–758

Roberts NJ, Evans SG (2013) The gigantic Seymareh (Saidmarreh) rock avalanche, Zagros Fold-Thrust Belt, Iran. J Geol Soc 170:685–700

Saito M (1969) Forecasting time of slope failure by tertiary creep. In: Proceedings of the 7th international conference on soil mechanics and foundation engineering, August 1969, Mexico City, Mexico, Vol 2, 677–683

Strahler AN (1957) Quantitative analysis of watershed geomorphology. Eos, Trans Am Geophys Union 38(6):913–920

Setudehnia A, Perry JTOB (1967) Dal Parri. 1:100000 geological Map. Iran oil operating companies, geological exploration division, Tehran, Iran

# Deformational Features of Deep-Seated Gravitational Slope Deformation of Slate Slopes in the Central Range, Taiwan

2020 Kyoto Japan

Ching-Ying Tsou, Masahiro Chigira, Yu-Chung Hsieh, Mien-Ming Chen, and Tai-Chieh He

**Abstract**

Catastrophic landslides are reported to be related to large amounts of unstable rocks preceded by deep-seated gravitational slope deformation (DGSD), where the slopes are underlain by Eocene to Miocene slate in the Central Range of Taiwan. To understand basic geological causes of the slope processes, the study was conducted in the Cifeng, Lixing, and Maliguan areas along the Lixing Industrial Road where many DGSDs threat important infrastructures and villages directly. Geological survey suggests that the beds beneath the slopes are mainly slate interlayered with thin metamorphosed sandstone. The slate along the river beds displays a distinct slaty cleavage which strikes NE–SW and dips 60–80° to the SE, but was observed to have gravitationally deformed to form buckling and toppling folds. DGSDs on overdip cataclinal and anaclinal slopes are subjected to buckling and toppling, respectively. Some of DGSDs of both types have led to deep-seated rockslides. Besides, sites with potential of future catastrophic landslides may be geomorphically identified at undercut slopes that have small head scarps upslope.

**Keywords**

Deep-seated gravitational slope deformation • Catastrophic landslide • Slaty cleavage • Toppling • Buckling

## Introduction

Deep-seated gravitational slope deformation, referred to as DGSD by Dramis and Sorriso-Valvo (1994), is a large-scale, voluminous, and long-lasting slope movement. DGSD has various types such as buckling, toppling, sliding, and dragging depending on the morphological relationship between bedding planes (foliation) and the slope (Chigira 2000). For example, toppling commonly develops when planar discontinuities, such as bedding, schistosity and cleavage, are vertical and subparallel to the slope. Buckling commonly occurs when planar discontinuities dip downslope more steeply than the slope angle. Thus, the deformation phase may be controlled by lithology and geologic structure.

The Central Range of Taiwan is widely underlain by Eocene to Miocene slate, which is brittle and easy to deform. Because of the fragile geological conditions and steep terrain, and weather conditions, many landslides were reported in this region, as could be seen, for example, after the 2009 Typhoon Morakot (Lin et al. 2011). Some are catastrophic cases occurred on slopes that had been preceded by DGSD beforehand in different geological conditions (Tsou et al. 2015; Chen et al. 2015). Therefore, DGSD may provide a clue in predicting potential sites of catastrophic landslides, and it must be interpreted on the basis of geological background.

The purpose of this study was therefore to clarify the geological features of landslide-prone areas in the Central

C.-Y. Tsou (✉)
Faculty of Agriculture and Life Science, Hirosaki University, 3 Bunkyo-cho, Hirosaki, Aomori, 036-8561, Japan
e-mail: tsou.chingying@hirosaki-u.ac.jp

M. Chigira
Disaster Prevention Research Institute, Kyoto University, Gokasho, Uji, Kyoto, 611-0011, Japan
e-mail: chigira@slope.dpri.kyoto-u.ac.jp

Y.-C. Hsieh · M.-M. Chen
Ministry of Economic Affairs, Central Geological Survey, No. 2, Lane 109, Huaxin Street, Zhonghe District, New Taipei City, 235, Taiwan
e-mail: hsiehyc@moeacgs.gov.tw

M.-M. Chen
e-mail: chenmm@moeacgs.gov.tw

T.-C. He
Department of Earth Science, National Cheng Kung University, No. 1, University Rd., Tainan, 701, Taiwan
e-mail: kikiroro@dprc.ncku.edu.tw


V. Vilímek et al. (eds.), *Understanding and Reducing Landslide Disaster Risk*,
ICL Contribution to Landslide Disaster Risk Reduction,
https://doi.org/10.1007/978-3-030-60319-9_19

Range. The study was conducted in a slate-rich area along the Lixing Industrial Road in the Beigang River catchment, central Taiwan (Fig. 1a), where DGSD and landslides commonly occur.

## Study Area and Methods

The study area is underlain by Lushan Formation of Miocene age with slate interlayered with thin metamorphosed sandstone (Lo and Yang 2002) (Fig. 1b). The geological structures in and around the study area are characterized by the NNE–SSW-trending Meihsi Fault and the Wushe Syncline, which run to the west and east of the study area, respectively. The beds on the west limb of the syncline strike NE–SW and dip 40° to the SE; the beds on the east limb strike NW–SE and dip 40° to the NW. The bedrock contains pervasive slaty cleavage which strike NE–SW dips 60° to 80° to the SE. The strikes of cleavage and bedding are approximately similar to the direction of the Lixing Industrial Road. Thus, many of the slopes along the road are cataclinal or anaclinal slopes, whereas attitudes of bedding and cleavage become the fundamental elements of DGSD.

We conducted field surveys to collect geological data, observe the geological structure, and investigate DGSD across the study area using a topographic map with a scale of 1:20,000. The topographic map was constructed from a 1 m digital elevation model (DEM) from LiDAR data published by Taiwan Central Geological Survey. A rock mass consists of deformed cleavage, brittle fractures and visible openings was assumed to be a DGSD, whereas fresh slate without the above features was assumed to be bedrock (Chigira 1992). DGSDs proceeded by buckling and toppling are observable in many places in the study area and 4 cases of them (Fig. 1a) will be described and discussed here along with geological structures. Others will be introduced in a separate paper.

## Results and Discussion

### Deformational Features (Buckling Type)

A deep-seated rockslide was investigated at Site B1 (Figs. 1a and 2). It is located on a SE-facing slope and the total fall height was 290 m from the top of the head scarp, at an

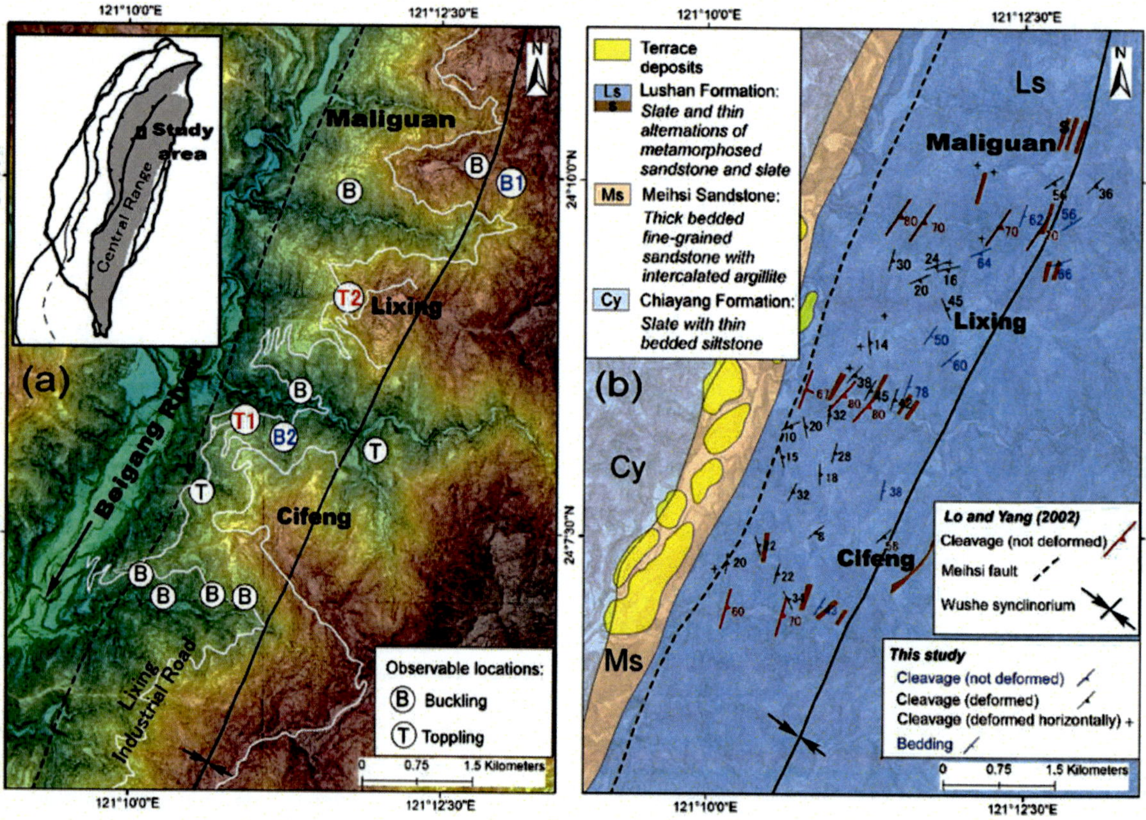


**Fig. 1** Maps of the study area. **a** A topographic map. B1 and B2 represent the example sites of buckling (see Figs. 2 and 3). T1 and T2 represent the example sites of toppling (see Figs. 4 and 5). T and B are observed sites of toppling and buckling, respectively. **b** Geologic map. B1 and T2 are cases of deep-seated rockslides preceded by DGSD. B2 and T1 are cases of DGSDs

elevation of 2030 m a.s.l., to the toe of the landslide deposit at 1740 m. Southwest of Site B1, at lower elevations, the slate bedrock is exposed with southeast-dipping slaty cleavage of 86°, suggesting Site B1 is at an underdip cataclinal slope. The steeply dipping cleavage had been gravitationally deformed to form buckling folds, which was observed on the lower right flank of Site B1 (Fig. 2b, c). The height of the deformed slate of the observed section is about 20 m, suggesting the landslide depth of >20 m. The observation suggests that the landslide was preceded by DGSD of buckling. In addition, the slope to the west of Site B1 shows a bulging topography (Fig. 2a), which may indicate that the subsurface geological structures have been deformed in the same manner.

Deformational feature of buckling type was also observed at Site B2, where slaty cleavage controls DGSD (Figs. 1a and 3). Slaty cleavage generally dips at 80° and has buckled, forming slope bulges and scarps at the tops. The geometry of the observed buckling fold is asymmetrical, with the downslope limb (dipping 76° NW) steeper than the upslope limb (dipping 30° SE) (Fig. 3c). The axial part of the fold was very fractured with many visible openings and consists of platy slate fragments that are parallel or oblique to the cleavage.

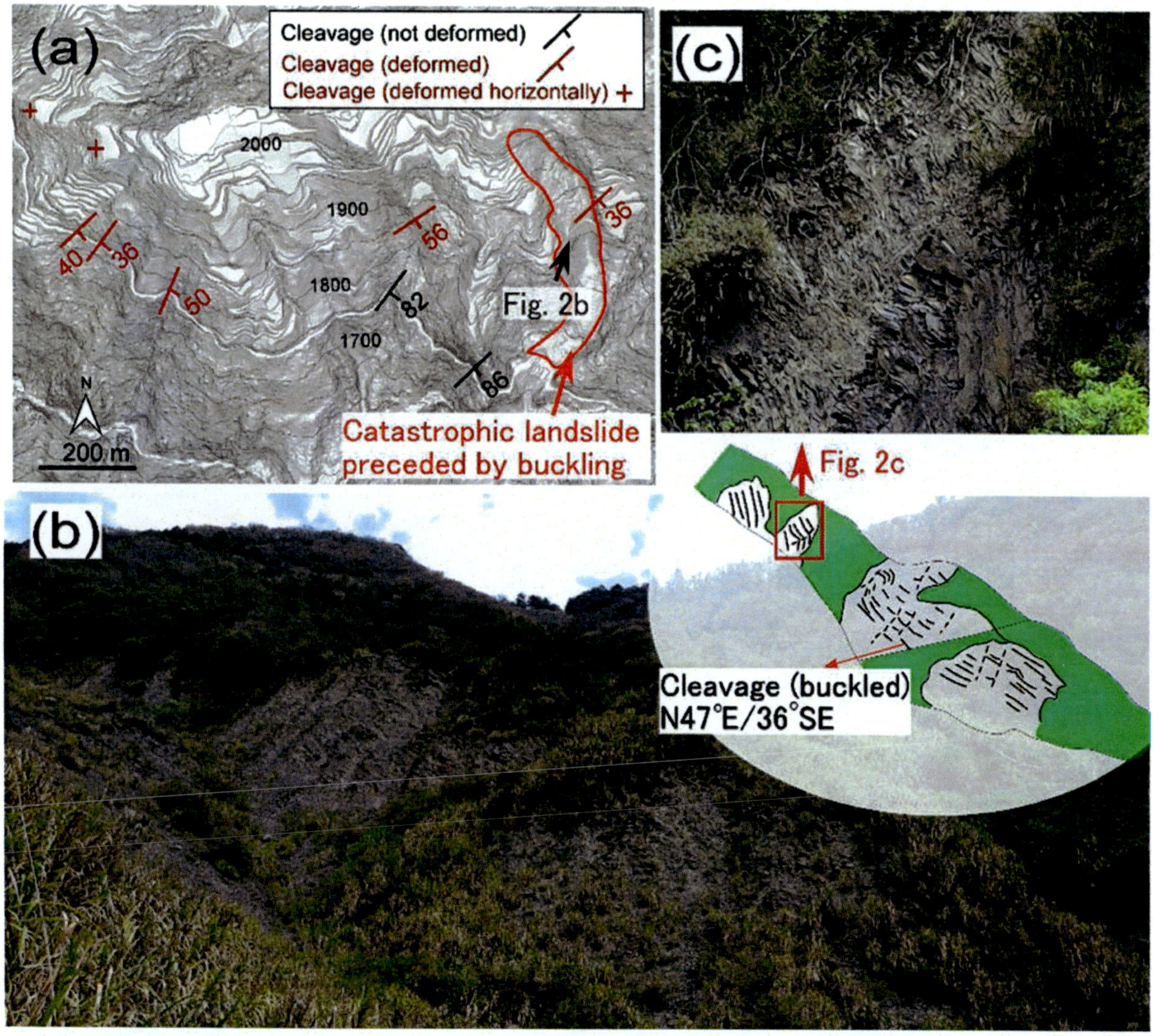


**Fig. 2** An example site (Site B1) of buckling. **a** A slope gradient image of Site B1 and surrounding slope. The lines are 50-m contours. **b** A close-up view of the lower right flank of the landslide and a sketch of the slope. **c** Gravitational deformation (a buckling fold) on the flank of (**b**)

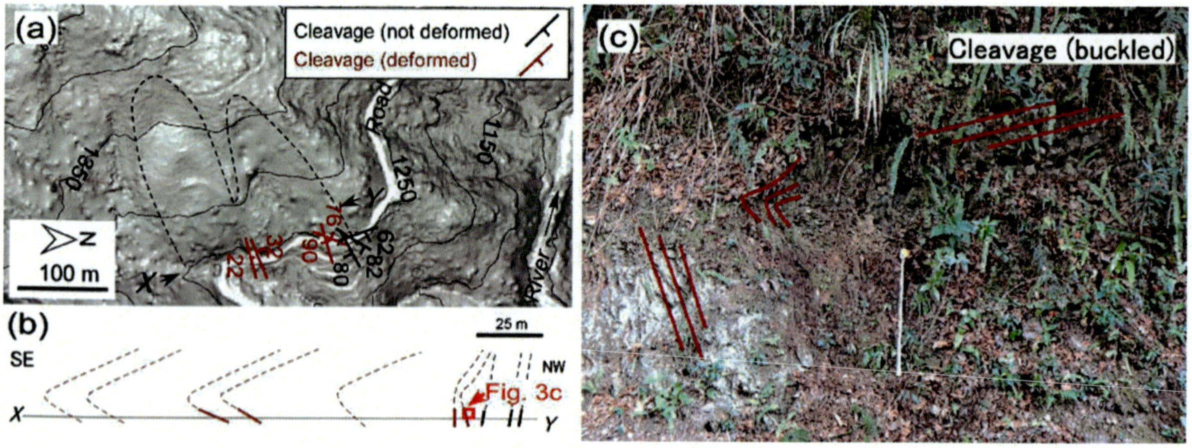


**Fig. 3** An example of buckling (Site B2). **a** A slope gradient image of Site B2. The lines are 50-m contours. **b** A geological cross section (XY, see Fig. 3a). **c** Gravitational deformation (a buckling fold) on the flank of (**b**)

## Deformational Features (Toppling Type)

Site T1 is located on a NE-facing slope with irregular shaped bulges and depressions (Figs. 1 and 4a). At lower flank, the slate bedrock has steeply dipping cleavage (dipping 80° SE), and slaty cleavage has toppled downslope, that can be observed at about 150 m above a tributary (Fig. 4a, b). The toppled cleavage can also be found at the lower left flank,

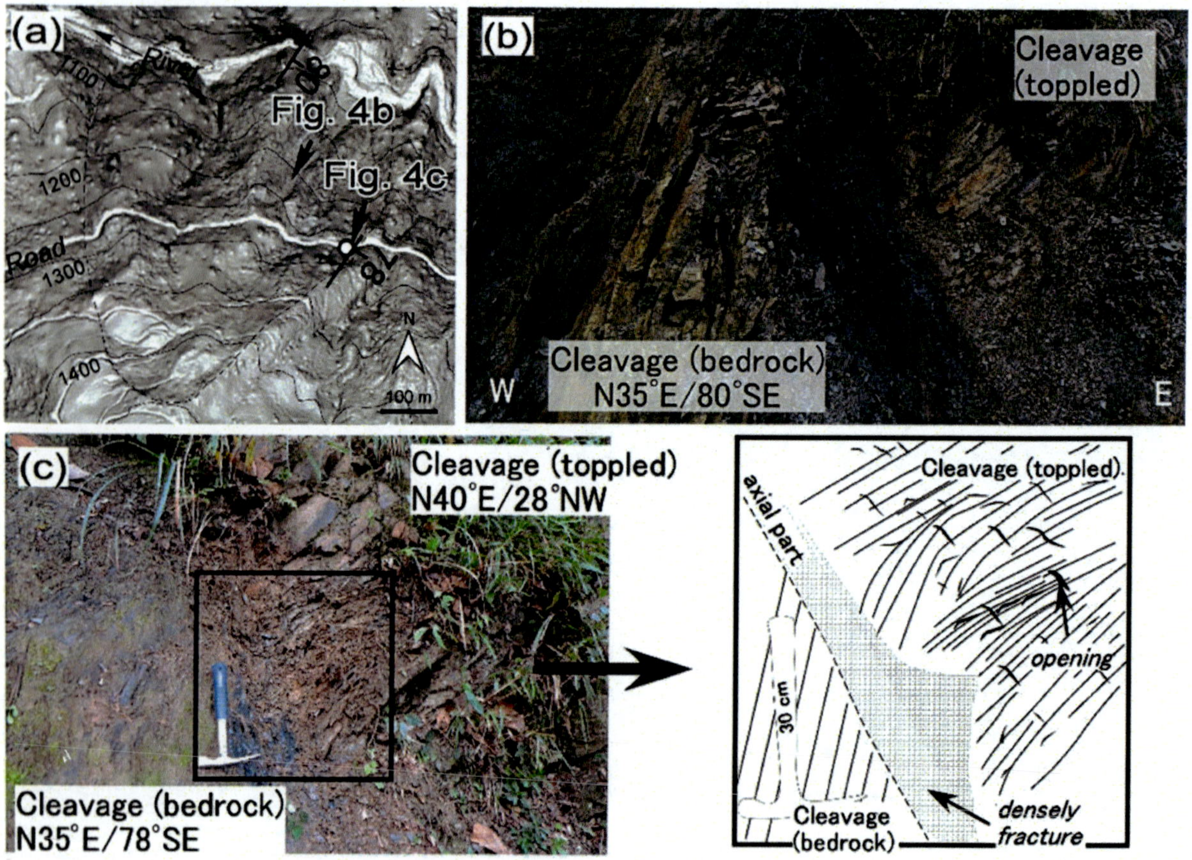


**Fig. 4** An example of toppling (Site T1). **a** A slope gradient image of Site T1. The lines are 50-m contours. **b** A close-up view of the lower flank of the landslide. **c** A close-up view of the lower left flank of the landslide and a sketch of the outcrop showing gravitational deformation of toppling

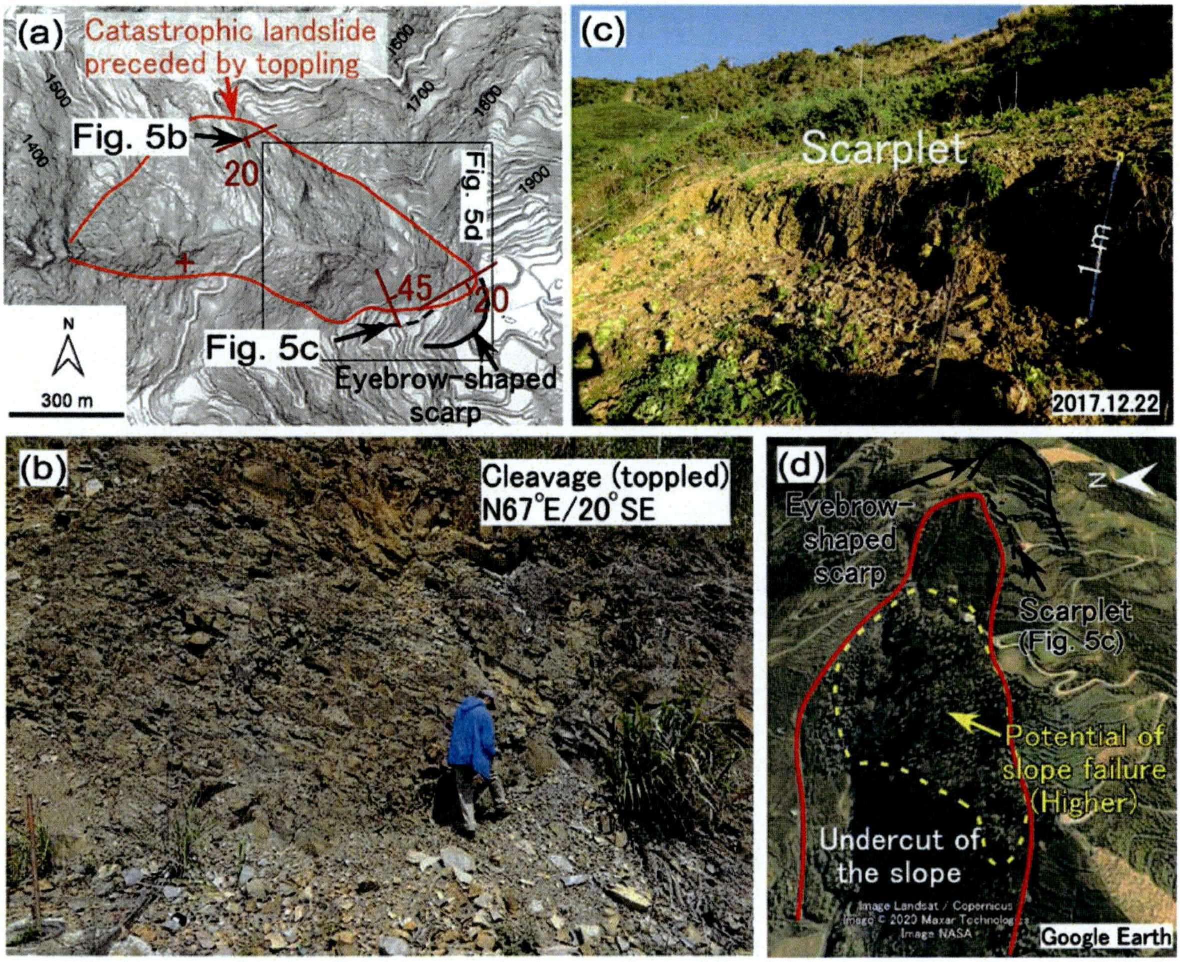


**Fig. 5** An example site of toppling (Site T2). **a** A slope gradient image of Site T2. The lines are 50-m contours. **b** A close-up view of the lower left flank of the landslide. **c** Fresh crack of about 1 m. **d** A close-up view of the source area of the landslide

dipping gently towards the upslope (dipping 28° NW) with bedrock beneath it (Fig. 4a, c). The deformation was accompanied by densely fracturing at the axial part, in contrast to the intact bedrock. The bedrock consists of dark, soft-weathering slate that the rock becomes easy to exfoliate (Fig. 4c), suggesting that the weathering may play an important role in the DGSD development.

Site T2, a 900 m-long and 400 m-wide deep-seated rockslide, is located on a W-facing slope facing to the Beigang River (Figs. 1 and 5a). The cleavage has toppled downslope in the landslide scar and within the deformed slope (Fig. 5b). The toppled cleavages were observed dipping horizontally and gently towards the upslope (dipping 20° SW) at the elevations of about 1550 and 1950 m a.s.l., respectively (Fig. 5a), suggesting the toppling mass was ca. 400 m deep. The deformation accompanied an eyebrow-shaped scarp along the ridge top (Fig. 5a), suggesting the formation of a continuous rupture surface separating the deformed mass and the underlying stable zone. Here the landslide had been preceded by toppling. Morphologically, Site 2 was failed within nearly complete coverage of the entire DGSD, there still remains some parts unstable (Fig. 5a). A scarplet up to 1 m high with a length of ca. 200 m in the plain (Fig. 5c) and numerous small cracks were visible beyond the landslide crown at a flat surface just below the eyebrow-shaped scarp, which appear to be newly formed after the landslide occurrence. These features could be the result of a secondary rupture surface developed within the deformation zone. In addition, toe of the toppled mass within upper part of the landslide scar has been undercut (Fig. 5d). This structure may result in the catastrophic failure of the displaced mass (Curden and Varnes 1992). Small head scarps upslope and undercut downslope suggest that potential of failure exists for additional landsliding on the same slope.

## Conclusions

To understand DGSD of slate slopes in the Central Range, Taiwan, we observed the geological structure in a slate-rich area along the Lixing Industrial Road and introduced 4 example sites in the Cifeng, Lixing, and Maliguan areas. The dominant geological structure is a distinct slaty cleavage which strikes NE–SW dips 60° to the SE. Steeply dipping cleavage controlled the DGSD development and was observed to have gravitationally deformed to toppling on anaclinal slopes and buckling on overdip cataclinal slopes. Both types of DGSD can let to deep-seated rockslides. In addition, DGSD with geomorphic features of eyebrow-shaped scarp upslope and undercut downslope may suggest a potential of catastrophic failure.

**Acknowledgements** This study was funded by Bilateral Programs of the Japan Society for the Promotion of Science and JSPS KAKENHI Grant Number 17H02973. We are indebted to Prof. Ching-Weei Lin of National Cheng Kung University, Prof. Rau-Fei Chen of Chinese Culture University, and Li-Yuan Fei, Hsi-Hung Lin, and Chung-Chi Chi of Central Geological Survey, Ministry of Economic Affairs, Taiwan for useful discussions. LiDAR DEM-derived maps were kindly supplied by Central Geological Survey, Ministry of Economic Affairs, Taiwan.

## References

Chigira M (1992) Long-term gravitational deformation of rocks by mass rock creep. Eng Geol 32(3):157–184

Chigira M (2000) Geological structures of large landslides in Japan. J Nepal Geol Soc 22(497–5):04

Chen RF, Lin CW, Chen YH et al (2015) Detecting and characterizing active thrust fault and deep-seated landslides in dense forest areas of Southern Taiwan using Airborne LiDAR DEM. Remote Sens 7 (11):15443–15466

Curden D M, Varnes D J, (1992) Landslides: Investigation and Mitigation. Turner A L, Shuster R L (eds). Transportation Research Board. (ISBN 030906208X). 673p.

Dramis F, Sorriso-Valvo M (1994) Deep-seated gravitational slope deformations, related landslides and tectonics. Eng Geol 38:231–243

Lin CW, Chang WS, Liu SH et al (2011) Landslides triggered by the 7 August 2009 Typhoon Morakot in southern Taiwan. Eng Geol 123:3–12

Lo W, Yang CN (2002) Explanatory text of the geologic map of Taiwan, Wushe sheet, scale 1/50, 000. Central Geological Survey, Ministry of Economic Affairs, Taiwan, p 59

Tsou CY, Chigira M, Matsushi Y, Chen SC (2015) Deep-seated gravitational deformation of mountain slopes caused by river incision in the Central Range, Taiwan: spatial distribution and geological characteristics. Eng Geol 196(28):126–138

# Bathymetric Analyses of Submarine Landslides on the Jan Mayen Ridge, Norwegian–Greenland Sea

2020 Kyoto Japan

Kiichiro Kawamura and Jan Sverre Laberg

**Abstract**

This paper considers the formation of submarine landslides on the Jan Mayen Ridge. A large submarine landslide, about 50 km wide, 40 km long and 500 m deep, has developed into an amphitheatre-shaped depression with subsequent retrogressive sliding along bedding planes to form secondary and tertiary landslide scars. This retrogressive process proceeded via small slope failures that generated erosive, confined turbidity currents. Consequently, long channels developed downslope of the secondary and tertiary landslide scars. Other small landslides are located on the deep-sea terraces at the foot of the southeast ridge. Internal normal faults on the terrace have generated an irregular bathymetry with many superimposed landslides. These small landslides formed on areas of relatively steep seafloor and are triggered by normal faulting.

**Keywords**

Retrogressive slide • Sliding from normal faulting • Channels • Turbidity currents • Norwegian sea

## Introduction

A large tsunamigenic submarine landslide, the Storegga Slide, occurred along the Atlantic continental margin of NW Europe in the Holocene (e.g., Bugge et al. 1987; Evans et al. 2005; Harbitz et al. 2006; Fig. 1). Along the northern Storegga Slide escarpment, a few minor slide/slump events have been dated to approximately 5700 and 2200–2800 cal. yrs BP, but the main Storegga Slide represents one retrogressive event that occurred at approximately 8100 ± 250 cal. yrs BP (Haflidason et al. 2005). Bondevik et al. (2005) showed that a tsunami of at least 20 m affected the Faeroe and Shetland islands following the Storegga Slide (Fig. 1a).

The triggering mechanism of these submarine landslides is closely related to the sedimentation history and/or tectonic history of the continental margin during the Quaternary (Owen et al. 2007). Evans et al. (1996) proposed a seismic trigger for the Storegga Slide. Bungum et al. (2005) argued that a recent detailed assessment of neotectonics in Fennoscandia revealed that several large-magnitude (>M7.0) earthquakes had occurred in northern Norway and northern Finland (Lapland) at that time. A possible explanation is that postglacial earthquakes developed on hidden thrusts (Bungum et al. 2005). Actually, the Storegga Slide is located on the Jan Mayen Fracture Zone (Fig. 1a). These recent studies suggested that submarine landslides might be the result of earthquakes induced by isostatic rebound during a postglacial period as discussed by Laberg and Vorren (2000) and Leynaud et al. (2009). The shaking generated by these earthquakes could have weakened the interbedded contouritic sediments that are commonly more porous than the intercalated glacial deposits and are considered to have lower strength (Bryn et al. 2005).

Several previous and recent possibly tsunamigenic submarine landslides have occurred along the Jan Mayen Ridge (Laberg et al. 2013; Fig. 1a, b), but the formation history of these landslides remains unknown. The trigger mechanism, sliding processes and history of these submarine landslides is possibly linked with those of the landslides on the NW European margin.

In this paper, we describe in detail the submarine landslides on the Jan Mayen Ridge using bathymetric and

K. Kawamura (✉)
Graduate School of Sciences and Technology for Innovation, Yamaguchi University, 1677-1 Yoshida, Yamaguchi, 753-8512, Japan
e-mail: kiichiro@yamaguchi-u.ac.jp

J. S. Laberg
Department of Geosciences, UiT The Arctic University of Norway, N-9037 Tromsø, Norway
e-mail: jan.laberg@uit.no



V. Vilímek et al. (eds.), *Understanding and Reducing Landslide Disaster Risk*, ICL Contribution to Landslide Disaster Risk Reduction,
https://doi.org/10.1007/978-3-030-60319-9_20

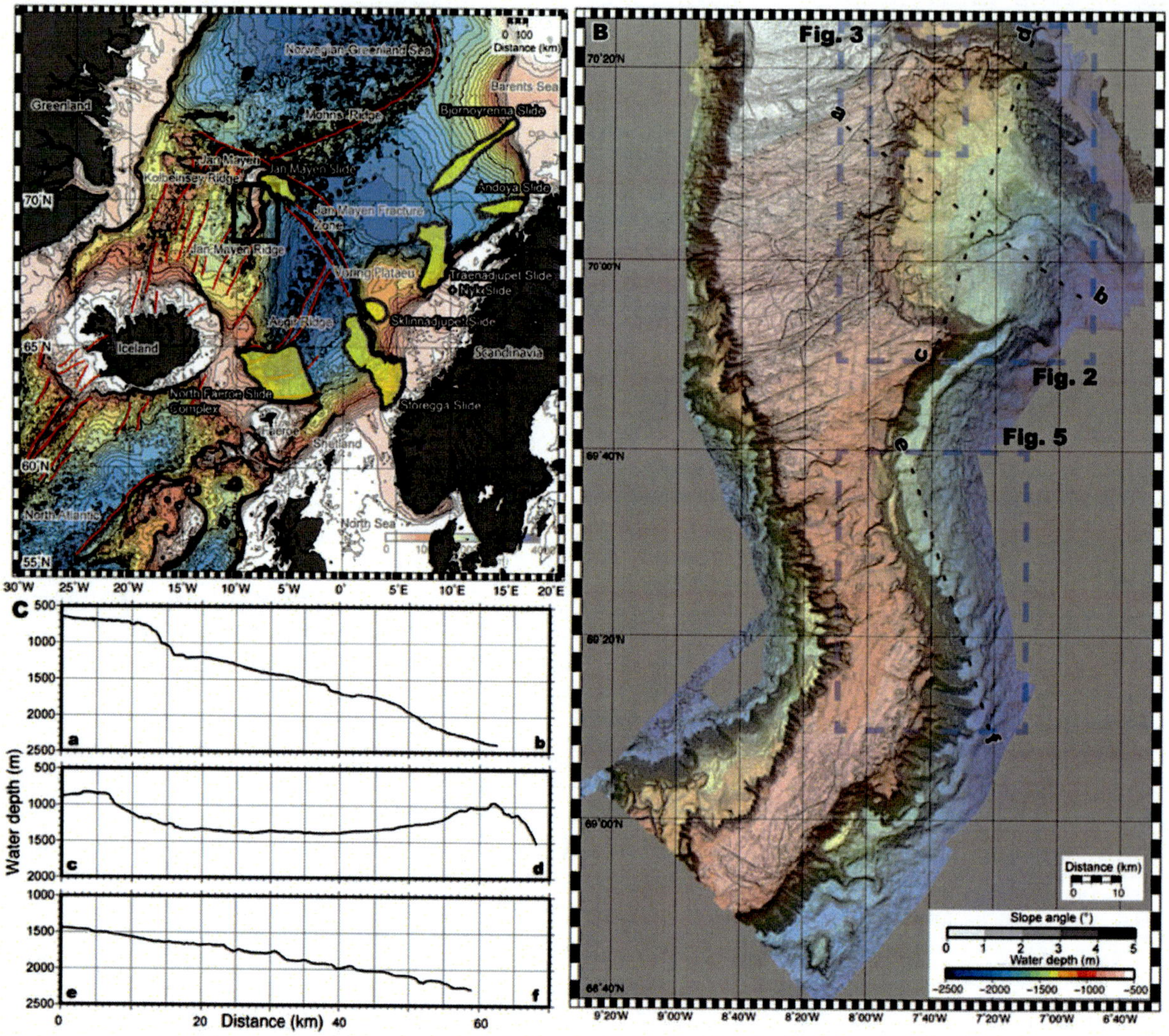


**Fig. 1** Detailed bathymetry. **a**: Bathymetry of the Norwegian–Greenland Sea. Black box outlines the Jan Mayen Ridge showing in Fig. 1b. **b**: Bathymetry of the Jan Mayen Ridge. Blue boxes indicate the location of Figs. 2, 3 and 4. Open circles indicate Deep Sea Drilling Project drill sites. **c** Bathymetric profiles along **a–b**, **c–d** and **e–f** showing in Fig. 1b

seismic data. We also discuss the formation processes of these submarine landslides.

## General Geology of the Norwegian–Greenland Sea

The Norwegian–Greenland Sea began to open with a pulse of breakup magmatism around 55–54 Ma in 24th reverse magnetic chron (hereafter C24 reverse) (Faleide et al. 2008) or 56.5–55 Ma in C25 normal–C24 reverse (Gernigon et al. 2012), during the early Eocene. The oceanic plate has since spread progressively along two spreading ridges. The Aegir Ridge (Fig. 1) was active from the breakup until around 28.5 Ma in C10, during the early Oligocene, when the spreading ridge switched to the Kolbeinsey Ridge (Fig. 1a) between C10 and C6. The Kolbeinsey Ridge began to open around 20 Ma in C6b (early Miocene), at which time the Jan Mayen Ridge (Fig. 1a) was separated from the east Greenland continental margin (Talwani and Eldholm 1977; Fig. 1a).

On the basis of the stratigraphic record of the lower Neogene, Stoker et al. (2005) proposed that compressional tectonism between the early and early-mid Miocene created an early Neogene syncline that became the Faroe Bank Channel (Fig. 1a). This channel connected the North Atlantic to the Nordic Seas and facilitated the formation of the present-day deep-water circulation system.

Furthermore, plate tectonic reconstructions indicate that the Fram Strait finally opened as a North Atlantic–Arctic Gateway between 20 and 10 Ma, during the Miocene (Jakobsson et al. 2007). Thus, the basal Neogene strata were deposited on top of the Oligocene strata, but the two are separated by a sharp unconformity caused by ocean current erosion (Stoker et al. 2005).

During the late Pliocene and Pleistocene epochs, glaciogenic sediments were widely deposited along the Norwegian continental shelf. These upper Neogene sediment layers collapsed following several large submarine landslides (e.g., the Storegga and Traenadjupet slides; Laberg et al. 2005; Fig. 1a) generated by a trigger associated with earthquake shaking and/or gas-hydrate decomposition (e.g., Laberg and Vorren 2000; Bryn et al. 2005; Laberg and Camerlenghi 2008).

## Characteristics of the Seabed Bathymetry Around the Jan Mayen Ridge

### Amphitheatre-Shaped Depression and Retrogressive Sliding Process

We analysed high-resolution bathymetric data obtained using a Kongsberg Simrad EM300 multi-beam echo sounder. The grid cell size is 50 × 50 m. The area around the Jan Mayen Ridge contains an amphitheatre-shaped depression

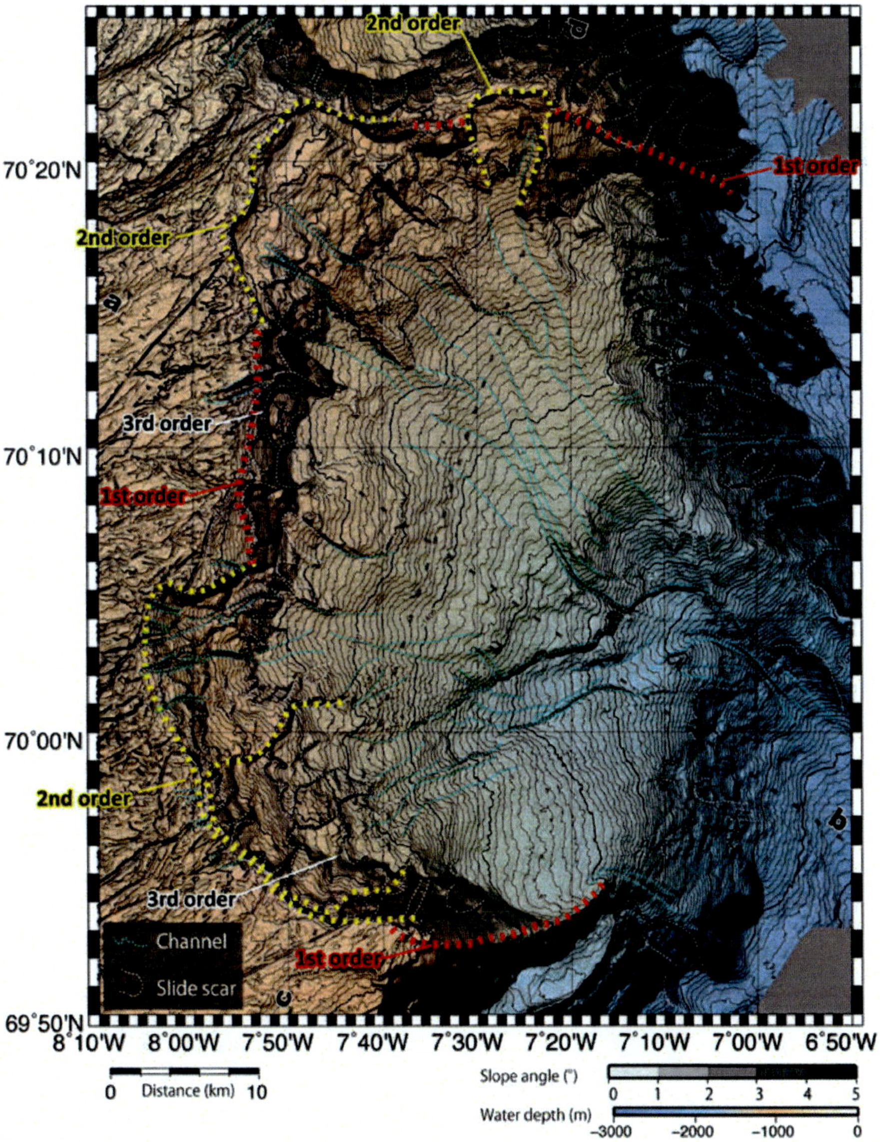


**Fig. 2** Detailed bathymetry of the amphitheatre-shaped depression. We identified first-, second- and third-order head scarps (red, yellow and white narrow dashed lines, respectively). Several long channels in the depression are connected to the second-order landslides, suggesting that turbidity currents continue to be generated by small collapses of third-order landslides on second-order head scarps. The current pathways of the third-order landslides merge with each other to build a long channel in the depression. Thus, the first-order landslide scarp is enlarged by the retrogressive small landslides. Dashed lines a–b and c–d indicate the locations of the bathymetric profiles shown in Fig. 1c

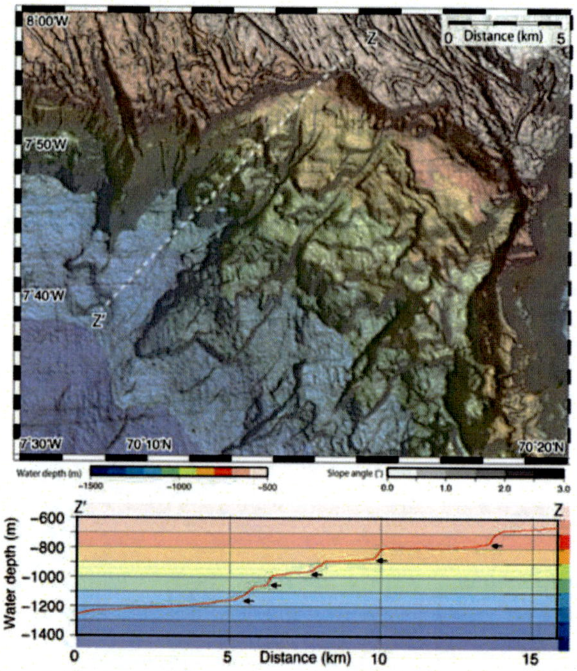


**Fig. 3** Detailed bathymetry of the second-order landslide scar (left) and its bathymetric profile along line Z–Z' (right)

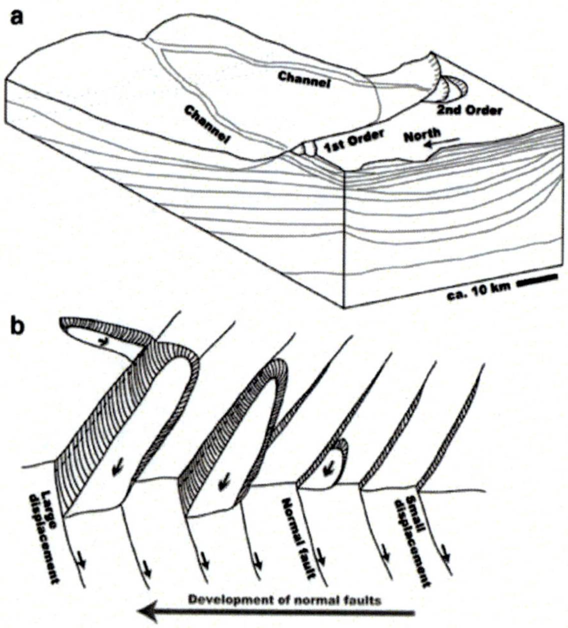


**Fig. 4** Schematic illustrations of **a** the formation process of the retrogressive slides and channels and **b** the irregularly shaped terrace and small submarine landslides

that is approximately 50 km wide, 40 km long and 500 m deep (Figs. 1b, c and 2), indicative of a large submarine landslide. The head scarp is about 500 m deep (Figs. 1b, c and 2). We have identified submarine landslides of at least three orders of magnitude. The first-order event corresponds to the outline of the amphitheatre-shaped depression, and the second-order is a landslide 10–20 km wide located within the first-order landslide scar (Figs. 2 and 3). The third-order features are small landslides a few kilometres wide, located within the scars of the first- and second-order events (Fig. 2).

Several channels, up to 30 km long, occur within the first-order landslide. These channels are connected with the second-order landslides (Fig. 2). The second-order slides contain step escarpments up to 100 m high. This result implies that the second-order landslides (and probably also the third-order slides) could have slipped down retrogressively along the layering of the sedimentary strata. This process would have widened the first-order scars as shown in Laberg et al. (2013), and also produced turbidity currents to form the long channels that incise the depression. Thus, the first-order scar would most likely have developed retrogressively (Fig. 4a).

Our analytical results suggest that it will be necessary to carefully revaluate our previous estimates of the volume of material associated with the specific slides that are the focus of this paper using bathymetric and seismic data. As shown in Løvholt et al. (2017), retrogressive sliding mechanisms are common in submarine landslides; consequently, the modelling of their associated tsunamis should incorporate the geological and sedimentological insights presented above.

## Irregularly Shaped Terrace and Related Small Landslides

The northern terrace has a smooth surface at almost the same water depth (Fig. 1b, c and 5), implying that the terraces might have been formed basically by recent bottom current systems.

However, the southern terraces have a unique bathymetry as an irregularly shaped terrace cut by lineations at intervals of approximately 4 km in a N–S to NE–SW direction (Figs. 1b, c and 5). These lineations are aligned sub-parallel to the lineations that were formed by the normal faults associated with the seafloor spreading at the Kolbeinsey and Aegir ridges, as shown by the red lines in Fig. 1a. We consider that this unique bathymetry was caused by the normal faults. Irregularly shaped terraces and related small landslides is closely related to the area of the normal faults, and small submarine landslides are present along the normal faults (Fig. 5).

Based on these observations, we suggest that the submarine sliding processes were controlled by the normal faulting as follows (Fig. 4a). Firstly, the normal faulting

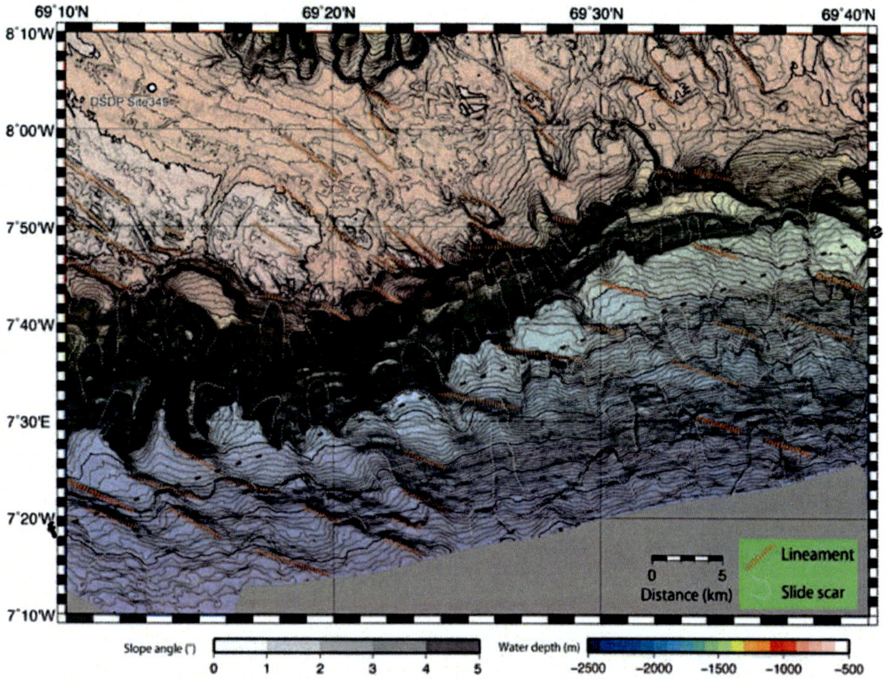


**Fig. 5** Detailed bathymetry of the irregularly shaped terrace. There are many small landslides on the southeast slopes and the deep-sea terrace. A dashed line e–f indicates the location of the bathymetric profile shown in Fig. 1c

built a step-like escarpment through rotation. At the foot of the normal faults, the sea floor then subsided. Secondly, the surface sediments appear to have been eroded along the normal fault cliffs, and the erosion progressed to form small valleys along the cliff. Finally, when the valley was extended and matured sufficiently, the valleys grew in the upslope direction. Later, small submarine landslides might have occurred in the valleys as they extended upslope.

## Conclusions

We summarize our results as follows.

(1) We studied a large submarine landslide characterized by an amphitheatre-shaped depression that was approximately 50 km wide, 40 km long and 500 m deep. The landslide developed into an amphitheatre-shaped depression as a result of retrogressive sliding that occurred along the layering structure and formed secondary landslide scars (second- and third-order landslides). This retrogressive process of small slope failures formed erosive, confined turbidity currents resulting in channels that developed downslope of the secondary landslide scars.

(2) We also studied other landslides that were located on deep-sea terraces at the foot of the southeast ridge. The terraces were formed by the bottom current systems and are cut by normal faults. Small submarine landslides formed on the terraces and are controlled by the normal faulting.

**Acknowledgments** We thank S. Kobayashi at Yamaguchi University for valuable input and discussions. We also acknowledge M. Sand at the Norwegian Petroleum Directorate for access to the multi-beam bathymetric data. We thank Dr. Hans-Balder Havenith and one anonymous reviewer for their constructive suggestions. This study was partly supported by the 2012 Exchange programme between Norway (The Research Council of Norway; RCN) and Japan (Japan Society for the Promotion of Science; JSPS; Program No. 219004).

## References

Bondevik S, Lovholt F, Harbitz C, Mangerud J, Dawson A, Svendsen JI (2005) The Storegga Slide tsunami——comparing field observations with numerical simulations. Mar Pet Geol 22:195–208

Bugge T, Befring S, Belderson RH, Eidvin T, Jansen E, Kenyon NH, Holtedahl H, Sejrup HP (1987) A giant three-stage submarine slide off Norway. Geo-Mar Lett 7:191–198

Bungum H, Lindholm C, Faleide JI (2005) Postglacial seismicity offshore mid-Norway with emphasis on spatio-temporal-magnitudal variations. Mar Pet Geol 22:137–148

Evans D, Harrison Z, Shannon PM, Laberg JS, Nielsen T, Ayers S, Holmes R, Hoult RJ, Lindberg B, Haflidason H, Long D, Kuijpers A, Andersen ES, Bryn P (2005) Paleoslides and other mass failures of Pliocene to Pleistocene age along the Atlantic continental margin of NW Europe. Mar Pet Geol 22:1131–1148

Evans D, King EL, Kenyon NH, Brett C, Wallis D (1996) Evidence for long-term instability in the Storegga Slide region off western Norway. Mar Geol 130:281–292

Gernigon L, Gaina C, Olesen O, Ball PJ, Peron-Pinvidic G, Yamazaki T (2012) The Norway Basin revisited: From continental breakup to spreading ridge extinction. Mar Pet Geol 35:1–19. https://doi.org/10.1016/j.marpetgeo.2012.02.015

Haflidason H, Lien R, Sejrup HP, Forsberg CF, Bryn P (2005) The dating and morphometry of the Storegga Slide. Mar Pet Geol 22:123–136

Harbitz CB, Lovholt F, Pedersen G, Masson DG (2006) Mechanisms of tsunami generation by submarine landslides: a short review. Norw J Geol 86:255–264

Jakobsson M, Backman J, Rudels B, Nycander J, Frank M, Mayer L, Jokat W, Sangiorgi F, O'Regan M, Brinkhuis H, King J, Moran K (2007) The early Miocene onset of a ventilated circulation regime in the Arctic Ocean. Nature 447:986–990. https://doi.org/10.1038/nature05924

Laberg JS, Camerlenghi A (2008) The significance of contourites for submarine slope stability. In: Rebesco M, Camerlenghi A (eds) Contourites, Development in Sedimentology 60. Elsevier Science, The Netherlands, pp 537–556

Laberg JS, Kawamura K, Amundsen H, Beaten N, Forwick M, Rydningen TA, Vorren TO (2013) A submarine landslide complex affecting the Jan Mayen ridge, Norwegian-Greenland Sea: slide scar morphology and processes of sediment evacuation. Geo-Mar Lett 34:51–58. https://doi.org/10.1007/s00367-013-0345-z

Laberg JS, Stoker MS, Dahlgren KIT, de Haas H, Haflidason H, Hjelstuen BO, Nielsen T, Shannon PM, Vorren TO, van Weering TCE, Ceramicola S (2005) Cenozoic alongslope processes and sedimentation on the NW European Atlantic margin. Mar Pet Geol 22:1069–1088. https://doi.org/10.1016/j.marpetgeo.2005.01.008

Laberg JS, Vorren TO (2000) The Traenadjupet Slide, offshore Norway —morphology, evacuation and triggering mechanisms. Mar Geol 171:95–114. https://doi.org/10.1016/S0025-3227(00)00112-2

Leynaud D, Mienert J, Vanneste M (2009) Submarine mass movements on glaciated and non–Glaciated European continental margins: A review of triggering mechanisms and preconditions to failure. Mar Pet Geol 26:618–632

Løvholt F, Bondevik S, Laberg JS, Kim J, Boylan N (2017) Some giant submarine landslides do not produce large tsunamis. Geophys Res Lett 44. https://doi.org/10.1002/2017GL074062

Owen M, Day S, Maslin M (2007) Late Pleistocene submarine mass movements: occurrence and causes. Quatern Sci Rev 26:958–978. https://doi.org/10.1016/j.quascirev.2006.12.011

Stoker MS, Hoult RJ, Nielsen T, Hjelstuen BO, Laberg JS, Shannon PM, Praeg D, Mathiesen A, van Weering TCE, McDonnell A (2005) Sedimentary and oceanographic responses to early Neogene compression on the NW European margin. Mar Pet Geol 22:1031–1044. https://doi.org/10.1016/j.marpetgeo.2005.01.009

Talwani M, Eldholm O (1977) Evolution of the Norwegian-Greenland Sea. Geol Soc Am Bull 88:969–999

# Forkastningsfjellet Rock Slide, Spitsbergen: State of Activity in a Changing Climate

2020 Kyoto Japan

Dirk Kuhn, Reginald L. Hermanns, Jewgenij Torizin, Michael Fuchs, Tim Redfield, Raymond Eilertsen, and Dirk Balzer

**Abstract**

The coastal ridge of Forkastningsfjellet comprises a $\sim$100 million $m^3$ rock rotational slide, which occurred in the hangingwall of a listric, northwest-dipping slide surface, probably in postglacial times. Active mass wasting and seacliff erosion, mainly controlled by the inherent discontinuities of the fractured and tilted rock masses, currently take place along the steep slopes of the coastal tilt blocks. A catastrophic reactivation of such slide blocks as very rapid rock slide could have severe consequences for the surrounding coastal regions and for Longyearbyen, the capital of Spitsbergen, which is situated nearby. This paper introduces the present knowledge about the rock slide and describes the current multidisciplinary investigation program that is conducted to better understand the nature and state of activity of the rock slide. The investigation is necessary to improve the hazard assessment of Forkastningsfjellet rock slide, which also serves as a showcase for the impact of climate change on landslide activity in the Longyearbyen region.

**Keywords**

Forkastningsfjellet rock slide • Spitsbergen • Natural hazard • Climate change

## Introduction

Climate change and the related permafrost degradation are thought to influence slope stability, landscape evolution, and the natural hazard potential in polar- and high mountain regions. On Svalbard, projections of climate change scenarios predict increased annual temperatures and rainfall, permafrost thaw and increased erosion and sediment transport, among others. As such meteorologic factors are important triggers for all types of mass movements, an increase of landslide activity is expectable (Hanssen-Bauer et al. 2019).

In this context, we investigate a large-scale rock slide affecting the coastal ridge of Forkastningsfjellet, situated at the eastern margin of Isfjorden. The rock slide is characterized by a number of rock slope instabilities of different kinematics, volumes and age.

Based on the affected rock volume, these mass movements can be divided into three classes. First, small scale translational landslides, so called "active layer detachment slides" (ALD), and debris flows affect the uppermost surface layer and occur in thawing soil overlying the permafrost. More obvious are coastal cliff failures, which comprise translational and rotational landslides affecting volumes in the order of $10^3$–$10^5$ $m^3$. The largest mass movement mapped so far is the huge Forkastningsfjellet rock slide itself that affects the entire mountain range and involves a displaced rock volume of > 0.1 $km^3$ (Kuhn et al. 2019). While the first two categories are active, ongoing processes that shape the slope morphologies and coast line, the state of activity of the rock slide is yet unknown. Such large- scale rock slope failures entering a fjord might generate large displacement

D. Kuhn (✉) · J. Torizin · M. Fuchs · D. Balzer
Federal Institute for Geosciences and Natural Resources, Stilleweg 2, 30655 Hannover, Germany
e-mail: dirk.kuhn@bgr.de

J. Torizin
e-mail: jewgenij.torizin@bgr.de

M. Fuchs
e-mail: michael.fuchs@bgr.de

D. Balzer
e-mail: dirk.balzer@bgr.de

R. L. Hermanns · T. Redfield · R. Eilertsen
Geological Survey of Norway, Torgarden, 6315 N–7491 Trondheim, Norway
e-mail: reginald.hermanns@ngu.no

T. Redfield
e-mail: tim.redfield@ngu.no

R. Eilertsen
e-mail: raymond.eilertsen@ngu.no


V. Vilímek et al. (eds.), *Understanding and Reducing Landslide Disaster Risk*,
ICL Contribution to Landslide Disaster Risk Reduction,
https://doi.org/10.1007/978-3-030-60319-9_21

waves that pose a threat to the marine traffic and inhabited surroundings in case that they are catastrophic (Hermanns et al. 2013; Gauthier et al. 2018).

Therefore, the Forkastningsfjellet rock slide serves as a pilot study area for the temporal development of an unstable rock slope in a changing climate. Moreover, the situation in the vicinity of Longyearbyen calls for a detailed investigation of the hazard potential in order to support hazard assessment.

In this paper, we present the results of a multi-disciplinary investigation of the Forkastningsfjellet rock slide, focussing on the active mass movements and specifically, the reactivation potential of large slide blocks that might have a decisive impact on hazard assessments.

## Location and Geological Setting

Forkastningsfjellet is a narrow, NE–SW-oriented mountain ridge to the north of Longyearbyen, the capital of Svalbard (Fig. 1).

The coastal mountain range reaches a maximum altitude of 490 m and is characterised by a rugged morphology, steep cliffs and a multitude of structural elements of tectonic origin. The area is characterized by the well-bedded sedimentary rocks of the Upper Jurassic to Cretaceous Adventdalen Group (Parker 1967; Dypvik et al. 1991; Dallmann et al. 2001). The sequence was differentiated into the lower Janusfjellet subgroup comprising mudstones and shales of the Agardhfjellet Formation (Jurassic) and Rurikfjellet Formation (Early Cretaceous). It is overlain by the deltaic Helvetiafjellet Formation with its lower cliff-forming sandstone unit of the Festningen Member. The latter is overlain by the heterolithic Carolinefjellet Formation comprising alternating thin sandstones and mudstones (Parker 1967).

At Forkastningsfjellet, the sequence was affected by two different deformation phases that formed the dominant morphostructural features of the ridge (Dallmann et al. 2001; Albertsen 2016; Kuhn et al. 2019). During the first deformation phase in the Eocene, the rocks were affected by thin-skinned deformation leading to NE- and SW-vergent folding and thrust faulting that is attributed to the development of the transpressional West Spitsbergen Fold-and-Thrust Belt along the western coast of Spitsbergen (Bergh et al. 1997; Tessensohn and Piepjohn 2000; Dallmann et al. 2001). The second deformation phase was a postglacial,

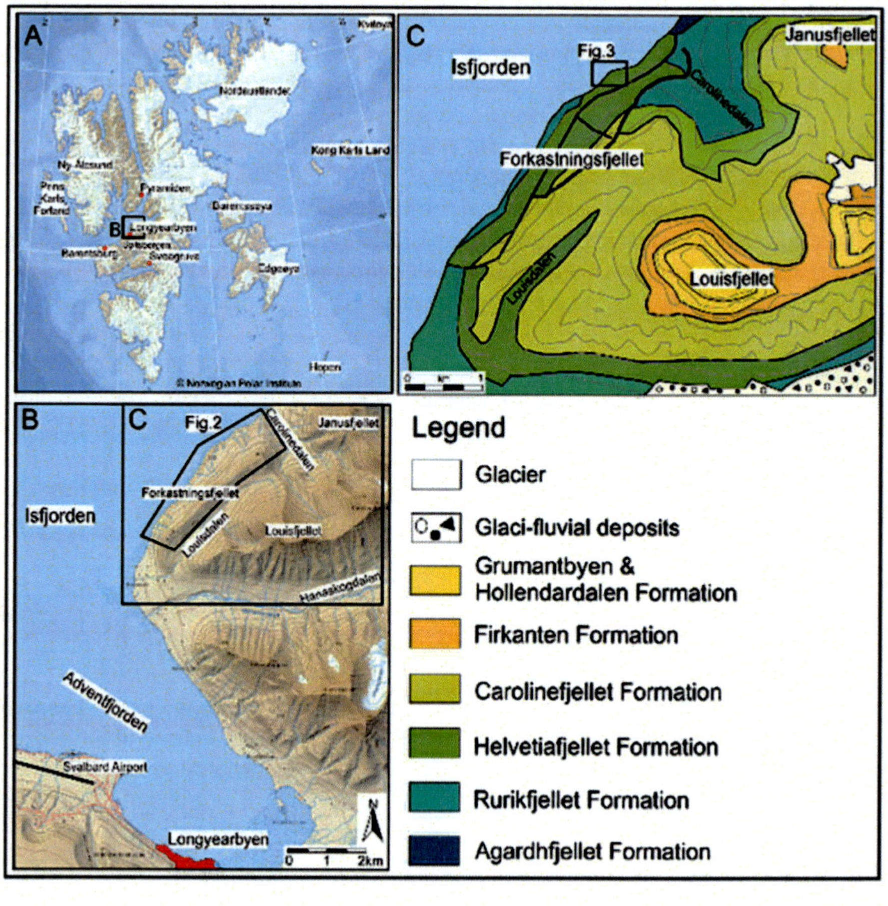


**Fig. 1** **a** Overview map of Svalbard, **b** regional over-view map of the working area, **c** geological map of the Forkastningsfjellet region (Norsk Polarinstitutt 2018)

gravitationally driven extensional deformation, initiated by a huge rock slide that led to the exceptional dissection of the coastal ridge (Kuhn et al. 2019).

## Methods

Detailed geomorphological and structural field mapping was conducted using an existing geological map at a 1:100,000 scale (Sheet C9G Adventdalen; Dallmann et al. 2001) and a digital elevation model (DEM) with a grid size of 5 m 5 m. Based on photo interpretation and DEM data, the cliff-forming marker horizon of the Festningen Member was mapped and used to construct cross-sections and estimate displacements in inaccessible or unstable terrain. To get better information about the timing of the rock slide, six rock samples of quarzitic sandstones from large boulders and the main scarp were taken to date the surface exposure based on cosmogenic nuclide dating of $^{10}$Be.

To monitor potential block movements, new dGPS stations were installed and an initial measurement was performed.

Offshore bathymetry adjacent to Forkastningsfjellet was surveyed with a 125 kHz and 250 kHz GeoSwath Sidescan Sonar system of GeoAcoustics Ltd (Geoacoustics 2020) that was attached to an expedition vessel.

During a flight campaign with an unmanned airborne vehicle (UAV) in summer 2019, high-resolution stereo images were acquired. We deployed a DJI Phantom 4 RTK drone and the images were processed with the DJI Terra Pro software (DJI 2019). The resulting DEM has a spatial resolution of 1 m.

## Forkastningsfjellet Rock Slide

### Morphostructural Description

The Forkastningsfjellet rock slide has been presented in detail by Kuhn et al. (2019); therefore only a summary is given here. The rock slide displays a rugged, stepped topography with steep scarps, plateaus and rotated fault blocks, dominated by the main scarp dipping to the NW. After failure initiation along a main rupture surface, the rock mass slid downward into the Isfjorden and disintegrated into 17 discrete slide/tilt blocks along transverse and en echelon cracks that developed into secondary sliding planes (Fig. 2).

With a measured footprint of >2.03 km$^2$ and an affected rock volume of >0.10 km$^3$ the rock slide is one of the largest described slope instabilities on Svalbard. Such large scale rock-slope failures are known as a principal type of mass movement in Norway and the Arctic region and have been termed as 'listric style complex field' type (Braathen et al. 2004). Ongoing investigations indicate that the rock slide might extend even farther to the North into the western slopes of Janusfjellet.

### Type, Kinematics and Age of the Rock Slide

The resulting complex landscape and deformation pattern are assigned to a multiple, rock rotational slide (Hungr et al. 2014) that occurred on a major west-dipping cylindrical rupture surface.

This listric sliding surface produces the steep major fault scarp at the eastern limit of the rock slide and probably flattens downward into a low-angle detachment zone situated in the flat-lying bedding planes of the less competent shales of the lower Rurikfjellet/Adventdalen Group sediments (Fig. 2, inset). The depth of the basal décollement is unknown but based on geometric considerations deduced from differential displacements of the rotated hangingwall blocks, it should daylight in the west-dipping offshore slope of Isfjorden. A minimal depth of 60 m below sea level is assumed. The initial failure was most likely related to the deglaciation of Isfjorden and the resulting instability of the weakened rock mass along the oversteepened slopes.

## Current State of Activity

Based on morphologic features and the horizontal extension, the rock slide can be differentiated into a northern, central, and southern segment. Currently, all segments are affected by active mass wasting and seacliff erosion that takes place along the steep slopes of the coastal tilt blocks (Fig. 3). In particular, a substantial increase in superficial active-layer detachments (ALD) and debris flows are observed. In contrast, uncertainty exists concerning the reactivation potential of large tilt blocks, the kinematics of potential movement (creep versus high-speed sliding), and triggering factors.

In this context, the potential activity of Block 5, situated in the northern segment, is an investigation priority (Fig. 4). Block 5 is the northernmost of a coastal block gallery with three adjoining smaller blocks 6, 7 and 8 to the south. It shows a stepped surface indicating active downslope displacement. A deeply scored block fault delimits Block 5 to the north; to the south, a clearly developed escarpment is assumed as the surface expression of a potential sliding plane. The stepped alignment of the adjoining blocks is assigned to an increasing vertical displacement from north to south, when the neighboring blocks 6, 7, and 8 slid down along a secondary sliding plane. While blocks 6, 7 and 8 appear to have reached a stable position, Block 5 is still

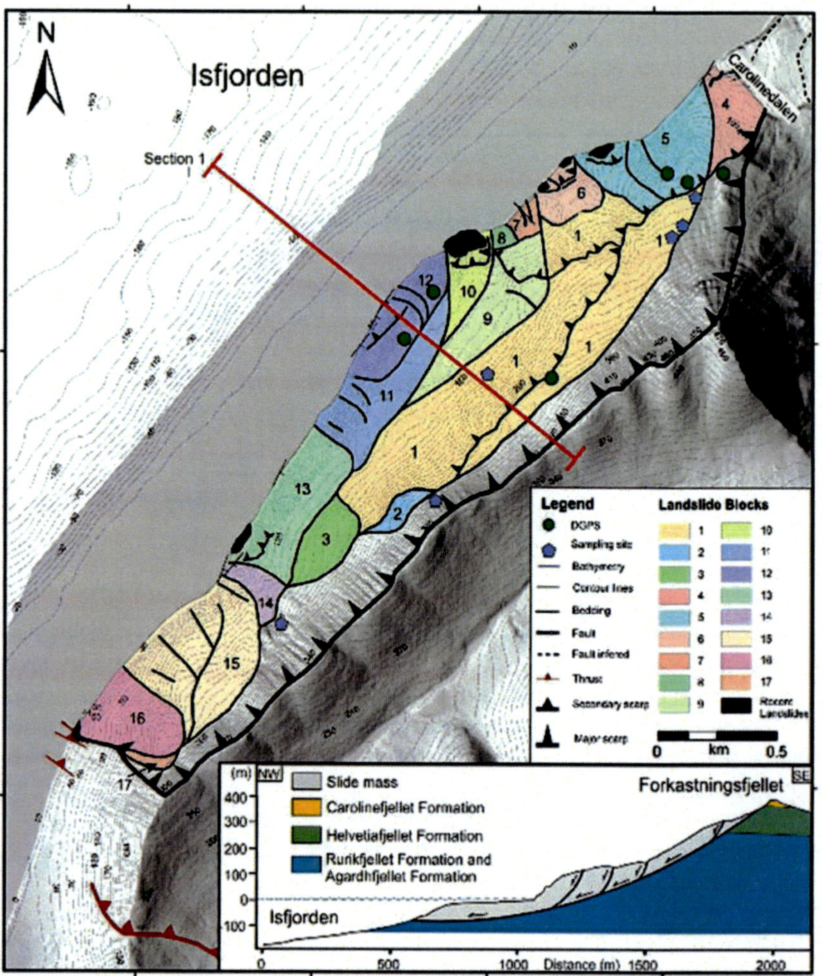


**Fig. 2** Morphostructural map of Forkastningsfjellet, showing main structures and numbered blocks of the rock slide. Inset shows NW–SE oriented cross-Sect. 1

**Fig. 3** Rock slope failure along adversely, NE-SW oriented fracture planes as a consequence of basal sea-cliff erosion and slope oversteepening

attached to northern prolongation of this secondary scarp/sliding plane. Block 5 is therefore assumed to be in a metastable condition and calls for further investigation.

## Follow Up Investigations

Due to the unknown state of activity of the large sliding blocks and its relevance to a hazard assessment, a detailed mapping and monitoring program is required. Therefore, a further investigation and survey program was established. This will also help to assess the impact of the warming climate on the different types of mass movements (number and affected volume) and will be used to quantify the mass transfer (mass balancing).

## Differential GPS Monitoring

To monitor potential horizontal and vertical displacements of the slide blocks, seven GPS stations were installed in the

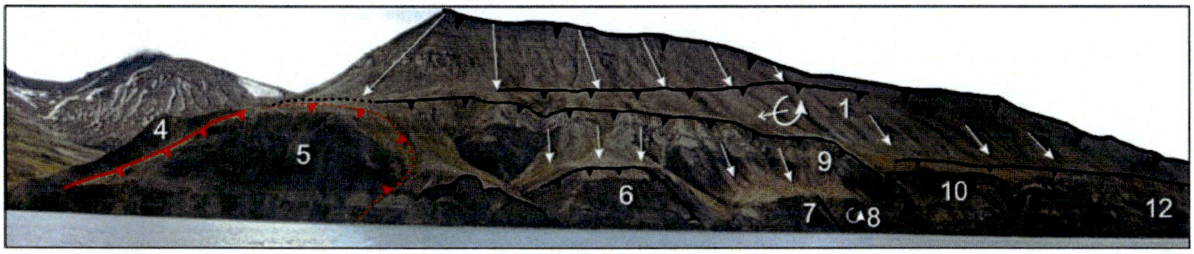


**Fig. 4** The northern segment of Forkastningsfjellet rock slide with the coastal Blocks 5–8, view to the east. In red, the trend of the potential sliding plane affecting Block 5

northern and central segments of the rock slide (Figs. 2, 5). Three of these stations were installed on the surface of Block 5 and the vicinity towards the east.

The network includes a reference station that was installed in stable bedrock outside and to the south of the Forkastningsfjellet rock slide. Repeated annual dGPS surveys will be performed during the upcoming summer field campaigns to evaluate potential surface deformations and related displacements of the slide blocks.

## Bathymetry

A bathymetric survey along the Forkastningsfjellet coastline was performed in order to study the shape of underwater features, evaluate the existence of further rock slide deposits in the offshore area and the maximum transport distance of the rock slide.

Parallel courses were driven along the shore with a GeoSwath Sidescan Sonar mounted beneath an expedition vessel. The resulting bathymetric map shows an undulating slope and a salient in front of the central segment that could be interpreted as basement high or a large slump (Fig. 6). Also, small slides and debris flows are visible to the north of the salient, but larger rock debris that might be linked with the rock slide were not identified.

**Fig. 5** Installed GPS Station to the east of Block 5, northern segment of the Forkastningsfjellet rock slide

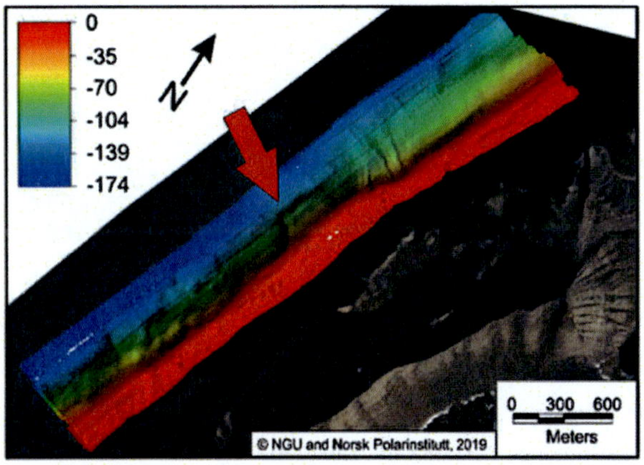


**Fig. 6** Bathymetric map of the coastal fringe of the rock slide. A salient (red arrow) of the slope and debris flows to the north are clearly visible

## UAV-Based Remote Sensing

Based on an UAV-based aerial survey, a high resolution DEM was elaborated (Fig. 7). The DEM covers the entire rock slide area (4.45 $km^2$) and will be used for site specific geotechnical modelling like stability analysis and run out modelling. In combination with geomorphological field mapping, this DEM is also used to map and quantify ALD and debris flows. This initial inventory will be successively extended by the annual UAV-surveys that are planned for next years field campaigns. Based on this inventory, an annual change detection will be performed and mass transport along the Forkastningsfjellet rock slide will be quantified.

## Discussion

Two distinct phases characterize the structural evolution of Forkastningsfjellet. Thin-skinned tectonics due to NE-SW shortening during the Eurekan deformation phase lead to NW–SE trending folds and thrusts. In addition, tectonic thickening and an intensive joint system with NE-SW

**Fig. 7** UAV-based DEM of the Forkastningsfjellet rock slide (ground resolution 1 × 1 m). Annual repeat surveys will be used for change detection (inset), mapping of surficial ALD's and the quantification of the mass transport

oriented fractures affected the incompetent weak mudstones and shales of the Jurassic and Early Cretaceous sequence.

In contrast, the recent tilt-block-structure of Forkastningsfjellet and the steep NE–SW-oriented releasing extensional structures that served as sliding surfaces have been assigned to rock slide activity in postglacial times (Major and Nagy 1972; Albertsen 2016). It is reasonable to assume that the inherited structural precondition played an essential role in the deformation history of Forkastningsfjellet.

This observation is also important when the landslide hazard is considered. While the majority of the mapped mass movements are of limited areal extend and involve only small volumes, the reactivation of a slide block along predefined extensional structures clearly would exert a major hazard to the surroundings.

In this connection, the global warming in the Arctic plays an important role. Projected scenarios for the Svalbard region expect an enhancement of geomorphic activity, landslide activity and mass transfer (De Haas et al. 2015; Hanssen-Bauer et al. 2019). Permafrost degradation because of increased summer temperatures and precipitation will probably enhance permafrost-controlled subaerial processes such as freeze and thaw cycles that further weaken the rock mass through weathering and seasonal crack propagation and widening. Consequently, a deeper infiltration of the melting water along the widening fractures and joints of the bedrock could lead to an increasing outward- directed joint water pressure and a reduced shear resistance on the potential sliding surfaces (Gruber and Haeberli 2007; McColl 2012; Krautblatter et al. 2013; Blikra and Christiansen 2014). Therefore, an increase of landslide frequency and a reactivation of block bounding sliding planes is likely at Forkastningsfjellet.

Based on the presented results, a long-term investigation program is warranted to improve the understanding of the past and present day rock slide behaviour and support a reliable hazard assessment. Thereby, the Forkastningsfjellet rock slide will serve as a showcase for a spatio-temporal assessment of multiscale mass movements (number and affected volume) in a changing climate.

## Conclusions

Mass movements of different type and magnitude are currently the main factors that control mass transport and shape the slope morphology of the Forkastningsfjellet rock slide. An increasing number of ALD and debris flows indicate a correlation with climate- controlled factors (e.g., permafrost thaw and frost-heave, meltwater infiltration).

Concerning the hazard potential, the shallow ALD and the sea cliff erosion along the coastal tilt blocks are only local phenomena. In contrast, a sliding reactivation of a large slide block (e.g., Block 5 in Fig. 4) and a related displacement wave might affect a larger area and could even reach the port of Longyearbyen. Therefore, a long-term monitoring program is necessary to unravel the nature and activity of the coastal slide blocks and the hazard potential of the

Forkastningsfjellet rock slide as a whole. Also, this monitoring program will help to model different climate-related mass movement processes and assess the sediment budget of the Forkastningsfjellet catchment in a warming climate.

**Acknowledgements** The authors wish to thank BGR and NGU for funding of the fieldwork. John Dahl is thanked for supporting the sonar installation. Terrapolaris and Hennigsen Transport and Guiding along with affiliated wildlife guards are thanked for supporting the field campaigns. The comments of an unknown reviewer on an earlier version helped to improve this paper.

## References

Albertsen A (2016) Cenozoic compressional tectonics and post-glacial landsliding at Forkastningsfjellet, Spitsbergen, Svalbard. MS thesis, Department of Geosciences Faculty of mathematics and natural sciences, University of Oslo, Norway

Bergh SG, Braathen A, Andresen A (1997) Interaction of basement-involved and thin-skinned tectonism in the Tertiary fold thrust belt of central Spitsbergen, Svalbard. Am Asso Petrol Geol Bull 81:637–661

Blikra LH, Christiansen HH (2014) A field-based model of permafrost-controlled rock slide deformation in northern Norway. Geomorphology 208:34–39

Braathen A, Blikra LH, Berg SS, Karlsen F (2004) Rock-slope failures in Norway; type, geometry, deformation mechanisms and stability. Norw J Geol 84:67–88

Dallmann W K, Kjaernet T, Nottvedt A, (2001) Geological map of Svalbard 1:100,000, Sheet C9G Adventdalen, Norsk Polarinstitutt

DJI Terra (2020) https://www.dji.com/de/dji-terra. Last accessed 06 Mar 2020

Dypvik H, Nagy J, Eikeland TA, Backer OK, Johansen H (1991) The depositional conditions of the Bathonian to Hauterivian Janusfjellet Subgroup. Spitsbergen; a Revised Lithostratigraphy. Sediment Geol 72:55–78

Gauthier D, Anderson S A, Fritz H M, Giachetti T, (2018) Karrat Fjord (Greenland) tsunamigenic landslide of 17 June 2017: initial 3D observations. Landslides 15:327–332

Gruber S, Haeberli W, (2007) Permafrost in steep bedrock slopes and its temperature-related destabilization following climate change. J Geophys Res 112:F02S18, 10 p

Hermanns R L, L'Heureux J-S, Blikra L.H, (2013) Landslide triggered tsunami, displacement wave. Bobrowsky, P.T. (ed.). Encyclopedia of natural hazards, Springer, Berlin, pp 611–615

Hungr O, Lerouei S, Picarelli L (2014) The Varnes classification of landslide types, an update. Landslides 11:167–194

Kongsberg GeoAcoustics Ltd, (2020). URL: https://geoacousticsltd.business.site/. Last accessed 6 Mar 2020

Krautblatter M, Funk D, Günzel FK (2013) Why permafrost rocks become unstable: a rock–ice-mechanical model in time and space. Earth Surf Proc Land 38:876–887

Kuhn D, Redfield TF, Hermanns RL, Fuchs M, Torizin J, Balzer D (2019) Anatomy of a mega-rock slide at Forkastningsfjellet, Spitsbergen and its implications for landslide hazard and risk considerations. Norw J Geol 99:41–61

Major H, Nagy J (1972) Geology of the Adventdalen map area. Skrifter 138, Norsk Polarinstitutt.

McColl ST (2012) Paraglacial rock-slope stability. Geomorphology 153–154:1–16

Norsk Polarinstitutt (2018) Geological maps of Svalbard. https://svalbardkartet.npolar.no/html5/index.html?viewer=svalbardkartet.html5. Accessed 5 May 2018

Parker JR (1967) The Jurassic and Cretaceous sequence in Spitsbergen. Geol Mag 104:487–505

Tessensohn F, Piepjohn K, (2000) Eocene compressive deformation in Arctic Canada and Svalbard and its plate tectonic causes. Roland N W, Tessensohn F, (eds.). ICAM III—III. In: International conference on arctic margins I, Polarforschung, vol 68, pp 121–124

# Catastrophic Landslides in Indian Sector of Himalaya

Vinod K. Sharma

**Abstract**

The Himalayan geo-environment has been experiencing slope instability hazards of catastrophic dimensions in historical past. Large rocks and debris avalanche have been recognized largely in higher domains in easternmost extremity covering Sikkim and Arunachal and Northwest Himalaya encompassing Kumaun, Garhwal, Himachal, Jammu and Kashmir and Ladakh Himalayas. These domains have witnessed societal and morphological impacts due to such extreme events. The catastrophic landslides are mainly triggered by earthquakes and climate change related intense rainfall, cloudbursts, Glacial Lake outburst Floods (GLOF) etc. in different sectors of Himalaya. These cataclysmic events are caused by various proportions of geological, geomorphologic and geotechnical characteristics of slopes, relief and tectonic conditions including anthropogenic activities. Each of the recorded landslides differs greatly in their causes, triggering agents, run-out distances and impacts. A case study of recent catastrophic landslide in Garhwal Himalaya has been presented. The paper highlights importance of the historical records of catastrophic landslides and attendant changes in terrain morphology to plan for strategy for sustainable regional development.

**Keywords**

Catastrophic landslides • Himalaya • Cataclysmic events

## Introduction

Catastrophic landslides in Himalayas generally occur in high relief conditions and their impact is extremely hazardous. Depending upon the volume and nature of displaced material, cascading run out, the consequential effects are variable. These extremely rapid (>25 $ms^{-1}$), large (>1 × $10^6$ $m^3$) mass movements, are potentially very destructive. In general, these rock or debris avalanches are high magnitude, low frequency events, triggered by hydro-meteorological, seismic, freezing and thawing conditions on higher slopes.

Causes of such large landslides include uplift, fluvial or glacial undercutting of slopes,over-steepening, heavy precipitation, snowmelt, and unfavorable orientation of rock discontinuities, sedimentary layering, joints, faults, or schistosity (Korup et al. 2007). Fluvial, glacial and mass movement processes modify topography, limit slope angles, and produce or destroy relief in the Himalaya (Zeitler et al. 2001). The nature of triggering mechanism of large landslides in high relief conditions, however, is not well understood. Extreme fluvial and glacial erosion commonly destroys the diagnostic morphologies of glacial and mass movement landforms, making their identification difficult (Dortch et al. 2009). The most likely triggers are seismic shaking and intense monsoon precipitation events (Barnard et al. 2004). Coseismic landsliding has been suggested to occur frequently in tectonically active mountain belts (Keefer 1994). Bookhagen et al. (2006), however, suggested that intense monsoon activity may be more significant than earthquakes in triggering large landslides.

The paper focuses on the nature and importance of large catastrophic landslides in the modifying landscape and their devastating impacts in Himalaya. An attempt has been made to summarize significant large landslides in the Indian Himalaya besides presenting a case study of a widespred and large landslide event in Garhwal Himalaya. The paper focuses on large catastrophic landslides, their impact and

V. K. Sharma (✉)
Engineering Geologist, Water Resource Department, Government of Bihar, Bihar Vikas Mission, Room No.14, Block-A, Jalsansadhan Bhawan, Anisabad, Patna, 800002, India
e-mail: vksharma_gsi@outlook.com


V. Vilímek et al. (eds.), *Understanding and Reducing Landslide Disaster Risk*, ICL Contribution to Landslide Disaster Risk Reduction,
https://doi.org/10.1007/978-3-030-60319-9_22

possible association with climate change scenario in Himalayan geo-environment.

## Geodynamics of Himalaya

### General Ecosystem

Himalayan mountain belt is about 2500 km long with variable width of 230–320 km, formed due to the convergent movement of two blocks of the earth's lithosphere the Indian and Asian plates that collided resulting lithosphere deformations and continued modification of the seismic activities of the region with the span of time. The young and rising Himalayas have immature topography, thus, slopes particularly in the Greater Himalayan part are highly vulnerable to slope failures.

The amount of precipitation varies widely in Himalayan ranges, some of the areas in the south of the main Himalayan ranges getting more rain while in Tibetan plateau, north of the Himalayan ranges is the largest rain shadow zone.

The Himalayas have more area under glaciers, than any other region of the world outside polar caps. These glaciers are source of the major river systems of the Indian subcontinent. Himalayas are also rich in glacial lakes. Glaciers are important components of the hydrological cycle in Himalaya and are quite sensitive to changes in the regional and global climate.

### High Vulnerability of the Terrain

Continued collision tectonics, high seismicity, high relief expressions and high-energy regime of Himalayas make the terrain conducive to landslide phenomenon. Himalayas are likely to witness more hydrological disturbances owing to changes in ecological and warming pattern. Warming is predicted to enhance hydrological cycle, more erosion and landslides are expected to occur in high Himalayas.

In brief, the Himalayan belt is likely to be affected by several factors including glaciers melt, enhanced hydrologic cycle, erosion and others mass movements. The Himalaya constitute an end member location of slope processes where events of high magnitude and low to high frequency can be expected as fairly common occurrences (Shroder and Bishop 1998).

In addition, Himalayan and north-eastern regions are potential sites where landslide dams have formed at many places in the past and the potentiality of such occurrences in the future is also high. The landscape in the Himalayan and north-eastern regions is most susceptible to landslides (Sharma 2020).

Indian part of Himalaya alone contains 9575 glaciers (Mukherjee and Sangewar 2001). However, climate change may adversely impact the Himalayan ecosystem through increased temperature, altered precipitation patterns and episodes of drought. Himalayan glaciers, along with other entities in the global cryosphere, may lose significant ice-mass, and thereby endanger river flows. In addition, the recession of the glaciers during the present interglacial (Holocene) has left behind a highly fragile glaciated terrain, which is susceptible to large-scale mass movement and other geomorphic hazards.

The Himalaya abounds in thrusts and faults which have profound effect on slope stability. Siwalik foot hills are separated from the Lesser Himalaya by the Main Boundary Thrust (MBT). Again, Greater Himalaya are separated from the Lesser Himalaya by the Main Central Thrust (MCT) (Fig. 1).

It has been postulated that the northern boundary of the Indian plate coincides with the southern margin of the Tien Shan-Nan Shan mobile fold belt (Carosi et al. 2018).

## Catastrophic Landslides in Himalaya

### Historical Events

Dynamics of change in the Himalaya give rise to the extensive mass movements of frightening proportions in the region. During last century, catastrophic landslides have been recorded from the Himalayan belt.

The slope instability around Nainital in Kumaun Himalaya (Fig. 2) where a disastrous landslide (1880, 1896) along a planar discontinuity inflicted colossal loss of life and property, has been well documented by T. Oldham of the Geological Survey of India and is probably one of the earliest geological investigations done for the landslide anywhere in the world. The area comprises calcareous, greyish to greenish slates, phyllites with thin bands of limestone belonging to Lower Krol Formation dipping 30–35° parallel to the slope. It is reported that 1880 landslide took place following a heavy rainfall for 3–4 days and an earth tremor. Large portions of hill including number of buildings were swept away in a matter of few minutes killing 120 people and the landslide debris filled a portion of the lake lying at the toe of the slope. The debris mass covered a part of the lake changing the landscape of the beautiful hill town. Presence of longitudinal fissure at the top of the *Sher-Ka-Danda* ridge, running almost over the full length is considered to be a unique tensional opening due to slow gliding of entire slope along deep seated cleavage planes or shear zones dipping towards the lake.

The catastrophic rock slide (1989) in Birehiganga valley in Garhwal Himalaya leading to creation of a huge reservoir

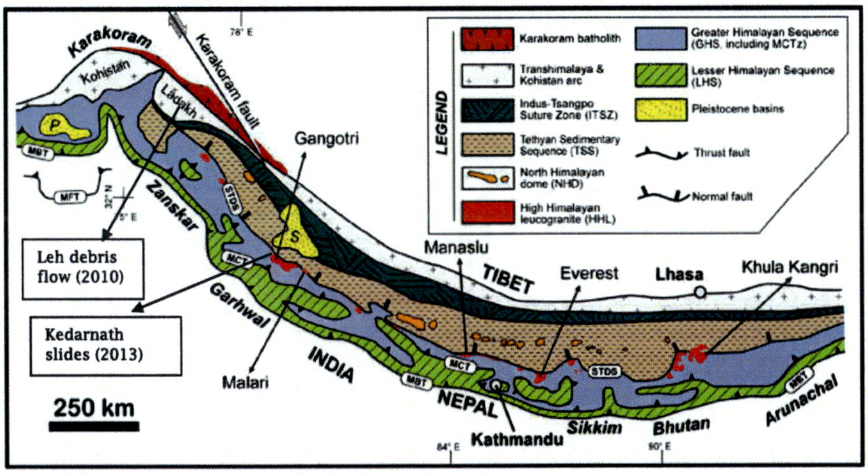


**Fig. 1** Geological map of Himalaya showing various tectonic features (Carosi et al. 2018)

**Fig. 2** Landslide in Nainital (1880), Kumaon Himalaya (British library archives)

had an impact on the geo-environment of the fragile mountain system.

At Kaliasaur, on the Rishikesh-Badrinath road in Garhwal Himalaya, number of devastating landslides occurred in 1952, 1963, 1965 and 1969 and continued even today. The critical spot located on left bank of the Alaknanda river had blocked three forth of the width of the river flowing about 100 m below the road level on 19th September, 1969. The crown portion of the slide extended nearly 120 m above the road level and is active even today. Kaliasaur landslide is essentially a multi-tier, retrogressive landslide in quartzite with clear evidences of fault planes revealing intense tectonic activity. The interface of quartzite and shale and slope undercut are responsible for initiating point of sliding.

A striking example of devastating landslides in eastern Himalaya is provided by Darjeeling floods of 1968.Towards the end of the monsoon, vast areas of Sikkim and West Bengal were destroyed by some 20,000 landslides inflicting huge loss of life and property. The landslides occurred over a three day period with precipitation ranging 500 to 1000 mm, considered a 100 year event in the area.Most of the landslides were released during the third day. The 60 km mountain highway to Darjeeling was cut off at 92 places leaving behind a trail of devastation and disruption to the communication system (Bhandari 1977). The National highway in Sikkim is the vulnerable corridor where numerous rock fall zones are controlled by intense shearing in lesser Himalayan lithology leading to numerous rock slides.

The disastrous rockslide at Malpa (1998) in Higher Himalaya close to Indo-Nepal border buried over 220 people in a flash with complete vanishing of the village settlements and camping pilgrims. The extreme rapidity of the mass movement was a unique example of rock fall–cum-rock avalanche with shooting, bouncing and rolling of rock blocks of quartzite and rock dust engulfing 1–1.5 km radial area. The kinematics of all the discrete rock falls has been favoured by several factors including the high relief, geological discontinuities, glacial environment, ongoing seismic activity, and hydro-meteorological conditions related to climatic change scenario (Sharma and Sanwal 1998).

The record of cases of typical rock fall hazards in northeast and northwest sector Himalaya bring out the predisposing kinematics of the phenomenon such as zone of maximum uplift rate within the crustal-scale ramp of the Himalaya (6–8 mm/year; Bilham et al. 1997); proximity to the Himalayan thrusts viz. Main Boundary Fault and Main Central Thrust, shear zones and tectonic deformations; local high relief; slope/discontinuity interplay and rock-mass characterization. Accordingly, there has been a phenomenal rise in rock fall events which have impacted the life of the people in the entire mountain belt.

Numbers of debris flow hazards have also increased in the Himalayas as per historical records of huge debris flows near Okhimath (1998), Pithoragarh and Almora (2010) in the Garhwal Himalaya; the debris flows movements around Nathpa Jhakri project and Parechu (2004) in Sutlej River catchments in Himachal Himalaya and; unprecedented mud flows/debris flows (2010) in Trans Himalayan region of Ladakh Himalaya, earthquake induced debris movements in Sikkim and rain induced failures in Arunanchal Himalaya which are conspicuous examples in historical past..

Records of spatial and temporal distribution of typical debris flow hazards in Himalayas is of significance as the relationships have been influenced by the occurrence of intense rainfall events, cloudbursts and earthquakes leading to the hypothesis of connect between temporal distribution of slope movements and dramatic impact of climatic changes in Himalayan geo-tectonic environment.

The lake formation caused due to the blockage of the river by a sudden rockslide (2004) in Parechu River, in Himachal Himalaya and Varnavrat slide (2003) in Garhwal Himalaya posed serious threat to urban areas.

The Catastrophic debris/mud flow events (Fig. 3) occurred in Leh (Ladakh) on the night of 5 and 6th August 2010, when an intensive rainfall followed by massive cloudbursts in high altitude terrain of Leh (3505 m), triggered numerous debris flows, mud flows (earth spreads) and sand flows. These widespread, composite flows due to excessive precipitation caused over 200 deaths including 3 foreign tourists, damaged 607 houses, breach of all major communication networks to Leh town besides an unaccountable loss of livestock and agriculture fields etc.

The disaster caused severe damage to the settlements in parts of Leh town, *Sabu* village, *Choklamsar*, Leh airport area, *Nimu, Ney, Bazgo* area around the town (Kumar et al. 2011). This extreme natural hazard occurred on a plateau at around 3500 m above sea level that receives very little rainfall (around 100 mm/yr). Described as a 'high altitude cold desert', the area has sparse rainfall.

The only weather station at Leh airport, recorded only 12.8 mm of total rainfall for the night of 6 August. However, precipitation intensities were of the order of atleast 150 mm/h over Leh during the most intense part of the storm. Some estimates of the maximum total rainfall in some places were as high as 250 mm. It is worth noting that 75 mm of rain over Leh, during the most intense part of the storm, is equivalent to around a year's worth of rain falling in 30 min. The Trans Himalayan region, known for its cold and desert type of climatic condition, had never witnessed such disaster scenario of debris movement (Fig. 4) and the incidence may be well be considered as climatic anomaly in the region.

## Case Study of a Recent Catastrophic Landslide

Widespread landslides of catastrophic magnitude occurred during 16–17th June 2013 around *Kedarnath Temple*, in Rudraprayag district, Uttarakhand in Garhwal Himalaya. The temple complex, at an approx. elevation of 3583 m is one of the major sites of Hindu pilgrimage and heritage structure, located on a glacial outwash plain (1 km × 0.5 km) bounded by R. Mandakini in the west and Saraswati River in the east. Mandakini river originates from Chorabari glaciers and Saraswati River from Companion glacier, both located about 1.5 km upslope of the temple. Augen gneiss and migmatites of Central Crystalline Group of rocks are exposed in the area. The glacier regime of the area is evidenced by 'U' shaped valley profile and presence of morainic materials and eratic boulders.

The intense rainfall and cloudbursts between 16 and 17th June, 2013 in the area were so severe that it inflicted 786 deaths, over 3000 people reported to be missing, damage to temple and razed other construction around the complex. The village houses and road network were severely damaged. The temple area was the epicenter of most of the damages in the district due to unprecedented debris/boulder movement. The intense rainfall event was so widespread that five districts of the state of Uttarakhand were badly affected and in all 2214 houses were completely damaged and deaths of 8231 livestock and severe damage to road network were reported.

The deluge at Kedarnath and other areas along Mandakini valley was the result of heavy and incessant rainfall and the breach or Glacial Lake Outburst Flood (GLOF), of Chorabari Lake.

The intensity, extent and pattern of rainfall in Mandakini valley, and Kedarnath area in particular, during the period from 15th to 17th June' 2013 is given in Fig. Figure 6 Damage by debris flows around Kedarnath temple, Uttarakhand, Garhwal Himalaya Fig. 5 indicated that maximum rainfall of 166.25 mm was recorded in Okhimath, a proximal location of Kedarnath, on 17-06-2013. The total rainfall recorded between 15th and 17th June' 2013 was also highest in in the area (337.50 mm) followed by Rudraprayag town

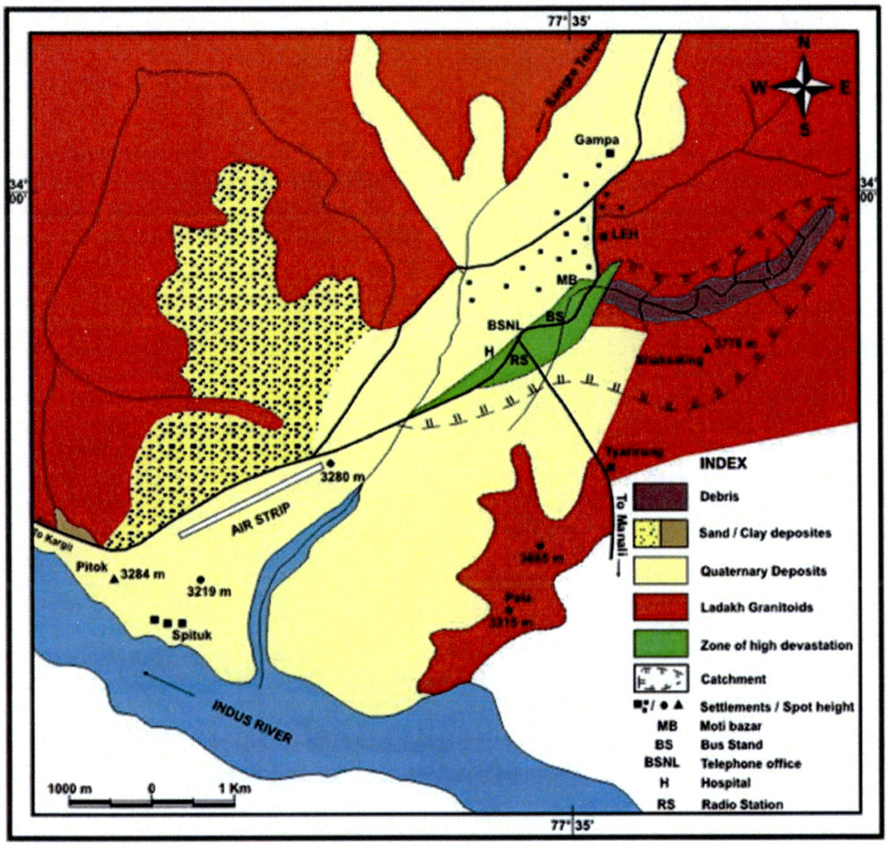


**Fig.3** Damages in run-out zone of catastrophic debris flow (2010) in Leh (Ladakh)

**Fig. 4** Pile of debris flow (2010) damaging a house in its run out zone in Leh area

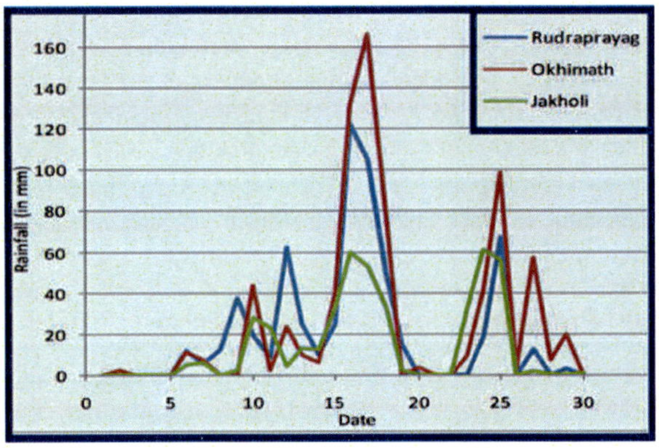


**Fig. 5** Rainfall data of at three observatories in nearby areas of Kedarnath township during in the month of June 2013

**Fig. 6** Damage by debris flows around *Kedarnath* temple, Uttarakhand, Garhwal Himalaya

(252.50 mm) and Jakholi area (146.40 mm).The Chorabari GLOF was reported on 17th June 2013. The moraine-dammed lake was defined by the right lateral moraine of Chorabari glacier.

A morphological impact of debris flows out of the GLOF was that the bed levels of both the channels of R. Mandakini have been raised due to huge flux of debris; consequently, the river changed its course to join flow into the course of R. Saraswati flowing east of the temple complex.

The intense rainfall induced a total of 98 mass wasting incidences in Rudraprayag district and have been categorized into three groups i.e. landslides, subsidences zones and bank erosion. Out of 98 incidences, there are 76 incidences of landslides, 3 incidences of subsidences and 19 instances of bank erosion. Analysis of 76 landslides incidences indicates that maximum 74% events are of debris slides followed by rock-cum-debris slides (22%) and rock slides (4%) indicating the debris movement were the most predominant phenomenon of the catastrophic event. The catastrophic event resulted due to an unprecedented floods, debris flows and Glacial Lake Outburst Flood (GLOF) in Garhwal Himalaya severely damaging the temple complex and its environs (Sharma and Rawat 2013).

## Discussions and Result

The historical accounts of occurrences of catastrophic landslides points towards randomness of such cataclysmic events that are generally governed by greatest sub- aerial relief, steep slopes, jointed/fractured rock-mass, interplay of discontinuities surfaces with slopes and deeply eroded, undercut slopes by glacier or river actions and valley profile. However, relationship between Himalayan tectonic uplift and landslide phenomenon is difficult to understand owing to the lack of information on spatial and temporal scales at which processes operate.

It has been observed that debris flows are occurring in response to greater rainfall anomalies, possibly due to climatic effects and breach of vulnerable moraine-dam lakes in higher domains. A highly vulnerable Himalayan ecosystem is also highly sensitive to slope failures even with a modest changes in climatic pattern.

The study of the historical records of catastrophic landslides, relict scars and cascading run-out of the debris provide a clue to the vulnerability profile of the area and records of the events help in planning the mitigation strategy for developmental activities in vulnerable domains of higher Himalaya.

## References


Barnard PL, Owen LA, Sharma MC, Finkel RC (2004) Late Quaternary landscape evolution of a monsoon-influenced high Himalayan valley, Gori Ganga, Nanda Devi, NE Garhwal. Geomorphology 61:91–110

Bhandari RK (1977) Some typical landslides in the Himalaya. In: Proceedings of second international conference on landslides. Tokyo, pp 1–23

Bilham R.K., Larson K., Freymuller J., Project IDHYLIM members (1997) GPS measurements of present-day convergence across the Nepal Himalaya. Nature 386:61–64

Bookhagen B, Thiede RC, Strecker MR (2006) Late Quaternary intensified monsoon phases control landscape evolution in the northwest Himalaya. Geol Soc Am 33:149–152

Carosi R, Chiara M, Salvatore T (2018) 20 years of geological mapping of metamorphic core across Central and Eastern Himalaya. Earth Syst Rev 177:114–118

Dortch JM, Owen LA, Haneberg WC, Caffee MW, Dietsch C, Kamp U (2009) Nature and timing of large landslides in the Himalaya and Trans-Himalaya of northern India. Quatern Sci Rev 28:1037–1054

Keefer DK (1994) The importance of earthquake-induced landslides to long-term slope erosion and slope-failure hazards in seismically active regions. Geomorphology 10:265–284

Korup O, Clague J, Hermanns RL, Hewitt K, Strom AL, Weidinger JT (2007) Giant landslides, topography, and erosion. Earth Planet Sci Lett 261:578–589

Kumar, Pankaj., Sharma,V.K., Pankaj, A. and Kumar, A., (2011). Landslide disaster and its mitigation-A case study of debris flow at Leh,Ladakh, J&K. In: proceedings recent advances in geosciences. Indian School of Mines, Dhanbad, p 138

Mukherjee S, Sangewar CV (2001) Glaciology of Indian Himalaya. GSI Special Publ 63:213

Sharma VK (2020) Landslides in India: Issues and perspective. J Geol Soc India 95:110

Sharma VK, Rawat PVS (2013) Post disaster slope stability evaluation of catastrophic events in Uttarakhand. Indian J Geosci 67(3–4): 337–346

Sharma VK, Sanwal R (1998) Disastrous Rock Avalanche at Malpa on 18 August 1998, Kumaon Himalaya. India J Eng Geol XXVII Nos. 1–4:71–74

Shroder JF, Jr., Michael P. Bishop, (1998) Mass movement in the Himalaya: new insights and research directions. Geomorphology 26:13–35

Zeitler PK, Meltzer AS, Koons PO, Craw D, Hallet B, Chamberlin CP, Kidd WSF, Park SK, Seeber L, Bishop MP, Shroder JF (2001) Erosion, Himalayan, geodynamics, and the geomorphology of metamorphism. Today 113:1443–1455

# Part II

# Frontiers of Landslide Science

# Enhancing Preparedness Against Impact of Climate Change on Slope Safety in Hong Kong

2020 Kyoto Japan

K. K. S. Ho, H. W. Sun, E. J. Lam, and F. L. C. Lo

**Abstract**

Over the past 40 years, a comprehensive slope safety system has been developed and implemented by the Geotechnical Engineering Office (GEO) to manage the landslide risk in Hong Kong to a reasonably low level. However, extreme rainfall events due to the effect of climate change could bring about widespread and large-scale landslides. Therefore, the GEO has made concerted efforts to enhance our preparedness for extreme landslide events. This paper highlights the holistic, multi-pronged approach adopted to manage the landslide risk associated with climate change.

**Keywords**

Landslide • Slope safety system • Extreme rainfall event • Rain-induced landslide • Emergency preparedness and response • Risk management

## Introduction

Hong Kong is a densely populated city with hilly terrain and seasonal torrential rainfall. All these have made Hong Kong prone to landslide hazards. Nevertheless, fatalities from landslides have dropped significantly since the establishment of the Geotechnical Engineering Office (GEO) in 1977, as a result of the implementation of a comprehensive slope safety system. However, extreme rainfall events are observed around the world, possibly exacerbated in recent years by climate change. More frequent and extreme rainfall events are expected, which could cause major landslide disasters with multiple casualties and economic losses. It is paramount to enhance our preparedness for the impact of climate change on slope safety in Hong Kong.

## Lessons Learnt from Record-Breaking Rainstorm in June 2008

Hong Kong experienced a very severe rainstorm on 7 June 2008, which struck Lantau Island (Fig. 1). This resulted in widespread debris flows and debris avalanches on natural terrain (some 2,400), many of which are large scale failures with high mobility. The maximum rolling 4 and 24 h rainfall was 384 mm and 623 mm respectively, which was the most severe rainstorm since rainfall record began in Hong Kong in 1885, as corroborated by the significant response of the natural hillsides.

The landslides caused havoc to the communities and major difficulties in respect of the response and recovery operations rendering Government's emergency response system on the verge of breakdown.

The serious landslides were studied extensively and the technical insights were summarised by Wong (2009). In essence, the number, scale and mobility of natural terrain landslides under such intense rainfall conditions were unprecedented.

K. K. S. Ho (✉)
Geotechnical Engineering Office, Civil Engineering and Development Department, The Government of the Hong Kong SAR, Civil Engineering and Development Building, 101 Princess Margaret Road, Hong Kong, China
e-mail: kenho@cedd.gov.hk

H. W. Sun · E. J. Lam · F. L. C. Lo
Geotechnical Engineering Office, Civil Engineering and Development Department, The Government of the Hong Kong SAR, Hong Kong, China
e-mail: hwsun@cedd.gov.hk

E. J. Lam
e-mail: elainejlam@cedd.gov.hk

F. L. C. Lo
e-mail: frankielclo@cedd.gov.hk


V. Vilímek et al. (eds.), *Understanding and Reducing Landslide Disaster Risk*, ICL Contribution to Landslide Disaster Risk Reduction,
https://doi.org/10.1007/978-3-030-60319-9_23

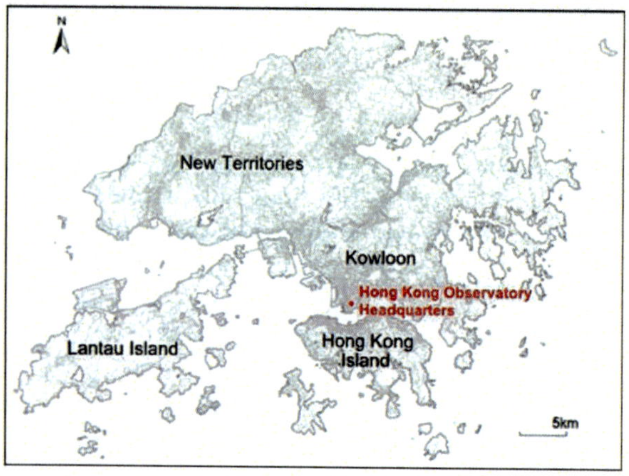


**Fig. 1** Location map of Hong Kong

Figure 2 illustrates the relationship between landslide density and rainfall based on the available data in Hong Kong. When an area is subjected to severe rainfall that exceeds the threshold (rolling 24 h corresponding to about 20% of the mean annual rainfall), natural terrain landslides would overtake man-made slope failures in terms of landslide density and become the principal landslide hazard under extreme rainfall. The corresponding shift in the risk profile associated with the occurrence of extreme rainfall has major implications on landslide risk management.

Apart from increased landslide density, observations suggest that the mobility of debris flows in a severe rainstorm could increase significantly. Such an increase is partly related to the increased volume due to significant entrainment, and partly attributed to watery debris in the transportation process. The following settings were diagnosed as being prone to the development of sizeable and mobile channelised debris flow: (a) sizeable debris flow at a major drainage line (e.g. site with a large catchment and a long flow path where a large amount of storm water and entrainable materials may be available for mixing with the landslide debris), (b) sizeable debris flow along a major drainage line into which many tributaries are feeding (i.e. possible sudden increase in the water content of the moving debris when the debris passes through a confluence point), and (c) the discharge of debris onto a pool of water on the drainage line or debris from a small drainage line onto a major drainage line, where there is a potential for a large amount of surface runoff.

The torrential rain in June 2008 fell largely on Lantau Island, which is one of the sparsely populated areas, hence it was a near-miss event.

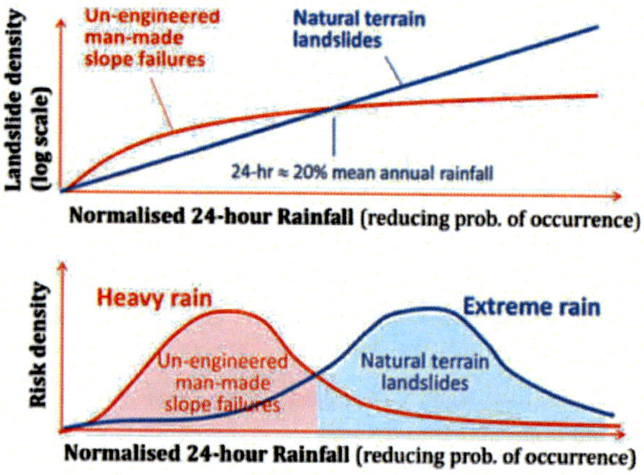


**Fig. 2** Distribution of landslide and risk density in response to extreme rainfall

## Consideration of Extreme Rainfall Events

### Climate Modelling and Rainfall Projections

The Hong Kong Observatory (HKO) has been conducting regional scale climate projections using statistical downscaling to project the rainfall changes in the twenty-first century. Under the high greenhouse gas concentration scenario RCP8.5, the number of extremely wet years (annual rainfall >3,168 mm) is expected to increase from three in 1885–2005 to about 12 in 2006–2100 (with a likely range of 5–20). In particular, it is projected that the occurrence of extreme rainfall events will experience increasing frequency and severity due to the effect of climate change.

The maximum hourly rainfall record of the HKO Headquarters (Fig. 1) has been broken several times since 1885 and the time interval between new records is getting shorter (Fig. 3). In particular, the record was broken three times since 1966. The latest record of 145.5 mm was set on 7 June 2008, breaking the previous one by a wide margin of 30 mm.

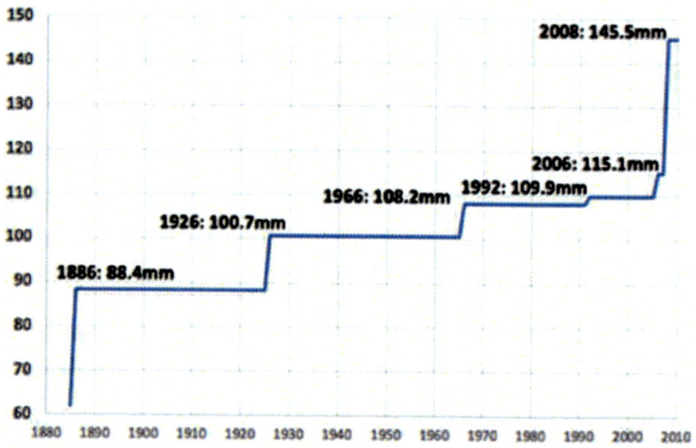


**Fig. 3** Maximum hourly rainfall recorded at Hong Kong Observatory

## Scenario-Based Assessment and Stress Testing

The GEO conducted a scenario-based landslide hazard assessment incorporating the Probable Maximum Precipitation (PMP) concept. PMP is the theoretical maximum precipitation for a given duration under modern meteorological conditions over a given area at a certain time of year (WMO 2009).

The extreme landslide scenarios under two credible extreme weather events were assessed: (a) Level-1 scenario: transposing the June 2008 rainstorm event to give a direct hit on the densely populated Hong Kong Island. The maximum daily rainfall of rainstorm was about 600 mm (corresponding to about 40% PMP); (b) Level-2 scenario: simulating a direct hit to Hong Kong Island by transposing the rainstorm caused by Typhoon Morakot that struck Taiwan in 2009, with suitable orographic corrections and extra rainfall added to account for the effect of climate change projected to the end of the twenty-first century. The estimated daily rainfall was about 840 mm to 1,040 mm (corresponding to about 70% PMP).

The results revealed that under Level-1 rainstorm, the current emergency response system of the GEO could barely cope with the landslide scenario, and streamlining and enhancements are called for. Under Level-2 rainstorm, the capacity of the emergency response system would be exceedingly overwhelmed. It would not be cost effective nor practicable to deal with the potential hazards solely by engineering measures. A more pragmatic approach would be to enhance our preparedness to extreme landslide events by a combination of adaptation strategies to lessen the impact of extreme landslide scenarios and resilience strategies to strengthen community's capability to cope with the landslide hazards.

## Adaptation Strategies

### Expanded Scope of Systematic Slope Retrofitting Programme

The GEO has been systematically retrofitting substandard high-risk man-made slopes since 1977 under the Landslip Preventive Measures Programme (LPMP), which was completed in 2010. With more urban encroachment to steep hillsides and the potential impacts of extreme weather events due to climate change, the GEO launched a long-term Landslip Prevention and Mitigation Programme (LPMitP) in 2010 with an expanded scope to include both vulnerable natural hillsides and the remaining man-made slopes that are not up to the current safety standard. Vulnerable natural hillsides are those with known historical landslide in close proximity to existing buildings and important transport corridors pursuant to the "react-to-known-hazard" principle. A risk-based priority ranking system has been developed to select the most deserving natural hillside catchments and man-made slopes for action under the LPMitP.

Unlike man-made slopes, it is often impractical, costly and environmentally undesirable to carry out extensive slope stabilisation works on natural hillsides to prevent their failure. The strategy for natural hillsides is to mitigate the risk of landsliding by reducing their consequences. Some common mitigation measures include construction of rigid and flexible debris-resisting barrier, together with other debris flow control and erosion control measures.

Taking cognisance of the effect of extreme rainfall on debris mobility, the GEO is expanding the existing selection criteria for vulnerable natural hillsides, (which are based on proximity of historical landslide to existing facilities) to include those catchments with a known history of failure (irrespective of their proximity to existing facilities) and a potential for developing mobile debris flows under extreme rainfall.

### Enhancing Slope Engineering Practice

Lessons learnt from systematic landslide studies in Hong Kong have emphasised the need to improve the reliability of slope design and detailing. This is done by means of robust engineering measures that will be less sensitive to unforeseen adverse geological and groundwater conditions (Ho and Lau 2010).

For slope drainage design, guidelines have been promulgated to account for the effects of climate change on rainfall intensity projected up to the end twenty-first century. The mean rainfall intensity given by the of intensity–duration–frequency curves is to be increased by about 15% for surface drainage design (GEO 2018). In addition, prescriptive slope surface protection and drainage measures have been promoted as contingency provisions (Wong et al. 1999). Guidance on improved detailing of and enhanced redundancy in drainage provisions has also been promulgated.

### Regular Slope Maintenance

All government-owned slopes are systematically maintained with annual routine maintenance and engineer inspection every 5–10 years to avoid deterioration and upkeep their functions. To reduce small-scale and washout failures, enhanced maintenance works would be carried out if necessary.

## Resilience Strategies

### Strengthened Public Education Strategy

Insights from the stress testing (see above) highlight the need for a new public education strategy to address the impact of climate change on slope safety. As a vast area and population will be affected by extreme rainfall, the government's emergency response system is likely to be overwhelmed and jeopardised by poor accessibility conditions. Hence a community-based approach is of the essence for emergency preparedness and response. The public communication strategy has been strengthened accordingly to improve community resilience to landslides with more emphasis on extreme rainfall events and to enhance public communications at times of landslide emergency. A key initiative is to promote self-protection and neighbourhood support in case of emergency by introducing updated sets of messages on "self-rescue measures when threatened by landslides" (Fig. 4). These simple messages include (i) stay at home or in a safe shelter and keep vigilant if already indoor, (ii) keep away from slopes if still outside, and (iii) pay attention to and follow evacuation notices. Partnering with non-government and quasi-government agencies in emergency management has also been strengthened.

In addition, the GEO has recently launched the Landslide Potential Index (LPI) (GEO 2020) to provide a simple measure of the severity of a rainstorm in terms of the corresponding risk of landslides causing casualties (Table 1). The LPI is correlated with the estimated number of landslides based on rainfall intensity, rainfall location and spatial distribution of slopes in Hong Kong. The risk description would help the public better understand the landslide risk associated with a particular rainstorm. It would also help to illustrate the key message that there is no room for complacency even though no fatal landslides have occurred in the past decade as slopes in Hong Kong had not faced the challenges of severe rainstorms of "very high" or "extremely high" categories during this period.

### Enhancing the Landslip Warning System

The GEO has been operating the Landslip Warning System to alert the public to take precautionary measures and reduce their exposure to risk posed by landslides during periods of heavy rainfall. A new generation of the system has been launched, which gives real-time prediction of the landslide distribution throughout the territory based on rainfall data collected from an extensive cloud-based network of automatic raingauges and rainfall forecasts from the HKO using state-of-the-art nowcasting models. The system is the backbone of the multi-tiered LPI mentioned above.

In addition, the Natural Terrain Landslip Alert (NTLA) was implemented as an enhancement of the landslip warning system to facilitate better emergency preparedness and response under extreme rainfall events. The NTLA was based on real-time prediction of natural terrain landslide distribution using the rainfall-based landslide susceptibility model (Ko and Lo 2016). It served as an internal alert system on the possible widespread occurrence of natural terrain landslides.

### Enhancing Emergency Preparedness and Response

The GEO operates its Emergency Control Centre when a high demand for emergency service and professional advice is anticipated, e.g. when a landslip warning has been issued. In view of the possible extreme landslide scenarios due to the effect of climate change, the GEO has implemented several initiatives to enhance the resilience of our landslide emergency service against any prolonged regional breakdown of communication networks and power grid. These include joining the Unified Digital Communication Platform for using radio services to maintain communications in case of mobile network breakdown, setting up a backup Emergency Control Centre and an Uninterruptible Power Supply system to enhance the continuity of emergency systems.

To enhance emergency response, the GEO has utilised handheld laser scanning equipment and unmanned aerial vehicle (UAV) for emergency inspections at landslide sites where access is difficult and unsafe. These enabling tools allow instant acquisition of the post-landslide terrain topography and enhance timely situation awareness under adverse site and weather conditions.

### Common Operational Picture (COP)

By virtue of the vulnerable and densely urbanised setting, it is perceivable that an extreme weather event hitting Hong Kong could result in concurrent multiple hazards, such as landslides, flooding, storm surges, etc., each of which could be of an unprecedented intensity. Overall situation awareness and efficient cross-functional communication are essential to emergency management. To gear up for potential climate change impact, a Common Operational Picture (COP) is being developed. This is a new information technology platform with geographic information system (GIS) functions for sharing real-time emergency information amongst the Works Departments and security/rescue

**Fig. 4** Leaflet of landslide self-help tips

**Table 1** Risk categories of Landslide Potential Index

| Landslide Potential Index | Risk description |
|---|---|
| >100 | Extremely high |
| 51–100 | Very high |
| 10–50 | High |
| <10 | Low |

agencies. It can provide a timely overall picture of emergency situations in multiple hazard scenarios, thereby facilitating decision-making and coordination of emergency response across the departments.

The GEO also organises regular desktop drills jointly with 30 key departments/agencies to walk through several hypothetical extreme weather events and assess the likely constraints in respect of emergency management.

## Smart Barrier System

Landslide debris-resisting barriers are common natural terrain mitigation measures adopted to deal with natural terrain landslide risk. These barriers are capable of retaining landslide debris of volume and mobility designed for severe rainfall scenarios. In case of exceptionally massive landslides triggered by extreme rainfall scenarios, the barriers may be overwhelmed and overflow may occur in the form of pulses. To provide timely emergency responses, a Smart Barrier System (LSCM 2020) has been developed by the GEO to detect landslide debris impacts onto debris-resisting barriers, as well as to disseminate warning messages to the public near the barriers.

The Smart Barrier System is a state-of-the-art information technology system which comprises debris impact detection modules, a signal transmission system, monitoring instrumentation and a power supply system (Fig. 5). It is designed to provide instant alerts to the GEO through a desktop/mobile application when a landslide debris-resisting barrier intercepts landslide debris. At the same time, flashing lights and moving message displays installed on landslide warning signboards would be triggered to alert the public to stay away from the natural hillside. At present, the Smart Barrier System has been installed on several landslide debris-resisting barriers for trial operations at selected critical facilities (e.g. hospital). An emergency response plan has also been established for the critical facilities in case the Smart Barrier System detects debris impact.

## Further Work

Some areas for further enhancement of preparedness to landslide hazards arising from extreme rainfall events due to climate change are discussed below.

### Further Update of Rainfall Projections for Hong Kong

Further to the statistical downscaling for Southern China based on CMIP5 models adopted in Fifth Assessment Report (AR5) of the Intergovernmental Panel on Climate Change

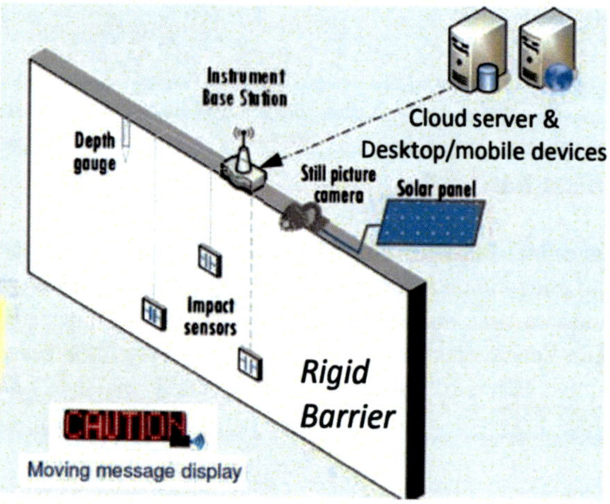


**Fig. 5** Schematic diagram of Smart Barrier System

(IPCC), a study using dynamical downscaling technique is now underway by the HKO. This would involve the use of a sophisticated regional climate model to simulate the physical effects of the output from the global climate models on a regional scale. Upon release of the AR6 by IPCC scheduled in 2022, further update of the downscaling study for the future CMIP6 models would be necessary.

### Low-Frequency Large-Magnitude Events

Catchments with major drainage lines have been identified as one of the adverse site settings that are prone to the development of channelised debris flows of high mobility. These drainage lines are associated with large catchments and long debris flow path where a large amount of storm water and entrainable materials may be available for mixing with the landslide debris resulting in large-magnitude channelised debris flows under extreme rainfall scenarios with long return periods.

A pilot study has been conducted to estimate the order of risk arising from these low-frequency high-magnitude debris flows that are not explicitly covered in the conventional study and design approach of mitigation measures. It calls for the development of a new strategy with innovative solutions that suit the urban setting in Hong Kong for addressing the associated landslide risk. It would be complementary to the prevailing approach with a view to enhancing the robustness of the slope safety system amidst the looming threats of extreme weather events.

The projection of low-frequency, large-magnitude landslides based on past experience will inevitably be subject to a high degree of uncertainty. The key is whether there would be a change in landslide mechanism under the effect of extreme rainfall. With a view to narrowing the knowledge gap, the GEO has embarked on a programme of systematically dating ancient debris fan deposits to study the potential relationship between landslide activity and climate conditions in the geological past. The results have been considered in the projection of the magnitude-frequency relationship of such events.

### Landslide Hazards on Critical Infrastructures

A study has been conducted to assess the resilience of the critical infrastructures in Hong Kong under the effects of climate change and extreme weather. These critical infrastructures of government ownership were identified based on three key selection principles: (a) long recovery time upon hazard impact; (b) non-substitutability, and (c) territory-/region-wide socio-economic impacts in case of service disruption.

The GEO is now undertaking a vulnerability and risk assessment of landslide affecting the continual functioning of the critical infrastructures. In the assessment, the exposure and sensitivity of the critical infrastructures to potential landslide hazards, as well as the likelihood and the consequence of the associated hazards are considered.

## Application of Innovation and Technology

To tackle the acute challenge of the impact of climate change on slope safety, the GEO endeavours to apply innovation and new technologies to enhance our slope safety system and services. The GEO has collaborated with industry partners and academia to explore the use of deep learning techniques in slope engineering work, e.g. pilot studies on applying Convolution Neural Network (CNN) for landslide mapping using aerial photographs and satellite remote sensing imageries as well as data-driven landslide prediction methodology and improved stress testing.

Studies are also underway to assess the potential of wider use of bio-engineering techniques and practicality of nature-based solutions for landslide risk mitigation. The possible use of robotics to carry out landslide inspection at dangerous locations, such as below a landslide scar, is being explored. The GEO will continue to strive for opportunities to bring innovation and technology into geotechnical practice to enhance our preparedness for extreme landslide scenarios.

## Discussion and Conclusion

The multi-pronged slope safety system implemented in Hong Kong is a role model for landslide risk management for densely-urbanised cities. Despite this, it is understood through observations from the impact of record-breaking rainfall and insights from projected extreme landslide scenarios and stress testing that the comprehensive slope safety system still falls far short of dealing with the risk of extreme rainfall events that could be exacerbated by climate change. Hence, considerable efforts have been devoted in the past decade in Hong Kong to gear up our preparedness for the impact of climate change on landslide risk through a wide range of adaptation and resilience strategies (Ho et al. 2017).

**Acknowledgements** This paper is published with the permission of the Head of the Geotechnical Engineering Office and the Director of Civil Engineering and Development, Hong Kong SAR Government.

## References

GEO (2018) Updated Intensity-Duration-Frequency curves with provision for climate change for slope drainage design Technical Guidance Geotechnical engineering office, Hong Kong, p 4

Ho KKS, Lau JWC (2010) Learning from slope failures to enhance landslide risk management. Q J Eng Geol Hydrol 43(1):33–68

Ho KKS, Sun HW, Wong ACW, Yam CF, Lee SM (2017) Enhancing slope safety preparedness for extreme rainfall and potential climate change impacts in Hong Kong. Ho, Lacasse, Picarelli (eds). Taylor & Francis Group, CRC Press. pp 105–150

Ko FWY, Lo FLC (2016) Rainfall-based landslide susceptibility analysis for natural terrain in Hong Kong—a direct stock-taking approach. Eng Geol 215:95–107

LSCM (2020). Smart Barrier. https://www.lscm.hk/eng/channel.php?channel=Case-Smart-Barrier. Last accessed 13 Mar 2020

WMO (2009) Manual on estimation of probable maximum precipitation (WMO-No. 1045). World Meteorological Organisation. Geneva. 257p.

Wong H N, (2009) Rising to the challenges of natural terrain landslides. Natural hillsides: Study and risk management measures. In: Proceedings of the 29th annual seminar of the HKIE geotechnical division. Hong Kong Institution of Engineers, Hong Kong, pp 15–53

Wong H N, Pang L S, Wong A C W, Pun W K, Yu Y F, (1999) Application of prescriptive measures in slopes and retaining walls (GEO report no. 56) 2nd edition. Geotechnical Engineering Office. Hong Kong. p 73

Check for updates

2020 Kyoto Japan

# Climate Change and Surface Deformation Characteristics in Degradation Area of Permafrost in Lesser Khingan Mountain, China

Wei Shan, Chengcheng Zhang, Ying Guo, Monan Shan, Xujing Zeng, and Chunjiao Wang

**Abstract**

Over the past 100 years, the process of global climate change has accelerated, and the air temperature has also appeared an increasing trend in the high latitude permafrost area of northeast of China, the permafrost degradation which dramatically changed the environmental geological conditions,resulting in the change of landform and the increase of geological disasters.However, there are few research results about this area. In this paper, we take the road area of Bei'an-Heihe highway crossing the Lesser Khingan Mountain as the study area. Using a total of 18 ALOS/PALSARSAR images, through the SBAS-InSAR technique to draw the surface deformation map of the study area from July 2007 to December 2010, combined with the local meteorological data and the permafrost distribution map in the study area obtained by our research group, the relationship between distribution characteristics of permafrost, permafrost degradation, and surface deformation were analyzed. The results show that the annual average deformation rate of surface in the study area is ± 70 mm/year, the surface deformation is closely related to the permafrost distribution, and the distribution law is consistent.Climate warming is the main factor leading to the permafrost degradation. In the context of continued warming and accelerating degradation of permafrost, we will face landslides, ground surface subsidence or more ecological and engineering problems, threaten the safety of local engineering structures. The results reveal the law of the permafrost degradation process under the background of climate warming, and provide a reference for future research.

**Keywords**

climate change • Permafrost degradation • Permafrost distribution • Surface deformation • SBAS-InSAR

W. Shan (✉) · C. Zhang · Y. Guo · M. Shan · X. Zeng · C. Wang
Northeast Forestry University, Harbin, Heilongjiang, 150040, China
e-mail: shanwei456@163.com

C. Zhang
e-mail: 93038478@qq.com

Y. Guo
e-mail: samesongs@163.com

M. Shan
e-mail: shanmonan@yeah.net

X. Zeng
e-mail: jingle1992@163.com

C. Wang
e-mail: zlj_832008@126.com

## Introduction

The fifth assessment report of Intergovernmental Panel on Climate Change(IPCC) pointed out that: from the end of the nineteenth century to the 1980s, the global air temperature rised by 0.6 °C, about 0.5 °C every hundred years. This warming trend is very obvious in the world, especially in the middle and high latitudes area, and all seasons are getting warmer, especially in winter Zhao et al. (2007), Zhao et al. (2018). There is no doubt about the warming trend also in China in the past 100 years, which is 0.19–0.72 °C/100 years, especially in the past 50 years the warming trend is more obvious about 0.64–0.92°C/50 years Ding et al. (2006).

The global permafrost is mainly in the northern hemisphere, including the Eurasia and North America, accounting for 63% of the total area of the global permafrost. In China, the permafrost is divided into high altitude permafrost and high latitude permafrost. The former is distributed in the western mountain plateau and some higher mountains in the east, and the latter in the northeast. Permafrost is a kind of soil medium which is very sensitive to temperature. In recent years, permafrost degradation is widespread because of the



V. Vilímek et al. (eds.), *Understanding and Reducing Landslide Disaster Risk*, ICL Contribution to Landslide Disaster Risk Reduction,
https://doi.org/10.1007/978-3-030-60319-9_24

climate warming and human activities. The permafrost degradation area in Alaska are greatly increased. Through the spectral analysis, it is shown that the degradation causes the lowland landform to be affected, and the tundra ecosystem has changed greatly Jorgenson et al. (2001), Jorgenson et al. (2006). Since the 1950s, most parts of the Qinghai Tibet Plateau have experienced obvious warming, and the warming trend in the cold season is greater than that in the warm season. Climate warming also causes a large number of permafrost degradation, which has an impact on the stability of the ecological environment and structures, and accelerates the process of surface desertification Yang et al. (2010), Wang et al. (2000).

The permafrost region in northeast is the second largest permafrost regions in China, which located in the south of the permafrost region in Eurasia. The geographic southern boundary is between 46°36 N and 49°24'N, which is a W-shaped line (in Fig. 1)swinging between the isolines of ±1 °C with the annual average temperature of 0 °C as the axis. The permafrost distribution has an obvious law of latitudinal zonality. From north to south, the distribution region decreases. The local factors such as snow cover, vegetation, water content, topography and atmospheric inversion in this area lead to significant thermal offset, thus forming the distribution characteristics of "Xing'an-Baikal" permafrost, which is quite different from polar permafrost and high-altitude permafrost. However, the wide permafrost region has a very unstable thermal state, forming island melting area or island permafrost area, extremely vulnerable to climate and external environmental changes Wei et al. (2011).

However, the range of temperature rise in northeast China is higher than the national average (in the Fig. 2) China Meteorological Administration National Climate Center (2009), which is one of the regions with the fastest temperature rise Chen et al. (2004); Gao and Wang (2008). The signs of permafrost degradation are obvious, and it is a trend of acceleration He et al. (2009). The degradation not only affects water resources and ecosystem, but also affects the stability and operation safety of highway, railway and other important infrastructure in permafrost area due to surface deformation, debris flow, landslide and other engineering geological problems caused by permafrost melting. With the rapid development of infrastructure construction, some important projects in Northeast China, such as Harbin-Yichun highway, Hegang-Yichun highway, Yichun-Jiayin highway, Bei'an-Heihe highway and Sino Russian crude oil pipeline, are facing the problems of thawing settlement and frost heave which is the main harm of engineering construction caused by the permafrost degradation.

Interferometric Synthetic Aperture Radar (InSAR), with all-time and all-weather capability, has a certain penetration ability for ground objects and strong penetration ability for clouds, rain and smoke. It is a potential technique of the earth observation from the space. In some areas with the complex terrain, the bad weather conditions, and it is not suitable for traditional geodetic methods, showing the advantages Rosen et al. (1996) in extracting the surface deformation caused by earthquake, volcano and subsidence.

Because of its high precision, high spatial resolution and high surface coverage of differential interferometric radar technique in the research of obtaining surface deformation, it has advantages over other traditional methods. Using D-InSAR methods to monitor the roadbed deformation along the Qinghai Tibet highway is basically consistent with the measured results. Monitoring results of the surface deformation along the Bei'an-Heihe highway is consistent

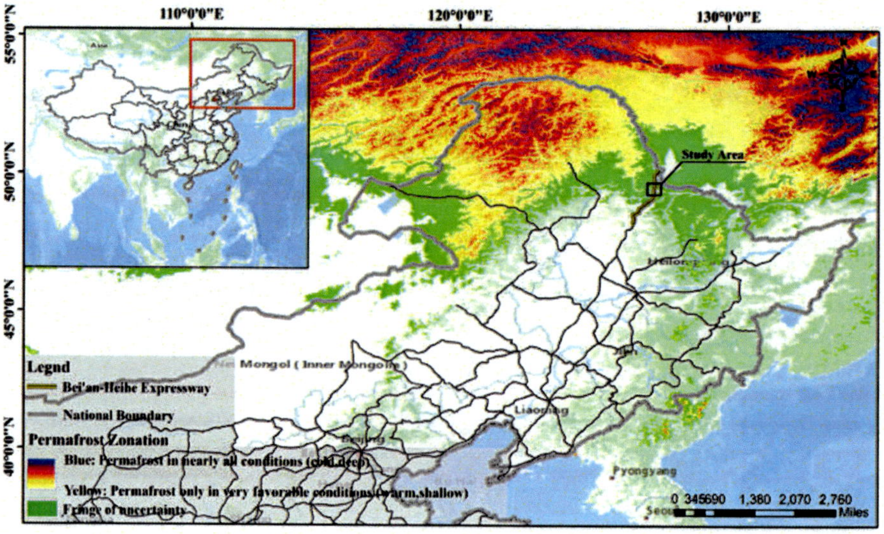


**Fig. 1** Permafrost distribution map in northeast China and location of the study

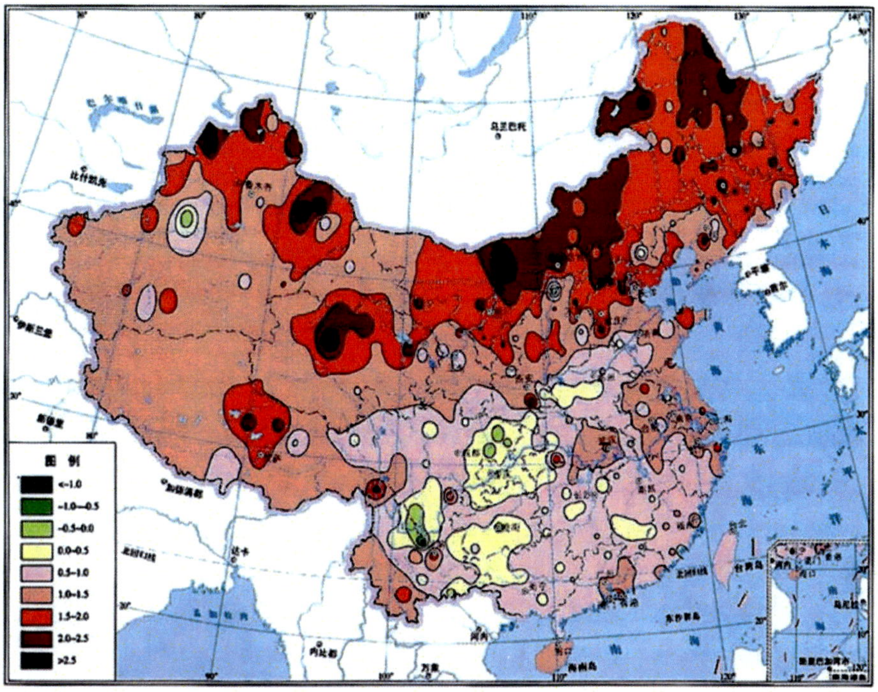


**Fig. 2** 1957–2007 spatial distribution of average temperature changes (°/10 years) in China

with the actual situation of subsidence, landslide and other disasters in the region Wang et al. (2015), Xie et al. (2008). It is proved that the Differential Interferometry Synthetic Aperture Radar (D-InSAR) method can also be applied to the monitoring of surface deformation in permafrost area.

However, the monitoring precision of D-InSAR technique is affected by temporal and spatial decorrelation and atmospheric delay, it is not possible to obtain the surface deformation change with the temporal variationa. In order to overcome these factors, some scholars have proposed time series InSAR technique on the basis of D-InSAR in recent years. Through the time series analysis of high-coherence point targets in a large number of data, the surface deformation rate and deformation time series are obtained. And the permanent scatterers InSAR (PS-InSAR) technique proposed by Ferretti et al. (2001) and small baseline subset InSAR (SBAS-InSAR) technique proposed by Berardino et al. (2002) is the most widely used.

In this paper, we take the road area of a section of the Bei'an-Heihe expressway crossing the Lesser Khingan Mountain which is located in the southern boundary the permafrost region as the study area. Apply the SBAS-InSAR technique to monitor the surface deformation of the permafrost area in the study area. The distribution law of permafrost and the characteristics of the surface deformation are statistically analyzed. Combined with the regional meteorological data, the correlation between surface deformation, permafrost distribution and degradation is analyzed.

## The Study Area

### Landform and Geological Structure

The study area is located in the road area of Bei'an-Beihe Expressway in China. It is at the junction of Sunwu county and Aihui district in Heilongjiang province and it intersects the northwest section of the Lesser Khingan Mountain. Its geographical location is between 127°17′31″–127°21′24″E and 49°30′57″–49°41′50″N, Yang et al. (2008) the southern edge of high latitude permafrost region in China (see Fig. 2), with typical periglacial landform. In recent years, the south boundary of permafrost and the boundary of discontinuous permafrost zones move northward because of climate change, resulting in a decrease in the total area, while the permafrost near the south boundary which belongs to the remains of ancient glacial deposition, is distributed in an island pattern, and in the stage of degradation with distribution the characteristics of lower altitudes, higher latitudes and instability. The study area belongs to the landform of lower mountains and hills, the elevation is between 200 and 530 m. The surface vegetation is mainly grassland and woodland with "Sabre Lin" and "Drunkard Lin". As a part of the Sunwu fault depression, the new fault depression first falls and then rises, under the long-term denudation and planation, the ground is covered with thick layered sediments. The thickness of the sediments on the mountain top and hillside is 1–2 m, and on the intermountain basin and valley is about 10 m, all sediments are sandy loam, clayey loam, gravelly sand, etc., and the surface layer has large thickness of peat and peat.

## Climate and Meteorology

The geographical location of the study area is the connecting part of the middle temperate zone and the cold temperate zone, which is the transitional zone between the two zones. It has a long, dry and cold winter and short, humid and hot summer, with continental monsoon climate (Zhou et al. 2000), the annual extreme maximum temperature of 38.6 °C, the annual extreme minimum temperature of −48.1 °C; the precipitation is mostly concentrated in June to September; the annual average sunshine duration to 2500 h, and the annual total radiation of 1148 kcal/cm$^2$. All the soil near surface can freeze and form seasonal permafrost, the maximum frozen depth is about 2.26–2.4 m from April to May. In mountainous areas, the maximum frozen depth is more than that. From April to September is the melting period, the seasonal frozen soil distributed in the dry area is completely melt in early July. In the swamped thick layer of peat and humus area, melt completely until the end of October. The regional seasonal frozen soil is very developed, and there is island permafrost in the valley section.

According to the data of Sunwu meteorological station (see in Fig. 4) provided by the National Meteorological Network of China, the change curves of the average temperature and precipitation over time in this area as shown in the Fig. 3, the average annual temperature is between −4 and 2 °C, the average annual precipitation is about 550 mm, and the maximum is about 800 mm. From 1954 to now, the precipitation has decreased slightly with little change, which has little effect on the long-term change of the soil thermal state; between 1954 and 1980, the annual average temperature has a small change, since 1980, the annual average temperature began to show a significant change with a increase trend in general, and the rate of temperature increase is relatively fast. It increased from −1.7 °C in 1980 to 1.95 °C in 2018, with a temperature rise of 3.65 °C whichis much higher than the other areas in Northeast China in the same period, and since 2000, the annual average temperature is reached above 0 °C, that will gradually change the permafrost storage conditions; in addition, climate change is not only reflected in the increase of annual average temperature, but also mainly in the change of average minimum temperature. The change trend of annual average minimum temperature changes from decrease to increase, and temperature rise range from 1980 to 2018 reachs to 5.26°, the rate is 0.1 °C/ year, almost three times of the annual maximum temperature. Therefore, under the background of climate warming, permafrost in northeast of China is gradually degenerating and melting.

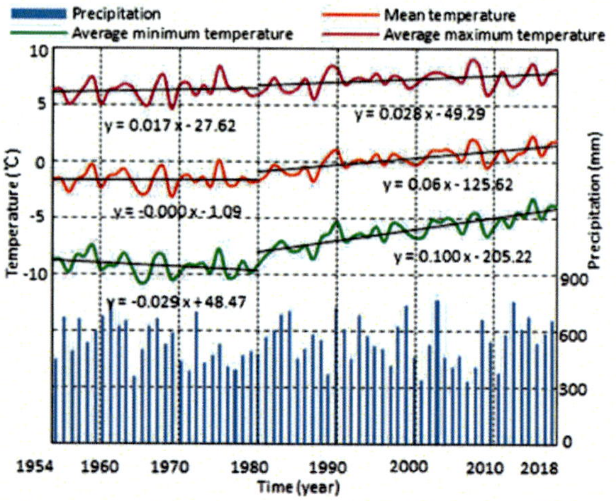


**Fig. 3** 1954–2018 annual average of temperature and precipitation at Sunwu meteorological station

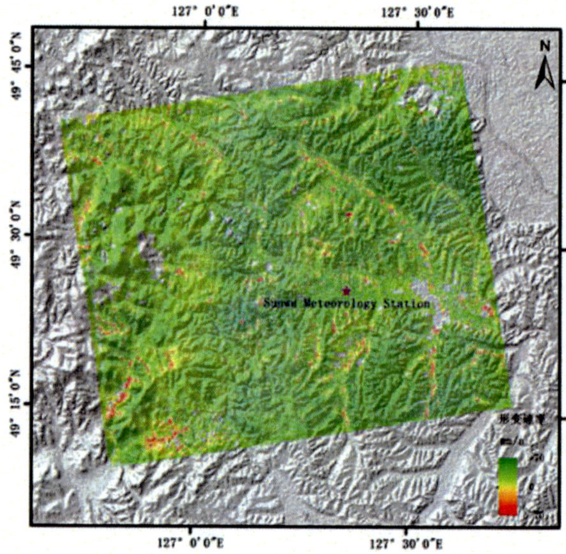


**Fig. 4** Average annual deformation rate along the line of sight direction

## Permafrost Distribution and Degradation

Because winter is controlled by Siberia Mongolia high pressure, inversion layer is widely distributed in the study area, which has an important influence on the development process and regional distribution of permafrost. In addition, the Greater Khingan Mountains and the Lesser Khingan Mountains are the areas with dense forest distribution in China. Low-lying wetland and forest vegetation is the unique ecological environment in forest zone, which can reduce the heat radiation from the sun to the ground surface, block or reduce the heat transfer from the atmosphere to the permafrost, and provide the storage conditions for the permafrost, permafrost and vegetation interact benefit coexists and restricts each other. The study area is located in the permafrost degeneration area, in which the permafrost is basically developed in the low-lying swamp wetland, valley and shade slope with surface water, dense grass growth, thick peat and peat accumulation, and has typical distribution characteristics of "Xing'an-Baikal" permafrost. Thus the permafrost is extremely sensitive to temperature. Due to the interference of climate warming and human activities, the total area of permafrost in the south boundary of permafrost and the boundary of discontinuous permafrost zones move northward, resulting in the reduction of the total area, while the permafrost near the south boundary is distributed in an island pattern.

The study area is located in the Lesser Khingan Mountain island permafrost area, it complex topography, landform and geological conditions. It is one of the most serious geological disaster areas in Northeast China. The Bei'an-Heihe secondary highway was built in 1997 and completed in 2000. In 2009, it was widened and expanded into a highway, island permafrost distribute in some section. Affected by climate warming and environmental change, the existing conditions of permafrost become more fragile. Some sections of Bei'an-Heihe highway with permafrost degenerated from 17 in 2000 to 6 in 2009. Because of the melting of permafrost, geological disasters such as iceing, debris flow, landslide and uneven settlement of foundation occur frequently.

## Data Source and Processing Results

The study area is mainly distributed in the cold temperate coniferous and broad-leaved mixed forest, there are also meadows, shrubs and marshes. The surface vegetation grow vigorously, covering a large area, the period of average snowfall is about 200 days. Moreover, the permafrost active layer is affected by seasonal changes, and the periodic change of soil moisture content is obvious, which has a great impact on the coherency between SAR image, is the main factor affects the accuracy and reliability of InSAR monitoring results in the study area. Therefore, the long wave L-band data with certain penetration ability are selected for differential interference processing, it is less affected by time decorrelation and can maintain good coherence. In this paper, a total of 18 L-band ALOS/PALSAR (path: 423, frame: 980, orbit direction: Descending) SAR images over the study area from July 2007 to December 2010 are used as data sources to retrieve the surface deformation caused by the permafrost degradation. The data set information used for inversion is shown in Table 1.

To improve coherence between SAR images, SBAS-InSAR technique is used to process the long time series data set, and the time series deformation and the average deformation rate are retrieved. The GAMMA software is used in the processing, including three parts: pre-processing, selection of target points and inversion of time series deformation. Selecte the image acquired on 31 July 2009 as the reference image for Single Look Complex coregistration by minimizing the sum decorrelation. Finally, the annual average deformation rate map along the line of sight direction from 2007 to 2010 is obtained, as shown in Fig. 4, and the background is DEM shaded relief. About 860,000 point targets are obtained through data processing, and the deformation rate range is $\pm$ 70 mm/year. We can be that the most of the point targets are relatively stable. The area with obvious deformation is mainly located in the relatively low-lying areas such as valleys and depressions, especially the point targets with annual deformation rate greater than $-35$ mm/year. The distribution characteristics of the deformation rate are consistent with that of the "Xing'an-Baikal" permafrost, which also shows that the permafrost in the study area is degenerating.

**Table 1** Information of data used

| No | Acquired time | No | Acquired time |
|---|---|---|---|
| 1 | 26/07/2007 | 10 | 31/07/2009 |
| 2 | 10/09/2007 | 11 | 15/09/2009 |
| 3 | 11/12/2007 | 12 | 16/12/2009 |
| 4 | 26/01/2008 | 13 | 31/01/2010 |
| 5 | 27/04/2008 | 14 | 18/03/2010 |
| 6 | 12/06/2008 | 15 | 03/05/2010 |
| 7 | 13/12/2008 | 16 | 18/06/2010 |
| 8 | 28/01/2009 | 17 | 18/09/2010 |
| 9 | 15/06/2009 | 18 | 19/12/2010 |

## Results Analysis and Discussion

### Mode of Occurrence of Permafrost

Nelson et al. proposed the ground freezing number model. Using the ground temperature freezing and thawing index and its relationship F-freezing number to determin the plane distribution boundary between permafrost and seasonal permafrost, between continuous and discontinuous permafrost in permafrost area. The simulation results of the model can basically reflect the current permafrost zoning conditions in the study area. The ground freezing number is calculated based on the surface temperature of the meteorological data from Sunwu National Meteorological Station in the study area. The results are shown in Fig. 5. The ground freezing/thawing index fluctuated steadily from 1960 to 1988, with no obvious change trend. The ground freezing number F basically remained between 0.5 and 0.666, indicating the discontinuous distribution of permafrost in the study area. Since 1989, the ground freezing/thawing index has an obvious trend of decreasing or increasing, and the ground freezing number F has been less than 0.5 all the time, indicating that there is no permafrost. The change of ground freezing number F shows that discontinuous permafrost distribution begins to transition to seasonal permafrost distribution in the study area, which is an index of regional permafrost degradation. Of course, the distribution of permafrost is also affected by terrain conditions, land cover, soil properties and other factors. Combined with the field survey, it can be more accurately applied to the division of permafrost distribution.

The 3 Landsat 7 ETM + image datas were choosen respectively on March, May, September in 2009. The infrared surface temperature using the radiation transfer

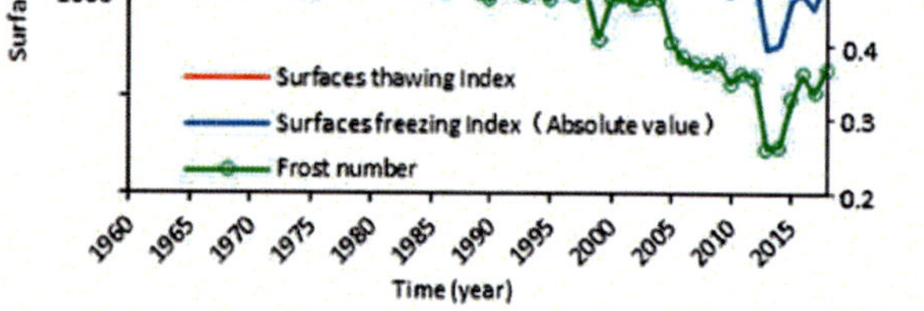


**Fig. 5** 1960 ~ 2018 ground melting/freezing index,freezing number in sunwu region, China

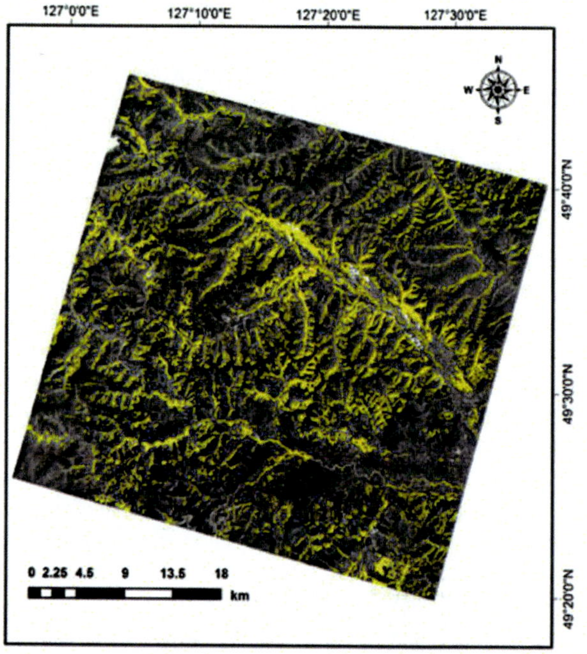


**Fig. 6** The permafrost distribution map in study area

equation method, the inversion results were classified using natural breakpoint classification, and obtained the low temperature areas distribution map of the corresponding moment.Permafrost distribution map was obtained in 2009 on the intersection operation of the low temperature zone, as shown in the yellow area in Fig. 6, the background is Landsat 7 single band image on September in 2009. It can be seen that the permafrost distributed in an island pattern, it is a manifestation of the degradation process of discontinuous permafrost. The permafrost distribution law will be analyzed and discussed in the following.

### Surface Deformation, Permafrost Distribution and Topographic Factors

Attributing to the combined influences of the climate, topography, vegetation and other factors, the permafrost in the high latitude of northeast China distribution is characterized by low-lying and shade slope, and the degradation is characterized by the law of from a high place to a low place, from a sunny slope to a shade slope. These special phenomena constitute the distribution and degradation characteristics of "Xing'an-Baikal" permafrost. Using the SRTM DEM with a resolution of 30 m provided by the U.S. Geological Survey, and taking the overlapping area of permafrost distribution map and deformation rate map as the study area of this paper. According to the accumulated deformation values of the last SAR image on December 19 2010, the point targets are further divided into 8 grades deformation of surface, namely, the accumulated deformation values are respectively <100, −100 ~ −50, −50 ~ −25, −25 ~ −0, 0 ~ 25, 25 ~ 50, 50 ~ 100 and >100 mm, and the three factors i.e. elevation values, slope and aspect of permafrost area and point targets with different grades deformation are extracted respectively, so as to further analysis the relationship between permafrost distribution, surface deformation and topographic factors. Based on the study area data, the statistical results are converted into the percentage of the different factors of permafrost area and surface deformation in the study area. The results are shown in Fig. 7a, according to the elevation distribution of different deformation grades represented by different color curves, we can see that, firstly, the point targets with the accumulated deformation values within ± 25 mm accounts for about 78% of the total point targets, and it is widely distributed in the area, which also menas that the most surface in the study area is relatively stable, and the deformation values is small; secondly, permafrost area accounts for about 14% of the total area of the study area, the elevation values of permafrost distribution is between 230 m and 400 m, and the main distribution range is between 230 and 320 m, which belongs to the relatively low terrain in the area. Moreover, the distribution curve of permafrost has three peaks at the elevation values of 240 m, 310 m and 360 m, which indicates that the permafrost distribution is the same as the topography of the study area with stepped distribution, and the terrain is higher in the west and lower in the east; and then, the accumulated deformation variables within the range of less than 100 mm and greater than 100 mm are less, indicating that with larger deformation variables is less distributed in the area; all the deformation with skewed positively distribution, mainly below the elevation values of 300 m, indicating that the deformation distribution is at location with the lower elevation, according to the analysis of Fig. 7b, permafrost and deformation are mainly distributed in the gentle area with a gradient of 0–6°.

In the same valley area, the slope received the most solar radiation is south slope, followed by the southeast slope and the southwest slope, followed by the east slope, the west slope, the northeast slope and the northwest slope, the least is the north slope. Therefore, the surface temperature of the sunny slope is higher than that of the shade slope, the surface temperature of the shade slope is slightly higher than that of the valley bottom. There is a temperature gradient from the bottom of the valley to the top of the slope. The lowest surface temperature is at the bottom of the valley. Furthermore, under the joint action of climate, topography, landform, surface vegetation and other factors, the high latitude permafrost in Northeast China is characterized by the distribution of low-lying, shade slope, and the degradation is

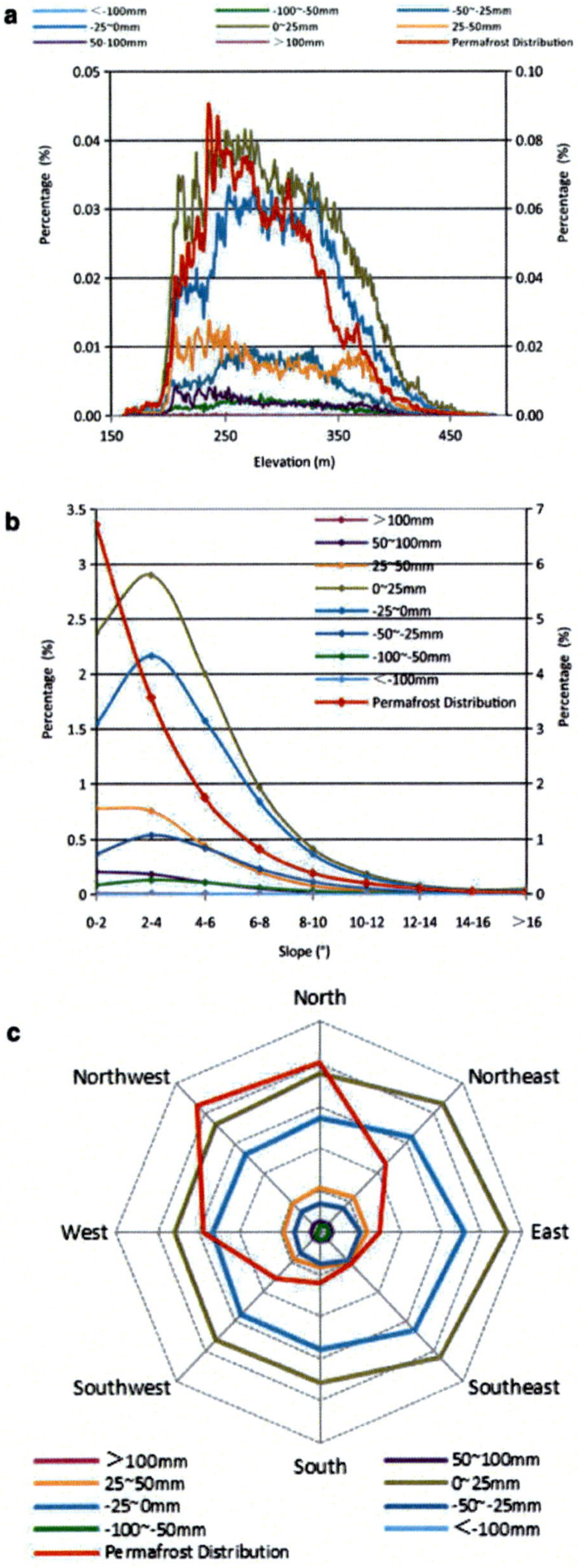


**Fig. 7** **a** Elevation distribution of permafrost and point targets with different deformation. **b** Slope distribution of permafrost and point targets with different deformation. **c** Aspect distribution of permafrost and point targets with different deformation

**Fig. 8** The superposition map of the surface distribution and different accumulated deformation

characterized by the law of from higher to lower, from sunny slope to shade slope, the permafrost can be preserved in low-lying wetlands and shady slopes at the bottom of the ditch Wang et al. (1996). We can see from Fig. 7c that the permafrost distribution is significantly different, with the most on the shade slope facing north and northwest, followed by northeast and west; the deformation distribution is mainly distributed on the slope facing east and southeast, followed by northeast and north. According to the elevation and slope distribution in Fig. 7a, b and the law of permafrost degradation, permafrost is mainly distributed on the shade slope in the gentle and low-lying position, particularlly in the slope range of 0–2°. These indicate that the deformation are mainly caused by the melting of the already degraded permafrost located at the higher place and the sunny slope, and there are a large number of deformation points on the shade slope, which proves that the permafrost still exists is also in the process of degradation.

This can also be proved by the distribution map of cumulative deformation variables. The superposition map of the surface distribution and accumulated deformation values with 25 ~50 mm and −50~−25 mm as shown in Fig. 8. Through comparison, it can be seen that the surface deformation in the gentle valley of permafrost area are mainly realized as uplift, in the boundary of permafrost area are mainly realized as subsidence. In combination with the above mentioned the comparative analysis of distribution situatons of slope, aspect and elevation, prove that the permafrost degradation is characterized by the law of from higher to lower, from sunny slope to shade slope. The soils on the slope fall and slide caused by the first degraded permafrost melting, and accumulate in low-lying valleys.

## Permafrost Degradation and Temperature Change

The thawing period of permafrost is from April to September of each year. Monthly average data from Sunwu meteorological station during the period of 1971 to 2000 as shown in the Fig. 9. The average temperature reach the peak value from June to August, the soil temperature of 40 cm below the surface reach the peak value in July and August, and the maximum frozen soil depth become significantly shallower from July, and change greatly in September. Ground temperature is the expression that the surface soil absorbs solar radiation and converts it into soil heat energy and then transmits it to the deep layer. The changes of ground temperature are lagged than that of air temperature. The maximum freezing depth and the maximum melting depth are obviously later than the maximum and the minimum air temperature. During the melting period, the heat absorbed by the soil is greater than the heat lost, the ground temperature rises and the permafrost melts. That is the time that the surface subsidence reach the largest. From October, the air temperature began to fall. When the temperature dropped below 0 °C, the water in the soil was frozen, and the

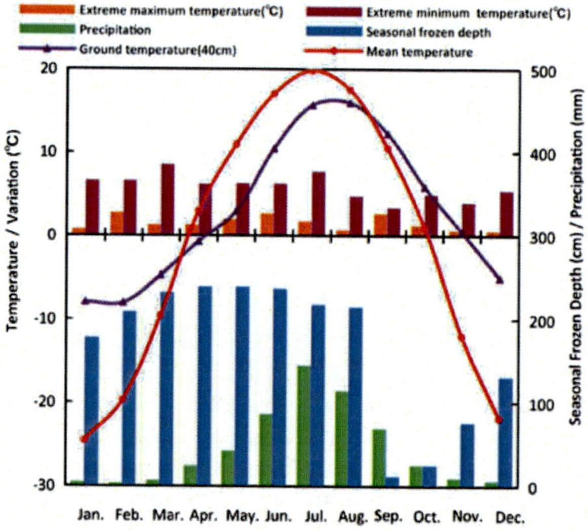


**Fig. 9** Monthly value changes for years in succession of different meteorological elements

surface frost heaved. We can see from the monthly average temperature variation during the period of 1954–2011, that the extreme maximum temperature and extreme minimum temperature in each month are rising, but the extreme minimum temperature is rising faster, and the total increase of extreme maximum temperature in the melting period is greater than the total increase of extreme minimum temperature, while that in the freezing period is the opposite. Except for September,the extreme minimum temperature variation is more than twice of the extreme maximum temperature variation, and the thawing depth of seasonal frozen soil in September is deepest which prove the distribution and degradation of frozen soil have a relationship with the climate change.

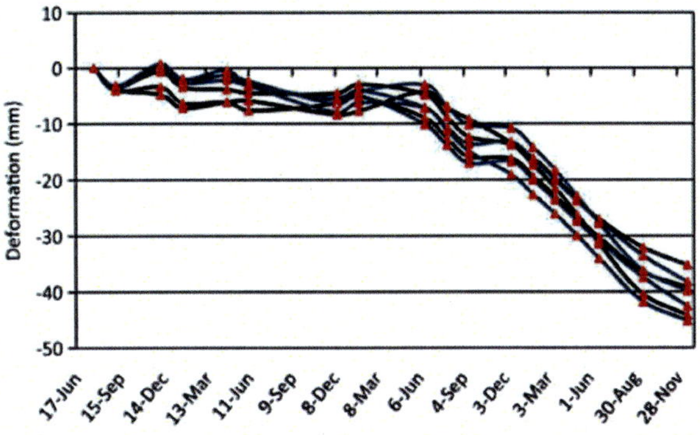


Fig. 11 Time series curves of pavement deformation of Bei'an-Heihe highway

## Permafrost Degradation and Geological Disaster

Climate change leads to the permafrost degradation, human activities further accelerates the degradation process of permafrost. The melting process of permafrost leads to the more landslides and surface subsidence. The permafrost distribution map in the road area of some sections of Beian-Heihe highway is shown in Fig. 10. Mark A and B were confirmed to be permafrost by drilling in 2009. During period of road construction, the permafrost melted in a large area, resulting in the subsidence of the foundation and subgrade and the obvious cracks in the pavement. Mark E was a landslide in 2000, resulting in the overall instability of the subgrade, and the permafrost had degenerated toward to the valley in 2009, during the degradation process, the soil on the slope slips and falls, and finally accumulates in the valley. Therefore, the surface of the valley area has obvious uplift, while the

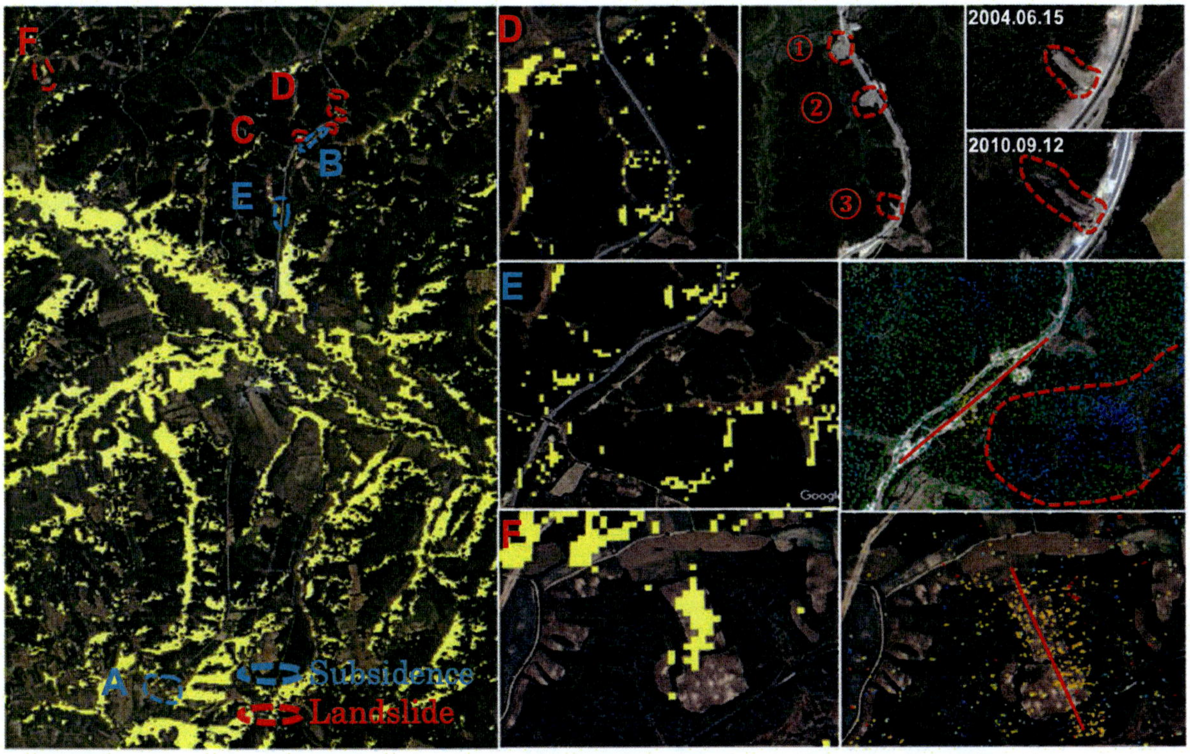


Fig. 10 Distribution map of geological hazards in the road area of Bei'an-Heihe highway

old road still has large surface deformation, and the time sequence curve of the surface deformation along the direction of the red marking line is shown in Fig. 11. After the old road is abandoned, one side of the road were widened and expanded into a highway in 2009, and the road in this section were rebuilted, while there is permafrost distributed on the left side of the road, the permafrost degradation also causes landslides. There are three landslides in mark D, with the continuous degradation of permafrost, new landslides are generated on the original landslides (mark as ③). In June 2010, a comprehensive survey was carried out on the ③ landslides and started monitoring of the landslide process. The results show that the landslide movement has a low angle, intermittent slip characteristics; the occurrence time and rate of sliding have a corresponding relationship with the pore water pressure of the soil at the back edge of the landslide. With the decrease of the pore water pressure, the landslide movement gradually stops, and the permafrost melting water seepage into the landslide body is the main cause of the landslide movement. Due to the lack of investigation and monitoring data before 2009, the initial sliding time of the landslide cannot be determined. From the satellite photos of June 15, 2004 and September 12, 2010, it was found that the straight-line distance between the front edge and the back edge of the landslide increased from 101.26 m to 145.05 m, the front edge of the landslide moved forward 43.79 m, and the position of the back edge basically did not change Shan et al. (2015); mark F has permafrost distribution are located outside the road area, the terrain around mark F is multi-level platform distribution, which means that the permafrost gradually degeneration and the surface creep. The deformation time series curve of the red line marked section in the figure is shown in Fig. 12. The linear trend change is related to the permafrost degradation, while the periodic change is related to the seasonal replacement and the squeezing and movement process of slope soil.

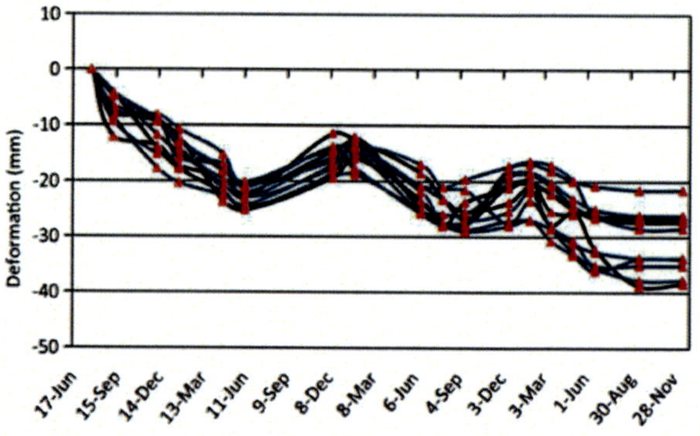


**Fig. 12** Change curves of slope body in permafrost degradation area of Bei'an-Heihe highway

## Potential Geological Environment Problems

Under the background of climate warming, permafrost of high latitude in China is in the process of general degradation. The permafrost degradation not only destroys the ecological environment in the cold region, but also affects the construction engineering. Permafrost is the cornerstone of structures in cold regions, which temperature state, soil texture and ice content affect the soil force and thermal properties of permafrost, and then the stability of structures. The Bei'an-Heihe expressway area is one of the regions with the highest temperature rise rate in China oner the past 50 years. Since it was built in 2000, 17 frozen soil sections along the route have degenerated decrease to 6 in 2009. The landslides, uneven settlement of foundation and other geological disasters caused by the melting of permafrost occur frequently. In addition, in recent years, due to the degradation of permafrost, it including hydrology, vegetation and greenhouse gas emission, wildfire and other ecological environmental problems have gradually attracted widespread attention.

## Conclusion

In summary, in the context of global warming, the southern boundary of permafrost is moving northward, and the degradation of permafrost is accelerating. This paper uses the time series InSAR technique to conduct large-scale monitoring of the surface deformation in the research area. Based on the permafrost distribution and meteorological monitoring data in the research area, the relationship between climate change and permafrost degradation, permafrost degradation and surface deformation was studied. The following conclusions were finally reached:

From 1954 to 2018, the temperature in the research area increased significantly, with an annual average temperature increase of 3.65 °C, which is much higher than the temperature change in other parts of Northeast China at the same time, especially the annual minimum temperature rise rate reached 0.1 °C/year. Permafrost is sensitive to climate change, and the increase in temperature is the main factor for the degradation of permafrost. The increase in temperature makes the permafrost conditions in the region more fragile and the degradation of permafrost accelerates. If the future temperature continues to rise, permafrost will accelerate degradation.

The surface deformation map of Bei'an-Heihe road area is obtained. The surface deformation rate of this area is ±70 mm/year during the time of 2007–2010. The surface deformation is closely related to the permafrost distribution, and the distribution law is consistent. Contrastive analysis the elevation values, slope and aspect of permafrost and deformation, we find that the permafrost in the study area forming the distribution characteristics of "Xing'an-Baikal" type, characterized by low-lying, gentle terrain and shady slope; surface deformation characteristics consistent with the law of permafrost distribution, mainly in the permafrost degraded area in sunny slope with relatively less solar radiation of east and southeast, the slope range of 2–4°, realized as uplift. The surface deformation at the boundary of the degenerated permafrost is realized as settlement, which proves that the permafrost degradation is characterized by the law of from a high place to a low place, frome a sunny slope to a shady slope, also shows that the permafrost the study area is in degradation process.

In the process of permafrost degradation, a series of environmental and engineering geological problems such as uneven ground settlement, seasonal landslides and spring slip will be caused, which have a great impact on the construction, safe operation and maintenance of the project, and are important factors of road diseases in cold regions. At the same time, permafrost degradation is accompanied by the succession of surface vegetation, the decline of groundwater level and the loss of soil water, which affect the regional surface hydrology and carbon cycle. Permafrost is also one of the main sources and sinks of greenhouse gases, with the melting of permafrost, a series of problems such as greenhouse effect and wildfire come one after another, so the monitoring of permafrost degradation area is very important. Beian-Heihe highway in the permafrost degradation area of lesser Khingan Mountains in Northeast China selected in this paper is only a representative of the permafrost degradation area. With the continuous warming of climate and the expansion of human activities, more permafrost degradation areas will appear, and there are many other factors affecting the process of permafrost degradation and deformation, which need to be combined with more abundant data to improve the deformation observation step by step, it is hoped that more colleagues and relevant people will pay more attention to it and join this study.

## References


Berardino P, Fornaro G, Lanari R et al (2002) New algorithm for surface deformation monitoring based on small baseline differential SAR interferograms. IEEE Trans Geosci Remote Sens 40 (11):2375–2383

Chen LX, Zhou XJ, Li WL et al (2004) Characteristics of the climate change and its formation mechanism in China in last 80 years. Acta Meteorologica Sinica 62(5):634–646

China Meteorological Administration National Climate Center (2009) China climate impact assessment 2008. China Meteorolgical Press, Beijing

Ding YH, Ren GY, Shi GY, et al (2006) National assessment report of climate change (I):climate change in China and its future trend. Adv Climate Change Res 2(1):3–8+50

Ferretti A, Prati C, Rocca F (2001) Permanent scatterers in SAR interferometry. IEEE Trans Geosci Remote Sens 39(1):0–20.

Gao F, Wang BS (2008) A study on global warming and thetemperature changes in the Northeast region. Marine Forecasts 25(1):25–30

He RX, Jin HJ, Lv LZ et al (2009) Recent changes of permafrost and cold regions environments in the northern part of northeastern China. J Glaciol Geocryol 31(3):525–531

Jorgenson MT, Racine CH, Walters JC et al (2001) Permafrost Degradation and Ecological Changes Associated with a WarmingClimate in Central Alaska. Climatic Change 48(4):551–579

Jorgenson MT, Shur YL, Pullman ER (2006) Abrupt increase in permafrost degradation in Arctic Alaska. Geophys Res Lett 33 (2):356–360

Rosen PA, Hensley S, Zebker HA et al (1996) Surface deformation and coherence measurements of Kilauea Volcano, Hawaii, from six month repeat pass interferometry with SIR-C radar. J Geophys Res Planets 101(E10):23109–23125

Shan W, Hu ZG, Guo Y et al (2015) The impact of climate change on landslides in southeastern of high-latitude permafrost regions of China. Frontiers Earth Sci 3

Wang CH, Zhang BL, Liu FT (1996) A Preliminary analysis on the regularity of permafrost degradation, its advantages and disadvantages in the Greater and Lesser Xing'an mountains. J Glaciol Geocryol S1:174–180

Wang S, Jin H, Li S et al (2000) Permafrost degradation on the qinghai-tibet plateau and its environmental impacts. Permafrost Periglac Process 11(1):43–53

Wei Z, Jin HJ, Zhang JM et al (2011) Prediction of permafrost changes in Northeast China under climate change conditions. Scientia Sinica Terrae 41(1):74–75

Wang CJ, Shan Mn, Hu ZG, Shan W (2015) Multi-spectral remote sensing based land surface temperature retrieval and isolated permafrost zone segmentation. Infrared Laser Eng 44(4)

Xie C, Li Z, Li XW (2008) A study of deformation in permafrost regions of Qinghai-Tibet Plateau based on ALOS/PALSAR D-InSAR interferometry. Remote Sens Land Resour 20(3):15–19

Yang YP, Cheng DX, Fu HX (2008) Engineering geological characteristics and evaluations of permafrost in Daxing'an Mountains. J Eng Geol 16(5):657–662

Yang M, Nelson FE, Shiklomanov NI et al (2010) Permafrost degradation and its environmental effects on the Tibetan Plateau: A review of recent research. Earth Sci Rev 103(1–2):31–44

Zhou YW, Guo DX, Qiu GQ (2000). Geocryology in China. China Science Press, Beijing; Zuo HC , Lü SH , Hu YQ (2004) Variations trend of yearly mean air temperature and precipitation in China in the last 50 years. Plateau Meteorology.

Zhao ZC, Wang SW, Luo Y (2007) Assessments and projections of temperature rising since the establishment of IPCC. Adv Climate Change Res 3(3):183–184

Zhao ZC, Luo Y, Huang JB (2018) Review IPCC 30 years (1988–2018). Adv Climate Change Res 14(05):540–546

2020 Kyoto Japan

# Climate Change Impact Evaluation on the Water Balance of the Koroška Bela Area, NW Slovenia

Nejc Bezak, Tina Peternel, Anže Medved, and Matjaž Mikoš

**Abstract**

Climate change is expected to affect the elements of the hydrological cycle, which are also related to the landslides and debris flows triggering mechanics. In the scope of this contribution, we evaluated climate change impact on the water balance components in the hinterland of the Koroška Bela (NW Slovenia), which is one of the endangered settlements in Slovenia. For this purpose, we focused on the representative concentration pathway (RCP) RCP4.5 scenario, which can be described as a midway scenario. Lumped conceptual hydrological model was used in order to model the relationship between the hydrological cycle elements. Moreover, climate change impact on the intensity-duration-frequency (IDF) curves was also estimated. The results indicate that total rainfall as well as effective rainfall could increase in the future. Mainly due to air temperature increase, we could see an evapotranspiration increase in the future. Moreover, hydrological model results indicate a possible surface runoff increase from the Bela stream catchment. However, these changes are mostly in the range of the 5% with the exception of the air temperature data where difference between future and past is larger. Furthermore, subsurface water storage included in the model indicates that only slight changes in the wetness of the catchment might be expected. Additionally, climate change could also affect the extreme rainfall characteristics, which can be seen as a change in the IDF curves. However, these changes are relatively small.

**Keywords**

Climate change • Water balance • Rainfall • Runoff • Koroška bela • Landslides

N. Bezak (✉) · M. Mikoš
Faculty of Civil and Geodetic Engineering, University of Ljubljana, Jamova Cesta 2, 1000 Ljubljana, Slovenia
e-mail: nejc.bezak@fgg.uni-lj.si

M. Mikoš
e-mail: matjaz.mikos@fgg.uni-lj.si

T. Peternel
Geological Survey of Slovenia, Dimičeva Ulica 14, 1000 Ljubljana, Slovenia
e-mail: Tina.Peternel@GEO-ZS.SI

A. Medved
Slovenian Environment Agency, Vojkova 1b, 1000 Ljubljana, Slovenia
e-mail: Anze.Medved@gov.si

## Introduction

Climate change is expected to affect the water cycle components such as rainfall, surface runoff, subsurface fluxes and their relationships. However, it is not clear how will the temperature increase impact on the rainfall-runoff dynamics (Blöschl et al. 2019), which is also related to the landslides and debris flows triggering mechanics (Ciervo et al. 2017). Therefore, site-specific investigations should be carried out in order to evaluate climate change impact in specific region or catchment. Slovenia, which is positioned in Central Europe and covers 20,273 km$^2$, is characterized by steep slopes and complex geological characteristics, which means that variety of different landslide forms can be found in the area (Jemec Auflič et al. 2017). One of the most endangered areas is the north-west part of the country. One of the endangered villages is the Koroška Bela that is located below the Potoška planina slow deep-seated landslides with the sliding mass composed of tectonically deformed and weathered Upper Carboniferous and Permian clastic rocks covered with a large amount of talus material (Peternel et al. 2017; Jemec Auflič et al. 2019). Due to the landslide characteristics, knowledge about hydrological (i.e. water) balance of the area is important because the landslide movement is driven by the subsurface flows. Moreover, multiple processes such as rainfall interception or evapotranspiration affect these flows. Therefore, longer duration



V. Vilímek et al. (eds.), *Understanding and Reducing Landslide Disaster Risk*, ICL Contribution to Landslide Disaster Risk Reduction,
https://doi.org/10.1007/978-3-030-60319-9_25

rainfall events or changed frequency of events (e.g. extreme events more frequently occur in autumn) could affect the landslide stability. In order to investigate characteristics of the future hydrological conditions in the study area, climate change investigation should also be carried out. Moreover, since past experiences have shown that landslide mass could transform into a debris flow as was the case in the Log pod Mangartom debris flow (Mikoš et al. 2004, 2007), knowledge about changes in the triggering mechanics is important. These can be investigated with the analysis of the changes in the relationship between empirical rainfall thresholds and intensity-duration-frequency (IDF) curves.

Therefore, the main aim of this contribution is to evaluate climate change impact on the water cycle in the Koroška Bela area where several potential landslides are located. More specifically, impact on the rainfall, evapotranspiration, air temperature, surface and subsurface runoff were investigated. Additionally, we also investigated changes in the intensity-duration-frequency (IDF) curves and their relationship to the empirical rainfall thresholds.

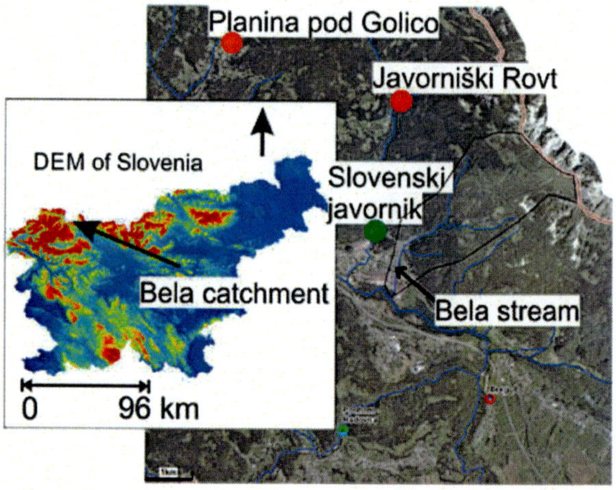


**Fig. 1** Location of the investigated area on the DEM of Slovenia (i.e. red, high elevation; blue, low elevation) and location of the discharge (green circle) and rainfall (red circle) measuring stations. The Bela catchment boundary is also shown with black line

## Data and Methods

### Koroška Bela Area

In the scope of this contribution, we investigated the Bela stream catchment that is located in the NW part of Slovenia. Main characteristics of the catchment are shown in Table 1. Moreover, Fig. 1 shows location of the Bela stream and the surrounding area. One can see that area is characterized by steep slopes, which lead to torrential response of the Bela stream. Since there is no discharge gauging station available in the catchment, the area is regarded as ungauged. Therefore, in order to calibrate the hydrological model, data from neighbour catchment was used (i.e. Javornik catchment where main characteristics are also shown in Table 1). The hinterland of Koroška Bela is known to be prone to several slope mass movements, which are especially common in saturated soils with low permeability. These landslides represent direct risk to settlement Koroška Bela (Peternel et al. 2017, 2018; Jemec Auflič et al. 2019).

### Hydrological Modelling and Empirical Rainfall Thresholds

In order to evaluate the climate change impact on the hydrological cycle lumped conceptual hydrological model named GR6J with included snow module CemaNeige was used (Perrin et al. 2003; Coron et al. 2017). In order to run the model, daily rainfall, potential evapotranspiration and air temperature data are needed. Moreover, for the snow module also catchment hypsometric curve is required, which was defined based on the 20 m DEM of Slovenia. For the rainfall data, we used Javorniški Rovt station (Fig. 1), for air temperature data, we used Planina pod Golico station because Javorniški Rovt station does not have air temperature data available. Smaller percent of missing air temperature data was calculated using data from another station (i.e. Lesce) and using linear regression where $R^2$ between two stations was 0.98. The Equation proposed by Oudin and included in airgr package was used for potential evapotranspiration calculation (Coron et al. 2017). In order to calibrate the model, discharge data from the Slovenski Javornik station on

**Table 1** Main characteristics of Bela and Javornik catchments

| Properties | Javornik | Bela |
|---|---|---|
| Catchment area [$km^2$] | 16.5 | 6.6 |
| Mean catchment slope [%] | Around 46 | Around 51 |
| Mean stream slope [%] | Around 20 | Around 27 |
| Three major land-use types | 82% forest, 12% pastures, 2% urban areas | 72% forest, 18% pastures, 5% urban areas |
| Mean annual precipitation [mm] | Around 1900 | Around 1900 |

the Javornik stream was used (Fig. 1). We used to ratio between catchment areas shown in Table 1 (i.e. Bela and Javornik catchments) in order to estimate the discharge values for the Bela stream catchment. A similar approach was for example used by Zhang and Han (2017) who also applied ratio of catchment areas. Moreover, catchments with similar properties show similar response to rainfall and can be modelled using similar model parameters (Bárdossy 2007). Since only discharge data from 6.2.2012 until 31.10.2017 was available, this period was used for model calibration. Year 2012 was used for model warm-up. Detailed description of the hydrological model used can be found in Perrin et al. (2003) or Coron et al. (2017).

Recently, Bezak et al. (2019) showed that production store level (i.e. underground reservoir that is part of the conceptual representation of the catchment in the scope of the GR6J model) can be used as a proxy of wetness conditions in the catchment, which are also related to the landslides triggering. If production store level is high for several days, this could have a negative effect on the landslides stability. Moreover, Bezak et al. (2019) showed that this variable could be used for the prediction of landslides and proposed methodology yielded better performance than empirical landslide threshold curves.

## Climate Change Impact

In order to evaluate the climate change impact on the hydrological conditions and potential connection to landslide and debris flow triggering mechanisms we used results of the study entitled "The estimation of the average and extreme meteorological and hydrological conditions in Slovenia over the 21st century", which was carried out by the Slovenian Environment Agency (Bertalanič et al. 2018). Therefore, detailed description about bias correction and downscaling methodology used can be found in Bertalanič et al. (2018). This data was also used in order to investigate climate change impact on the high and low-flows of the karst catchments in Slovenia (Sapač et al. 2019). For the purpose of this study, we used five different combinations of global climate models (GCM) and regional climate models (RCM), namely (GCM/RCM): CNRM-CM5-LR/CCLM4-8-17 (model 1), MPI-ESM-LR/CCLM4-8-17 (model 2), MPI-ESM-LR/RCA4 (model 3), EC-EARTH/HIRHAM5 (model 4), and IPSL-CM5A-MR/WRF331F (model 5) (i.e. RCP4.5 scenario was used). The data is now available with a 1 km resolution. We used bias-corrected model data from the project described in Bertalanič et al. (2018) and then downscaled it to 1 km grid. We also used the observed data at 1 km resolution. We applied bias-correction with non-parametric quantile mapping using empirical quantiles (Gudmundsson et al. 2012) with 61-day moving window for each grid cell separately. For the purpose of this study, we additionally derived the bias-corrected downscaled point rainfall for the Javorniški Rovt rainfall station. This bias-correction was made a bit differently as described above. In this casse, we used raw model data in 12 km resolution. We applied bias-correction for the model cell in which Javorniški Rovt rainfall station is located with quantile delta mapping (Cannon 2018) with the same moving window as above. This method is more complicated and time-consuming, so we used it just for this point and not for the whole grid. Data period from 1981 to 2100 was investigated. Period from 1981-2020 was used as past (i.e. reference) period and periods 2021-2060 and 2061-2100 were used as future periods. The contribution shows relative difference of future conditions in comparison to past conditions (i.e. ((X2021-2060 – X1981-2020)/X1981-2020)*100)). The exception is the air temperature data, where we used temperature anomalies with respect to the reference period.

## Results and Discussion

### Hydrological Cycle

In the first step of the study, we carried out calibration of the hydrological model GR6J CemaNeige. Calibration results are shown in Fig. 2. Nash-Sutcliffe coefficient was 0.75 where a value of 1 would indicate a perfect fit between measured and modelled data and this coefficient ranges between $-\infty$ and 1. Based on graphical comparison of results and this coefficient we can assume that hydrological model provides satisfactory results. Moreover, Fig. 3 shows catchment boundary and 1 km grid cells used. Since the correlation between grid cell and point data is very high (i.e. $R^2$ was 0.98) we used the 1 km grid cell data as an input to the hydrological model.

Information about rainfall, air temperature and evapotranspiration were used as an input to the calibrated hydrological model. Model was run for the 1981-2100 period. Based on the model results, we compared future periods and the reference period. Tables 2, 3, 4, 5, 6, 7 and 8 show the climate change impact on the hydrological conditions in the area. One can notice that climate change could lead to more rainfall, much higher air temperature values and consequently also higher evapotranspiration values in the Bela stream catchment (Tables 2, 3 and 4). The largest differences can be detected for the air temperature variable.

Moreover, Tables 5, 6, 7 and 8 show results of the hydrological model. Similarly as total rainfall, also effective rainfall could increase in the future (Table 5). Furthermore,

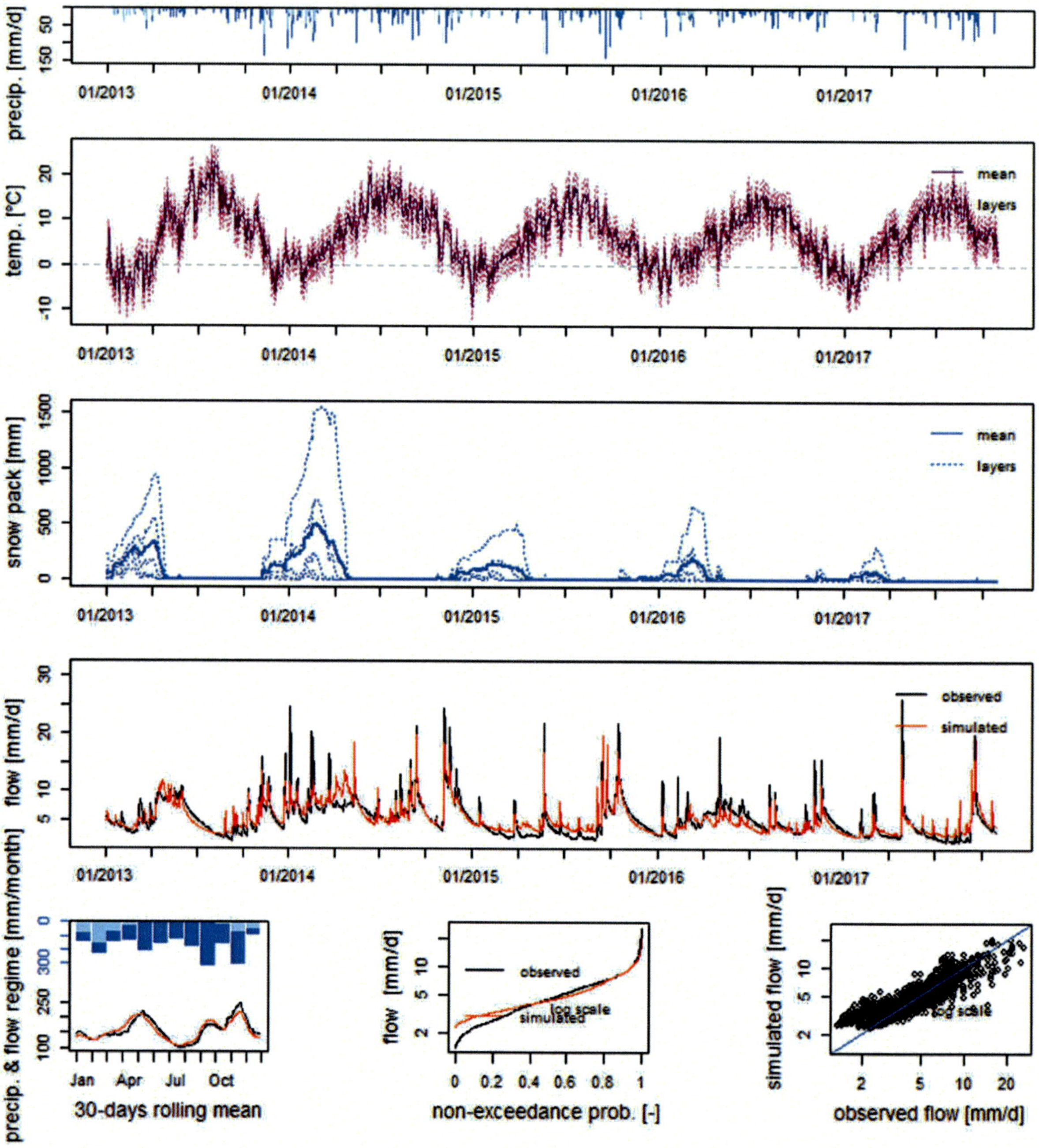


**Fig. 2** GR6J CemaNeige calibration results. Measured rainfall, measured air temperature, modelled snow pack, measured and modelled flows are shown in the above four panels. Snow pack and air temperature data for 5 different elevation layers used in the CemaNeige snow module are also shown. The lowest panel shows 30-days rolling mean, flow duration curve and scatter plot between measured and observed data. Orange and black lines indicate simulated and observed values, respectively

runoff from the Bela stream catchment could also increase (Table 6). Table 7 shows climate change impact on the production store (i.e. underground reservoir) included in the hydrological model that can be used as proxy of wetness conditions. One can see that climate change models predict only slight changes in this variable (i.e. less than 1%).

**Fig. 3** Bela stream catchment area and 1 km climate change grid cells with Gauss-Kruger Slovenia coordinate system shown

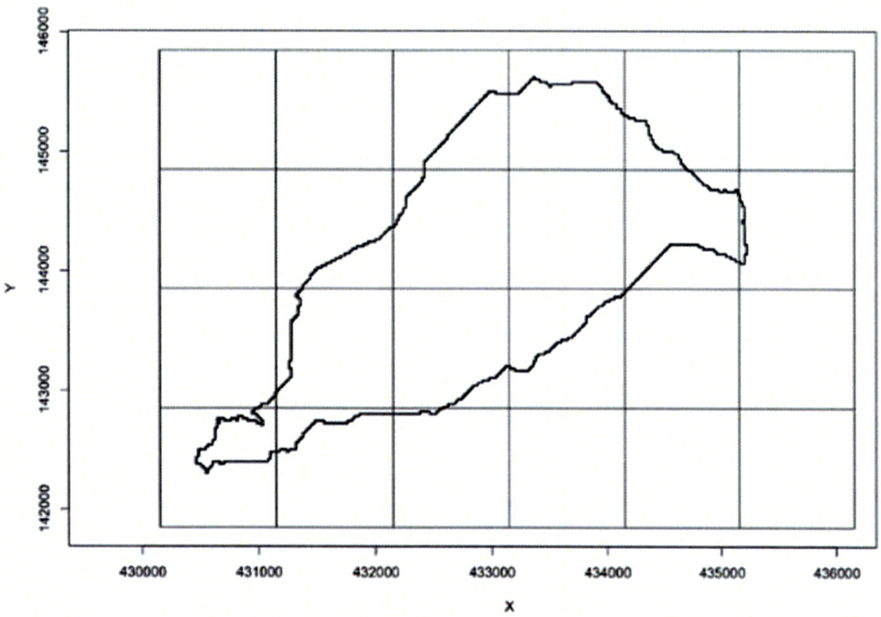


**Table 2** Climate change impact on the rainfall

| Rel. difference [%] | 2021–2060 in comparison to 1981–2020 | | | 2061–2100 in comparison to 1981–2020 | | |
|---|---|---|---|---|---|---|
| Model | 1st quartile | Mean | 3rd quartile | 1st quartile | Mean | 3rd quartile |
| 1 | 0.0 | −6.8 | -12.9 | 0.0 | 6.9 | 0.8 |
| 2 | 0.0 | 1.9 | -2.7 | 0.0 | 6.5 | 6.8 |
| 3 | 0.0 | 4.9 | 2.4 | 0.0 | 18.2 | 25.3 |
| 4 | 0.0 | 5.9 | 4.8 | 0.0 | 2.7 | −14.7 |
| 5 | 0.0 | 8.2 | 6.5 | 0.0 | 3.8 | −13.3 |
| Average | 0.0 | 2.8 | −0.4 | 0.0 | 7.6 | 1.0 |

**Table 3** Climate change impact on the average air temperature

| Temp. anomaly [°C] | 2021–2060 in comparison to 1981–2020 | | | 2061–2100 in comparison to 1981–2020 | | |
|---|---|---|---|---|---|---|
| Model | 1st quartile | Mean | 3rd quartile | 1st quartile | Mean | 3rd quartile |
| 1 | 0.9 | 0.7 | 0.5 | 1.7 | 1.6 | 1.4 |
| 2 | 1.0 | 1.1 | 1.1 | 1.6 | 1.5 | 1.3 |
| 3 | 1.0 | 1.0 | 0.9 | 2.0 | 1.9 | 1.6 |
| 4 | 0.7 | 0.8 | 0.8 | 1.0 | 1.4 | 1.4 |
| 5 | 1.3 | 1.3 | 0.9 | 1.9 | 2.1 | 1.6 |
| Average | 1.0 | 1.0 | 0.8 | 1.6 | 1.7 | 1.5 |

**Table 4** Climate change impact on the average evapotranspiration

| Rel. difference [%] | 2021–2060 in comparison to 1981–2020 | | | 2061–2100 in comparison to 1981–2020 | | |
|---|---|---|---|---|---|---|
| Model | 1st quartile | Mean | 3rd quartile | 1st quartile | Mean | 3rd quartile |
| 1 | 9.7 | 3.1 | 1.9 | 8.5 | 4.6 | 4.6 |
| 2 | 4.5 | 4.3 | 4.6 | 7.1 | 3.6 | 2.4 |
| 3 | 9.7 | 5.3 | 5.3 | 7.1 | 5.1 | 4.6 |
| 4 | 5.0 | 3.3 | 3.0 | 7.7 | 7.4 | 6.6 |
| 5 | 7.2 | 7.0 | 7.1 | 18.0 | 11.3 | 11.0 |
| Average | 7.2 | 4.6 | 4.4 | 9.7 | 6.4 | 5.8 |

**Table 5** Climate change impact on the effective rainfall

| Rel. difference [%] | 2021–2060 in comparison to 1981–2020 | | | 2061–2100 in comparison to 1981–2020 | | |
|---|---|---|---|---|---|---|
| Model | 1st quartile | Mean | 3rd quartile | 1st quartile | Mean | 3rd quartile |
| 1 | 0.0 | -6.5 | −16.1 | 0.0 | 8.8 | −1.7 |
| 2 | 0.0 | 0.7 | −11.0 | 0.0 | 6.5 | −3.0 |
| 3 | 0.0 | 4.8 | 2.0 | 0.0 | 19.7 | 22.3 |
| 4 | 0.0 | 4.3 | −0.5 | 0.0 | 1.1 | −17.6 |
| 5 | 0.0 | 7.3 | 0.7 | 0.0 | 2.7 | −16.7 |
| Average | 0.0 | 2.1 | −5.0 | 0.0 | 7.8 | −3.3 |

**Table 6** Climate change impact on the runoff from the Bela stream catchment

| Rel. difference [%] | 2021–2060 in comparison to 1981–2020 | | | 2061–2100 in comparison to 1981–2020 | | |
|---|---|---|---|---|---|---|
| Model | 1st quartile | Mean | 3rd quartile | 1st quartile | Mean | 3rd quartile |
| 1 | −9.8 | −8.6 | −8.2 | 5.4 | 7.7 | 9.1 |
| 2 | −8.6 | −0.4 | 3.3 | 0.2 | 5.4 | 10.3 |
| 3 | 2.0 | 3.3 | 4.9 | 17.1 | 20.2 | 21.6 |
| 4 | 2.6 | 4.0 | 4.2 | −4.1 | −0.4 | 2.1 |
| 5 | 3.8 | 6.8 | 9.4 | −0.1 | 0.5 | −0.9 |
| Average | −2.0 | 1.0 | 2.7 | 3.7 | 6.7 | 8.5 |

**Table 7** Climate change impact on the production store level (i.e. conceptual underground reservoir)

| Rel. difference [%] | 2021–2060 in comparison to 1981–2020 | | | 2061–2100 in comparison to 1981–2020 | | |
|---|---|---|---|---|---|---|
| Model | 1st quartile | Mean | 3rd quartile | 1st quartile | Mean | 3rd quartile |
| 1 | −2.5 | −1.9 | −1.5 | 0.6 | 1.1 | 1.6 |
| 2 | −2.2 | −0.9 | 0.4 | −0.2 | 0.6 | 1.8 |
| 3 | −0.4 | −0.3 | 1.0 | 3.0 | 3.2 | 3.5 |
| 4 | 0.1 | 0.5 | 0.9 | −1.7 | −0.7 | 0.3 |
| 5 | 0.2 | 0.8 | 1.8 | −0.7 | −0.5 | 0.1 |
| Average | −1.0 | −0.4 | 0.5 | 0.2 | 0.7 | 1.5 |

**Table 8** Climate change impact on the percolation from production store level (i.e. conceptual underground reservoir)

| Rel. difference [%] | 2021–2060 in comparison to 1981–2020 | | | 2061–2100 in comparison to 1981–2020 | | |
|---|---|---|---|---|---|---|
| Model | 1st quartile | Mean | 3rd quartile | 1st quartile | Mean | 3rd quartile |
| 1 | −11.8 | −8.7 | −7.4 | 3.0 | 6.1 | 8.3 |
| 2 | −10.4 | −2.2 | 1.8 | −0.8 | 4.1 | 9.6 |
| 3 | −2.0 | 1.0 | 5.2 | 15.9 | 17.3 | 18.9 |
| 4 | 0.3 | 2.6 | 4.6 | −8.0 | −2.5 | 1.3 |
| 5 | 1.0 | 5.0 | 9.4 | −3.3 | −1.3 | 0.7 |
| Average | −4.6 | −0.5 | 2.7 | 1.4 | 4.7 | 7.7 |

## Empirical Rainfall Thresholds

Moreover, we also evaluated climate change impact on the IDF curves where bias-corrected downscaled point data was used for this purpose. The analysis was carried for rainfall durations of 0.5, 1, 3 and 5 days.

The comparison was made for 2, 5, 10, 25, 50, 100 and 250-years return period. The results indicated that design rainfall values in the future will be more or less the same as in the reference period (i.e. 1981-2020). For example, for the 100-year return period and for the 24 h rainfall duration, the design rainfall should increase for 2% and 4% for the 2021-2060 and 2061-2100 periods, respectively. Differences of the same range were also observed for other return periods (Fig. 4). This also means that future relationship between IDF curves and empirical rainfall thresholds should remain more or less the same (Fig. 4). However, it should be noted that large uncertainties could affect the results for sub-daily rainfall durations and large return periods.

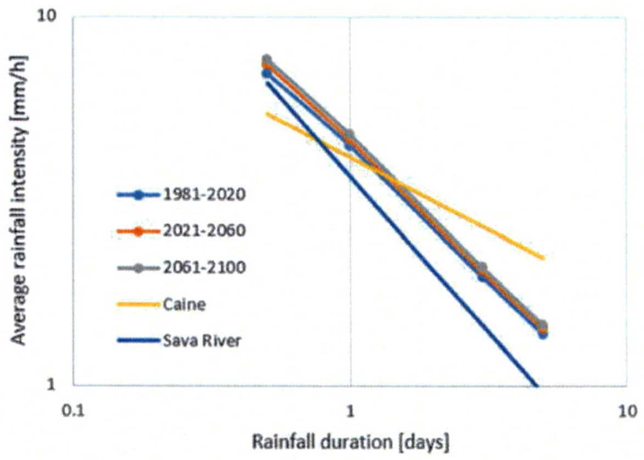


**Fig. 4** Comparison between IDF curves for the 2-year return period and empirical rainfall thresholds defined by Caine (1980) and Rosi et al. (2016) for the Sava River catchment

## Conclusions

Based on the presented results next conclusions can be made

Total and effective rainfall, air temperature, evapotranspiration and runoff from the Bela stream catchment might increase in the future. For most of the variables, the average differences calculated using five difference GCM/RCMs are in the range of 5% with the exception of the air temperature where temperature anomalies are in the range of 1–2 °C.

The production store level that can be understand as a proxy of the catchment wetness conditions is projected to be more or less the same as in the past (i.e. relative difference less than 1%). This could indicate that landslides activity in the future could be similar to the situation in the past.

Moreover, future changes in the IDF curves are also mostly in the range of 5%.

However, it should be noted that there exist relatively large differences among five tested GCM/RCMs and that some models predict more significant future changes. Moreover, it should be noted that we have evaluated the RCP4.5 scenario and not the RCP2.6 and RCP8.5 scenarios, which could lead to different dynamics of the water cycle processes. Bertalanič et al. (2018) provided detailed comparison among the three RCP scenarios for the air temperature and evapotranspiration data. Moreover, the RCP4.5 can be regarded as a midway scenario positioned between the optimistic (i.e. RCP2.6) and pessimistic (i.e. RCP8.5) scenarios.

**Acknowledgments** The results of this study are part of the research projects J7-8273 "Recognition of Potentially Hazardous Torrential Fans Using Geomorphometric Methods and Simulating Fan Formation" and J1-8153 "Studying Landslide Movements from Source Areas to Zone of Deposition using a Deterministic Approach" that are financed by the Slovenian Research Agency. Both projects were approved by the International Programme on Landslides (IPL) as the IPL-225 and IPL-226 Project, respectively. Moreover, climate change investigation was also financially supported by the Ministry of the Environment and Spatial Planning.

## References


Bárdossy A (2007) Calibration of hydrological model parameters for ungauged catchments. Hydrol Earth Syst Sci 11:703–710. https://doi.org/10.5194/hess-11-703-2007

Bertalanič R, Mojca D, Andrej D, et al (2018) Ocena podnebnih sprememb v Sloveniji do konca 21. stoletja, 1st edn. Slovenian Environment Agency, Ljubljana

Bezak N, Jemec Auflič M, Mikoš M (2019) Application of hydrological modelling for temporal prediction of rainfall-induced shallow landslides. Landslides 16:1273–1283. https://doi.org/10.1007/s10346-019-01169-9

Blöschl G, Hall J, Viglione A et al (2019) Changing climate both increases and decreases European river floods. Nature 573:108–111. https://doi.org/10.1038/s41586-019-1495-6

Caine N (1980) The rainfall intensity-duration control of shallow landslides and debris flows. Geogr Ann Ser A 62:23–27. https://doi.org/10.1080/04353676.1980.11879996

Cannon AJ (2018) Multivariate bias correction of climate model output: Matching marginal distributions and intervariable dependence structure. MBC Packag

Ciervo F, Rianna G, Mercogliano P, Papa MN (2017) Effects of climate change on shallow landslides in a small coastal catchment in southern Italy. Landslides 14:1043–1055. https://doi.org/10.1007/s10346-016-0743-1

Coron L, Thirel G, Delaigue O et al (2017) The suite of lumped GR hydrological models in an R package. Environ Model Softw 94:166–171. https://doi.org/10.1016/j.envsoft.2017.05.002

Gudmundsson L, Bremnes JB, Haugen JE, Engen-Skaugen T (2012) Technical Note: Downscaling RCM precipitation to the station scale using statistical transformations—a comparison of methods. Hydrol Earth Syst Sci 16:3383–3390. https://doi.org/10.5194/hess-16-3383-2012

Jemec Auflič M, Jež J, Popit T et al (2017) The variety of landslide forms in Slovenia and its immediate NW surroundings. Landslides 14:1537–1546. https://doi.org/10.1007/s10346-017-0848-1

Jemec Auflič M, Kumelj Š, Peternel T, Jež J (2019) Understanding of landslide risk through learning by doing: case study of Koroška Bela community, Slovenia. Landslides 16:1681–1690. https://doi.org/10.1007/s10346-018-1110-1

Mikoš M, Četina M, Brilly M (2004) Hydrologic conditions responsible for triggering the Stože landslide, Slovenia. Eng Geol 73:193–213. https://doi.org/10.1016/j.enggeo.2004.01.011

Mikoš M, Fazarinc R, Majes B (2007) Delineation of risk area in Log pod Mangartom due to debris flows from the Stože landslide| Določitev ogroženega območja v Logu pod Mangartom zaradi drobirskih tokov s plazu Stože. Acta Geogr Slov 47:171–198. https://doi.org/10.3986/AGS47202

Perrin C, Michel C, Andréassian V (2003) Improvement of a parsimonious model for streamflow simulation. J Hydrol 279:275–289. https://doi.org/10.1016/S0022-1694(03)00225-7

Peternel T, Jež J, Milanič B et al (2018) Engineering-geological conditions of landslides above the settlement of Koroška Bela (NW Slovenia)| Inženirskogeološke značilnosti plazov v zaledju naselja Koroška Bela (SZ Slovenija). Geologija 61:177–189. https://doi.org/10.5474/geologija.2018.012

Peternel T, Kumelj Š, Oštir K, Komac M (2017) Monitoring the Potoška planina landslide (NW Slovenia) using UAV photogrammetry and tachymetric measurements. Landslides 14:395–406. https://doi.org/10.1007/s10346-016-0759-6

Rosi A, Peternel T, Jemec-Auflič M et al (2016) Rainfall thresholds for rainfall-induced landslides in Slovenia. Landslides 13:1571–1577. https://doi.org/10.1007/s10346-016-0733-3

Sapač K, Medved A, Rusjan S, Bezak N (2019) Investigation of low- and high-flow characteristics of karst catchments under climate change. Water (Switzerland) 11:925. https://doi.org/10.3390/w11050925

Zhang J, Han D (2017) Catchment Morphing (CM): A Novel Approach for Runoff Modeling in Ungauged Catchments. Water Resour Res 53:10899–10907. https://doi.org/10.1002/2017WR021403

Check for updates

2020 Kyoto Japan

# Global Warming as a Predisposing Factor for Landslides in Glacial and Periglacial Areas: An Example from Western Alps (Aosta Valley, Italy)

Jessica Maria Chicco, Marco Frasca, Giuseppe Mandrone, Damiano Vacha, and Laurie Jayne Kurilla

**Abstract**

Signals of global climate change are increasingly evident in the Alps where, in recent years, a growing number of landslides occurred in glacial and periglacial areas. In order to document such a case, a landslide is described and analysed, which affected the SE side of a rocky ridge and the Trajo Glacier below in the Gran Paradiso Massif (NW Italy) during the warm summer of 2017. It was a multiple rockfall of about ½ million $m^2$, with vertical drop of ≈300 m, a runout distance of ≈900 m, and 17° of travel angle. Prompt field surveys, interpretation of Sentinel-2 satellite images, and an analysis of data from several weather stations, are used to reconstruct the phenomenon and its causes. This study highlights the geological structure of the area (which reflects the morphology and geo-mechanical characteristics of the slope) and the meteorological conditions during the months before the main failure. Moreover it reveals that the landslide was not a single event but developed over time through at least five failures. According to available information, several predisposing factors seem to have played an important such as the degradation of permafrost (probably affecting rock masses at depth), the alternation of freeze-thaw cycles, and the availability of a considerable amount of water from rainfall and snowmelt, infiltrating the rock mass.

**Keywords**

Global warming • Landslides • Permafrost • Remote sensing • Western alps

J. M. Chicco · G. Mandrone · D. Vacha (✉) · L. J. Kurilla
Dipartimento di Scienze Della Terra, Università Di Torino, Via Valperga Caluso 35, 10125 Turin, Italy
e-mail: damiano.vacha@unito.it

J. M. Chicco
e-mail: jessica.chicco@unito.it

G. Mandrone
e-mail: giuseppe.mandrone@unito.it

L. J. Kurilla
e-mail: lauriejayne.kurilla@unito.it

M. Frasca
Fondazione Montagna Sicura, Loc. Villard de la Palud 1, 11013 Courmayeur, AO, Italy
e-mail: marcofrasca@hotmail.it

## Introduction

Summer 2017 was the second hottest in Italy since 1800 (about +2.5 °C), as well as the fourth driest (ISPRA 2018). During the annual summer campaign of the Italian Glaciological Committee, a new landslide (Figs. 1 and 2) was observed (Frasca et al. 2020) in an arduous area near Cogne (Aosta Valley region, NW-Italy). The landslide occurred in the Mount Grivola area (Fig. 2), on the northern side of the Gran Paradiso Massif (4061 m a.s.l.). The slope failure is at approximately 3500 m a.s.l., and the deposit overlaps a mountain glacier characterized by a single SW-NE oriented tongue.

Landslides in high mountains are reported more frequently and the scientific literature is growing accordingly (Fischer et al. 2012; Hewitt et al. 2008). Typically, landslides initially run very fast on steep slopes and can travel for long distances over snow and ice. As a result, the deposition zone can be very far from the source area.

Landslides occurring on glaciers propagate with a different mechanical behaviour than those occurring on rocky ground. This is due to a basal friction coefficient of the moving mass on ice, which is approximately 75% less than that of a rock surface (Bottino et al. 2002; Sosio et al. 2012). Landslides on glacial surfaces are characterized by a low H/L ratio or "overall slope" where H and L are the horizontal and vertical distances, respectively, between the headscarp and the toe of the landslide (Varnes 1978). The presence of ice and snow in the moving mass can play an important role


V. Vilímek et al. (eds.), *Understanding and Reducing Landslide Disaster Risk*,
ICL Contribution to Landslide Disaster Risk Reduction,
https://doi.org/10.1007/978-3-030-60319-9_26

**Fig. 1** The study area viewed from a nearby peak (Fondazione Montagna Sicura, Sept. 2017): in red the detachment zone, in purple the deposit on the glacier

in this anomalous elongation. All these factors lead to a higher average mobility than non-glacial landslides of comparable magnitude (Schneider et al. 2011).

The impact of such landslides on tourism in mountains and at downstream villages is well known, and the effects must be studied to minimize the risk. Some authors highlight the possibility of attributing landslide events at high elevations to increasing impacts from global warming (e.g., Beniston 2003; Fischer et al. 2013; Patton et al. 2019).

## The Landslide

The top of the landslide is on a very steep (about 60°) rocky slope with a S-SE exposure at approximately 3500 m a.s.l.; while the bottom of the scarp is at 3300 m a.s.l. where it reaches the glacier (Fig. 3). The depletion zone covers an area of about 200 × 200 m. The head is asymmetric and directly related to the arrangement of the tectonic lineaments and discontinuities. The deposition zone follows the very flat surface of the glacier (dipping ≈8°) for approximately 900 m, where the front stops at about 3200 m a.s.l. The avalanche track reaches a maximum width of 320 m.

The bedrock consists of metabasites characterized by a massive facies. They present a structural setting and secondary joints controlling the slope aspect. The resulting rock mass is composed of portions and fragments of intact rock separated by open discontinuities. Joint surfaces are smooth, with evidence of extensive water circulation. At a larger scale, the ridge is crossed by a series of pervasive faults and discontinuities at a spacing of tens to hundreds of meters.

The landslide has a thin average thickness of approximately 1.5 m, in agreement with the literature on similar landslides (Shugar and Clague 2011) with a heterogeneous grain size, from boulders (the largest with a volume of approximately 2000 $m^3$) to centimetre-sized pebbles of angular rock fragments.

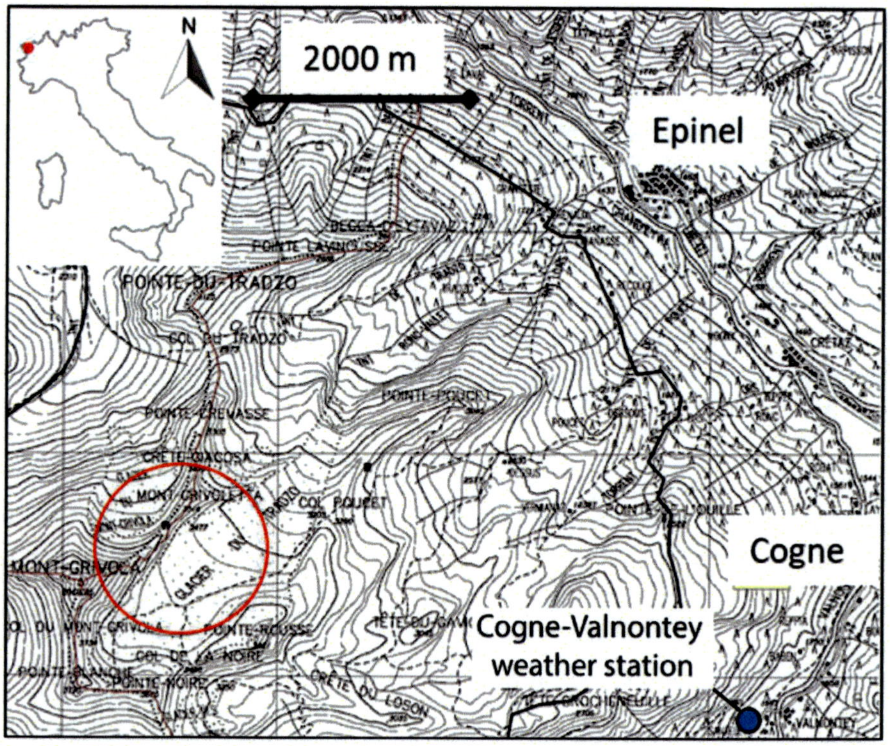


**Fig. 2** In the red circle the study area (WGS84/ UTM zone 32 N 365353 E, 5051071 N). The blue dot represents the Cogne-Valnontey weather station (WGS84/UTM zone 32 N 370277 E, 5049336 N)

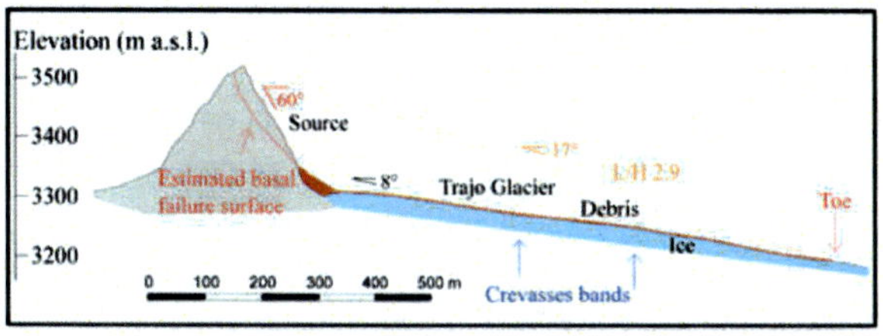


**Fig. 3** Sketch of the landslide. In colours: bedrock (grey), debris (brown) and ice (blue); red line indicates the surface of rupture; angles indicate the inclination of the slope, of the glacier and the travel angle of the 2017 landslide

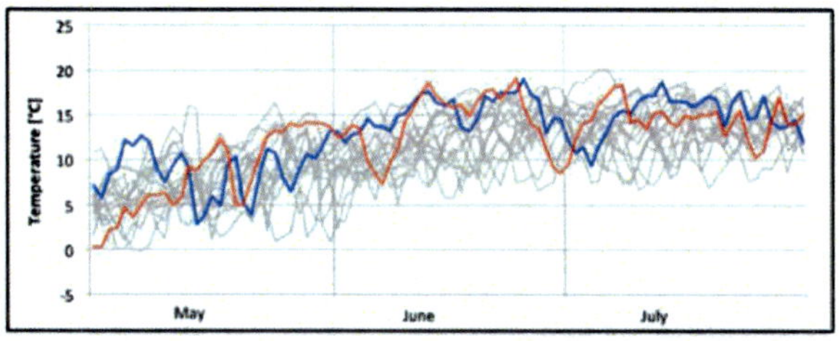


**Fig. 4** Cogne-Valnontey weather station: average daily temperatures for May, June and July from 2002 to 2017; 2017 trend is red, 2003 is blue and other years are grey

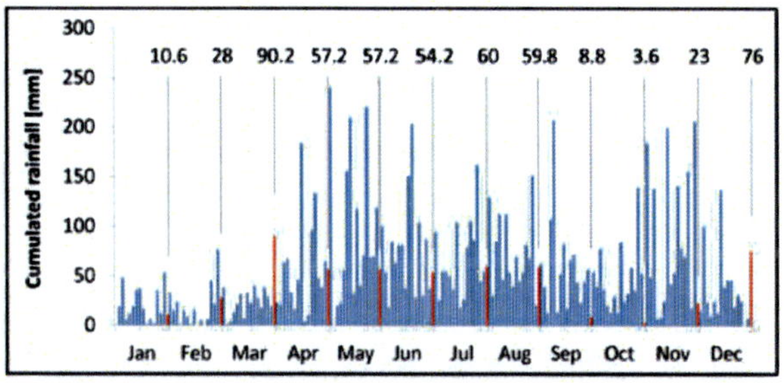


**Fig. 5** Cogne-Valnontey weather station: monthly cumulative rainfall (mm) from 2002 to 2017; 2017 is highlighted in red, other years are blue. 2017 values are reported on top of the graph

European Space Agency Sentinel-2 multispectral satellite images were used to reconstruct the evolution of the landslide over time. At least five events were recognized between 10/05/2017 and 14/07/2017. They can be described as follow:

(1) the incipient failure can be dated between 10/05/2017 and 30/05/2017, highlighted by a scarp in the slope and new small debris deposits on the glacier surface;
(2) between 31/05/17 and 09/06/17, the debris deposit at the bottom of the slope increased slightly;
(3) between 10/06/17 and 19/06/17, an increase in the size of both the landslide scarp and deposit occurred;
(4) between 20/06/17 and 04/07/17 again the scarp increased in area and the deposit too;
(5) the last and largest collapse occured in the period between 05/07/17 and 14/07/17.

According to Varnes (1978), the landslide can be classified as a multiple rockfall, evolving (in space and time) into a rock avalanche of approximately 500 000 m$^3$.

## Climate and Meteorological Conditions Before 2017 Event

Raw meteorological data were collected from weather stations in a range of 10 km from the Trajo Glacier. Two of the stations were in other watersheds and three were at elevations much lower than the landslide (less than 1000 m a.s.l.). As a result, the temperature and rainfall data from these stations show anomalous trends compared to evidence found at the study area. A geostatistical approach did not improve our understanding of the events. Only Cogne-Valnontey station data (WGS84/UTM zone 32 N 370277 E, 5049336 N; ≈6 km from landslide and at 1682 m a.s.l.) from May, June and July of 2002 to 2017 were used.

Average daily temperatures from 2002 to 2017 are plotted in Fig. 4. Average daily temperatures in 2017 (red line) are compared to 2003 (blue line) which is considered the hottest years in Italy since 1800 (Regione Valle d'Aosta 2006). Temperatures progressively increase in May 2017, particularly during the second half of the month, reaching values higher than those of 2003. June 2017 temperatures are extremely high and comparable to June 2003. Temperatures recorded in the first 8 days of July 2017 were higher than those in July 2003. The 2017 data show a generally increasing trend over time, with five temperature troughs (lower average daily temperatures) corresponding to five rainfall events.

Overall, 2017 was characterized by a dry autumn and a wet spring and summer with a constant rainfall of approximately 60 mm per month (Fig. 5).

## Discussion on Predisposing and Triggering Factors

The landslide involved a steeply dipping slope of hard rocks fractured by a high-density network of tectonic discontinuities, increasing secondary porosity and causing high water infiltration. Cryoclasts, induced by freeze-thaw cycles, deeply affect rock mass, but freeze-thaw cycles cause reductions in fracture openings during the summer due to thermal expansion. This led to an increase of fracture openings during the winter, due to rock contraction (Weber et al. 2019).

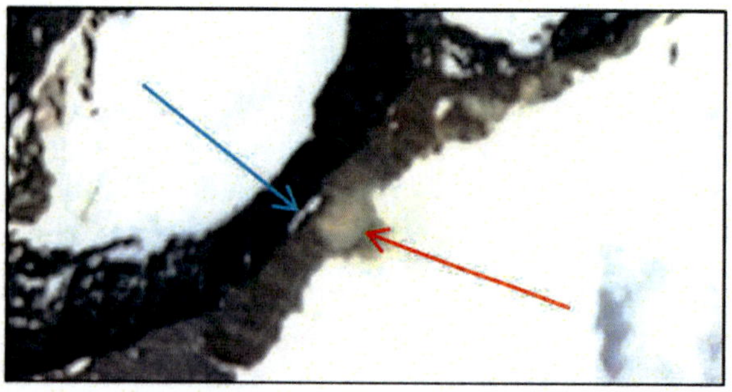

**Fig. 6** Satellite image (ESA Sentinel-2, June 2017): the elongated trench full of snow on the divide and the landslide, highlighted by blue and red arrows, respectively. (https://apps.sentinelhub.com/sentinel-playground)

On the ridge divide corresponding to the landslide crown, satellite images from June 2017, show the presence of a trench approximately 120 m in length (Fig. 6). It is clearly marked by the persistence of snow during periods when the rest of the slope had already melted. This depression may act as a water source for infiltration during summer. Double crests often mean deep-seated movements and are usually referred to as unstable areas.

Investigating links between climatic variations and a high-elevation instability phenomenon is complex from both a logistical and technical perspective (Huggel et al. 2012a). To justify the conjecture of a linkage, air temperature at the crown should be known as well as the temperature at depth within the rock mass, but these parameters are difficult to acquire, and measurements are rare, discontinuous and scattered around the world. In Italy, at ≈50 km from the study area, the reference is the Plateau Rosà weather station (3488 m a.s.l.) which records the annual average air temperature at approximately −3 °C. Moreover, direct measurements on permafrost and active layers at Mt. Cervino (60 km from the study area) show an active layer between −2 and −5 m (http://www.permanet-alpinespace.eu/home.html).

The temperature at the crown of the landslide was reconstructed using data from the Cogne-Valnontey weather station and corrected with a standard thermal gradient of 0.6 °C for every 100 m of elevation (Regione Valle d'Aosta 2016). This assumption does not take into consideration multiple factors, such as the energy balance of the Earth's surface, reflected longwave radiation, the stability of the boundary layer (Nigrelli et al. 2017), and the fact that solar rays reflected from the mountain slopes may have a considerable thermal effect on the surrounding atmosphere.

Figure 7a presents the trend of maximum and minimum daily temperatures (T) at the crown for the five events. During the three major ones (10–19 June, 20 June-04 July and 05–14 July), maximum T were considerably above 0°C (10–15 °C). For the event on 10–30/05, maximum T were also higher than 0 °C. Subsequently, minimum T progressively increased, with values close to 0 °C (−5 to 0 °C) for several consecutive days, especially for the three major events (10–19 June, 20 June-04 July and 05–14 July). Particularly high minimum T occurred for several days before Event 5. Moreover, the period immediately preceding Event 1 was characterized by daily T always substantially below 0 °C, so the rock mass was probably "frozen". T for Events 1 and 2 show daily maximum and minimum widely above and below the thermal zero, respectively, so the activation of freeze-thaw cycles probably occurred over a relatively long period. High T for several days during Events 3, 4 and 5 indicate a prevailing thaw. Figure 7b shows daily elevation of the thermal zero from the Regional Functional Centre's daily meteorological bulletins. These values were obtained from a few direct measurements and a physical-mathematical model for the rest of the valley. The thermal zero stands at 3000 m or lower until the 15th of May, then it rises above 3500 m for the entire period of the five rockfalls.

Figure 7c presents the cumulative daily rainfall for each reactivation, which does not exceed 25 mm, a normal quantity for the area (https://cf.regione.vda.it/precipitazioni). As a result, rainfall cannot be claimed as a major triggering factor.

## Did Permafrost Play a Role in This Rockfall?

Permafrost is defined as a thickness of soil or bedrock in which the temperature has remained at or below the freezing point of water (0 °C) for 2 or more consecutive years. It is exclusively defined by temperature and duration, independent of texture, water content and lithology. The importance of permafrost as an indicator to assess the impacts of global climate change is universally recognized, such that the Global Climate Observing System has included it among the essential climate variables (ECV). Permafrost is basically influenced by climate, but characteristics of topographic surface and of substratum play a fundamental role (Noetzli et al. 2007), especially for Alpine permafrost.

The existing literature highlights degradation of permafrost as a likely cause of rockfall occurrences in many areas of the world (Harris et al. 2009; Ravanel and Deline 2011; Krautblatter et al. 2013). However, attribution of specific events to permafrost degradation is difficult, especially without monitoring field data (Allen et al. 2011; Huggel et al. 2012b).

Progressive degradation of permafrost in the Alps is well documented (Noetzli and Gruber 2009). It can favour water infiltration in fractures causing deep defrosting (tens of

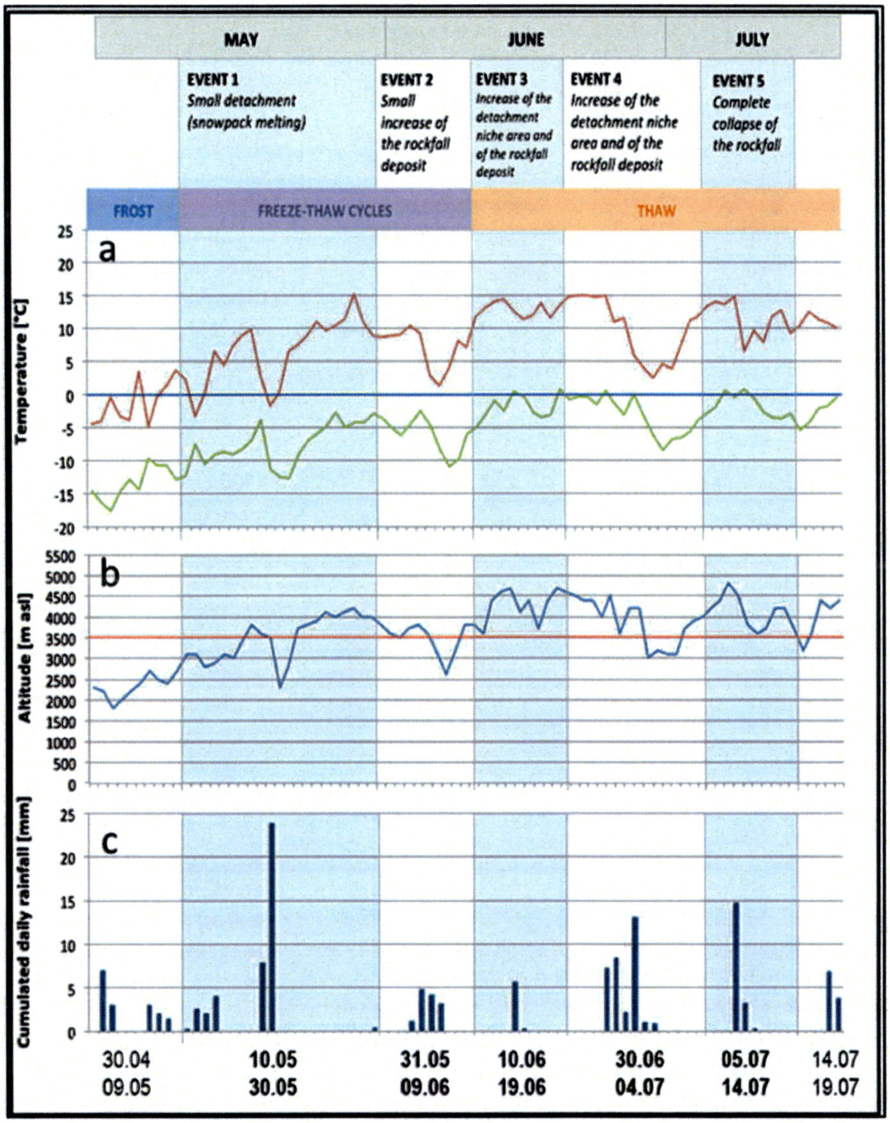


**Fig. 7** Weather conditions for the five landslide activation events: **a** maximum (red) and minimum (green) daily air temperature at 3500 m a.s.l.; **b** elevation of thermal zero; **c** cumulative daily rainfall

metres). This process leads to an increase in the stress on the joints surfaces, a reduction in the strength of bedrock, and a destabilization of large rock volumes. According to the Alpine Permafrost Index Map (Fig. 8), the study area is likely to be affected by permafrost. This map presents an index to estimate the potential presence of permafrost in the Alps (Boeckli et al. 2012). In addition, using the schematic diagram of Haeberli (1975) this presence is "probable". The abundant infiltration and circulation of meteoric and snow-melt waters in the rock mass discontinuities, together with the circulation of warmer air, may have favoured the heating of the rocky permafrost by conduction and advection, contributing to the melting of subsurface ice. Moreover, hydrostatic overpressure due to infiltration can contribute to instability. Overpressures can be easily induced through temporary obstructions of drainage pathways in open fractures and joints inside the rock mass (Davies et al. 2001). On this basis, it can be reasoned that permafrost affected the slope in the past and that the hot period in spring-summer 2017 could have contributed to its degradation. This can be presumed to be one of the predisposing factors for the landslides.

Recent literature confirms that the morphology of slope can favour more rapid degradation of the permafrost over time because, on a narrow ridge, the increase in temperature acts simultaneously from both sides of the relief (Fig. 9). In fact, the configuration of slopes can affect the thermal regime of rock at depth, creating great differential thermal dilatations (Noetzli et al. 2007). In terms of temperature-related instabilities, the migration of the permafrost threshold (0 °

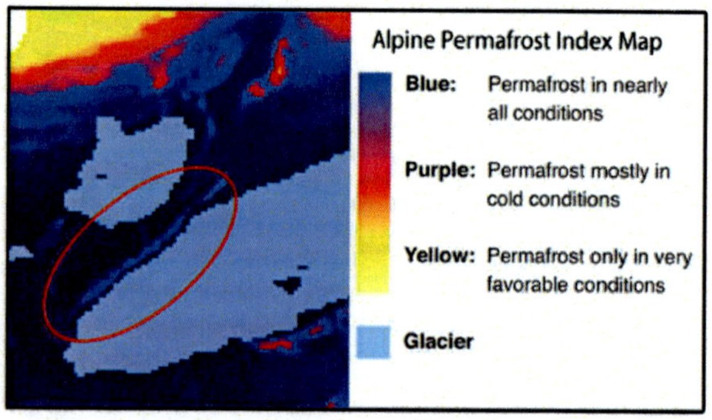


**Fig. 8** Alpine Permafrost Index Map (Boeckli et al. 2012) of the study area (red circle) and the surroundings

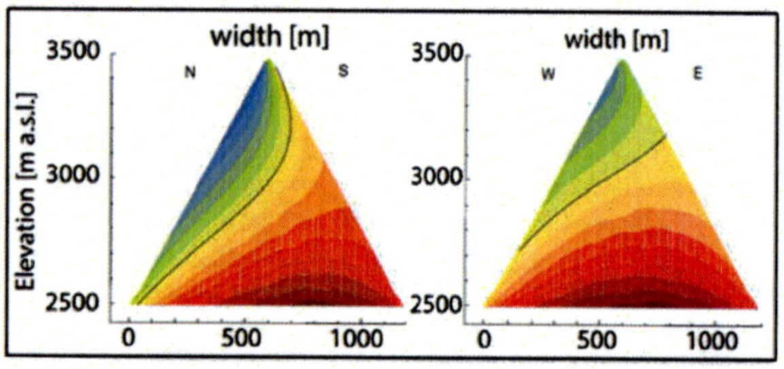


**Fig. 9** Subsurface temperature conditions in steady-state for two idealized ridges (black line = 0 °C isotherm—lower limit of the permafrost). N-S and E-W exposed slopes on the left and right, respectively (Noetzli et al. 2007)

C), due to defrosting, can lead to rockfalls and collapses on the warmer or "permafrost-free" side of a ridge. In the case of the Trajo event, the SE aspect is exposed to direct solar irradiation plus solar radiation reflected by the underlying glacier surface with its high albedo. These two effects can substantially increase thermal energy affecting the slope, especially during more favourable hours of the day.

In summary, it seems reasonable to hypothesize that the following climatic factors played an important role in the Trajo landslide: (i) intense activity of freeze-thaw cycles preceding the first failure, (ii) high minimum temperatures recorded for several consecutive days before the landslides and iii) rainfall precipitation, rather than snow, at elevations higher than 3500 m.

## Summary and Conclusions

In summer 2017, a landslide occurred in the central sector of the steep SE slope of a rocky ridge at approximately 3500 m a.s.l. in the Gran Paradiso Massif (Cogne, Aosta Valley Region, NW Italy). It was a multiple rockfall involving about ½ million $m^2$ of "green stones" that reached the Trajo Glacier at the bottom of the slope. The displaced material ran out for more than 900 m over the glacier. The landslide developed over time in several events from 10/05/17 to 14/07/17 and was surveyed for the first time during a glaciological expedition in September 2017.

No field information was available. In fact, direct measures in such remote places require expensive instrumentation which was not available in this study. Data from satellite images, low-elevation weather stations, and a few expeditious surveys served to reconstruct the dynamics of the landslides and the predisposing triggers.

May, June and July 2017 were characterized by anomalously high temperatures, and thermal zero remained above 3500 m for many weeks. These conditions made available great quantities of water (from snowmelt and rain) which favoured, by infiltrating at depth, the heating of the rock mass causing permafrost degradation. So, groundwater overpressure within joints and faults of rock bridges, due to freeze-thaw cycles, can be recognized as a possible additional cause of slope failures.

Geological and climatic studies highlight several other elements that contributed to the development of the instability, such as the tectonic structure of the area and climatic trend of high temperatures for decades before the event. Field surveys and borehole temperatures, reported in literature, document that slopes at the same elevation of the Trajo landslide in the western Alps contain relatively warm permafrost reaching to considerable depths inside the rock mass. Permafrost degradation is unlikely to be the only decisive factor in triggering a rockfall, but it may play a major role.

Probably this failure was caused by the interaction of several factors, including (i) degradation of permafrost, (ii) alteration of the freeze-thaw cycles, (iii) availability of a considerable amount of rainfall water, and iv) high rate of infiltration due to the opening of joints caused by the heating of rock masses.

Landslides studies at high elevation are important because periglacial areas are among the most sensitive with respect to global climate changes. Moreover, an increasing number of people are hiking such zones and risks connected to high-mountain landslides can also have impacts at much lower elevations, where nature and humans have a close relationship (Haeberli et al. 2017).

**Acknowledgments** Many thanks to dott. F. Troilo of "Fondazione Montagna sicura" and to dott. D. Bertolo of Aosta Valley Region for supporting the study.

## References

Allen SK, Cox SC, Owens IF (2011) Rock avalanche and other landslides in the central Southern Alps of New Zealand: a regional study considering possible climate change impacts. Landslides 8 (1):33–48

Beniston M (2003) Climatic change in mountain regions. A review of possible impacts. Clim Chang 59:5–31

Boeckli L, Brenning A, Gruber S, Noetzli J (2012) Permafrost distribution in the European Alps: calculation and evaluation of an index map and summary statistics. Cryosph Discuss 6(4):807–820

Bottino G, Chiarle M, Joly A, Mortara G (2002) Modelling rock avalanches and their relation to permafrost degradation in glacial environments. Permafrost Periglac Process 13(4):283–288

Davies MCR, Hamza O, Harris C (2001) The effect of rise in mean annual temperature on the stability of rock slopes containing ice-filled discontinuities. Permafrost Periglac Process 12(1):137–144

Fischer L, Purves RS, Huggel C, Noetzli J, Haeberli W (2012) On the influence of topographic, geological and cryospheric factors on rock avalanches and rockfalls in high-mountain areas. Nat Hazards Earth Syst Sci 12:241–254

Fischer L, Huggel C, Kääb A, Haberli W (2013) Slope failures and erosion rates on a glacierized high mountain face under climatic changes. Earth Surf Process Landforms 38(8):836–846

Frasca M, Vacha D, Chicco J, Troilo F, Bertolo D (2020) Landslide on glaciers: an example from Western Alps (Cogne–Italy). J Mountain Sci 17

Haeberli W (1975) Untersuchungen zur Verbreitung von Permafrost zwischen Flüelapass und Piz Grialetsch (GR). Mitteilung der Versuchsanstalt für, Wasserbau, Hydrologie und Glaziologie, 17

Haeberli W, Schaub Y, Huggel C (2017) Increasing risks related to landslides from degrading permafrost into new lakes in de-glaciating mountain ranges. Geomorphology 293:405–417

Harris C, Arenson LU, Christiansen HH, Etzelmüller B, Frauenfelder R, Gruber S, Haeberli W, Hauck C, Hölzle M, Humlun O, Isaksen K, Kääb A, Kern-Lütschg MA, Lehning M, Matsuoka N, Murton JB, Nötzli J, Philips M, Ross N, Seppälä M, Springman SM, Mühll DV (2009) Permafrost and climate in Europe: Monitoring and modelling thermal, geomorphological and geotechnical responses. Earth-Sci Rev 92(3–4):117–171

Hewitt K, Clague JJ, Orwin JF (2008) Legacies of catastrophic rock slope failures in mountain landscapes. Earth-Sci Rev 87(1–2):1–38

Huggel C, Allen S, Deline P, Fisher L, Noetzli J, Ravanel L (2012a) Ice thawing, mountains falling are alpine rock slope failures increasing? The Geologists' Association & The Geological Society of London, Geology Today, Vol. 28, No. 3, May–June 2012

Huggel C, Clague JJ, Korup O (2012b) Is climate change responsible for changing landslide activity in high mountains? Earth Surf Process Landforms 37(1):77–91

Krautblatter M, Funk D, Guenzel FK (2013) Why permafrost rocks become unstable: a rock-ice-mechanical model in time and space. Earth Surf Process Landforms 38:876–887

ISPRA (2018) Gli indicatori del clima in Italia nel 2017, anno XIII. Stato dell'ambiente, 80/2018, 75 p

Nigrelli G, Fratianni S, Zampollo A, Turconi L, Chiarle M (2017) The altitudinal temperature lapse rates applied to high elevation rockfalls studies in the Western European Alps. Theor Appl Climatol 131:1479–1491

Noetzli JS, Gruber T, Kohl N, Salzmann, Haeberli W (2007) Three-dimensional distribution and evolution of permafrost temperatures in idealized high-mountain topography. J Geophys Res 112, F02S13, 1:14

Noetzli J, Gruber S (2009) Transient thermal effects in Alpine permafrost. Cryosphere 3(1):85–99

Patton AI, Rathburn SL, Capps DM (2019) Landslide response to climate change in permafrost regions. Geomorphology 340 (2019):116–128

Ravanel L, Deline P (2011) Climate influence on rockfalls in High-Alpine steep rockwalls: the north side of the Aiguilles de Chamonix (Mont Blanc massif) since the end of the 'Little Ice Age'. The Holocene, 21, 2:357–365. ISBN 88-900099-9-3. In Italian

Regione Valle d'Aosta (2006) Cambiamenti climatici in Valle d'Aosta. Opportunità e strategie di risposta, Società Meteorologica Subalpina, Bussoleno

Regione Valle d'Aosta (2016) Rendiconto idro-meteorologico 2016 Pioggia, temperatura, neve e dissesti.http://cf.regione.vda.it/download.php?fileID=152&lang=ita

Schneider D, Huggel C, Haeberli W, Kaitna R (2011) Unraveling driving factors for large rock-ice avalanche mobility. Earth Surf Process Landforms 36:1948–1966

Shugar H, Clague J (2011) The sedimentology and geomorphology of rock avalanche deposits on glaciers. Sedimentology 58(7):1762–1783

Sosio R, Crosta GB, Chen JH, Hungr O (2012) Modelling rock avalanche propagation onto glaciers. Quat Sci Rev 47:23–40

Varnes DJ (1978) Slope movements types and processes. In: Schuster RL, Krizek RJ (eds) Landslides: analysis and control. Transportation Research Board Spec. Rep. 176. National Academy of Science, Washington, pp 11–33

Weber S, Beutel J, Da Forno R, Geiger A, Gruber S, Gsell T, Hasler A, Keller M, Lim R, Limpach P, Meyer M, Talzi I, Thiele L, Tschudin C, Vieli A, Mühll D, Yücel M (2019) A decade of detailed observations (2008–2018) in steep bedrock permafrost at the Matterhorn Hörnligrat (Zermatt, CH). Earth Syst Sci Data 11:1203–1237

# Characteristics and Causes of the Debris Flow in Shelong Gully, China

2020 Kyoto Japan

Qiang Zou, Peng Cui, Hu Jiang, Yanguo Liu, Cong Li, Sheng Hu, and Bin Zhou

**Abstract**

A rainfall-induced debris flow disaster occurred in Shelong Gully on the 22th June, 2019. More than 26,300 $m^2$ of farmland and 14 houses were buried and transportation and power supply broke down in the worst catastrophe in the history of Shelong area. To investigate this disaster, a detailed interpretation was carried out using (Unmanned Aerial Vehicle) UAV-based images. Analysis of disaster characteristics based on images and field survey revealed that the disaster could be identified as a consequence of compound mountain hazards including mass movements and mountain torrent in Shelong Gully. Moreover, the dynamic process of debris flow was analysed by using the finite difference method. Its dynamic characteristics indicated that the structure of debris materials changed during the movement process of debris flow. The results reveal that the loose materials, steep terrain condition and abundant rainfall are the main causal factors of this disaster event. This research on the analysis of hazard formation conditions and dynamic process for debris flows is of guiding significance to the formulation of disaster risk assessment and disaster mitigation plans.

**Keywords**

Debris flow • Dynamic process • Hazard formation • Disaster risk reduction • Rainfall

Q. Zou (✉) · C. Li · B. Zhou
Key Laboratory of Mountain Hazards and Earth Surface Process, Institute of Mountain Hazards and Environment (IMHE), Chinese Academy of Sciences (CAS), Chengdu, 610041, China
e-mail: zouqiang@imde.ac.cn

Q. Zou · P. Cui
The Centre for Excellence in Tibetan Plateau Earth Sciences (CETES),, Chinese Academy of Sciences (CAS) of the Beijing, Beijing, 100101, China

P. Cui
Institute of Geographic Sciences and Natural Resources Research, Chinese Academy of Sciences, Beijing, 100101, China

H. Jiang · Y. Liu
School of Environment and Resource, Southwest University of Science and Technology, Mianyang, 621010, China

S. Hu
Shaanxi Key Laboratory of Earth Surface System and Environmental Carrying Capacity, Northwest University, Xi'an, 710127, China

B. Zhou
University of Chinese Academy of Sciences, Beijing, China

## Introduction

The Shelong Gully is located in Jinchuan County, Sichuan Province, China. It is situated on the left bank of the Carla feet River, which is an anabranch of the upper reaches of the Dadu River (Fig. 1). At 11:02 pm on the 22th June, 2019, a strong debris flow destroyed Kalajiao village in Jinchuan County, Sichuan Province, China. About 170 thousand $m^3$ of debris and mud piled up at the gully exit. Although no one was killed or injured in this event, it still caused substantial losses, including nearly 26, 300 $m^2$ of farms and 14 houses buried, 50 goats missed, transportation and power system broken (Fig. 2).

## Watershed Background

Due to the active crustal movement at the eastern margin of Qinghai-Tibet Plateau, the study site has complex geological and geomorphological environments. Affected by the Ganzi-Kangding-Shimian deep fault and Longmenshan fault, strong earthquakes frequently hit the Shelong area (Luo and Liu 2010) Accordingly, geologic hazards including slope failures, collapses and soil slips are prone to occur on both sides of the Shelong Gully. These hazards generate abundant loose materials to promote the forming of debris



V. Vilímek et al. (eds.), *Understanding and Reducing Landslide Disaster Risk*, ICL Contribution to Landslide Disaster Risk Reduction,
https://doi.org/10.1007/978-3-030-60319-9_27

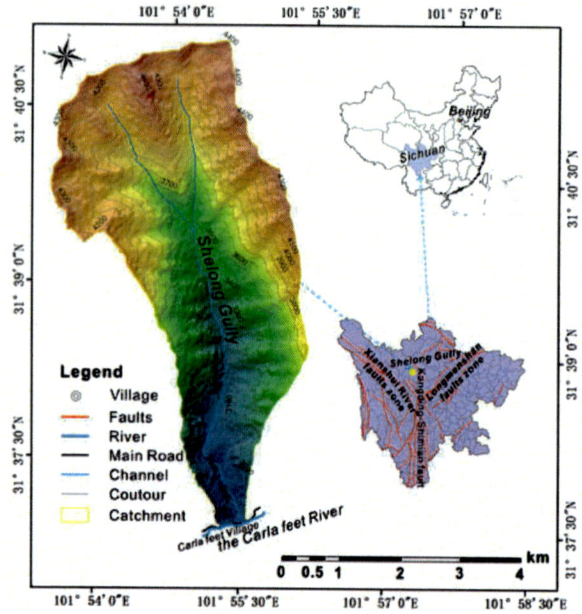


Fig. 1 Location of Shelong Gully

flows (Hungr et al. 2014). Moreover, the high elevation difference and straight channel of the Shelong Gully provides the favourable terrain and energy condition to transport the debris.

The intensive rainfall due to the monsoon climate is one of the important triggers of debris flow in China (Cui et al. 2018). According to data from Maerkang meteorological observatory, the average annual rainfall in the Shelong Gully in the past ten years is 858.8 mm, the maximum annual precipitation is 988.2 mm and the maximum daily precipitation is 48.5 mm (Fig. 3). The annual rainfall is concentrated in summer, accounting for 87.3%, during which the debris flows are highly prevalent.

## Disaster Characteristics

### Hazard Distribution

Surveying and mapping accuracy and resolution are important in the topographic analysis of the debris flow. A high-resolution topographic model and orthoimage are imperative to analyse the hazard distribution and movement process. Since the aerial photo was initially taken on a balloon by the Frenchman Nadar, photogrammetry and (Unmanned Aerial Vehicle) UAV photography technology have been developed (Chen 2003; Lambert et al. 2010). Niethammer et al. (2010) successfully developed methods to obtain terrain data and scene models by using from the images taken by small UAVs. Then, scholars have applied it widely to topographic analysis (Marx et al. 2017; Lazzari and Gioia 2017; Pineux et al. 2017; Hu et al. 2018; Boreggio et al. 2018).

In this article, we used UAV-based method to acquire the terrain and image data including 17.1 cm/pixel DEM and 8.5 cm/pixel DOM (Digital Orthophoto Map). According to the actural topography conditions in the Shelong Gully, the controlled locations of the UAV and the camera parameters were determined. Table 1 shows the flight parameters of UAV. Figure 4 shows the UAV flight strips and camera posture in data collection. Consequently, 229 images were captured for Shelong Gully.

### Dynamic Process

The Reynolds-averaged Navier-Stokes formula written in the differential conservation form shown as Eqs. (1) and (2) is reduced to a two-dimensional shallow water form (Eq. (3))—by integrating for various dynamic parameters on

Fig. 2 Debris flow buried the roads and buildings

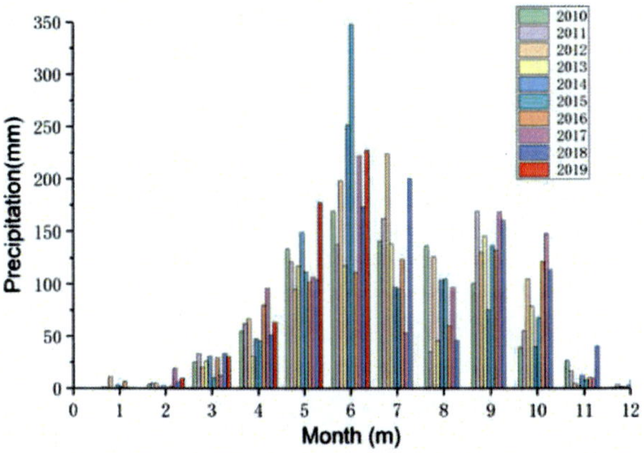


**Fig. 3** Average annual precipitations in the Shelong Gully from 2010 to 2019

**Fig. 4** UAV flight strips and camera posture in data collection

the vertical direction (Denlinger and Iverson 2004). Then, the finite difference numerical method is proposed to simulate the dynamic process of debris flow.

$$\frac{\partial \rho}{\partial t} + \nabla \cdot \rho \tilde{u} = 0 \quad (1)$$

$$\frac{\partial \rho \tilde{u}}{\partial t} + \nabla \cdot \rho \tilde{u} \tilde{u} = \nabla \cdot \tilde{\tau} + \rho g \quad (2)$$

$$\frac{\partial A}{\partial t} + \frac{\partial F}{\partial x} + \frac{\partial G}{\partial y} = S + T \quad (3)$$

Where, ρ is mass density; g is gravitational acceleration; u is velocity vector of mass in the three axes of the global Cartesian coordinates x, y, and z. The symbol (~) represents the parameters varying with vertical direction; A is the vector of conserved flow variables; F and G are the vectors describing the flux of mass and momentum in the x- and y-directions, respectively; S and T are the vectors containing source terms gravitational driving and resulting from internal stresses and boundary forces in the x- and y-directions on x-y plane normal to z, respectively.

Based on the finite difference method, an analytical solution is developed for governing equations of the hazard dynamic process. Then, the movement process of debris flow is simulated in Shelong Gully based on the 12.5 m/pixel terrain data generated from ALOS/PALSAR imagery (Huang et al. 2015) before this disaster event. Accordingly, the depth and velocity at each time point are obtained, and the accumulation pattern of debris flow is shown in the Fig. 5.

Through the field survey, the simulated debris-flow deposition was in good accordance with the actual situation, which verified our numerical method on this debris flow simulation. By using the spatial analysis tools (David 2001; Wheaton et al. 2010), the debris flow accumulation area and volume were 5.79 × 104 $m^2$, 17.37 × 104 $m^3$, respectively.

## Damage Features

The pre-disaster image (Fig. 6a) was obtained from Google Maps, and the post-disaster image (Fig. 6b) was acquired by UAV in Shelong Gully. The pre-disaster image shows that the channel of Shelong Gully was narrow and smooth. Compared with the pre-disaster image, the channel of Shelong Gully in the post-disaster image was widened and silted by a lot of debris deposits. By interpreting and overlaying the pre- and post-disaster image, 500 m of the main road at the exit of Shelong Gully was damaged or buried by the debris flow, approximately 16 houses were situated within areas that exposed to debris flows, and nearly 26300 $m^2$ of farmland and 14 houses was buried.

**Table 1** UAV flight parameters in the Shelong Gully(DOM: Digital Orthophoto map)

| Longitude | Latitude | Flight elevation (m) | (Digital Orthophoto Map) DOM resolution (cm/pixel) | DEM resolution (cm/pixel) | Overlap (%) | Sidelap (%) | Tilt angle (°) |
|---|---|---|---|---|---|---|---|
| 101.92°E | 31.62° N | 3076–3425 | 8.5 | 17.1 | 75 | 75 | −90 |

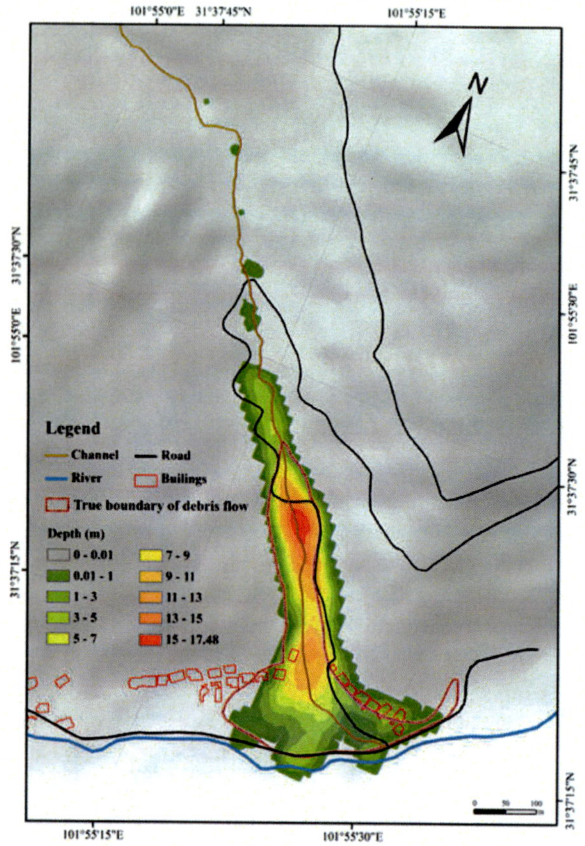


**Fig. 5** Flow depth of debris flow in the Shelong Gully

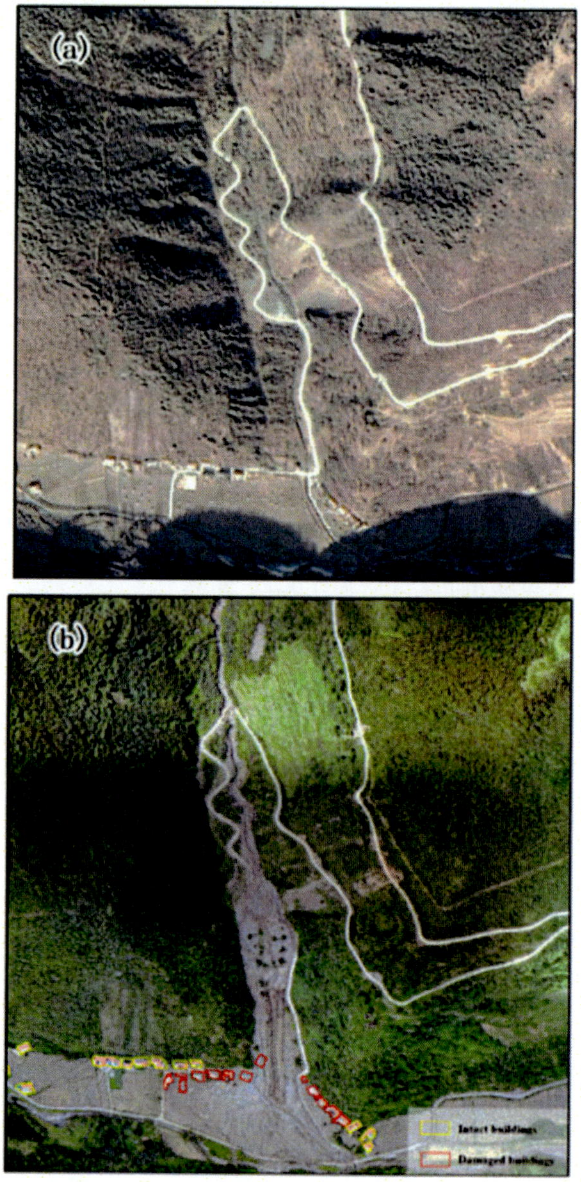


**Fig. 6** Remote sensing images of a pre- downloaded from Google Maps and **b** post-disaster imagery taken by UAV

## Causal Factors

### Geologic Conditions and Loose Solid Material

The Shelong Gully is located at fault zone with active geologic tectonic structures. The rock strata in this area are heavily fragmented by active tectonic activity. Fault and fracture development are considerably extensive. As a result, rocks are especially susceptible to collapse from steep slope conditions (Fig. 7a). The main lithology in this area includes muddy limestone with inclusion sandy conglomerate, occasionally mixed up with volcanic and limestone. A variety of loose accumulations from the Triassic period are distributed on the inclinations and riverbanks of the upper reaches of the gully. Furthermore, accumulation and alluvium are mainly distributed on the channel, the riverbed and the entrance of branches gullies. In addition, a large number of landslides and collapses occur on both sides of the main channel of the gully. And the detritus suffering from seasonal congelifraction is a primary source for subsequent debris flow (Fig. 7b). The loose solid materials are available for debris-flow in the Shelong Gully.

### Topographic Conditions

The Shelong Gully has characteristics with high mountains, deep-cutting rivers and steep slopes. The gradient of slopes in the upper reaches (debris-flow source areas) ranges from 30° to 60°. The main channel length is 5.22 km, and the catchment area is 17.71 km$^2$. The lowest point within the basin is located at the gully entrance at 2652 m a.s.l. The highest point is located at the north-eastern watershed with an elevation of 4424 m a.s.l. By using field surveys and GIS technology, the terrain parameters of the main and tributary

**Fig. 7** Sources of loose solid materials for debris flow: **a** slope failure; **b** moraine deposit

channels are calculated, and are shown in Tables 2 and 3. Furthermore, the cross sections of gully channels are shown as U-shape and the longitudinal gradients currently range from 8‰ to 993‰. These terrain conditions are especially conducive to loose solid mass movement and debris-flow formation.

## Rainfall

The antecedent rainfall and the critical rainfall are interrelated and seriously affect the debris flow formation (Qiao 1997). Due to the sufficient antecedent rainfall, the water content of slope soil increases, and the critical rainfall descends. According to the meteorological data in Shelong Gully (Fig. 8a), abundant precipitation was in June among which only eight days had no rains. The precipitation reached its peak of 22.5 mm on June 22. The rainstorm process in Fig. 8b shows the whole process of rain lasted about 8 h from 14:00 to 24:00 on June 22. Especially, the rainfall increased from 15:00, with an hourly rainfall intensity of 7.7 mm/h, and reached the maximum rainfall intensity of 16 mm/h at 16:00, and stopped intermittently at 23:00. Accordingly, the high-intensity precipitation reached the threshold to trigger and form debris flow at around 23:05 on June 22. Therefore, accumulated antecedent rainfall and the heavy rainfall on June 22 are the main causes of the debris flow in Shelong Gully.

## Discussion and Conclusion

Through the interpretation of remote sensing images, this study obtained rich information about mountain hazard distributions, dynamic process, and damage features. This research shows that the Shelong disaster was the comprehensive result of mass movements and mountain torrent. As a typical compound disaster, the Shelong disaster caused great damage. The finite difference method by analysing the various dynamic parameters on the vertical direction was developed to simulate and analyse the process of debris flow. The dynamic characteristics indicated that the debris-flow volume was amplified by the abundant loose materials and heavy rainfall during the movement process of debris flow. Moreover, although the topography, geology, and loose

**Table 2** Terrain parameters of the main and tributary channels

| Channels | Area ($km^2$) | Length (km) | Longitudinal profile (‰) | Elevation at entrance (m) | Elevation of highest point (m) | Largest relative elevation difference (m) |
|---|---|---|---|---|---|---|
| Main channel | 17.71 | 5.22 | 292 | 2652 | 4424 | 1772 |
| Tributary channel on left | 3.51 | 2.51 | 292 | 3472 | 4424 | 952 |
| Tributary channel on right | 4.70 | 2.58 | 280 | 3472 | 4407 | 935 |

**Table 3** statistical results of areas in each slope classification

| Channels | Items | Slope (°) | | | | Summation |
|---|---|---|---|---|---|---|
| | | [0,15] | [15,25] | [25,35] | [35,90] | |
| Main channel | Area (km$^2$) | 1.42 | 3.92 | 8.15 | 4.22 | 17.71 |
| | Percentage (%) | 8.04 | 22.13 | 46.00 | 23.83 | 100.00 |
| Tributary channel on left | Area (km$^2$) | 0.29 | 1.06 | 1.55 | 0.61 | 3.51 |
| | Percentage (%) | 8.26 | 30.29 | 44.17 | 17.28 | 100.00 |
| Tributary channel on right | Area (km$^2$) | 0.60 | 1.61 | 1.72 | 0.77 | 4.70 |
| | Percentage (%) | 12.82 | 34.24 | 36.61 | 16.33 | 100.00 |

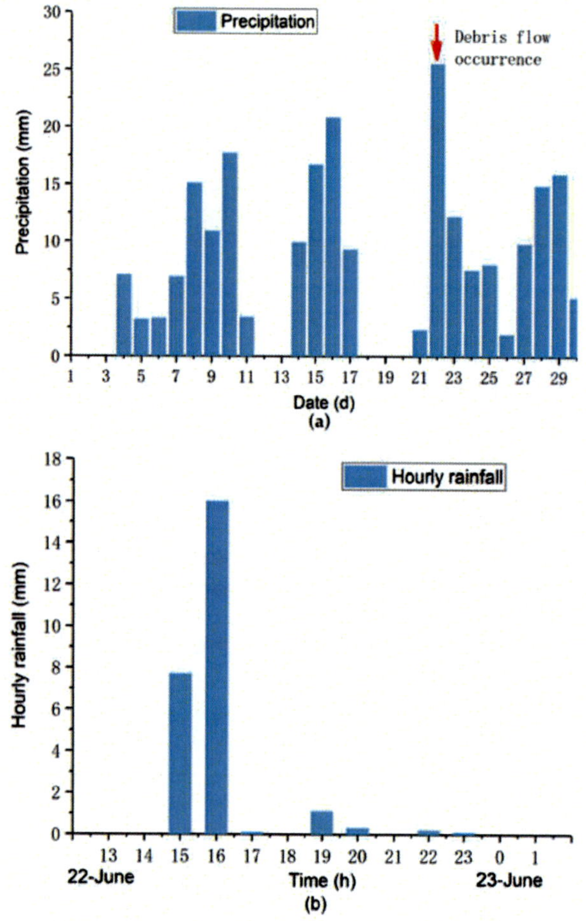


**Fig. 8** Rainstorm process in Shelong Gully in June: **a** daily precipitation 1st to 30th, June; **b** hour precipitation from 12:00 on June 22 ~ 01:00 on June 23

solid material conditions are very favorable for this disaster, it should be realized that previous accumulative rainfall and the critical rainfall triggering the debris flow.

Large-scale debris flows caused by heavy rainfall commonly lead to a complex process of damage accompanied by multiple hazards. Moreover, hydrothermal change has enhanced the complexity of the landslide failure mechanism under changing climate conditions. With the increase of extreme weather events, compound disaster events increase in the adequate-rainfall mountainous areas (Cui and Jia 2015; Zou et al. 2017). Thus, it is necessary to pay attention to the intensive rainfall and the research on the consequent risks of debris flows. The research on the analysis of hazard formation conditions and dynamic process for debris flows is of significance to disaster risk reduction.

**Acknowledgments** This work is supported by the National Nature Science Foundation of China (Grant No. 41790432), the Sichuan Science and Technology Planning Program (Grant No. 19YYJC0660), the Key Research Program of Chinese Academy of Sciences (Grant No. KFZD-SW-425), and the Scientific Research Program of Institute of Mountain Hazards and Environment, Chinese Academy of Sciences (Grant No. SDS-QN-1916).

## References

Boreggio M, Bernard M, Gregoretti C (2018) Evaluating the differences of gridding techniques for digital elevation models generation and their influence on the modeling of stony debris flows routing: a case study From Rovina di Cancia Basin (North-Eastern Italian Alps). Front Earth Sci 6:1–26

Cui P, Jia Y (2015) Mountain hazards in the Tibetan Plateau: research status and prospects. Natl Sci Rev 2(4):397–399

Cui P, Guo XJ, Yan Y, Li Y, Ge YG (2018) Real-time observation of an active debris flow watershed in the Wenchuan Earthquake area. Geomorphology 321:153–166

Chen YM (2003) Aerial photographic survey. China Architecture & Building Press

David P (2001) MapScript: a map algebra programming language incorporating neighborhood analysis. GeoInformatica 5(2):145–163

Denlinger RP, Iverson RM (2004) Granular avalanches across irregular three dimensional terrain: 1. Theory and computation. J Geophys Res 109:F01014

Hu S, Qiu HJ, Wang XG (2018) Acquiring high-resolution topography and performing spatial analysis of loess landslides by using low-cost UAVs. Landslides 15(3):593–612

Hungr O, Leroueil S, Picarelli L (2014) The Varnes classification of landslide types, an update. Landslides 11:167–194

Surhone LM, Tennoe MT, Henssonow SF (2010) Nadar (photographer). Betascript Publishing

Lazzari M, Gioia D (2017) UAV images and historical aerial-photos for geomorphological analysis and hillslope evolution of the Uggiano

medieval archaeological site (Basilicata, southern Italy). Geomat Nat Hazards Risk 8(1):104–119
Luo G, Liu M (2010) Stress evolution and fault interactions before and after the 2008 Great Wenchuan earthquake. Tectonophysics 491(1–4):127–140
Marx A, Mcfarlane D, Alzahrani A (2017) UAV data for multi-temporal landsat analysis of historic reforestation: a case study in costa rica. Int J Remote Sens 38(8–10):2331–2348
Niethammer U, Rothmund S, James MR, Travelletti J, Joswig M (2010) UAV-based remote sensing of landslides. In: Mills JP, Barber DM, Miller PE, Newton I (eds) Proceedings of the ISPRS Commission V mid-term symposium close range image measurement techniques. Gottingen: Copernicus Gesellschaft Mbh, pp 496–501
Pineux N, Lisein J, Swerts G, Bielders CL, Lejeune P, Colinet G, Degré A (2017) Can DEM time series produced by UAV be used to quantify diffuse erosion in an agricultural watershed? Geomorphology 280:122–136
Qiao JP (1997) Theories and Applications of Landslide Mitigation. Beijing:Science Press. pp 95–96. (In Chinese)
Wheaton JM, Brasington J, Darby SE, Sear DA (2010) Accounting for uncertainty in DEMs from repeat topographic surveys: improved sediment budgets. Earth Surf Process Landforms 35:136–156
Zou Q, Zhou GD, Li SS, Ouyang CJ, Tang JB (2017) Dynamic process analysis and hazard prediction of debris flow in eastern Qinghai-Tibet Plateau area—a case study at Ridi Gully. Arctic Antarctic Alpine Res 49(3):373–390

# MPM Modelling of Buildings Impacted by Landslides

2020 Kyoto Japan

Sabatino Cuomo, Angela Di Perna, and Mario Martinelli

**Abstract**

The paper discusses the topic of dynamic impact of fast-moving flow-like landslides against structures such as masonry walls and buildings. A set of numerical simulations is developed through the Material Point Method (MPM) to investigate the landslide-structure interaction. First, some experimental results available in the literature for masonry walls made of clay bricks and mortar joints are simulated in 2D and 3D conditions appropriately reproducing the overall stiffness, resistance and displacement of the wall in out-of-plane loading, until plastic hinges are formed and complete collapse occurs. In these cases, a known external pressure is applied to the wall. Then, realistic flow-like landslide scenarios are considered for analysing the impact on reinforced concrete buildings with unreinforced masonry infilled walls. Particularly, the impacting mass is modelled as a frictional material composed of a solid skeleton saturated with water. MPM simulations are developed in 2D conditions, focusing the attention on the failure of non-structural elements and damage to the different floors. MPM simulations are varied again over a range of properties to assess the role of key parameters, among impact velocity, volume, and soil properties such as unit weight and internal friction angle. The potential of a unitary approach for simulating the fast propagation of a saturated soil and the stress-strain response of a structural element is discussed.

**Keywords**

MPM • Displacement • Interaction • Failure • Building • Out-of-plane

## Introduction

Landslides of the flow type have been attracting the attention of scholars for (i) the threat to populations and structures and (ii) the scientific challenges related to their understanding and modelling.

The need to tackle the fundamental and sometimes urgent issue to protect populations has favoured a number of theories and approaches to be proposed, tested and applied. Landslide risk zoning demonstrated to be a formidable tool to deal with technical, economical and social aspects within a single unitary framework, applicable from site-specific cases to regional scale. About this topic, research projects have been developed in Italy and Europe (e.g. Safeland EU FP7 project, www.ngi.no/eng/Projects/SafeLand, 2010–2012), and long-term dissemination initiatives have been launched (e.g. LARAM School, www.laram.unisa.it, from 2006). Related to landslide risk zoning, either land use planning, restrictions or regulations for the areas at risk, warning systems and communication strategies have been used to manage the threat of potential catastrophic landslides impacting urban areas.

In each of the previous tools, the topics of landslide propagation and impact process on structures (Fig. 1) must be necessarily addressed and a model (simplified or sophisticated) for landslide impact is always adopted.

For the sake of the simplicity, one may consider that any structure impacted by a flow-like landslide will be destroyed. In other term, the vulnerability of the impacted structure is assumed equal to one (i.e. total loss of the exposed element). This safe approach is widely used for risk analysis over large areas.

S. Cuomo (✉) · A. Di Perna
Department of Civil Engineering, University of Salerno, Fisciano, 84084, Italy
e-mail: scuomo@unisa.it

A. Di Perna
e-mail: adiperna@unisa.it

M. Martinelli
Deltares, P.O. Box 177 2600 MH Delft, The Netherlands
e-mail: Mario.Martinelli@deltares.nl


V. Vilímek et al. (eds.), *Understanding and Reducing Landslide Disaster Risk*,
ICL Contribution to Landslide Disaster Risk Reduction,
https://doi.org/10.1007/978-3-030-60319-9_28

**Fig. 1** Example of damage caused by the impact of flow-like landslides in the Sarno area, Campania, southern Italy, 1998 (From Mavrouli et al. 2014)

However, it is frequently observed that the impacted structures suffer different types of damage: (i) complete collapse, (ii) partial destruction, (iii) only damage to non-structural elements. The occurrence of any of these three cases depends at least on the urban area (dense or sparse), construction features (type, materials, age, etc.), exposure (direct or limited) and impact (frontal or lateral), and other site- or case-specific features. Thus, the analysis (and zoning) of such issues must be addressed at detailed scale.

Examples are reported in Figs. 1, 2 and 3, with the different type of construction and the diverse amount of damage easily recognizable. In all the cases of those figures, flow-like landslides were induced by rainfall according to different triggering mechanisms (Cascini et al. 2008). Then, the unstable masses propagated downslope as debris flows, debris avalanches or combined phenomena (Cuomo et al. 2016). Independent on the type, all the flows recorded in 1998, 1999, and 2005 (and as already occurred in the previous centuries) arrived at high velocity and impacted the inhabited areas causing destruction or damage to structures.

What is really challenging is to find a mathematical framework to properly describe the so different materials and behaviour of both (i) the flow-like landslide and (ii) the impacted structure.

(i) A flow-like landslide, for its inner nature, consists of a multi-phase material travelling long runout distances at high velocity, with a total volume depending on source area, hillslope features, entrainment, etc.; in addition, soils are non-linear materials with irreversible deformations depending on specific stress path. Thus, staying within reasonable computational time usually requires depth-averaged approaches or meshless numerical techniques with some mandatory assumptions about water pressure, effective stress distribution or rheological behaviour.

**Fig. 2** Example of a building partially destroyed by a flow-like landslide in the Cervinara area, Campania, southern Italy, 1999 (From Cascini et al. 2011a)

(ii) The impacted structure is usually made of bricks or concrete elements eventually reinforced with steel bars. It means that linear and planar structural elements are connected in complex 3D configurations. And, while the behaviour of each single material can be nicely reproduced in continuous mechanics framework, the combined materials (e.g. bricks and mortar, or concrete and steel) requires a structural scheme to be defined,

**Fig. 3** Example of a building destroyed by a debris avalanche occurred in the Nocera Inferiore area, Campania, southern Italy, 2005 (From Cascini et al. 2011b)

where geometrical items (e.g. pillars, beams) and connections (hinges, joints) are modelled.

One relatively new method, namely Material Point Method (MPM), allows simulating large deformations and displacements (with no restriction), that are typical features of both a flow-like landslide and a building structure under severe external loads such an impulsive impact.

In the paper, after a review of the methods available for landslide propagation modelling and for structural damage analysis, the MPM fundamentals are presented. Then, the case of an Unreinforced Masonry (URM) wall is analyzed in 3D and 2D conditions in response to a known external pressure applied during an experimental campaign. Later, the case of the impact of a fast-moving landslide on a reinforced concrete building is analyzed, with the structural frame considered fixed and the masonry walls exposed to the impact of the landslide, whose features are varied over a range of values.

## Literature Review

### Modelling of Flow-like Landslide Propagation

Despite their potential especially for the back-analysis of case histories, the so-called empirical methods (Corominas 1996; Cascini et al. 2011a), which relate field observations of landslide volume, morphology, etc. to the landslide run-out distance, they may quantify landslide mobility but cannot provide enough accurate information about landslide dynamic features, such as propagation height, velocity and pressure inside the moving mass. The knowledge of those features is fundamental to assess the impact forces on structures.

Therefore, analytical methods are required as they can simulate landslide propagation using physical-based equations coming from fluid dynamics, soil mechanics or mixture theories (Pastor et al. 2009; Hungr and McDougall 2009). Among them, the Eulerian and Lagrangian approaches use a different strategy to represent the movement of a highly deformable and mobile body such as a landslide.

The Eulerian approach is common if not standard in fluid mechanics applications. The computational nodes are fixed, organized into a computational mesh, regular or not, and from there the phenomenon is described by means of the governing equations. In a sense, this is the most intuitive way of describing the motion of fluids, flows or similar phenomena such a fluid-like material while it exhibits very large displacements. There are examples of models using Finite Difference Method (FDM) to compute height and velocity of different types of flows in 2D or 3D conditions (O'Brien and Julien Fullerton 1993; Cascini et al. 2011b). But, there are also Finite Element Method (FEM) models with an Eulerian formulation (Pastor et al. 2002). However, the material properties are advected across the fixed computational grid. Such a procedure causes a spurious (numerical) diffusion of history variables (e.g. plastic strains) and interfaces of heterogeneous material setting are smoothed in space through time. In addition, different than fluids, a flow-like landslide has a non-negligible amount of solid, about 50% or more as so specific requirements are needed in those models. For instance, rheological formulations including any feature of the solid particle, or definition of pore water pressure inside the moving mixture must be included.

In a Lagrangian framework, while the material is deforming, there are the computation points which are linked to it. This class of method includes the Finite Element Method (FEM), the Discrete Element Method (DEM) and Smoothed Particles Hydrodynamics (SPH).

FEM has been extensively used in Lagrangian formulation for applications in solid- and soil-mechanics, especially to simulate small strains accumulated prior of failure (pre-failure stage) and eventually also during the failure stage. The strength of such models is that accurate assessment can be done by using complex constitutive laws capable to take into account the stress path of the soil (Zienkiewicz et al. 1999; Pastor et al. 2004) and strain history including accumulation of irreversible strains. However, there is the tendency of the computational mesh to be distorted until inaccuracy and numerical instability may prevent FEM to proceed further. With such premises the description of a soil flowing at high velocity becomes demanding.

On the other hand, DEM has been conceived for modelling the motion of granular materials. Many applications are reported in the literature about modelling of granular flows compare to experimental evidences (Favier et al. 2009, Faug et al. 2011). However, large domains and long time series are difficult to manage due to the high number of unknowns and large time for computing the particle connectivity. However, landslides can be schematized in most of the cases and continuous media and so the discontinuous description used by DEM can be inconvenient.

SPH, although originally proposed for describing clusters of individual points/bodies, has been later used for geophysical and continuum-mechanics applications. Since many years applications to landslide propagation problems have been proposed including a number of special features such as pore water pressure (Pastor et al. 2009), bed entrainment (Cuomo et al. 2014), special boundary conditions (Dai et al. 2016, Lin et al. 2019).

As common feature, in the Lagrangian FEM, SPH and DEM, the computational points coincide with material points (Fig. 4). However, this correspondence could be also avoided as proposed in some alternative methods. Among these, the Finite Element Method with Lagrangian Integration Point (FEMLIP) (Moresi et al. 2002), which has been derived from the Particle In Cell method (Sulsky et al., 1994), and the Material Point Method (MPM), which has been applied also to a number of different slope stability and landslide cases (Wang et al. 2016; Ghasemi et al. 2018; Cuomo et al. 2019a, b, 2020a).

Both methods, among others similar, are based on material points dissociated from the computational nodes of the Eulerian finite element mesh. The material points are used as integration points at a given material configuration. The resolution of the equilibrium equation at the mesh nodes gives a velocity field. At the end of each step, the velocity is interpolated from the nodes to the material points which are moved accordingly throughout the fixed mesh up to a new configuration. Internal variables and all material properties are stored at the material points, so that they are accurately tracked during the advection process. Actually, thanks to the distinction between mesh nodes and material points, such approach benefits both from the ability of the Eulerian FEM (the mesh is kept fixed) to support large transformations, and from the possibility of the Lagrangian FEM to track internal variables during the material movement.

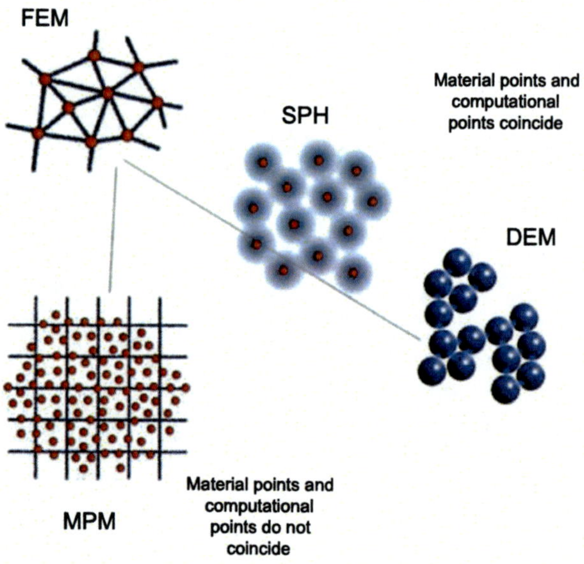


**Fig. 4** Different methods for modelling soil deformation and landslides (Modified from Cuomo 2020)

Hence, MPM is one of the most suitable method to deal with: i) static equilibrium of elasto-plastic materials in the pre-failure stage, ii) large deformations upon failure, iii) large displacements during the propagation stage while still tracking the history of material properties. A more exhaustive literature review on these topics is provided by Cuomo (2020) and other similar contributions, while more specific details on MPM can be found later on and in Fern et al. (2019).

## Collapse Mechanisms of Infill Panels in RC Buildings

Once a physically-based modeling of landslide has been completed, one should address the effects and damage of buildings in consequence of the impact of such fast-moving mass. Based on the observations of damage on Reinforced Concrete (RC) buildings with URM walls caused by the impact of flow-like landslides, some collapse mechanisms have been pointed out for both structural and non-structural elements. With reference to such collapse mechanisms, the ultimate bearing capacity of both the overall building and its constituent members could be computed (Faella and Nigro 2003a). Particularly, the load bearing capacity of a RC frame cannot be affected by the destruction of the external URM walls, as building collapse occurs only in the case of impact-induced formation of plastic hinges in the pillars of the ground floor. Anyway, the collapse of non-structural elements represents a serious danger to the safety of people in their own home.

The failure mechanisms (Figs. 5 and 6) of masonry infill can be divided into two groups: (i) In-Plane failure (IP, Fig. 5a) due to the interaction with the frame structural members (Parisi and Sabella 2016), and (ii) Out-Of-Plane failure (OOP, Fig. 5b) under a pressure perpendicular to the wall face (Di Trapani et al. 2018).

The frame-wall interactions have been largely investigated under IP and OOP seismic loads, through the adoption of theoretical (e.g. equivalent diagonal struts method as in Di Trapani et al. 2018), numerical models (Pantò et al. 2017; Abdulla et al. 2017), or also through experimental tests (Griffith and Vaculik 2007; Ricci et al. 2018; Valera-Rivera et al. (2012); Graziotti et al., 2019). In general, the proper modelling of the OOP response of an URM wall would require considering the effect of the IP loading resistance to accurately assess the performance of a three-dimensional structure. Anyway, in the case of buildings impacted by flow-like landslides, one could assume that the OOP response is more relevant than the IP behaviour.

Under OOP loads an infill wall can exhibit a significant resistance thanks to the formation of an arching mechanism. The latter is generated by (i) two-way bending mechanism if the panels are well confined in both directions by the surrounding RC structure (Fig. 6b), or by (ii) one-way bending if the opposite edges are supported (Fig. 6a). The occurrence of a two-way rather than a one-way mechanism strongly depends on support conditions (Fig. 6), as highlighted by numerous studies (Varela-Rivera et al. 2012; Vaculik and Griffith 2017; Tomassetti et al. 2018; Graziotti et al. 2019). The wall with top and bottom or the two lateral sides supported is characterized by one-way spanning and uniaxial bending (vertical for horizontal span and vice versa), as shown in Fig. 6a. In this situation, the cracking has the same direction of the supported edges. Two-way spanning walls supported on at least two adjacent edges (Fig. 6b) have more complex behavior, (Derakhshan et al. 2018), since they are subjected to biaxial bending, in both the horizontal and vertical directions. In this case, the wall cracking becomes a combination of vertical, horizontal and diagonal crack lines. In turn, the internal moments along the different types of crack lines can consist of a combination of flexure and torsion (Samarasinghe and Lawrence (1994), Graziotti et al. 2019).

The ultimate horizontal pressure bearable from a URM wall in OOP conditions can be determined through the load-displacement curve (Fig. 7).

An uncracked URM wall subjected to an OOP pressure shows first a linear-elastic behavior until its flexural tensile strength is achieved and consequently initial cracks start forms. As the deformations increase, cracking also spread along the mortar-joints of the panel, leading to a reduction of the wall stiffness.

Many experimental tests carried out on URM walls under seismic loads have demonstrated that the ultimate strength is generally reached with displacements approximately in the range 20–40 mm, since the cracks are such to lead to a

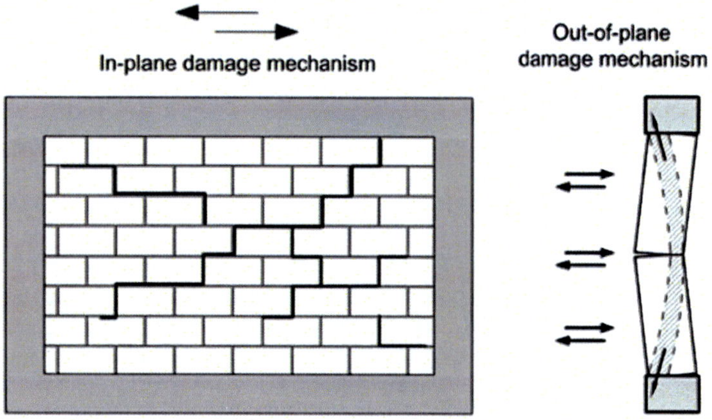


**Fig. 5** Damage mechanisms of URM infill walls: **a** In-Plane (IP), and **b** Out-Of-Plane (OOP), from Di Trapani et al. (2018)

Fig. 6 Crack patterns for different kinds of wall supports (From Vaculik 2012)

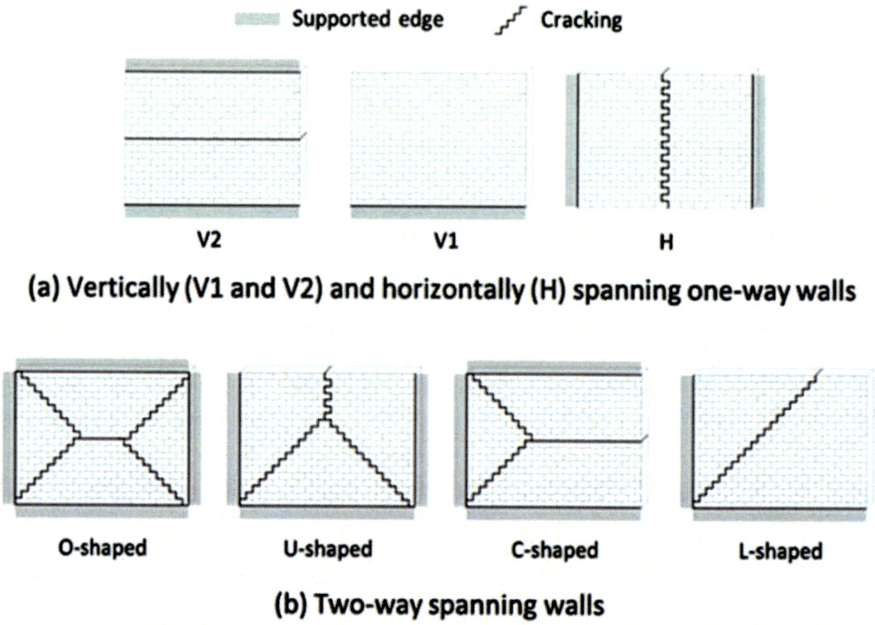


Fig. 7 Load-displacement curve for an URM wall subjected to OOP bending (Vaculik 2012)

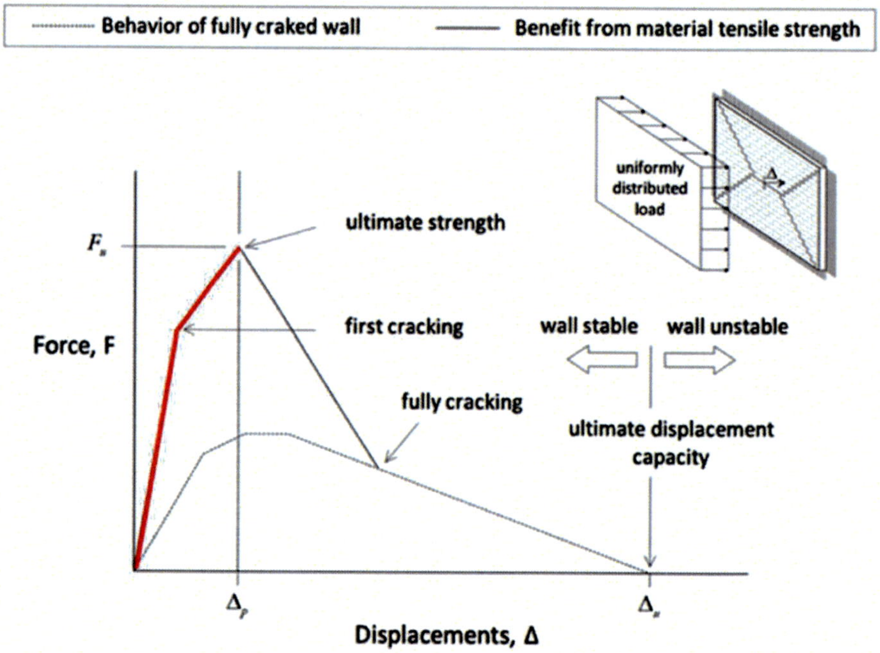


collapse mechanism (red line in Fig. 7). Both strength and stiffness continue to decrease with the increasing of deformations until the wall becomes fully cracked (rocking mechanism). However, the wall is characterized by a residual load capacity due to gravity effects. The ultimate displacement capacity is reached with the balance between resisting and destabilizing moments and it occurs when the wall reaches a deformation approximately equal to its thickness (Vaculik 2012). Beyond this point, the wall is unstable.

Focusing now the attention on the attainment of the ultimate strength and considering a vertical cross-section of the wall 1 m wide, the collapse mechanism is part of the one-way typology. Even the force-displacement curve

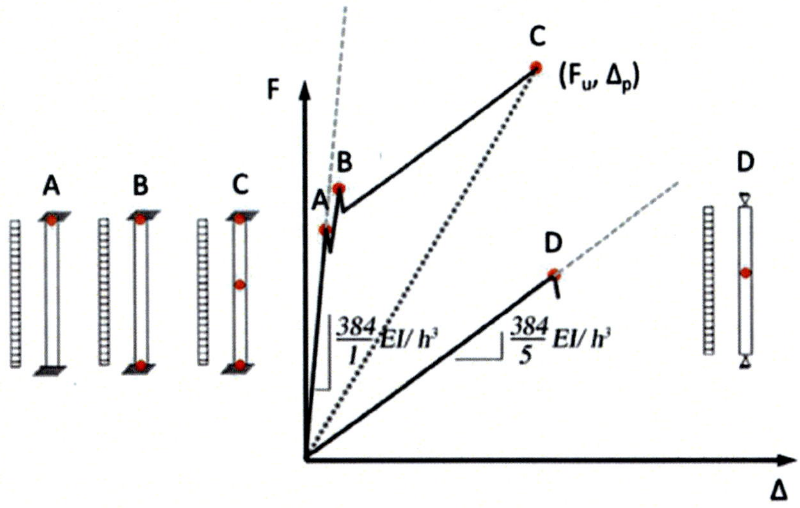


**Fig. 8** Load-displacement curve for different support conditions (Tomassetti et al. 2018)

strongly depends on the type of restraints at the top and bottom extremities (Fig. 8), which can be fixed-fixed or pinned-pinned.

In the fixed-fixed case, the first cracking arises at the top of the wall (point A), since the axial load due to the wall weight is lower around that point. When the external load increases, the second plastic hinge occurs at the bottom (point B) and finally at the mid-height of the wall (point C), generating a collapse mechanism.

In the pinned-pinned case, failure occurs quickly at the formation of just one plastic hinge located at the wall mid-height (point D).

Looking at the force-displacement curve, it emerges that the ultimate load capacity of the URM wall is higher in the case of fixed-fixed supports than the pinned-pinned case and is reached with three different levels of stiffness (Tomassetti et al. 2018).

In order to simulate such complex mechanisms, large displacements of the wall material must be accurately simulated and accurate landslide propagation modelling must be also possible in the same mathematical approach.

## The Material Point Method

As recalled before, MPM can be considered as an enhancement of FEM and it is very well suited for large deformation problems. The continuum body consists of a number of Lagrangian points (named material points, MPs), which carry all the physical properties and stress-strain values. The MPs move across a background fixed mesh which allows modeling large deformations of the body. The background mesh is used to solve the governing equations without storing any permanent information (Fig. 9).

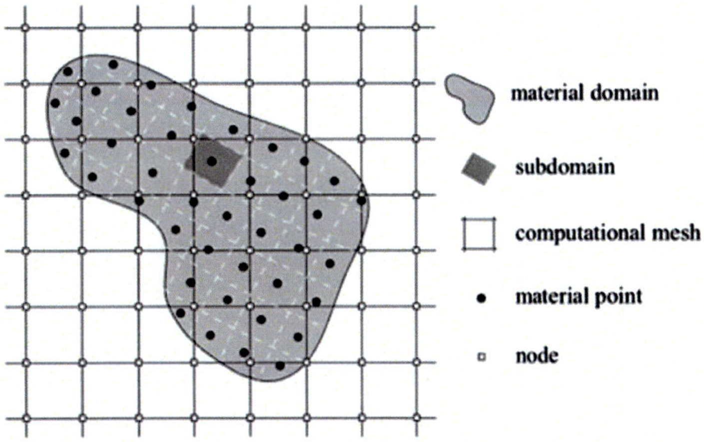


**Fig. 9** MPM discretization of space (computational mesh) and of material domain (filled with MPs), from Yerro (2015)

To simulate the interaction between phases (solid and liquid in a saturated soil), it is possible to use the so-called *two-phase single-point* formulation (Esch et al. 2011). This formulation describes the behaviour of the saturated porous media using only one set of MPs,, but containing all the information about both the solid and liquid phases.

The liquid flow is considered laminar hence the interaction force between solid and liquid phases is governed by the Darcy's law, with a constant permeability. Liquid fluxes due to spatial variations of liquid mass are also neglected. This assumption is acceptable when gradients of porosity are sufficiently small ($\nabla \approx 0$), or when an average value of porosity is assumed as representative for the whole simulation at stake.

The MPM algorithm for a single calculation step of a time increment is illustrated in Fig. 10. At the beginning of each step, the components of the momentum balance equations are defined by mapping information from the MPs to the nodes of the computational mesh by means of interpolation or shape functions (Fig. 10a). The equations of motion are solved for the primary unknown variables, i.e. the nodal accelerations (Fig. 10b). These nodal values are used to update acceleration, velocity and position of MPs as well as to compute strains and stresses at the MPs (Fig. 10c). At that point, no permanent information is stored any more at the mesh. Thus, it can be freely redefined at the end of each time step, but commonly it is kept fixed. The MPs assignment of MPs to elements is updated after mesh adjustment (Fig. 10d).

The *two-phase single-point* formulation is based on the assumption that the behavior of the saturated porous medium can be studied with the kinematics of the solid skeleton. Each MP represents a volume of the mixture $V$, given by the sum of the solid $V_S$ and liquid $V_L$ phase volumes. The movement of the MPs, representing the solid skeleton, is described in a Lagrangian framework, while the liquid motion can be obtained in respect to the solid one in an Eulerian specification (Figs. 11 and 12).

The governing equations are represented by the dynamic momentum balance of the liquid phase, the dynamic momentum balance of the mixture, the mass balances and the constitutive relationships of solid and liquid phases. The accelerations of the two phases are the primary unknown: the liquid acceleration $\boldsymbol{a}_\mathrm{L}$ is obtained by solving the dynamic momentum balance of the liquid phase (Eq. 1), while the solid one $\boldsymbol{a}_\mathrm{S}$ is calculated from the dynamic momentum balance of the mixture once known $\boldsymbol{a}_\mathrm{L}$ (Eq. 2). These equations are solved at grid nodes and their solutions are used to update the MPs velocities and momentum of each phase. The strain rate $\dot{\varepsilon}$ of MPs can be computed from the nodal displacements after nodal velocities have been obtained from the nodal momentum.

The resolution of solid and liquid constitutive laws (Eqs. 3 and 4, respectively) allows to calculate the increment of effective stress $\mathbf{d\sigma'}$ and excess pore pressure $dp_L$, respectively. The mass balance equation of the solid skeleton is then used to update the porosity of each MP (Eq. 5), while

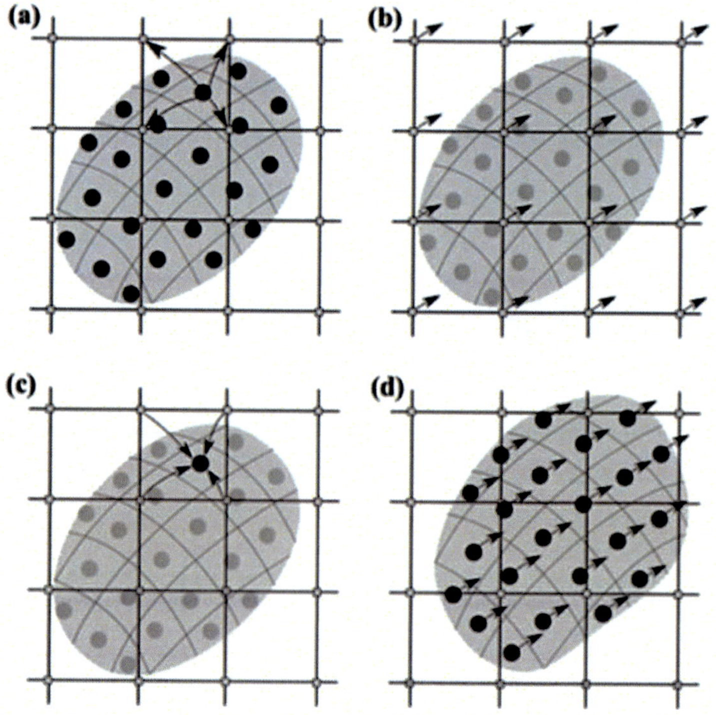


**Fig. 10** Schematic of Material Point Method integration scheme (at mesh nodes) and displacement computed at the MPs, from Fern (2016)

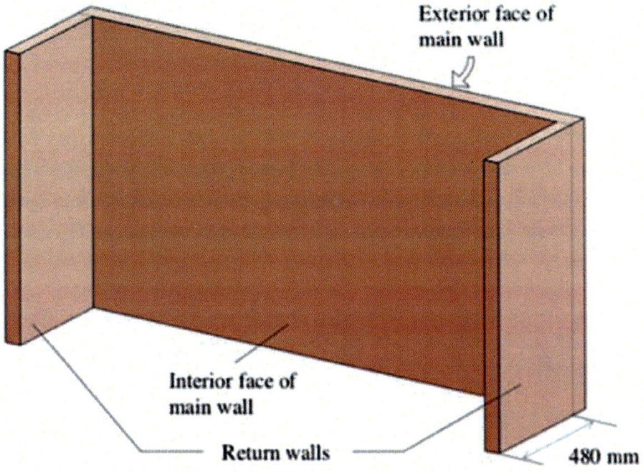


**Fig. 11** Schematic of the wall used in the experiments of Vaculik (2012)

**Fig. 12** Schematic of the interior face of the wall as cracked during the experiments of Vaculik (2012)

the total mass balance represent the volumetric strain rate of the liquid phase (Eq. 6) since fluxes due to spatial variations of liquid mass are neglected ($\nabla n\rho_L = 0$).

$$\rho_L \mathbf{a}_L = \nabla p_L - \frac{n\mu_L}{k_L}(\mathbf{v}_L - \mathbf{v}_S) + \rho_L \mathbf{b} \quad (1)$$

$$(1-n)\rho_S \mathbf{a}_S + n\rho_L \mathbf{a}_L = \nabla \cdot \sigma + \rho_m \mathbf{b} \quad (2)$$

$$d\boldsymbol{\sigma}' = \mathbf{D} \cdot d\varepsilon \quad (3)$$

$$dp_L = K_L \cdot d\varepsilon_{vol} \quad (4)$$

$$\frac{Dn}{Dt} = (1-n)\nabla \cdot \mathbf{v}_S \quad (5)$$

$$\frac{D\varepsilon_{vol}}{Dt} = \frac{(1-n)}{n}\nabla \cdot \mathbf{v}_S + \nabla \cdot \mathbf{v}_L \quad (6)$$

In the Eqs. 1–6, $\rho_L$ is the liquid density, $p_L$ is the liquid pressure, $\mu_L$ is the liquid dynamic viscosity, $k_L$ is the intrinsic permeability of the liquid, $\mathbf{v}_S$ and $\mathbf{v}_L$ are the solid and liquid velocity, **b** is the body force vector, *n* is the porosity, $\rho_S$ is the solid density, $\rho_m$ is the density of the mixture $\rho_m = (1-n)\rho_S + n\rho_L$, $\boldsymbol{\sigma}$ is the total stress tensor of the mixture, **D** is the tangent stiffness matrix ($\mathbf{K_L}$) is the elastic bulk modulus of the liquid.

In case of mono-phase materials (e.g. completely dry soils), the *one-phase single-point* formulation can be adopted in MPM. The formulation is based on the assumption that solid grains are incompressible, and the spatial variation of density and porosity is neglected. The motion of the continuum body is governed by the conservation of mass (Eq. 7) and momentum (Eq. 8). In addition, constitutive equation (Eq. 9) is also needed in order to describe the stress-strain relationship of the material. The equation refers to the Jaumann stress rate (Eq. 10), which is particularly suitable in simulating large deformations problem.

$$\frac{d\rho}{dt} + \rho\nabla \cdot \mathbf{v} = 0 \quad (7)$$

$$\rho\frac{d\mathbf{v}}{dt} = \nabla\boldsymbol{\sigma}^T + \rho\mathbf{g} \quad (8)$$

$$d\dot{\sigma}^J = \mathbf{D} \cdot d\dot{\varepsilon} \quad (9)$$

$$\dot{\sigma}^J = \dot{\sigma} + \boldsymbol{\sigma} \cdot \boldsymbol{\omega} - \boldsymbol{\omega} \cdot \boldsymbol{\sigma} \quad (10)$$

In Eqs. 7–9, ρ is the material density, **V** is the velocity vector, **g** is the gravity acceleration, $\dot{\varepsilon}$ is the strain rate matrix, $\dot{\sigma}^J$ is the Jaumann stress rate matrix, **D** is the tangent stiffness matrix, $\dot{\sigma}$ is the Cauchy stress rate and $\boldsymbol{\omega}$ is the rate-of-rotation matrix.

To model the propagation of the landslide and its impact on the retaining structure, a contact algorithm that allows relative motion at the interface between the interacting bodies is required. The frictional contact algorithm proposed by Bardenhagen et al. (2000) is here used. Such algorithm predicts the nodal velocity from the solution of the governing equation of each body separately and then it obtains the corrected value using the velocity of the combined system and the contact law.

## Modelling the Damage of URM Walls

### Input and Methods

Aimed to test the potential of MPM in properly simulating the response of a URM (Unreinforced Masonry) wall, some literature experiments are here considered. Particularly, the

**Table 1** Some material properties from literature

| | (*) E (MPa) | (*) ν | c' (kPa) | ϕ' (°) | $f_t$ (kPa) |
|---|---|---|---|---|---|
| Brick | 52700 | 0.2 | 1500(**) | 15(**) | 686§ |
| Mortar bed joints | 442 | 0.2 | 180(***)<br>81 ÷ 206§ | 30(*)<br>÷32(***) | 110 ÷ 140§§<br>170 ÷ 500§§§ |

(*)Vaculik (2012); (**)Hulse et al. 1982); (***)Graziotti et al. (2019); §Magenes et al. (1992); §§Varela-Rivera et al. (2012); §§§Wei and Stewart (2010)

URM walls used in the experimental tests of Vaculik (2012) were considered in the numerical simulations.

The walls were made of two components: standard Australian clay bricks with nominal dimensions 230 mm 110 mm × 76 mm, perforated with two rows of five holes and mortar joints with standard thickness of 10 mm using a composition of 1:2:9 (portland cement, lime and sand). The dimensions of the wall are $h$ = 2494 mm and $B$ = 4080 mm, with the lateral boundary confied in embedded return wall, each 480 mm long (Fig. 5). The mechanical properties of these materials and other similar are shown in Table 1.

The MPM computational domain has been set both in 3D and 2D for different purposes. The 3D scheme is used to properly reproduce the mechanical response of the wall including under out-of-plane two-way bending and traction at the central part of the wall, as well as propagation of extension mode and cracking inside the wall plane. In such model all the boundary conditions can be considered, either as the fixities or as mortar joints.

As seen before, masonry elements subjected to out-of-plane loading resist the load by flexural action. The load capacity of unreinforced masonry wall panels depends upon the dimensions and support conditions, the level of compressive stress in the wall and the tensile strength of the masonry (Griffith and Vaculik 2005).

As main advantage, such model could reproduce generalized deformation mechanisms and limit states including complex patterns of deformation, failure and post-failure of the wall. It is worth reminding that MPs have no restirction to move within the computational domain.

This feature of MPM is very well suited when the impact of a flow-like landslide is considered as external action against the wall, as very large displacements are expected.

On the other hand, the 2D model is here coinceived as a simplified scheme to be validated for later extensive application to design-oriented analyses. The minimum requirement for such 2D model is to properly reproduced out-of-plane bending, formation of plastic hinges at the top and bottom of the wall, simulation of tension cracking in the wall associated to the accurate evaluation of the load-displacement curve.

In both the 3D and 2D model, the URM wall is schematized as an equivalent uniform continuos frictional-cohesive material, with the mechanical parameters (Table 2), selected as follows. The density ($\rho$ = 1936 kg/m$^3$) and the Poisson ratio (ν = 0.2) are those mesaured for the wall by Vaculik (2012). The Young's modulus ($E$) is taken from the experimental evidence for the wall ($E$ = 2240 kPa) reported by Vaculik (2012), as for combo p1-p3 of Table 2 or assumed as equal to that of the mortar ($E$ = 442 kPa), in combo p4. The cohesion of the wall material (c′ = 180 kPa) is that reported by Graziotti et al. (2019) in combo p4 or is varied between 100 kPa and 130 kPa to get numerical results similar to the experimetal ones for combos p1–p3. The internal friction angle ($\varphi$') is set equal to the average value reported by Vaculik (2012) and Graziotti (2019) for the mortar bed-joints. The $K0$ value is determined as 1-sen$\varphi$' using the friction angle of the brick since it is plausible that the horizontal and vertical stress distribution is similar to that of a brick column. The tensile strength ($f_t$) of the wall is simply obtained as the ratio between cohesion ($c'$) and friction angle (tg$\varphi'$), assuming that the friction angle is the same both in compression and extension.

From there, 6 combos of material properties were individuated (Table 2).

**Table 2** Sets of masonry properties used for modelling

| | (*) ρ (kg/m³) | (*) ν | (*) E (MPa) | c' (kPa) | ϕ' (°) | $K_0$ | $f_t$ (kPa) |
|---|---|---|---|---|---|---|---|
| p1 | 1936 | 0.2 | 2240 | 100 | 31 | 0.74 | 166 |
| p2 | 1936 | 0.2 | 2240 | 130 | 31 | 0.74 | 216 |
| p3 | 1936 | 0.2 | 2240 | 115 | 31 | 0.74 | 191 |
| p4 | 1936 | 0.2 | 442 | 180 | 31 | 0.74 | 300 |
| p5 | 1936 | 0.2 | 442 | 115 | 31 | 0.74 | 191 |
| p6 | 1936 | 0.2 | 442 | 80 | 31 | 0.74 | 133 |

(*)Vaculik (2012)

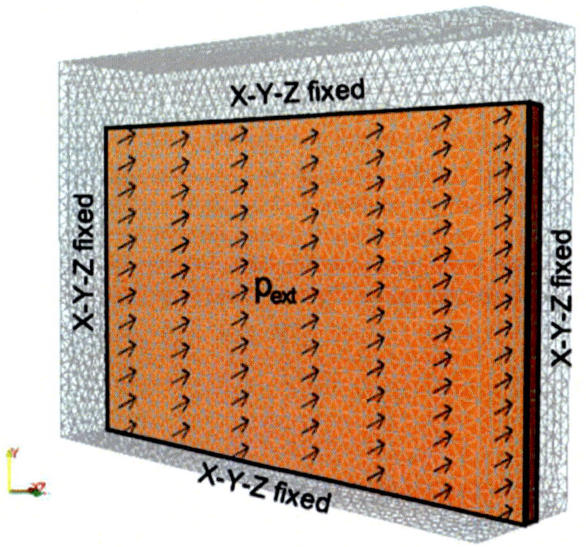


**Fig. 13** Computational 3D domain used to simulate the experiments of Vaculik (2012)

## Inspection of Full 3D Wall Response

In this case, the domain is 1.00 m × 4.08 m × 3.00 m and inside there is a fixed Eulerinan 3D mesh made of 116325 4-noded tethraedal elements with average size of about 0.1 m inside and outside the wall (Fig. 13).

Some of the results achieved are reported in Fig. 14, with reference to the deviatoric strain defined as $\varepsilon_d = \frac{2}{3}\sqrt{(\varepsilon_x - \varepsilon_y)^2 + (\varepsilon_y - \varepsilon_z)^2 + (\varepsilon_z - \varepsilon_x)^2}$, for combo *p6* of material properties allowing the fest-fitting of the experimental evidence. Before focusing on such satisfactory comparison, which is not the most important outcome, it is worth evidencing what are the physical processes correctly reproduced or not. As seen before, masonry elements subjected to out-of-plane loading resist the load by flexural action. The load capacity of unreinforced masonry wall panels depends upon the dimensions and support conditions, the level of compressive stress in the wall and the tensile strength of the masonry. Particularly, the crack patterns depend on the different kinds of wall supports (Fig. 6) and the load-displacement curve also depends on the different support conditions (Fig. 8). In this specific case, being fixed all the boundaries (Fig. 13) yielding occurs as follows: (i) first at the top of the wall, (ii) then at the bottom of the wall, (iii) at the side boundaries of the wall. From now on, yielding zones enlarged along the boundaries where they appeared but (iv) a plastic hinge forms at the mid-height of the wall, and (v) yielding also appears along inclined directions as expected for a two-way spanning wall (Fig. 6) and as observed by Vaculik (2012) in the experiments.

The formation of the first two plastic hinges at the vertical extremities of the wall gives starting to the collapse mechanism by reducing the overall stiffness of the panel (Fig. 14). Later on, for similar increments of the external pressure, the deviatoric strains and the horizontal displacements increase more and more.

Also, the out-of-plane displacements are correctly simulated (Fig. 15). Interestingly, the activation of the first plastic

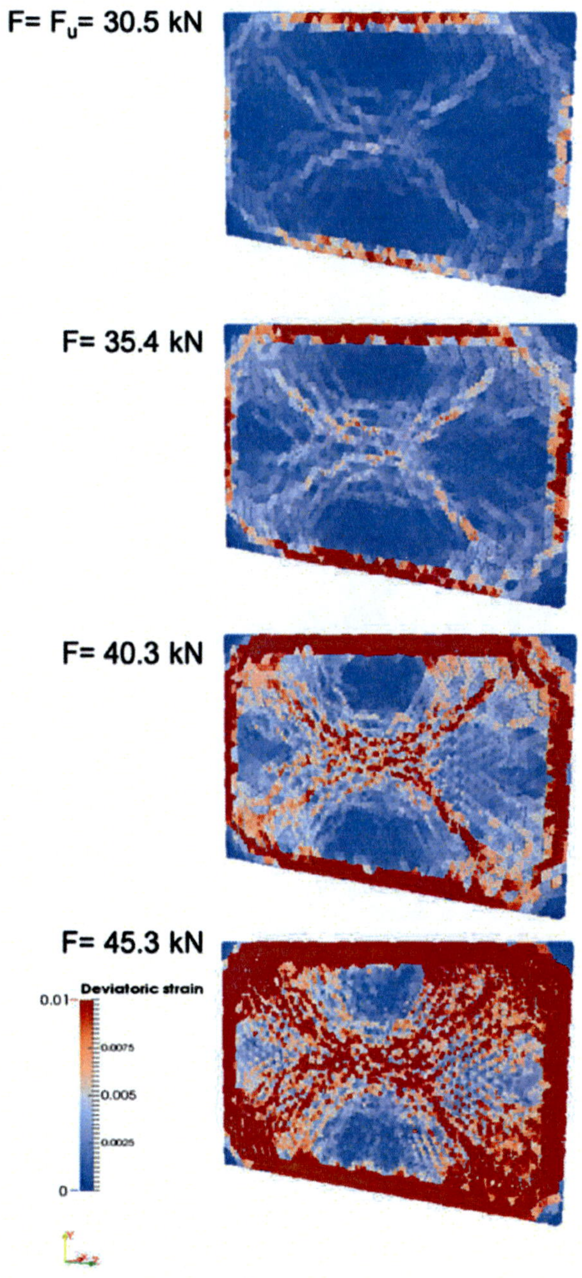


**Fig. 14** Deviatoric strain computed in the 3D domain for different external force and combo p6 of materials

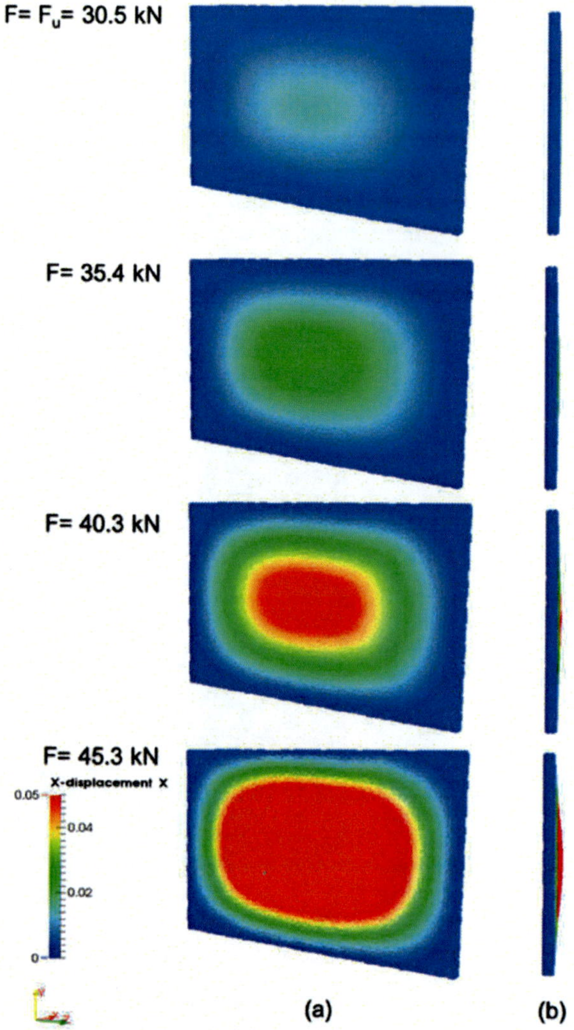


**Fig. 15** Computed horizontal displacements in the 3D domain at different values of the external force for combo p6 of materials

hinges and the corresponding reduction in wall stiffness (for Fu = 30.5 kN) does not correspond to the collapse, because most of the stress redistribution and strains occur inside the plane. Later on, the out-of-plane deformations and displacements start and collapse become unavoidable. Quantitative comparison of the horizontal displacements computed along different vertical sections versus the measurements is provided in Fig. 16, where it is shown that the 3D deformation mode is well captured by the model. Indeed, during the experiment (dashed line in Fig. 17), once the maximum horizontal displacement reached about 15 mm, then the external load was decreased so that the maximum displacement was about 30 mm with a final recovery of about 15 mm.

In Fig. 17 we can also assess how the stress-independent aliquot of the wall strength (cohesion) is capable to influence: (i) the stiffness of the wall, not at the origin of the plot but after that some deformation has already occurred, i.e. when yielding starts, and (ii) the ultimate load that the wall can sustain, as expected. While passing from combo p4 to p5 and p6, the decrement of cohesion is almost linear the response in the F-Δ plot is non-linear both for reduction in the ultimate load and decrease in the wall stiffness.

## Modelling in a Simplified 2D Scheme

While the 3D model has the evident advantage to carefully reproduce the deformation mechanisms, there is still an issue of long computational time especially when large structures are considered. And, on the other hand, in many applications and research investigations cross-section will be always a valuable schematization of the reality. Thus, it makes sense to try interpreting the same experiments seen before by means of a 2D model, whose parameters and performances must be evaluated again.

**Fig. 16** Displacements computed via 3D modelling for the combo p6 of material properties of Table 2 compared to those measured at the ultimate strength of the wall ($F_u$ = 30.5 kN) by Vaculik (2012)

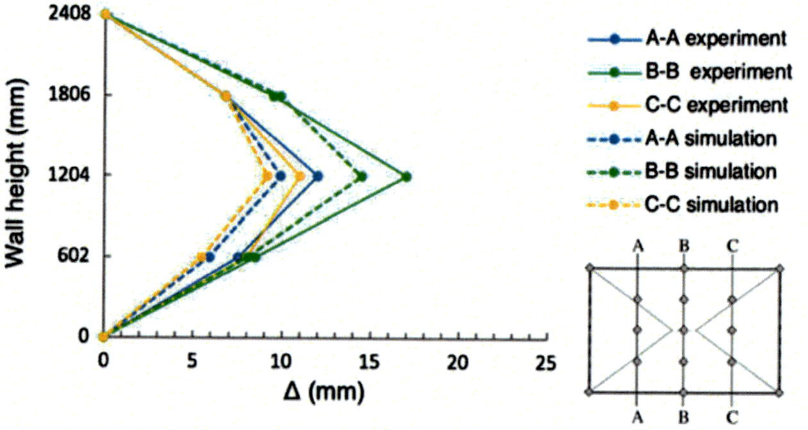

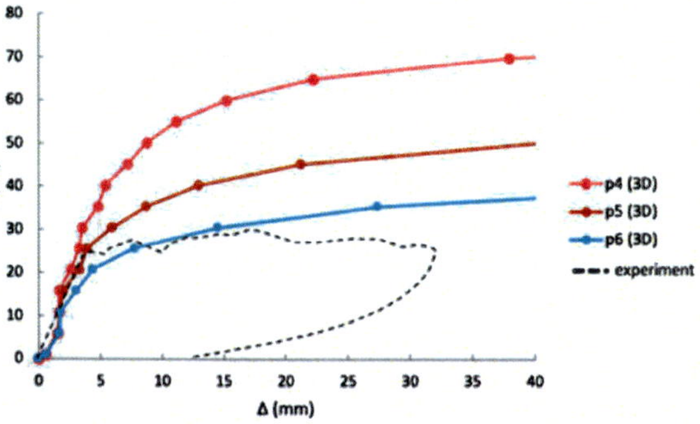


**Fig. 17** Load-displacement curves computed via 3D modelling for different combos of material properties of Table 2 compared to those measured by Vaculik (2012)

In the 2D case, the domain is 3.50 m × 3.00 m large, and made of 2605 3-noded triangles, 0.08 m sized inside (consistent with the 3D mesh size) and around the wall while 0.50 m sized more far (Fig. 18). The wall is considered fixed at top and bottom sides. Also in this case, a number of simulations were performed arriving to assess the combo p3 of Table 2 as the best-fitting set of parameters for simulating the experiments of Vaculik (2012).

Interestingly, when the external load reach about F = 30.5 kN, two hinges are simualted, then for F = 34.5 kN the yielding at mid-height of the wall, and finally the progression of yielding at the central zone of wall for F = 40.3 kN (Fig. 19). Such results are consistent with 3D model and it entails that also a 2D simplified model scheme could be used in real applications.

A quantitative comparison of the computed horizontal displacement versus the experimental values is provided in

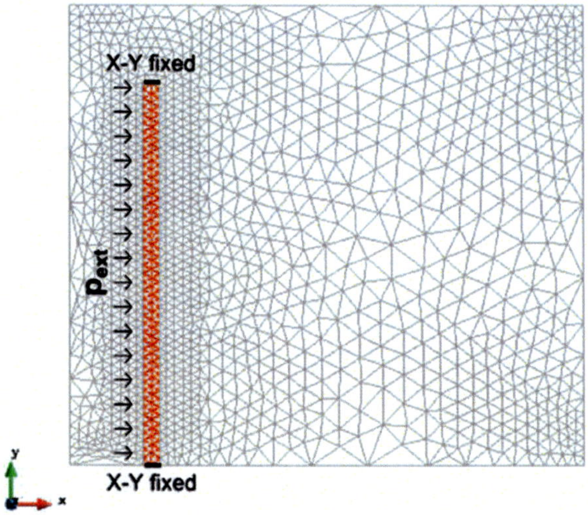


**Fig. 18** Computational 2D domain (barycentral vertical cross-section of the 3D domain)

Fig. 20, clearly identifying the combo p3 of materials as the best-fitting the experimental evidence. Other numerical results in terms of *F*-Δ curve are reported in Fig. 21, where the higher cohesion and related traction resistance (from p1 to p3 and then p2) result in higher strength and higher stiffness after the first cracking point.

## Modelling the Damage of RC Buildings

### Input Data

The previous results clearly show that the damage and the collapse of a single masonry wall, made of bricks and mortar joints, can be accurately simulated assuming the wall was as homogeneous frictional-cohesive material. Its properties, however, must be calibrated also in relation to the conditions (3D or 2D) referred in the modelling. Particularly, the 2D model proved to be simple and accurate enough, and with a reduced computational cost compared to 3D. For these reasons, a 2D model is here used to simulate the interaction of masonry walls with another complex (i.e. multi-phase) material such as a flow-like landslide.

The single wall case is in fact extended to a cross-section of a reinforced concrete building, where the framework is assumed fixed, while the non-structural external and inner masonry walls are impacted, deformed and eventually destroyed and transportated due to the impact of the landslide. The fundamental hypothesis beneath such modelling is that the framework is able to resist to the forces exerted by the flow on the pillars and the other structural elements. This is the case sometimes observed at some locations even for huge landslides (see Fig. 1). Fig 22 reports the geometry of the building and a zoom of the computational mesh characterized by elements with average size of 0.1 m for the smaller structural elements and of 0.5 m elsewhere.

The external walls are 0.3 m thick, with two panels 0.1 m thick separated by a 0.1 m internal cavity. The height of all the walls excluding the thickness of the exposed beam is equal to 2.4 m. For both walls, external and internal, the constitutive model is the same of the previous section and the mechanical properties are those of the combo p3 of Table 2, which proved to be the best set in the 2D modelling of the URM wall. Each wall is assumed to be connected to the structure by means of mortar joints, whose properties are obtained from literature values of Table 1, whereby a frictional and adhesive contact (e.g. tan$\varphi'$ = 0.6 and c' = 180 kPa) was assigned to the top and bottom ends of the wall. Moreover, the extremity of each wall is fixed in x-direction, so the wall can be considered pinned-pinned.

On the other hand, the lanslide of the flow-type (with volume ranging from hundreds to thousands of cubic meters) must be simulated. Here it is hypthotised that only the front

**Fig. 19** Deformed configuration and deviatoric strains simulated at different times for combo p3 of Table 2

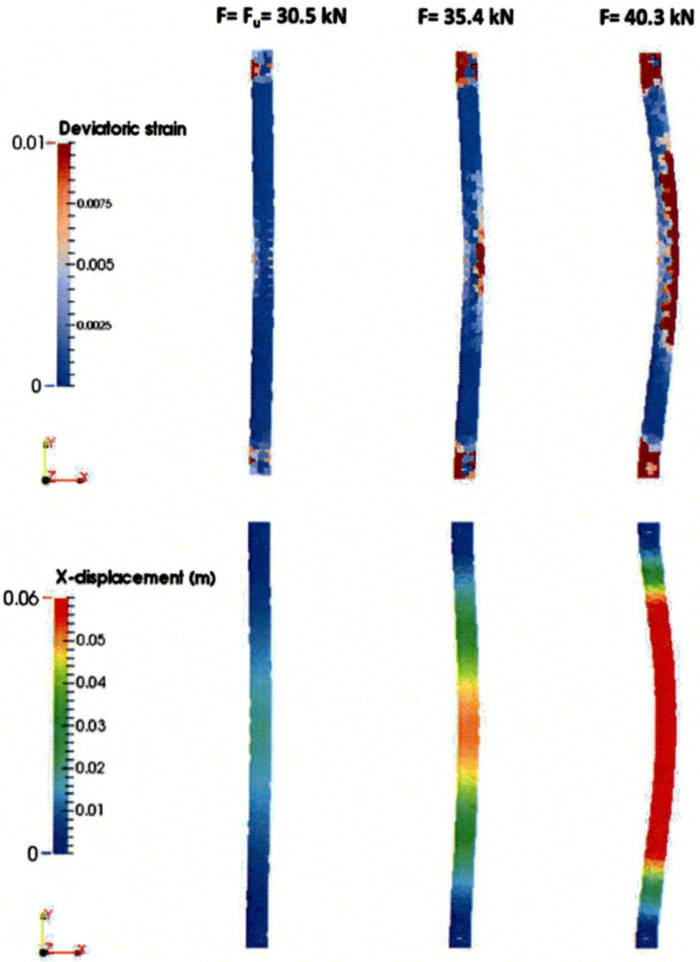


**Fig. 20** Horizontal displacements versus measurement

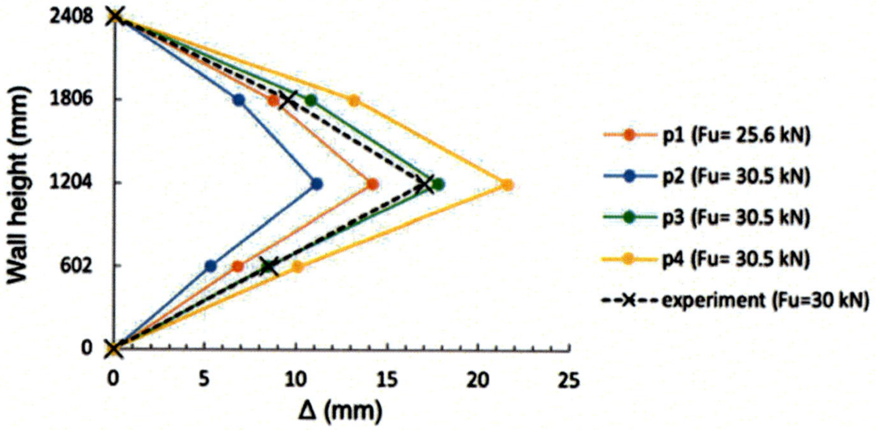

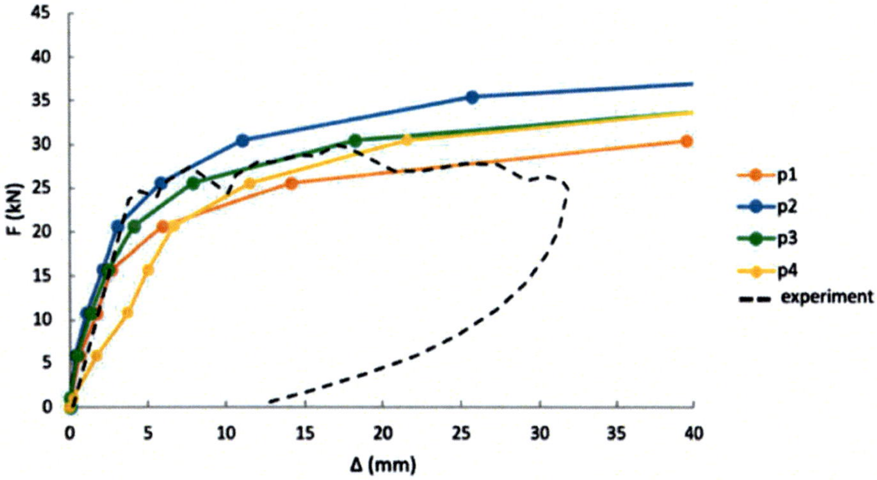


**Fig. 21** Load-displacement curves simulated for the wall via 2D modelling versus the measurements of Vaculik (2012)

part of the lanslide, which are typically very elongated, effectively interact with the structure during the impact process. The height of the impacting mass has been assumed generally equal to the first floor height on the basis of the surveys (Faella and Nigro 2003a, 2003b), hence in this case equal to 3 m, while the length is 15–30 m (i.e. 5 to 10 times the height).

It entails that, for instance for a width of 30 m, the volume here considered ranges from 700 to 1400 $m^3$. The computational discretization of such mass comprises a mesh of 3-noded triangular elements with average size of 0.5 m (Fig. 23).

The landslide material is assumed as a mixture of solid skeleton saturated with water and an elastic perfectly-plastic Mohr-Coulomb constitutive model is considered. The initial

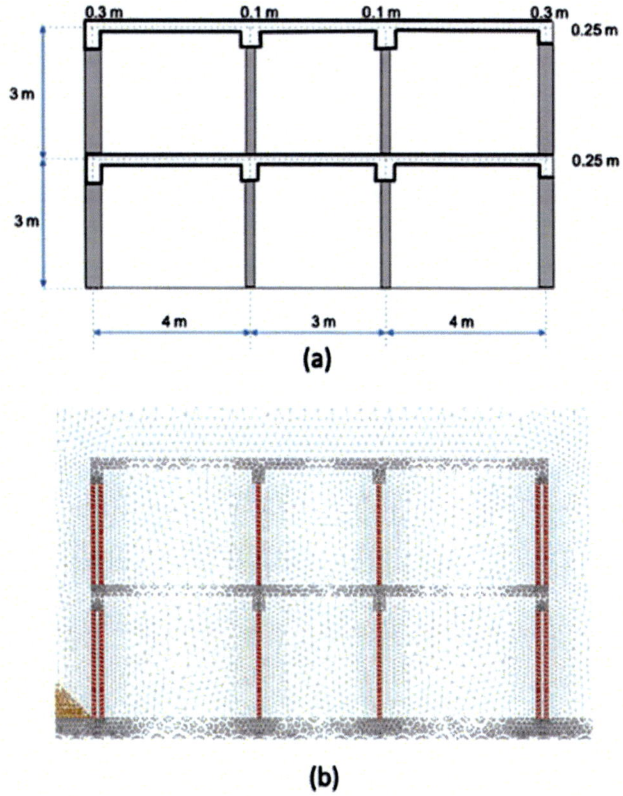


**Fig. 22** Geometry and computational discretization of a reinforced concrete building used as input for 2D modelling

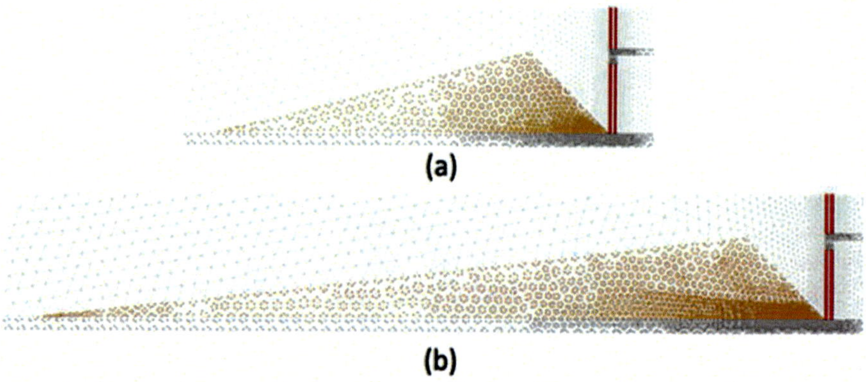


**Fig. 23** Discretization of the soil impacting mass with length of: **a** 15 m or **b** 30 m

stress distribution is set through gravity loading. Based on literature values and other related researches (Cuomo et al. 2020a), the mechanical properties are selected as follows: density ($\gamma$) of 18 kN/m$^3$, Poisson ratio ($\nu$) of 0.25, Young's moduls (E) equal to 2000 kPa, nil cohesion (c′), internal friction angle ($\varphi$') is 20°, nil dilatancy angle, hydraulic conductivity (k) equal to $10^{-3}$ m/s, liquid viscosity ($\mu_L$) of $10^{-3}$ Pa·s, liquid bulk modulus ($K_L$) of 200 MPa. At the base of the soil a frictional contact is considered with friction angle equal to that of the soil.

An overview of the whole computational domain including both the landslide and the builiding is provided in Fig. 24, where the impressive size of the lanslide can be appreciated compared to that of the building. In Table 3 the list of some selected simulations, which will be discussed in the next section.

## Numerical Results

The impact of a 30 m long soil mass impacting at 10 m/s against the building is firstly discussed with reference to some snapshots reported in Fig. 25. At the impact stage, the front of the lanslide is compressed against the external wall (Fig. 25), with increase in pore-water pressure and reduction of soil shear strength and consequent increment of total earth pressure on the wall itself, which of course fails with no chance to resist to such large external action. Indeed, the peak impact pressure averaged along the vertical could be assessed based on kinematic features of the impacting mass (h = 3 m, $\gamma$ = 18 kN/m$^3$, v = 10 m/s), as in Cuomo et al. (2020b). Here, the peak pressure would be about 921 kPa.

However, it is evident that the impact pressure is not only related to the impacting mass, but it is the outcome of the

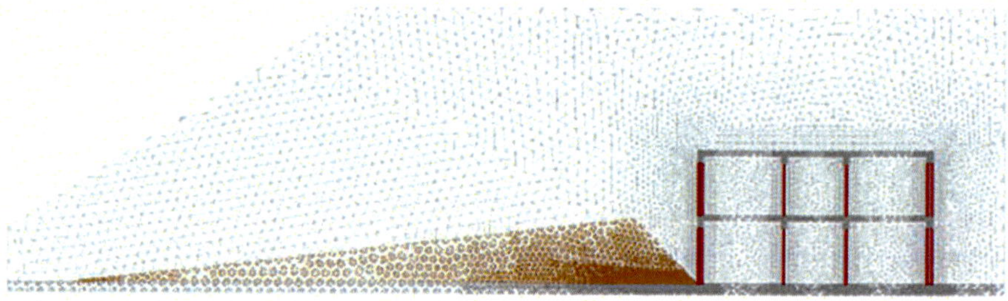

**Fig. 24** Computational domain used for 2D modelling

**Table 3** List of selected numerical simulations

| Case | L (m) | v (m/s) | $\varphi$‘ (°) |
|---|---|---|---|
| 1 | 15 | 10 | 30 |
| 2 | 30 | 10 | 30 |
| 3 | 30 | 10 | 20 |
| 4 | 30 | 5 | 20 |

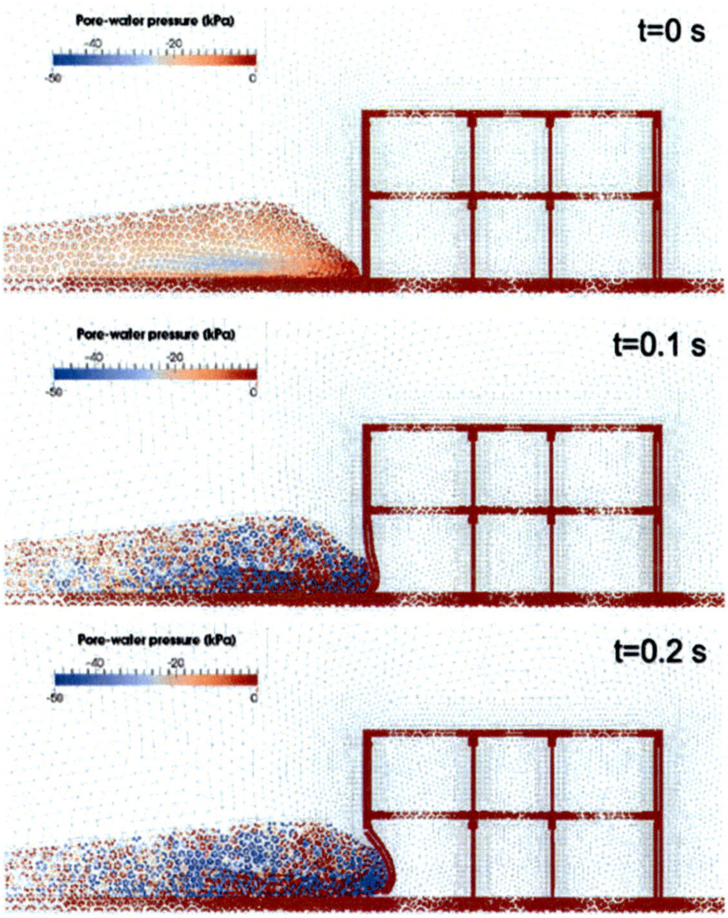


**Fig. 25** Increasing in pore-water pressure after the impact of a 10 m/s fast landslide on a building (Case 3 in Table 3)

internal forces arising from the landslide-wall interaction. In this sense, such mutual actions (force or pressure) depends on geometry, stiffness, resistance, ductilty of both the bodies involved into the impact. The latter is, in addition, impulsive, and the rate effects on deformations and mechanical behaviour should be also considered. From the results of the MPM modelling we inferred that the maximum impact pressure suffered by the ground floor wall was about 800 kPa in the zone impacted by the flow (t = 0.1 s in Fig. 25). Also the time trend of impact pressure is north of more discussion. If one compares the MPM numerical results to the simplified triangular time trend achievable through the impulse-theorem-based procedure proposed by Cuomo et al. (2020a). In this case, the time of peak pressure is about 0.5 s from the simplified procedure and 0.1 s from modelling.

Due to the impact, the landslide material runs up along the bailing, and also the wall of the first floor is damaged (Figs. 26 and 27). Such behaviour of the flow mostly depends on flow velocity (or related quantities), soil mechanical behavior and geometry of the "obstacle". Experimental results, theoretical studies and real case applications about such topic have been reported so far. Interested reader may refer to Faug (2015) and Cuomo et al. (2019a). The run-up and the effects on the first floor of the building (i.e. at elevation larger than the flow height) are more documented in Figs. 26 and 27, where the upper external wall is hit by the splash of the flow and fails. In the meantime, the flow propagates fast and inexorably destroys whatever it hits. If the flow arrives at slower velocity (i.e. v = 5 m/s) the consequence will be much worse (Fig. 26). In this case, the external wall suffers a cascade effect of the impact of the flow on the building, since the flow has much time to run up the building. Nevertheless, an inhabitants present at the first floor behind such wall would be dramatically endangered.

It is also true that in the same time lapse the pillars of the ground floor are overloaded laterally. But, we are assuming that the pillars do not fail, and so the frame can resist overall (like the real case in Fig. 1). Such issue must be more investigated, and the 3D model of the whole structure would be required, and a proper schematization of the mechanical

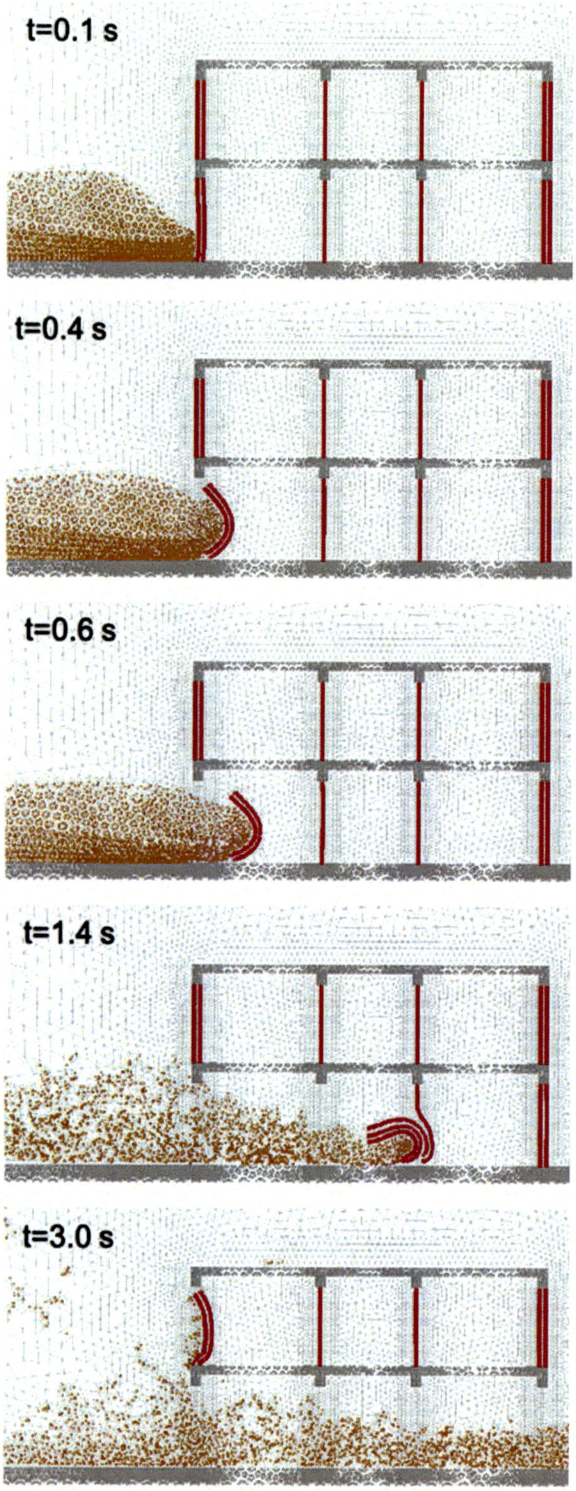


**Fig. 26** Impact of a 5 m/s fast landslide against the building simulated via 2D MPM modelling (case 4 in Table 3)

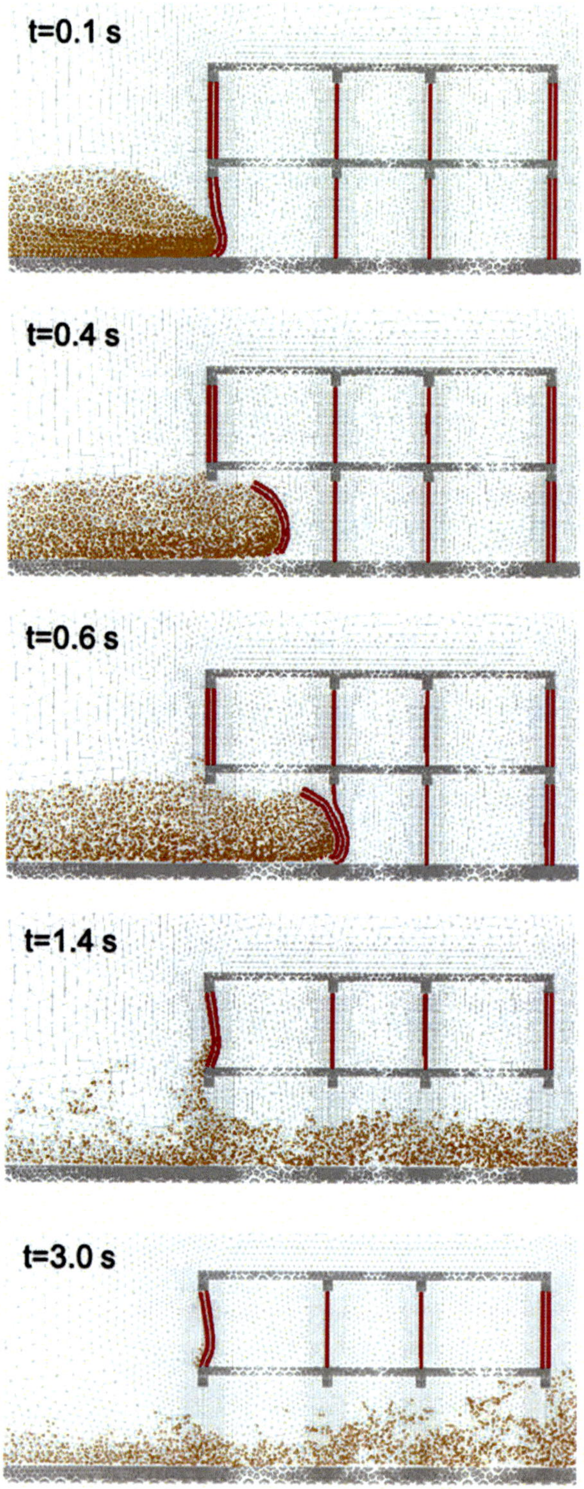


**Fig. 27** Consequence scenario during the impact of a 10 m/s fast landslide on a building as modelled through 2D MPM (case 3 in Table 3)

behaviour of the structural elements (pillars and beams) subjected to generalised loading conditions during the impact and flow-structure interaction. Some of these models already exist and have been used (as mentioned in the literature review) to analyse how a huge landslide of the flow-type may destroy (all or a part of) a urbanised area (or a cluster of buildings), and eventually stopping there or proceeding further. But this is not the topic we want to address here, which instead refers to landslides still catastrophic but not able to destroy a whole buildings. Such kind of model hence calls for a preventive assessment of the zones of the urban areas where height and velocity of the flows expected in a given return period (*T*) are below a threshold to safely assert that the building will not collapse. In such scenario, the attention must be focused on the type and amount of damage to indicate proper reaction behaviours to inhabitants in emergency cases and eventually individuated landslide migitation options or works for protection and reinforcement of the building.

As mentioned above, even the soil mechanical characteristics can modify the impact mechanism of the flow against the structure. For instance, comparing Figs. 27 and 28, it can be seen that different friction angle of the flowing mass can generate different impact scenarios against the structure.

What emerges is that for higher friction angle (Fig. 28) the flow doesn't reach the first floor, because the addictional strength allows the flow mass moving as a "rigid-like" body rathen than in the case represented in Fig. 27. Even the volume of the landslide has a crucial role. In fact, comparing Fig. 28 with Fig. 29, it is clear that smaller is the mobilized landslide volume lowest is the damage to the structure, since the flow doesn't impact the exposed beam. Anyway, velocity and friction angle of the approaching flow seems to be the most relevant factors in a 2D-analyzed impact problem.

## Concluding Remarks

The paper investigated how Material Point Method (MPM) modelling could be used to simulate the dynamic impact of fast-moving flow-like landslides against structures such as single masonry wall or groups of non-structural walls in a reinforced concrete building.

The present study even furnished a contribution to the modeling of the entire flow-structure dynamic interaction in a context of large deformations for both saturated flow and infill frame panels. As an enhancement of the previous

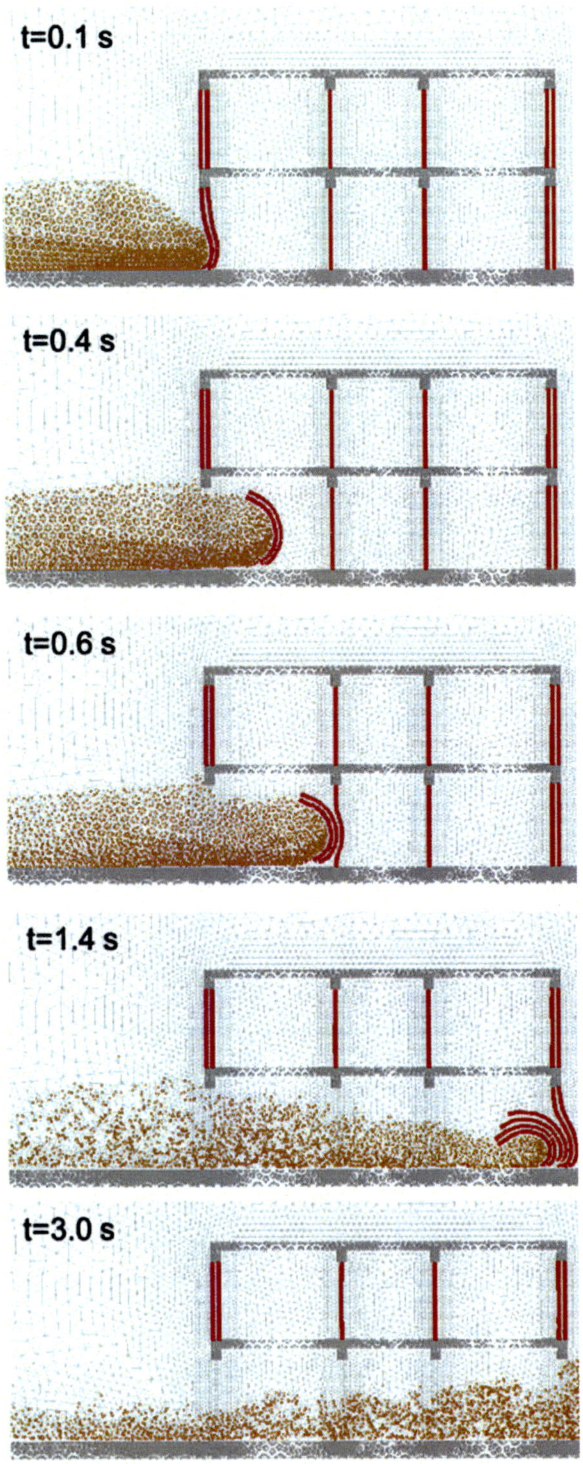


**Fig. 28** Consequence scenario during the impact of a 10 m/s and 30 m long fast landslide on a building (case 2 in Table 3)

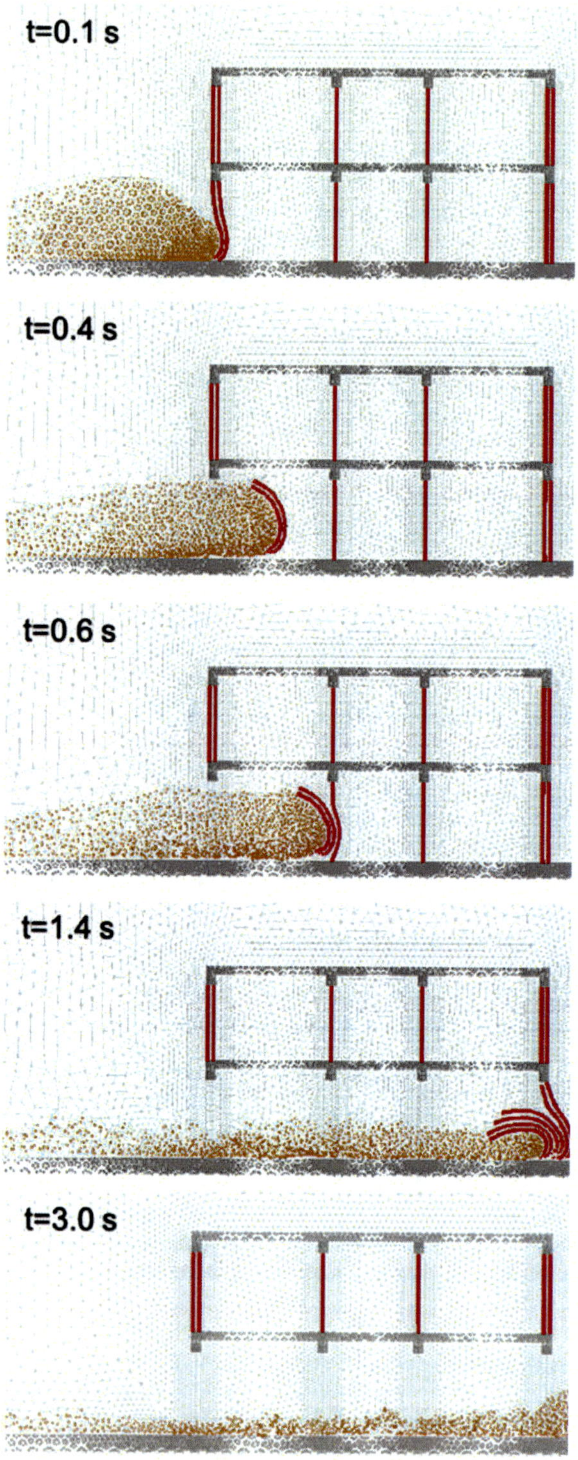


Fig. 29 Consequence scenario during the impact of a 10 m/s and 15 m long fast landslide on a building (case 1 in Table 3)

literature which safely calculated the impact pressure distribution by means of conservative empirical formulations, here it is shown that using the Material Point Method (MPM) it is possible to track accurately the spatial and temporal distribution of the stresses in both flow and structure.

The potentiality of MPM was first tested in simulating literature experimental results available for masonry walls, made of clay bricks and mortar joints, under a known uniform external pressure. The simulations in 2D and 3D conditions appropriately reproduced the overall stiffness, resistance and displacement of the wall in out-of-plane condition, until the formation of plastic hinges and complete collapse.

Then, the impact of a saturated soil on a realistic reinforced concrete building was simulated with reference to a 2D cross-section. The attention was focused on the failure of non-structural elements and damage to the different floors. MPM simulations were varied again over a range of properties to assess the role of the key parameters, such as volume, velocity and friction angle of the flow.

The potential of using a unitary approach for simulating the landslide propagation of a saturated soil and the stress-strain response of a structural element was highlighted and also the limitations and possible future development were discussed.

**Acknowledgements** This research was developed within the framework of Industrial Partnership PhD Course (Dottorato Industriale Regione Campania, Italia) and the Anura3D MPM Research Community. All the MPM simulations were performed using the Anura3D 2019 v2 code (http://www.mpm-dredge.eu/).

## References

Abdulla KF, Cunningham LS, Gillie M (2017) Simulating masonry wall behaviour using a simplified micro-model approach. Eng Struct 151:349–365

Bardenhagen SG, Brackbill JU, Sulsky D (2000) The material-point method for granular materials. Comput Methods Appl Mech Eng 187(3–4):529–541

Cascini L, Cuomo S, De Santis A (2011b) Numerical modelling of the December 1999 Cervinara flow-like mass movements (Southern Italy). In: Proceedings of 5th international conference on Debris-Flow Hazards mitigation: mechanics, prediction and assessment. Italian Journal of Engineering Geology and Environment, pp 635–644

Cascini L, Cuomo S, Della Sala M (2011a) Spatial and temporal occurrence of rainfall-induced shallow landslides of flow type: a case of Sarno-Quindici, Italy. Geomorphology 126(1–2):148–158

Cascini L, Cuomo S, Guida D (2008) Typical source areas of May 1998 flow-like mass movements in the Campania region, Southern Italy. Eng Geol 96(3–4):107–125

Corominas J (1996) The angle of reach as a mobility index for small and large landslides. Can Geotech J 33(2):260–271

Cuomo S, Di Perna A, Martinelli M (2020) Impact mechanism of granular saturated flows on protection structures. In: XXVII Convegno Nazionale di Geotecnica, June 24–26, 2020, Reggio Calabria, Italy (accepted)

Cuomo S (2020) Modelling of flowslides and debris avalanches in natural and engineered slopes: a review. Geoenviron Disast 7(1):1–25. https://doi.org/10.1186/s40677-019-0133-9

Cuomo S, Di Perna A, Ghasemi P, Martinelli M, Calvello M (2019a) Combined LEM and MPM analyses for the simulation of a fast moving landslide in Hong Kong. In: Proceedings of II international conference on the material point method for modelling soil–water–structure interaction. 8–10 January 2019, University of Cambridge, UK

Cuomo S, Ghasemi P, Martinelli M, Calvello M (2019b) Simulation of liquefaction and retrogressive slope failure in loose coarse-grained material. Int J Geomech 19(10):04019116

Cuomo S, Moretti S, Aversa S (2019c) Effects of artificial barriers on the propagation of debris avalanches. Landslides 16(6):1077–1087

Cuomo S, Moretti S, Frigo L, Aversa S (2020b) Deformation mechanisms of deformable geosynthetics-reinforced barriers (DGRB) impacted by debris avalanches. Bull Eng Geol Env 79:659–672

Cuomo S, Pastor M, Capobianco V, Cascini L (2016) Modelling the space–time evolution of bed entrainment for flow-like landslides. Eng Geol 212:10–20

Cuomo S, Pastor M, Cascini L, Castorino GC (2014) Interplay of rheology and entrainment in debris avalanches: a numerical study. Can Geotech J 51(11):1318–1330

Dai Z, Wang F, Huang Y, Song K, Lio A (2016) SPH-based numerical modeling for the post-failure behavior of the landslides triggered by the 2016 Kumamoto earthquake. Geoenviron Disast 3(1):1–14

Derakhshan H, Lucas W, Visintin P, Griffith MC (2018) Out-of-plane strength of existing two-way spanning solid and cavity unreinforced masonry walls. Structures (Elsevier) 13:88–101. https://doi.org/10.1016/j.istruc.2017.11.002

Di Trapani F, Shing PB, Cavaleri L (2018) Macroelement model for in-plane and out-of-plane responses of masonry infills in frame structures. J Struct Eng 144(2):04017198

Faella C, Nigro E (2003a) Dynamic impact of the debris flows on the constructions during the hydrogeological disaster in campania-1998: description and analysis of the damages. In: Proceedings of international conference on fast slope movements-prediction and prevention for risk mitigation (FSM2003), Napoli (Italy), pp 11–13

Faella C, Nigro E (2003b) Dynamic impact of the debris flows on the constructions during the hydrogeological disaster in Campania-1998: failure mechanical models and evaluation of the impact velocity. In: Proceedings of the international conference on "fast slope movements-prediction and prevention for risk mitigation", Napoli, vol 1. Pàtron Editore, pp 179–186

Faug T (2015) Macroscopic force experienced by extended objects in granular flows over a very broad Froude-number range. Eur Phys J E 38(5):1–10

Faug T, Caccamo P, Chanut B (2011) Equation for the force experienced by a wall overflowed by a granular avalanche: experimental verification. Phys Rev E 84(051301):1–18

Favier L, Daudon D, Donzé FV, Mazars J (2009) Predicting the drag coefficient of a granular flow using the discrete element method. J Statis Mech Theor Exper. https://doi.org/10.1088/1742-5468/2009/06/p 06012

Fern EJ (2016) Constitutive modelling of unsaturated sand and its application to large deformation modelling. PhD thesis, University of Cambridge

Fern J, Rohe A, Soga K, Alonso E (eds) (2019) The material point method for geotechnical engineering: a practical guide. CRC Press

Ghasemi P, Martinelli M, Cuomo S, Calvello M (2018) MPM modelling of static liquefaction in reduced-scale slope. Numer Methods Geotech Eng 9(2):1041–1046

Graziotti F, Tomassetti U, Sharma S, Grottoli L, Magenes G (2019) Experimental response of URM single leaf and cavity walls in out-of-plane two-way bending generated by seismic excitation. Constr Build Mater 195:650–670

Griffith MC, Vaculik J (2005) Flexural strength of unreinforced clay brick masonry walls. In: 10th Canadian masonry symposium

Griffith M, Vaculik J (2007) Out-of-plane flexural strength of unreinforced clay brick masonry walls. TMS J 25(1):53–68

Hulse R, Ambrose RJ, Lumbard P (1982) The shear strength of bricks and brickwork. In: 6th International brick/block masonry conference, pp 16–32

Hungr O, McDougall S (2009) Two numerical models for landslide dynamic analysis. Comput Geosci 35(5):978–992

Lin C, Pastor M, Yague A, Tayyebi SM, Stickle MM, Manzanal D, Liu X (2019) A depth-integrated SPH model for debris floods: application to Lo Wai (Hong Kong) debris flood of August 2005. Géotechnique 69(12):1035–1055

Magenes G, Calvi GM (1992) Cyclic behaviour of brick masonry walls. In: Proceedings of the 10th world conference on earthquake engineering, pp 3517–3522

Mavrouli O, Fotopoulou S, Pitilakis K, Zuccaro G, Corominas J, Santo A, Ulrich T (2014) Vulnerability assessment for reinforced concrete buildings exposed to landslides. Bull Eng Geol Environ 73 (2):265–289

Moresi LN, Dufour F, Muhlhaus HB (2002) Mantle convection modeling with viscoelastic/brittle lithosphere: numerical methodology and plate tectonic modeling. Pure appl Geophys 159(10):2335–2356

O'Brien JS, Julien Fullerton WT (1993) Two-dimensional water flood and mudflow simulation. J Hydrol Eng 119:244–261

Pantò B, Cannizzaro F, Caliò I, Lourenço PB (2017) Numerical and experimental validation of a 3D macro-model for the in-plane and out-of-plane behavior of unreinforced masonry walls. Int J Architect Herit 11(7):946–964

Parisi F, Sabella G (2016) Flow-type landslide fragility of reinforced concrete framed buildings. Eng Struct. https://doi.org/10.1016/j.engstruct.2016

Pastor M, Haddad B, Sorbino G, Cuomo S, Drempetic V (2009) A depth-integrated, coupled SPH model for flow-like landslides and related phenomena. Int J Numer Anal Methods Geomech 33 (2):143–172

Pastor M, Merodo JF, Gonzalez, E, Mira P, Li T, Liu X (2004) Modelling of landslides: (I) failure mechanisms. In: Degradations and instabilities in geomaterials, Springer, Vienna, pp 287–317

Pastor M, Quecedo M, Fernández Merodo JA, Herreros MI, González E, Mira P (2002) Modelling tailing dams and mine waste dumps failures. Geotechnique 52(8):579–592

Ricci P, Di Domenico M, Verderame GM (2018) Experimental investigation of the influence of slenderness ratio and of the in-plane/out-of-plane interaction on the out-of-plane strength of URM infill walls. Constr Build Mater 191:507–522

Samarasinghe W, Lawrence SJ (1994) Behaviour of masonry under combined torsion and compression. CSIRO. Division of Building, Construction and Engineering

Sulsky D, Chen Z, Schreyer HL (1994) A particle method for history-dependent materials. Comput Methods Appl Mech Eng 118 (1–2):179–196

Sulsky D, Zhou SJ, Schreyer HL (1995) Application of a particle-in-cell method to solid mechanics. Comput Phys Comm 87:236–252

Tomassetti U, Graziotti F, Penna A, Magenes G (2018) Modelling one-way out-of-plane response of single-leaf and cavity walls. Eng Struct 167:241–255
Vaculik J (2012) Unreinforced masonry walls subjected to out-of-plane seismic actions (Doctoral dissertation)
Vaculik J, Griffith MC (2017) Out-of-plane load–displacement model for two-way spanning masonry walls. Eng Struct 141:328–343
Van Esch J, Stolle D, Jassim I (2011) Finite element method for coupled dynamic flow-deformation simulation. In: 2nd International Symposium on Computational Geomechanics
Varela-Rivera J, Moreno-Herrera J, Lopez-Gutierrez I, Fernandez-Baqueiro L (2012) Out-of-plane strength of confined masonry walls. J Struct Eng 138(11):1331–1341
Wang B, Vardon PJ, Hicks MA (2016) Investigation of retrogressive and progressive slope failure mechanisms using the material point method. Comput Geotech 78:88–98
Wei X, Stewart MG (2010) Model validation and parametric study on the blast response of unreinforced brick masonry walls. Int J Impact Eng 37(11):1150–1159
Yerro A (2015) MPM modelling of landslides in brittle and unsaturated soils. PhD thesis, Universitat Politecnica de Catalunya
Zienkiewicz OC, Chan AHC, Pastor M, Shrefler BA, Shiomi T (1999) Computational geomechanics. Wiley

# Accelerating Landslide Hazard at Kandersteg, Swiss Alps; Combining 28 Years of Satellite InSAR and Single Campaign Terrestrial Radar Data

Rafael Caduff, Tazio Strozzi, Nils Hählen, and Jörg Häberle

**Abstract**

In summer 2018, in an area above lake Oeschinen in Kandersteg (Bernese Alps, Switzerland), significant terrain changes with indication of fast ground movements were observed. The NW dipping rock and debris slope named "Bim Spitze Stei" had been known to be under constant movement before. However, the rapid acceleration from a maximum volume prone to failure of about 20 $mm^3$ prompted the authorities to undertake a thorough analysis of the situation and analyse primary (rock avalanche) and secondary (floods and debris-flows out of the rock avalanche debris) hazard processes and the risk they pose to the nearby Village of Kandersteg. A first assessment of the most recent Sentinel-1 satellite InSAR data confirmed rapid ground movement in the order of several mm/d up to cm/d and a rapid acceleration of the west-flank of "Bim Spitze Stei" landslide from initially 7 mm/d to few cm/d within 2 weeks in July 2018. In addition, different sectors with different kinematics could be identified by interpretation of single interferograms. In a second step, an archive analysis of historical InSAR data reaching back to 1991 clearly showed that an acceleration trend from initially sub-stable conditions up to several m/a. Finally, based on the findings from the satellite InSAR analysis, a survey campaign with a terrestrial radar interferometer was performed in order to define the current state and location of the potentially outcropping glide plane in the west-flank. The successful campaign led to the observation of the presence of two active glide planes with the lowermost encompassing the maximum estimated volume of the mass in movement thus helping for the definition of potential failure scenarios thus helping in the selection of enhanced monitoring systems and increasing the preparedness for the runout-areas.

**Keywords**

Landslide hazard assessment • Landslide acceleration • Satellite InSAR • Terrestrial radar interferometry • Glide plane detection

## Introduction

A large part of the small mountain peak "Bim Spitze Stei" located above Lake Oeschinen at Kandersteg (Bernese Alps, Switzerland) has been known to be in constant movement for decades. Fractured bedrock in the upper part and freshly re-mobilized sediments at the middle part of the slope are the clearest signs of movement. Frequent observations at the peak area were taken by local mountaineering guides. During summer 2018 drastic changes in the area were reported, with new and unusually fast opening cracks around the peak.

The local authorities were requested to assess the current hazard and risk posed to the inhabited areas below. The volume of the unstable rock mass was estimated to be about 20 $mm^3$. While the slope itself does not bear any infrastructure that needs to be protected, high hazard is expected in the area of potential rock avalanche trajectory and more importantly in the area of potentially triggered secondary processes such as debris flows from sediment accumulations. In particular the village of Kandersteg, which lies downstream of the Oeschibach, needed further hazard and

R. Caduff (✉) · T. Strozzi
Gamma Remote Sensing AG, Worbstrasse 225, 3073 Gümligen, Switzerland
e-mail: caduff@gamma-rs.ch

T. Strozzi
e-mail: strozzi@gamma-rs.ch

N. Hählen · J. Häberle
Amt für Wald Und Naturgefahren AWN, Abteilung Naturgefahren, Schloss 2, 3800 Interlaken, Switzerland
e-mail: nils.haehlen@be.ch

J. Häberle
e-mail: joerg.haeberle@be.ch





V. Vilímek et al. (eds.), *Understanding and Reducing Landslide Disaster Risk*,
ICL Contribution to Landslide Disaster Risk Reduction,
https://doi.org/10.1007/978-3-030-60319-9_29

subsequently risk assessment for potential damage by floods and debris-flows that might occur after a large rock avalanche. Recent rock avalanches triggering secondary processes that occurred in the Swiss Alps (e.g. Pizzo Cengalo (Walter et al. 2020), Ritzlihorn/Spreitgraben (Frank et al. 2019), Val Selva, Ghirone (Caduff and Strozzi 2017)) suggested an exhaustive analysis of the situation.

As consequence, a monitoring system using multiple technologies was installed to assess the current displacements of the area (RT-Tachymeter, RT GNSS, Time Lapse Photography, Geophones). For an in-depth analysis of predisposition and triggering factors, geological, structural and hydrological analyses were mandated together with a historical analysis of the surface displacements in the area (Pixel Tracking on historical aerial imagery, satellite InSAR).

This paper focuses on the use of radar interferometry for the hazard assessment of a potentially large rock avalanche on different scales and platforms. After a geographical overview, we demonstrate (i) the use of near real-time satellite InSAR with Sentinel-1 data for the assessment of the current surface displacement, (ii) the use of historical InSAR data since 1991 with different sensors (ERS-1/2, ENVISAT ASAR, ALOS-1/2 PALSAR-1/2, RADARSAT-2, TerraSAR-X, Cosmo-Sky MED) to assess the evolution of the slope deformation in the past 28 years and (iii) the benefit of short campaign terrestrial radar-interferometry for assessing important structural parameters as the location of (multiple) active glide planes in the rugged parts of the target area. The results were used later for a hazard assessment mandated by the local authorities.

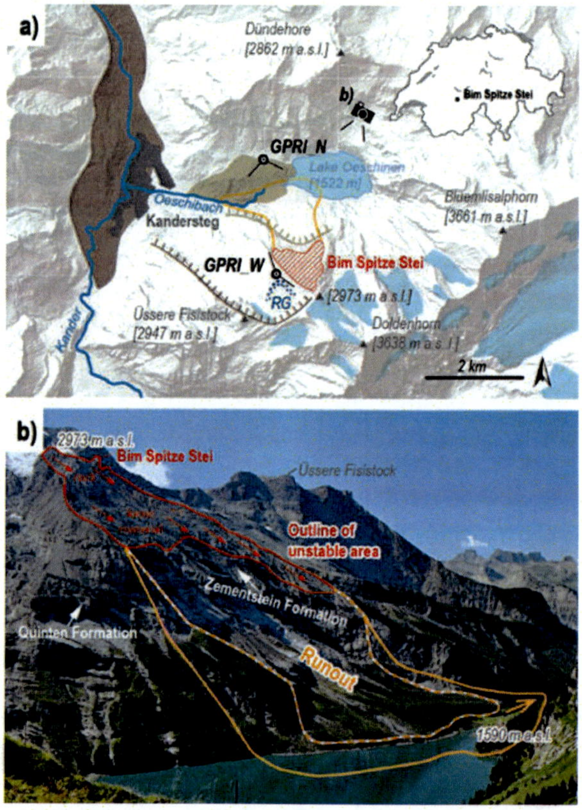


**Fig. 1** Geographical overview **a** with the outline of the active landslide (red), **b** lateral view of the target area with active slide (red) and potential runout zone (orange). GPRI_E and GPRI_W are the two positions of the terrestrial radar measurements, RG is a rock glacier

## Geographical and Geological Setting

### Location

The "Bim Spitze Stei" area is located in the central part of the Bernese Alps, Switzerland. It is part of the ~40 km$^2$ large Oeschinen sub catchment that is encompassed by the mountain chain including Fisistock—Doldenhorn—Blüemlisalp—Dündehore (Fig. 1).

In the centre of the catchment lies the touristically attractive landslide-dammed lake Oeschinen. The lake drains partly in an underground-stream and then at the surface as creek "Oeschibach" towards the river Kander. On the river-fan at the outlet of the Oeschinen catchment, the village of Kandersteg is located.

### Geology

From the geological point of view, the "SpitzeStei" is built from a sequence of rock formations of the infrahelvetic Doldenhorn-Nappe. Its main lithologies, from bottom to top, contain (i) light grey micritic limestone of the Quinten Formation (ii), dark limestone and marly limestones of the Zementstein Formation and (iii) dark marlstones and light grey bioclast bearing limestone, both composing the Öhrli Formation, at the "SpitzeStei" area itself. The main bedding is oriented towards NW with a dip-angle of 30°.

Several very large rock avalanches took place in the mid Holocene and characterized the main morphology of the current day Kander Valley. One very large rock-avalanche with an estimated volume of 0.8 km$^3$ originated from the northern face of the current day "Üsserer Fisistock" (Tinner et al. 2005). New research from Singeisen et al. (2020) suggests that this failure occurred about 3,200 years BP. The detachment zone is in direct vicinity to "Bim Spitze Stei" and consists of similar structural and geological predisposition.

Another large rock avalanche occurred ~2,300 years BP from the lower part of the target study area that led to the formation of Lake Oeschinen (Köpfli et al. 2018). Further investigations showed, that at least 6 other significant rockslide/avalanche events took place (Knapp et al. 2018).

The area that is moving spans from an elevation beginning at 2200 up to 2900 m a.s.l. This area and the slope exposition are prone to an ice-rich permafrost situation (Hauck et al. 2017). Rockglaciers close and in the target-area are a clear indication of a temperature regime supporting the formation of permafrost.

## Assessment of the Current Displacements with Sentinel-1 InSAR Data

### Method

As first order assessment of the spatial extent of the ground movements in the area, we considered data from the European Sentinel-1 satellite. Sentinel-1 has a 6-day observation period on the single interferometric orbits and multiple revisit orbits in the two acquisition geometries (ascending path: ENE looking; descending path WNW looking). The sensor itself has a nominal resolution in IW mode of 5 × 20 m, which was resampled for the interferometric processing to 20 × 20 m (multi-looking factor of 4 × 1) and then again resampled for geocoding to 5 × 5 m. An overview of the methodology using Sentinel-1 interferograms for the assessment of a large slope instability over the Swiss Alps is given by Manconi et al. 2018. Major limitations to the use of satellite SAR interferometry to assess landslide motion are due to the viewing geometry, the acquisition time interval, and the landcover. Around "Bim Spitze Stei", the north facing slope and the west-shoulder of the area are both covered without significant gaps caused by layover and shadow effects. However, the view on the uppermost areas of the slope instability is largely blocked because of this effect (black areas in Fig. 2a, b).

Another limitation of satellite SAR interferometry is given by the velocity of the landslide, more precisely by the total displacements along the line-of-sight of the sensor between a revisit-cycle. The feasibility for a successful determination of surface displacement values in the interferograms is a function of the sensor wavelength (Sentinel-1: 5.6 cm [1/2 $\lambda$ = 2.8 cm]), the revisit time, the spatial resolution and the extent and kinematics of the area in movement. In other words, the movement must not be too fast in order to be quantified unambiguously.

Regarding the surface cover, scenes with vegetation, snow and rockfall/avalanche within the area of interest are excluded from the analysis.

Our analysis was not restricted to archived data. An automated near-real-time interferometric processing chain was set-up to obtain the information of the current state of the movement hours after the acquisition of the newest Sentinel-1 scene.

### Results

A comprehensive product from the interpretation of the Sentinel-1 interferograms of the years 2015-2018 is shown in Fig. 2. The focus of the analysis was set on the separation of areas with similar temporal behaviour. Several zones with different average deformation rates and/or the temporal signatures could be identified. Significant changes could be observed in the western part of "Bim Spitze Stei", where an increase of the deformation started mid-summer 2018 and soon reached values that led to complete decorrelation in the 6 d interferograms (meaning >> than 2.8 cm movement).

A comparison with acquisitions from the same season in the years before showed that in this time significant movements were present as well, but the recorded surface deformations did not reach the same levels as 2018.

Additionally, the boundaries of the area affected from the movements could be mapped. However, the pixel resolution and projection errors in the steeper part of the area lead to a significant uncertainty in the estimation of the volume in movement.

## Historical Analysis Using Satellite InSAR Data

### Method

Similar to the above outlined assessment of the radar interferograms generated from Sentinel-1, data from various other SAR sensors were interferometrically processed. We could process X-Band data from TerraSAR-X and COSMO SkyMed, C-Band data from ERS-1/2, ENVISAT ASAR, RADARSAT-2, and L-Band Data from JERS-1 and ALOS-1/2 PALSAR-1/2, mostly along the descending geometry. After interferometric processing of the data over the subset of the test area, coherent pairs were selected and qualitatively interpreted for potential movements in the area similarly to the study performed by Strozzi et al. 2010. Wherever possible, the absolute displacement in line of sight was extracted at locations where optical targets were installed and continuously measured with an automated tachymeter starting on 13. August 2018. The displacement rates within the time intervals of the interferograms were normalized to yearly rates and plotted as time series (Fig. 3).

### Results

The results of the historical analysis using satellite InSAR data are shown in Fig. 3. First measurements from the ERS-1 mission in 1991 with a 3-day repeat interval showed movement in the eastern part of the area. This patch is identified as an ice rich sediment body (i.e. a rock glacier),

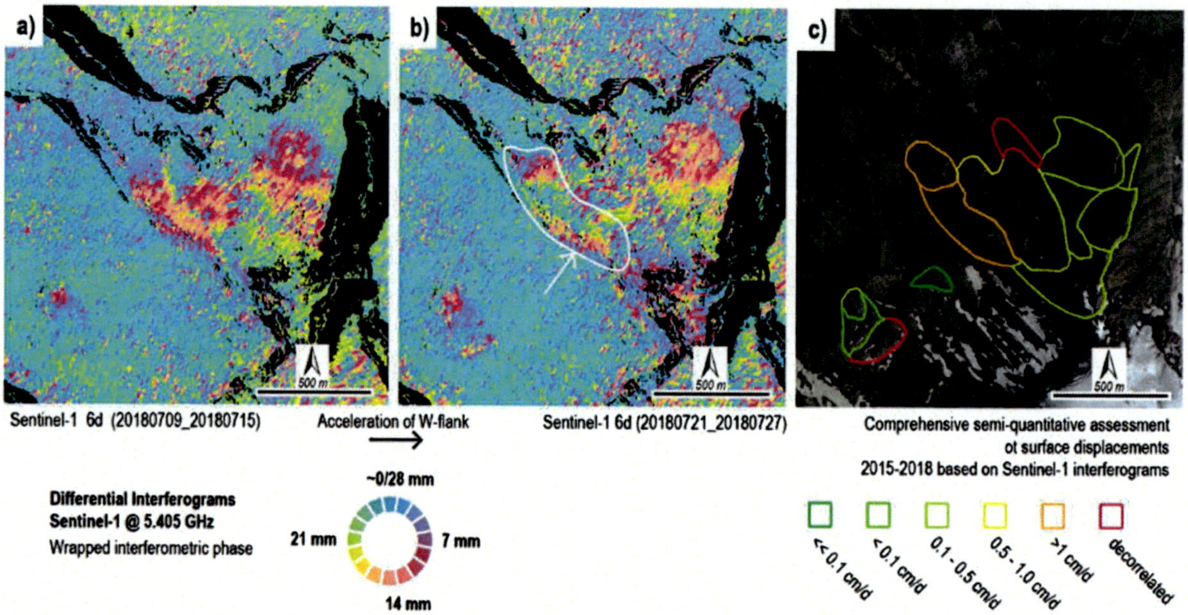


**Fig. 2** Comparison of two short time (6d) Sentinel-1 interferograms **a** before and **b** after initiation of a fast acceleration of the W-flank (white outline) in mid-July 2018. A synthetic analysis of multiple Sentinel-1 interferograms from 2015 to 2018 led to the delineation of different sectors with different kinematics (**c**)

which is in continuous movement also in recent times. Even though only short revisits (1–3 d) are coherent in 1991, it is clear that the movement rates by "Bim Spitze Stei" were significantly lower than in recent observations. The western part of the area is more or less stable or slower than 5 mm/month in LOS. An increase of the deformation rate can be observed only since 2007. at Point 52 and upward. The actual onset of this movement cannot be determined from the satellite data since there is a large data gap of coherent observations between 2000 and 2005.

From 2014 onwards, the area under significant movement was increasing at the lower western boundary, including point 63.

## Assessment of the Current Situation with Terrestrial Radar Interferometry

### Method

In order to verify the presence and location of the main basal glide plane in the west-lateral area of the instability, a terrestrial radar interferometric campaign was executed at the beginning of August 2019. Terrestrial radar interferometry works similarly to satellite-based interferometry and is described in detail in Caduff et al. (2015). In order to plan a terrestrial radar campaign, it is important to have a rough idea of the current movement rates. Since the accessibility with the instrument to the observation point was only possible with a helicopter transport, the measurement intervals have arranged to lie within the sensitivity range of the system, i.e. more than ~1 mm and less than about a phase cycle.

On 6. August 2019 first measurements with 2 min intervals were taken during 45 min from an inactive or transient rock glacier lobe (GPRI_W in Fig. 1). Since the Sentinel-1 analysis did not show significant movement of the rock glacier in the years and weeks before the campaign and no long term observations were planned, the risk of a moving observation platform was very small. Following the forecast of a thunderstorm, the instrument was disassembled and stored at 2470 m a.s.l. for two days.

After the rain- and thunderstorm, the instrument was set-up again on 8. August 2019 and a continued monitoring the rock slope for 26 h at acquisition intervals of first 2 min and later 5 min.

Before and after the campaign described above, additional terrestrial radar measurements were taken from Point GPRI_N located on the opposite slope of the "Bim Spitze Stei" (Fig. 1).

### Results

The results of the campaign at the west-lateral area is shown in Fig. 4. A selected interferogram is projected onto a set of

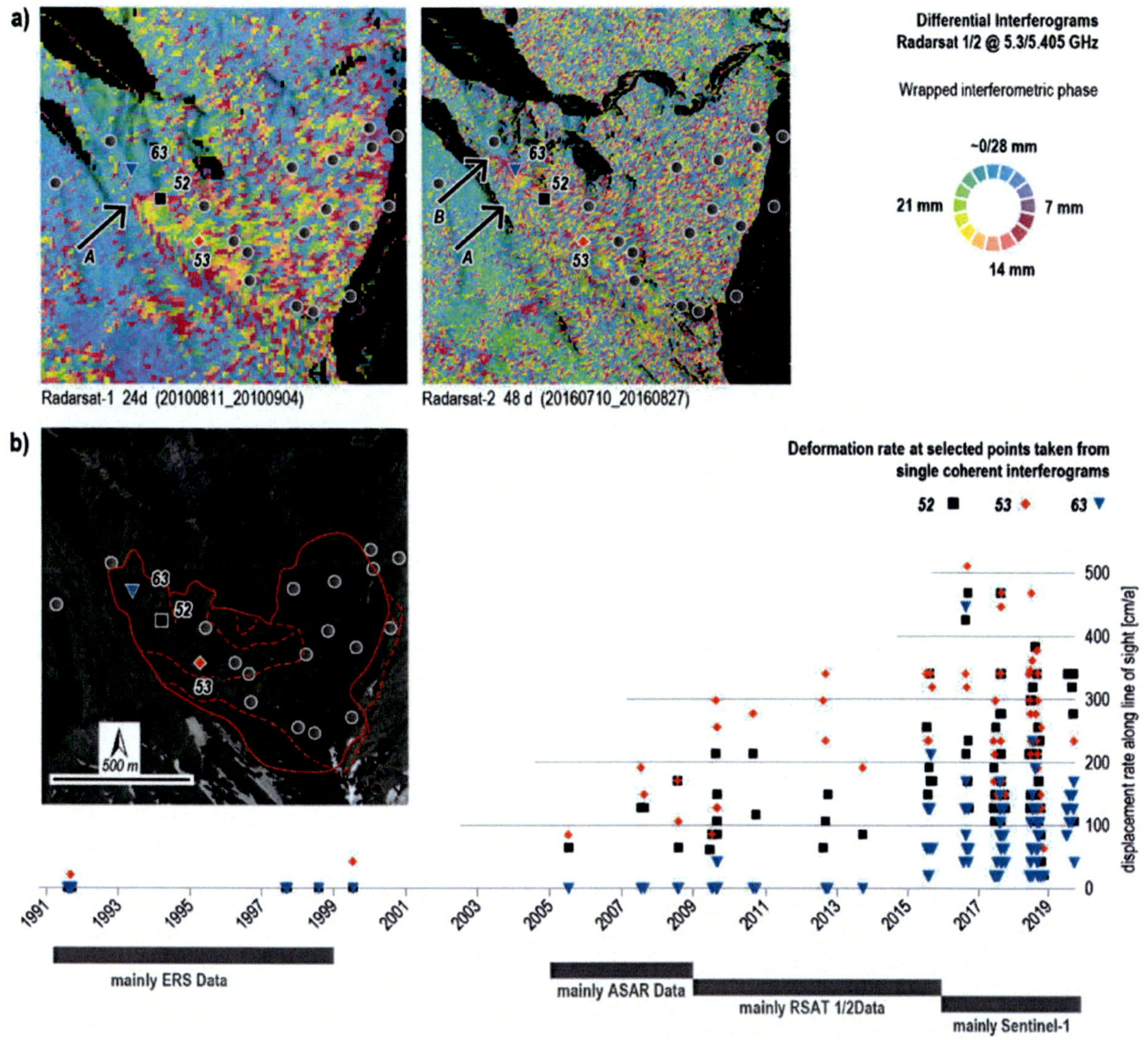


**Fig. 3** Historical satellite InSAR analysis for "Bim Spitze Stei". **a** Radarsat-2 interferograms indicating a lowering of the lowest limit of deformation from location A to location B. Note the different temporal baselines of interferograms. **b** Time series of line of sight displacement rates extracted for locations 52, 53 and 63 currently measured with tachymetry

terrestrial photographs. The data clearly show the presence of several discontinuities that can be followed over the entire W-lateral area. Isolated patches showing faster movement are identified as superficial movements (settlements) of the debris or ice rich debris (white outlines in Fig. 3c). The more extended structures in the upper part of the slope (red outlines in Fig. 3c) mark the outcropping line of the basal glide planes. A lower glide plane is starting right behind the highly fractured peak area of "Bim Spitze Stei" and then steeply dips downward until its path turns and follows more or less the bedding plane of the bedrock. A second, more superficial glide plane with significantly faster displacement rates can be traced several tens of meters above the basal glide plane. This separates the peak area from the lower compartment and ultimately joining the basal glide plane in the peak area.

This observation is in very good agreement to the Sentinel-1 interferograms, but with the good spatial resolution of the terrestrial radar, the basal gliding plan can now be detected at sub-meter scale.

## Conclusions and Outlook

Our study based on satellite and terrestrial radar interferometric data showed that the area of "Bim Spitze Stei" was affected from an unprecedented acceleration of the slope

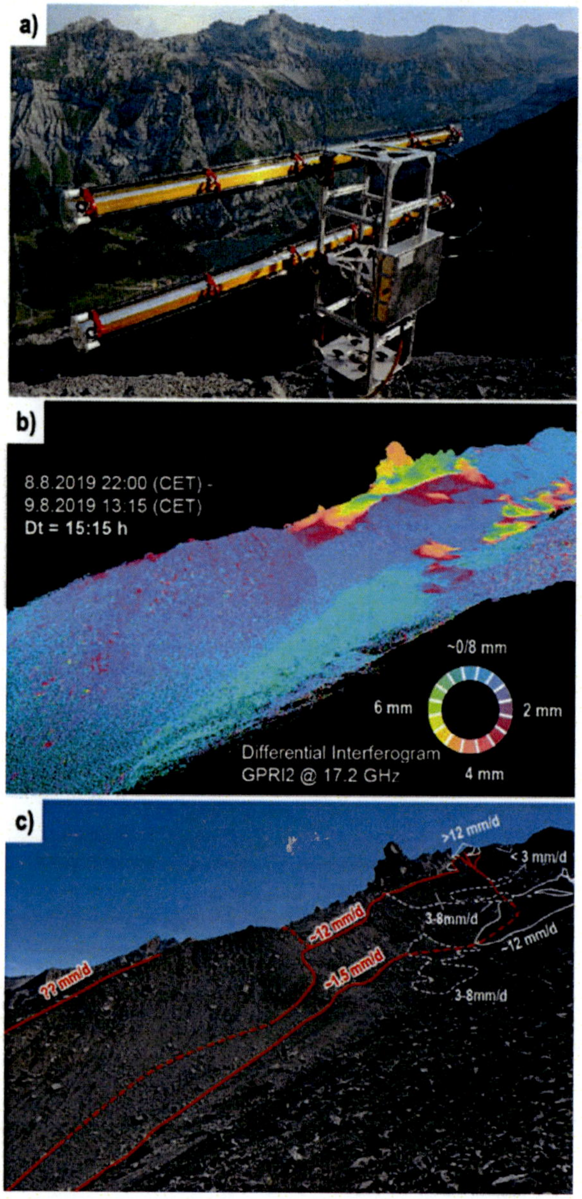


**Fig. 4** **a** Setup of the terrestrial radar interferometer (GPRI) at location W. **b** 15 h Ku-Band interferogram from the west-flank in precise photo geometry. Different sectors with different kinematics could be delineated (**c**) and identified as superficial movement (white) and most importantly as set of outcropping basal glide planes (red)

movements from several cm per year to cm per day in 2018. Different zones with different times of activation and velocities could be identified using Sentinel-1. A historic analysis going back until 1991 with satellite SAR data showed that a cascading progression of the unstable areas on the western side took place between 1999 and 2005. Unfortunately, a gap in usable data hindered a more precise delineation of the events during this time.

Ground based observations delivered valuable information on the current structural situation. Two stacked glide planes could be identified and mapped along a large part of the western flank. The results from terrestrial radar interferometry, satellite SAR interferometry and other conventional observation methods are in good agreement to each other and are in accordance to the results of a drilling campaign conducted in winter 2019/20. Currently, all available observations are used for an in-depth hazard analysis and risk assessment of the down-stream areas.

Our study demonstrated the possibility of delineating and quantifying with satellite and terrestrial radar interferometry the surface motion of different sectors with different kinematics of the "Bim Spitze Stei" area. On the other hand, our analysis is also clearly stating the technological and mission specific limitations related e.g. to sensor specific parameters or the acquisition time interval. However, even though a single observation method may not be enough to get the full picture including geological modelling, influence from triggering factors, and estimation of the timing and volume of potential catastrophic slope failures, the knowledge gain can be still significant. This results in a better understanding of the mechanisms and driving forces of the slope instabilities and thus to an enhanced hazard management.

**Acknowledgments** Data used for this study: JERS-1 SAR, ALOS-1 PALSAR-1 and ALOS-2 PALSAR-2 data are copyright JAXA. ERS-1/2 and ENVISAT data are copyright ESA. TERRASAR-X data are copyright DLR. Cosmo-SkyMED data are copyright ASI. Radarsat-2 data are copyright MDA. Sentinel-1 images available from Copernicus. Part of the work was funded by Eureka and Innosuisse in the frame of the EUROSTARS Project E! 113220 RAMON.

## References

Caduff R, Schlunegger F, Kos A, Wiesmann A (2015) A review of terrestrial radar interferometry for measuring surface change in the geosciences. Earth Surf Process Landforms 40(2):208–228

Caduff R, Strozzi T (2017) Terrestrial radar interferometry monitoring during a landslide emergency 2016, Ghirone, Switzerland. In: Workshop on world landslide forum, pp 301–309

Frank F, Huggel C, McArdell BW, Vieli A (2019) Landslides and increased debris-flow activity: a systematic comparison of six catchments in Switzerland. Earth Surf Process Landforms 44 (3):699–712

Hauck C, Hilbich C, Mollaret C (2017) Geophysikalische Untersuchungen zur Permafrost-Verbreitung im Berner Oberland, 2015–2016, Report 1. Departement Geowissenschaften Universität Fribourg

Knapp S, Gilli A, Anselmetti F S, Hajdas I (2018) Multistage rock-slope failures revealed in lake sediments in a seismically active alpine region (Lake Oeschinen, Switzerland). J Geophys Res Earth Surf 123(4)

Köpfli P, Grämiger LM, Moore J R, Vockenhuber C, Ivy-Ochs S (2018) The Oeschinensee rock avalanche, Bernese Alps, Switzerland: a co-seismic failure 2300 years ago? Swiss J Geosci

Manconi A, Kourkouli P, Caduff R, Strozzi T, Loew S (2018) Monitoring surface deformation over a failing rock slope with the ESA Sentinels: insights from Moosfluh instability Swiss Alps. Remote Sens 10(5):672

Singeisen C, Ivy-Ochs S, Wolter A, Steinemann O, Akçar N, Yesilyurt S, Vockenhuber C (2020) The Kandersteg rock avalanche (Switzerland): integrated analysis of a late Holocene catastrophic event. Landslides

Strozzi T, Delaloye R, Kääb A, Ambrosi C, Perruchoud E, Wegmüller U (2010) Combined observations of rock mass movements using satellite SAR interferometry, differential GPS, airborne digital photogrammetry, and airborne photography interpretation. J Geophys Res Earth Surf 115(F1)

Tinner W, Kaltenrieder P, Soom M, Zwahlen P, Schmidhalter M, Boschetti A, Schlüchter C (2005) Der nacheiszeitliche Bergsturz im Kandertal (Schweiz): Alter und Auswirkungen auf die damalige Umwelt. Eclogae Geol Helv 98:83–95. https://doi.org/10.1007/s00015-005-1147-8

Walter F, Amann F, Kos A, Kenner R, Phillips M, de Preux A, Bonanomi Y (2020) Direct observations of a three million cubic meter rock-slope collapse with almost immediate initiation of ensuing debris flows. Geomorphology 351:106933

# Identification Old Landslides in Permafrost Degradation Area in Northeast China by Difference Distribution of Surface Trees

2020 Kyoto Japan

Ying Guo, Wei Shan, Zhichao Xu, Chunjiao Wang, and Shuanglin Wang

**Abstract**

With the development of global warming, the permafrost in Northeast China, which is located in the south margin of permafrost area in Eurasia, has gradually degenerated, resulting in an increased incidence of landslides. At present, research on the occurrence mechanism, movement process characteristics and evolution trend of landslides in permafrost area, is limited and lacking in detail. It is difficult to identify landslides that have cured within the last hundred years as movements have been relatively small during this time period. In addition, the lack of written records, the unusual geological environment and landforms and the surface coverage, as well as the landslide mechanisms mean that accurate identification is difficult using existing technology. In this paper, Beian-Heihe expressway, which is located in the permafrost degradation area of Northeast China, is selected as the research area. Field investigation, 3D modeling of UAV image, high-density electrical survey, geological drilling and other methods are used to study the distribution of new and old landslides in the area. The study found that the distribution of tree populations at the boundary of the old landslide and some parts on the landslide body was significantly different from the distribution of tree populations outside the landslide body. In combination with tree core sampling, with the help of the relevant theories and methods of dendrochronology, Statistical analysis of tree species distribution and tree ring differences was used to estimate the occurrence time of old landslides. Combined with the previous research results, the relationship between the movement process of landslides and the degradation of permafrost in the study area is analyzed. This method can effectively realize the positioning and model reconstruction of old landslides in a short time scale, which will have important theoretical and practical significance for the early warning and treatment of landslides in permafrost degradation area.

**Keywords**

Eurasia permafrost • Climate change • Old landslides • Trees population • Tree ring

## Introduction

In the past 100 years, affected by climate change and human activities, the permafrost in the eastern Siberia region of Russia and the Northeast permafrost region of China has degenerated violently, resulting in a large number of geological problems such as landslides, debris flow, thermokarst lake and uneven land subsidence. At the same time, the change of environmental geological conditions has also led to the change of landform (Shan et al. 2014; Morgenstern et al. 2013). However, so far, there are relatively few research results on the occurrence conditions of landslides in permafrost regions, the relationship between the sliding process and the landforms and regional environmental changes. In the study of geomorphic change, it is very important to study the geochronology of geomorphic body and other surface features. Accurate chronological data can help to determine the history and rate of surface processes (Peng et al. 2007). For the study of the time of natural disasters in the past, in addition to the conventional field investigation, Carbon-14

Y. Guo · W. Shan (✉) · Z. Xu · C. Wang · S. Wang
Institute of Cold Regions Science and Engineering, Northeast Forestry University, Harbin, 150040 Heilongjiang, China
e-mail: shanwei456@163.com

Y. Guo
e-mail: samesongs@163.com

Z. Xu
e-mail: 976214528@qq.com

C. Wang
e-mail: wangchunjiao@iem.ac.cn

S. Wang
e-mail: 821457173@qq.com



V. Vilímek et al. (eds.), *Understanding and Reducing Landslide Disaster Risk*,
ICL Contribution to Landslide Disaster Risk Reduction,
https://doi.org/10.1007/978-3-030-60319-9_30

dating, optical stimulated luminescence dating and cosmogenic Nuclide dating methods are also more commonly used, but these methods have a large error in the dating of events with a relatively recent age, especially the centennial scale and less than centennial time (Jacoby et al. 1997; Geertsema and Clague 2006; Yu and Lai 2014; Yi et al. 2006). Therefore, using the environmental change information recorded by trees' growth process to study the time, location, scope and other aspects of mountain disasters shows great potential (Tang et al. 2018). A large number of studies have confirmed that the width of tree ring is restricted by environmental factors, and there is a good correlation between its width and environmental factors. It can record the growth status of trees and the environmental changes that affect the growth of trees (D'Arrigo et al. 1992), which has gradually become an important means to study environmental changes (Wimmer 2002). At present, some scholars have applied the theory and method of dendrochronology to study and analyze the avalanche and debris flow disasters that happened many times in the same area, accurately determine the time of disaster occurrence, and predict the trend of disaster evolution (Šilhán and Tichavský 2017); using the tree ring data, analyze and evaluate the stability of large-scale ancient landslides in the non frozen area (Migoń et al. 2014). In apuseni mountain of Carpathian Mountains in Romania, the lower boundary of sporadic permafrost in basalt glacial system of Detunata site was investigated by using ground and air temperature monitoring, geophysical exploration and tree ring study (Răzvan Popescu et al. 2017). Based on the tree ring width of 15 deciduous pine trees in the Sphagnum swamp in Central Siberia, the feedback cycle between the above ground and underground ecosystems since the last wildfire in 1899 was reconstructed, and then the change process of the active layer thickness of permafrost affected by the fire was inferred (Knorre et al. 2019). In the same section of 100 m long river bank, riparian zone and terrace in the permafrost area of Northern Siberia, the same regional climate parameters of tree ring reconstruction at different sampling points are used (Kirdyanov et al. 2013).In the context of global change, tree ring is an important data to study the relationship between tree growth and climate change, and it is also the key to accurately assess the climate change situation, response strategies and adaptability of species (Fonti et al. 2010). However, at present, there are few studies on the identification of old landslides in permafrost area by tree ring on the surface. Due to the differences of site conditions and thermal and hydrological conditions of the soil in different locations, as well as the changes of environmental factors controlling the growth of tree ring in different locations, tree ring differences in different sampling locations are large, which makes the climatic parameters of the same region reconstructed by tree ring in different locations different Quite different (Kirdyanov et al. 2013). Therefore, it is of great significance to further improve the old landslide identification method by surface trees in permafrost degradation area for reconstructing the geomorphic evolution process of permafrost area and preventing geological disasters such as landslides in advance.

## Distribution of Permafrost

The occurrence and distribution characteristics of high latitude permafrost in Northeast China are similar to those in East Siberia and Khangan Mountains in Russia. It is quite different from those in high mountains and plateaus in the Western China. It is mainly distributed in the low place and the shady slope. The degradation order is from sunny slope to shady slope, from higher place to lower place, from top of the hill to bottom of the valley. This kind of regional permafrost distribution is called " Khangan Baikal " permafrost (Zhou et al. 2000). The distribution of permafrost in Northeast China is shown in Fig. 1a.

In this paper, Beian–Heihe expressway, which is located in the permafrost degradation area of Northeast China (49° 30′57″N ~ 49°41′50″N, 127°17′31″E ~ 127°21′24″E), is selected as the study area. The geographical location of the study area is the connecting part of the middle temperate zone and the cold temperate zone, which is the transitional zone between them. The winter is long and the summer is short. In recent years, the average air temperature is −0.5 ~ 1.0 °C, the annual maximum air temperature is 38.6 °C, and the annual minimum air temperature is −48.1 °C; The initial snow is generally in mid October, and the final snow is mostly in the late March to early April of the next year. Within the depth of survey control in the study area, from top to bottom there are Quaternary loose layer, Tertiary sand conglomerate and sand mudstone, Cretaceous mudstone, monzonitic granite and Granodiorite in Indosinian and Yanshanian. The genetic types of Quaternary strata in study area are mainly eluvial Deluvial, Deluvial, diluvial, proluvial and alluvial, and patchy permafrost is distributed in the loose layer. The distribution of patchy permafrost in the study area obtained from Landsat 7 single band image (Wang et al. 2015) is shown in Fig. 1b.

According to the atmospheric temperature change curve of Sun Wu national meteorological station from 1960 to 2018 near the study area, the overall atmospheric temperature in the study area shows an upward trend, with an increase of 0.57 °C every 10 years, as shown in Fig. 1c.

Based on the ground surface temperature data of Sun Wu National Meteorological Station, the freezing index, melting index and freezing number of the ground surface in study area were calculated, and the curve of them was plotted as shown in Fig. 1d. Since 1989, both the freezing index and the melting index have significantly decreased or increased,

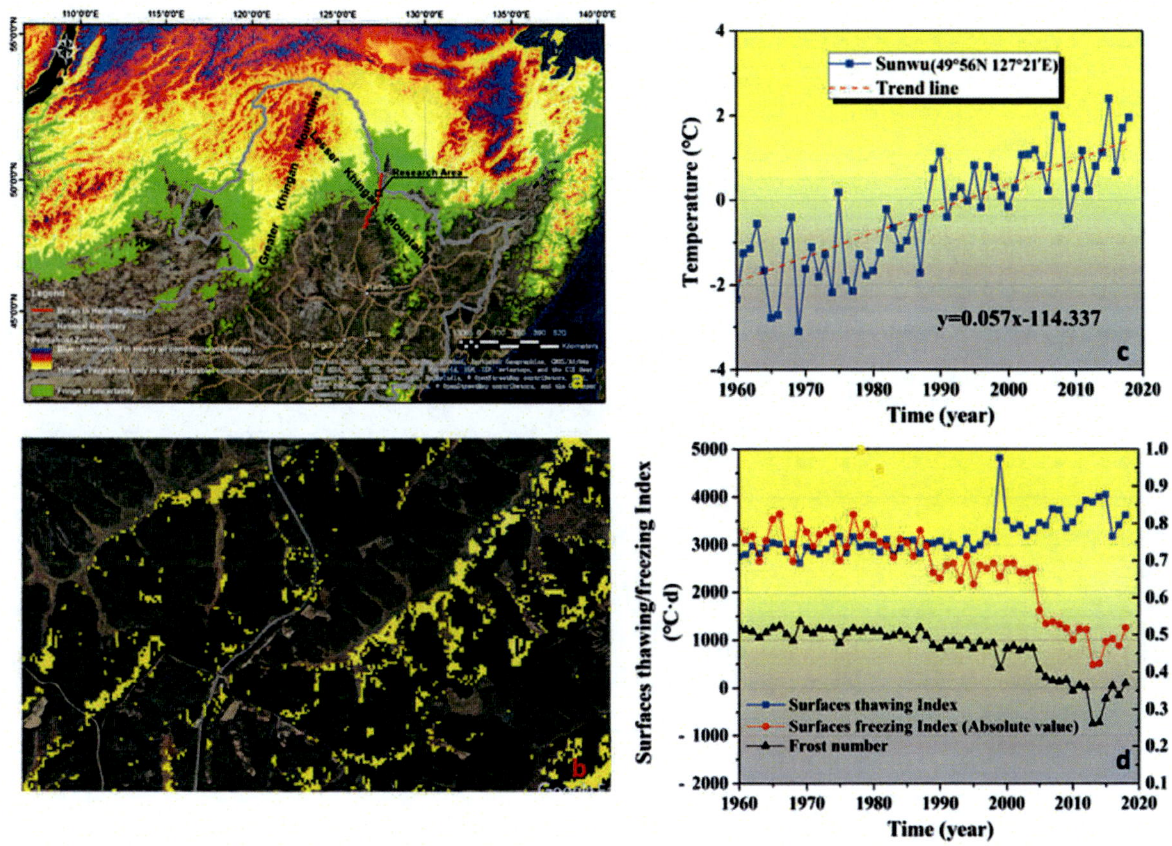


**Fig. 1** Distribution of permafrost and temperature change in Northeast China. **a** Distribution of permafrost in Northeast China and location of research area. **b** the distribution map of island permafrost in study area obtained from Landsat 7 single band image(the yellow area represents permafrost) **c** the average atmospheric temperature change curve from 1960 to 2018 obtained by Sun Wu national meteorological station near the study area. **d** Freezing index, melting index and freezing number curve of round surface calculated based on Nelson model and the data from Sun Wu National Meteorological Station

and the freezing number has started to be less than 0.5. According to the judgment limit of the Nelson model (Lv et al. 2008), the environmental conditions in the study area are no longer suitable for permafrost Exist. It means that the permafrost in the study area transitioned from discontinuous or island-like distribution to gradual disappearance.

## Distribution of Landslides

With the gradual degradation of permafrost in the study area, the process of permafrost thawing and phase transformation leads to the gradual increase of landslides (Shan et al. 2015). In the road area of section k178-k180 of Bei'an-Heihe Expressway in the study area, there are densely distributed landslides (as shown in Fig. 2a). The occurrence time and development of these landslides are different, but they are closely related to the distribution and degradation of permafrost. The distribution of these landslides and their relationship with the environment can be seen more clearly by the results of 3D modeling of UAV image (Fig. 2b).

The special geological environment of the study area provides material conditions for the formation of landslide. The superposition of permafrost thawing process and atmospheric precipitation provides dynamic conditions for landslide movement. Taking the k178 + 530 landslide (landslide ⑤ in Fig. 2) of Bei'an-Heihe Expressway in the study area as an example (Shan et al. 2015), it is found that through detailed investigation the soil layer distribution from top to bottom of the landslide profile is: Quaternary loose layer, Tertiary gravelly sandstone and siltstone, and Cretaceous mudstone. Quaternary loose layer: mainly composed of silty clay, gravelly sand and completely weathered mudstone, with a small amount of organic matter. Tertiary gravelly sandstone: completely weathered, mainly composed of quartz and sand minerals, well graded and highly permeable; Tertiary completely weathered siltstone: yellow, sandy structure, layered joint structure, poor water

**Fig. 2** Distribution of landslides in study area. **a** The position and development of several landslides in study area; ①–⑥Landslides site photos, **b** 3D modeling of UAV image of several landslides in study area

permeability. Cretaceous mudstone: completely weathered mudstone, yellow or grayish green, argillaceous structure, layered structure, poor water permeability. The loose soil layer on the hillside surface provides conditions for the infiltration of atmospheric precipitation, ice and snow thawing water and permafrost thawing water. During the infiltration process, the water is blocked by the lower frozen soil or dense soil layer, and infiltrates along the water barrier to the lower part, forming a potential slip zone. There are permafrost with thickness of 2-4 m on the shady slopes of both sides of the landslide mass. Due to the construction disturbance and the influence of climate warming, the phase change of permafrost results in the gradual infiltration of water into the lower part to form a sliding surface. The process of landslide movement changes with seasons, with obvious annual periodicity and seasonal activity characteristics; the movement of landslide has low angle and intermittent characteristics, and the occurrence time and rate of sliding have corresponding relationship with the pore water pressure of the soil at the back edge of the landslide. The thawing water of permafrost flows into the landslide body, affecting the change of pore water pressure of the soil at the back edge of the landslide, which also is the main reason of the landslide.

Affected by the occurrence of patchy permafrost, the occurrence time of different landslides in the study area is also different. According to the surface vegetation condition and the investigation and observation, the landslide group in Fig. 2 is divided into three categories: stable old landslide, stable landslide in recent years and unstable landslide with slow movement, which are respectively represented by blue, orange and green. Based on the distribution of surface tree species and tree ring differences in the region, the occurrence time of the old landslide was identified by statistical analysis.

## Old Landslides and Differences of Surface Trees

Site conditions and physiological characteristics of trees determine the distribution of forest populations, and the differences in forest population distribution are also a reflection of differences in site conditions. Through UAV investigation, it was found that there are many White birch (Betula platyphylla Suk.) in the Mongolican oak (Quercus mongolica Fischer ex Ledebour) community in a band-like, sheet-like distribution. And the White birch grows mostly on the edge of landslides and on landslide bodies in study area.

Mongolican oak enjoys a warm and humid climate, and can withstand certain cold and drought. The soil is not strict, and it can grow on acidic, neutral or limestone alkaline soil, and it is resistant to barren but not to water and humidity. The Mongolican oak forest is mostly secondary vegetation formed after the primary forest was damaged, and it mainly grows on high dry hillsides with plenty of sunshine. The White birch is highly adaptable to soil, resistant to barren, and resistant to severe cold. It likes acid soil, especially wet soil, and can grow in marshland, dry sun slope and wet shady slope. It has deep roots and is pioneer tree species of secondary forest. Laboratory experiments have shown that the maximum net photosynthetic rate of White birch is slightly higher than that of Mongolican oak, and its water use efficiency is much lower than that of Mongolican oak (Zhang et al. 2008). This indicates that white birch is resistant to extreme cold and fast growth, while Mongolican oak is resistant to cold and drought but not wet and humid (Zhang et al. 2019).

The biggest difference in physiological characteristics between White birch and Mongolican oak is water resistance. In the study area, the main driving force for landslides is the water produced during the melting and phase transformation process of permafrost. After the sliding occurs, the side and leading edges of the landslide are also the water-rich area, we speculate that When a landslide occurs in the place where the Mongolican oak forest is located, the soil mass at the edge of the landslide causes the death of the original Mongolican oak, and the birch grows rapidly because of snow accumulation or melt water in the cracks. Therefore, the distribution of birch population in mixed forests in the study area is closely related to landslides, and birch may be an indicator for landslides.

By using the high-density electrical method (WGMD) to detect the landslide⑥in two direction (vertical to the sliding direction and parallel to the sliding direction), it was found that There is a U-shaped low-resistance band on the cross section perpendicular to the sliding direction. The low-resistance is caused by high water content (Fig. 3b). It can be judged as a sliding belt (Hu and Shan 2016). It can be seen from the cross section parallel to the sliding direction that the permafrost exists only at a later position of the leading edge, and no high water content and low resistance sliding belt is formed along the sliding direction, resulting in insufficient landslide power. So the side edge did not continue to slide after cracking (Fig. 3a).The distribution of permafrost obtained by Landsat7 single-band image data also shows that there is permafrost distribution in the old landslide location, as shown in Fig. 3c, which is consistent with WGMD results.

In order to further verify the relationship between the development of the old landslide and the distribution of surface plants, we combined with the theory of dendrochronology and used the CO250 tree core sampler to take tree core samples of the trees on the landslide⑥ and outside the landslide. The sampling location is shown in Fig. 4. 72 birch core samples were taken along the edge of the landslide, 35 Mongolian oak core samples were taken inside and outside the landslide, combined with statistical theory to removal of failure samples and conduct tree age statistics and analysis. The size of the dots in the figure represents the age of the tree.

The age analysis results of White birch and Mongolican oak tree at landslide ⑥ is shown in Fig. 5. The white birch age on the edge of the landslide is 28-68 years, while the Mongolian oak age on the slope is 38-82 years, and the growth age of white birch is lower than Mongolian oak. However, it can be seen from Fig. 5a that the DBH of white birch with shorter growth time is larger than that of Mongolian oak with longer growth time, so the growth speed of white birch is higher than Mongolian oak. In addition, the main distribution areas of tree age obtained by calculating the upper and lower quartiles are: birch age is 38–48 years, Mongolian oak age is 49–74 years, birch average age is 42 years, Mongolian oak average age is 61 years. The average age of birch on the edge of the landslide is 19 years younger than that of Mongolian oak on the slope, as shown in Fig. 5b. The difference of tree age comes from the vegetation replacement on the edge of the landslide during the landslide occurrence. Therefore, the location and model reconstruction of old landslides in a short time scale can be effectively realized through the difference distribution of surface tree population.

## Conclusion

This paper studies the distribution of new and old landslides in the study area by means of field investigation, 3D modeling of UAV image, high-density electrical method survey, geological drilling, etc., and makes statistical analysis on the distribution of surface trees and tree ring differences in the area with the help of dendrochronology theory and method, combining with the tree core sampling on the landslide body and outside the landslide Ming: the soil instability caused by the gradual melting of permafrost has become an important factor of the landslide in permafrost area, and there are obvious differences in the types and rings of trees on the landslide slope due to the influence of the landslide. In view of the above characteristics, this paper proposes a new method to identify the old landslide in the permafrost degradation area of Northeast China by the distribution of population differences of trees on the ground surface, this method can effectively realize the location and model reconstruction of old landslides in a short time scale, which will have important theoretical and practical significance for the early warning and treatment of landslides in permafrost degradation area.

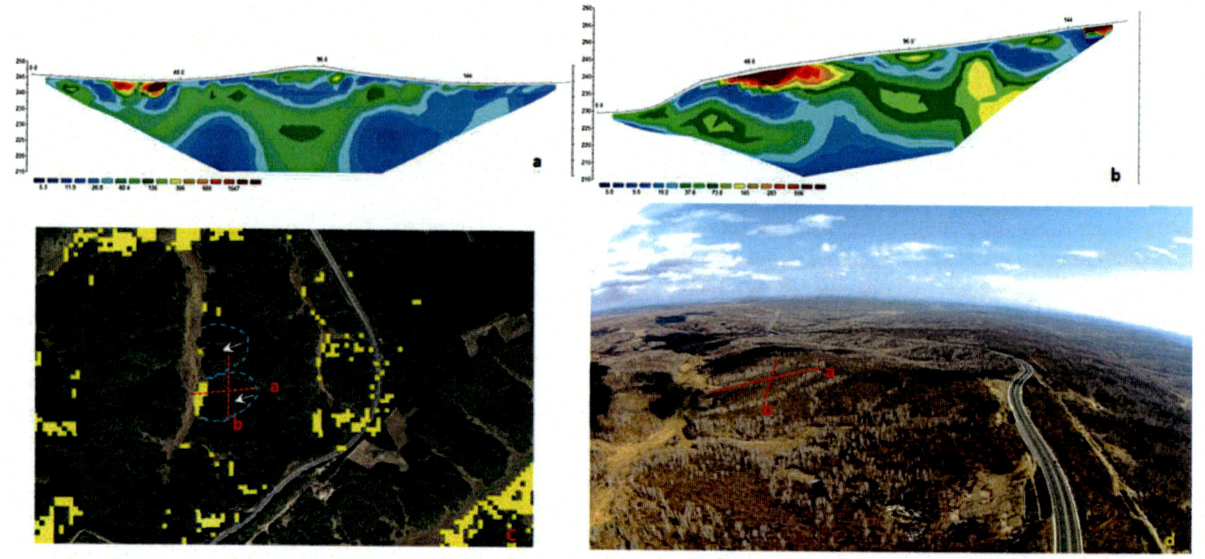

**Fig. 3** Geological distribution and tree species difference of landslide ⑥in study area. **a** The geological conditions of the landslide ⑥ obtained by the high-density resistivity method (WGMD) perpendicular to the sliding direction **b** The geological conditions of the landslide ⑥ obtained by the high-density resistivity method (WGMD) parallel to the sliding direction **c** The distribution of the permafrost obtained by Landsat 7 single band image data. **d** The tree species differences at the landslide ⑥

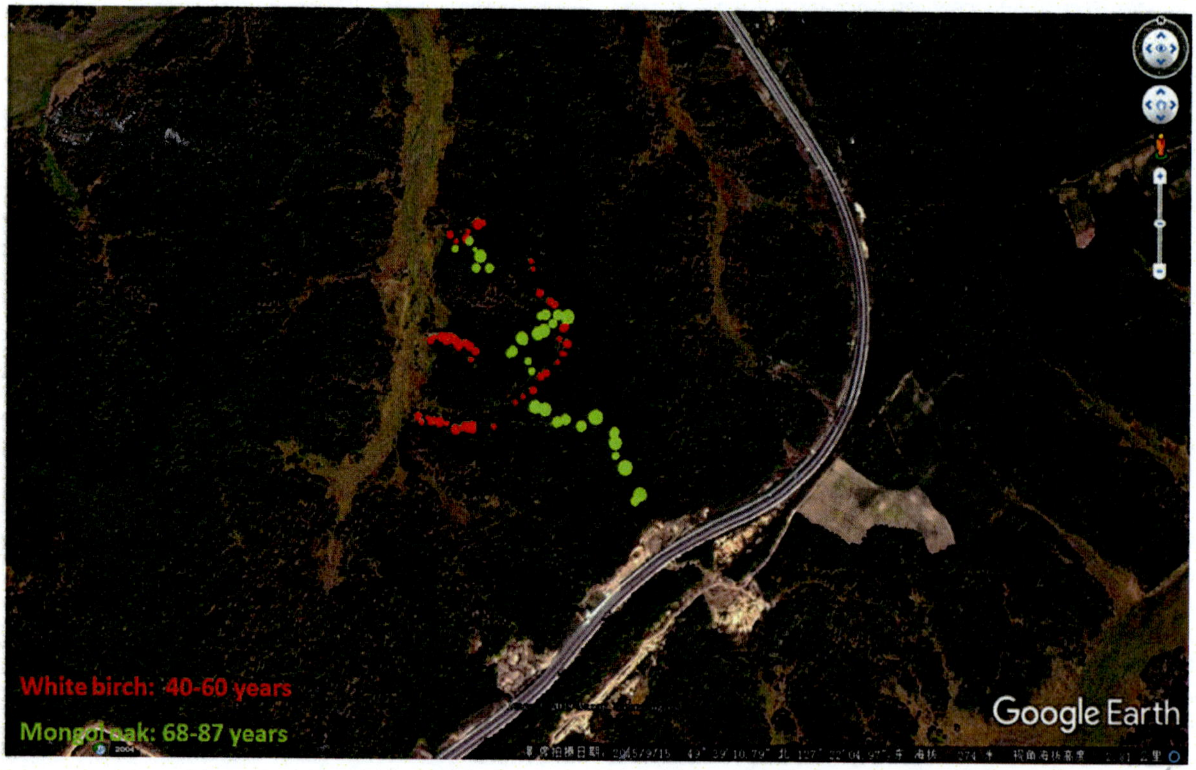


**Fig. 4** Sampling position of tree core and the corresponding tree ring at landslide⑥. **a** Landslide ⑥ tree core sampling location

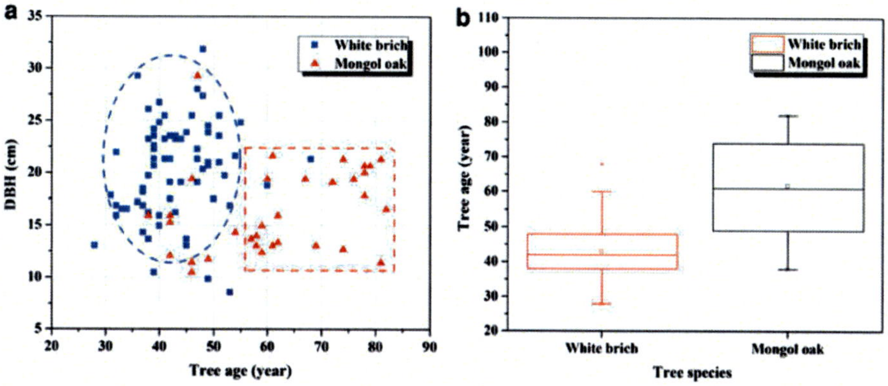


**Fig. 5** Trees age distribution of White birch and Mongolican oak tree at landslide⑥ **a** Scatter diagram of tree age and DBH of white birch and Mongolian oak **b** Box chart of tree age distribution of birch and Mongolian oak

**Acknowledgements** Thanks for the support of projects: Heilongjiang postdoctoral fund project and ICL- IPL203.

## References

D'Arrigo D, Jacoby GC, Free RM (1992) Tree-ring width and maximum latewood density at the North American tree line: parameters of climatic change. Can J Forest Res 22(9):1290–1296

Fonti P, Arx GV, García-González I et al (2010) Studying global change through investigation of the plastic responses of xylem anatomy in tree rings. New Phytol 185(1):42–53

Geertsema M, Clague JJ (2006) 1000-year record of landslide dams at Halden Creek, northeastern British Columbia. Landslide 3(1): 217–227. Jacoby GC (1997) Application of tree-ring analysis to paleoseismology. Rev Geophys 35(2):399–416

Hu ZG, Shan W (2016) Landslide investigations in the northwest section of the lesser Khingan range in China using combined HDR and GPR methods. Bull Eng Geol Environ 75(2):591–603

Kirdyanov AV, Prokushkin AS, Tabakova MA (2013) Tree-ring growth of Gmelin larch under contrasting local conditions in the north of Central Siberia. Dendrochronologia 31(1):114–119

Knorre AA et al (2019) Tree ring-based reconstruction of the long-term influence of wildfires on permafrost active layer dynamics in Central Siberia. Sci Total Environ 65(2):314–319

Lv JJ, Li XZ et al (2008) Application of frost number model in Northeast China permafrost region. Chin J Appl Ecol 19(10):2271–2276 (In chinese)

Migoń P, Kacprzak A, Malik I et al (2014) Geomorphological, pedological and dendrochronological signatures of a relict landslide terrain, Mt Garbatka (Kamienne Mts) SW Poland. Geomorphology 219(1):213–231

Morgenstern A, Ulrich M, Günther F et al (2013) Evolution of thermokarst in East Siberian ice-rich permafrost: a case study. Geomorphology 201(1):363–379

Peng JB, Fan W, Li XA et al (2007) Some key questions in the formation of ground fissures in the Fen-Weibasin. J Eng Geol 15 (4):434–440 (In Chinese)

Popescu R, Vespremeanu-Stroe A et al (2017) Low-altitude permafrost research in an overcooled talus slope–rock glacier system in the Romanian Carpathians (Detunata Goală, Apuseni Mountains). Geomorphology 29(5):840–854

Shan W, Guo Y, Wang FW et al (2014) Landslides in cold regions in the context of climate change. Springer International Publishing, Switzerland. (9783319008660), 310p

Shan W, Hu ZG, Guo Y et al (2015) The impact of climate change on landslides in southeastern of high-latitude permafrost regions of China. Front Earth Sci 3(7):1–11

Šilhán K, Tichavský R (2017) Snow avalanche and debris flow activity in the High Tatras Mountains: new data from using dendrogeomorphic survey. Cold Regions Sci Technol 134(1):45–53

Tang Y, Xie JH, Li PB et al (2018) Potential applications of dendrochronology in the research of mountain natural disasters. Adv Eng Sci 50(3):24–32 (In chinese)

Wang CJ (2015) Land surface deformation research of permafrost degradation area in northeast China based on D-INSAR. PhD thesis, Northeast Forestry University, Harbin, China

Wimmer R (2002) Wood anatomical features in tree-rings as indicators of environmental change. Dendrochronologia 20(1):21–36

Yi CL, Zhu L, Seong YB et al (2006) A late glacial rock avalanche event, Tianchi Lake, Tien Shan, Xinjiang. Quat Int 154/155(1):26–31

Yu LP, Lai ZP (2014) Holocene climate change inferred from stratigraphy and OSL chronology of Aeolian sediments in the Qaidam Basin, northeastern Qinghai-Tibetan Plateau. Quat Res 81 (1):488–499

Zhang W, Wang YJ, Mao ZJ et al (2008) Comparison of physiological characteristics of six greening tree species in Harbin City. Forest Eng 24(3):6–9. (In chinese)

Zhang QL, Liang Y, He HS et al (2019) Changes in species-level biomass and its relationship with climate and forest disturbances in the Great Xing'an Mountains. Acta Ecol Sinica 39(12):4442–4454. (In Chinese)

Zhou YW, Guo DX, Qiu GQ et al (2000) Permafrost in China. Science Press, Beijing. (7030082850), 480p. (In chinese)

# Forensic Geotechnical Investigation of the Skjeggestad Quick Clay Landslide, Norway

Thi Minh Hue Le, Vidar Gjelsvik, Suzanne Lacasse, Stein-Are Strand, Eirik Traae, and Vikas Thakur

**Abstract**

On February 2 2015, a quick clay landslide occurred adjacent to Mofjellbekken's two parallel bridges (referred to as the Skjeggestad landslide) on the E18 motorway from Oslo to the southeastern coast of Norway. The landslide damaged one bridge foundation which led to partial collapse of the southbound bridge. The damaged bridge had to be demolished and rebuilt. NVE, the Norwegian Water Resources and Energy Directorate, established an independent investigation commission to study the cause and the failure mechanism of the landslide. The Commission examined evidence from the soil investigation, available topographic data, observations in situ, eye witness statements and photographs taken by drones. These evidences were used to reconstruct the most probable sequence of events during the landslide. Both natural causes (erosion, rainfall, snow/ice melting) and man-made causes (vibration, fill construction or other disturbance) were investigated using the available data and stability analyses. The Commission concluded that fill placement on top of the slope in the period 1998–2006 reduced the safety margin considerably. The last fill placed shortly before the landslide occurred in 2015 was the trigger of the landslide which had by then a marginal factor of safety. This paper presents the forensic investigative process by the NVE investigation commission.

**Keywords**

Quick clay • Skjeggestad • Mofjellbekken bridge • Failure mechanism • Fill materials

T. M. H. Le (✉) · V. Gjelsvik · S. Lacasse
Norwegian Geotechnical Institute, Oslo, Norway
e-mail: thi.le@ngi.no

V. Gjelsvik
e-mail: vidar.gjelsvik@ngi.no

S. Lacasse
e-mail: suzanne.lacasse@ngi.no

S.-A. Strand
Norges Vassdrags-Og Energidirektorat (NVE), Trondheim, Norway
e-mail: sas@nve.no

E. Traae
Norges Vassdrags-Og Energidirektorat (NVE), Tønsberg, Norway
e-mail: eit@nve.no

V. Thakur
Norwegian University of Science and Technology, Trondheim, Norway
e-mail: vikas.thakur@ntnu.no

## Introduction

On Monday, 2 February 2015, a quick clay landslide occurred in Holmestrand in Vestfold county near the southeastern coast of Norway (Nord: 6,593,078, East: 573,700, EU89-UTM32) (Fig. 1). The landslide started partially under two parallel bridges crossing Mofjellbekken Creek, near the Skjeggestad farmhouse. The bridges are part of the E18 (European route 18), one of Norwegian national roads carrying the main traffic from Oslo to the southeastern coast of Norway. The landslide, often referred to as the Skjeggestad landslide, spanned an area of about 80 m long by 100 m wide (Fig. 2). Approximately 100 000 $m^3$ of soil were displaced under the sliding event (Fig. 3).

The landslide occurred in the plateau on the east side of Mofjellbekken Creek running under the bridges. The foundation of the southbound bridge was damaged and the bridge partially collapsed. The incidence led to closing of the bridge causing serious traffic disruptions, but fortunately no fatalities. The southbound bridge had to be demolished and rebuilt. The rehabilitation work cost almost 170 million



V. Vilímek et al. (eds.), *Understanding and Reducing Landslide Disaster Risk*, ICL Contribution to Landslide Disaster Risk Reduction,
https://doi.org/10.1007/978-3-030-60319-9_31

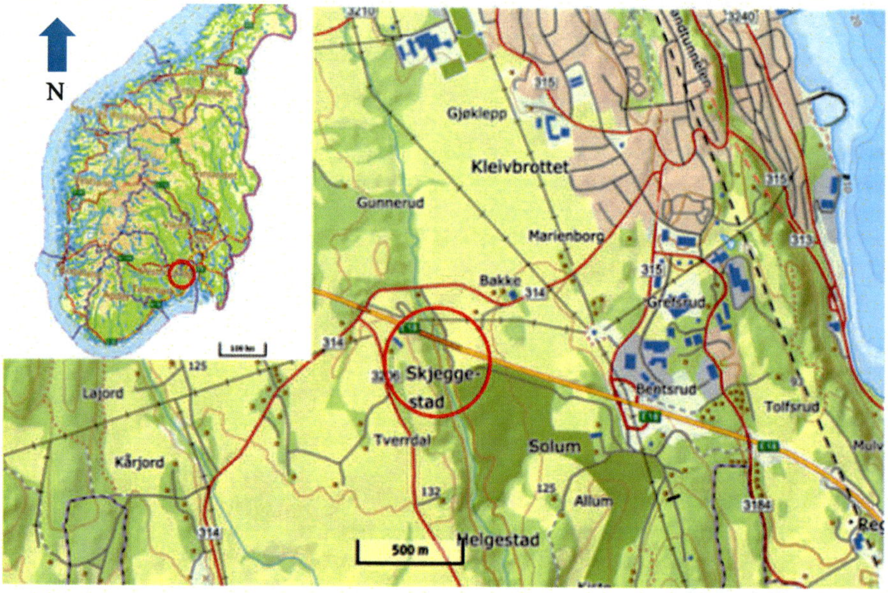


**Fig. 1** Location of Skjeggestad landslide in Holmestrand County, and location of Holmestrand in Norway (inset)

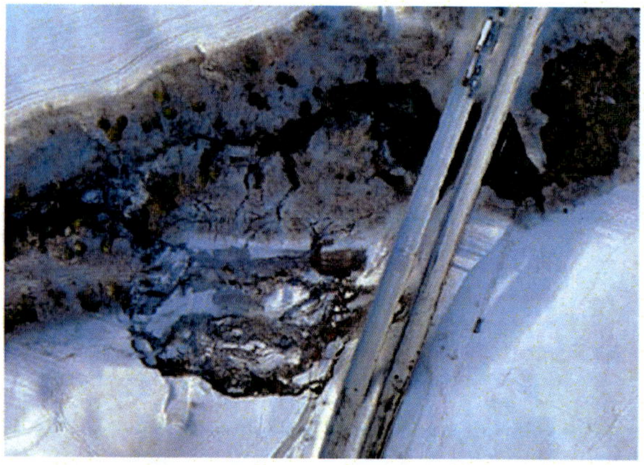

**Fig. 2** Drone photograph of landslide on Feb. 4 2015 (Orbiton)

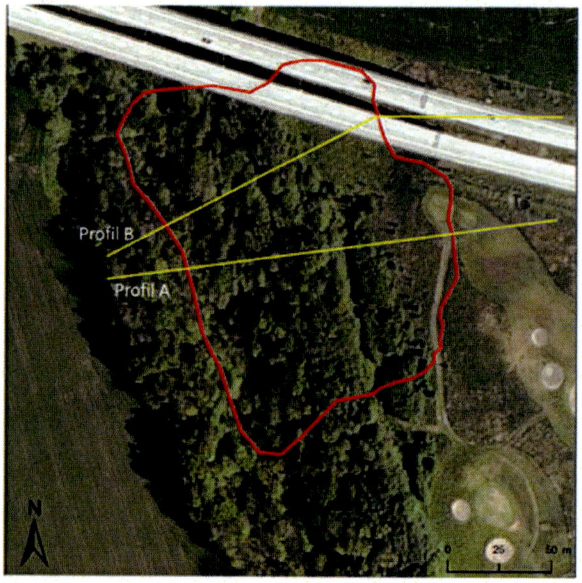


**Fig. 3** Landslide boundaries (Aerial photo *norgeibilde.no*)

NOK (19 million US dollar) (Skolund 2017). Traffic was reopened 18 months later. The incidence also prompted the safety inspection of 17 500 bridges in the country by the Norwegian Road Supervisory Authority.

To investigate the triggering and the propagation of the slide, the Norwegian Water Resources and Energy Directorate (NVE) established an independent investigation commission (referred to as the Commission in this paper), including members from NVE, the Norwegian Geotechnical Institute (NGI), the Norwegian University of Science and Technology (NTNU) and the Norwegian Railway Authority. The Commission examined the following documentation, among others:

- Ground investigation in or near landslide area (Fig. 4);
- Observations and data collected from site visits;
- Terrain analyses by GIS (Geographic Information System) of LIDAR data and data collected by drone-UAV (Unmanned Industrial Vehicle);
- Archived aerial photos before the landslide, and drone photos after the landslide;

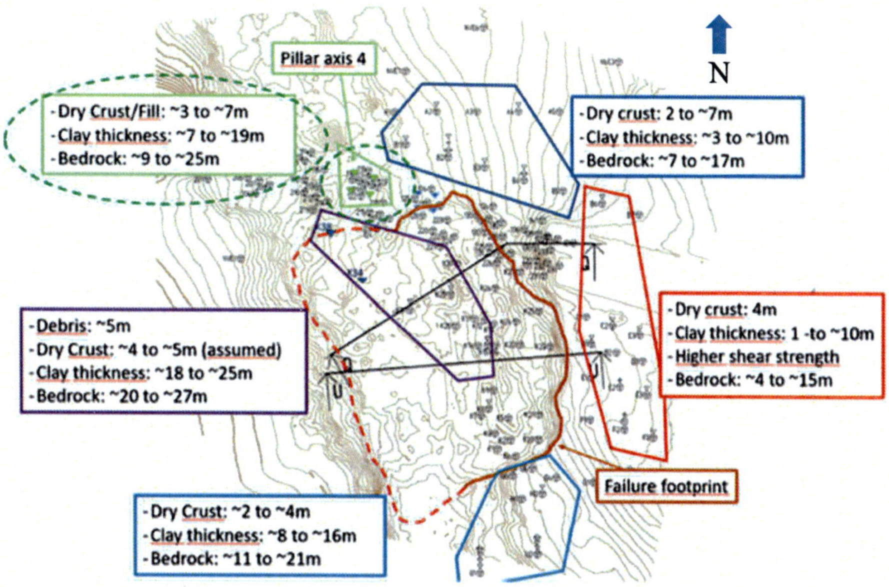


**Fig. 4** Ground investigation and soil layering in landslide area and in areas close to the landslide

- Rainfall and temperature data from nearby weather stations;
- Traffic vibration data from a similar bridge;
- Stability analyses for different situations.

With the permission of NVE and the members of the Commission, the authors summarize the investigations and the findings of the Commission.

## Norwegian Quick Clay

Quick clay is the term used for a strain-softening clay which collapses and flows like a fluid when overloaded. Quick clay is found primarily in Norway and Sweden, but also in Finland, Russia, Canada and Alaska. In Scandinavia, quick clay is found below the marine border (i.e. the elevation of the sea level at the end of the last ice age, about 14000 years ago). The marine border varies between 0 and 220 m higher than the mean sea level in present day in Norway, depending on location.

The Scandinavian Peninsula (Norway and Sweden) was covered by an ice layer of about 3000 m, 20000 years ago. When the ice started to melt, the fine clay particles flowed with the water and sedimented in the saltwater marine environment along the seashore. Due to the electrical charges from the saline environment, the clay particles formed a "card-house" structure. As the ice melted, the land formation slowly rose. Parts of the area with clay was elevated above the sea water level. After the land rise, the salt in the clay was washed out over time due to fresh groundwater flow. When the salt, which formed electrostatic binding in the clay, was leached, the strength characteristics of the clay drastically changed. When such clay is overloaded, the clay structure (similar to a house of cards) suddenly collapses. Such clay is denoted as "quick". If the clay is subjected to sustained loading or remoulding due to natural or man-made actions, the clay masses suddenly transform into a running fluid.

A typical quick clay slide starts with an initial local slide which causes unstable back-scarp. The intial slide can be triggered by natural causes such as erosion and/or pore water pressure rise due to rainfall/snowmelt. Man-made causes such as vibrations or overloading (e.g. fill construction) can also initiate quick clay landslide. The slide can propagate very quickly backwards (retrogressively) and/or forward over large areas, and the debris will flow over considerable distances.

## Topography, Geology and Geotechnical Conditions

The bridges cross a valley characterised with the Mofjellbekken Creek meandering in the north–south direction and two plateaus on the east and west of the creek (Fig. 3). The original ground in the sliding area was composed of 3 to 5 m of a hard clay dry crust, under which there was a layer of medium to hard clay and then a soft silty quick clay deposited about 10000 years ago (Figs. 4 and 5). The sea level then was almost 185–200 m higher than the present sea level. The soil particles in the seawater deposited and filled up the terrain depressions with alternating layers of clay and silt directly over the bedrock or over thin till (or moraine)

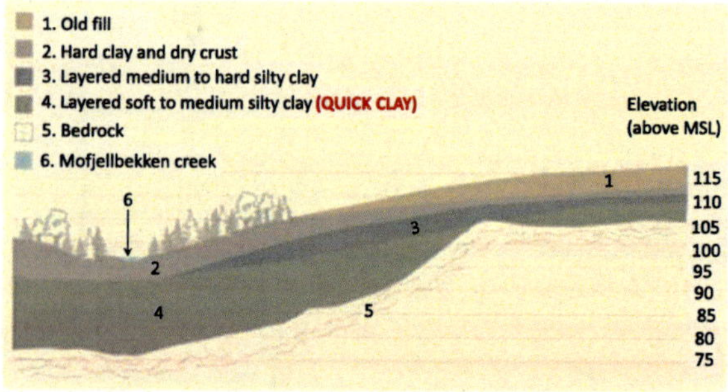


**Fig. 5** Soil layers and topography along Profile A (Fig. 3)

over bedrock. Ground investigations indicated a deep rock graben (or trench) under the soils (Fig. 4). The bedrock slopes down towards Mofjellbekken Creek from both the east and west sides. The depth to bedrock is less on the eastern side of the area than the western side. Rotary soundings show 38–45 m depth to bedrock in the valley, but less than 10 m depth on the eastern plateau (Figs. 4 and 5). The original ground has been modified over the years with cut and fills.

After post-glacial uplift, Mofjellbekken Creek eroded down 15–20 m and meandered through the layers of clay and silt. The erosion increased the gradient between groundwater level in the plateaus and the water level of the creek. This gradient caused groundwater flow from the plateaus through the soils and the bedrock into the creek. This groundwater flow, through thousands of years, leached salt from the ground leading to the formation of quick clay pockets in the area.

The interpretation of rotary soundings, CPTU (Cone Penetration Test) and soil sampling done before construction (NPRA 1997) indicated the presence of quick clay. Site investigation after the landslide, including CPTU, confirmed a continuous layer of quick clay under the bridge, increasing in thickness westward. During visits after the landslide, traces of quick clay being squeezed out and flowing down Mofjellbekken Creek were observed (Fig. 6).

**Fig. 6** Observation after the landslide of traces of quick clay downstream from the landslide

Piezometers installed in the bottom of the valley showed artesian hydraulic pressure. Those installed near the top of the slope show pore pressures that are lower than hydrostatic condition which indicates downward groundwater flow. As the soils are thinly layered, it is most likely that the groundwater followed the more permeable layer in a slope. During the installation of many of the piles for the bridge, water flowing up the casing was observed. This indicates artesian pressure in the area near the bedrock.

## Analyses of the Failure

The failure mechanism was investigated by the Commission using ground investigation, topographic data, archived photos, Lidar and drone images, site visits and accounts from eyewitnesses. The topography of the area before 1990 was reconstructed using archived maps, while LIDAR data (Light Detection And Ranging) was available for topographic analysis for the period 1998–2006. The topography after the landslide was analysed based on data collected by automatic UAV (Unmanned Industrial Vehicle). Many site visits were conducted to inspect the conditions of the sliding masses after the event as well as any evidence of erosion in the Mofjellbekken Creek. Interviews were conducted with eyewitnesses and residents living nearby to gather as much information as possible on the sequence of the landslide.

The Commission, after combining all of the data and information available, concluded that the evolution of the landslide can be divided into three phases (Phases I, II and III in Figs. 7 and 8). Phase I started in the area where the ground has been loaded with a considerable amount of fills over a long period between 1998 and 2015. As part of Phase I, the terrain in the sliding area sank down by as much as 10 m and moved approximately 20 m in the westward direction to Mofjellbekken Creek (Figs. 7 and 8). The northern part of the plateau moved in the south-west direction, but the movement was limited by the sliding masses from the east. The sliding surface penetrated several meters into the quick clay layer and followed this layer almost horizontal to the other side of the creek. The landslide progressed almost immediately backwards and sideways to Phase II as the steep unstable back-scarp of almost 10 m height, created by the sliding in Phase I, failed. The transition from Phase I to Phase II was within a few seconds according to eyewitnesses. The north-eastern sliding mass from Phase II torn out the backfill supporting the bridge foundation on one side, causing partial collapse of one of the pillars of the southbound bridge. The foundation of steel core piles to bedrock prevented a total collapse. Phase III,

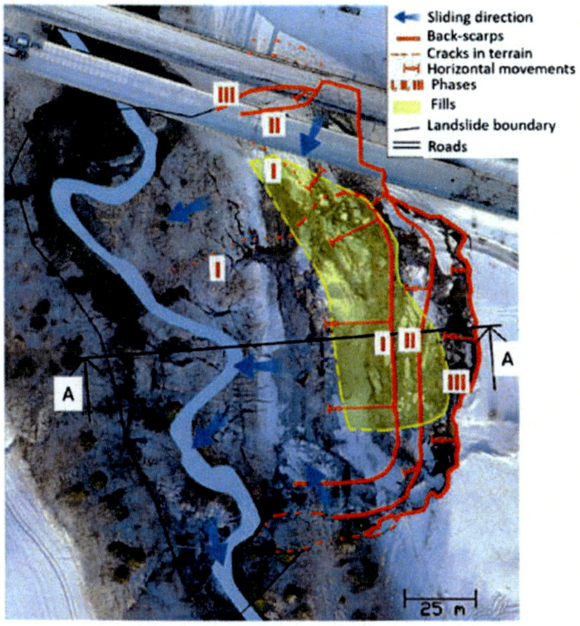


**Fig. 7** The three phases of Skjeggestad landslide

**Fig. 8** Backward propagation of Skjeggestad landslide along Profile A in Fig. 4

consisting of multiple slides in the steep scarps resulting from Phase II, extended over a few days.

The spatial propagation of the landslide was limited by the ground conditions at the site of the bridges: the shallow bedrock ridge in the eastern plateau stopped propagation eastward (Fig. 8). The slope in the northern part next to the bridges was gentle, and limited the movements in the northward direction. Propagation of the landslide in the south direction was limited partly by the sliding masses from the east. These masses deposited in Mofjellbekken Creek and contributed to stabilize the slopes in the south.

The thickness of the quick clay layer played an important role in limiting the extent of the landslide. Borehole data show that the quick clay layer was thicker in the lower part of the valley, and thinner in the upper part (Figs. 4 and 5). The thinner quick clay layer limited the propagation of the slide in the upper part of the plateau. In addition, the 3–4 m thick dry crust formed a solid flake which stopped the landslide to flow further down Mofjellbekken Creek. Therefore, the sliding mass deposited upon reaching the creek and contributed to stabilize the north and south slopes.

The failure was therefore a true three-dimensional case. Two-dimensional, plane strain or limit equilibrium analyses can provide useful indication on relative change in safety factor. This method is adopted in this study. A correction for 3D effects have also been investigated in Jostad et al. (2020) investigated the 3D effects on the stability analysis for the Skjeggestad landslide.

## Cause of the Landslide

To investigate the trigger of the landslide, ground investigation data from before the slide as well as supplementary ground investigations by the National Road Authorities after the landslide were analysed to establish the soil layering and the ground conditions. Temperature and rainfall data from two nearest weather stations were used to investigate if rainfall or snowmelt could have caused a pore pressure increase. The impact from traffic vibration or other activities was also examined. Vibration data measured on a nearby bridge on the same expressway were used. Vibration from heavy machinery (bulldozer) used to place the last fill on top of the slope was based on vibration values for similar types of construction equipment.

Analysis of slope stability in the short-term (undrained conditions) and on the long-term (fully drained conditions) played a central role in the determination of the causes of the landslide. Two profiles were analysed by NGI (NVE 2015): Profile A in the east–west direction through the steepest slope between the plateau and the creek; and Profile B near the bridge foundation in the northeast-southwest direction. Profile A was considered the more critical because it runs across the initial sliding mass from the eastern plateau (Phase I). Profile B stability analysis was done with the Limit Equilibrium Method (LEM) with the GeoSuite Stability software, version 14.1.1.0 *(geofuture.se)*; Haugen et al. (2016). The soil parameters used in stability analysis were based on the ground investigation conducted in nearby area before and after the landslide. Additional stability analyses were also conducted by (Karlsrud et al. 2015; Rødvand et al. 2019; Jostad et al. 2020). This paper focuses on the results obtained by the Independent Investigation Commission which were the basis for the conclusion of the forensic geotechnical investigation.

## Natural Causes

Natural causes often linked to quick clay landslides are erosion. Strong groundwater flow or high porewater pressure (due to rainfall and/or snowmelt or other causes) can also contribute to initiate an initial slide that develop further into a quick clay landslide.

Rainfall and temperature measurements were available from two weather stations nearest Mofjellbekken Creek over a ten-year period from 2005 to 2015 (Ramnes-Berg and Sande-Galleberg). Daily rainfall and temperature measurements immediately before the landslide were compared with data from the same calendar period over the last 10 years. The rainfall in early 2015 was slightly over the average. A heavy rainfall on 16.01.2015 recorded 30–60 mm rainfall in one day. The temperature fluctuated between −5 and 2 °C, which would caused cycles of freeze–thaw in the ground.

Stability analyses under fully drained conditions were done for Profiles A and B (Fig. 3). The analyses focused on the stability of the slope before the fill was placed to separate the effect of increased porewater pressure from rainfall/snow melt from the effect of fill placement. The soil parameters and groundwater table were used conservative values (Haugen et al. 2016). The analyses showed that the factor of safety (F) for sliding surfaces into the quick clay layer was high (F = 1.45 for sliding surface 1 in Fig. 9). The results indicate that the eastern slopes had good stability before fill construction (F over 1.4 in Fig. 9), even with higher porewater pressure due to rainfall and snowmelt. The results support the conclusion that porewater pressure increase is unlikely to be the main cause initiating the Skjeggestad landslide. It is however possible that high porewater pressures could have contributed to lowering the safety factor at the time of the Skjeggestad landslide in 2015.

Little erosion was observed in Mofjellbekken Creek during site inspection after the landslide. Erosion caused by runoff and groundwater seeping out from the plateaus is visible locally. The slopes adjacent to the creek were gentle and less than 3-m high, with factor of safety larger than 1.4. The Commission concluded that erosion did not trigger the Skjeggestad landslide.

## Man-made Actions

The analyses of man-made actions focused on changes in terrain due to filling and on vibrations on the eastern plateau. The Commission considered the vibration from traffic on the bridges and from heavy construction machinery used just before the landslide occurred.

## Vibrations

In Norway, there have been recent landslides initiated in quick clays by vibration. Strong vibration can lead to cyclic loading of the clay which can cause irreversible strain and possibly collapse of the structure of the quick clay. Vibrations from two sources were investigated: traffic on the bridge and the bulldozer used to place the fill when the slide occurred. Measurements of traffic vibrations were conducted in the ground near the foundation of two bridges over Gutufossen Creek. These bridges were chosen because they lie on the same E18 Expressway, about 20 km north of Skjeggestad Bridge. The two locations, with same speed limit, have therefore comparable quantity of traffic, and the same type of bridge foundations. The clay beneath the Gutufossen Bridge is slightly stiffer than the clay beneath the Mofjellbekken Bridge. The difference in shear strength was taken into account in the analysis of potential dynamic shear strain due to vibrations.

Vibrations were measured continuously from 20 to 23 May 2015. Approximately 250 vibration incidences were registered. Fewer than 10 incidences with the strongest vibration measures 0.5–0.6 mm/s occurring over about 100 cycles, approximately 20 Hz. In terms of dynamic shear strain in the ground, the strongest vibration incidence would give a maximum shear strain of $12 \times 10^{-4}$%. This value is one to two orders of magnitude lower than the threshold for initiating a landslide in the quick clay at the Skjeggestad Bridge location. Such low frequency will not cause any irreversible changes in the clay, regardless of the number of cycles. Therefore, it can be concluded that vibration from traffic on the bridge was not the cause for the of the Skjeggestad landslide.

The fill on top of the slope when the landslide occurred was constructed by a CATEPILLAR D6P LPG bulldozer which weighs around 200 kN. Maximum measured vibration

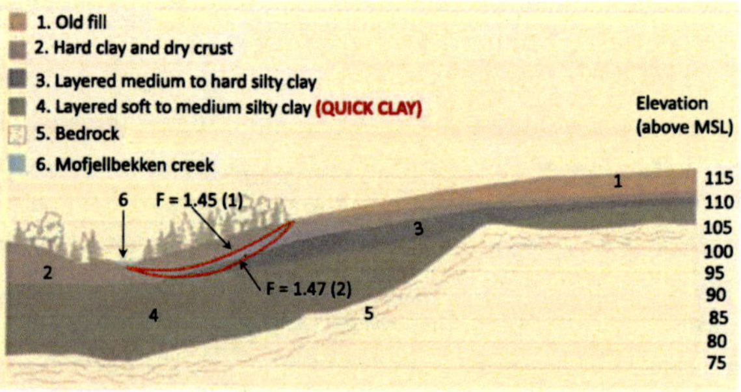


**Fig. 9** Sliding surfaces and factor of safety (F) from long-term (drained) stability analysis

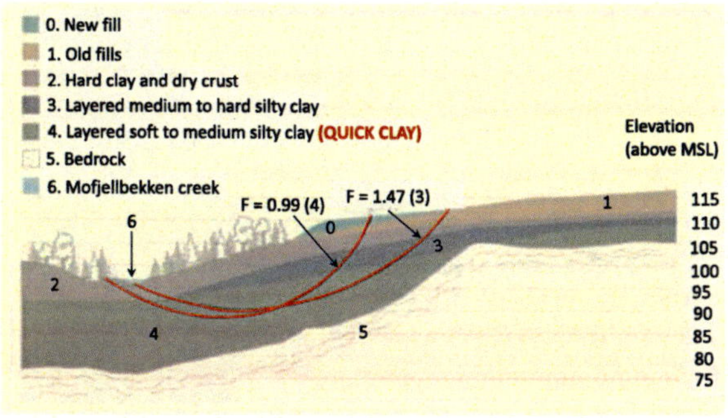


**Fig. 10** Sliding surfaces and factor of safety from short-term (undrained) stability analysis

from a similar type of heavy machine on quick clay is around 20–25 mm/s. The Norwegian Standard for vibration and shock (SN 2014) gives a maximum allowable vibration of 25 mm/s for quick clay. The Standard Norway value is set lower than the actual threshold that can trigger a quick clay landslide. Thus, vibration from the bulldozer was excluded as a trigger of the landslide.

### Changes in Terrain (Fill)

The analyses of terrain modifications used the GIS images showing that fills had been placed in the area in the periods 1998–2006 and 2006–2015. The fills in period 1998–2006 (denoted as "old fills" in Figs. 9 and 10) were mostly for the construction of the Mofjellbekken bridges. In this period, 3 m of fill were placed in the area under the bridge and up to 8 m for the eastern bridge foundation. Undrained stability analysis showed that these fills reduced the safety factor of the slope along Profile A by about 14% (from F = 1.36 in 1998 to F = 1.17 in 2006 (Sliding Surface 3, Fig. 10).

In the period 2006–2015, terrain modifications were made in the area south of the bridge. The most significant filling was the one immediately before the landslides. The fill had a total volume ca. 3500 $m^3$, and an average thickness of about 1.5–1.6 m. The highest part of the fill could be up to 3 m (Fig. 10). Results from 2D LEM undrained stability analyses showed that the factor of safety reduced to below 1 due to the additional load from this fill (Sliding Surface 4, Fig. 10). The Commission concluded that the fills in 1998–2006 reduced considerably the safety margin along Profile A, while the additional fill in 2015 triggered the landslide of a slope that was already only marginally stable.

Stability analyses for Profile B showed that the northeast slope had a high factor of safety (F > 1.4). It is therefore believed that it is unlikely that the landslide would have started in this direction. These slopes could have been loaded more before a landslide in the north-eastern part under the bridges could have been triggered. Thus, the soil masses from the northeast under the bridge foundation were likely "torn out" by the first sliding mass in the east (i.e. along Profile A).

## Summary and Conclusions

The forensic geotechnical investigation of the Skjeggstad quick clay landslide was conducted to reconstruct the failure mechanism and examine the cause of the landslide. The independent investigation commission reviewed the available information and data, and in particular LIDAR data and drone photographs to study the failure mechanism. Rainfall and temperature collected at nearby weather station were used to assess porewater pressure increase. Site visits were carried out to examine erosion and other physical characteristics. Vibration measurements were conducted to examine vibration a possible trigger. The stability was analysed with the LEM under both fully drained and fully undrained conditions.

The investigation shows that the landslide propagated in three stages. Phase I involved first a large sliding mass from the eastern plateau and then from the northeast part near the bridge foundation. Steep unstable back-scarps generated in Phase I triggered the propagation of the sliding activity backwards. Phase II started within seconds after Phase I. Movements of sliding masses in Phase II from the northeast torn out the backfill supporting the bridge foundation causing the partial collapse of the bridge pillar. Phase III involved local sliding of the back-scarps from Phase II, and spanned over a few days. The investigation also pinpointed that the shallow bedrock in the east and the layering of the soils played an important role in limiting further propagation of the landslide.

The Commission concluded from the observations of the failure and the stability analyses that the quick clay landslide was triggered by a fill which was placed a few days before the landslide, but that the stability at the location was already marginal due to the addition of fill on the slope in the period 1998-2006. The pore water pressure had increased due to rainfall and snowmelt in the months before the landslide and can have contributed to the landslide but was not the main cause of the movements. The investigation excluded erosion along the Mofjellbekken Creek as a trigger because little erosion was observed. Vibration as triggering cause was also excluded as a trigger the strongest vibration from both traffic and construction machinery near the site were lower than the expected threshold values that can initiate landslide in quick clay.

The case of the Skjeggestad landslide illustrates the importance of careful and continuous management of quick clay areas after the completion of an infrastructure project.

**Acknowledgements** The authors wish to thank the Norwegian Water Resources and Energy Directorate (NVE) and the member of the Independent Investigation Commission for the permission to publish the results, and the Norwegian Public Road Administration for access to the data at the site of the Skjeggestad landslide.

## References

Haugen SB, Henderson LA, Amdal ÅMW (2016) Case-study of a quick clay landslide that caused the partial collapse of Mofjellbekken bridges in Norway (Under review). In: Stefano Aversa LC, Luciano Picarelli, Claudio Scavia (ed) Proceedings of the 12th International Symposium on Landslides, Napoli, Italy, 12–19 June 2016. Taylor & Francis Group

Jostad HP, Sivasithamparam N, Le TMH, Lacasse S, Giese S, Åkershult AR a, Johansen T, Aabøe R (2020) 3D stability analyses of Skjeggestad landslide. Paper presented at the Nordic Geotechnical Meeting - NGM2020, Helsinki, Findland, 18–20 January, 2021

Karlsrud K, Kim Y, Hendersen L (2015) Sikringstiltak for-og refundamentering av Skjeggestad bruene etter skredet 2 februar 2015. Paper presented at the Geoteknikkdagen 2015, Oslo, Norway

NPRA-Norwegian Public Roads Administration (1997) ZD95C-1-E18 Nordre Vestfold parsell 5B Mofjellbekken bru profil 30152–30370. Norwegian Public Roads Administration

NVE-Norwegian Water Resources and Enery Directorate (2015) Skredet ved Mofjellbekken bruer (Skjeggestadskredet). Utredning av teknisk årsakssammenheng, Rapport, p 53

Rødvand LA, Andresen L, Grimstad G (2019) Case study of a road bridge hit by a landslide in highly sensitive clay. In: Proceedings of the XVII ECSMGE 2019 : Geotechnical Engineering foundation of the future: European Conference on Soil Mechanics and Geotechnical Engineering., Reykjavik, Iceland, 1–6 September 2019

SD-Standard Norge (2014) NS 8141–3:2014−Vibrasjoner og støt - Veiledende grenseverdier for bygge- og anleggsvirksomhet, bergverk og trafikk - Del 3: Virkning av vibrasjoner fra sprengning på utløsning av skred i kvikkleire

Skoglund J (2017) Alt Ved Det Gamle I Skjeggstad. doi:veier24.no

# Accuracy Assessment of Unmanned Aerial Vehicle (UAV) Structure from Motion Photogrammetry Compared with Total Station for a Deformed Slope

Vera Hui Loo and Chou Khong Wong

**Abstract**

Unmanned Aerial Vehicle (UAV) has been widely used for slope stability analysis. The objective of this research is to test the digital surface model (DSM) results generated from UAV images with the data acquired from total station for a deformed slope. A slope along the Pan Borneo Highway was selected for the study. The UAV survey was undertaken by utilizing DJI Inspire 1 with Zenmuse X3 Gimbal. A total of 10 ground control points (GCPs) were marked during the surveying for validation purposes. Structure from motion (SfM) technique adopted Pix4D enterprise version 4.3.33 to stitch the images for the production of orthophotos and DSM. The root mean square error (RMSE) of the GCPs were checked, where the horizontal RMSE in x direction and y direction are 1.4 cm and 1.8 cm respectively while RMSE in z direction is 2.6 cm. The total station surveying was taken at various locations of slope, which include slope surface with slight to moderate deformation, slope surface with severe deformation, surface channel and at the edge of surface channel. The elevations of DSM results were tested with those surveying data acquired from site. The results show that for slope with slight to moderate deformation, the accuracy of the RMSE in elevation of 4.2 cm can be achieved. Similar RSME accuracy can be attained for surface channel which is 5 cm. However, the RMSE for slope portion with severe deformation is 10.6 cm. From this research, it is found that the UAV-based DSM lower accuracy will be attained for locations of sharp changes in elevation.

**Keywords**

Unmanned aerial vehicle (UAV) • Total station • Photogrammetry • Slope with deformation

## Introduction

Slope instability issues are either natural or man-made in nature, which cause heavy damage to infrastructures as well as human life. In mountainous regions, slope instability problems heavily depend on developmental activities which includes construction of road, rail networks, hydroelectric projects and other human settlement developments. Monitoring slope instability and deformation will improve the understanding of the mechanism of such slope and will further facilitate in identifying the causes of failures. Slope instability analysis in traditional way includes the identification and mapping of susceptible areas with the help of historic instability information. The conventional methods of identification of slope deformation utilise aerial photographs and topography maps by on site surveying. For the past decades, UAV has been rapidly developed. This technique can have distinct advantages over the conventional methods in producing aerial photographs and topography maps. The UAV can provide high resolution images with the advantage of requiring less time and cost.

Several local and international studies used UAV imagery and SfM techniques to produce 3D model, dense point cloud, orthophotos and DSM/DEM. This includes geohazard reconnaissance mapping (Sharan Kumar et al. 2018), landslide displacement mapping (Lucieer et al. 2014), landslide evolution (Mozas-Calvache et al. 2017), characterization of reservoir landslides (Yang et al. 2019), modelling of landslide topography (Yu et al. 2017), construction of extreme topography (Agüera-Vega et al. 2018), and many more. In most of the studies, researcher have compared the results of DEM generated using UAV images with the data acquired

V. H. Loo (✉)
Curtin University Malaysia, Miri, Malaysia
e-mail: vera.loo@curtin.edu.my

C. K. Wong
Sanudra UAS Sdn Bhd, Miri, Malaysia
e-mail: michael@avetics.com


V. Vilímek et al. (eds.), *Understanding and Reducing Landslide Disaster Risk*,
ICL Contribution to Landslide Disaster Risk Reduction,
https://doi.org/10.1007/978-3-030-60319-9_32

from global positioning system (GPS), global navigation satellite system (GNSS), total station (TS) or laser scanning. Nagendran et al. (2018) used 10 number of GCPs to obtain a horizontal and vertical accuracy of 10 cm and 33 cm respectively while comparing with GPS measurements and UAV photogrammetric output. In mapping of landslide displacements, Lucieer et al. (2014) tested his UAV-based DEM results with coordinates obtained by geodetic GPS receivers and a horizontal accuracy of 7 cm and vertical accuracy of 6 cm can be achieved.

Due to the increasing development of infrastructure in Sarawak along with the intense tropical rainfall, slope failures are exacerbated and becoming increasingly common. Several slope stability problems were observed in slope cuttings where shale is present. Those failures are not instant; however, the continuing deformations bring significant hazard to the road users. Shale has been well recognized as a problematic deposit for slope stability (Stead 2016). Shale is easily disintegrated due to its laminated and fissile characteristics and its exposure to the weather accelerates disintegration process.

The objective of this research is to test the DSM results generated from UAV images with the data acquired from total station for a deformed slope in Sarawak. The testing of DSM results include the accuracy of DSM output as compared to 10 nos. of ground control points, the accuracy of the elevation of DSM for various areas on slope and the accuracy of cracks of surface channel.

## Site Investigation and Surveying

### Site Description

A cut slope with deformation along the highway under construction was selected for the study. It is located at about 240 km away from Miri in Sarawak (Fig. 1). The study area is about 3720 $m^2$. Geologically the area is covered with sedimentary formation. The lithology is mainly shale, siltstone and sandstone. This cut slope strikes S40°E, and the slope angle is 34°. The strata of the shale layer strikes S20°E to S30°E and dips 50° to 70° into the slope. The width of the front part of slope is 110 m and the length of the slope is 40 m. The slope height is 17.8 m, and its elevation at toe and crest is 58.6 m above principal datum (mPD) and 76.4 mPD respectively. The slope has 4 batters and the first 3 batters are about 5 m high while the top batter 3 m high. Figure 2 shows the general view of the slope.

The failure mechanism of this shale slope is complex. Figure 3 shows the surface channels which were broken into few parts on top of the 2nd batter slope. The slope deformed seriously at the 2nd batter of the slope (Fig. 4). The slope failure could be the result of the surface runoff which cannot be discharged effectively away by the surface channel from the slope during heavy rainfall. The runoff accumulated at the middle section of the channel ran over the slope surface, and seeped into slope which softened the shale. Gradually, this causes the settlement and dislocation of the surface channel.

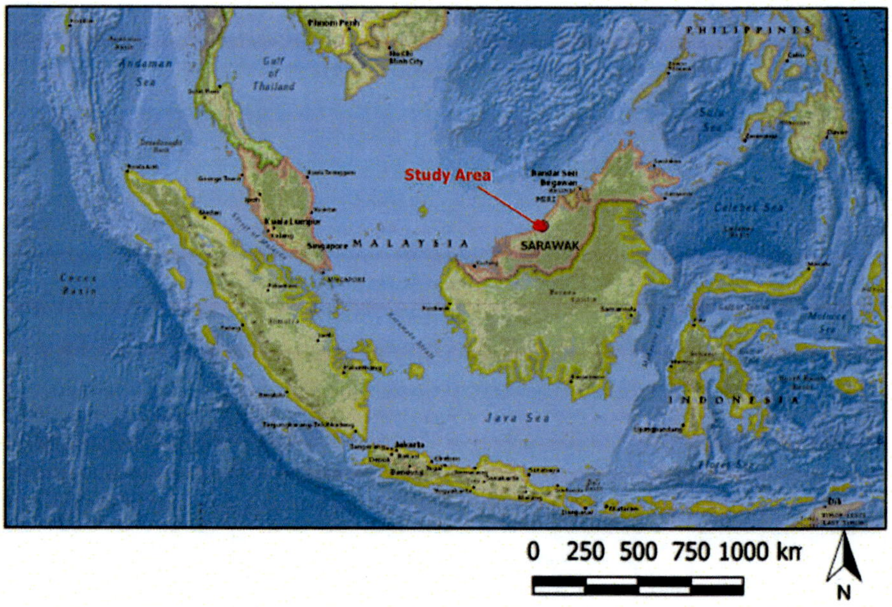


**Fig. 1** Location plan of study area

**Fig. 2** General view of the study area

**Fig. 3** Surface channel of the 2nd batter of slope

**Fig. 4** Close view of slope deformation at 2nd batter of the slope

## Capturing of Aerial Images

The UAV survey was undertaken in April 2018, utilizing DJI Inspire 1 with Zenmuse X3 Gimbal. The camera model is DJI FC350 with the resolution of 4000 × 3000 pixels. The flight height is 52.9 m above ground level (AGL) and a total of 278 photos were gathered for the generation of DEM/DSM, producing an average ground sampling distance (GSD) of 2.33 cm/pixel.

## Surveying of Ground Control Points

During the capturing of aerial images, a total of 10 GCPs as shown in Fig. 5 were marked on the slope. The coordinates and elevation surveyed by total station were listed in Table 1. These points are set for validation purposes during the photogrammetry process. The coordinates and reference systems of Timbalai 1948/RSO Borneo (m) are referred.

## Photogrammetry Data Processing and Outputs

Photogrammetric process adopted Sfm technique by using Pix4Denterprise version 4.3.33 to stitch the images for the production of digital surface model (DSM) and orthophotos. A total of 152 images were selected for the photogrammetric process. The workflow of photogrammetric includes (i) adding images, (ii) aligning images, (iii) placing GCPs markers, (iv) realign and reoptimize image tie points, (v) building the dense cloud, (vi) building the mesh and texture, and (vii) generating the orthophoto and DSM. The output of orthophotos and DSM will then import into QGIS software for further analysis.

## Accuracy of Digital Surface Model (DSM)

The accuracy of the DSM results were tested with coordinates and elevation of GCPs obtained from total station. The RMSE of the GCPs were checked. The horizontal RMSE in x direction and y direction are 1.4 cm and 1.8 cm respectively while RMSE in z direction is 2.6 cm.

## Results and Discussion

This study tests the elevation of DSM results as compared with total station (i) at slope surface with slight deformation, (ii) slope with significant deformation, (iii) surface channel, and (v) edge of surface channel. In addition, the cracks of surface channel obtained by DSM results are tested with site measurement.

### Elevation at the First Batter of Slope

The first batter of slope has slight deformation especially at the right portion of the slope. A total of 18 points were

**Fig. 5** Location of 10 ground control points

chosen at the first batter of the slope to compare the difference in elevations between DSM with TS (Fig. 6). When the difference is in positive value, this indicates that DSM produces a higher elevation as compared with total station; otherwise, DSM has a lower elevation as compared with TS. The difference in elevations for the first batter of slope ranges from −8.4 cm to 2.8 cm, with an average of −2.5 cm. The RMSE in z direction is 4.2 cm.

## Elevation at the Second Batter of Slope

The 2nd batter of the slope is where the deformation significantly observed. A total of 36 points were chosen at the 2nd batter of the slope to compare the difference in elevations from DSM and the TS (Fig. 7). The points with red colour are those locations with differences more than 10 cm. The differences range from −23 cm to 15.9 cm, with an average of −7.1 cm. The RMSE in z direction is 10.6 cm. Out of 36 points, 31 points show the negative value, indicating the elevation of DSM is lower than that measured by TS.

The close view of the deformation of the slope of front view (Fig. 8) and side view (Fig. 3) shows that the elevation changes abruptly from one location to the other location.

## Elevation at the Middle of Surface Channel

A total of 16 points were chosen along the middle point of surface channel for comparison (Fig. 9). The results show that except one particular point located at the cracking joint of the channel which show the difference of 12.7 cm, the difference of all other points is less than 5.8 cm. The difference ranges from -5.8 cm to 12.7 cm, with an average of -0.5 cm. The RMSE in z direction is 5.0 cm.

## Elevation at the Edge of the Surface Channel

Figure 10 shows the results of 42 points located on the slope surface along the edge of the surface channel. The points of red colour are those locations with difference more than 10 cm. It is observed that those points are located at the edge of surface changes with detachment from the soil surface. A clearer view of the detachment of surface channel can be viewed from 3D model in Fig. 11. On the other hand, all the points in red colour show the value of the difference in positive value, which indicates that the elevation acquired from DSM is higher than those surveyed by total station.

**Table 1** Coordinates and elevation of ground control points

| GCP | Easting (X) | Northing (Y) | Elevation (mPD) |
|---|---|---|---|
| 1 | 404,854.1460 | 363,281.5782 | 77.5031 |
| 2 | 404,864.7896 | 363,282.2931 | 73.6596 |
| 3 | 404,887.4614 | 363,286.1751 | 68.7034 |
| 4 | 404,866.7496 | 363,270.7318 | 67.2299 |
| 5 | 404,842.7379 | 363,251.6561 | 67.1361 |
| 6 | 404,893.9400 | 363,281.8071 | 62.7993 |
| 7 | 404,883.8900 | 363,270.2800 | 58.9751 |
| 8 | 404,852.5953 | 363,246.1565 | 58.1756 |
| 9 | 404,906.1079 | 363,276.0409 | 77.5031 |
| 10 | 404,853.5720 | 363,236.2098 | 73.6596 |

**Fig. 6** Difference in cm of the elevation from DSM and TS at the 1st batter of slope

**Fig. 7** Difference (in cm) of the elevation between DSM results and total station at the 2nd batter of slope

**Fig. 8** Front view of surveying points at the deformation area located at the 2nd batter of the slope

This may due to the rendering process of photogrammetry process which will give the average value at the interface of surface channel and slope surface. The difference between the surveyed point and DSM ranges –9.6 cm to 68.1 cm, with an average of 3.7 cm. The RMSE in z direction for 42 points is 16.3 cm. The low accuracy is due to 3 points where the differences are more than 30 cm. If those points are not considered, the RMSE for 39 points out of 42 points will be 7.4 cm.

Figure 12 shows the results of all the points. Those points with higher discrepancy (red/light red colour) are mainly located at 2nd batter and the edge of the channel.

Table 2 summarises the results for 1st batter and 2nd batter of slope, and the mid-point and the edge points of surface channel. The accuracy of 1st batter of slope is the

**Fig. 9** Difference (in cm) of the elevations between DSM and TS at the mid-point of surface channel

**Fig. 10** Difference (in cm) of the elevations between DSM and the TS along the edge of the surface channel

best, followed by the midpoint of surface channel, the 2nd batter of slope then the edge of the surface channel.

## Horizontal Displacement of Surface Channel

Figure 13 shows the three locations (H1–H2, H3–H4 and H5–H6) of the crack of surface channel at the top of 2$^{nd}$ batter of slope from DSM. Figure 14 shows the close view of the crack on site. The crack width from DSM results is compared to that measured on site and the results are listed in Table 3. The accuracy of less than 1.9 cm can be achieved.

## Conclusion

This research applies the combination of UAV with SfM techniques in order to produce DSM for use to assess the deformation of slope. The model is validated by 10 GCPs

**Fig. 11** 3D model of DSM results

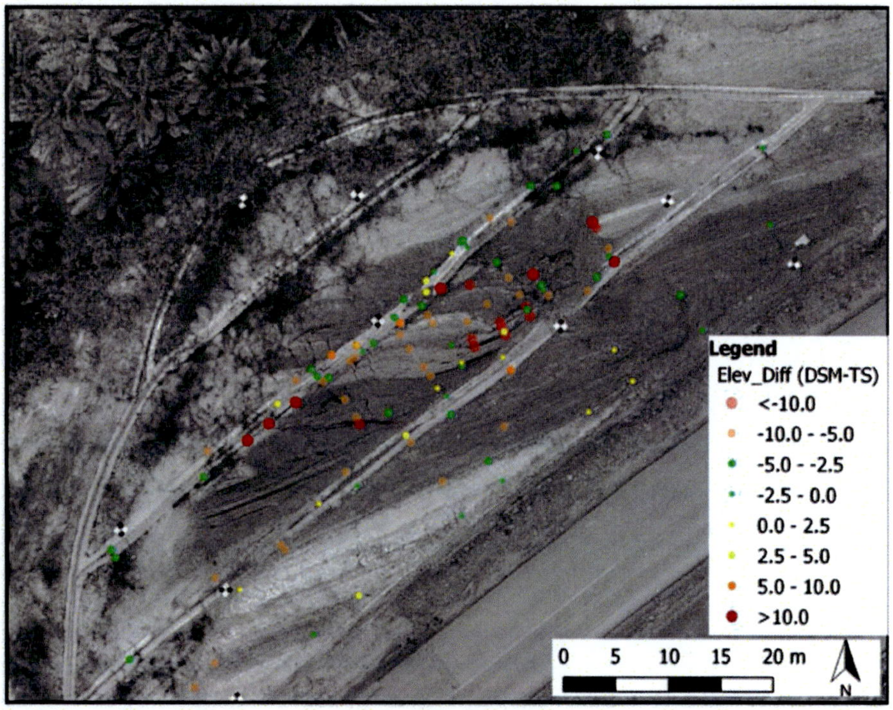


**Fig. 12** Difference (in cm) of the elevations between DSM and the TS for all points on slope

measured by total station which gives the horizontal of 1.4 cm and 1.8 cm respectively while the vertical RMSE is 2.6 cm. It is found that a higher accuracy in horizontal RMSE is achieved as compared to the elevation's RMSE.

The total station surveying was taken at various location of slope, which include slope surface with slight to moderate deformation, slope surface with severe deformation, surface channel and at the edge of surface channel. The surveying data were compared with the DSM results.

For slope with slight to moderate deformation, the accuracy of RMSE of 4.2 cm can be achieved. For slope with severe deformation, the accuracy of RMSE of 10.6 cm can be achieved.

For the surface channel, the accuracy of RMSE of 5.0 cm can be achieved. However, the accuracy for the surveying point taken at the interface of surface channel and the slope is the lowest, which is 16.3 cm. This is due to the detachment of surface channel from the slope in which the

**Fig. 13** Crack at the surface channel

rendering process of photogrammetry cannot reflect the actual abrupt changes of the elevation.

The crack width of the surface channel measured on site is checked against the DSM result. The accuracy of horizontal displacement is within 2 cm.

From this research, it is found that at most locations, the accuracy of the RMSE in elevation of 4 cm to 5 cm can be reasonably achieved. However, UAV-based DSM can give erroneous information for location of sharp changes in elevation. On the other hand, the accuracy of UAV imagery will be affected by the vegetation on the slope. In addition, weather condition is critical during the images capturing.

The vertical accuracy of this research is comparable to the accuracy (~6 cm) obtained by UAV-based model carried out by Lucieer et al. (2014) tested with GCPs based on geodetic GPS receivers and that carried out by Agüera-Vega et al. (2018) tested against TS and GNSS.

UAV imagery in combination of photogrammetry method can be a cost effective method in measuring the deformation of slope failure. It can be served as a means of long term monitoring for slope along the roadside, as well as a means to acquire fast and actionable data in the field of disaster management.

**Table 2** Summary table for accuracy comparision at different locations of slope

| Area | DSM-TS (cm) | Average (cm) | RSME, Z (cm) |
|---|---|---|---|
| 1st batter of slope | −8.4 to 2.8 | −2.5 | 4.2 |
| 2nd batter of slope | −23 to 15.9 | −7.1 | 10.6 |
| Mid-point of surface channel | −5.8 to 12.7 | −0.5 | 5.0 |
| Edge of surface channel | −9.6 to 68.1 | 3.7 | 16.3 |

**Table 3** Accuracy of crack width of surface channel measured from DSM compared with the site measurement

| Location | DSM (cm) | Site measurement (cm) | Difference \|DSM-Site\| (cm) |
|---|---|---|---|
| H1 to H2 | 22.6 | 23 | 0.4 |
| H3 to H4 | 8.9 | 7 | 1.9 |
| H5 to H6 | 11.8 | 10 | 1.8 |

**Fig. 14** Crack measured on site

**Acknowledgements** The project is supported by start-up funds by Office of Research and Development of Curtin University Malaysia. Special thanks to Mr Fum from JKR to allow the survey to be carried out on the site. The author would also like to thank Tiu Chin Yong, Chieng Yiew Chee and Chung Ming Yong who helped to take the survey data of using total station of the site. On the other hand, the author would like to express gratitude to Dr. Vijith H for your kind review on the paper.

## References

Agüera-Vega F, Carvajal-Ramírez F, Martínez-Carricondo P, Sánchez-Hermosilla López J, Mesas-Carrascosa FJ, García-Ferrer A, Pérez-Porras FJ (2018) Reconstruction of extreme topography from UAV structure from motion photogrammetry

Lucieer A, Jong SMD, Turner D (2014) Mapping landslide displacements using Structure from Motion (SfM) and image correlation of multi-temporal UAV photography. Prog Phys Geogr 38(1):97–116

Mozas-Calvache AT, Pérez-García JL, Fernández-del Castillo T (2017) Monitoring of landslide displacements using UAS and control methods based on lines. Landslides 14(6):2115–2128

Nagendran SK, Tung WY, Mohamad Ismail MA (2018) Accuracy assessment on low altitude UAV-borne photogrammetry outputs influenced by ground control point at different altitude. IOP Conference Series 169(1):012031

Sharan Kumar N, Ismail MAM, Sukor NSA, Cheang W (2018) Geohazard reconnaissance mapping for potential rock boulder fall using low altitude UAV photogrammetry. In IOP Conference Series: Materials Science and Engineering

Stead D (2016) The Influence of shales on slope instability. rock mechanics and rock engineering 49(2): 635–651

Yang Y, Wang X, Jin W, Cao J, Cheng B, Xiong M, Zhou S, Zhang C (2019) Characteristics analysis of the reservoir landslides base on unmanned aerial vehicle (UAV) scanning technology at the Maoergai Hydropower Station, Southwest China. In IOP Conference Series: Earth and Environmental Science 349

Yu M, Huang Y, Zhou J, Mao L (2017) Modeling of landslide topography based on micro-unmanned aerial vehicle photography and structure-from-motion. Environmental earth sciences 76(15)

2020 Kyoto Japan

# ARAS: A Web-Based Landslide Susceptibility and Hazard Mapping System

Murat Ercanoglu, Mehmet Balcılar, Fatih Aydın, Sedat Aydemir, Güler Deveci, and Bilgekağan Çintimur

**Abstract**

As in the case throughout the world, landslides in Turkey have accounted for significant amount of economic losses and caused damage to properties as well as the environment and inhabitants. For example, considering the last 70 years' landslide records in Turkey, it was revealed that more than 60,000 people were affected due to landslides. Thus, in order to minimize the undesired landslide consequences, some measures and initiatives had to be taken. Particularly in the last 10 years, many projects have been initiated by the governmental agencies in Turkey. Of these, a web-based landslide susceptibility and hazard mapping system project, namely Disaster and Risk Reduction System (ARAS), has been initiated in 2016. The most important objective of ARAS is to establish a spatial decision support and analysis system to reduce the mass movement risk for mitigation efforts using today's technologies and data processing techniques. In this study, it was aimed at introducing ARAS and its applications on producing landslide susceptibility and hazard maps in a web-based platform. For this purpose, whole stages of ARAS in its current version were explained and landslide susceptibility and hazard maps of Kastamonu city, located in Middle Black Sea region of Turkey, were produced for this study. In ARAS, risk assessment stage is an ongoing process at the moment, and will be completed in a few years. When completed, decision-makers, planners and local authorities will benefit from the advantages of ARAS, which will provide significant gains for sustainable risk management in Turkey.

**Keywords**

Landslide • Susceptibility • Hazard • Web-based mapping

## Introduction

Turkey with its geological, geomorphological and climatic characteristics is one of the most affected countries with respect to natural disasters on a global scale. In terms of consequences of these events, earthquakes were the most destructive one causing the loss of lives and properties in Turkey. The country has also been experienced major disasters such as landslides, floods and avalanches. When examined the disaster records of the last 70 years, direct and indirect economic losses due to the above mentioned disasters were about 3% with respect to Gross Domestic Product of Turkey (https://www.afad.gov.tr).

One of the most important studies to be done in order to minimize the damages caused by the disasters is to produce inventory, susceptibility, hazard and risk maps. The maps to be prepared for this purpose are extremely important in order to transmit accurate, fast and reliable information to the authorities such as decision makers, planners and local

M. Ercanoglu (✉)
Geological Engineering Department, Hacettepe University, Ankara, Turkey
e-mail: murate@hacettepe.edu.tr

M. Balcılar
Department of Economics, Eastern Mediterranean University, Famagusta Mersin, Turkey
e-mail: mehmet.balcilar@emu.edu.tr

F. Aydın · S. Aydemir · G. Deveci
Ministry of Interior, Disaster and Emergency Management Authority, Üniversiteler Mahallesi, Ankara, Turkey
e-mail: fatih.aydin@afad.gov.tr

S. Aydemir
e-mail: sedat.aydemir@afad.gov.tr

G. Deveci
e-mail: guler.deveci@afad.gov.tr

B. Çintimur
Globetech Geographical Information Technologies, Hacettepe University Technopark, Üniversiteler Mahallesi, Ankara, Turkey
e-mail: kagan.cintimur@globetech.com.tr


V. Vilímek et al. (eds.), *Understanding and Reducing Landslide Disaster Risk*, ICL Contribution to Landslide Disaster Risk Reduction,
https://doi.org/10.1007/978-3-030-60319-9_33

authorities. Utilization of these maps will enable authorities to make rational and healthier regional development and land-use plans as well as to protect the inhabitants and the environment.

Although Turkey's disaster policy dates back to the aftermath of the 1939 Erzincan earthquake (which claimed nearly 33.000 lives and left at least 100.000 injured), the 1999 Marmara earthquake marked the turning point in disaster management and coordination. This devastating disaster clearly demonstrated the need to reform disaster management and compelled the country to establish a single government institution to single-handedly coordinate and exercise legal authority in cases of disaster and emergencies (https://www.afad.gov.tr). In this regard, the Disaster and Emergency Management Authority, namely AFAD, was established in 2009. After this date, significant progress has been achieved in the area of sustainable disaster management in the last decade in Turkey. There were many projects initiated by AFAD to reach the global and scientific standards with the goal of minimizing the effects of natural disasters. For example, AYDES (abbreviation of "Disaster Management and Decision Support System" in Turkish) is a sustainable disaster management and decision support system (https://aydes.afad.gov.tr). With AYDES, the decision makers are able to access the data they need in the disaster management process in a fast and accurate fashion. It helps user to reach the databases, spatial maps related to any type of disaster in Turkey. ARAS (abbreviation of "Disaster Risk Reduction System" in Turkish) project was initiated in 2016 and it has been working on a web-based platform together with AYDES (or with the user's own database and inventory map) to analyse the mass movements such as landslides and avalanches in any selected region in Turkey (https://aras.afad.gov.tr). Main purpose of this study is to introduce ARAS that works on a web-based platform for producing landslide susceptibility and hazard maps.

In ARAS, the user can analyse landslide susceptibility and hazard using AYDES or his/her own database and inventory map in a short time ranging between a few minutes to a few hours depending on the selected assessment algorithm and the size of the area. However, the system is only open to AFAD people and its provincial users at the moment. The main goal of ARAS is to assess the mass movement risk in any part of the country. The risk assessment module has not been completed at the moment, but in the next years, after completing the related database records, it will be implemented to assess landslide risk in any region/part of Turkey. The details related to landslide module of ARAS were presented in the following lines.

## General Characteristics of ARAS

One of the main goals of ARAS landslide module is to establish a spatial decision support system which aids the determination of landslide risk zones throughout Turkey to minimize the possible undesired effects of landslides. As mentioned before, risk assessment stage is still an ongoing process in the system. However, landslide susceptibility and hazard assessments are currently in use by head of AFAD and its 81 provincial directorates. ARAS could be reached by https://aras.afad.gov.tr link, but it works via intranet at the moment. The language used in the system interfaces is in Turkish.

Technological background of ARAS is based on a 3 layered architecture and works on many web-based components/services developed by ArcGIS Server, Geoprocessing Services, Java REST services and Angular Javascript Framework. All components of the system are integrated into a unified framework and the system is developed through a cloud based architecture. The users reach the system via a web browser without needing a third party software. ARAS comprises of three main interfaces such as administrative, mapping and analysis on the web-based GIS (Geographical Information System) platform. In administrative interface, the authorized personnel at the head of AFAD can see the number of performed analyses, types and areal extents of landslide inventories, considered parameters and methods, top analysts etc. Map and analysis interfaces work interchangeably to produce landslide susceptibility and hazard maps of any selected region. A general flow chart of these operations was shown in Fig. 1.

When the system is initiated, the user is asked to enter the username and password since the authorized users can only use the system both in headquarter of AFAD and in its 81 province of directorates throughout Turkey. At the first stage, the user should select the study area to be analysed. The selected study area could be a city, a province or a district boundary, a neighbourhood, or a basin. The other options provided for the users are manual digitizing of any interested area on the system or loading the files with "*.kml" or "*.shp" extensions. After completing this stage, landslide inventory map is asked to the users whether they will use their own maps or AYDES database. If the user will use his/her own inventory map, it can be loaded to the system by a file with "*.kml" or "*.shp" extension. The next stage is to select the statistical analysis model. The model could be an inventory based one or not. Inventory based models fed into ARAS are Artificial Neural Network

**Fig. 1** Operational flow-chart of ARAS system for landslide susceptibility and hazard mapping

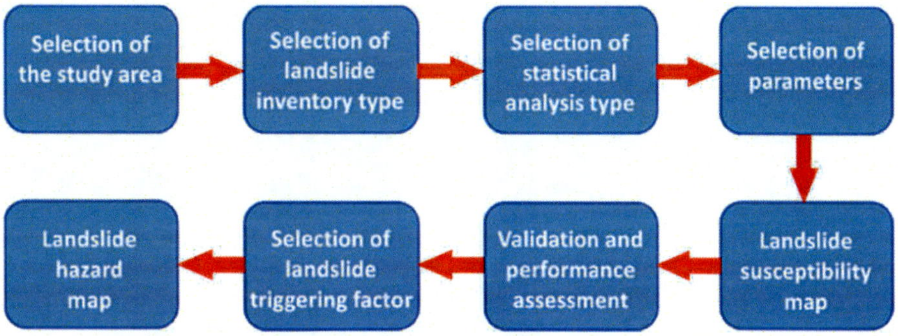


(ANN), Logistic Regression Model (LRM), Frequency Ratio Approach (FRA), Linear Discriminant Analysis (LDA), Weights of Evidence Model (WEM), Support Vector Machines (SVM) and Bayesian Deep Learning (BDL), while Analytical Hierarchy Process (AHP) and Simple Additive Weighting (SAW) models are grouped as non-inventory based ones. These methods are known as the most commonly used ones in the literature. Of these, FRA, WEM, AHP and SAW model algorithms are developed by the AFAD project team using Python and ArcPy libraries (https://pro.arcgis.com/en/pro-app/arcpy/get-started/what-is-arcpy-.htm). The other model algorithms are developed by using the open source SciPy (https://scikit-learn.org/stable/index.html) library. In addition, these developed models are published via ArcGIS Server as an asynchronous GeoProcessing Service (https://pro.arcgis.com/en/pro-app/help/analysis/geoprocessing/share-analysis/what-is-a-geoprocessing-service.htm). Through these services, the users can easily run the analyses with the web-based components and interfaces and have option to follow the analyses asynchronously. The user may choose one of the above mentioned statistical methods to produce a landslide susceptibility map on the web platform with the selected conditioning parameters, which is the next stage following the statistical model selection phase. There are three main landslide conditioning parameter groups in ARAS such as geological (e.g. lithology, fault, distance to faults, lineaments etc.) topographical (elevation, slope, aspect, curvature, distance to drainage, topographical wetness index etc.) and environmental (land-use, land cover, distance to roads, vegetation indices etc.) ones, constituting a total of 26 different parametric maps that cover overall Turkey. These parameters were fed into the system in digital format with 10m x 10m resolution, which were the most abundantly used conditioning parameters in the landslide literature. Instead of using certain or pre-determined parameters in the system, it is left to the user's choice which parameters will be used in landslide susceptibility analyses. The main rationale herein is to allow the user to better define the landslide characteristics and regional conditions of the area to be analysed. Before the landslide susceptibility analyses, the users are also asked to select how many points will be used for training and validation stages both from landslide and non-landslide areas if an inventory-based model is selected. The number of points are adjusted to at least 2000, but the user has option to use more than 2000 points. If non-inventory map based model such as AHP is selected, the user is asked to fill the weight matrix of AHP. In addition, the users have option to see the relationship between the considered parameter(s) and the landslide locations with the response curves. It is certain that the duration of the analyses mainly depends on many factors such as the number of points and parameters, size of the area, complexity of the selected algorithm etc. However, regardless of the number of parameters, points and the complexity of the statistical method etc., landslide susceptibility maps can be produced in a short time (within a few hours at most), without needing a third party software, by ARAS. An application by ARAS from Kastamonu city (Fig. 2a) is performed for this study. Landslide susceptibility map of Kastamonu city produced by ARAS was also shown in Fig. 2b. This analysis is performed for overall Kastamonu city, located in the Middle Black Sea Region of Northern Turkey, covering approximately 13,064 $km^2$. Details related to this application of ARAS are given in the results section.

In the literature, many confusing definitions could be encountered for landslide, susceptibility, hazard and risk terms. However, there were some studies, which could be used as the guidelines for any landslide study (e.g.: Varnes and IAEG 1984; Aleotti and Chowdhury 1999; Fell et al. 2008a, b). In ARAS, it was tried to perform the analyses based on the definitions (i.e. the terms landslide susceptibility and hazard) given in these guidelines as much as possible.

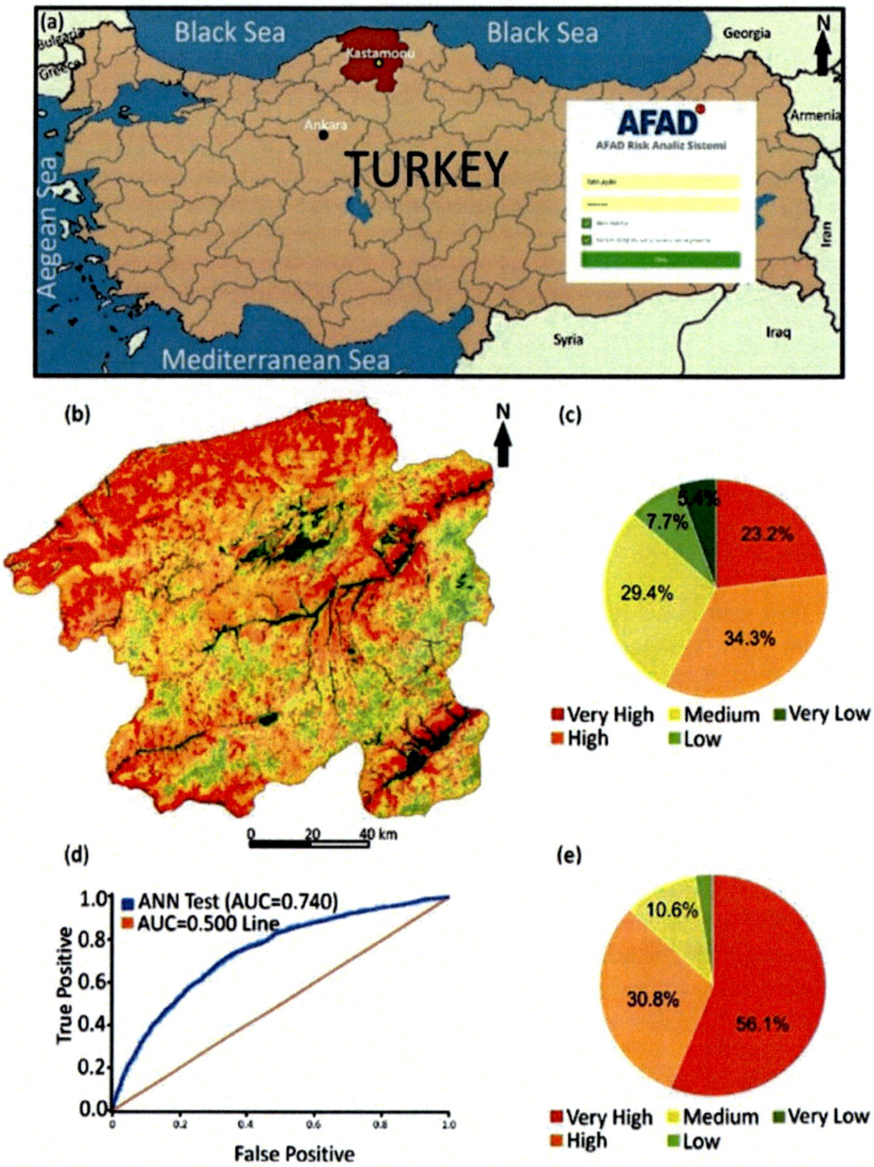


**Fig. 2** An example of production of landslide susceptibility map with ARAS system: **a** location map of the Kastamonu city; **b** landslide susceptibility map of the selected area; **c** areal % distributions of different susceptibility levels throughout the selected area; **d** ROC curve of the analysis and **e** % areal distributions of different susceptibility levels throughout the landslided areas

## Results

ANN method was selected to produce landslide susceptibility map of Kastamonu city for the current study. Based on the information gathered from local experts in Kastamonu directorate of AFAD, slope, land-use, lithology, aspect, curvature and topographical elevation parameters were selected as input parameters. Multilayer Perceptron (MLP) classifier model was used for the ANN analyses in ARAS. The ANN architecture (e.g.: number of hidden layers, activation function etc.) in ARAS can be run by defaults (https://scikit-earn.org/stable/modulesgenerated/sklearn.neural_network.MLPClassifier.html), but these could be changed based on the user demands. For example, in this study, the ANN architecture was comprised of 6 inputs, 4 hidden layers and 1 output (i.e. 6*4*1 architecture). The landslide susceptibility map produced by ANN model was based on randomly selected 3000 points from landslide locations (i.e. output value: 1) and 3000 points from non-landslide locations (i.e. output value: 0). A total of 1529 rotational earth slide type landslide locations were transmitted from AYDES system to ARAS for training and testing phases. These locations were previously mapped in the field by the local landslide experts of Kastamonu directorate of AFAD.

All algorithms for landslide susceptibility mapping modules in ARAS produce some useful outputs. For example, Relative Operating Characteristics (ROC) curve, percentage distribution of landslide susceptibility in the selected area and in the landslided areas (see Fig. 2c, d, e). The performance of the so-produced landslide susceptibility maps was assessed by ROC curve, which was one of the most commonly used approaches in the landslide literature. For landslide susceptibility map of Kastamonu city carried out for this study, the Area Under Curve (AUC) value was calculated from ROC approach was 0.740 and was considered satisfactory. The susceptibility classes are divided into five groups considering natural breaks (jenks) approach.

The next step, after completing the landslide susceptibility mapping, was to assess the landslide hazard in the system. Based on the definition of landslide hazard given by Fell et al. (2008a), it consists of three main components such as spatial, size (or magnitude) and temporal probabilities. Of these, spatial component was evaluated by landslide susceptibility stage. For temporal component, the triggering factor such as rainfall or earthquake analysis is needed. The magnitude and size probability is perhaps by far the most challenging task because of the incompleteness of the landslide inventories (Guzzetti et al. 1999) as in the case of this study. Since the complete landslide inventory is lacking for the selected area in this study, the landslide hazard is evaluated by multiplying the landslide susceptibility and spatial landslide triggering factor probability. Based on the landslide database of Kastamonu city, it was revealed that most of the landslides were triggered by heavy rainfall. Thus, rainfall was selected as a main triggering factor for this study. In addition, the users have option to assess the earthquake probability for any selected area in ARAS based on Earthquake Hazard Map of Turkey, produced by AFAD.

The theoretical background of calculation of probabilities of rainfall events is mainly based on the dates of both landslide occurrences and rainfall amount. However, the dates of landslide events are also lacking in the study area. The studies related to complete landslide inventory and to determine the dates of landslides are still continuing in the country. Thus, in ARAS, it was not possible to calculate the threshold values and their probabilities for certain periods as performed in the literature (e.g. Crozier and Eyles 1980; Glade et al. 2000; Zêzere et al. 2005). In order to overcome this problem, it was preferred to define different rainfall thresholds (u) by the users and to calculate their probabilities for exact return periods in ARAS. The daily rainfall data for Turkey is available for many weather stations operated by the Turkish State Meteorological Service. The Turkish State Meteorological Service is currently operating 1171 weather stations throughout Turkey. Although weather station rainfall data are available in all provinces of Turkey, it is not sufficiently fine gridded for the analysis. Thus, the rainfall distribution above a given threshold level (u), using station level data and based on kernel regression, was estimated, which allowed the users to interpolate rainfall probabilities to grids (location) of interest. In other words, it was almost impossible for the users to connect the landslide dates and rainfall thresholds. Therefore, instead of identifying thresholds between landslide dates and rainfall data, it was left to the users to make a choice from the specified return period. In this way, decision makers or users can set the specific level of relevant daily rainfall thresholds and obtain the exceedance probabilities for the relevant return period. This approach is based on Extreme Value Theory (EVT) (Embrechts et al. 2013; Beirlant et al. 2004; De Haan et al. 2006). EVT offers a rich base for modelling extreme maximum, which fits well to estimating probability of rainfall extreme maximums. Modelling exceedance over a threshold, which is the interest in ARAS, can be based on the generalized error distribution (Subbotin 1923). In ARAS, the maximum series are derived using the partial duration series approach of Langbein (1949). Following Pickands (1975), it was possible to model the distribution of the exceedance Generalized Pareto (GP) distribution, which asymptotically approximates Generalized Extreme Value (GEV) distribution. For an identically and independently distributed random variable X, the GEV distribution can be parametrized as shown in Eq. 1:

$$F(x) = \frac{1}{2\sigma\xi^{1/\xi}\Gamma(1+1/\xi)}\exp\left\{-\frac{1}{\xi}\left|\frac{x-\mu}{\sigma}\right|^{\xi}\right\} \tag{1}$$

where $\mu = E(X)$ is the location parameter, $\sigma = [E|X-\mu|\hat{}\xi]\hat{}(1/\xi) > 0$ is the scale parameter, $\xi > 0$ is the shape parameter, and $\Gamma$ is the Euler Gamma function. The function in Eq. 1 represents a generalization of a large set of distributions, allowing the users to model the rainfall data for arbitrary distribution shapes. Details related to these statistical terms and distributions could be found in above-mentioned literature.

In ARAS, it was approached to estimate the distribution function F(u) of values of X that exceed a predefined rainfall threshold u. Thus, it is needed to compute the conditional probability that represents the probability of the rainfall value X exceeds the threshold u by at most the amount x given that condition X > u holds. This probability is given in Eq. 2 by Davison et al. (2012):

$$\Pr(X > x + u | X > u) = \frac{1 - F(x+u)}{F(u)} \tag{2}$$

Considering all these statistical approaches in ARAS, the user was first asked to determine the rainfall threshold and to select the return period of time. Then, the landslide hazard map could be produced based on the statistical approaches mentioned above and the selected thresholds and return periods, if the landslide susceptibility map was produced.

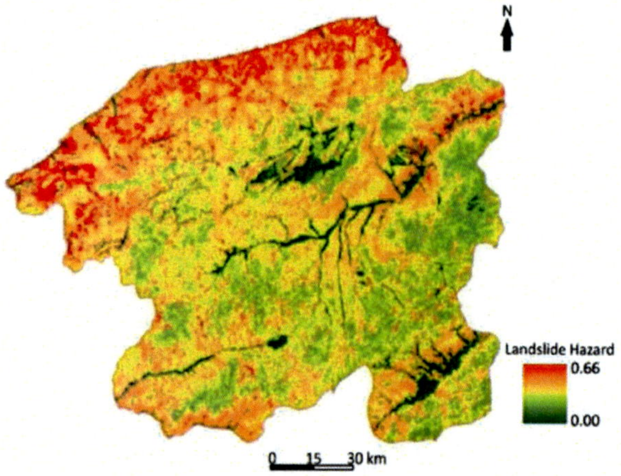


**Fig. 3** Landslide hazard map of Kastamonu city with 100 mm rainfall threshold and 25 year return period produced by ARAS

The analyses in ARAS can be performed by selecting 50 mm, 100 mm or 150 mm rainfall thresholds with 5, 10, 25, 50 and 100 years return periods, respectively. An example of such a landslide hazard map with 100 mm rainfall threshold and 25 year return period was represented in Fig. 3.

## Discussion and Conclusions

In this study, a web-based landslide susceptibility and hazard mapping system, namely ARAS, was represented with its general characteristics. In addition, an application of landslide susceptibility and hazard mapping procedure in ARAS was performed for this study. This system is mainly established for a governmental agency, namely AFAD, and its province directorates in Turkey. The main goal of ARAS is to produce risk maps for mass in overall country. However, risk assessment stage has not been reached yet, and mass movement inventories and their update process are still under construction. Nonetheless, the system works in its current state and users can perform their susceptibility and/or hazard analyses. The most powerful feature of the system is that it has a flexible structure and can operate in a short time without the need for a third party software. For example, selection of input parameters for landslide susceptibility assessment was left to the users. The main idea herein is that analyses are also allowed to be performed by local users (province directorates of AFAD users) instead of performing them from the central head of AFAD because they know the regional characteristics better. Another advantage of the utilization of the system is that the users can use their own inventories to produce a landslide susceptibility and/or hazard map with one of the statistical algorithms fed into the system or any triggering factor, respectively.

The system is planned to operate completely in a few years. The most important deficit of ARAS is the incompleteness of the inventories and the uncertainty in date of occurrences of landslides and the other mass movements. These issues will be attempted to be overcome as soon as possible during the update processes. It is also a useful platform where the produced maps can be digitally served to the other central or local public institutions. When ARAS comes fully operational, decision-makers, planners and local authorities will have the chance to make much healthier and more reliable plans for future works such as safer urbanization, regional development plans and sustainable risk management.

## References

Aleotti P, Chowdhury R (1999) Landslide hazard assessment: Summary review and new perspectives. Bull Eng Geol Env 58 (1):21–44

Beirlant J, Goegebeur Y, Segers J, Teugels JL (2004) Statistics of extremes: Theory and applications. John Wiley & Sons, England. (ISBN: 978-0-471-97647- 9) pp 514

Crozier M, Eyles R (1980) Assessing the probability of rapid mass movement. In: Proceedings of Third Australia-New Zealand conference on Geomechanics. Wellington, New Zealand, pp 2–47

Davison AC, Padoan SA, Ribatet M (2012) Statistical modeling of spatial extremes. Statistical Science. 27(2):161–186

De Haan L, Ferreira A, Ana F (2006) Extreme value theory: An introduction. Springer-Verlag, New York. (ISBN: 978-0-387-23946-0) pp 418

Embrechts P, Klüppelberg C, Mikosch T (2013) Modelling extremal events: For insurance and finance. Springer-Verlag, Berlin. (ISBN: 978-3-540-60931-5) pp 648

Fell R, Corominas J, Bonnard C, Cascini L, Leroi E, Savage WZ (2008a) Guidelines for landslide susceptibility, hazard and risk zoning for land-use planning. Eng Geol 102(3–4):85–98

Fell R, Corominas J, Bonnard C, Cascini L, Leroi E, Savage WZ (2008b) Guidelines for landslide susceptibility, hazard and risk zoning for land-use planning. Eng Geol 102(3–4):99–111

Glade T, Crozier M, Smith P (2000) Applying probability determination to refine landslide-triggering rainfall thresholds using an empirical antecedent daily rainfall model. Pure Appl Geophys 157:1059–1079

Guzzetti F, Carrara A, Cardinali M, Reichenbach P (1999) Landslide hazard evaluation: A review of current techniques and their application in a multi-scale study, Central Italy. Geomorphology 31:181–216

https://www.afad.gov.tr [Last accessed: 30.12.2019].

https://aras.afad.gov.tr [Last accessed: 30.12.2019].

https://aydes.afad.gov.tr [Last accessed: 30.12.2019].

https://pro.arcgis.com/en/pro-app/arcpy/get-started/what-is-arcpy-.htm [Last accessed: 30.12.2019].

https://pro.arcgis.com/en/pro-app/help/analysis/geoprocessing/share-analysis/what-is-a-geoprocessing-service.htm [Last accessed: 30.12.2019].

https://scikit-earn.org/stable/modulesgenerated/sklearn.neural_network.MLPClassifier.html [Last accessed: 30.12.2019].
https://scikit-learn.org/stable/index.html [Last accessed: 30.12.2019].
Langbein WB (1949) Annual floods and the partial-duration flood series. Trans Am Geophys Union 30(6):879–881
Pickands J (1975) Statistical inference using extreme order statistics. Ann Stat 3(1):119–131
Subbotin MT (1923) On the law of frequency of error. Mathematicheskii Sbornik 31(2):296–301
Varnes DJ, (1984) IAEG commission on landslide and other mass movement on slopes landslide hazard zonation: A review of principles and practice. UNESCO, Paris, pp 63
Zêzere JL, Trigo RM, Trigo IF (2005) Shallow and deep landslides induced by rainfall in the Lisbon region (Portugal): Assessment of relationships with the North Atlantic Oscillation. Nat Hazards Earth Syst Sci 5:331–344

2020 Kyoto Japan

# A Landform Evolution Model for the Mannen Area in Romsdal Valley, Norway

Paula Hilger, Reginald L. Hermanns, and Bernd Etzelmüller

**Abstract**

The Quaternary geology of western Norway's landscape is the result of glacial and post-glacial sedimentation and erosional processes, a significant sea-level drop and high rock-slope failure activity. All these processes are represented within a small valley section below the Mannen rock-slope instability in Romsdal valley, western Norway. Here, exposure ages, Quaternary geological mapping and geophysical investigations permit the development of a paraglacial landscape evolution model. The model contextualises at least six catastrophic rock-slope failure events within the overall sequence of fjord-valley infilling following deglaciation. A transition from a wide basin-like valley into a strongly confined valley section led to the build-up of more than 40 m thick stratified drift, which was at least partly deposited within a marine environment. The morphology of these sediments features two distinct erosional levels, which are interpreted to be connected to tidal currents during post-glacial sea-level drop. The landform evolution model illustrates the importance of catastrophic rock-slope failures and the impact of strong tidal currents on the typical sediment fill in narrow, high-relief fjord valleys.

**Keywords**

Fjord-valley fill • Catastrophic rock-slope failures • $^{10}$Be dating • Quaternary geology • Paraglacial landscape evolution

## Introduction

The unique landscape of Norway, with its deep fjords and valleys is the result of several glaciations, interglacial isostatic rebound and connected sedimentation and erosion processes. The oversteepened slopes along the valleys are characterised by a high mass-wasting activity, including rock falls, rock slides and rock avalanches. These processes add significantly to the paraglacial landscape evolution within the valleys. The importance of large rock-slope failures for landscape responses has been shown on examples in the Himalaya, the Tien-Shan, the European Alps and the Southern Alps in New Zealand (Korup et al. 2006; Korup and Schlunegger 2007; Larsen et al. 2010). However, well accepted evolution models of Norwegian fjord-valley fill, do often not account for the impact and evolutional effect of large catastrophic rock-slope failures (e.g. Corner 2006; Eilertsen et al. 2006).

This conference paper is an extension to a previously published study by Hilger et al. (2018). The aim is to present a landform evolution model of an area in a glacially eroded valley that has been subjected to both glacial and deglacial sedimentation processes, significant sea-level fall and colluvial and rock-slope failure activity within a time period of a few thousand years. We present an interpretation of previously mapped Quaternary stratigraphy and Holocene landforms, contextualising local rock-slope failure activity within the framework of a fjord-valley fill succession. This novel model may be relevant for many fjord-valley settings in Norway.

P. Hilger (✉)
Department for Environmental Sciences, Western Norway University of Applied Sciences, Sogndal, Norway
e-mail: paula.hilger@hvl.no

R. L. Hermanns
Geohazards and Earth observation, Geological Survey of Norway, Trondheim, Norway
e-mail: reginald.hermanns@ngu.no

R. L. Hermanns
Department of Geoscience and Petroleum, Norwegian University of Science and Technology, Trondheim, Norway

B. Etzelmüller
Department of Geosciences, University of Oslo, Oslo, Norway
e-mail: bernd.etzelmuller@geo.uio.no


V. Vilímek et al. (eds.), *Understanding and Reducing Landslide Disaster Risk*, ICL Contribution to Landslide Disaster Risk Reduction,
https://doi.org/10.1007/978-3-030-60319-9_34

## Setting

Romsdal valley is a glacially formed, sediment-filled palaeofjord. The valley cuts through the Western Gneiss Region and is characterised by a high rock-slope failure activity due to inherent critical weaknesses along the over-steepened rock slopes (Blikra et al. 2006; Saintot et al. 2012). In the lower reach of the valley the slopes are between 800 and 1400 m high and have an average gradient of >45° (Fig. 1).

The area of interest of this paper is a valley section below the rock-slope instability of Mannen, where the valley width decreases significantly. A basin-like low-gradient section is followed by a strongly confined cross section and a knick-point in the longitudinal profile (Figs. 1 and 2).

During late deglaciation (ca. 13–11.7 ka; Hughes et al. 2016; Stroeven et al. 2016; Hermanns et al. 2017) this part of the Romsdal valley was inundated by seawater, reaching a marine limit of 120 m above modern sea level (Høgaas et al. 2012). Sediments covering the valley floor are thus the product of processes connected to the Pleistocene glaciation, post-glacial sea-level fall and paraglacial slope processes.

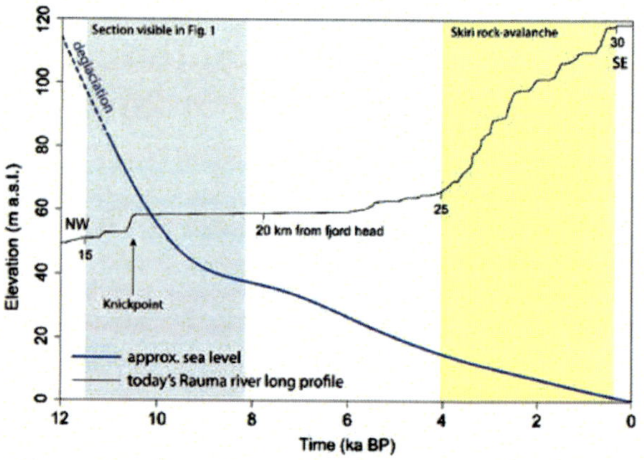


**Fig. 2** Sea-level curve approximated after Svendsen & Mangerud (1987) and long profile of today's Rauma river. Numbers along the river profile represent the distance to the fjord head in kilometres. The area of interest of this paper lies between km 15 and 19 and is marked in grey. In the upper reach, deposits of the Skiri rock avalanche form a dam and thus a significant knickpoint in the river profile (yellow section)

## Brief Summary of Methods

### Quaternary Geological Mapping

We combined GIS-based mapping and relief analyses using high-resolution digital elevation models (DEM) and aerial photographs with field work. This allowed us to map and interpret the stratigraphy in the context of the valleys' morphology. Quaternary geological mapping included geo-physical surveys using Electrical Resistivity Tomography (ERT) and Ground Penetrating Radar (GPR) in order to derive the stratigraphical sequences of defined sedimentary facies (Hilger et al. 2018).

### Geochronology

In order to establish apparent exposure ages of pre-historical rock-slope failures at Mannen, Hilger et al. (2018) analysed

**Fig. 1** Photograph of the study site taken in 2016 from the top of the Mannen rock-slope instability. The photo features the transition from a wide cross section to a strongly confined valley section. A knickpoint in the river's long profile, visible due to rapids, was created by a rock-slope failure event. Numbers 1–6 represent rock-slope failure deposits hidden under the forest

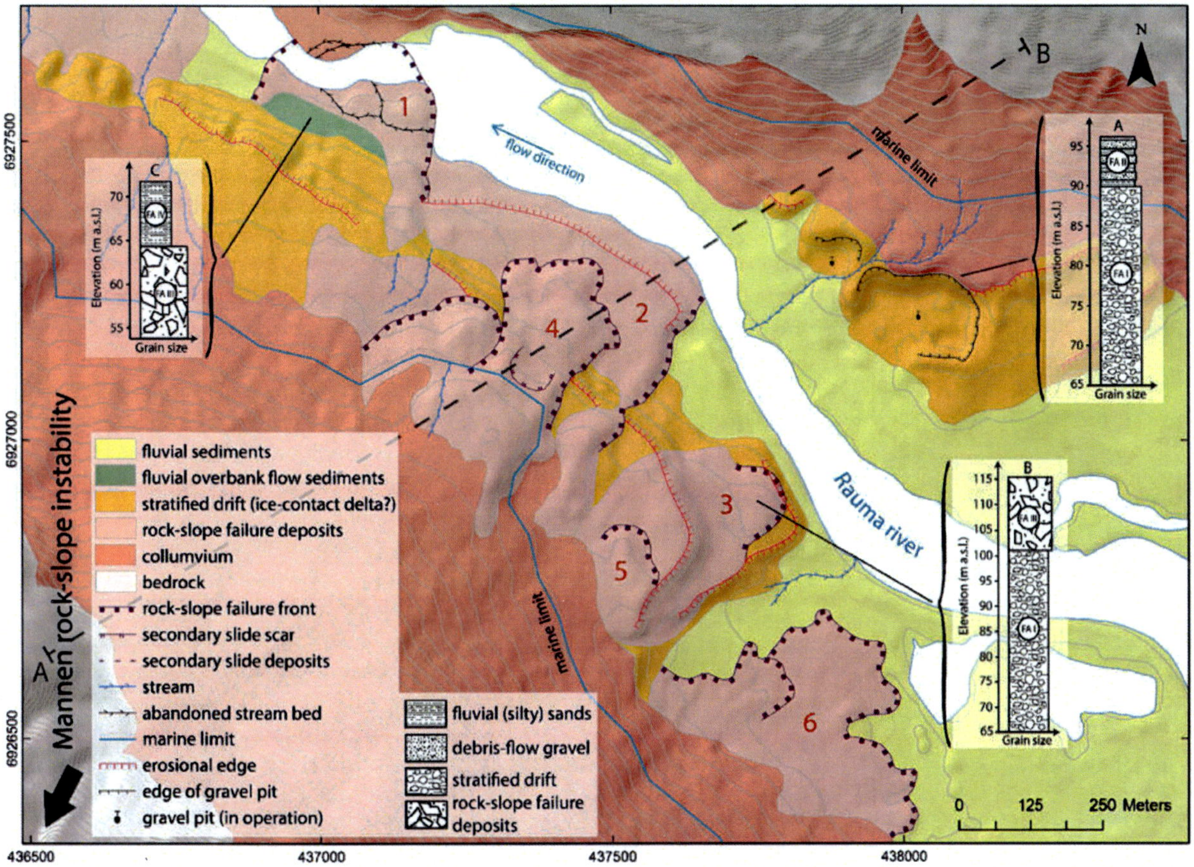


**Fig. 3** Quaternary geological map of the Romsdal valley section below the Mannen rock-slope instability. Deposits from catastrophic rock-slope failure events are numbered from 1 to 6 for easier referencing. The stratigraphy for three locations is included. The dashed gray line indicates the approximate location of profile A–B in Fig. 4. (Adapted from Hilger et al. 2018)

samples from 13 rockslide boulders for cosmogenic $^{10}$Be. The sampled boulders represent at least four catastrophic rock-slope failure events with two to five samples per deposit. Except for the deposits of rock avalanche N°2 (Figs. 1 and 3) the samples are well distributed (cf. Hilger et al. 2018).

Due to expected old ages and the likeliness of failure into water, samples for $^{10}$Be dating were neither collected from the lower part of event N°2 nor from the deposits of events N °1 and 3. The failure timing of these three events was obtained through relative correlation of the mapped stratigraphy and erosional features.

## Summary of Results

### Quaternary Geology

One of the main results of the mapping is a simplified Quaternary geological map of the study site (Fig. 3). Hilger et al. (2018) present three composite sediment profiles from the local stratigraphy featuring the following sedimentary facies: stratified drift (FA I; sand and gravel with blocks and boulders), distal fluvial/debris flow deposits from the slopes (FA II; mostly sand), catastrophic rock-slope failure deposits (FA III; gravel with chaotic boulder fields on top) and a small section of overbank fluvial deposits (FA IV; silt and fine sand). The central part of the valley, closest to the Rauma river, is characterised by recent fluvial sediments.

The superficial deposits are dominated by large talus slopes and several deposits from catastrophic rock-slope failures reaching beyond the foot of the colluvium.

The sands and gravels of the stratified drift dips c. 5° into downstream direction and form two rather flat (<5°) elevated surfaces c. 40 and 20 m above today's river level, respectively. The surfaces are 10–20 m higher than any glacio-fluvial or fluvial terrace mapped along Romsdal valley (Hilger et al. 2018). These surfaces relate to two well-defined topographical steps with a distinct transition from 35° steep erosional slopes to the flat surfaces (Fig. 3).

There is an almost abrupt transition from a cross-valley direction into a valley parallel direction of these topographical steps (Fig. 3).

## Age Constraints

$^{10}$Be-dating results suggest that ~5000 years ago three to six catastrophic rock-slope failures took place within a few hundred years (N°4–6, Table 1, Hilger et al. 2018). Based on the error weighted mean of two samples, an additional event (N°2, Fig. 3) has an apparent exposure age of 9.39 ± 0.64 ka. The geomorphological appearance of the deposit however suggests that the real failure timing lies within the upper range of this result, or beyond. The lower part of this deposit shows clear erosional signs up to 20 m above todays river level, which indicates failure into a higher water level and thus earlier than 10 ka (Fig. 2). This is also true for deposits N°1 and 3, where the erosional edge of the stratified drift runs very distinctly through the deposits (Fig. 3).

## A Landform Evolution Model

Our Quaternary geological and geomorphological observations invite to reconstruct the landscape development in such a typical but also specific fjord-valley setting. The Quaternary geology in this type of valleys is dominated by glacial and marine sediments modified by a high mass-wasting activity and fluvial erosion and deposition.

The several tens of meter thick stratified gravels and sands below Mannen are considered to be older than the failure timing of a large rock-avalanche in the upper reach of Romsdal valley (10.5–11.7 ka; Hermanns et al. 2017) and underlie deposit N° 3 (Fig. 3). The rock masses of the Skiri rock avalanche (Fig. 2) dammed the river and must have retained at least large parts of the later, coarse glacio-fluvial sediments. The dam was never eroded and still represents a barrier in the Rauma valley (Hermanns et al. 2013). Based on the failure timing of the Skiri rock-avalanche, the stratified drift was deposited in a marine environment (Fig. 4A). Deposition processes could have been similar to those of a submarine fan or ice-contact underwater fan (Lønne 1995; Winsemann et al. 2009). This is supported by the abrupt appearance of the glacial drift within the basin-like valley opening, with a steep slope running orthogonally to the valley (Fig. 3).

The confined valley section on the downstream side probably led to a quick build-up of 20–30 m thick layers of glacio-marine till with a sharp ice-contact line valley upwards. While fine sediments were evacuated, coarser till deposits persisted behind the chokepoint. An observed sharp transition from glacio-marine till to fine-grained sediments on top could be an indicator for the timing of the Skiri rock-avalanche, retaining the coarse grain size fraction (Fig. 4B). The top surface of these sediments reaches an elevation of 90 m a.s.l. in the north-west to 100 m a.s.l. in the south-east. These elevations coincide with the sea level between 11 and 12 ka, when wave and tidal activity must have produced flat erosional surfaces and steep longitudinal erosional edges (Fig. 4B). The ~5° inclination of the upper elevated surface indicates the formation at an early stage of isostatic rebound, when the sea level was still high. It is likely that velocities of the currents, connected to the tide reversals were increased by the strong convergence in valley width in this section (Ross et al. 2017) and led to incision and strong lateral erosion in the central part.

The deposits of three early (12–10 ka) catastrophic rock-slope failures reached, at least partly, into the fjord environment (Fig. 4C). Erosional modification of the deposits can be connected to the continued strong tidal activity and valley-parallel currents during the ongoing rapid sea-level drop. The height of the erosional edges suggests that tidal currents have been effective from deglaciation to about 10 ka when the sea level dropped to the recent river level (Figs. 3 and 4D).

**Table 1** Age constraints for a minimum of six catastrophic rock-slope failure events from Mannen

| CRSF *N°* | Approximate failure timing *ka* ($\pm 1\sigma$) | Method |
|---|---|---|
| 1 | 10.5–12 | Stratigraphy |
| 2 | 9–12 | $^{10}$Be (2 samples) + stratigraphy |
| 3 | 10–12 | Stratigraphy |
| 4 | 4.93 ± 0.31 | $^{10}$Be (5 samples) |
| 5 | 4.96 ± 0.33 | $^{10}$Be (2 samples) |
| 6 | 4.95 ± 0.31 | $^{10}$Be (3 samples) |

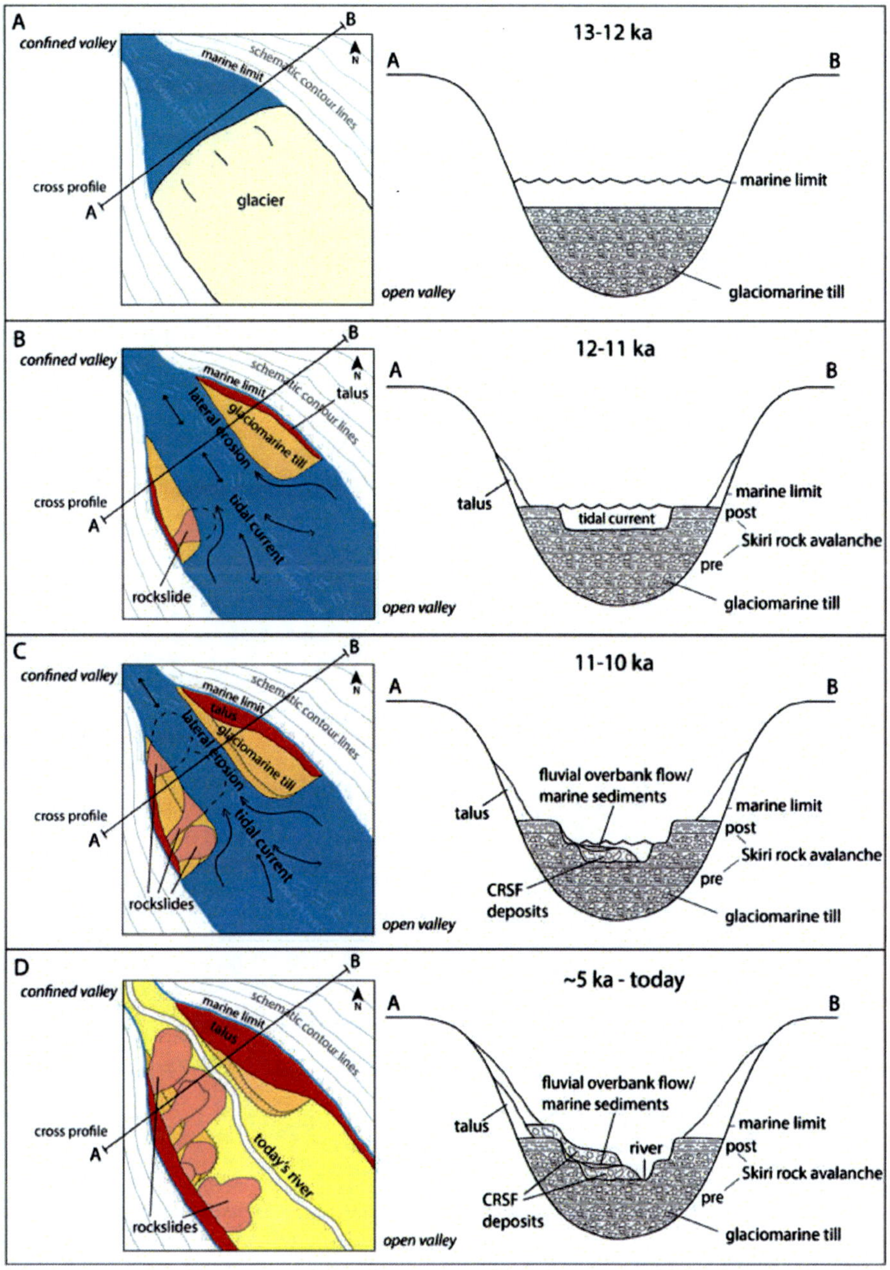


**Fig. 4** Schematic illustration of a Holocene landscape evolution model for the section of the Romsdal valley below Mannen. The section is characterised by an increasing valley confinement which enhanced currents during tidal reversals while the sea level was dropping. The cross-section A-B indicates the sedimentology with depth for each time step. Catastrophic rock-slope failures are abbreviated with CRSF

## Discussion and Conclusion

The presented model is plausible and demonstrates the erosional force tidal currents can have in such settings. Interesting within this context are the two distinct surface levels, which may be connected to changes in the velocity of isostatic rebound, and thus the elevation of the shoreline.

The valley section below the Mannen instability in Romsdal valley stands out from the typical fjord-valley stratigraphy and morphology (cf. Corner 2006). This is caused by a short transition from a wide u-shaped cross section to a confined valley section and a high activity of catastrophic rock-slope failures. On one hand, the valley constriction acted as chokepoint and led to a quick build-up of thick stratified drift, likely deposited as glaciomarine contact-fan or -delta. On the other hand, it intensified tidal currents during post-glacial sea-level drop (cf. Ross et al. 2017), increasing lateral erosion and forming two 20 m high steps above today's riverbanks.

Because of the particular topography in an otherwise typical fjord-valley setting, we expect similar features in several valleys with alike topographical chokepoints, sustaining our model.

This study shows the significance of catastrophic rock-slope failures on landscape development in valleys. The Skiri rock avalanche has dammed the Rauma river and thus retained coarse glaciofluvial sediments. More than 10,000 years later, the deposits still form a significant knickpoint in the valley's long profile (Fig. 2). The effect of the landslide dam is thus overruling forcing factors for river adjustment (e.g. Korup et al. 2006; Ouimet et al. 2007).

Another minor but distinct knickpoint was formed at a similar time in the area of interest of this study. Here, catastrophic rock-slope failures, reaching far beyond large talus slopes along the valley sides, are significant components of the paraglacial landscape development and the Quaternary geology of this valley section. The deposits of three large rock-slope failures are affected by the early tidal erosion, and a failure timing before 10 ka is suggested by stratigraphic relations. Three distinct deposits, which represent up to six rock-slope failure events are dated to 5 ka. The deposits are well preserved and will most likely survive and dominate the landscape for thousands of years to come.

**Acknowledgements** This study is part of the project 'CryoWALL - Permafrost slopes in Norway' (243784/CLE) funded by the Research Council of Norway (RCN).We thank our project partners and co-authors from Hilger et al. (2018) Benjamin Jacobs, Michael Krautblatter and John C. Gosse for their cooperation and support. Sample processing for $^{10}$Be dating was conducted by the first author at Cosmic Ray Isotope Sciences at Dalhousie University (CRISDal) with great help of John C. Gosse and G. Yang.

## References

Blikra LH, Longva O, Braathen A, Anda E, Dehls JF, Stalsberg K (2006) Rock slope failures in Norwegian fjord areas: Examples, spatial distribution and temporal pattern. In: Evans SG, Mugnozza GS, Strom A, Hermanns RL (eds) Landslides from Massive Rock Slope Failure (NATO Science Series, vol. 49). Springer, Dodrecht. pp 475–496

Corner GD (2006) A transgressive-regressive model of Fjord-Valley fill: Stratigraphy, facies and depositional controls. In: Dalrymple RW, Leckie DA and Tillman RW (eds) Incised Valleys in Time and Space. SEPM Society for Sedimentary Geology (Special Publication no. 85, ISBN-10: 1565761227) pp 161–178

Eilertsen RS, Corner GD, Aasheim O, Andreassen K, Kristofferson Y, Ystborg H (2006) Valley-fill stratigraphy and evolution of the Malselv Fjord Valley, Northern Norway. In: Dalrymple RW, Leckie DA and Tillman RW (eds) Incised Valleys in Time and Space. SEPM Society for Sedimentary Geology (Special Publication no. 85, ISBN-10: 1565761227) pp 179–195

Hermanns RL, Dahle H, Bjerke PL, Crosta GB, Anda E, Blikra LH, Saintot A, Longva O (2013) Rockslide Dams in Møre og Romsdal County, Norway. In: Margottini C, Canuti P, Sassa K (eds) Landslide Science and Practice - Volume 6: Risk Assessment, Management and Mitigation. Springer, Berlin, Heidelberg. (ISBN: 978–3–642–31319–6) pp 3–12

Hermanns RL, Schleier M, Bohme M, Blikra LH, Gosse JC, Ivy-Ochs S, Hilger P (2017) Rock-avalanche activity in W and S Norway peaks after the retreat of the Scandinavian Ice Sheet. In: Mikoš M, Vilimek V, Yin Y, Sassa K (eds) Advancing Culture of Living with Landslides. Springer, Cham, pp 331–338

Hilger P, Hermanns RL, Gosse JC, Jacobs B, Etzelmüller B, Krautblatter M (2018) Multiple rock-slope failures at Mannen in Romsdal Valley, western Norway, revealed from Quaternary geological mapping and $^{10}$Be exposure dating. The Holocene 28 (12):1841–1854

Hughes ALC, Gyllencreutz R, Lohne OS, Mangerud J, Svendsen JI (2016) The last Eurasian ice sheets - A chronological database and time-slice reconstruction, DATED-1. Boreas 45:1–45

Høgaas F, Hansen L, Rinstad BI, Sveian H, Olsen L (2012). Database for registrering av marin grense (MG) i Norge. NGU-report 2012.063, Geological Survey of Norway, Trondheim. (ISSN: 0800–3416)

Korup O, Strom AL, Weidinger JT (2006) Fluvial response to large rock-slope failures: Examples from the Himalayas, The Tien Shan, and the Southern Alps in New Zealand. Geomorphol 78(1–2):3–21

Korup O, Schlunegger F (2007) Bedrock landsliding, river incision, and transience of geomorphic hillslope-channel coupling: Evidence from inner gorges in the Swiss Alps. J Geophys Res: Earth Surf 112 (F3)

Larsen IJ, Montgommery DR, Korup O (2010) Landslide erosion controlled by hillslope material. Nat Geosci 3(4):247–251

Lønne I (1995) Sedimentary facies and depositional architecture of ice-contact glaciomarine systems. Sed Geol 98(1–4):13–43

Ouimet WB, Whipple KX, Royden LH, Sun Z, Chen Z (2007) The influence of large landslides on river incision in a transient landscape: Eastern margin of the Tibetan Plateau (Sichuan, China). Geol Soc Am Bull 119(11–12):1462–1467

Ross L, de Swart H, Ensing E, Valle-Levinson A (2017) Three-dimensional tidal flow in a fjord-like basin with converging width: An analytical model. J Geophys Res: Oceans 122(9):7558–7576

Saintot A, Dahle H, Derron M-H, Henderson I, Oppikofer T (2012) Large gravitational rock slope deformation in Romsdalen valley (Western Norway). Revista De La Asociación Geológica Argentina 69:354–371

Stroeven AP, Hättestrand C, Kleman J, Heyman J, Fabel D, Fredin O, Goodfellow BW, Harbor JM, Jansen JD, Olsen L, Caffee MW, Fink D, Lundqvist J, Rosqvist GC, Strömberg B, Jansson KN (2016) Deglaciation of Fennoscandia. Quatern Sci Rev 147:91–121

Svendsen JI, Mangerud J (1987) Late Weichselian and Holocenesea-level history for a cross-section of western Norway. J Quat Sci 2:113–132

Winsemann J, Hornung JJ, Meinsen J, Asprion U, Polom U, Brandes C, Bußmann M, Weber C (2009) Anatomy of a subaqueous ice-contact fan and delta complex, Middle Pleistocene, North-West Germany. Sedimentology 56(4):1041–1076

# Multimethodological Study of Non-linear Strain Effects Induced by Thermal Stresses on Jointed Rock Masses

Guglielmo Grechi and Salvatore Martino

**Abstract**

A multimethodological method based on environmental, stress–strain, microseismic, and ambient seismic noise monitoring is here presented, with a view to identifying non-linearity of thermally-induced deformation of jointed rock masses at different dimensional scales. Rock masses experience non-negligible deformation cycles due to the continuous fluctuations of their surficial temperatures. However, the interpretation of such strain effects, in terms of the ratio between elastic and inelastic percentages, is still debated. In particular, the relation between microseismic emissions, considered as primary indicators of crack-growth related energy release, and resonant frequencies fluctuations of rock structures, witnesses of the thermally-induced effect at the macro- or structure-scale, have not been yet studied within a coupled framework. The combination of different approaches able to investigate the behavior of rock masses from micro- to macro-scale, then from fracture-scale to joint-isolated rock blocks up to rock structures, could provide new insights and perspectives on the effects related to shallow thermal stresses fluctuations. This paper presents the preliminary outcomes from two case studies, the Acuto experimental test-site (Italy) and the Wied Il-Mielaħ sea arch (Malta), where multiparametric monitoring surveys were conducted and are still ongoing, aiming at the assessment of the cause-to-effect relation between near-surface thermal stresses and induced strains. Data analysis was carried out following different approaches, with a particular emphasis on the Acuto test-site dataset recorded so far, allowing to establish a well-constrained correlation among temperature fluctuations and rock mass deformation both at the daily and seasonal scale.

**Keywords**

Multiparametric monitoring • Thermal stress • Rock masses

## Introduction

In the last few decades, the scientific interest has increased with regards to the analysis of thermally-induced strain and progressive mechanical weathering of rock masses caused by daily and seasonal near-surface temperature fluctuations. In fact, the study of the deformative response to cyclical thermal stresses of rock masses is considered crucial in the field of geological risk mitigation relatively to those instabilities that, due to their impulsiveness and high frequency of occurrence, are able to configure high hazard slope instability scenarios. Under specific climatic conditions, the effects related to the superposition of heating–cooling cycles, negligible if considered in short- to middle-term, can influence rock mass mechanical behavior acting as a thermal fatigue process (Hall 1999; Gunzburger et al. 2005): the intensity of thermomechanical effects, cumulated over time, cannot be then omitted from all those factors actively contributing to rock mass progressive damaging and mechanical weathering (Collins and Stock 2016; Greif et al. 2016). Temperature fluctuations are able to exert slight perturbations of stress fields, resulting in a day-to-day cumulative effect, which contributes over wide-time-scales to lead rock masses toward conditions of instability. As a direct consequence of the superposition of thermal expansion–contraction cycles, the rock mass stress field undergoes such perturbations capable of inducing both the genesis of new cracks and the growth of preexisting ones, i.e. *subcritical crack-growth* (Eppes et al. 2016). These processes can

G. Grechi (✉) · S. Martino
Department of Earth Sciences, University of Rome La Sapienza, Rome, Italy
e-mail: guglielmo.grechi@uniroma1.it

S. Martino
e-mail: salvatore.martino@uniroma1.it



V. Vilímek et al. (eds.), *Understanding and Reducing Landslide Disaster Risk*, ICL Contribution to Landslide Disaster Risk Reduction,
https://doi.org/10.1007/978-3-030-60319-9_35

induce inelastic deformations consisting of triggering thresholds of shallow slope instabilities, e.g. rock-falls and rock-topples (Bakun-Mazor et al. 2013; Collins and Stock 2016). The study of thermomechanical effects on jointed rock masses has been primarily approached through comparative analyses (statistical and observational-based approaches) between thermal and stress–strain time-series derived by long-term multiparametric monitoring systems (Fiorucci et al. 2018; Alcaino-olivares and Perras 2019). Therefore, it was possible to highlight a strong correlation between near-surface temperature fluctuations and rock masses deformative response, as well as the existence of inelastic deformation trends even in the absence of brief and violent phenomena such as heavy rainfalls, strong winds or local seismicity, commonly considered as main triggering factors of unrecovered strain (Gunzburger et al. 2005). Over recent years the exploitation of high-sensitivity seismic sensors on rock masses affected by gravitational instabilities has well constrained the existence of a cause-to-effect relation between temperature trends and microseismic events, proving that the genesis or the progressive growth of pre-existing cracks, potentially leading to instabilities, is followed by a sudden release of energy propagating in the form of seismic waves (Coviello et al. 2015; Arosio et al. 2018).

Similarly, through the acquisition of ambient seismic noise, it was observed the existence of cyclic thermally-driven shifts of fundamental resonant frequencies of rock slopes (Burjánek et al. 2012; Bottelin et al. 2013; Colombero et al. 2018) and rock structures (Starr et al. 2015). The wandering of natural frequencies, which was proven to be strongly correlated with an in-phase relation with yearly and daily temperature fluctuations, is attributed to the influence of temperature fields on both fractures and rock matrix (Bottelin et al. 2013; Starr et al. 2015; Colombero et al. 2018). The continuous superposition of stiffening and relaxation stages can potentially cause a loss in the contact stiffness along fracture and microfracture patterns, causing the irreversible detachment of sub-volumes or compartments from the stable rock mass, which induces a significant drop in resonant frequencies values (Lévy et al. 2011; Colombero et al. 2018). The comprehension of the thermomechanical behavior of rock masses, where the role of joints is not negligible, requires the implementation of a multimethodological approach in order to identify micro- and macro-scale thermally-driven effects, such as subcritical crack-growth and resonant frequencies wandering. The integration of widespread geotechnical and geophysical monitoring techniques could then provide a significant enhancement in the recognition of non-linear strain in intensely stressed systems such as jointed rock masses.

## Case Studies

### Acuto Test-Site

The Acuto test-site (Central Italy) is located within the western sector of the Mt. Simbruini-Ernici ridge, inside an abandoned quarry where a succession of intensely tectonized Mesozoic limestones outcrops (Accordi et al. 1986). After a preliminary stage of geomechanical surveying, which allowed the reconstruction of an engineering-geological model of the quarry wall and led to the identification of five main joint sets, a 20 $m^3$ rock block was identified as the optimal target of the multisensor monitoring system (Fig. 1a, d).

The investigated rock block is 6 m high, whereas the height of the quarry wall ranges between 15 and 50 m, and is characterized by a protruded geometry with respect to the quarry yard in virtue of an open joint separating it from the rock mass behind. The multisensor monitoring system is operating since April 2016 and it was specifically designed in the framework of multiple experimental activities managed by the Research Centre for Geological Risks (CERI) of La Sapienza University of Rome, with the aim of understanding how natural and anthropogenic stressors can act as conditioning factors jointed rock mass long-term stability.

### Wied Il-Mielaħ Sea Arch

The Wied Il-Mielaħ sea arch is located in the north-western sector of the Island of Gozo (Maltese archipelago) and, since the collapse of the famous Azure Window sea arch in March 2017, it has become one of the major touristic attraction of the island. The arch is approximately 25 m high with a span of 15 m and thickness of 5 m, 0°-10° N orientated, and is anchored to the adjacent vertical cliff (Fig. 1e, f). The structure is entirely composed of Oligo-Miocenic limestones ascribable to the Lower Coralline Limestone Fm. (Pedley 2011). Moreover, due to the presence of several open joints, the arch results to be partially dislodged into multiple compartments. Geomechanical surveys were carried out in order to achieve a preliminary engineering-geological model of the rock arch and four main joint sets were identified. Due to the impossibility of predisposing a permanent multiparametric monitoring system, experimental activities at this case study consisted of a preliminary characterization of the thermal and vibrational behavior of this structure, through respectively IRT surveys and ambient seismic noise measurements.

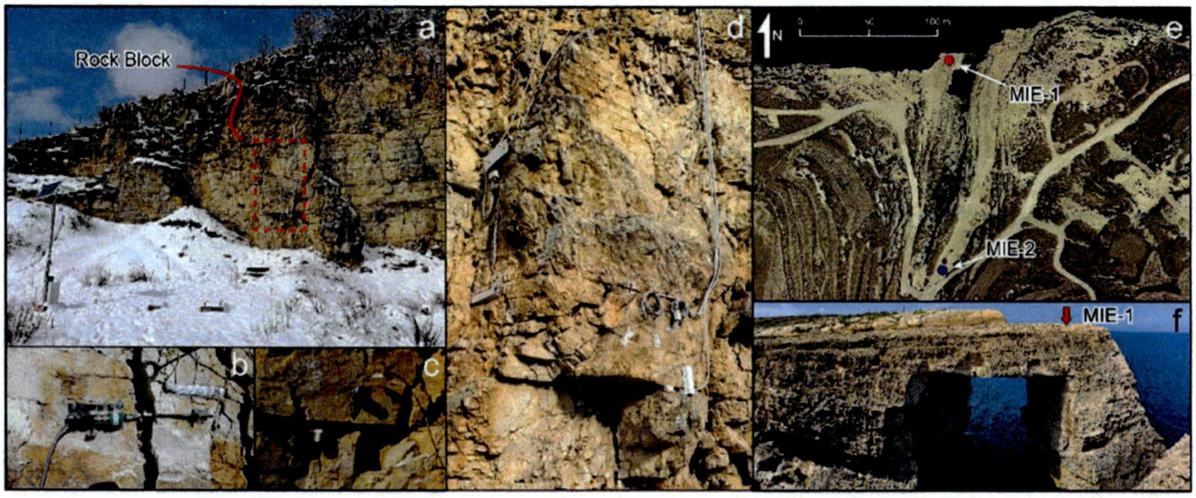


**Fig. 1** View of the Acuto test-site **a** and of the monitored rock block **d**; Example of one extensometer installed in correspondence of an open joint **b** and of two high-sensitivity accelerometers installed at the Acuto test-site **c**; Aerial **e** and panoramic view **f** of the Wied Il-Mielaħ sea arch, where the location of seismic noise stations is reported (MIE-1, MIE-2)

## Methods

The Acuto test-site monitoring system was firstly designed with geotechnical devices in order to provide advancement in the comprehension of the cause-to-effect relations between environmental factors and jointed rock masses response. The rock block was instrumented with different typologies of devices, including contact sensors and remote sensing equipment. The monitoring system includes (i) two weather stations, located at the top and the bottom of the quarry, equipped with air thermometer, hygrometer, rain gage, and anemometer; (ii) one Type-K thermocouple for rock mass temperature monitoring; (iii) six strain gauges and four extensometers (Fig. 1b), respectively installed in correspondence of microfractures and open joints. All the exploited devices are cable-connected to a CR100 Campbell Scientific datalogger with 1-min sampling interval. In order to investigate the spatial distribution and the temporal evolution of surficial temperatures of both the rock block in the Acuto quarry and the Wied Il-Mielaħ sea arch, pondering the effect of fracture networks and surface irregularities in the heat propagation process, a seasonal recurrence InfraRed Thermography (IRT) monitoring is ongoing. IRT surveys consist in the acquisition of radiometric images with a sampling interval of 1 h, allowing to reconstruct the evolution of surficial temperatures through days and different seasons. The image collection is carried out by exploiting a Testo® 885–1 thermal camera. Starting from November 2018, the multisensor monitoring system was implemented with a high-sensitivity microseismic array composed of 7 BKSV® type-8344 monoaxial piezoelectric accelerometers (frequency range 0.2–3000 Hz). The microseismic array is cable-connected to a modular digital acquisition system consisting of an HBM® Somat CX23-R data recorder and a Somat XR MX1601 B-R amplifier set to a sampling frequency of 2400 Hz. Microseismic sensors (Fig. 1c) were directly installed on the rock mass surface using predisposed steel studs, and a double-triaxial configuration (Up-Down, North–South, East–West) was designed in order to investigate both the rock block and the rock mass dynamic response to environmental forcings. The main objective of this microseismic monitoring has to be seen in relation to the integration of geophysical and geotechnical approaches, which could lead to a deeper understanding of the role played by near-surface temperature fluctuations in the dynamic of the progressive failure process involving jointed rock masses.

Seismic noise measurements were performed at the Wied Il-Mielaħ sea arch to preliminarily characterize the vibrational behavior of this natural structure and to retrieve key-information for the future design of a microseismic monitoring system. Ambient seismic noise was recorded for 3 days using two seismometers in a site-to-reference configuration: one placed on the outermost sector of the arch (MIE-1), in correspondence of the freestanding pillar, and the other on horizontal bedrock 150 m away as reference (MIE-2) (Fig. 1d, e). Two SARA® three-component portable seismometers were used with a sampling frequency set to 200 Hz. Each 1-h record was divided into 16 windows of 409.16-s with 50% overlap and 10% cosine taper, then power spectral densities (PSD) were computed for every segment by FFT (Fast Fourier Transform) and averaged over all windows in order to reduce the variance of the outcome.

The resulting PSDs were smoothed and used to compute spectral ratios between the two sites, aiming at the preliminary identification of resonance frequencies of the sea arch and at investigating their relation with temperature fluctuations.

## Results

### Multiparametric Monitoring

In order to detect seasonal and daily fluctuations in the rock mass mechanical behavior, eventually highlighting the existence of irreversible trends of deformation, the whole multiparametric monitoring time-series was analyzed. Starting from the earliest stages of the Acuto test-site, thermal and strain monitoring time-series allowed to constrain a strong correlation between near-surface temperature fluctuations (both air and rock mass temperatures) and strain cycles of fractures and microfractures. Thermally-induced deformations are characterized by a cyclical behavior that appears in phase with both daily and seasonal thermal variations (Fig. 2). The resulting intensity of strain cycles, in terms of contraction and expansion magnitude, is directly proportional to thermal ranges, evidencing how temperature can actively control rock mass strain through the continuous superposition of slight, but non-negligible, stresses. From the observational and statistical analysis of strain time-series, no irreversible trends of deformation were recognized during the monitoring period. This result suggests either the need for longer time-windows of analysis (years or decades) in order to appreciate the insurgence of inelastic effects, and the elastic nature of thermally-induced deformations in the short-term. Concerning the analysis of temperature-related effects, IRT allowed to evaluate the role of different factors affecting the heat-exchange between jointed rock masses and the environment, such as differential exposure of surfaces to direct solar radiation and the influence of open joints on the heat propagation in depth. In particular, through the execution of multitemporal surveys, it was observed that main open joints could markedly influence both the spatial distribution and the evolution of temperatures, in virtue of the air mass circulation that can be hosted within fracture networks (Fig. 3). As a consequence of that, high contrasts of temperature are induced in jointed rock masses, forcing differently stressed sub-volumes to experience non-homogeneous strain magnitudes. Then, when dealing with rock masses, open joints can significantly influence the thermal behavior of these systems, especially in case of high temperature ranges at the daily and seasonal scale. From the analysis of microseismic monitoring data, it was possible to highlight the existence of inelastic effects attributable to a process of progressive deformation involving the investigated rock block. In particular, the rate of microseismic events can be increased by the occurrence of rainfall events and high thermal stresses within the thermal active layer. The plot of Fig. 4 shows exemplary multiparametric monitoring time-series comprehensive of microseismic events, and where the yellow box highlights a 4-day monitoring period when a sharp temperature drop and rainfalls were registered. As it can be observed, the occurrence of intense rainfalls and a net decrease of temperatures below 0 °C, thus determining freezing conditions, induced a substantial increase in microseismic emissions. The microseismic activity related to this 4-day monitoring period is characterized by a marked increase in both the total number and the associated peak ground acceleration (PGA) of the detected events. Minor concentrations of less energetic microseismic events were recorded in periods when no rainfalls or freezing conditions occurred, testifying how the combination of these factors can represent a thread for rock mass stability. Nevertheless, in order to strengthen the cause-to-effect relation between temperature fluctuations and non-linear strain effects, a statistical analysis has to be performed, taking into account microseismic events clustering during heating and cooling phases of the rock mass, both in presence and absence of other external disturbing factors.

### Ambient Seismic Noise at Wied Il-Mielaħ Sea Arch

The computation of PSD for each recording station provided a preliminary comparison between spectral content of seismic noise in correspondence of the arch (MIE-1) and on horizontal bedrock considered as reference (MIE-2). In the investigated frequency range (0.5–20 Hz) different sharp spectral peaks were observed, in particular between 4 and 15 Hz. The identified peaks are visible in all spectra of MIE-1, and are interpreted as resonance frequencies of the Wied Il-Mielaħ arch (Fig. 4a–c).

To strengthen this evidence, the same spectral peaks emerge from the computation of site-to-reference spectral ratios (Fig. 4d–f): for all three components, the observed frequencies show high values of the spectral ratio, testifying a significant amplification of seismic noise realistically related to a structural resonance. Hourly averaged PSDs of the EW component of MIE-1 are represented in three frequency bands, corresponding to the observed spectral peaks within 20 Hz (Fig. 4g–l). Despite the discontinuity of the recorded data, a clear wandering of resonance frequencies can be observed and appears to be directly correlated with rock temperature variations. The existence of an in-phase relation between temperature and resonance frequencies has already been observed according to other authors (Starr et al. 2015; Bottelin et al. 2017) and interpreted as the result of the thermal stiffening of the rock mass, mainly deriving from the

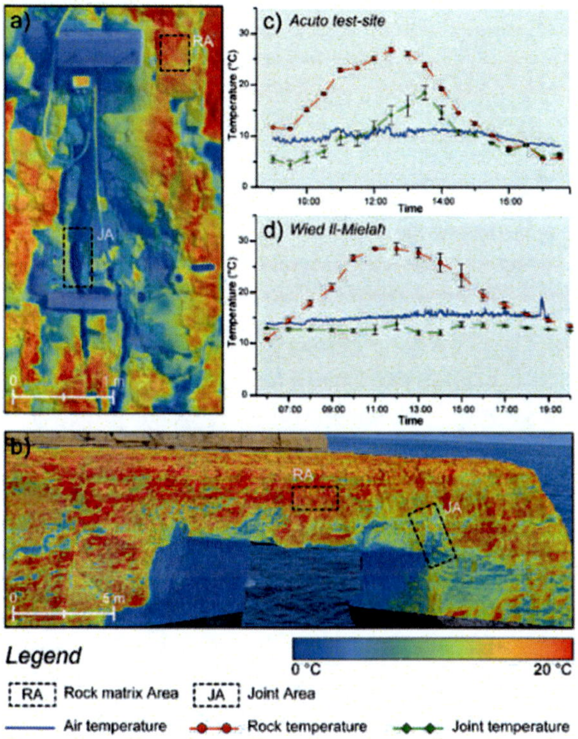


**Fig. 2** Radiometric images of thermal ranges recorded during 1-day monitoring survey at the Acuto test-site **a** and Wied Il-Mielaħ arch **b**. The two plots **c**, **d** highlight the relation between the evolution of rock, joints and air temperature. Rock and joint temperatures were extracted and averaged over the pixels contained within RA and JA respectively

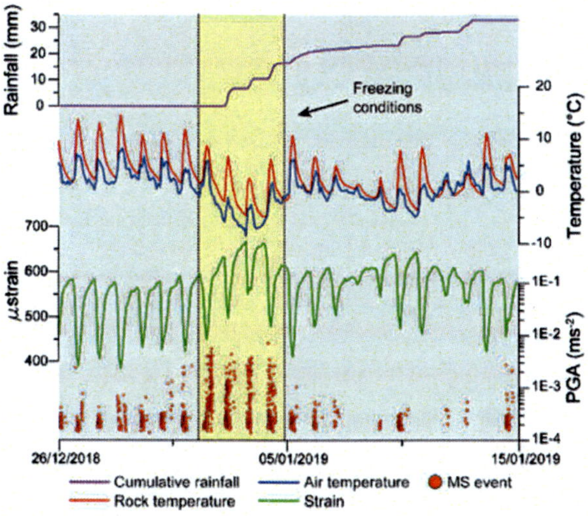


**Fig. 3** Plot of multiparametric time-series recorded by the Acuto test-site monitoring system during the coldest period of Winter 2019

cyclic thermally-driven opening-closing dynamics of joints. These evidences are relevant for inferring how thermal stresses could play a non-negligible role in creating structural damage, as witnessed by irreversible frequency shifts. Seismic noise measurements at the Wied Il-Mielaħ arch must be intended as a preliminary attempt to achieve useful information on the vibrational behavior of this natural structure, in the perspective of a future implementation of a coupled microseismic and seismic noise monitoring system aimed at investigating non-linear deformation induced by near-surface thermal fluctuations both at micro- and structure-scale.

## Conclusions

The analysis of the thermomechanical behavior of jointed rock masses represents a major objective in the perspective of understanding how near-surface temperature fluctuations can drive these natural systems toward prone-to-failure conditions. In fact, rock masses continuously experience temperature variations whose influence, if considered over both short and wide time-scales, is able to induce significant stresses. The study of thermally-induced effects has to focus on the evaluation and identification of those natural thresholds able to highlight inelastic deformation trends attributable to a phenomenon of rock mass progressive damaging. The here presented results from the multiparametric monitoring system in the Acuto test-site allowed to directly observe cause-to-effect relations between thermal stresses and strain response of a fractured rock block.

In particular, through microseismic monitoring it was possible to observe the occurrence of microseismic events, interpreted as indicators of inelastic deformation at fracture-scale. This research is still in progress, and further activities have already been planned. Concerning the Wied Il-Mielaħ sea arch, based on the results obtained from ambient seismic noise, which preliminarily highlighted not only the existence of several vibrational modes of this structure, but also their in-phase fluctuation with temperature variations, the installation of a discontinuous monitoring system coupling microseismic and seismic noise monitoring is planned. In this framework, the possibility to couple different techniques able to provide information on rock mass mechanical damaging at different dimensional scales, such ambient seismic noise and microseismic monitoring, could represent a useful tool to further investigate subsisting relations between thermal fluctuations and stress–strain response of jointed rock masses.

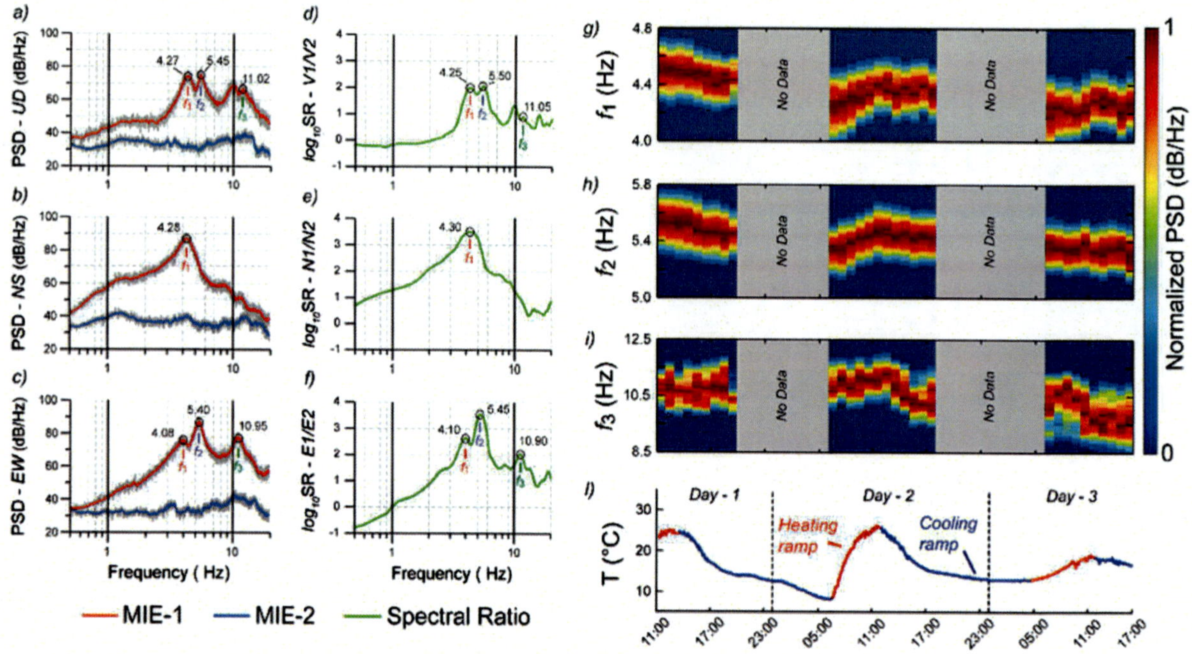


**Fig. 4** Comparison between PSD of active (MIE-1) and reference sensor (MIE-2) **a-c**. Spectral ratios for all three components **d–f**. Hourly averaged normalized PSD in three frequency bands centered on resonance frequencies **g–i**. Rock temperature measured on the arch (l)

**Acknowledgements** This research is part of the Ph.D. research of Guglielmo Grechi. Activities were carried out in the framework of the project "Rock failures in cliff slopes: from back- to forward-analysis of processes through monitoring and multi-modelling approaches" (University of Rome La Sapienza—Year 2016, P.I. Prof. Salvatore Martino).

## References

Accordi G, Carbone F, Civitelli G et al (1986) Lithofacies map of Latium- Abruzzi and neighbouring areas. Quaderno C.N.R. "La Ricerca Scientifica", Roma, 114(5), pp 223

Alcaino-olivares R, Perras MA (2019) Cliff stability at tomb KV42 in the Valley of the Kings, Egypt: A first approach to numerical modelling and site investigation. 53rd US ROCK Mech/Geomech Symp 1–10

Arosio D, Longoni L, Papini M et al (2018) Analysis of microseismic signals collected on an unstable rock face in the Italian Prealps. Geophys J Int 213:475–488

Bakun-Mazor D, Hatzor YH, Glaser SD, Carlos Santamarina J (2013) Thermally vs. seismically induced block displacements in Masada rock slopes. Int J Rock Mech Min Sci 61:196–211

Bottelin P, Lévy C, Baillet L et al (2013) Modal and thermal analysis of les arches unstable rock column (vercors massif, french alps). Geophys J Int 194:849–858

Bottelin P, Baillet L, Larose E et al (2017) Monitoring rock reinforcement works with ambient vibrations: La Bourne case study (Vercors, France). Eng Geol 226:136–145

Burjánek J, Moore JR, Yugsi Molina FX, Fäh D (2012) Instrumental evidence of normal mode rock slope vibration. Geophys J Int 188:559–569

Collins BD, Stock GM (2016) Rockfall triggering by cyclic thermal stressing of exfoliation fractures. Nat Geosci 9:395–400

Colombero C, Baillet L, Comina C et al (2018) Integration of ambient seismic noise monitoring, displacement and meteorological measurements to infer the temperature-controlled long-term evolution of a complex prone-to-fall cliff. Geophys J Int 213:1876–1897

Coviello V, Pogliotti P, Chiarle M, et al (2015) Monitoring Rock Wall Temperatures and Microseismic Activity for Slope Stability Investigation at J.A. Carrel Hut, Matterhorn. In: Engineering Geology for Society and Territory—Vol 1: Climate Change and Engineering Geology. Springer, pp 305–309

Eppes MC, Magi B, Hallet B et al (2016) Deciphering the role of solar-induced thermal stresses in rock weathering. Bull Geol Soc Am 128:1315–1338

Greif V, Brcek M, Vlcko J et al (2016) Thermomechanical behavior of Pravcicka Brana Rock Arch (Czech Republic). Landslides 14:1441–1455

Gunzburger Y, Merrien-Soukatchoff V, Guglielmi Y (2005) Influence of daily surface temperature fluctuations on rock slope stability: Case study of the Rochers de Valabres slope (France). Int J Rock Mech Min Sci 42:331–349

Fiorucci M, Marmoni GM, Martino S, Mazzanti P (2018) Thermal response of jointed rock masses inferred from infrared thermographic surveying (Acuto test-site, italy). Sensors 18

Hall K (1999) The role of thermal stress fatigue in the breakdown of rock in cold regions. Geomorphology 31:47–63

Levy C, Jongmans D, Baillet L (2011) Analysis of seismic signals recorded on a prone-to-fall rock column (Vercors massif, French Alps). Geophys J Int 186:296–310

Starr AM, Moore JR, Thorne MS (2015) Ambient resonance of Mesa Arch, Canyonlands National Park, Utah. Geophys Res Lett 42:6696–6702

Pedley M (2011) The Calabrian Stage, Pleistocene highstand in Malta: a new marker for unravelling the Late Neogene and Quaternary history of the islands. J Geol Soc Lond 168(4):913–926

# Extreme Rainfall Induced Landslide Susceptibility Assessment Using Autoencoder Combined with Random Forest

Kounghoon Nam and Fawu Wang

**Abstract**

The purpose of this study is to evaluate Autoencoder-based landslide susceptibility areas triggered by extreme rainfall in Shimane Prefecture, Japan. Deep learning methods can take advantage of the high-level representation and reconstruction of information from landslide-affecting factors. In this paper, a novel deep learning-based algorithm that combine classifiers of both deep learning and machine learning is proposed for landslide susceptibility assessment. A stacked autoencoder (StAE) and a sparse autoencoder (SpAE) both consist of an input layer for raw data, hidden layer for feature extraction, and output layer for classification and prediction. As a study case, Oda City and Gotsu City in Shimane Prefecture, southwestern Japan, were used for susceptibility assessment and prediction of landslides triggered by extreme rainfall. The prediction performance was compared by analyzing real landslide and non-landslide data. The prediction performance of random forest (RF) was evaluated as better than that of a support vector machine (SVM) in traditional machine learning, so RF was combined with both StAE and SpAE. The results show that the prediction ratio of the combined classifiers was 93.2% for StAE combined with RF model and 92.5% for SpAE combined with RF model, which were higher than those of the SVM (87.4%), RF (89.7%), StAE (84.2%), and SpAE (88.2%). This study provides an example of combined classifiers giving a better predictive ratio than a single classifier. The asymmetric and unsupervised autoencoder combined with RF can exploit optimal non-linear features from landslide-affecting factors successfully, outperforms some conventional machine learning methods, and is promising for landslide susceptibility assessment.

**Keywords**

Autoencoder • Support vector machine • Random forest • Landslide susceptibility • Deep learning

## Introduction

### Background

Landslide susceptibility mapping is to evaluate the spatial probability of landslide reoccurrence landslides continuously result in damages and casualties worldwide (Nam and Wang 2019a). It is necessary for decision makers to fast recognize large areas where landslides are expected to result in land use planning and disaster control. Landslide susceptibility prediction based on statistical approaches can achieve this goal efficiently. Most of the quantitative methods for producing landslide susceptibility maps refer to regression or classification approaches between real landslide data and artificially created non-landslide data. Recently, deep learning algorithms have made a series of revolutions in the field of machine learning since the classification capability of a neural network to fit a decision boundary plane has become significantly more reliable which can successfully learn and extract patterns and unique features from big data. Deep learning also can effectively avoid local optimization and eliminates the need to set model parameters because of autonomous processes. At the moment, the core techniques of deep learning are neural networks that have two or more hidden layers, including the following techniques: the adaptive neuro-fuzzy inference system; recurrent neural networks; deep belief networks; long short-term memory; and convolutional neural networks (Nam and Wang 2019b). Deep learning-based autoencoder is a semi-unsupervised

K. Nam · F. Wang (✉)
College of Civil Engineering, Tongji University, Shanghai, China
e-mail: wangfw@tongji.edu.cn

K. Nam
e-mail: geonamsoil@gmail.com


V. Vilímek et al. (eds.), *Understanding and Reducing Landslide Disaster Risk*,
ICL Contribution to Landslide Disaster Risk Reduction,
https://doi.org/10.1007/978-3-030-60319-9_36

learning method with no prior knowledge, such as landslide inventory, which means that landslide and non-landslide labels and linear and non-linear correlation assumptions are not needed (Huang et al. 2019). For landslide susceptibility assessment, traditional methods for de-correlation are based on the prior assumption that there are linear correlations between landslides and non-landslides. However, landslide-affecting factors are usually non-linear in practical applications. The autoencoder driven by data rather than prior knowledge can transform raw data into non-linear correlated features. In this paper, novel deep learning algorithms, namely, both stacked autoencoder and sparse autoencoder combined with traditional machine learning, are proposed for landslide susceptibility prediction. StAE and SpAE are unsupervised learning as it does not require external labels on landslides information. The encoding and decoding process all happen in the dataset. The input and output data have the same number of dimensions, and the hidden layer has fewer dimensions. Autoencoders are learned automatically from dataset, which is easy to train specialized instances of the algorithm that will perform well on a specific type of landslide-affecting factors. The autoencoder technique takes advantage of dimension reduction by stacked autoencoder and dropout by sparse autoencoder for non-linear correlations of the landslide-affecting factors and gives better feature descriptions than the original data. It does not require any additional methods which are required for appropriate training data. In summary, this study proposes the combined method of the advantage of deep learning and the benefits of machine learning for landslide susceptibility assessment. The landslides in Oda City and Gotsu City in Shimane Prefecture, southwestern Japan, are used as case study. A stacked autoencoder and sparse autoencoder are combined with random forest acquired from the results of a better predictive performance between support vector machine and random forest.

## Spatial Mapping

### Landslides in Shimane Prefecture

The study area is located in Oda City and Gotsu City, Shimane Prefecture, southwestern Japan (Fig. 1). The elevation varies from sea level to 1,123 m. The average annual rainfall recorded from the rainfall stations at Fukumitsu, Oda, and Sakurae are 1,657 mm, 1,786 mm, and 2,011 mm from 2008 to 2018. The cumulative rainfall for 2013 recorded from the rainfall stations at Fukumitsu, Oda, and Sakurae are 2,270 mm, 2,102 mm, and 2,656 mm, respectively. In this study, a total of 90 landslides were caused by extreme rainfall from May to October 2013, and 69 of the landslides were triggered by extreme rainfall in August 2013. These landslides can be described as shallow landslides that were determined based on field investigation.

### Landslide Affecting Factors

Landslide susceptibility prediction can be evaluated as a binary classification problem between landslides and non-landslides. A spatial database setting including landslide pixel grid, non-landslide pixel grid, and related landslide-affecting factors is needed for statistical analysis. This spatial database was divided into a training dataset and a validation dataset. These real 90 of landslides and 90 of non-landslides artificially generated from ArcGIS software were randomly split into two parts with a ratio of 70% and 30%. Seventy percent of the landslide and non-landslide grid cells were selected for the training model, and the remaining 30% were used for the validation model. Furthermore, the landslide (event) and non-landslide (non—event) grid cells were set to 1 and 0, respectively, and the values of 1 and 0 were used for classification and prediction as the output variables of the landslide susceptibility prediction models. Thereafter, the calculated frequency ratio (FR) values were considered as numeric input variables of landslide susceptibility prediction models. The landslide-affecting factors in study area are complex, and it is difficult to confirm which affecting factors are the most important and necessary among the topographic, geological, hydrological, distance to stream and distance to road. In landslide susceptibility modeling, landslides may reoccur under conditions similar to those of past landslides. A total of 14 affecting factors were acquired and chosen as input variables for landslide

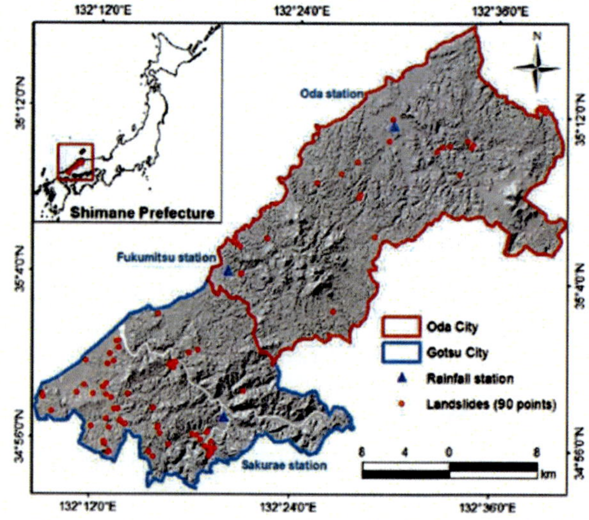


Fig. 1 Study area and landslide inventory

susceptibility models. The topographic factors were acquired and calculated based on the digital elevation model (DEM), with a spatial resolution of 10 m, including altitude, slope angle, plan curvature, profile curvature, distance to stream, stream power index (SPI), and topographic wetness index (TWI). The distance to road was acquired from Geospatial Information Authority of Japan. Normalized difference vegetation index (NDVI), normalized difference water index (NDWI), and bare soil index (BI) were derived from the Landsat TM 8 image data, resampled with a 10 m resolution. The geological factors were derived from the 1:200,000 scale geological map, which was obtained from the Geological Survey of Japan, AIST. These landslide-affecting factors were reflected using the raster format with a spatial resolution of 10 × 10 m, which results in raster format that has the advantages of regular shape, quick subdivision, and high modeling efficiency. For continuous affecting factors, the Jenks natural break method was used to divide each continuous affecting factor into five classes. Then the frequency ratio of all subclasses of each landslide affecting factor was calculated. The frequency ratio allows that all 14 landslide-affecting factors have significant influences on landslide occurrence. Some studies have suggested that the correlations between affecting factors should be eliminated to reduce model noise for the landslide susceptibility assessment. However, the number of input variables of the deep learning algorithm is generally hundreds or thousands due to their strong feature extraction ability, and 14 input variables will not result in information redundancy. On the other hand, some collinearity phenomena between landslide-affecting factors can be tolerated by the fast-developed machine learning models. These 14 landslide-affecting factors provide valuable information for producing landslide susceptibility maps, as quantitative measurement determined by frequency ratio. Therefore, all 14 landslide-affecting factors are utilized as input variables in the model to evaluate their capabilities in performance and feature extraction for the landslide susceptibility assessment.

## Methodology

### Support Vector Machine (SVM)

The purpose of optimal sorting hyperplanes is to accurately distinguish the two types of samples between landslides and non-landslides while maximizing the sorting margin. Determining kernel function and optimal parameters are critical for evaluating landslide susceptibility using SVM. Polynomial kernels and radial basis function are the most commonly used kernels in the literature. To optimize two parameters, both penalty coefficient C and kernel function parameters are needed in the SVM model (Huang and Zhao 2018).

### Random Forest (RF)

The random forest, a classification tree algorithm with repeated dichotomy data, can significantly reduce the computations required for classification and regression. In RF algorithms, predictive models are established by utilizing many decision trees. Based on randomly selected variables and samples, these trees and their decisions are generated. Once the model is established, the samples are first sorted individually according to all decision trees in the model, and then by all trees (Huang and Zhao 2018). The proportion of decision tree estimates and generates landslide susceptibility indexes, which can predict landslide occurrence between all decision trees in the RF model (Huang and Zhao 2018).

### Stacked Autoencoder (StAE)

The StAE is an artificial neural network, which is a special type of multi-layer perceptron. It is a type of unsupervised learning algorithm with an asymmetric structure, in which the middle layer represents the encoding of the input data in the bottleneck layer. The bottleneck constrains the amount of information that can traverse the full network, forcing the learned compression of the input data. The StAE is trained to reconstruct the input of landslide-affecting factors onto the output layer for feature representation, which prevents the simple copying of the data and the network. The middle layer has a lower dimension to avoid overfitting, which can either select a subset of features with the highest importance or apply some dimension reduction techniques. In this study, the StAE combined with back propagation neural network was processed for a lower dimension of features than the input data have, which can be used for learning the most important features of the data.

### Sparse Autoencoder (SpAE)

The SpAE consists of an input layer, hidden layers, and an output layer. Each layer in this neural network contains a sufficient number of neurons. Dropout can randomly classify the weight of some implicit layer nodes and reduce the mutual dependence between nodes to realize the normalization of neural networks. Additionally, dropout can effectively prevent overfitting and gradient disappearance. To initially achieve de-correlation among the 14 landslide-affecting factors, dropout was added to the input layer.

The process of StAE is as follows. First, some of the neurons in the network are randomly dropped in the mini-batch training samples and the remaining neurons are fed to the next layer. After obtaining this mini-batch training sample, the deleted neurons are recovered and some neurons in the network are randomly deleted once again. The corresponding parameters are updated based on the stochastic gradient descent method, performed on the neurons that have not been removed (Huang et al. 2019).

## Results and Discussion

### Landslide Susceptibility Modelling

All models based on the deep learning and machine learning were coded in R language on RStudio. For the SVM model and RF model, parameters were determined using a tenfold cross-validation approach. With radial basis function, SVM model was acquired from grid search for SVM parameter tuning. For RF model, it was composed of 'mtry' and 'tree', which were 3 and 300, respectively. The autoencoder models based on the deep neural network were coded in R language on RStudio using H2O packages. These algorithms were performed using hyperbolic tangent function (i.e., the tanh function) in every hidden layer which was used to encode and decode the input to the output in the undercomplete autoencoder. In the H2O library, five hidden layers with encoders and decoders were designed by using the tanh activation function in each layer. Stacked autoencoders (StAE) were constructed by organizing autoencoder on top of each other also known as deep autoencoder. StAE consists of multiple autoencoder stacked into multiple layers where the output of each layer was wired to the inputs of the successive layers, which was composed of 80–50-2–50-80. To obtain good parameters, StAE employed greedy layer-wise training. The benefit of StAE was that it can evaluate the benefits of deep network, which has greater expressive power. Furthermore, it usually can capture useful hierarchical grouping of the input. Finally, model construction was determined by the majority vote among all trees using RF models. The aim of sparse autoencoder (SpAE) was to make a large number of neurons to have low average output so that neurons may be inactive most of the time. The limitation of autoencoders to have only small numbers of hidden units can be overcome by adding a sparsity constraint, where a large number of hidden units can be utilized usually more than one input. Three hidden layers with encoders and decoders were designed by using the tanh activation function in each layer in the H2O library. Sparsity can be achieved by introducing a loss function during training or manually zeroing few strongest hidden unit activations, which was composed of 200–200-200. For classification, model class was constructed by RF model by means of the majority vote among all trees. Reconstruction error value employing mean square error was used by means of anomaly detection in both StAE and SpAE, which were 0.068 and 0.088, respectively.

### Landslide Susceptibility Mapping

The landslide susceptibility maps were derived from SVM (Fig. 2), RF (Fig. 3), StAE (Fig. 4), SpAE (Fig. 5), StAE with RF (Fig. 6), and SpAE with RF (Fig. 7) in the ArcGIS 10.6 software. For better visualization and comparison, the indices were reclassified into five classes using the equal interval function: very low (0–0.2), low (0.2-0.4), moderate (0.4-0.6), high (0.6-0.8), and very high (0.8-1). The susceptibility class area of the StAE model as the best performance (Table 1) were 6.31%, 13.58%, 33.04%, 36.81%, and 10.26%, respectively. The susceptibility class area of the RF model and StAE model has very high value. The susceptibility index value of the SVM model and StAE models were prominent near the road SpAE and SpAE with RF have lower values of class area percentage for a very high index of the susceptibility map. RF and StAE have lower values of class area percentage for a moderate index of the susceptibility map. StAE with RF and SpAE with RF have lower values of class area percentage for a very low (0.0–0.2) index of the susceptibility map.

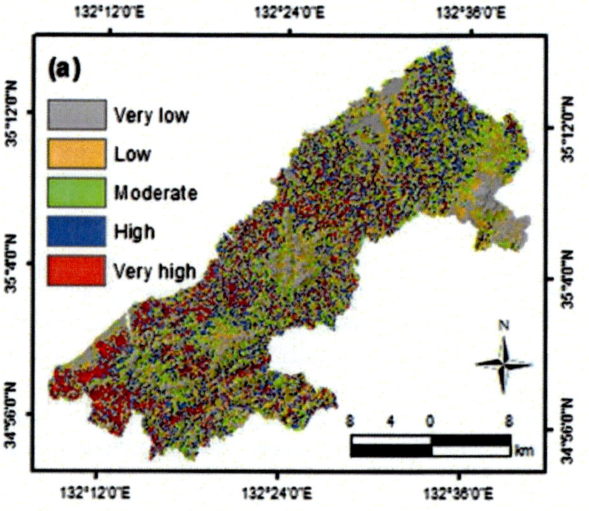


Fig. 2 Landslide susceptibility assessment by SVM

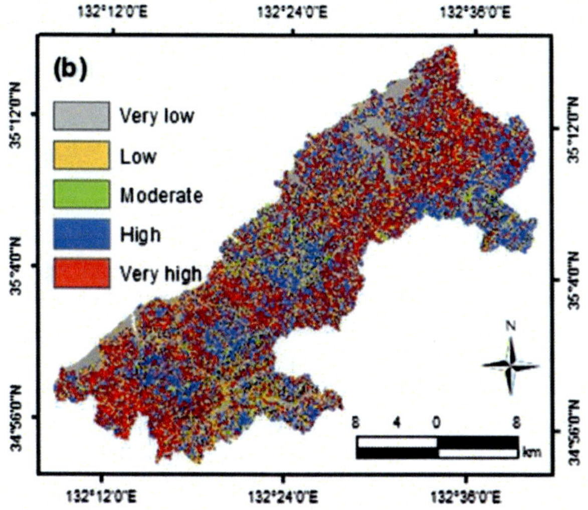


**Fig. 3** Landslide susceptibility assessment by RF

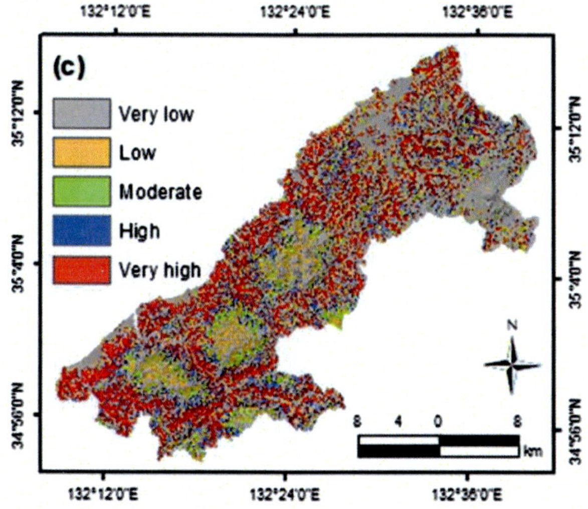


**Fig. 4** Landslide susceptibility assessment by StAE

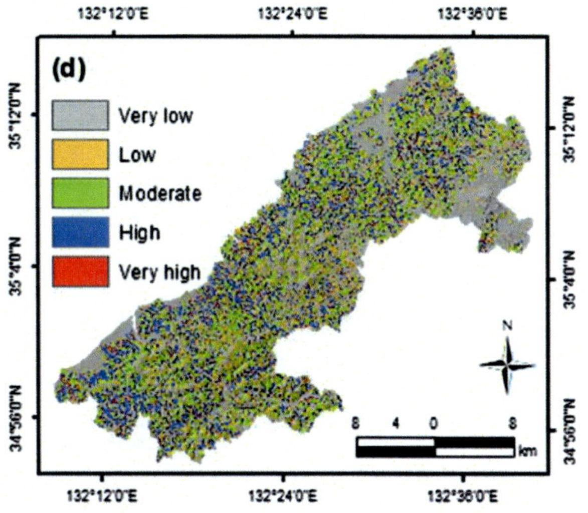


**Fig. 5** Landslide susceptibility assessment by SpAE

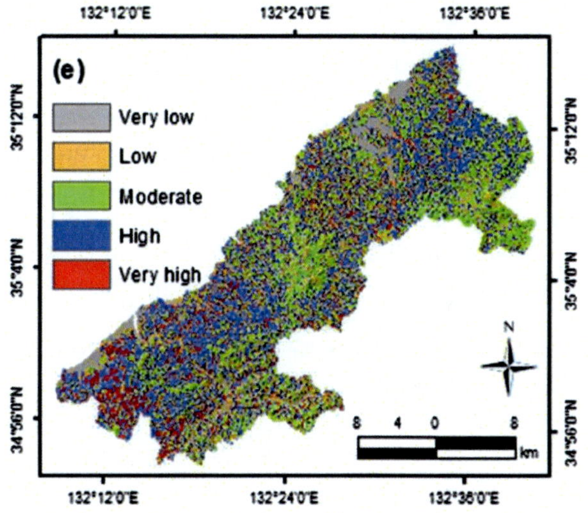


**Fig. 6** Landslide susceptibility assessment by StAE with RF

## Validation of Prediction Performance

The landslide susceptibility assessment was verified using the area under the curve on the validation dataset for six models. The predictive ratio for landslide susceptibility assessment is mainly calculated by confusion matrix. The true positive rate (TPR) is defined as the ratio of true positive to the sum of true positive and false negative, and the false positive rate (FPR) is defined as the ratio of false positive to the sum of false positive and true negative to the number of validation samples. In general, the true positive defines the landslide grid cells that are predictive as landslides, true negative means non-landslide grid cells that are predictive as non-landslides, false-positive reflects non-landslide grid cells that are predictive as landslides, and false negative means landslide grid cells that are predictive as non-landslides. The area under the curve was applied to assess the prediction performance of landslide susceptibility index values on the validation dataset. The prediction rate values of SVM, RF, StAE, SpAE, StAE with RF and SpAE with RF model are obtained by calculating the area under the prediction rate curves. The StAE with RF and SpAE model of combined classifier have relatively higher prediction rates than using SVM, RF, StAE, and SpAE model of single classifier. This means that the classifiers combined with both autoencoder and traditional machine learning are better than using a single classifier. Autoencoder is unsupervised learning as it does not require external labels on landslide information. The encoding and decoding process all happen within the dataset. The input and output data have the

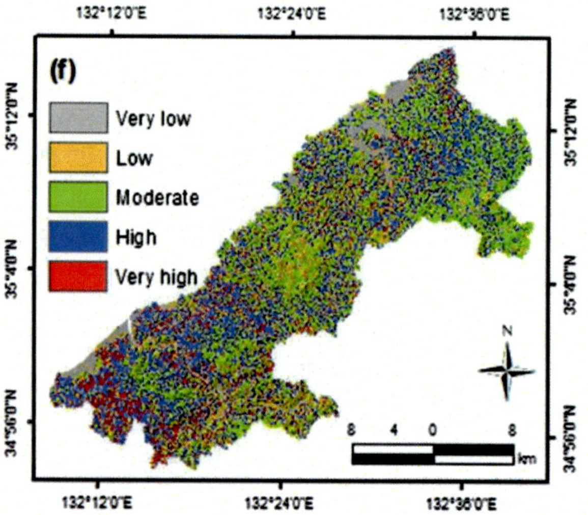


**Fig. 7** Landslide susceptibility assessment by SpAE with RF

same number of dimensions, and the hidden layer has fewer dimensions. Thus, it contains compressed information of the input layer, which is why it acts as a dimension reduction for the original input layer. From the hidden layer, the neural network is able to decode the information to its original dimensions. Autoencoders are learned automatically from data examples, which is a useful property. It means that it is easy to train specialized instances of the algorithm that will perform well on a specific type of input. It does not require any additional methods which are required for appropriate training data.

## Conclusion

In this study, the classifiers combined with both deep learning and traditional machine learning, StAE with RF and SpAE with RF models, are proposed for landslide susceptibility prediction. The autoencoder consists of input layers for raw data, hidden layers for feature extraction, and output layers for landslide susceptibility prediction. The combined classifiers have the advantage of both machine learning and deep learning, i.e., dimension reduction of the StAE model and dropout of the SpAE model for feature extraction.

The six models were applied in Oda City and Gotsu City, Shimane Prefecture, southwestern Japan. The correlation between landslides and landslide-affecting factors using frequency ratio was high in NDVI, distance to road, and altitude. Performance assessment was carried out with the SVM, RF, StAE, SpAE, StAE with RF, and SpAE with RF models. The results show that the proposed StAE with RF and SpAE with RF models have a relatively better prediction rate than a single classifier such as SVM, RF, StAE and SpAE models. In conclusion, the proposed combined classifier is promising for classification between landslide and non-landslide following landslide susceptibility prediction because it can overcome the limitations of

**Table 1** Number of landslides occurred and percentage of landslide susceptibility class area

| Susceptibility class (index value) | SVM | | RF | | StAE | | SpAE | | StAE with RF | | SpAE with RF | |
|---|---|---|---|---|---|---|---|---|---|---|---|---|
| | No | % | No | % | No | % | No | % | No | % | No | % |
| Very low (0–0.2) | 15 | 17.00 | 1 | 11.80 | 5 | 35.65 | 7 | 30.16 | 2 | 6.31 | 3 | 5.82 |
| Low (0.2–0.4) | 4 | 19.61 | 2 | 10.45 | 1 | 11.09 | 10 | 19.07 | 6 | 13.58 | 8 | 14.38 |
| Moderate (0.4–0.6) | 7 | 25.83 | 4 | 8.43 | 3 | 10.52 | 15 | 24.73 | 16 | 33.04 | 13 | 35.96 |
| High (0.6–0.8) | 17 | 24.94 | 13 | 33.43 | 4 | 13.07 | 44 | 22.60 | 66 | 36.81 | 66 | 33.25 |
| Very high (0.8–1.0) | 47 | 12.62 | 70 | 35.89 | 77 | 29.67 | 14 | 3.44 | 0 | 10.26 | 0 | 10.59 |
| Sum | 90 | 100 | 90 | 100 | 90 | 100 | 90 | 100 | 90 | 100 | 90 | 100 |

conventional machine learning algorithms, extract features and pattern recognition, reduce computations, and improve performance.

## References

Huang Y, Zhao L (2018) Review on landslide susceptibility mapping using support vector machines. CATENA 165:520–529

Huang F, Zhang J, Zhou C, Wang Y, Huang J, Zhu L (2019) A deep learning algorithm using a fully connected sparse autoencoder neural network for landslide susceptibility prediction. Landslides 17:1–13

Nam KH, Wang FW (2019a) An extreme rainfall-induced landslide susceptibility assessment using autoencoder combined with random forest in Shimane Prefecture. Japan. Geoenvironmental Disasters 7 (6):1–16

Nam KH, Wang FW (2019b) The performance of using an autoencoder for prediction and susceptibility assessment of landslides: A case study on landslides triggered by the 2018 Hokkaido Eastern Iburi earthquake in Japan. Geoenvironmental Disasters 6(19):1–14

# Economizing Soil Nailing Design by Drainage Improvement—Case History at Ginigathhena

2020 Kyoto Japan

S. O. A. D. Mihira Lakruwan and S. A. S. Kulathilaka

**Abstract**

As a tropical country which faces two monsoons with heavy precipitation, rainfall is identified as the major triggering factor for landslides in Sri Lanka. Reactivation of ancient landslides is a major challenge faced by the engineers in infrastructure development in the hilly terrain of the country. An ancient landslide at Bridge No. 48/2 in Avissawella—Hatton—Nuwaraeliya road was reactivated by excavation at the toe for the proposed widening of the bridge. This propagated further by the extensive rainfall that occurred subsequently. The site is a sloping ground with undulating topography towards upper slope in the form of a valley formed by an ancient landslide. Mitigation of the landslide was done to accommodate the widening of the bridge and to minimize the risk of further activation by rainfall. Mitigation measures adopted were drainage improvement, slope modification and soil reinforcement. Soil nailing was used as the reinforcement technique. Surface and subsurface drainage improvement was used for economizing the nailing design. This study presents how the nailing design was optimized by using different drainage arrangements. GeoStudio SEEP/W and SLOPE/W software were used for seepage and slope stability analyses, respectively.

**Keywords**

Drainage • Landslide • Slope stability • Soil nailing • Rainfall

## Introduction

### Background

Sri Lankan landform has very significant elevation differences over a short distance from the coastal area to the central regions. The elevation difference between the capital Colombo which is at around the sea level, to the city at the highest elevation Nuwara Eliya at 2,000 m above MSL, is within a road length of only 150 km. Central region of the country has a rugged terrain with many steep slopes formed of colluvial soils, residual soils and parent metamorphic rock at different degrees of weathering.

Sri Lanka's extensive road network connects the capital city Colombo and all regional cities in the country. Mostly, roads are taken along the contour lines but invariably there are many stretches of steep gradients surrounded by sloping terrain. There are many requirements to widen and improve the current road network to accommodate the increasing demands of traffic flow due to economic development and population growth. Such road widening demands excavations into the hilly terrain.

Since the Bridge No. 48/2 on Avissawella—Hatton—NuwaraEliya Road near Ginigathhena area is narrow, there is a need for widening that bridge to accommodate the increasing flow of traffic. Anancient landslide at this location got activated during an attempt to excavate at the toe of the slope for the widening of the bridge approach. Considering the social, economic and environmental importance, this Ginigathhena landslide was stabilized by extensive mitigation measures.

S. O. A. D. M. Lakruwan (✉)
Graduate School of Engineering, Tohoku University, Sendai, Japan
e-mail: samarasekara.okandapala.arachchige.don.mihira.lakruwan.q8@dc.tohoku.ac.jp; mihiralakruwan@gmail.com

National Building Research Organisation, Colombo, Sri Lanka

S. A. S. Kulathilaka
Department of Civil Engineering, University of Moratuwa, Moratuwa, Sri Lanka


V. Vilímek et al. (eds.), *Understanding and Reducing Landslide Disaster Risk*, ICL Contribution to Landslide Disaster Risk Reduction,
https://doi.org/10.1007/978-3-030-60319-9_37

Mitigation measures were designed to stabilize the slope while allocating space necessary for the widening. Drainage improvement, slope geometry modification and slope reinforcement were used as mitigation measures. A separate study was carried out to identify the importance of combined mitigation measures and necessity of staged construction. This study done as a continuation and the paper focuses on the economizing of the soil nailing design by application of different drainage improvement measures. Detailed analysis associated with the soil nailing design is presented in Lakruwan (2019).

## Landslides in Sri Lanka

There are many reasons for activation of landslides and, among them, rainfall is identified as the major triggering factor of landslides in Sri Lanka. Landslides triggered by rainfall are termed rain induced landslides. Other factors that contribute to the activation of rain induced landslides include human activities and natural weathering processes. Rain induced landslides and cut slope failures have become a major natural disaster in Sri Lanka, during the past few decades (e.g. Ratnayake and Herath 2005).

## Background

Sloping terrain in the hilly country of Sri Lanka is formed by residual soils, colluvial soils and rock at different states of weathering. Residual soils, which are formed by weathering of parent rock, have significant variations within short distances due to different degrees of weathering caused by the mineralogical composition of the parent rock and weathering under the conditions of high rainfall and humidity. Colluvium soils are loose, unconsolidated sediments formed by downslope movements of the residual soils and rock fragments under the force of previous landslide events. These colluvium soils also have variations in formation. Shear strength properties of these soils are significantly low and variable due to variations in soil formation. Process of rain induced landslides is quite complex due to these variable soil properties.

Slopes which have tendencies to fail during rainy periods are generally quite stable during dry periods due to prevailing matric suction under low ground water table conditions. However, with the infiltration during heavy rains, matric suction would be diminished and positive pore pressures may develop in the slope. Eventually, the slope becomes unstable due to elevated pore pressures. If the slope is found to be unstable, even under extreme conditions, mitigation measures need to be applied to enhance the stability.

To assess rain induced landslides two types of analyses are required:

- seepage analysis to analyse the ground water regime variations; and
- stability analysis to evaluate the safety margins of the slope under those conditions'.

These two analyses have to be coupled together to get realistic results. Semi coupled GeoStudio SEEP/W (GEO-SLOPE International Ltd. 2017a) and SLOPE/W (GEO-SLOPE International Ltd. 2017b) software are generally used for seepage and slope stability analyses, respectively.

Drainage improvement along with other structural mitigation measures are used for strengthening susceptible slopes. Improvement of drainage may be catogorized as related to surface drainage and subsurface drainage. Surface drainage improvement is done mainly to reduce infiltration while subsurface drains have capacity to lower the ground water table.

Retaining walls, ground modification and soil reinforcement are some other types of slope strengthening measures.

Reinforcement by soil nailing is used for stabilization of most steep high cut slopes. Soil nailing is a technique of reinforcing in-situ soil by using passive intrusions from the facing, proceeding in a top down construction sequence.

Soil nailing systems are generally assessed using a two-zone model, namely the active zone and the resistant zone, which are separated by a potential failure surface as shown in Fig. 1. The active zone is the region in front of the potential failure surface, where it tends to be detached from the main slope. The resistant zone is the region behind the potential failure surface, where it remains more or less intact. The soil nails act to tie the active zone to the resistant zone (GEO, Hong Kong 2008).

Soil nails are activated when the slope experiences some movement. Tensile stress developed in a soil nail reduces the shear stress to be mobilized on the slip surface and increases the shear resistance force by increasing the normal stress on the failure surface. Both these phenomena increase the safety margin of a slope. For this to be effective, nails should extend to a sufficient distance beyond the failure surface (into the resistant zone) and should be capable of mobilizing a sufficiently large tensile force without getting pulled out.

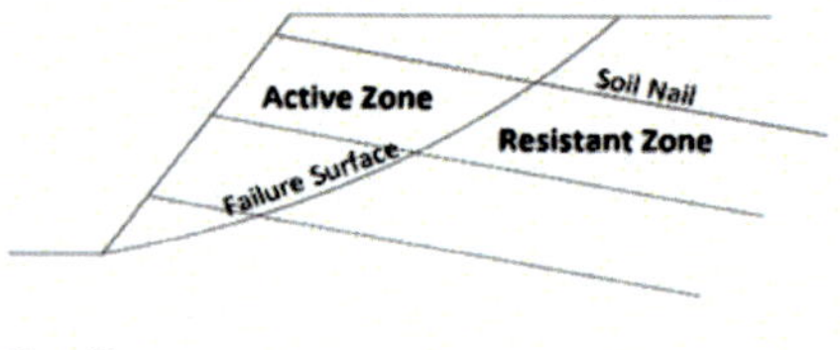


**Fig. 1** Soil nailing system

Drainage improvement is considered as the primary mitigation measure in stabilization of rain induced landslides. It is a very important component in a soil nailing system. GEO, Hong Kong (2008) and Federal Highway Administration (2006) highlight the requirements of drainage improvement in soil nailing systems in terms of improving the performance of the soil nailing system, preventing local instabilities and overcoming difficulties due to high ground water table during construction. With the enhancements achieved by drainage improvement, amount of soil nails required to maintain the stability of a particular slope would be less. In general, cost of drainage improvement is less than cost of soil nailing. Therefore, considerable amount of cost can be saved by application of drainage improvement measures in soil nailing systems.

Economizing the soil nailing design of Ginigathhena landslide by drainage improvement was studied in depth in this paper.

## Ginigathhena Landslide

### Background

Avissavella—Nuwara Eliya road is a major road which connects the capital Colombo to central region of the country. Bridge No. 48/2 on this road is very narrow. The traffic demand on the road is increasing rapidly with economic and population growth in the area and the bridge was proposed to be widened. Some excavation was done at the toe of upper slope of the bridge to widen the approach to the bridge. An ancient landslide above the bridge activated during this excavation and a failure occurred propagating upslope. It was reported that there was no rain at the time of the initial failure. With the occurrence of rainfall in the area later, failure propagated further. Figures 2 and 3 show the location map of the area and failure just after the rainfall event, respectively.

**Fig. 3** Failure after toe excavation (image by Mihira Lakruwan)

### Geology and Hydrology of the Area

The landslide area is thickly covered with colluvium with inclusion of boulders transported from the ancient landslide. This is underlain by highly crystalline metamorphic bed rock. Main landslide area is a sloping ground with an undulating topography and just above the crest of the landslide there is a water logged marshy area which is the main reason for the prevailing high water table in the area. Water collected in the upper slope flows through the water stream at a side of the landslide. Figure 4a and b show the waterlogged marshy area at the top and the stream flowing at the side of the landslide, respectively.

### Mechanism of Failure and Justification for Mitigation

With the high ground water table of the area, slope was already at a critical stable condition and with the removal of the toe support by the excavation, slope became unstable

**Fig. 2** Location map of the area

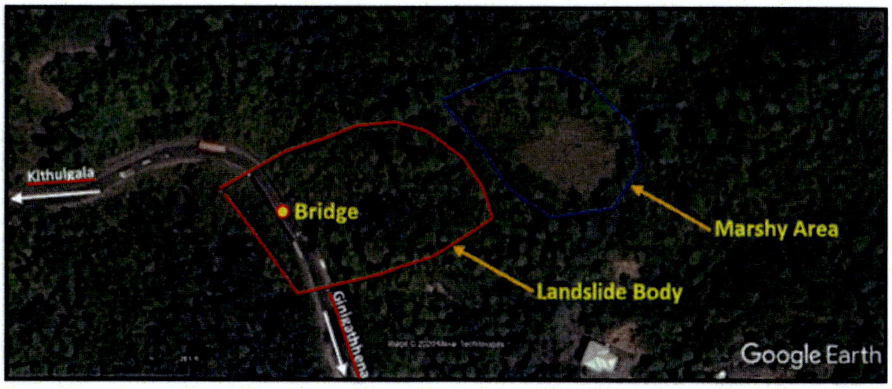

(a) Marshy area at the top (b) Stream beside the landslide

**Fig. 4** Landslide at Ginigathhena (images by Mihira Lakruwan)

leading to reactivation of the landslide. With the subsequent rainfall that occurred in the area, landslide propagated further. After the failure, the unstable soil mass was resting against the bridge structure and the slope was at a critical stable condition. There was a possibility of further activation of this landslide in the form of a deep-seated failure with heavy rainfall. Such activation would have led to a catastrophic failure damaging the bridge and the road. In addition to mitigation, there was also the requirement for road widening. Therefore, it was necessary to mitigate this landslide while accommodating the widening requirements.

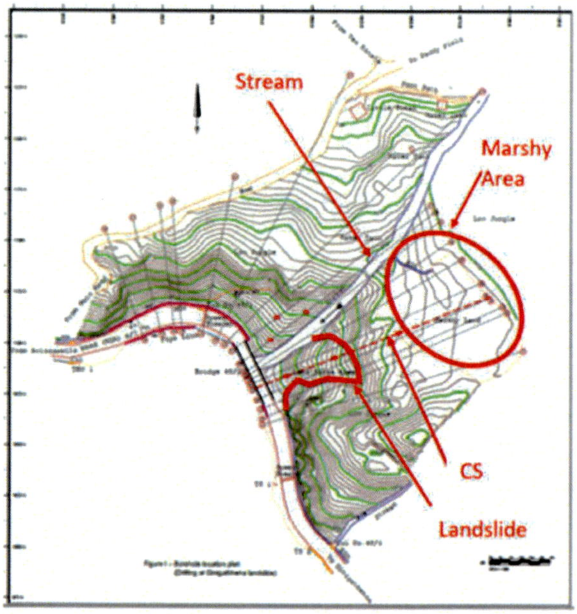


**Fig. 5** Survey map (Lakruwan 2019)

## Investigation

A detailed investigation of the landslide area was carried out, including a topographical survey and a geotechnical investigation. Topographical survey produced a contour map with 1 m contour intervals and slope cross sections (Fig. 5). During the geotechnical investigation, three boreholes were advanced with SPT's, followed by laboratory testing on recovered undisturbed samples. Subsoil profile from the geotechnical investigation is summarised and presented in Fig. 6.

## Mitigation Measures

Based on the investigation results, mitigation measures for the Ginigathhena landslide were designed focusing on three main categories:

- drainage improvement;
- ground modification; and
- reinforcement.

Under drainage improvement, surface drainage improvement was achieved by cut-off drains, berm drains, trench drains & slope cover by vegetation. The subsurface drainage improvement was achieved by sub horizontal gravity drains.

The proposed road widening required cutting into the slope at the toe region which reduces the safety margin of the slope. Therefore, slope reshaping was done to accommodate the required road widening. Since the required safety margins of this reshaped slope could not be achieved only by drainage improvement, reinforcement in the form of soil nailing was used for enhancing the safety margin of the slope. After several trials, stability of the slope was achieved by the nailing pattern presented in Fig. 7. This include:

- 5 nos. of 16 m long soil nails;
- 2 nos. of 12 m long soil nails; and
- 4 nos. of 8 m long soil nails.

All soil nails were installed in a 2.0 m × 2.5 m pattern, in horizontal and vertical directions, respectively.

## Monitoring

During the application of mitigation measures, a field monitoring scheme was established to monitor the behaviour of the slope. This monitoring system consisted of three parts, viz. rainfall monitoring by a manual rain gauge, ground water level monitoring by two observation wells and slope movement monitoring by a number of fixed points established on the slope surface.

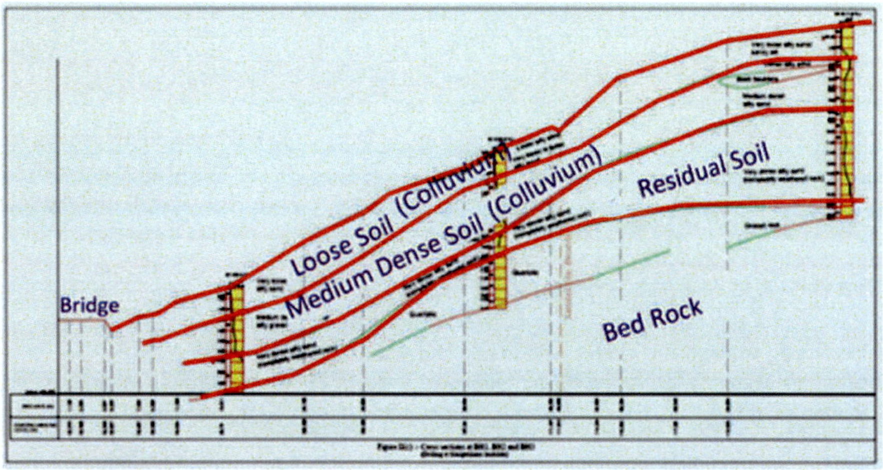


**Fig. 6** Subsoil profile (Lakruwan 2019)

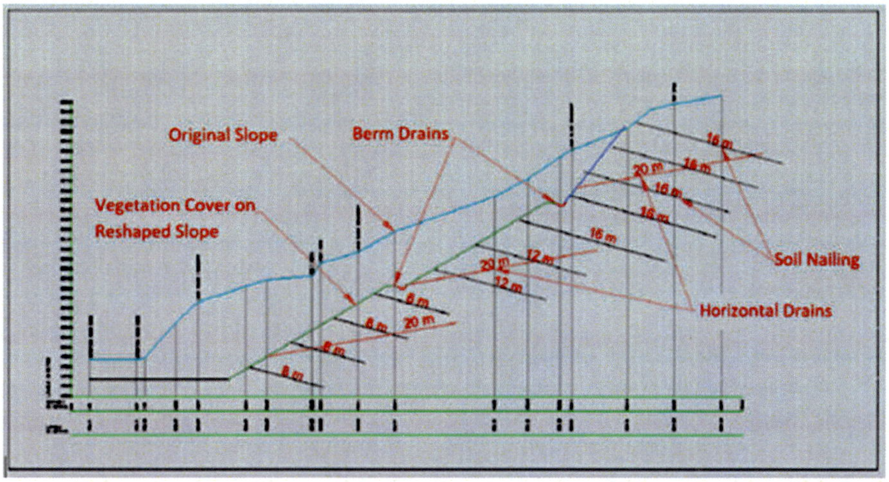


**Fig. 7** Proposed mitigation measure—cross sectional view (Lakruwan 2019)

## Economizing the Soil Nailing Design

### Overview

Performance of the designed nailing pattern was assessed for different drainage conditions under the critical design rainfall condition and nailing pattern was revised if the required safety margins were not achieved. Following steps were followed in this study:

- Development of SLOPE/W & SEEP/W models;
- Definition of critical rainfall event;
- Analyses for different drainage conditions and revision of the nailing patterns; and
- Costcomparison of different nailing patterns.

Development of geotechnical computer models are not discussed here in depth. To give an overview, results of the geotechnical investigation was used to develop both SEEP/W & SLOPE/W models. Verification of the SLOPE/W model was done by simulating the failure that occurred after excavation at the toe. SEEP/W model was verified by simulating an actual rainfall event on the model and comparing the results with actual ground water levels. Surface drainage improvement was modeled on the SEEP/W analysis by applying a thin layer of very low permeability on the slope surface. In the modeling of subsurface drains, sub horizontal drains installed were idealized to plain strain condition by a zero pressure boundary condition applied onto a reprenstative line of actual drains.

### Drainage Conditions

Three drainage conditions were considered in the analyses as "*No Drains*", "*Only Surface Drains*" and "*Both Surface & Subsurface Drains*" conditions. Original design was carried out for "*Both Surface & Subsurface Drains*" condition. Therefore, conditions of "*No Drains*" and "*Only Surface Drains*" were considered in this study and nailing patterns to maintain the required safety margin for these two drainage conditions were found.

## Design Factor of Safety (FoS)

The slope is located on a major road and any failure of the slope would affect the lives and properties of users of this busy road, damage the bridge and road, and disturb the traffic. Considering all these facts, the design FoS was selected as 1.3.

## Design Critical Rainfall Event

During the monitoring period, a peak rainfall of 300 mm/day was observed on 20 May 2018 at the landslide location. Therefore, two days of 300 mm/day rainfall was selected as the critical rainfall followed by a 50 mm/day residual rainfall for seven days. Five days of no rain period was considered prior to the critical rainfall event (Fig. 8) (Lakruwan 2019).

## Analysis

### Analysis for "No Drains" Condition

Critical design rainfall event was simulated on the SEEP/W model and it was analyzed for "*No Drains*" condition. Results from seepage analysis were used as input to the slope stability analysis. SLOPE/W analysis with the original design showed that the minimum FoS value will occur just after the peak rainfall of design rainfall event. The resultant FoS was 1.042 (Fig. 9). Therefore, soil nailing pattern was modified to achieve the required FoS of 1.3 against the critical rainfall (Fig. 10). Since increasing the nail length beyond 16 m was not practical and effective, the nail grid was densified while increasing length of some 12 m nails to 16 m. After several trials, the following nail pattern was obtained:

- 12 nos. of 16 m long nails in 1.5 m horizontal and 1.5 m vertical spacing; and
- 4 nos. of 8 m long nails in 2.0 m horizontal and 2.5 m vertical spacing.

Nailing arrangement in the lower part of the slope was not changed since the critical slip surface was not passing through that area.

The above indicates that, if no drainage improvement is done, the total number of nails and nail lengths will increase significantly. The, total nail length was increased to 3,112 m from 1,624 m of the original design. Further, facing arrangement of soil nailing system had to be modified due to densification of the grid and the increased number of nails.

### Analysis for "Only Surface Drains" Condition

As similar to the previous case, analysis was carried out for "*Only Surface Drains*" condition and the resulting FoS of the original design after peak rainfall was 1.124 (Fig. 11). Therefore, the soil nailing pattern was modified to achieve the design FoS (Fig. 12). After several trials, the following nail pattern was obtained:

- 9 nos. of 16 m long nails in 1.5 m horizontal and 2.0 m vertical spacing; and
- 4 nos. of 8 m long nails in 2.0 m horizontal and 2.5 m vertical spacing.

Bottom nailing arrangement was not changed since the critical slip surface was not passing through that area.

The results show that the total nail length was increased to 2,624 m for "*Only Surface Drains*" condition. Further, facing arrangement of soil nailing system had to be increased due to densification of the grid and increased number of nails.

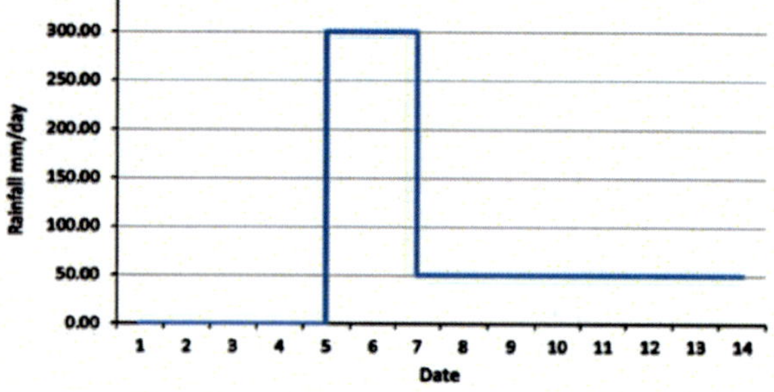


**Fig. 8** Design rainfall event

## Cost Comparison

Construction cost for drainage measures were calculated based on the average rates currently prevailing in the Sri Lankan industry. Table 1 presents the comparison of construction cost under different drainage conditions.

Cost of soil nailing also includes the cost of facing structure with nailheads connecting beams, high tensile mesh, coir mesh and vegetation. With the densification of the soil nailing arrangement, cost of facing structure also increased.

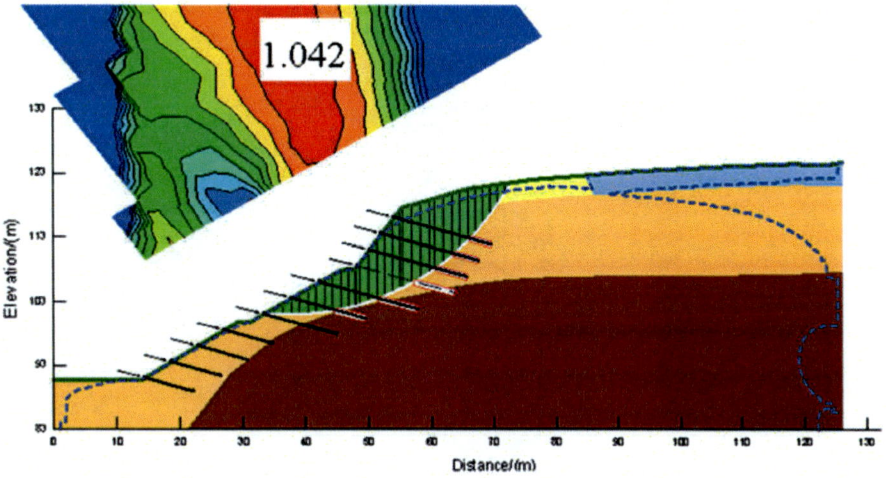


**Fig. 9** "No Drains"—Original design

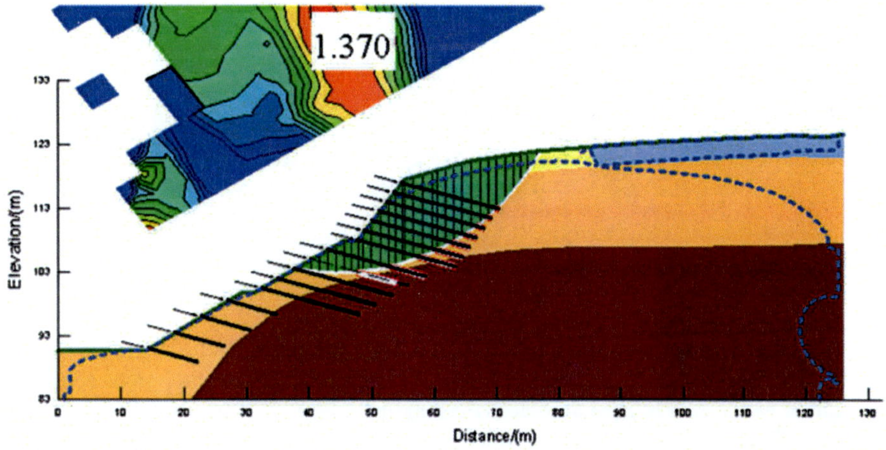


**Fig. 10** "No Drains"—Modified design

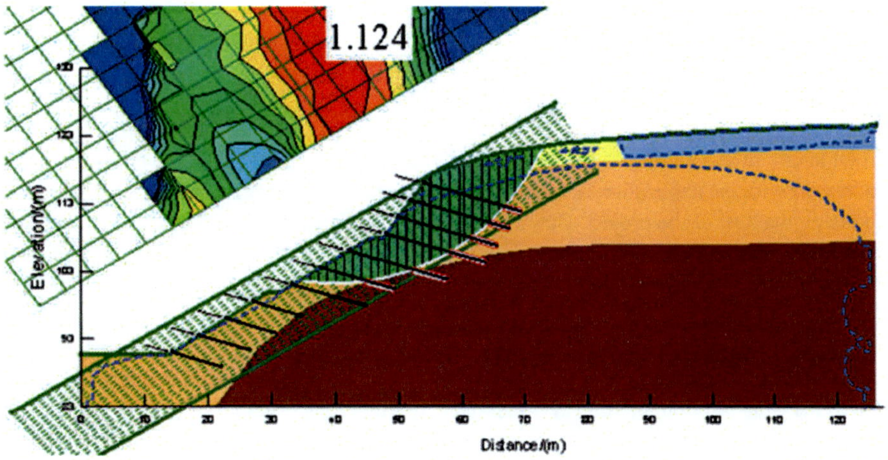


**Fig. 11** "Only Surface Drains"—Original design

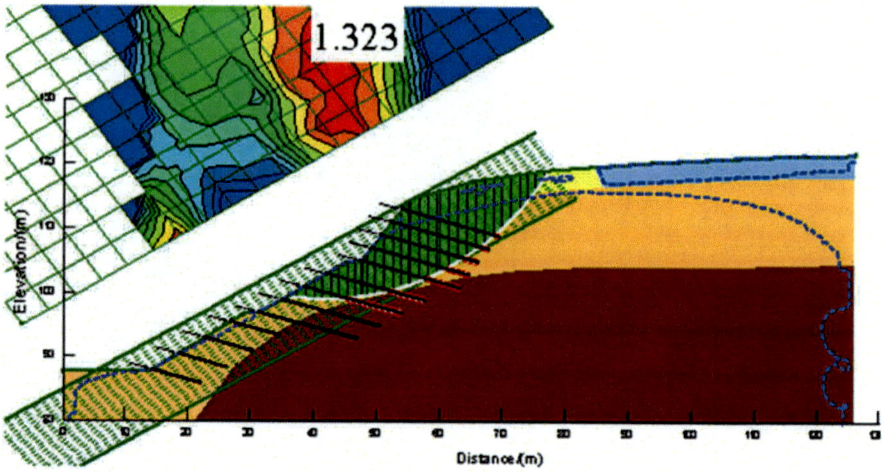


**Fig. 12** "Only Surface Drains"—Modified design

**Table 1** Construction cost comparison for different drainge measures

| Drainage condition | Cost of nailing and associated works (Rs. Mn) | Cost of drainage improvement (Rs. Mn) | Cost of excavation and other works (Rs. Mn) | Total cost (Rs. Mn) | Cost saving (%) |
|---|---|---|---|---|---|
| No Drains | 41.6 | 0 | 7.4 | **49.0** | 0.00 |
| Only surface drains | 34.9 | 5.4 | 7.4 | **47.7** | 2.65 |
| Surface & subsurface drains | 25.4 | 9.1 | 7.4 | **41.9** | 14.49 |

When the drainage measures are reduced a greater amount of nails had to be used to achieve the required safety margins. This caused a significant cost increase.

## Summary and Conclusions

### Summary

Total length of soil nailing was reduced and nail spacing was increased with the application of full drainage measures thus reducing the cost of the project. But, drainage measures also come with a cost. However, this cost is less than the cost of nailing. As an example, in general, cost of nailing is about Rs. 8,500 per meter while the horizontal drains cost only about Rs. 4,500 per meter. Cost of excavation and associated works remain unchanged. Therefore, total construction cost reduces by application of drainage measures. Further, the process of nailing would be less complicated during construction due to the lowered ground water table.

Construction cost without any drainage measures is Rs. 49.0 Mn. Cost saving by application of surface drainage alone is Rs. 1.3 Mn and it is a 2.65% saving. Cost saving by application of both surface and subsurface drainage is Rs. 7.1 Mn and it is a saving of 14.49%.

Above results illustrate how a soil nailing design can be economized by application of drainage measures. Therefore, it is important to have both surface and subsurface drainage improvement to economize the nailing design.

### Conclusions

Soil nailing technique can be used for stabilization of high steep slopes where other retaining structures are not feasible due to stability, economics and space constraints. Generally, soil nailing is applied when required safety margins could not be achieved only through drainage improvement. However, drainage measures are very effective in optimizing the cost and minimizing problems during installation of nails.

In this research, the reactivated rain induced landslide at Ginigathhena area was selected as a case study. Results of the analysis highlighted the significant construction cost reduction that can be achieved by application of drainage measures in soil nailing designs.

## References

Federal Highway Administration (2006) Soils and foundations, reference manual. 1st ed. National Highway Institute, U.S. Department of Transportation, Federal Highway Administration. Washington, DC

GEO, Hong Kong (2008) Guide to soil nail design and construction. 1st ed. Civil Engineering and Development Department, The Government of the Hong Kong Special Administrative Region, Hong Kong

GEO-SLOPE International Ltd. (2017a) Heat and mass transfer modeling with geo studio 2016. 2nd ed. GEO-SLOPE International Ltd., Calgary, Alberta, Canada

GEO-SLOPE International Ltd. (2017b) Stability modeling with geo studio. GEO-SLOPE International Ltd., Calgary, Alberta, Canada

Lakruwan SOADM (2019) Enhancement of the stability at the site of an ancient landslide in a road cutting with drainage and reinforcement—Case history at Ginigathhena, M.Eng. thesis, Department of Civil Engineering, University of Moratuwa, Moratuwa, Sri Lanka

Ratnayake U, Herath S (2005) Changing rainfall and its impact on landslides in Sri Lanka. J Mountain Sci 2(3):218–224

# Performances of Geosynthetics-Reinforced Barriers for Protection Against Debris Avalanches

Sabatino Cuomo, Sabrina Moretti, Lorenzo Frigo, and Stefano Aversa

**Abstract**

Flow-like landslides, such as debris avalanches, are often dangerous for both structures and human life due to the high velocities and large runout distances. Artificial barriers are proposed to mitigate the landslide threats by reducing the flow velocity and the runout distance as well as diverting the flow towards lateral zones constrained by the barriers. A procedure is here proposed based on multiple steps. The first step focuses on the propagation of debris avalanche along a real topography through the meshless GeoFlow_SPH model in presence of one or more barriers. The second step deals with the behaviour of Deformable Geosynthetics Reinforced Barriers (DGRBs) under horizontal pressures, calculated with reference to literature formulations which include a dynamic component (dependent on landslide velocity at the impact stage), and a static component (height dependent). The time trend of impact pressure is computed from landslide volume and impulse theorem. The third step concerns the evaluation of the performances of the barrier under the impact. The main goal is to present such procedure and to evaluate the performances of DGRBs in: changing the area affected by landslide and dissipating impact energy through deformation and displacement mechanisms.

**Keywords**

Propagation • Impact • Modeling • SPH • FEM • FDM

## Introduction

Mitigation works are very important in landslide management, particularly for fast landslides. Among these, debris avalanches typically occur in open slopes with initial low propagation heights and then they greatly increase in volume along the path (Cuomo et al. 2014a, 2016).

Several dramatic case histories occurred all over the world and some of them are carefully reported in the scientific literature. Just to mention: fast landslides of Rio de Janeiro (Avelar et al. 2013); shallow fast-moving landslides recorded in the South-West of China (Xu et al. 2009); debris avalanches occurred in Southern Italy (Revellino et al. 2013; Cuomo et al. 2014a; Cascini et al. 2016). Furthermore, several studies showed that obstacles along the slopes could be a valid option for protection against debris avalanches. Cuomo et al. (2019a,b) and Moretti et al. (2019) show that one or more barriers positioned at the toe of a slope are effective in reducing or confining the affected areas.

Barriers reinforced with geosynthetics have been used in several fields especially due to their highly satisfactory seismic performance. In particular, they have been used against rockfall, tsunami, avalanches (Brandl 2011; Lambert and Bourrier 2013; Yang et al. 2016).

Deformable Geosynthetics Reinforced Barriers (*DGRBs*) are here proposed as a mitigation option against debris avalanches. The *DGRBs* are composed of granular soil reinforced with polyester geogrids (*PET*) that provide tensile resistance to the soil through the soil-geogrid frictional contact. The wrap-around technique is here selected being associated to higher tensile stresses in the geosynthetics compared to simple runout anchorage. The barrier is also free to slide along the ground surface under lateral external

S. Cuomo (✉)
University of Salerno, Via Giovanni Paolo II, 132, 84084 Fisciano, Italy
e-mail: scuomo@unisa.it

S. Moretti · S. Aversa
Centro Direzionale, University of Naples Parthenope", Isola C4, 80133 Naples, Italy
e-mail: sabrina.moretti@uniparthenope.it

S. Aversa
e-mail: stefano.aversa@uniparthenope.it

L. Frigo
Geosintex S.R.L, Via Leonardo da Vinci, 12, 36066 Sandrigo, Vicenza, Italy
e-mail: Lfrigo@geosintex.com


V. Vilímek et al. (eds.), *Understanding and Reducing Landslide Disaster Risk*, ICL Contribution to Landslide Disaster Risk Reduction,
https://doi.org/10.1007/978-3-030-60319-9_38

pressure. In the paper, dynamic analyses are developed for several impact scenarios and different geometries to analyse the behaviour and performances of such barriers.

## Proposed Methodology

### Propagation Analysis

The propagation analyses are carried out through the GeoFlow_SPH model (Pastor et al. 2009) which is time-effective in simulating the propagation of flow-like landslides over long distances.

This is a depth-integrated hydromechanical coupled model that schematizes the propagating mass into a mixture of solid skeleton and pore water. The Smoothed Particle Hydrodynamics (SPH) numerical technique is used and is able to ensure good accuracy in reasonable computational times. More details are provided in Pastor et al. (2009) and Cuomo et al. (2014a, 2016). One or two barriers are positioned in the foothill area and the heights and velocities of the flow are computed; the involved areas are compared with those eventually affected without any barrier.

### Deformation Analysis

The behavior of DGRBs is evaluated through two different numerical models implemented in through two commercial codes. The first one, namely Plaxis v. 8 (Brinkgreve 2007), is based on the Finite Element Method, which relies on the discretization of complex objects into small elements where the differential equations are discretized and solved. The results of FEM analyses are compared whether possible to those given by the second commercial code, i.e. Flac v. 8 (Itasca 2016), which is based on the Finite Difference Method. The latter is the oldest numerical technique used for the solution of problems described through a set of differential equations, with given initial values and boundary conditions. In this method, every derivative equation is replaced by an algebraic expression written in terms of the field variables, stress or displacement, at discrete points in space; these variables are undefined within elements. The large displacement conditions are generally better simulated through FDM analyses than FEM analyses, which instead suffers in case of high distortion of the deformed elements. This is the reason why we used two different approaches for impact analyses.

The impact pressure exerted by a debris avalanche on the barrier is evaluated over time as proposed by Gioffrè et al. (2017), given by the sum of a static component (height-dependent) (Eq. 1) and a dynamic component (velocity-dependent) (Eq. 2).

$$ILP_{peak}{}^{stat} = 0.5 \cdot K_p \cdot \gamma \cdot h \quad (1)$$

$$ILP_{peak}{}^{dyn} = k \cdot v^2 \cdot \gamma / g \quad (2)$$

where $K_p$ is the passive earth pressure coefficient of landslide derived from Rankine theory; $h$ is the landslide height and here considered constant during the impact, $\gamma$ is the unit weight of debris flow, $k$ is an impact coefficient assumed equal to 5 by Gioffrè et al. (2017), and $v$ the landslide velocity at the impact stage.

Several impact pressures are applied against the barriers to evaluate their performances in terms of displacements and deformations.

### Performance Analysis

The performance analyses are focused on two different issues: (i) the efficacy of the barrier into constraining propagation area within a target zone; (ii) the amount of deformations and displacements of the barrier during the impact stage.

As the debris avalanche interacts with the barrier, three main Scenarios ($SC$) may occur: (i) the debris avalanche is stopped ($SC_1$); (ii) the flow overcomes the barrier with reduction in runout at valley-side ($SC_2$); (iii) the flow overcomes the barrier that practically does not reduce the runout at valley-side ($SC_3$).

Two empirical coefficients are defined (Cuomo et al. 2019a) for the propagation stage: the Index of Piedmont Runout Reduction ($I_{PRR}$) and the Index of Lateral Spreading ($I_{LS}$). Both indexes are equal to one in case of any barrier exists. Once one or more barriers are installed those indexes may change desirably as follows: $I_{PRR} < 1$, which corresponds to a reduction of the runout, while $I_{LS} > 1$ is as side effect of the runout reduction.

On the other hand, three deformation mechanisms and three correspondent Limit States ($LS$) are individuated for the barriers due to the impact pressure, namely: (i) local deformation ($LS_1$); (ii) sliding of the barrier over the base ($LS_2$); (iii) combination of previous mechanisms ($LS_3$).

## Application Case

### Input Data

The area study of Cervinara (Southern Italy) is chosen (Cascini et al. 2011) to evaluate the features of potential debris avalanches as far as the propagation areas and realistic impact pressures.

The input data for the case study are: the 2 m × 2 m Digital Terrain Model (*DTM*) of Cervinara slopes and downstream areas taken from technical dataset of a local territorial authority (Fig. 1); the location and the extent of some hypothetic landslide source areas, (Fig. 1a); the rheological features of the propagating masses taken from literature back-analyses of flows occurred in Cervinara and other similar sites (Cuomo et al. 2014a, b). The friction angle $\varphi$' is equal to 22°; $h_w^{rel}$, i.e. the height of the water table relative to the soil thickness in the triggering area, is equal to 0.4; $p_w^{rel}$, namely the base pore water pressure divided by the soil liquefaction value, is equal to 1.0; $c_v$ is the soil consolidation coefficient equal to 0.01 $m^2/s$; and $E_r$ is the bed entrainment coefficient regulating the increase of landslide volume equal to 0.007 $m^{-1}$.

Particularly, there are four source areas with an overall extent of 271,463 $m^2$ where 5,139 computational points are considered, each with an initial soil height ($h_{trig}$) of 2.0 m. All the volumes are released at the same time and propagate together in what we can consider the worst-case scenario. The installed barrier is 6 m high with a trapezoidal cross-section and it is here simply modelled as a modification of the topography, then, it is not deformable.

Two control points are selected at the mountain side of the barrier and respectively named $B_1$ and $B_2$ to estimate the impact pressure starting by the pairs of the computed propagation height and propagation velocity. Two specific cross sections are taken into account to account to estimate the effectiveness of the barrier and named *Section DX* and *Section SX*.

In the deformation analyses, three different barriers are analysed, namely *Type A*, *Type B* and *Type C*, all 6 m high, and equipped with horizontal geogrid layers spaced 0.6 m along the vertical. In each barrier, several control points are defined in the structure for tracking the computed displacements (Fig. 1b). The base computational domain is assumed rectangular and large 50 m × 20 m. The *Type A* barrier has a trapezoidal shape with a top width of 4 m, with the external walls inclined 60°, and the base 11 m wide.

The *Type B* barrier is similar to *Type A*, but equipped also with two geosynthetic-reinforced layers below the ground surface. The *Type C* has a trapezoidal shape with the mountain-side (i.e. impact side) inclined at 80°, the valley-side (rear) inclined 60°, the top width equal to 6.40 m and the base 11 m large.

The barriers and the eventual reinforced layers below the ground surface are assumed as built according the wrap around technique, which is largely used inside reinforced walls or slope (Lajevardi et al. 2015). The reinforcements are bidirectional geogrids, simulated as elasto-plastic elements reacting only to traction axial forces. The iron formworks are schematised as beam elements, resistant to both traction and bending moments.

The numerical analyses adopt a non-associative elastic-perfectly plastic Druger-Prager constitutive model for the granular soil inside the barrier. A negligible cohesion is considered only to avoid numerical instabilities along the barrier facing. The mechanical properties are listed in Table 1. The Rayleigh approach is used to model the damping effect, here considered as a linear combination of those related to mass and stiffness, with the coefficients $\alpha$ and $\beta$ derived from Aversa et al. (2007) and equal to 0.01 and 0.001, respectively. Elasto-plastic frictional interfaces are inserted between the granular soil and the geogrids, with strength properties reduced to 80% of those of nearby soil.

The analyses are performed for a unitary width of the barrier, in plane strain conditions.

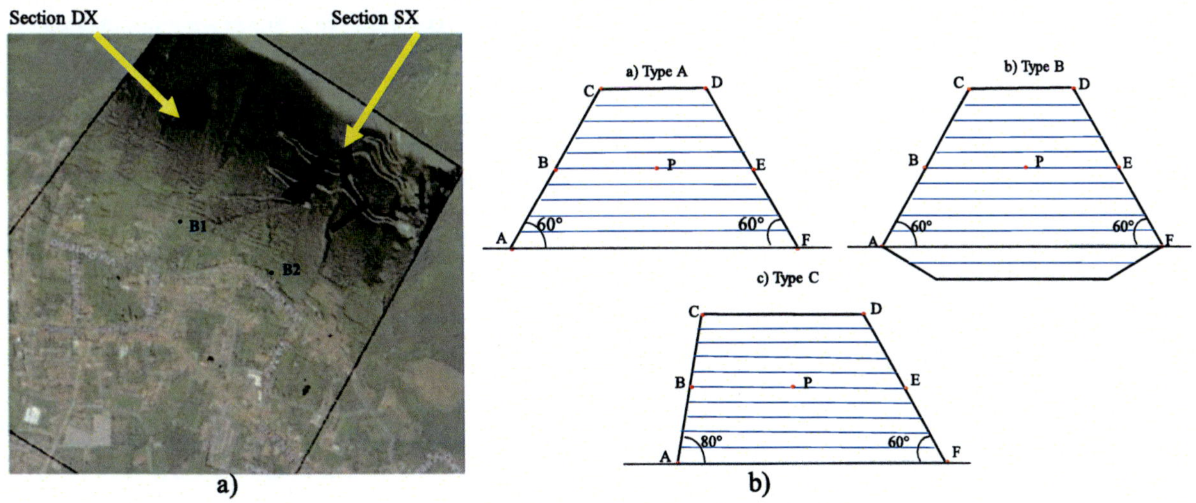


**Fig. 1** a Cervinara study area (Italy) with the indication of the control points and b cross section of the barriers

**Table 1** Mechanical properties of the materials

| | $\gamma_{dry}$ | c' | $\varphi$ | $k_0$ | N | E | Tu | EA | EI |
|---|---|---|---|---|---|---|---|---|---|
| | (kN/m$^3$) | (kPa) | (°) | (–) | (–) | (kPa) | (kN/m) | (kN/m) | (kN/m$^2$) |
| Soil | 18 | 0.1 | 33 | 0.45 | 0.25 | 2.0e + 4 | | | |
| Subsoil | 18 | 0.1 | 30 | 0.5 | 0.30 | 1.0e + 5 | | | |
| Geogrid | | | | | | | 50 | 500 | |
| Anchors | | | | | | | | 6.34e + 4 | 35 |
| Formwork | | | | | | | | 2.11e + 4 | |

## Propagation Analysis

The numerical cases are named as follows: the two-digit number after the letter indicates the triggering height expressed in decimetres; the number after the underscore refers to the barrier height in meters; the last letter indicates the distance from the source area, namely nearest (*n*) or farthest (*f*), or the case of two barriers (*t*). Conversely, the last term could be the digit 0, which indicates the natural slope (absence of the barrier). The four triggering areas pertain to two distinct propagation masses that affect the above mentioned *Section DX* and *Section SX*.

Figure 2 shows the involved areas in the case of natural slope (Fig. 2a), or with one barrier installed (Fig. 2b, c) or with two barriers (Fig. 2d). The largest reduction of the involved area is obtained in the case of two barriers. In fact, the first barrier contributes in reducing the volumes that may overcome the barrier and which are conversely almost totally stopped by the second barrier. The case *F20_6_n* and case *F20_6_f* are representative of the propagation scenario named $SC_2$; while, in the case of two barriers, the scenario is the $SC_1$.

The *ILP* values are calculated versus time in correspondence of the two control points previously individuated. The maximum value is for point B2 and equal to about 900 kPa. This value agrees those values already calculated by Cuomo et al. (2019b, c); while the pressure evaluated in correspondence of the point B1 is lower (about 400 kPa). The deformation analyses are carried out considering several values of $ILP_{peak}$ to show the possible different cases (Figs. 3 and 4).

## Deformation Analysis

The results of the deformation analyses include both the individuation of the deformation mechanisms of the impacted barriers and the assessment of the maximum displacements of the barriers, through the two commercial softwares mentioned before.

It can be seen that both FEM and FDM analyses are able to simulate the impact against geosynthetics-reinforced barriers and its components, with some limitations in the case of FEM. The *Type B* barrier is here chosen to show the results of FEM analyses (Plaxis) compared to those of FDM analyses (Flac). The *Type B* barrier presents a final displacement of about 1.30 m for an impact peak pressure of 246 kPa. Furthermore, the numerical results outline both a local deformation and a translation along the base (Limit State $LS_2$) with a good agreement between the two methods. The application of pressures higher than 246 kPa was not possible in the FEM analyses due to loss of convergence, as usual in geometrically nonlinear problems. Thus, FDM must be used.

The *Type B* and *Type A* barriers present deformations and a limit state, identifiable as $LS_3$. The *Type C* barrier has the lowest displacements without local deformations ($LS_2$); this is due to the highest mass of *Type C* barrier. The deformation mechanisms of the barriers are further highlighted considering the time-displacement curves (Fig. 5).

The different displacements simulated at the selected control points indicate that the barrier undergoes a large internal deformation. In the *Type A* barrier, the cumulative displacement depends on both translation and local deformation. The *Type B* barrier exhibits a similar behaviour to *Type A*; however, the *Type B* presents the maximum difference among the displacements simulated at the point "F" and the upper points "C" and "D", which indicates a large compression of the *Type B* barrier.

In *Type C* barrier, the displacement curves of all the control points almost completely overlap, which is representative of a global translation along the base and very limited local deformation ($LS_2$).

## Performance Analyses

The non-dimensional indexes IPRR and ILS are reported in Fig. 6 for the two reference cross sections. In all the cases of one barrier used, the landslide runout is diminished of about

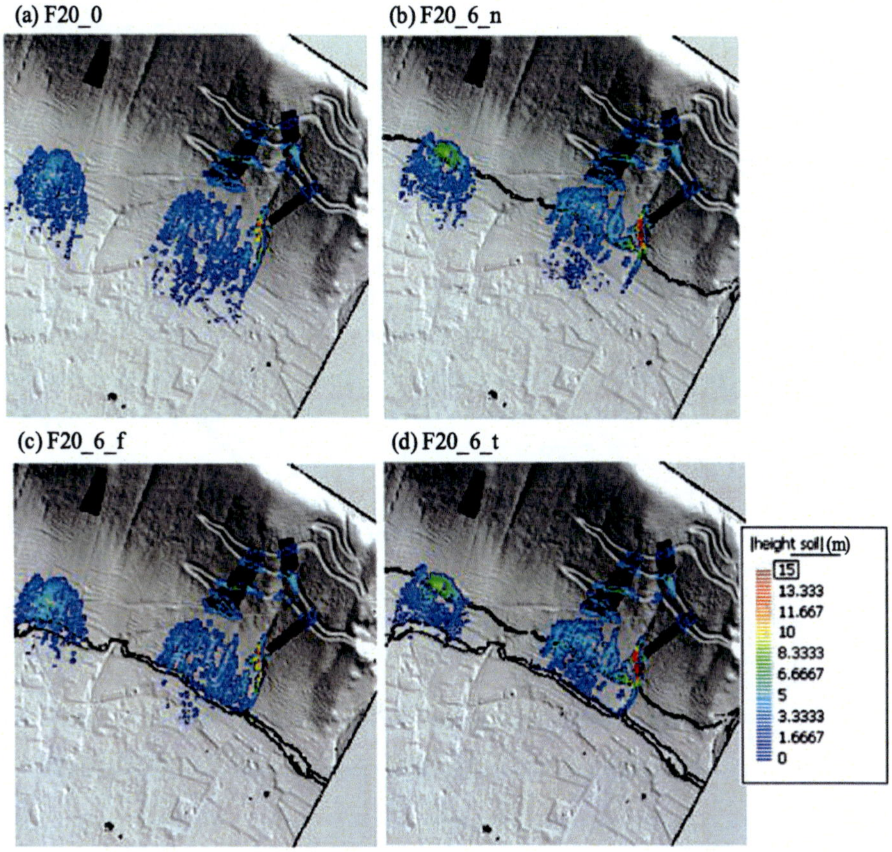


**Fig. 2** Final deposits of debris avalanches for the cases F20: **a** without barrier; **b** with the barrier nearest to the source areas; **c** with the barrier located at the foothill area; **d** with both the barriers

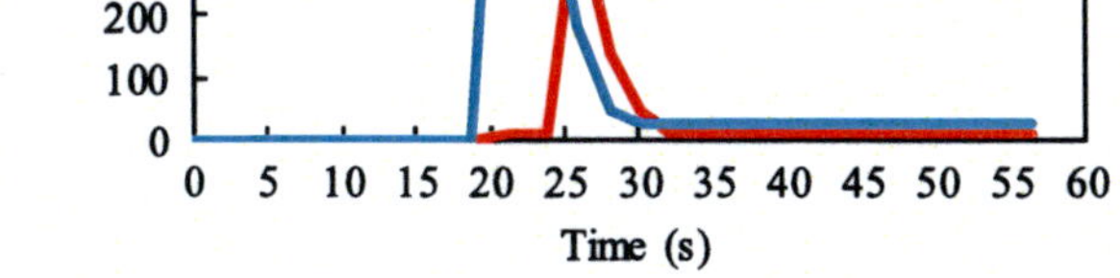


**Fig. 3** Maximum Impact Loading Pressure (*ILP*) of the flow versus time at control points ($B_1$ and $B_2$)

10%; though the percentage is small, the reduction of runout could be significant due to the high runout distances. Conversely, the a low or no increase of lateral spreading is computed. With two barriers used, the runout is reduced up

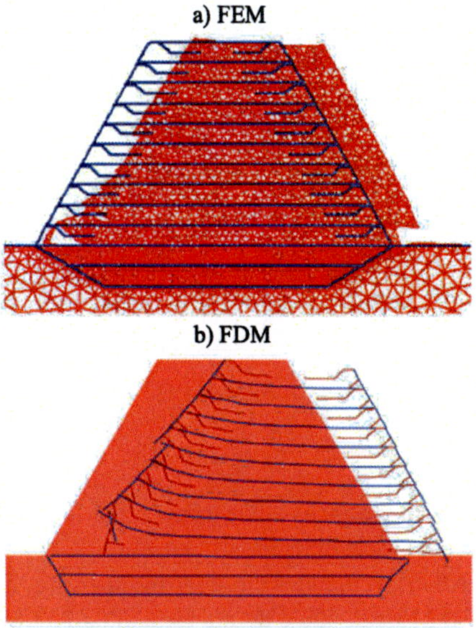


**Fig. 4** Deformation of the barrier *Type B*, with a peak impact equal to 246 kPa (at t = 0.56 s) and impact energy equal to about 1.0 MJ: **a** FEM analysis; **b** FDM analysis

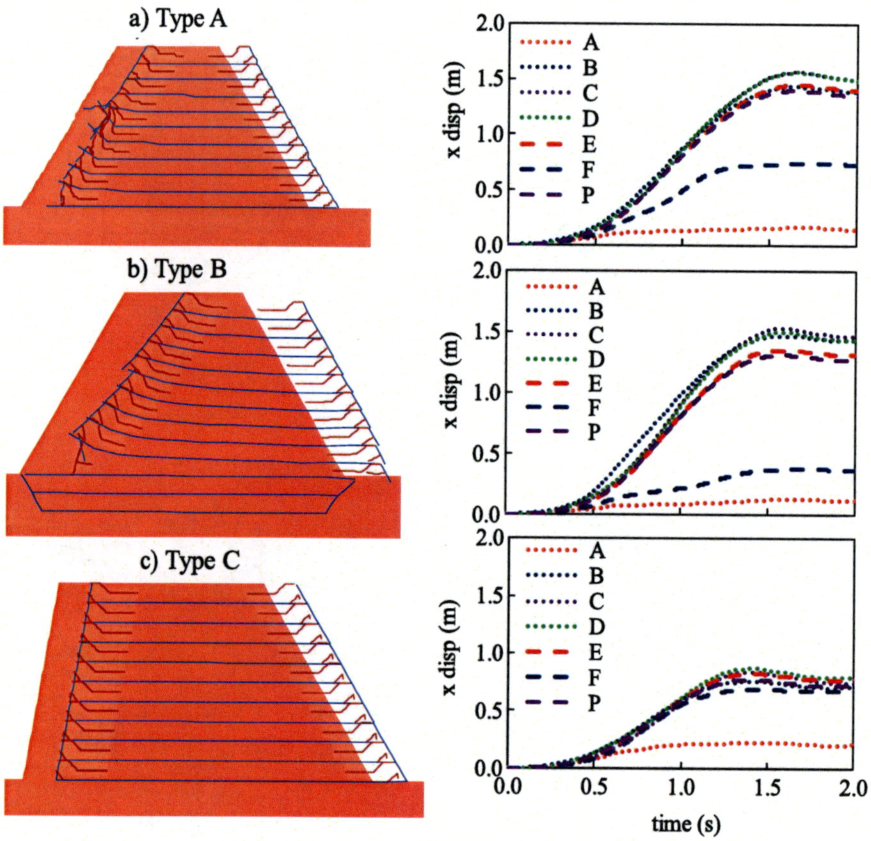


**Fig. 5** Deformation mechanism of several barriers in FDM analysis with a peak impact pressure equal to 246 kPa (at t = 0.56 s) and energy equal to about 1.0 MJ

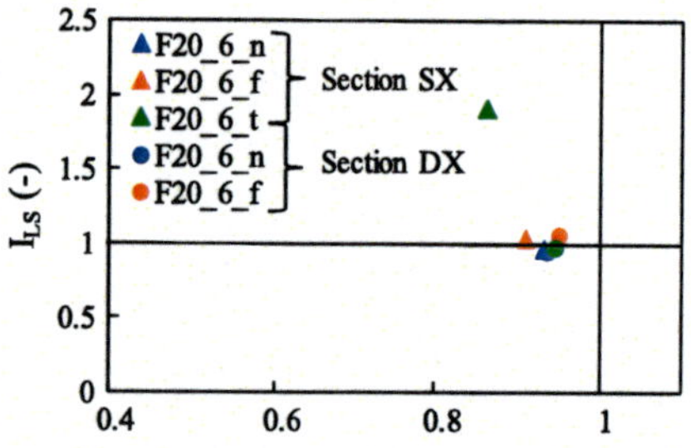


**Fig. 6** Indexes $I_{PRR}$ and $I_{LS}$ computed for different positions of the barrier

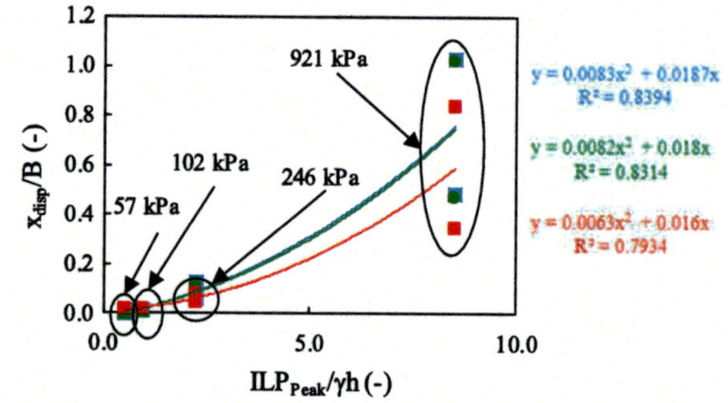


**Fig. 7** Dimensionless x-displacements of control point P versus $ILP_{peak}$ divided by the stress at the base of the barriers; in blue *Type A*; in green *Type B*; in red *Type C*

to 15% with a significant increment of the lateral spreading (case F20_6_t, Section SX).

In relation to the deformation mechanisms and Limit States of the barrier, FEM allows simulating *LS*1 but present some numerical instability in the cases of displacements higher than 2 m, recorded in particular for $LS_2$ and $LS_3$. However, the deformation mechanism $LS_1$ is never recorded for the analysed barriers. Instead, through the FDM analyses it is possible considering any impact pressure *ILP*, and so also 900 kPa, that is the maximum value estimated for the real case study, considering volumes of 27–45 m$^3$ impacting the barrier with velocity of 2–10 m/s, and with impact energy comprised between 0.1 MJ and 4.1 MJ.

Aimed at generalising the results, the non-dimensional values of horizontal displacement ($x_{disp}/B$) and impact pressure ($ILP_{peak}/\gamma \cdot h$) are plotted for different impact scenarios in Fig. 7, where $B$ is the base of the barrier while the product of $\gamma$ (soil unit weight) and $h$ (height of the barrier) represents the vertical stress at the base of the barrier. The trend is highly non-linear already for low values of $ILP_{peak}$, but reasonably well interpreted by second-order polynomial curves, with correlation coefficient $R^2$ always higher than 0.8, which could be used for preliminary assessment of barrier performance.

## Conclusions

This paper investigates two aspects of the barrier used as mitigation work against landslides. The first principal result is that the propagation mechanism scenario here named $SC_3$ is not recorded; indeed, all the geometrical configurations of the barrier are able at least to reduce the runout of the debris avalanches. On the other hand, the deformation mechanism and Limit State $LS_1$ is not simulated, but only a combination of local deformation with the global translation, namely $LS_3$. Non-dimensional displacement of the barriers increases more than linearly with the impact pressure and depends on barrier geometry. Furthermore, the trend of dimensionless displacement versus peak pressure could be well interpreted through polynomial curves.

**Acknowledgements** This research was developed within the framework of different projects: (i) Industrial Partnership PhD Course (Dottorato Industriale Regione Campania, Italia); (ii) project FARB 2017 "Modellazione numerica e analisi inversa per frane tipo flusso" funded by the Italian Education and Research Ministry. The authors would like to thank Geosintex s.r.l. for support to the research. The authors wish to thank the Autorità di Bacino Liri, Garigliano e Volturno for the input data necessary to the landslide propagation analysis.

## References

Avelar AS, Netto ALC, Lacerda WA, Becker LB, Mendonça MB (2013) Mechanisms of the recent catastrophic landslides in the mountainous range of Rio de Janeiro, Brazil. Landslide science and practice. Springer, Berlin, pp 265–270

Aversa S, Maiorano RMS, Tamagnini C (2007) Influence of damping and soil model on the computed seismic response of flexible retaining structures. In: 14th European conference on soil mechanics and geotechnical engineering

Brandl H (2011) Geosynthetics applications for the mitigation of natural disasters and for environmental protection. Geosynth Int 18 (6):340–390

Brinkgreve R (2007) PLAXIS 2D. Version 8.5 finite-element code for soil and rock analyses: complete set of manuals. Balkema, Rotterdam

Cascini L Cuomo S De Santis A (2011) Numerical modelling of the December 1999 Cervinara flow-like mass movements (Southern Italy). In: Proceedings of 5th international conference on debris-flow hazards mitigation: mechanics, prediction and assessment. Italian Journal of Engineering Geology and Environment, pp 635–644

Cascini L, Cuomo S, Pastor M, Rendina I (2016) SPH-FDM propagation and pore water pressure modelling for debris flows in flume tests. Eng Geol 213:74–83

Cuomo S, Pastor M, Cascini L, Castorino GC (2014a) Interplay of rheology and entrainment in debris avalanches: a numerical study. Can Geotech J 51(11):1318–1330

Cuomo S, Frigo L, Manzo M (2014b) Reinforcement of fluvial levees: a case study of Tevere River (Italy). In: Proceedings of 10th international conference on Geosynthetics, Berlin, 21–25 September 2014, pp 1–8

Cuomo S, Pastor M, Capobianco V, Cascini L (2016) Modelling the space–time evolution of bed entrainment for flow-like landslides. Eng Geol 212:10–20

Cuomo S, Moretti S, Aversa S (2019) Effects of artificial barriers on the propagation of debris avalanches. Landslides 16(6):1077–1087

Cuomo S, Moretti S, Lanza A, Aversa S (2019b). Geosynthetics reinforced barriers impacted by flow-like landslides. In: Proceedings of European conference on soil mechanics and geotechnical engineering, 1–6 September 2019

Cuomo S, Moretti S, Petrosino S, Aversa S (2019c). Dynamic impact of debris avalanches on structures. Proceedings overview. COMPDYN 2019, 24–26 June 2019

Gioffrè D, Mandaglio MC, Di Prisco C, Moraci N (2017) Evaluation of rapid landslide impact forces against sheltering structures. Rivista Italiana Di Geotecnica 51(3):64–76

Itasca Consulting Group (2016) FLAC Fast Lagrangian Analysis of Continua, vol 8

Lambert S, Bourrier F (2013) Design of rockfall protection embankments: a review. Eng Geol 154:77–88

Lajevardi SH, Silvani C, Dias D, Briançon L, Villard P (2015) Geosynthetics anchorage with wrap around: experimental and numerical studies. Geosynth Int 22(4):273–287

Moretti S, Cuomo S, Aversa S (2019) Feasibility of foothill barriers to reduce the propagation of debris avalanches. National Conference of the Researchers of Geotechnical Engineering. Springer, Cham, pp 309–317

Pastor M, Haddad B, Sorbino G, Cuomo S, Drempetic V (2009) A depth-integrated, coupled SPH model for flow-like landslides and related phenomena. Int J Numer Anal Meth Geomech 33(2):143–172

Revellino P, Guerriero L, Grelle G, Hungr O, Fiorillo F, Esposito L, Guadagno FM (2013) Initiation and propagation of the 2005 debris avalanche at Nocera Inferiore (southern Italy). Ital J Geosci 132 (3):366–379

Yang KH, Wu JT, Chen RH, Chen YS (2016) Lateral bearing capacity and failure mode of geosynthetic-reinforced soil barriers subject to lateral loadings. Geotext Geomembr 44(6):799–812

Xu Q, Fan XM, Huang RQ, Van Westen C (2009) Landslide dams triggered by the Wenchuan earthquake, Sichuan Province, south West China. Bull Eng Geol Environ 68(3):373–386

Check for updates

WLF
2020 Kyoto Japan

# Large and Small Scale Multi-Sensors Remote Sensing for Landslide Characterisation and Monitoring

Carlo Tacconi Stefanelli, Teresa Gracchi, Guglielmo Rossi, and Sandro Moretti

**Abstract**

In the last years, the use of Unmanned Aerial Vehicles (UAVs) has developed rapidly across several field of earth sciences application, including landslides characterisation and monitoring, therefore providing a strong support for hazard and risk management activities especially with the introduction and advances in the miniaturization of traditional and new generation sensors. The flexibility, low cost, easy operability, and rapidness of intervention in emergency situation gives to these instruments, a strong potential in opening up a vast new area of opportunities in remote sensing for observation, measuring, mapping, monitoring, and management in various landslide environment. Unless initially only air photography was the main application for UAVs, recently new sensors, both passive and active, are being increasingly used. This paper, through some case studies on landslide investigations, aims at giving an overview on several sensors and techniques using UAVs platform addressed to landslide detection, characterization and monitoring.

**Keywords**

Remote sensing • Drone • Landslide

C. T. Stefanelli · T. Gracchi · G. Rossi · S. Moretti (✉)
Earth Sciences Department, University of Florence, Via La Pira 4, 50121 Firenze, Italy
e-mail: sandro.moretti@unifi.it

C. T. Stefanelli
e-mail: carlo.tacconistefanelli@unifi.it

T. Gracchi
e-mail: teresa.gracchi@unifi.it

G. Rossi
e-mail: guglielmo.rossi@unifi.it


V. Vilímek et al. (eds.), *Understanding and Reducing Landslide Disaster Risk*,
ICL Contribution to Landslide Disaster Risk Reduction,
https://doi.org/10.1007/978-3-030-60319-9_39

## Introduction

Displacement monitoring of unstable slopes is a crucial tool for the prevention of hazards. It is often the only solution for the survey and the early warning of large landslides that cannot be stabilized or that may accelerate suddenly (Travelletti et al. 2012). The techniques to monitor displacement are broadly subdivided into two main groups: ground based and remote sensing techniques. The firs group needs installation of sensors in and/or out the target area measuring data in time nevertheless they suffer of their punctual nature and relatively high cost of installation and maintenance. The Remote sensing techniques are basic tools for applying spatially distributed information of unstable slopes (Delacourt et al. 2007). The main advantage of using UAVs remote sensing payload sensors for landslide investigations is the capability to provide spatially continuous data with very high spatial resolution, especially if integrated with the point-wise measurements typically acquired by ground-based techniques. In this work a resume of drone and associated typology of sensors application is described for landslide characterization and monitoring.

## UAV Classification

Unmanned aircraft vehicles are classified mainly into 3 categories: fixed wing, rotary wing and aerostatics. They have different characteristics, advantages and drawbacks.

A fixed wing UAV is basically an airplane, where the lift is produced by an aerodynamical appendix that moves fast in the air but is fixed with respect to the aircraft axes. Fixed wings need high relative airspeed to generate the lift then take-off and landing maneuvers are critical. More space or the catapult method is needed to gain speed during take-off; landings are often uncontrolled because they are accomplished by releasing a parachute. During the acquisition, the high flight speed, usually more than 20 m/s, can blurry data:

high level of sensibility sensors are needed to obtain sharp data when high resolution is the key point. Also, data georeferencing can be affected by speed: 100 ms of delay in the triggering at cruise speed can add metrical error in the geotagging chain of data. High speed GPS and precise detection of acquisition time stamps are crucial. Speed can be useful in long distance cruise and big areas to cover, combined with a high efficiency of fixed wings leads to long flight time which make them more suitable for long survey campaign.

Rotary wings UAV have a propeller that produces lift rotating in the air. This category is populated by helicopters and multi rotors. Multi rotors or multicopters are the most common rotary wing UAV because the mechanical design of the rotors is simplified due to the lack of cyclic and collective to control the helicopter movement. Change in direction, altitude, rotation and speed are operated only changing the speed of the motors: a mechanically safe and simpler method that relies on the electronics and software that operate a synchronized change of speed of the propeller. The drawback is that multi rotors cannot be maneuverable without a software to manage the motors: it is not possible to override with manual control if the software fails. Helicopters, like fixed wings can still be maneuverable and flyable only from a pilot input. The main advantages of rotary wings are the very fast deployment on field, vertical take-off and landing and high mass useful load respect dimensions. The possibility of hovering in a fixed point or acquiring at low speed can be very useful to increase quality of data in all conditions and the georeferencing accuracy. Rotary wings are energy hungry and do not excel in flight endurance.

Aerostatics UAV are the most common because of the dimensions and the limited operational weather conditions. Dirigible are the most used in this category for low altitude flight and they excel in flight time.

Fixed wings and rotary wings can merge in VTOL airplanes (Vertical Take Off and Landing) to overcome some reciprocal drawbacks and obtain a better flight endurance and easier take offs and landings. The merging is not a definitive solution because disadvantages can add up resulting in an aircraft made for specialized operations and tasks.

## Sensors Classification

UAVs can carry different kinds of sensors, one or even more at time, according to the payload capacity and connections with the control station. Their usage depends on several factors as the type of investigation, the topography of the area, the required products and the economics resources. The most commonly used in the field of landslides mapping and monitoring are listed in Table 1, together with their characteristics and applicability.

They can be divided into two main categories, active and passive sensors. The first one can be defined as sensors that actively send a pulse and measure the signals backscattered to the sensor itself. Conversely, passive sensors detect some type of input from the physical environment, always receiving and not transmitting signals.

Among them, optical imagery is certainly the most spread technology, due to its cost effectiveness, easiness of use and variety of achievable products such as DEMs, DSMs and Orthophotos (Colomina and Molina 2014; Kršák et al. 2016; Niethammer et al. 2012; Remondino et al. 2011; Rossi et al. 2018). The basic principle is to reconstruct objects in 3D through the process of recording, measuring and interpreting photographic images. UAV photogrammetry can be considered a low-cost alternative to the classic aerial manned photogrammetry, with respect to whom it allows more detailed surveys but on smaller areas. The resolution is the main variable of this kind of survey, and it is strictly connected to the resolution of the camera itself as well as the flight height.

The same principle applies for the thermal cameras producing thermograms, i.e. images of the radiation in the long-infrared range that is emitted by the investigated objects. The presence of fractures, moisture and voids influences the heat transfer of materials generating thermal anomaly in the observed thermal imagery (Casagli et al. 2017; Frodella et al. 2017; Pappalardo et al. 2018; Zhou et al. 2020). These can give fundamental information about the fracturing of rock walls as well as drainage systems or more in general of weak shallow surfaces. As in the case of photogrammetry, this technique has some limitation connected to the boundary circumstances at the time of the survey. Both require indeed an optimal atmospheric conditions and sun exposition (Wierzbicki et al. 2018; Yu et al. 2017), limits that could be however overcome when considering the low-cost effectiveness of the sensors.

Multispectral and hyperspectral sensors are passive instruments capable of recording the amount of reflected energy of objects on the earth's surface in the different wavelengths of the electromagnetic spectrum. The outputs are therefore multiband images that allow to extract spatial information and produce accurate thematic maps with the use of classifiers (Baccani et al. 2018; Natesan et al. 2018; Lin et al. 2017). The NDVI (Normalized Difference Vegetation Index) is one of the most used indicators achievable through these technologies, helpful to assess whether the observed target contains live green vegetation (Carlson and Ripley 1997).

As previously explained, passive sensors (like all those mentioned so far) are limited in observing and detecting external inputs. For this reason, the obtained information (such as images or thermograms) need to be carefully post-processed. On the other hand, active sensors acquire

**Table 1** UAV sensors and their applicability

| | | | | |
|---|---|---|---|---|
| Optical camera | Passive | 3D point cloud, DEM, DSM, Orthophoto | Landslides volume estimation, deformation maps trough multi-temporal surveys, georeferenced layer for further elaborations, emergency plan | Low |
| Thermal camera | Passive | Thermogram | Detection of fractures, subsoil voids and moisture | Middle-Low |
| Multispectral | Passive | Spectral Images | Extraction of spectral index, Soil characteristics (mineralogy), land use | Middle |
| Hyperspectral | Passive | Spectral Images | Spectral signatures, extraction of spectral index, soil characteristics (mineralogy), land use | Very High |
| LiDAR | Active | High resolution 3D point cloud, DEM, DSM, Orthophoto | Landslides volume estimation, deformation maps trough multi-temporal surveys, georeferenced layer for further elaborations | Very High |
| Ground Penetration Radar | Active | Profiles of the signals reflected from the subsurface | Detection of non-conductive materials, voids, sand and gravel deposits, fractures and faults | High |

measures, gathering usually the required information without the need of further elaborations, saving thus processing time.

An example of active sensor that can be adaptable to drones is the Ground Penetrating Radar (GPR). In this case, a transmitting antenna sends high frequency pulses of electromagnetic energy into the ground, while another antenna receives the reflections that bounce back up from layers and targets. The depth of the targets can be therefore evaluated if the Relative Dielectric Permittivity is known (Lanzarone et al. 2016). The information is therefore recorded by the instrument and the user has only to interpret them. Depth penetration is a function of the antenna frequency and the electrical conductivity of the soils in the survey area. Lower the frequency of the antennas, greater the penetration depth is. GPR is best suited for low conductivity or high resistivity materials like sandy soils, limestone and granite and work well in most saturated environments (Benson 1995; Mellett 1995). GPR on drones are a quite recent technique, but these kinds of radar have been already used several times in the field of landslides mapping, characterization and monitoring (Bichler et al. 2004; Hu and Shan 2016; Sass et al. 2008).

Thanks to the increasing reliability of drones, also heavy and costly active instruments can be safely transported with low risk of damage. It is the case of LiDAR (Light Detection And Ranging), a consolidated remote sensing technique used to obtain high resolution 3D point clouds of the topographic surface through the emission of highly collimated, directional, coherent and in-phase electromagnetic radiation (Jaboyedoff et al. 2012). The 3D point cloud can be then easily transformed in DEM, DSM and orthophotos, that are the same typical outcomes of a photogrammetric survey. However, while optical imagery requires optimal weather conditions, illumination rates and sun exposition, laser scanners can properly work even in low-light conditions. Moreover, in a wooded terrain, LiDAR can penetrate the canopy, allowing to detect and map landslides in forested areas, that is a competitive advantage over the use of optical cameras not able to compute feasible surveys in this kind of areas, thus justifying the higher costs of the LiDAR technique (Guzzetti et al. 2012). Because of these aspects, and considering also the higher precision obtainable, UAV laser scanners is becoming a very widespread technique (Karantanellis et al. 2018; Kwon et al. 2019; Pellicani et al. 2019).

## Materials and Methods

Battery life, payload capacity and flight safety are key and critical characteristics for a UAV. According to these points, a new multicopter concept has been thought by Earth Sciences Department and Center of Civil Protection of the University of Florence (Italy) to overcome some lack of traditional drone configuration. The patented Saturn drones are designed with an innovative ring perimeter airframe that allows the weight reduction in multi-rotor configurations improving the load capacity or flight autonomy.

Furthermore, the rigid connection between the motors, crucial for a correct flight dynamic, take place along the ring perimeter frame avoiding the central structure and cutting the vibrations spreading to the sensors and the flight instruments. Flight safety of the drones is enhanced through the use of an automatic engine positioning system along the chassis that allows a reconfiguration of the engines even during the flight: in the event of an engine failure, it is possible to rebalance the flight attitude in order to complete an emergency landing without payload damage or uncontrolled fall with harm of people.

**Fig. 1** The Saturn Mini X-21 drone

## The Multi-Copter Saturn Family

The ring perimeter frame patent led to the development of two drones each specialized in different tasks: the Saturn mini and the Saturn M2. Both drone editions, designed and built by the Department of Earth Sciences of the University of Florence, have a configuration with six motors, whose propulsive redundancy reduce the falling chances during emergency situations or limiting the ground impact in case of failure up to 2 engines. The extremely rigid frame in flight and the aerodynamics of the Saturn drones allow remarkable performance both uphill and downhill. The descent rate is a critical parameter of the multi-rotors which offer the capacity to handle rapid altitude losses without threatening aerodynamic phenomena that can lead to an uncontrolled fall. The control radio frequencies are redundant and differentiated to increase operational safety even in case of strong radio interference. A third frequency is dedicated to the pilot camera placed in the bow of the drone with very low latency. The dedicated pilot camera allows to operate safely even during flights that require the aiming of the sensors in different directions to those of the vehicle's progress. This configuration is extremely practical during nadiral acquisitions in confined spaces within gorges or with vertical walls. The landing gear is a tripod that guarantees stability on takeoff but above all on landing even in rough terrain and in case of sustained winds and gusts.

The Saturn mini (Fig. 1) has a diameter of 55 cm, maximum payload of 1.5 kg and take-off operating mass under 4 kg, with a maximum autonomy of 30 min. The frame is entirely produced by 3D printing and offers a high integration with instruments and sensors. The maximum take-off weight under 4 kg lets it to be classified in a less restrictive category and together with the payload capacity of up to 1.5 kg make it suitable for a wide range of scenarios.

The maximum speed of the Saturn mini is about 90 km/h giving it an excellent resistance to wind and possible gusts. The extreme maneuverability and the reduced flight restrictions make it particularly suitable for photogrammetric survey during rapid intervention for emergency or in areas with limited accessibility. The Saturn M2 drone (Fig. 2) has a 100 cm diameter frame, a maximum payload of 10 kg and a take-off operating mass of 22 kg. This take-off mass puts it in a category with some restriction of use. The maximum flight autonomy is approximately 30 min, which may vary depending on the flight conditions and the payload carried. The Saturn M2 can be used in emergency scenarios even in adverse weather and climatic conditions (strong winds or rain). The drone can be used safely in critical areas as it is equipped with appropriate failsafe, independent flight termination system and parachute. The characteristics of Saturn M2 are ideal for performing flight activities with multi-sensor platforms thanks to the high stability and flexibility in the implementation of loads of various shapes, sizes, weights and positions.

**Fig. 2** The Saturn M2 drone

## The Multi Sensor Family

Drones are used to provide all kind of data required for the analysis of hydrogeological instability, even in emergency conditions when the rapid acquisition becomes the priority, minimizing risks for operators. Within the sensors that can be mounted on Saturn drones there are: Haigh resolution

Panchromatic camera, Thermal camera, Hyperspectral camera, Lidar, Ground-Penetrating Radar, Radar.

With height resolution panchromatic camera three products can be achieved (Remondino et al. 2011): orthophotos, Digital Surface Model (DSM) and Digital Terrain Model (DTM). The orthophoto is a product created from aerial photos that are ortho-rectified and geo-referenced. The resolution on the ground can be increased to cover a large area by composing mosaics of images without signs of discontinuity through the photogrammetric technique. The orthorectification requires a digital model of the DSM, which is normally produced together. The DSM is a three-dimensional rendering of what is visible to the drone's camera through a programmed acquisition of images and subsequent processing on the fly. The DTM is the three-dimensional representation of the geodesic surface mapped. It is an interpolated product, which provides, through automated algorithms, the removal of vegetation and anthropogenic artifacts, fixed or not (buildings, vehicles, poles, etc. ...).

In recent years, infrared thermography (TIR = Terrestrial Infrared Thermography) has seen its use grow in the field of identification and characterization of landslides (Frodella et al. 2017; Morello 2018). Interesting studies have been carried out in the identification of critical elements in rock slopes, such as sub-superficial hollows, humidity areas linked to fractures and intensely fractured or detensioned sectors of rock mass. The dependence of physical thermal parameters linked to the density and composition of geological materials, such as thermal inertia, suggests the use of TIR in order to characterize the physiography of very heterogeneous deposits such as landslides. Thus, TIR constitutes a valid ancillary technique, which integrated with other remote sensing techniques (ex: Lidar, GBInSAR), can provide an important contribution for the purpose of a detailed characterization of slope-level instability phenomena.

Hydracam (HYperspecral DRone Advanced CAMera) is a hyperspectral camera for drone environmental monitoring by using an electro optical liquid crystal technology combined with a commercial camera, which allows to continuously space in a wide range of wavelengths (Baccani et al. 2018). The sensor works between 650 and 1100 nm and has a resolution of about 10–20 cm from a flying elevation between 30 and 150 m with a field of view of 40° (Fig. 3). Compared to a mechanical device with classic mobile filters, with this sensor it is possible to have high spectral resolution images in a short time. Moreover, the high integration of electronics, combined with the computing power of modern processors, allows to develop integrated and compact hyperspectral systems reducing the mass, making it a drone payload, and costs.

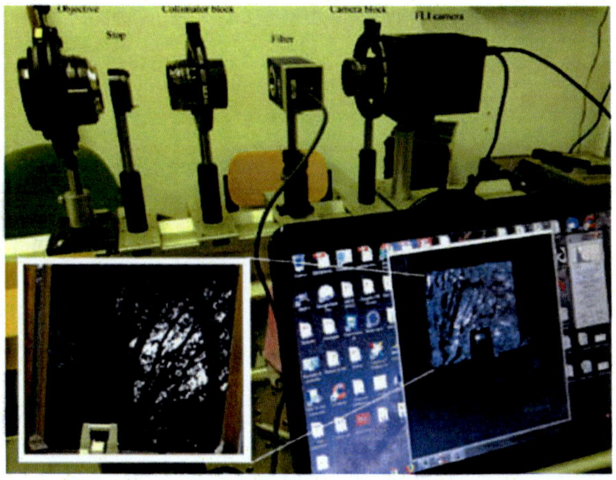


**Fig. 3** Laboratory setup aligned using self-centering jaw clamps and a theodolite. A FLI camera in place of XIMEA GmbH sensor has been used. The actual acquisition at 800 nm fixed by the tunable filter is enlarged in the box at the bottom left

The Ground-Penetrating Radar (GPR) is a geophysical method that uses electromagnetic radiation pulses in the microwave band (UHF/VHF frequencies) to reconstruct the subsurface detecting the reflected signals from underground surfaces and structures. GPR can be used through rock, soil, water and concrete and can detect buried objects, material properties changes, voids and cracks (Altdorff et al. 2014; Neal 2004). The GPR sensors for drone platform use high-frequency radio waves (from 10 MHz to 2.6 GHz) that allow, with a flying elevation between 30 and 50 m, to explore the subsoil from a depth of a few meters to a maximum depth of a few tens of meters.

For the radar system, an experimental lightweight Frequency Modulated Continuous Wave (FMCW) radar system has been developed to produce low resolution DEMs from long distances (Tarchi et al. 2017). This experimenting technique has great potential that still have to be improved with the opportunity to pierce into snow and low vegetation coverage and a wide range of applications.

## Applications

In soil sustainable exploitation planning and land resource evaluation, modern and accurate up-to-date scale maps of soil properties is one of the most important need such as inexpensive and non-invasive mapping methods. In that respect, geophysical sensing techniques allow to acquire rapidly a large amount of data on soil conditions with respect to conventional direct field sampling and traditional analyses which are generally invasive, time-consuming, expensive and often require a large number of samples to capture variability of the soil properties (Brevik et al. 2003; Kweon 2012). Among existing geophysical techniques, hyperspectral (Hs), results as particularly interesting as it allows characterization of relatively large areas with fine spatial and/or temporal resolutions. For soil and landslide characterization, clay content is one of the most important data to investigate. Ciampalini et al. (2015), Garfagnoli et al. (2013), Ben-Dor et al. (2009) applied hyperspectral analysis using a push broom aerial hyperspectral sensor to investigate the soil clay content and clay minerals. In their research the authors produce a map that represents the different clay minerals and their total amount in percentage (Fig. 4). In this map brown corresponds to the higher percentages and blue corresponds to the lower percentages. A mask was applied to the neighboring grassland. The maps of hillite and montmorillonite (Fig. 4a, b) show a similar trend. They assume that the lower values clustered in the northern part of the parcel, is due to a topographically more elevated area, while an increasing trend towards the south (i.e., parallel to the flow direction, towards the bottom) can be observed for both minerals. This tendency can be interpreted as a normal effect due to soil erosion by rainwater and the consequent removal and concentration of finer material downhill. On the contrary, the map of kaolinite content (Fig. 4c) displays a completely inverted trend, with higher concentrations located to the northern and south-eastern parts of the field. As shown in Fig. 4d, the distribution of the total clay content closely resembles that of illite and montmorillonite. A comparison among the maps shows that kaolinite scarcely affects the total clay areal distribution.

In this section, some examples of landslide characterization and monitoring using different UAV sensors are presented. The first cases are related to the application of optical cameras and therefore photogrammetric processes, focusing in particular on surveys in emergency conditions and multitemporal analysis. Concerning the latter, the comparison of terrain 3D models acquired in different times allows indeed to evaluate changes occurred on the surfaces. This technique founds application in several environments such as river dynamics, erosion processes and several geomorphological processes (James et al. 2013; Riquelme et al. 2019), but it is widely used to reconstruct volume changes due to the occurrence of a landslide.

An example on landform reconstruction at different stages and estimation of volumes has been performed in the village of Ricasoli, an area strongly affected by diffuse slope instability in the Upper Arno river Valley (Tuscany, Italy). The substrate of the basin where the village is located consists in flysh-type formations constituted by sandstones interlayered with siltstone, overlain with fluvial-lacustrine sediments. Several landslides affect this area, mainly made by sands and sandy silts and characterized by high slope angles. The study here presented focuses on two shallow landslides occurred respectively on March 1st (Landslide 1, LS1) and March 30th, 2016 (Landslide 2, LS2) after a period of heavy rainfall (Fig. 5) (Rossi et al. 2018).

Three aerial photogrammetric surveys were performed on the study area with the Saturn Drone, respectively on July 30th, 2015, March 2nd, 2016, and April 6th, 2016, using a Sony digital RGB camera with 8-MP resolution and flying at a constant altitude of approximately 70 m a.g.l.. Data have been handled in Agisoft Photoscan Professional software and the acquired 3D point clouds have been processed and appropriately filtered removing all the points corresponding to buildings, unwanted elements, and high vegetation, to obtain high-resolution DTMs (0.05 m/pix). The DTMs were

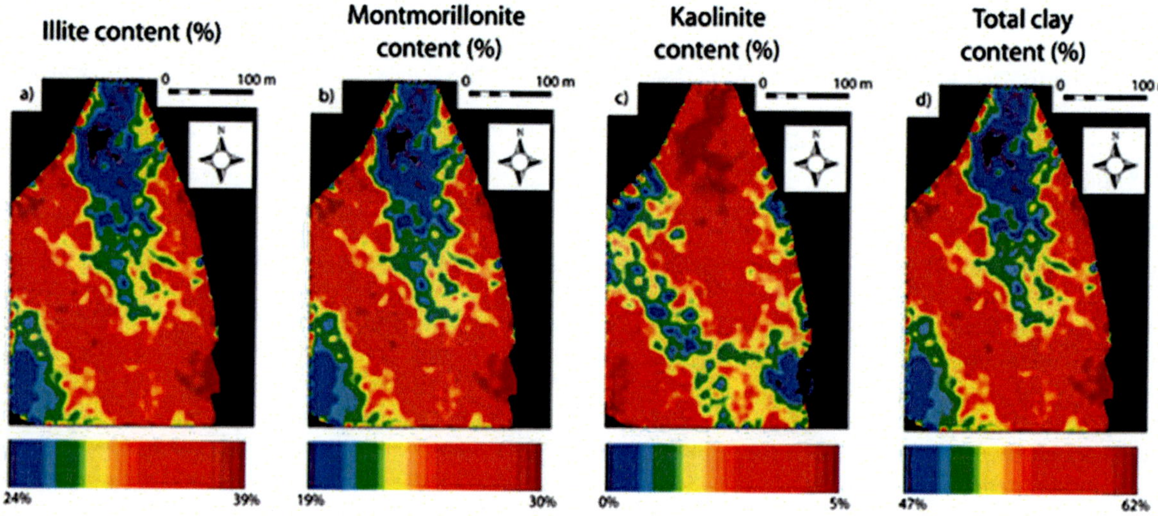


**Fig. 4** Hyperspectral-derived clay map for illite (**a**), montorillonite (**b**), kaolinite (**c**) and total clay (**d**)

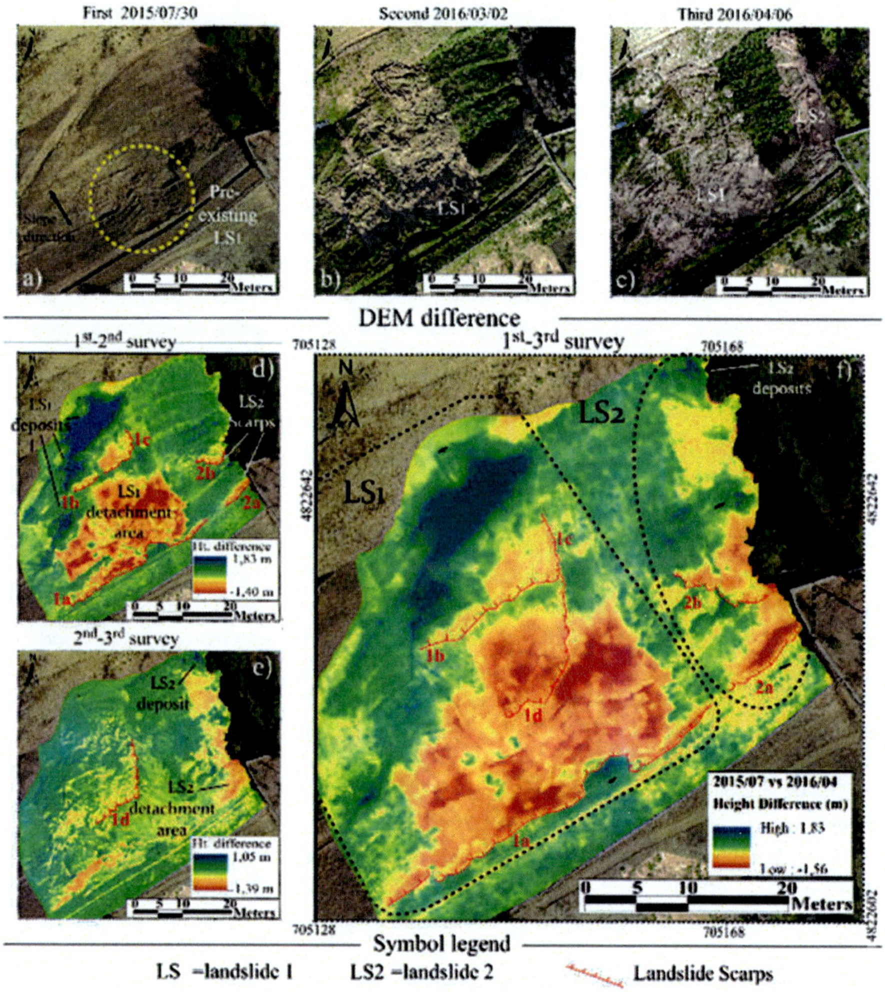


**Fig. 5** Orthophotos of the area affected by the landslides (**a**, **b**, **c**) and DEM differences among different acquisitions (**d**, **e**, **f**) (Modified after Rossi et al. 2018)

then compared to detect any morphological change between the three acquisitions, permitting to characterize the landslide and, in addition, to precisely point out geomorphological features of landslide-prone areas on the slope (Fig. 5).

In particular, Fig. 5 shows the orthophotos of the three surveys (Fig. 5a–c) together with the DTMs comparisons (Fig. 5d–f).

The results of the first survey carried out in July 2015 showed a deformation on the surface, clearly visible in the yellow dotted circle of Fig. 5, as the only sign of instability. That area was the first signal of a shallow landslide that occurred the 1st of March 2016 after intense rainfall, identifiable in the second drone survey, i.e. the one made in March 2016 as the LS1 area. The comparison of the two DTMs (obtained in 2016 and 2015) allowed to characterize the landslide, making it possible to recognize the detachment, transportation and deposit areas and to estimate an unstable area with an extension of 1250 $m^2$ (Fig. 5d). Moreover, the DTMs comparison pointed out the formation of two new scarps, called 2a and 2b in Fig. 5d. Once again, these were the precursor of a new landslide (called LS2 in Fig. 5c, f) verified in between the second and the third surveys, i.e. the one of March and April 2016, respectively. The phenomenon was characterized thanks to the comparison of the DTM obtained from the 3rd survey with the ones obtained in the 1st and 2nd surveys (Fig. 5e, f). The LS2 ground deformation was hence classified as a planar translational slide (according to Varnes 1978) with an extension of 300 $m^2$ (Rossi et al. 2018).

The Ricasoli area example shows how the multitemporal analysis not only allows to characterize and monitor existing landslides but can also be a powerful instrument for landslides recognition at a preliminary stage. Furthermore, one of the main advantages is the potential repeatability of the surveys in a relatively short time and with high resolution, especially when compared to other techniques such as terrestrial laser scanning, as well as the low cost.

Another fundamental aspect of the UAVs is the opportunity to make surveys in emergency condition, when the impossibility to directly reach the area affected by the landslide, time and low-cost effectiveness are basics requirements.

It is the case of the Marano landslide, a roto-translational slide located in the Province of Bologna, in Central Italy. The landslide, mainly composed of clayey material, occurred in March 2018, covering an area of 140.000 $m^2$ and threatening the total obstruction of the Reno river. For this reason, on March 18th, 2018 (14 days after the slide occurrence) a drone survey using the optical camera Canon IXUS 240 has been carried out on request of the Italian National Civil Protection Department. The urgency to carry out the flight has led to perform the survey in non-optimal conditions and therefore to adopt some precautionary measures. Due to difficult climatic conditions linked to the strong wind which reached gusts of more than 15 m/sec, two distinct flights starting from two different points of take-off, respectively near the crown and foot of the landslide were performed (Fig. 6a). A precautionary approach has been adopted also concerning the images superimposition, thus considering an overlap and a sidelap respectively never less than 70% and 55%. The flight height was set from 70 to 80 m a.g.l..

The photogrammetric processing has been done as previously using the Agisoft Photoscan Professional software, obtaining the following georeferenced outcomes: a 3D point cloud with a density of 1500 points/$m^2$; a 35 million polygons mesh; a DTM and an orthophoto both with a resolution of 4 cm/pixel. All these products allowed a detailed characterization of the landslide that, due to the emergency condition and to the inaccessibility of the area, was of fundamental importance (Fig. 6).

Moreover, the DTM and the orthophoto were essential to have a spatial reference for all data acquired by the in-situ monitoring instrumentations (such as data acquired from a GBInSAR) as well as for emergency activities planning.

As mentioned in Sect. 3, whereas the optical surveys focus on landslide mapping and on the estimation of surface changes, the multispectral survey allows to characterization of the vegetation cover and its partially burned canopy.

In the following case study, the Saturn drone, equipped with a Near Infrared (NIR) sensor, was used to evaluate the wildfire affected area extent and to preliminary assess the effects on landslide susceptibility. The key factor that allows the distinction among the vegetation and burned area is the relevant higher reflectance of green vegetation (Pereira et al. 1999). Furthermore, the combination of NIR-green-blue spectrums can be used to analyse the health condition of vegetation. This special combination results in the NDVI index (Normalized Difference Vegetation Index) (Eq. 1).

$$NDVI = \frac{(NIR - B)}{(NIR + B)} \tag{1}$$

When green vegetation is burned NDVI decreases owing to a rise of the reflectance in the blue band and a decrease in the NIR (Fraser and Cihlar 2000).

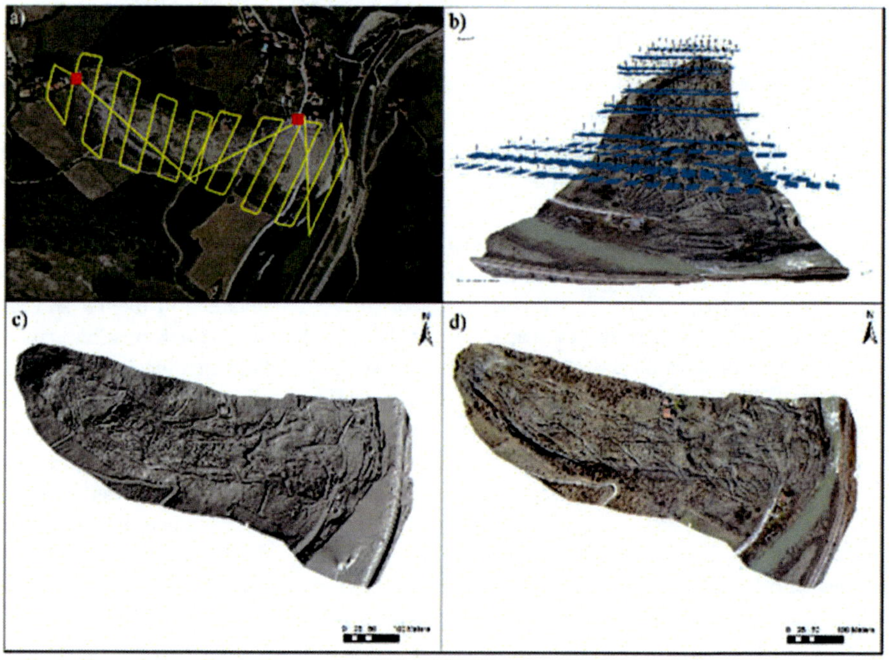


**Fig. 6** Photogrammetric survey of the Marano landslide. **a** GPS track of the flight plan. Red squares represent the two take-off points; **b** 3D model of the landslide in real colors. In blue the nadiral acquisition geometry; **c** Digital Elevation Model (DEM); **d** Orthophoto in real colors

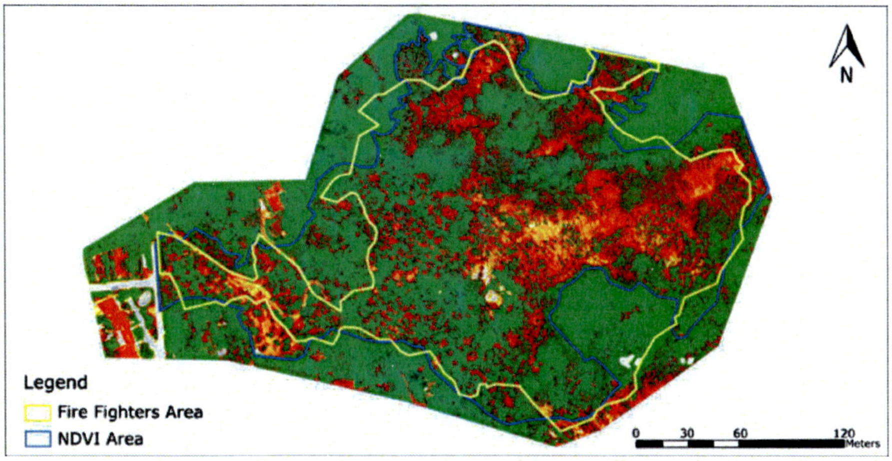


**Fig. 7** Comparison of the burned area mapped by the field GPS mapping of the Italian Fire Fighters Department (yellow polygon), and by the NDVI analysis (blue polygon)

The study area is located near the Vinchiana village, in the Lucca municipality, Italy. From a geological point of view, it is characterized by Holocenic eluvial-colluvial deposits overlying a bedrock composed by massive silicoclastic sandstones stratifications with variable granulometry, ranging from fine to coarse classified as greywackes, and interbedded with silts and clays. The study area, particularly prone to instability phenomena due to the presence of a dormant landslide, was affected by a wildfire on August 8th, 2018.

The UAV survey was performed on November 13th, 2018 using a Canon Powershot S100 digital camera, customized with a filter for the collection of NIR radiation. The flight was carried out with a constant height of 60 m (a.g.l.). Interesting results were achieved by merging false-colour image of the Normalized Difference Vegetation Index (NDVI) on the DTM of the area, obtaining a 3D NDVI map which provided information about the vegetation health conditions and the extension of the burned area.

Results pointed out the complexity of the area affected by the wildfire (Fig. 7). The Comparison between the NDVI mapped burned area and the one mapped by field GPS by the Italian National Fire Corps personnel highlighted the potential of this technique as a tool to remotely and rapidly recognize burned areas, refining the field mapping and providing a wider coverage while granting at the same time the safety of the involved personnel (Fig. 7).

The data quality is affected by the time elapsed from the moment of the extinction of the fire and the data collection (Chuvieco et al. 2002): the longer the time span, the weaker the traces related to the fire will be. Using a multispectral sensing technique with a drone over the area right after the fire, instead of waiting for the satellite revisiting time, can provide better and unaltered data.

The adopted method provides details not achievable with optical cameras, representing therefore a useful low-cost tool to evaluate the terrain conditions as a preliminary input parameter for landslide susceptibility in areas affected by wildfires.

## Conclusions

Recently the combination of rapid development of low-cost small unmanned aerial vehicles (UAVs), and new and miniaturized sensors (Active and Passive) have increase the possibility to be used in terms of cost and dimensions, leading to new opportunities in landslides monitoring and classification.

The aim of this work was to make a review of drone and sensors characteristics able to test their use and applicability in landslide characterization and monitoring. Several combinations of drone and payload were considered to better understand the terrain condition in a critical area. In all the examples the drone survey has proven to be an easy and effective approach for landslide monitoring and surveying and thanks to these potentialities and to its repeatability, in many cases it has become an integral part of the monitoring system. The use of multi-sensors (Active and Passive) is increasing in scientific research and this is becoming more and more of common use in commercial activities too. For that reason, the development of drone with a strong reliability both in terms of safety and flight stability is crucial, in this terms the Department of Earth Sciences and the Civil Protection Centre at the University of Florence have developed a new drone airframe that overcomes some critical issues for scientific and relatively heavy payload or long flight applications.

**Acknowledgements** We would like to thank University of Firenze – Centre of Civil Protection for logistic support as well as for providing computing infrastructure, and Italian Civil Protection Department for financial support.

**Author Contribution** Carlo Tacconi Stefanelli and Teresa Gracchi contribute to writing the part of data acquisition and analysis and the review part, Guglielmo Rossi contribute to writing the UAVs sensors analysis and the review part, Sandro Moretti provided guidance and support throughout the research process to develop research work and did the final editing.

## References

Altdorff D, Schliffke N, Riedel M, Schmidt V, van der Kruk J, Vereecken H, Stoll JB (2014) UAV-borne electromagnetic induction and ground-penetrating radar measurements: a feasibility test. Water Resour Res 42:W11403

Baccani C, Rossi G, Landini F, Salvatici T, Romoli M, Pancrazzi M, Focardi M, Nove V, Moretti S, Casagli N (2018) Optical design of a hyperspectral drone advanced camera for soil monitoring using an electro-optical liquid crystal technology. In: Optical Design and Engineering VII. International Society for Optics and Photonics, Vol 10690, pp 106900P

Ben-Dor E, Chabrillat S, Dematte JAM, Taylor GR, Hill J, Whiting ML, Sommer S (2009) Using imaging spectrscopy to study soil properties. Remote Sens Environ 113:38–55

Benson AK (1995) Applications of ground penetrating radar in assessing some geological hazards: examples of groundwater contamination, faults, cavities. J Appl Geophys 33(1–3):177–193

Bichler A, Bobrowsky P, Best M, Douma M, Hunter J, Calvert T, Burns R (2004) Three-dimensional mapping of a landslide using a multi-geophysical approach: the Quesnel Forks landslide. Landslides 1(1):29–40

Carlson TN, Ripley DA (1997) On the relation between NDVI, fractional vegetation cover, and leaf area index. Remote Sens Environ 62(3):241–252

Casagli N, Frodella W, Morelli S, Tofani V, Ciampalini A, Intrieri E, Raspini F, Rossi G, Tanteri L, Lu P (2017) Spaceborne, UAV and ground-based remote sensing techniques for landslide mapping, monitoring and early warning. Geoenviron Disast 4(1):9

Colomina I, Molina P (2014) Unmanned aerial systems for photogrammetry and remote sensing: a review. ISPRS J Photogram Remote Sens 92:79–97

Delacourt C, Allemand P, Berthier E, Raucoules D, Casson B, Grandjean P, Pambrun C, Varel E (2007) Remote-sensing techniques for analysing landslide kinematics: a review. Bull de Societe Geologique 178(2):89–100

Fiolleau S, Borgniet L, Jongmans D, Bièvre G, Chambon G (2019) Using UAV's imagery and LiDAR to accurately monitor Harmalière (France) landslide evolution. Geophys Res Abstracts 21

Fraser RH, Li Z, Cihlar J (2000) Hotspot and NDVI differencing synergy (HANDS): a new technique for burned area mapping over boreal forest. Remote Sens Environ 74(3):362–376

Frodella W, Gigli G, Morelli S, Lombardi L, Casagli N (2017) Landslide mapping and characterization through infrared thermography (IRT): suggestions for a methodological approach from some case studies. Remote Sens 9(12):1281

Garfagnoli F, Ciampalini A, Moretti S, Chiarantini L, Vettori S (2013) Quantitative mapping of clay minerals using airborne imaging spectroscopy: new data on Mugello (Italy) from SIM-GA prototypal sensor. Europ J Remote Sens 46:1–17

Guzzetti F, Mondini AC, Cardinali M, Fiorucci F, Santangelo M, Chang KT (2012) Landslide inventory maps: New tools for an old problem. Earth Sci Rev 112(1–2):42–66

Hu Z, Shan W (2016) Landslide investigations in the northwest section of the lesser Khingan range in China using combined HDR and GPR methods. Bull Eng Geol Env 75(2):591–603

Jaboyedoff M, Oppikofer T, Abellán A, Derron MH, Loye A, Metzger R, Pedrazzini A (2012) Use of LIDAR in landslide investigations: a review. Nat Hazards 61(1):5–28

James MR, Ilic S, Ruzic I (2013) Measuring 3D coastal change with a digital camera. In Proceedings of the 7th international conference on coastal dynamics, pp 24–28

Karantanellis E, Marinos V, Papathanassiou G (2019) multitemporal landslide mapping and quantification of mass movement in red beach, Santorini Island Using Lidar and UAV platform. In IAEG/AEG annual meeting proceedings, San Francisco, California, Vol 1, pp 163–169

Kwon TH, Choi SK, Myung H, Jung S, Kim YT (2019) Topographic data acquisition and reduction for landslides modeling using UAV-LiDAR system. Geophys Resh Abstracts 21

Kršák B, Blišťan P, Pauliková A, Puškárová P, Kovanič Ľ, Palková J, Zelizňaková V (2016) Use of low-cost UAV photogrammetry to analyze the accuracy of a digital elevation model in a case study. Measurement 91:276–287

Lanzarone P, Garrison E, Bobe R, Getahun A (2016) Examining fluvial stratigraphic architecture using ground-penetrating radar at the Fanta Stream fossil and archaeological site, central Ethiopia. Geoarchaeology 31(6):577–591

Lin J, Wang M, Yang J, Yang Q (2017) Landslide identification and information extraction based on optical and multispectral uav remote sensing imagery. IOP Conf Series: Earth and Environ Sci 57 (1):012017

Mellett JS (1995) Ground penetrating radar applications in engineering, environmental management, and geology. J Appl Geophys 33(1–3):157–166

Morello R (2018) Potentialities and limitations of thermography to assess landslide risk. Measurement 116:658–668

Natesan S, Armenakis C, Benari G, Lee R (2018) Use of UAV-borne spectrometer for land cover classification. Drones 2(2):16

Neal A (2004) Ground-penetrating radar and its use in sedimentology: principles, problems and progress. Earth Sci Rev 66(3–4):261–330

Niethammer U, James MR, Rothmund S, Travelletti J, Joswig M (2012) UAV-based remote sensing of the Super-Sauze landslide: Evaluation and results. Eng Geol 128:2–11

Pappalardo G, Mineo S, Angrisani AC, Di Martire D, Calcaterra D (2018) Combining field data with infrared thermography and DInSAR surveys to evaluate the activity of landslides: the case study of Randazzo Landslide (NE Sicily). Landslides 15(11):2173–2193

Pellicani R, Argentiero I, Manzari P, Spilotro G, Marzo C, Ermini R, Apollonio C (2019) UAV and airborne LiDAR data for interpreting kinematic evolution of landslide movements: the case study of the Montescaglioso landslide (Southern Italy). Geosciences 9(6):248

Pereira JM, Sá AC, Sousa AM, Silva JM, Santos TN, Carreiras JM (1999) Spectral characterisation and discrimination of burned areas. Remote Sens Large Wildfires 123–138

Remondino F, Barazzetti L, Nex F, Scaioni M, Sarazzi D (2011) UAV photogrammetry for mapping and 3d modeling–current status and future perspectives. Int Archives Photogram Remote Sens Spatial Inf Sci 38(1):C22

Riquelme A, Del Soldato M, Tomàs R, Cano M, Baldeehore LMJ, Moretti S, (2019) Digital landform reconstruction using old and recent open access digital aerial photos. Geomorphology 329:206–233

Rossi G, Tanteri L, Tofani V, Vannocci P, Moretti S, Casagli N (2018) Multitemporal UAV surveys for landslide mapping and characterization. Landslides 15(5):1045–1052

Sass O, Bell R, Glade T (2008) Comparison of GPR, 2D-resistivity and traditional techniques for the subsurface exploration of the Öschingen landslide, Swabian Alb (Germany). Geomorphology 93(1–2):89–103

Tarchi D, Guglieri G, Vespe M, Gioia C, Sermi F, Kyovtorov V (2017, June) Mini-radar system for flying platforms. In: 2017 IEEE international workshop on metrology for AeroSpace (MetroAeroSpace, pp 40–44. IEEE

Travelletti J, Delacourt C, Allemand P, Malet JP, Schmittbuhl J, Toussaint R, Bastard M (2012) Correlation of multi-temporal round-based optical images for landslide monitoring: Application, potential and limitations. ISPRS J Photogram Remote Sens 70:39–55

Varnes DJ (1978) Slope movement types and processes. Special Report 176:11–33

Wierzbicki D, Kedzierski M, Fryskowska A, Jasinski J (2018) Quality assessment of the bidirectional reflectance distribution function for NIR imagery sequences from UAV. Remote Sens 10(9):1348

Yu X, Liu Q, Liu X, Liu X, Wang Y (2017) A physical-based atmospheric correction algorithm of unmanned aerial vehicles images and its utility analysis. Int J Remote Sens 38(8–10):3101–3112

Zhou C, Ying C, Hu X, Xu C, Wang Q (2020) Thermal infrared imagery integrated with multi-field information for characterization of pile-reinforced landslide deformation. Sensors 20(4):1170

# Novel Cosmogenic Datings in Landslide Deposits, San Juan, Argentina

2020 Kyoto Japan

Pilar Jeanneret, Stella Maris Moreiras, Silke Merchel, Andreas Gärtner, Steven Binnie, Maria Julia Orgeira, G. Aumaître, D. Bourlès, and K. Keddadouche

**Abstract**

High-mountain environments in an active tectonic setting are prone to landsliding. The triggering mechanisms can vary, as these areas are influenced by several pre-conditioning factors coupled with active seismicity and climatic forcings. Understanding the intrinsic and external mechanisms by which these events are influenced would help to establish better constraints onto their timing and periodicity and, eventually, hazard assessment and prediction. Glacially eroded valleys are especially prone as they deeply incise mountain ranges leaving unstable slopes once they retreat. Establishing the timing of such events enables better understanding of the triggering and pre-conditioning factors of landslides. To this aim, 10Be and 26Al cosmogenic age determinations were performed in three landslide deposits in a poorly studied area of San Juan province, all of which are novel to the area. Coupled with remote sensing techniques, field observations and detailed stratigraphic and sedimentological studies, these new large landslides represent a first approach to understand this dynamic environment. The three landslides were categorized as rock avalanches found in the middle and lower reaches of the Blanco River, sourced from the Choiyoi Group with evidence of hydrothermal alteration and including/deforming moraine deposits during their fall. Ages are 20.9 ± 1.4, 10.8 ± 0.7 and 12.8 ± 0.9 ka from the lowermost deposit to the highest, respectively. Even though one sample per deposit is not enough to have statistically significant exposure ages, these values, along with the established chronostratigraphy, allow first order interpretations regarding the links between deglaciation processes and readjustment of the slopes via large landslide events.

**Keywords**

Landslide • Chronostratigraphy • Central andes • Argentina

## Introduction

The abrupt topography of the Central Andes is prone to landsliding as the valleys were deeply incised by past glacial advances which also have left unstable and un-supported glacigenic landforms and/or glaci-fluvial terraces (Cossart et al. 2008; McColl 2012). These erosional processes are key aspects of landscape evolution in this high-relief and arid environments (Bookhagen and Strecker 2012). The studied area comprises high relief, heavily glaciated and tectonically dynamic valleys surrounding the Mercedario Peak (6770 m asl) in the Frontal Cordillera of San Juan, Argentina (32° S, 70° W). This normally leads to natural impoundments such

P. Jeanneret · S. M. Moreiras (✉)
Instituto de Nivologia, Glaciologia y Ciencias Ambientales, CONICET-Mendoza, Argentina
e-mail: moreiras@mendoza-conicet.gob.ar

P. Jeanneret
e-mail: pilar.jean87@gmail.com

S. M. Moreiras
Universidad Nacional de Cuyo, Mendoza, Argentina

S. Merchel
Helmholtz Zentrum Dresden-Rossendorf, Dresden, Germany
e-mail: s.merchel@hzdr.de

A. Gärtner
Senckenberg Naturhistorische Sammlungen Dresden, Museum für Mineralogie Und Geologie, Sektion Geochronologie, Germany
e-mail: Andreas.Gaertner@senckenberg.de

S. Binnie · M. J. Orgeira
Department of Geology and Mineralogy, University of Cologne, Cologne, Germany
e-mail: sbinnie@uni-koeln.de

M. J. Orgeira
e-mail: mjorgeira@yahoo.com.ar

G. Aumaître · D. Bourlès · K. Keddadouche
Aix-Marseille University, CNRS-IRD-Collège de France, INRAE, UMR34 CEREGE Aix-en-Provence, France



V. Vilímek et al. (eds.), *Understanding and Reducing Landslide Disaster Risk*,
ICL Contribution to Landslide Disaster Risk Reduction,
https://doi.org/10.1007/978-3-030-60319-9_40

as the case of Blanca Lake, which formed after a rock avalanche that blocked the river (Jeanneret et al. 2017).

Many landslides in the Central Andes have been linked to certain lithologies (Moreiras 2006, 2009; Moreiras et al. 2015; Junquera et al. 2017; Junquera Torrado 2019), with a seismic triggering factor. However, huge landslides have been linked to unconsolidated Quaternary deposits (Moreiras et al. 2017), which are resting un-stable laterally to over-steepened slopes. Another destabilizing factor is the periglacial environment, which expands and shrinks accordingly to glaciations. These changes in freeze-thaw action affect rock massifs and permafrost contained in these deposits adding other conditions to slope stability (Ballantyne 2002; Ballantyne and Stone 2013). This was observed in the studied area in previous works (Jeanneret and Moreiras 2018).

Even though landslides are very frequent in the Andean mountains, in the Central Andes of San Juan, there are only a few studies that try to establish a exposure history and the influence of global climatic changes in slope stability. The timing and chronology of glacial advances in the Mendoza River, an adjacent basin within Central Andes, has been investigated by the use of terrestrial cosmogenic radionuclides (TCN) over moraines (Hermanns et al. 2015). In Mendoza River basin, those age estimations were also constrained by numerical age determinations on landslide by detailed chronostratigraphy (Moreiras et al. 2015).

Recent studies in the Central Andes suggest a close link between warmer and more humid interglacial stages with a higher frequency of large landslides (Moreiras et al. 2018). This interpretation is based on a compilation of different dating techniques used across the mountain range, and supported by observations from high-mountain environments around the world (Beniston et al. 2007; Haeberli and Gruber 2009; Huggel et al. 2012; Slaymaker 2009; Zech et al. 2008, 2009; Zemp et al. 2015) and in tropical areas of the Andes (Trauth et al. 2001, 2003). Even though past landslides have been proven to be proxies of past climatic changes (Borgatti and Soldati 2010), more numerical age estimations should be made in order to better constraint these assumptions.

## Study Area

The analyzed area is located in the Central Andes of Argentina (31° S), south San Juan province (Fig. 1a). It comprises the last tributary rivers and creeks of the Blanco River in the district of Calingasta. It includes the Laguna Blanca stream (which contains the homonymous lake) and other minor creeks. This area comprises several high peaks of La Ramada and Ansilta's mountain ranges such as Mt. Mercedario (6720 m asl) and Peak #7 Ansilta (5721 m asl). This section of the Andes is characterized by its aridity caused by the high topographic barrier to the west that prevents moisture transfer from the Pacific Ocean. Precipitation occurs mostly in the Chilean side or as snow in the higher peaks. Rainfalls are more common during summer months and can reach a mean value of 300 mm/year, decreasing to the east (Minetti et al. 1986). The limit between periglacial and non-periglacial areas is defined by the 0°C isotherm of medium annual air temperature that was estimated to be located at around 3500 m asl (Tapia Baldis and Trombotto Liaudat 2015).

## Geological Setting

The study area is located in the Frontal Cordillera geological province (Ramos 1999) (Fig. 1b). This morphotectonic unit is characterized by a Permian to Triassic basement of rhyolitic and more intermediate composition lava flows along with pyroclastic flows and tuffs all extruded during the same extensional phase. Several granitic to granodioritic batholiths intrude these sequences, in most cases leaving a ring of hydrothermally altered host rock from the Choiyoi Group (Caminos 1965). Its composition variability responds to an orogeny formation phase that collapsed (Heredia et al. 2002) settling an extensional phase that continues until the early Jurassic, filled with La Ramada basin (Alvarez and Ramos 1999). The Miocene volcanic stage in this area is manifested as dacitic to andesitic dykes within the Choiyoi. The cease of the volcanic record is the clear evidence of the shift of the subduction angle towards 5–10° (Fernández et al. 1997), responsible for the highest elevations to this point. This uplift shaped and filled the Manantiales basin with synorogenic continental sediments with minor marine intrusions (Pérez 2001), called the Chinches Formation. This is dominated by conglomerate, sandstone, silt and some shale. The migration of the tectonic front tilted the Manantiales basin exposing it to erosion which led to an angular unconformity with the subsequent Quaternary deposits.

The latter is dominated mainly by glacigenic and glacial deposits, piedmont type deposits such as coalescent alluvial terraces, lower fluvial/glacifluvial/alluvial terraces and large mass removal processes that will be analyzed in this study. Glacigenic and glacial deposits as form of moraines are found at latitudes of 2000 m asl indicating a larger extension of the ice fields in the past. These are lateral moraines and may be of Pleistocene age because of its extension and correlation to nearby glacial extension maps (Espizúa 1993). The piedmont deposits include all proximal to medium alluvial accumulations on the foothill of La Ramada range that are now terraced by the lowering of the baseline. The other terraces are fluvial, glacifluvial, alluvial or a mixture between two or the three of these components that were also left hanging on the valley sides by successive lowering of

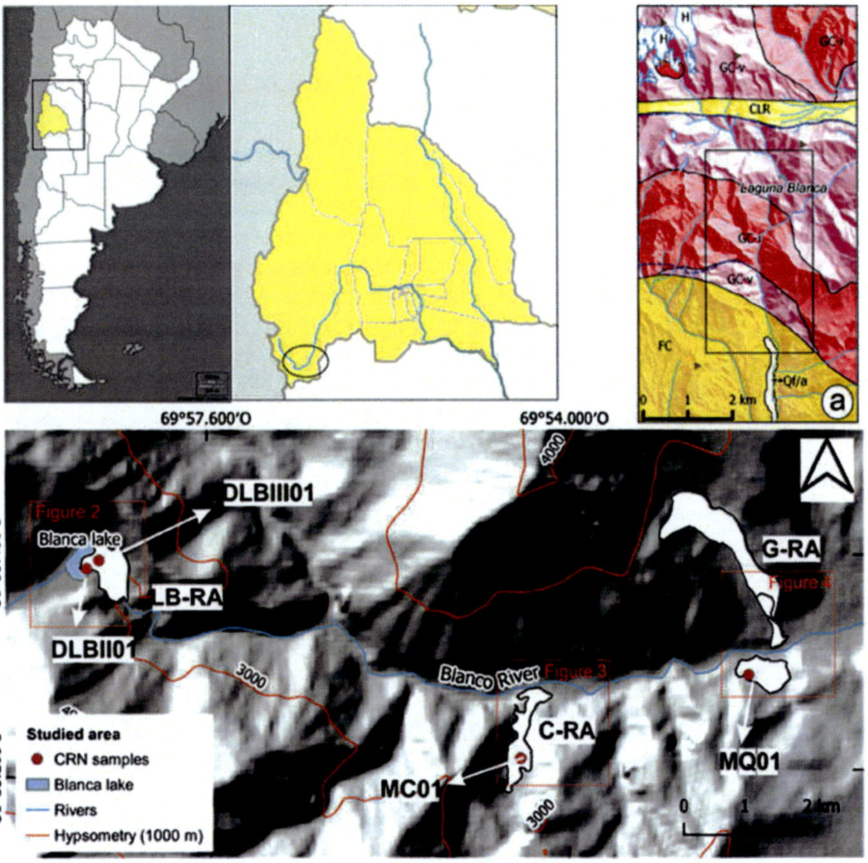


**Fig. 1** Rock avalanches within the lower reaches of the Blanco River, San Juan province. The geological map is also presented even though the main lithology involved is composed of Choiyoi Group rhyolites and granitoids. H: ice cover; GC-v: Choiyoi Group volcanic; GC-I: Choiyoi Group intrusive; CLR: La Ramada basin; FC: Chinches Formation; Qf/a: Quaternary (Fluvial/alluvial)

the baseline after the formation of the piedmont terraces and after the largest glaciation.

## Methodology

### Remote Sensing Analysis

A first approach to landslide identification was done with GoogleEarth TM images taking advantage of the three-dimensional views. An inventory map was generated in GIS environment (QGIS) using georeferenced LANDSAT images. After detecting these landforms, their source area and maximum extent was estimated using the geometric tools provided by QGIS.

### Field Works

Field campaigns were conducted during the summers of 2016 and 2018. During these sessions, the landslide distribution was checked. Surficial boulders of large landslides were sampled for TCN dating (Table 1).

### Sedimentology and Chrono-Stratigraphy

To complement the age estimations, a detailed sedimentological and stratigraphic study was performed. Fortunately, a mining company cut fresh faces of all the deposits to build a road which enabled to do detailed sedimentological profiles of the distal phases. The lithological facies were described along with modal clast size. Stratigraphic relations with the surrounding Quaternary deposits were conducted to better understand the geomorphological evolution of the Blanco River valley.

### TCN Dating

Due to scarcity of organic matter in this dry area, this technique was implemented, taking advantage of the availability of large boulders at the highest elevation of each deposit (where available). Within the limitations of the technique, this is a valid approachto obtain first order interpretations of the landslide timing in the Central Andes. This method has been widely used to date surface exposure events, including the timing of deep seated, rapid mass

**Table 1** TCN samples information. DLBIII01 did not contain enough quartz for TCN dating

| Lat S | Long W | Alt m asl | Landf | Lithology | Sam |
|---|---|---|---|---|---|
| 31,891 | 69,980 | 3148.9 | Lag. Blanca RA | Granitic | DLBII01 |
| 31,889 | 69,978 | 3118.1 | Lag. Blanca RA | Rhyolitic | DLBIII01 |
| 31,918 | 69,908 | 2464.7 | Chinches RA | Granitic | MC01 |
| 31,906 | 69,869 | 2184.9 | Granitic RA | Granitic | MQ01 |

movements (Dunai 2010). Typically, minimum ages for the formation of the landslide deposit are obtained. For the studied area, samples from exposed boulder over 1 m height above the surface, on top of landslides deposits were extracted by sampling the upper 2 cm of the surface using chisel and hammer. Preference was given to sample sites that were flat, although measurements of the average surface inclination and surrounding topographic shielding with which to correct production rates were taken on site with a Brunton geologic compass. Latitude, longitude and altitude were measured on site with a GPS (Garmin ETREX) in order to derive production rate scaling factors (Balco et al. 2008). Snow cover was considered to have negligible influence on production rates as snowfalls are currently extremely uncommon at this altitude (<3500 m asl).

Four samples were taken from three rock avalanches, two boulders exposed in the Laguna Blanca rock avalanche (LB-RA), one in the Chinches rock avalanche (C-RA) and 1 in the Granitic rock avalanche (G-RA). Even though it is highly recommended to sample multiple boulders of each deposit for a more robust analysis, it was extremely difficult to find suitable boulders. Furthermore, one of the boulders from LB-RA had a rhyolitic composition containing too little amounts of TCN useable crystalline quartz (<20% of total volume) to make it suitable for measurement. The other three samples were granites and granodiorites with a higher content of crystalline quartz, allowing a 10Be and 26Al extraction and determination.

A first physical extraction of the quartz was done at the University of Cologne. A density separation, chemical cleaning (Brown et al. 1991) and radiochemical separation of the radionuclides, a method further described by Merchel and Herpers (1999), were performed at the DREsden Accelerator Mass Spectrometer (DREAMS) at HZDR based on Merchel et al (2008, 2019). The stable isotope carriers used for AMS sample preparation were a dedicated low radionuclide one for 9Be ("Phenakite EA") (Merchel et al. 2013) and a commercial one from the company ROTH (lot E03624) for 27Al. After preparation as oxides (BeO and Al2O3), and mixed with Nb and Ag powder, respectively, they were pressed into Cu cathodes and sent to ASTER in Aix-en-Provence for AMS measurements (Arnold et al. 2010). All AMS data has been normalized to secondary standards from HZDR, i.e. SMD-Be-12 (Akhmadaliev et al. 2013) and SMD-Al-11 (Rugel et al. 2016), respectively. Details about AMS measurements at ASTER are given by Arnold et al. (2010). The determination of native 27Al in liquid aliquots of the dissolved quartz samples was performed by inductively coupled plasma mass spectrometry (ICP-MS) at HZDR. One blank sample with a known concentration of 9Be and 27Al was performed along with this batch to estimate the uncertainties within the chemical procedure.

The obtained results, in units of radionuclide atoms per gram quartz, are calculated from AMS isotopic ratios and, in the case of 26Al, also from ICP-MS results. Ages were derived using the online tool CRONUS-EARTH V.3 based on Balco et al. (2008) (https://hess.ess.washington.edu). This tool enables non-specialized users to calculate the exposure ages using different scaling factors schemes and corrections. Here, we report ages using the scaling scheme of LSDn (Lifton, Sato, Dunai) with a 1 mm/ka erosion rate. This erosion rate was selected as a middle ground between a null erosion rate, which is most likely erroneous because of the dynamic nature of the area, and a 5 mm/ka erosion rate, which "saturated" the older boulders. Both ages (10Be and 26Al) and their corresponding uncertainties (internal errors) were averaged in order to get a final, more robust age.

## Results

### Sedimentology and Chronostatigraphy of Rock Avalanche

#### Laguna Blanca Rock Avalanche (LB-RA)

This rock avalanche is located at coordinates −31.89116 and −69.9764 at around 3100 m asl damming the Laguna Blanca creek and forming an upstream lake with the same name (Fig. 2). The estimated area of this deposit is 0.42 km$^2$ with a maximum height of 180 m. The deposit could have travelled 4.5 km to the intersection with the Blanco River based on hanging relict avalanche deposits further downstream. Most of the deposit is concentrated within 1.3 km from the source scarp on the southern slope of the valley (Jeanneret and Moreiras 2017).

**Fig. 2** Laguna Blanca rock avalanche. The white dotted line delimits the lithological difference between the granitic (to the southwest) and more rhyolitic (to the northeast) dominance within the rock avalanche with some mixed compositions towards the middle. Some photos taken in the field around the circled areas are schematic of the studied sections (a, b, c). 1: granitic composition; 2: rhyolitic composition; 3: dacitic-andesitic composition. The light blue arrow shows the flow's direction

It involves rhyolites and granitoids from the Choiyoi Group that were intruded by andesitic-dacitic dykes. These are recognized as red, light pink and dark green facies, respectively. The rhyolitic facies is preserved adjacent to the lake and was described as a poorly sorted, monolithological deposit, clast-supported with a 70% of the total volume consisting in angular blocks of maximum diameter of 50 cm and in average of 15 cm. The matrix is sandy to gravelly also rhyolitic in composition. There is a subtle but clear tendency of increase in proportion of large blocks towards the top of the sequence.

The run-up was recognized 90 m above the original deposit over the northeast side of the valley. In a terrace from the run-up the three facies could be described (Fig. 2a). All of them are poorly sorted, clast supported with angular blocks and sandy to gravelly matrix. Towards the top, the granitic facies has a minimum height of 1 m but it is estimated to have been eroded and part of the original deposit is missing. Blocks larger than 30 cm in diameter are extremely infrequent, blocks between 15 and 20 cm are frequent and matrix is abundant (<40%). The rhyolitic facies in the middle seems to have the highest proportion of matrix and smaller sized blocks (<15 cm). The dacitic-andesitic facies at the bottom has a lower content of matrix (~30%) and blocks of 15 cm average width with very infrequent occurrences of larger boulders. There was some evidence of pulverization by friction in the matrix with soap-type consistency which increase in proportion towards the bottom. Downriver, this facies is covered by alluvial and colluvial deposits.

Downstream, moving away from the lake, a larger proportion of large granitic boulders were found and some smaller andesitic-dacitic boulders. Next to the road, the same rhyolitic facies was described but with higher abundance of blocks having 15 cm average diameter and occasional blocks of more than 40 cm wide. Matrix seems to be more abundant (40%) but it remains being clast-supported. There were also some dark green layers, predominantly fine grained, belonging to the andesitic-dacitic facies.

## Chinches Rock Avalanche (C-RA)

This rock avalanche is downstream Blanco River with coordinates −31.92531, −69.9092 at an average altitude of 2500 m asl (Fig. 3). The source scar was described to be

3 km to the south covering a valley. This valley has evidence of being glacially eroded and some lateral moraines are still preserved.

The rock avalanche covered the entire valley floor and the sedimentological evidence indicates that some of the moraine deposits were included and re-transported during the event. Because of the height of such a deposit (>10 m) it is assumed that it impounded temporally the Blanco River but no evidence of a run-up on the opposite side or of the impoundment itself were found.

The deposit is poorly sorted with two distinctive facies: one comprises the Chinches Formation basal lithology of volcanic-rhyolitic breccias with sub-rounded boulders of up to 2 m wide (Fig. 3a) in a clayey to sandy matrix that supports smaller boulders of 7–10 cm in diameter. The matrix is abundant (>60%) and it is matrix supported. The other facies was found on top of the rhyolitic breccia with an irregular but clear-cut contact and consists of a granitic facies from Choiyoi Group. This has a lower percentage of matrix (50%) with larger boulders of up to 40 cm in diameter and a matrix of gravel less than 10 cm in diameter including sand. Mixed in this facies, dark green bands were described but could not be linked to a possible source or be analyzed for their composition as they are very thin with a fine-grained matrix. It is estimated they these belong to the dacitic-andesitic dykes within Choiyoi Group. These green layers were also found as lobes and as superficial patches all around the rock avalanche. Downriver, following the deposit and towards its eastern limit, the granitic facies was found also under the Chinches Fm. facies with a less clear-cut edge indicating some mixture between the both.

According to the evidence, it was assumed that the rock avalanche sourced in a steep outcrop from Chinches Fm. It dragged older moraines and included them into the deposit. Some of the original lateral moraines were affected by the rock avalanche's path. That is why the granitic (moraine) facies is found under and over the rhyolitic breccia's facies. This could be deduced from the predominance of clay in the matrix of the bottom granitic facies.

## Granitico Rock Avalanche (G-RA)

This rock avalanche was a challenge to describe because the deposit is covered by colluvial taluses (Fig. 4). It is predominantly of granitic composition with some layers of a dark green/blue facies that is recognized in the field and via remote sensing. It was sourced in a Choiyoi Group outcrop

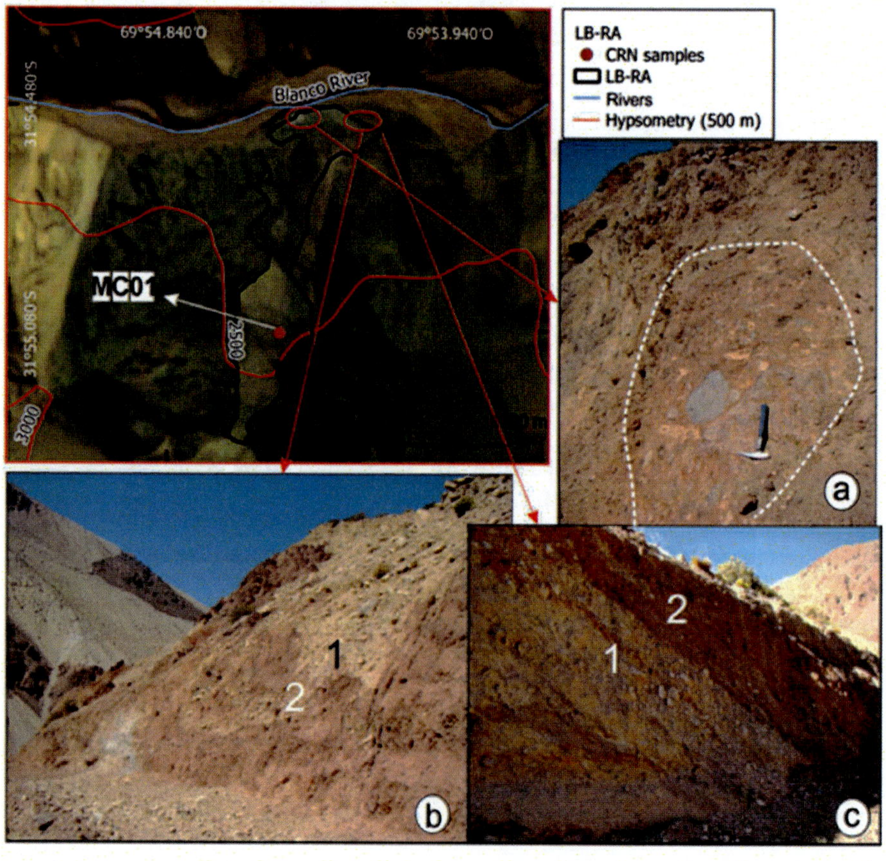


**Fig. 3** Chinches rock avalanche. Only the distal section (**b**, **c**) could be studied in detail. The photos show the boulder's size from the Chinches Fm (**a**) and the mixed nature of the deposit which incorporates moraine deposits in their downstream movement (**b**, **c**). 1: moraine-sourced; 2: Chinches Fm-sourced. The light blue arrow shows the flow's direction

in the north-side valley and travelled 3.5 km crossing the Blanco River causing a run-up and an impoundment. The medium point coordinates are −31.89117, −69.86907 with an average altitude of 2500 m asl.

The estimated height of the deposit is 125 m and the run-up covered the lateral moraines on the opposite side of the valley. The impoundment is evidenced by a subtle alluvial layer adjacent to the deposit on the southern side of the valley with a similar lithology as an upstream (younger) rock avalanche that could not be dated. It also sourced a younger rock avalanche through the same creek.

## Geochronology: TCN Age Estimations

Four samples were extracted from top boulders of the three rock avalanches and their location and extracted parameters are given in Fig. 5 and Table 2. According to the scaling factor and corrections, the estimated ages are: 12.8 ± 0.87 ka for sample DLBII01, 10.8 ± 0.7 ka for sample MC01 and 20.9 ± 1.4 ka for sample MQ01. The first two correspond to a late Pleistocene-early Holocene age and the third is within the last glacial maximum.

## Discussion

Even though more dating is required, the few samples here give first-order estimates with which to understand the timing of development of a glacially eroded valley. Both C and LB-RA are now placed within the Holocene-Pleistocene limit in agreement with other large scale mass removal processes around the Central Andes of Argentina (Hermanns et al. 2015; Moreiras et al. 2015; Trauth et al. 2000).

G-RA was firstly described as a moraine but with further studies it was re-interpreted as part of the run-up deposit of a rock avalanche sourced in the northern side of Blanco River valley. Its age is coincident with the LGM but according to glacial advances' interpretations (Jeanneret et al. 2017), the valley was already ice-free at this stage. Further evidence is given by the fact that this rock avalanche partially covers older moraine deposits on the southern side of the valley. The same can applied for C-RA, as this event incorporate older moraine deposits and, in the ridge of the valley, halted over the moraine with a high degree of mixture. Lastly, LB-RA is more complicated to understand but there is clear evidence that the event included not only outcrops from the surrounding Choiyoi Group but also moraine deposits that

**Fig. 4** Granitic rock avalanche. Also the distal section of this rock avalanche could be defined (G-RA) and a younger rock avalanche (Y-RA) involving the original deposit can be seen on the north size valley (green–blue lithology). The original deposit is seen in the distal section of the photo a and a former lake-shore deposit caused by the original impoundment marked by the white dotted line in b. RA: another rock avalanche upstream from Blanco River that was not sampled but it is assumed to be younger. The light blue arrow shows the flow's direction

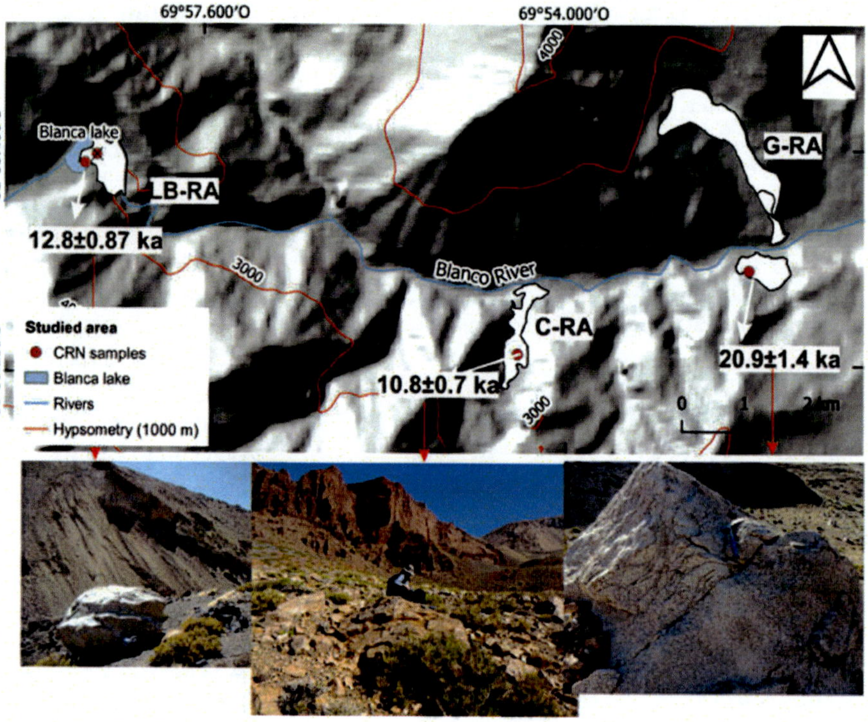


**Fig. 5** TCN samples including exposure ages with schematic photos of the sampled boulders. The light blue arrow shows the flow's direction

**Table 2** CRONUS EARTH data to estimate ages. The LSDn scaling scheme (Lifton, Sato, Dunai) was used without snow correction because of the altitude and three erosion rates that were considered relevant. The 1 mm/ka erosion rate was the selected for the age's estimation as the null erosion rate is not realistic for high mountain environments and more than 5 mm/ka tend to 'saturate' older ages, which was considered too exaggerated for this area. Blue shading shows glacial stages and red shading shows interglacial/warm stages

| Samples | Isotopes | LSDn (Erosion rate = 0) | | | | LSDn (Erosion rate = 1 mm/ka) | | | | LSDn (Erosion rate = 5 mm/ka) | | | |
|---|---|---|---|---|---|---|---|---|---|---|---|---|---|
| | | Age (years) | Internal error (years) | External error (years) | Average error (years) | Age (years) | Internal error (years) | External error (years) | Average error (years) | Age (years) | Internal error (years) | External error (years) | Average error (years) |
| DLBII01 | Be-10 (qtz) | 9606 | 468 | 736 | 602 | 12,397 | 611 | 960 | 785,5 | 12,893 | 664 | 1044 | 854 |
| DLBII01 | Al-26 (qtz) | 10,283 | 474 | 1008 | 741 | 13,247 | 619 | 1314 | 966,5 | 13,794 | 675 | 1434 | 1054,5 |
| MC01 | Be-10 (qtz) | 8138 | 391 | 620 | 505,5 | 10,909 | 529 | 839 | 684 | 11,163 | 562 | 892 | 727 |
| MC01 | Al-26 (qtz) | 8045 | 388 | 796 | 592 | 10,802 | 526 | 1079 | 802,5 | 11,081 | 560 | 1149 | 854,5 |
| MQ01 | Be-10 (qtz) | 19,695 | 937 | 1497 | 1217 | 21,694 | 1052 | 1680 | 1366 | 23,228 | 1220 | 1949 | 1584,5 |
| MQ01 | Al-26 (qtz) | 18,122 | 859 | 1792 | 1325,5 | 20,102 | 970 | 2023 | 1496,5 | 21,361 | 1110 | 2315 | 1712,5 |

crowned such outcrops. It is possible that some original moraine deposits from younger glacial advances are still preserved within or covered by the rock avalanche but a differentiation between moraine-original and re-worked deposits is difficult even with detailed sedimentological studies. The fact that the granitic (moraine) facies is covering the other avalanche facies towards the lowermost section of the deposit indicates that this is avalanche-sourced, re-worked moraine deposit. However, the same assumption could not be made in the uppermost sections were a clean vertical cut of the deposit was not found. Another assumption that has to be made on the sampled boulder is that, in some way, the horizontal face was exposed during the event and the estimated age corresponds to the event's age. It is assumed that, if the horizontal face of the boulder was previously exposed, it would have an inherited age that would over-estimate the actual age. With a calculated numerical exposure of around 10 ka, we can make the assumption that there are no inherited isotopic counts, meaning from a previous exposure, and that this is in fact the age of the rock avalanche's collapse. Of course, more samples have to be taken and analyzed in order to have a more accurate exposure age.

The fact that the three mapped and dated rock avalanches incorporate moraine deposits during the event is a significant observation towards understanding how glacial erosion affects the hill slopes which turn to be more prone to landsliding. Integrating sedimentological and straigraphic studies can help mitigate some of the uncertainties in the TCN dating.

## Conclusion and Outlook

In this study we present preliminary results of the first TCN exposure ages in a poorly-studied area of the arid Andes. These are first indicators for constraining the timing of occurrence of large landslides. The results are in accordance with the estimation from adjacent mountain basins (Moreiras et al. 2015; Hermanns et al. 2015). There are hints for a correlation between humid and warmer interglacial stages (or late glacial stages in non-glaciered lower catchments) and landslide formation. It is estimated that they are linked to the physical and thermal effects that rapid climatic changes cause to the hill slopes. Other types of dating techniques could also be implemented to complement the TCN data (i.e. OSL). In this study we used detailed sedimentological mapping and chronostratigraphy to complement for the lack of further numerical data and make first assumptions on the timing of these large events.

**Acknowledgements** This article was based on previous studies and findings of running researches supported by PIP 11220150100191CO project, the ANLAC (Natural Hazards in the Central Andes: analysis, prediction and valuation) Program from University of Cuyo, and SeCTyP 2016-2018 (A06/669-Res.3820/2016) project lead all of them by Prof. Moreiras. We would like to thank the team in the Institute of Geology and Mineralogy in the University of Cologne for lending the crush lab and magnetic separator used for the physical preparation of the samples, specially to Tomasz Goral for his assistance. We also would like to thank the DREAMS team in HZDR, especially Georg Rugel for helping with AMS data treatment, Johannes von Borany for the official invitation to HZDR, Sabina Beutner (HZDR) for the ICP-MS measurements of 27Al and the ASTER-CEREGE team for taking special care of the AMS samples sent to France. The ASTER AMS national facility (CEREGE, Aix-en-Provence) is supported by the INSU/CNRS, the ANR through the "Projets thématiques d'excellence" program for the "Equipements d'excellence" ASTER-CEREGE action and IRD.

## References


Akhmadaliev S, Heller R, Hanf D, Rugel G, Merchel S (2013) The new 6 MV AMS-facility DREAMS at Dresden. Nucl Instr Meth Phys Res B 294:5–10

Alvarez PP, Ramos VA (1999) The Mercedario rift system in the principal Cordillera of Argentina and Chile (32 SL). J South Ame Earth Sci 12:17–31. https://doi.org/10.1016/S0895-9811(99)00004-8

Arnold M, Merchel S, Bourlès DL, Braucher R, Benedetti L, Finkel RC, Aumaître G, Gottdang A, Klein M (2010) The French accelerator mass spectrometry facility ASTER: improved performance and developments. Nucl Inst Meth Phys Res B 268:1954–1959. https://doi.org/10.1016/j.nimb.2010.02.107

Balco G, Stone JO, Lifton NA, Dunai TJ (2008) A complete and easily accessible means of calculating surface exposure ages or erosion rates from 10Be and 26Al measurements. Quat Geochronol 3:174–195

Ballantyne CK (2002) Paraglacial geomorphology. Quater Sci Rev 21:1935–2017. https://doi.org/10.1016/S0277-3791(02)00005-7

Ballantyne CK, Stone JO (2013) Timing and periodicity of paraglacial rock-slope failures in the Scottish Highlands. Geomorphology 186:150–161. https://doi.org/10.1016/j.geomorph.2012.12.030

Beniston M, Stephenson DB, Christensen OB, Ferro CAT, Frei C, Goyette S, Halsnaes K, Holt T, Jylhae K, Koffi B, Palutikov J, Schoell R, Semmier T, Woth K (2007) Future extreme events in European climate: an exploration of regional climate model projections. Clim Change 81(suppl. 1):71–95

Bookhagen B, Strecker MR (2012) Spatiotemporal trends in erosion rates across a pronounced rainfall gradient: examples from the southern Central Andes. Earth Planet Sci Lett 327:97–110. https://doi.org/10.1016/j.epsl.2012.02.005

Borgatti L, Soldati M (2010) Landslides as a geomorphological proxy for climate change: a record from the Dolomites (northern Italy). Geomorphology 120(1-2):56–64

Brown ET, Edmond JM, Raisbeck GM, Yiou F, Kurz MD, Brook EJ (1991) Examination of surface exposure ages of Antarctic moraines using in situ produced 10Be and 26Al. Geochimica et Cosmochimica Acta 55(8):2269–2283

Caminos R (1965) Geología de la vertiente oriental del Cordón del Plata, Cor-dillera Frontal de Mendoza. Revista de la Asociación Geológica Argentina 20:351–392

Cossart E, Braucher R, Fort M, Bourlès DL, Carcaillet J (2008) Slope instability in relation to glacial debuttressing in alpine areas (Upper Durance catchment, southeastern France): evidence from field data and 10 Be cosmic ray exposure ages. Geomorphology 95:3–26. https://doi.org/10.1016/j.geomorph.2006.12.022

Dunai TJ (2010) Cosmogenic Nuclides: Principles, concepts and applications in the Earth surface sciences. Cambridge University Press

Espizua LE (1993) Quaternary glaciations in the río Mendoza valley, Argentine Andes. Quatern Res 40:150–162. https://doi.org/10.1006/qres.1993.1067

Fernández LR, Hereida N, Espina RG, Cegarra MI (1997) Estratigrafía y es-tructura de los Andes Centrales Argentinos entre los 30° y 31° de Latitud Sur. Acta geológica hispánica 32:51–76

Haeberli W, Gruber S (2009) Global warming and mountain permafrost. In: Margesin R (eds) Permafrost Soils. Soil Biology, 16st edn. Springer, Berlin, Heidelberg

Heredia N, Fernández LR, Gallastegui G, Busquets P, Colombo F (2002) Geo-logical setting of the Argentine Frontal Cordillera in the flat-slab segment (30° 00′–31° 30′ S latitude). J S Am Earth Sci 15:79–99. https://doi.org/10.1016/S0895-9811(02)00007-X

Hermanns R, Fauqué F, Wilson C (2015): 36Cl terrestrial cosmogenic nuclide dating suggests late pleistocene to early holocene mass movements on the south Face of Aconcagua Mountain and in the Las Cuevas-Horcones Valleys, Central Andes, Argentina. In: Geological Society of London, Special Publication, 399, Geodynamic Processes in the Andes of Central Chile and Argentina. https://doi.org/10.1144/SP399.19

Huggel C, Clague JJ, Korup O (2012) Is climate change responsible for changing landslide activity in high mountains? Earth Surf Proc Land 37(1):77–91. https://doi.org/10.1002/esp.2223

Jeanneret P, Moreiras SM, Orgeira MJ, Korup O (2017) Nuevos aportes de la evolución geomorfológica de la cuenca del Arroyo Laguna Blanca, Cerro Mercedario, San Juan. Sesión Técnica 15: Riesgo Geológico y Geología Ambiental y Urbana. Presented at the XX Congreso Geológico Argentino, Asociación Geoló-gica Argentina, San Miguel de Tucumán, pp 62–63

Jeanneret P, Moreiras SM (2018) Inventario de procesos de remoción en masa en la cuenca baja del Río Blanco (31 S), Andes Centrales Argentinos. Revista Mexicana de Ciencias Geológicas 35(3):215–227

Junquera Torrado S, Moreiras SM, Sepúlveda SA (2019) Distribution of land-slides along the Andean active orogenic front (Argentinean Precordillera 31–33° S). Quatern Int 512:18–34. https://doi.org/10.1016/j.quaint.2019.01.030

McColl ST (2012) Paraglacial rock-slope stability. Geomorphology 153–154:1–16. https://doi.org/10.1016/j.geomorph.2012.02.015

Merchel S, Herpers U (1999) An update on radiochemical separation techniques for the determination of long-lived radionuclides via accelerator mass spectrometry. Radiochim Acta 84:215–220

Merchel S, Arnold M, Aumaître G, Benedetti L, Bourlès DL, Braucher R, Alfimov V, Freeman SPHT, Steier P, Wallner A (2008) Towards more precise 10Be and 36Cl data from measurements at the 10–14 level: Influence of sample preparation. Nucl Instr Meth Phys Res B 266:4921–4926. https://doi.org/10.1016/j.nimb.2008.07.031

Merchel S, Bremser W, Bourlès DL, Czeslik U, Erzinger J, Kummer NA, Leanni L, Merkel B, Recknagel S, Schaefer U (2013) Accuracy of 9Be-data and its influence on 10Be cosmogenic nuclide data. J. Radioanal Nucl Chem 298:1871–1878

Merchel S, Gärtner A, Beutner S, Bookhagen B, Chabilan A (2019) Attempts to understand potential deficiencies in chemical procedures for AMS: Cleaning and dissolving quartz for 10Be and 26Al analysis. Nucl Instrum Methods Phys Res, Sect B 455:293–299

Minetti J, Barbieri PM, Carletto MC, Poblete AG, Sierra EM (1986) El régimen de precipitaciones de la provincia de San Juan y su entorno. San Juan, Centro de Investigaciones Regionales, Informe Técnico 8:47

Moreiras SM (2006) Chronology of a probable neotectonic Pleistocene rock avalanche, Cordon del Plata (Central Andes), Mendoza, Argentina. Quatern Int 148:138–148. https://doi.org/10.1007/978-3-319-53483-1_42

Moreiras SM (2009) Análisis de las variables que condicionan la inestabilidad de las laderas en los valles de los ríos Las Cuevas y Mendoza. Revista de la Asociación Geológica Argentina 65 (4):780–790

Moreiras SM, Páez MS, Lauro C, Jeanneret P (2017) First cosmogenic ages for glacial deposits from the Plata range (33° S): new inferences for quaternary landscape evolution in the Central Andes. Special Volume. Interactions between Quaternary climatic forcing, tectonics and volcanism along some different tectonic settings of South America. Quatern Int 438: 50–64. https://doi.org/10.1016/j.quaint.2016.08.041

Moreiras SM (2017) Coupled slope collapse—cryogenic processes in De-glaciated valleys of the aconcagua region, central andes. In: Workshop on world landslide forum. Springer, pp 355–359

Moreiras SM, Hermanns RL, Fauqué L (2015) Cosmogenic dating of rock avalanches constraining Quaternary stratigraphy and regional neotectonics in the Argentine Central Andes (32° S). Quat Sci Rev 112:45–58

Moreiras SM, Benavente CA, Bufarini CD (2018) Análisis de régimen sedimentario del paleolago de altura pleistoceno donde se hallaron los primeros restos dentarios del équido Hippidion Devillei, Mendoza-Argentina. Lat Am J Sedimentol Basin Anal 25(1):55–67

Pérez DJ (2001) Tectonic and unroofing history of Neogene Manantiales fore-land basin deposits, Cordillera Frontal (32° 30′ S), San Juan Province, Argentina. J S Am Earth Sci 14:693–705. https://doi.org/10.1016/S0895-9811(01)00071-2

Ramos VA (1999) Las provincias geológicas del territorio argentino. Geología Argentina 29:41–96

Rugel G, Pavetich S, Akhmadaliev S, Enamorado Baez SM, Scharf A, Ziegenrücker R, Merchel S (2016) The first four years of the AMS-facility DREAMS: status and developments for more accurate radionuclide data. Nucl Instr Meth Phys Res B 370:94–100

Slaymaker O (2009) Proglacial, periglacial or paraglacial? Geological Society, London, Special Publications 320(1):71–84

Tapia Baldis CC, Trombotto Liaudat D (2015) Cinturones altitudinales criogé-nicos en la cuenca del río Bramadero, San Juan, Argentina. Acta Geológica Lil-Loana 27:146–158

Torrado SJ, Moreiras SM, Sepúlveda SA (2017) Huge slope collapses flashing the andean active orogenic front (Argentinean Precordillera 31–33° S). In: Workshop on World Landslide Forum, Springer, pp 535–541

Trauth MH, Alonso RA, Haselton KR, Hermanns RL, Strecker MR (2000) Climate change and mass movements in the NW Argentine Andes. Earth Planet Sci Lett 179:243–256

Trauth MH, Bookhagen B, Marwan N, Strecker MR (2003) Multiple landslide clusters record Quaternary climate changes in the northwestern Argentine Andes. Palaeogeogr Palaeoclimatol Palaeoecol 194:109–121. https://doi.org/10.1016/S0031-0182(03)00273-6

Zech J, Zech R, Kubik PW, Veit H (2009) Glacier and climate reconstruction at Tres Lagunas, NW Argentina, based on 10Be surface exposure dating and lake sediment analyses. Palaeogeogr Palaeoclim Palaeoecol 284:180–190

Zech R, Kull C, Ilgner J, Kubik PW, Veit H (2008) Timing of the Late Quaternary Glaciation in the Andes from ~15 to ~40°S. -. J. Quaternary Sci. 23:635–647

Zemp M, Frey H, Gärtner-Roer I, Nussbaumer SU, Hoelzle M, Paul F, Haeberli W, Denzinger F, Ahlstrom AP, Anderson B, Balracharya S, Baroni C, Braun LN, Cáceres BE, Casassa G, Cobos G, Dávila LR, Delgado Granados H, Demuth MN, Espizua L, Fischer A, Fujita K, Gadek B, Ghazanfar A, Hagen JO, Holmlund P, Karimi N, Li Z, Pelto M, Pitte P, Popovnin VV, Portocarrero CA, Prinz R, Sangewar CV, Severskiy I, Sigurdsson O, Soruco A, Usubaliev R, Vincent C (2015) Historically unprecedented global glacier decline in the early 21st century. J Glaciol 61(228):745–761

# Modeling Landslide Volumes: A Case Study in Whatcom County, Washington, USA

Gabriel Legorreta Paulin, Trevor A. Contreras, Katherine A. Mickelson, Kara E. Jacobacci, and William Gallin

**Abstract**

In mountainous terrains, landslides and their deposits form a major natural hazard for human settlements and economic activities. The characterization and analysis of landslide volumes is important to understanding their behaviors and the effects of their potential propagation distances. The Washington Geological Survey (WGS) created a new protocol to map landslides with consistent data in Washington State using Light Detection and Ranging (LiDAR). The protocol uses the surface area and the average depth of landslides to estimate their initial volumes. The aim of this paper is to present the ongoing research project from the Institute of Geography at the National Autonomous University of Mexico (UNAM) and WGS to estimate sediment production and distribution by taking full advantage of LiDAR and standardized landslide volume calculation in a Geographic Information System (GIS). We develop two landslide volume models by using Python scripts. These models are a systematic methodology for modeling volume of shallow and deep-seated landslides. The methodology and its implementation in the GIS-based technology is presented and discussed.

**Keywords**

GIS • Landslide inventory map • Landslide volume • Washington state

## Introduction

Worldwide, landslide hazard scientists and mitigation workers seek to determine landslide distribution, susceptibility, hazard, and landslide sediment production at multiple scales to build risk models and hazard zonation maps. To that end, these workers compile landslide inventories and model landslide susceptibility by using a GIS. (Washington State Department of Natural Resources (DNR), Forest Practices Division 2006; Hervás and Bobrowsky 2009; Blahut et al. 2010; Slaughter et al. 2017). In order to estimate landslide volumes, the frequency, density, and geometry of landslides are recorded using historical or multi-temporal landslide inventory maps and their related geodatabases (Hervás and Bobrowsky 2009; Guzzetti et al. 2009; Wenkey et al. 2011; Slaughter et al. 2017). Techniques for calculating the landslide volume are based on the geometric shape (conical, elliptical, circular, planar, irregular, etc.) of the current physical form of a landslide (rock falls, deep-seated landslide, debris flow, etc.), on analytical methods for resolution (geometric calculation, differential calculus, topographic change detection, etc.) (Šíma and Seidlová 2014; Tang et al. 2018; Di Traglia, et al. 2020), and on statistical methods that link geometric measurements of landslide area to landslide volume (Guzzetti et al. 2009; Wenkey et al. 2011; Catani et al. 2016). The physical form of the landslide and landform units are obtained by using detailed Digital Elevation Models (DEMs) through the use of differential GPS (Travelletti, and Malet 2012; Young 2015) and imaging equipment (digital cameras, multispectral sensors, LiDAR instruments and Synthetic Aperture Radar (SAR)) carried by

G. L. Paulin (✉)
Instituto de Geografía, Universidad Nacional Autónoma de México, Circuito Exterior, Ciudad Universitaria, Coyoacán, 04510 Ciudad de México, México
e-mail: legorretag@hotmail.com

T. A. Contreras · K. A. Mickelson · K. E. Jacobacci · W. Gallin
Washington Geological Survey, Washington State Department of Natural Resources, Olympia, WA, USA
e-mail: Trevor.Contreras@dnr.wa.gov

K. A. Mickelson
e-mail: Kate.Mickelson@dnr.wa.gov

K. E. Jacobacci
e-mail: Kara.Jacobacci@dnr.wa.gov

W. Gallin
e-mail: William.Gallin@dnr.wa.gov


V. Vilímek et al. (eds.), *Understanding and Reducing Landslide Disaster Risk*,
ICL Contribution to Landslide Disaster Risk Reduction,
https://doi.org/10.1007/978-3-030-60319-9_41

airplanes or unmanned aerial vehicles (UAVs) (UNEP 2013; Chen et al. 2014).

In Washington, several GIS applications have been used to represent and assess slope susceptibility and hazard (Shaw and Vaugeois 1999; Sarikhan et al. 2008; Slaughter et al. 2017). In a similar way, the role of landslides as source of sediment that impact fish habitat, infrastructure, and private property has been conducted by heuristic, statistical, or deterministic approaches at a local or regional scale (McKenzie-Johnson 2004; Wartman et al. 2014; Iverson et al. 2015; Perkins et al. 2017; Slaughter et al. 2017). In spite of this, little work has been done on the systematic comparison of modeling landslide volumes, in order to outline advantages and limitations of the methods.

## Method

The main goal of the model is to provide a systematic methodology for modeling and mapping volumes of shallow and deep-seated landslides. A case study of landslide volume calculation is conducted in Whatcom County, Washington, USA. The study area was selected in order to model landslide volumes delivered to stream systems. The study area has pre-historic and recent hazardous deep-seated and shallow landslides. In the study area, two inventories of more than 200 shallow and deep-seated landslides were mapped to assess and describe landslide distribution from interpretation and local field surveys (Fig. 1).

Two landslide volume models are implemented to calculate volumes for shallow and deep-seated landslides. The method and the implementation of each model in a GIS-based technology is as follows.

Model-1 uses the landslide inventory headscarps (and/or deposit) and LiDAR elevation data to calculate the pre-sliding surface. The model overlays the pre-sliding surface and the current landslide surface to obtain the elevation differences which are multiplied by the pixel size to obtain the potential landslide volume and its distribution (Fig. 2).

The model works for shallow and deep-seated landslides. The analysis for shallow landslides is performed only in the headscarp polygon area by considering the pre-slide surface and the current surface only in the evacuation (initialization) area. The analysis for deep-seated landslides can be performed in the headscarp and/or deposit polygon area where the landslide deposit is not separated from the evacuation (initialization) area by a bypass (runout) zone (Fig. 3). In the latter case of deep-seated landslides, separate headscarp, flank, and deposit polygons are merged before calculating the landslide volume.

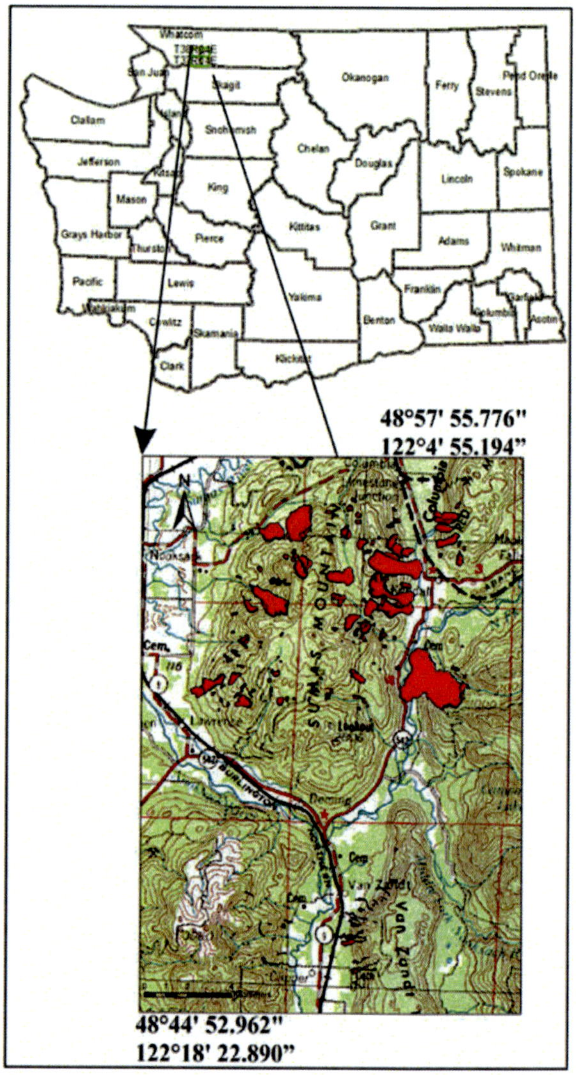


Fig. 1 Study area

To execute the model, the ArcMap toolbox called VolumeUNAM-DNR.tbx should be accessed. The toolbox contains model-1 (VolumenbyInterpolationMainMenu) and model-2 (VolumebyOverlayMainMenu) (Fig. 4).

The model extracts headscarps from the inventory and saves them as polygons. Then the model extracts the nodes from polygons to create a point shape file which in turn is used to extract the elevation data from the LiDAR DEM (Fig. 5).

The nodes are interpolated in ArcMap to create the pre-sliding surface. The landslide volume is calculated by subtracting the real current landslide surface from the

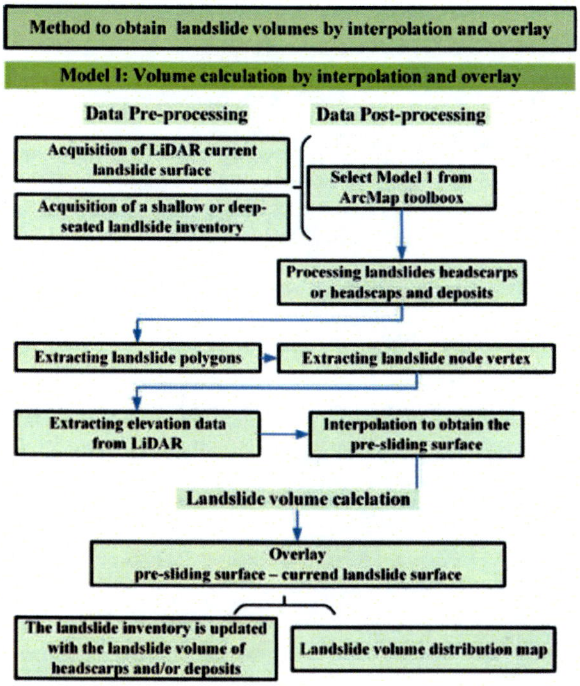


**Fig. 2** General procedure for conducting landslide volumes calculation by interpolation and overlay

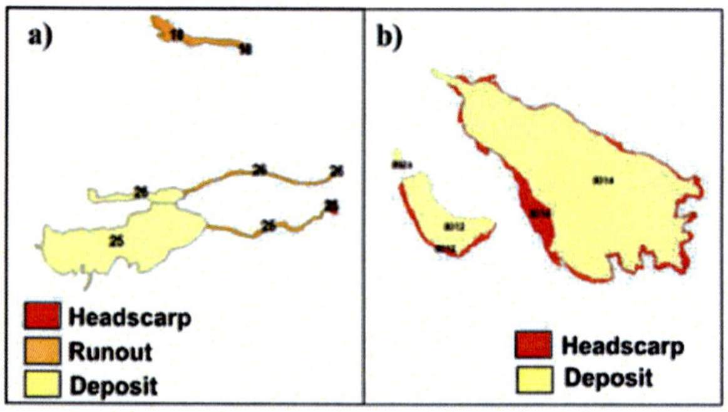


**Fig. 3** Complex landslide polygons. a) shallow landslide inventory; b) deep seated landslide inventory

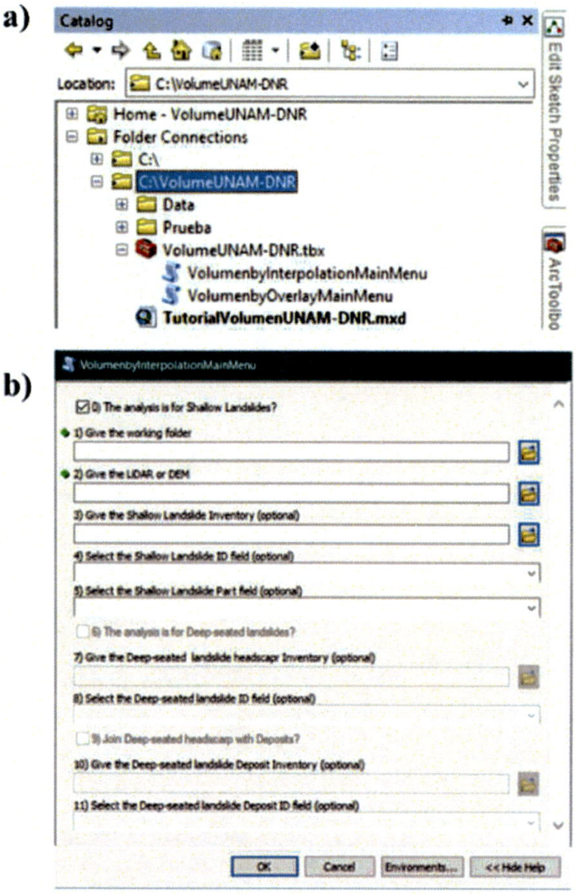


**Fig. 4** **a** Toolbox containing the landslide volume models; **b** main menu of model-1

pre-sliding landslide surface and multiplying by pixel size (Fig. 6).

The landslide volume for each headscarp will have negative or positive values. Negative values represent accumulation of material in the headscarp while positive values represent loss of material.

Model-2 does the same but with the real pre-sliding and current surfaces (Figs. 7 and 8). The model works for shallow and deep-seated landslides.

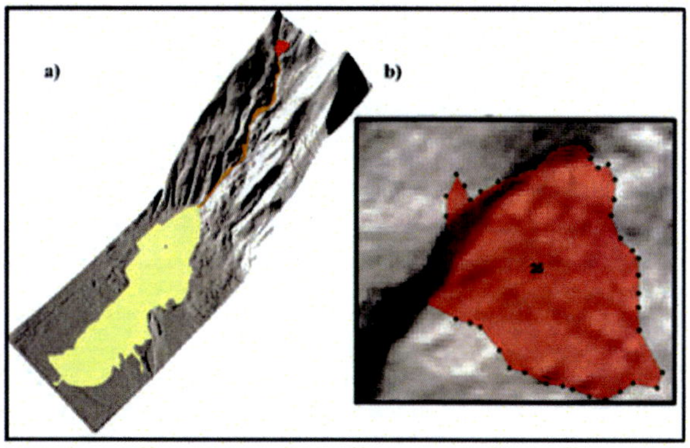


**Fig. 5** **a** Shallow landslide; **b** Shallow landslide headscarp (red polygon) and polygon nodes (green dots)

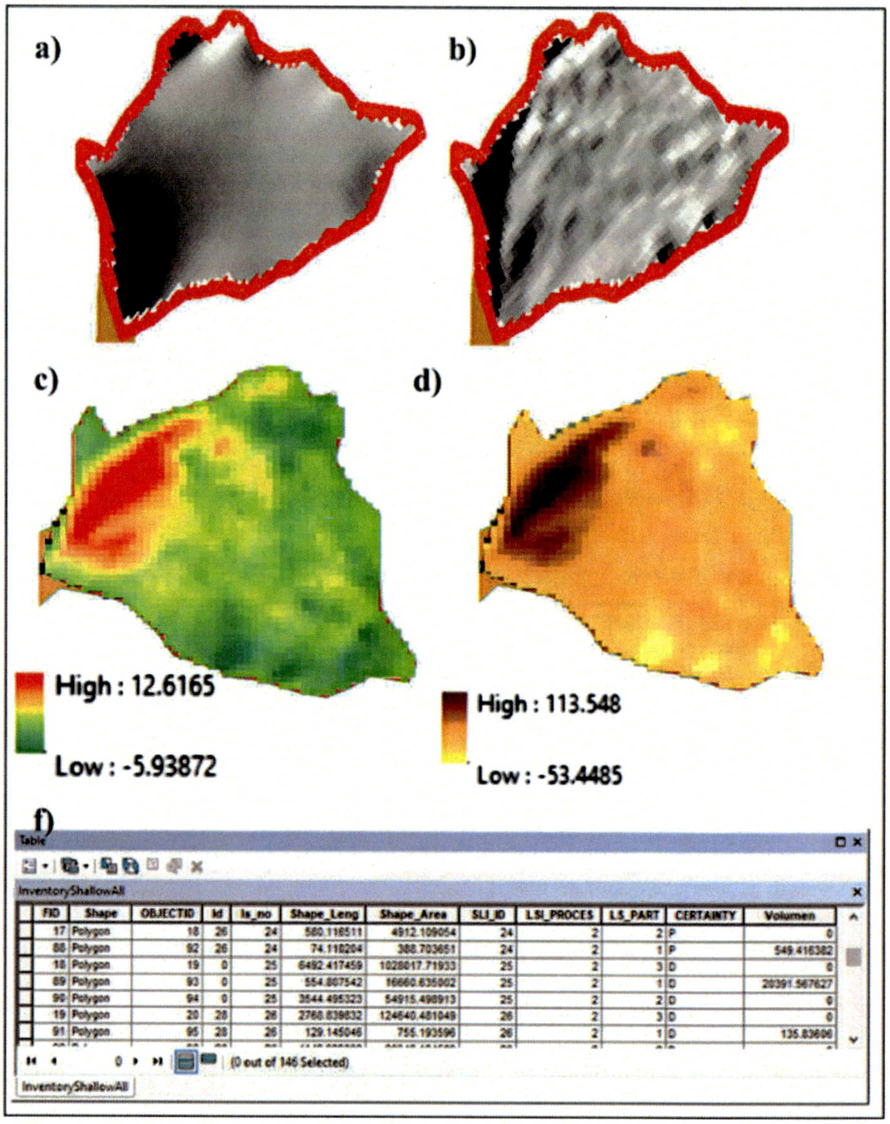


| FID | Shape | OBJECTID | Id | ls_no | Shape_Leng | Shape_Area | SLI_ID | LSI_PROCES | LS_PART | CERTAINTY | Volumen |
|---|---|---|---|---|---|---|---|---|---|---|---|
| 17 | Polygon | 18 | 26 | 24 | 580.118511 | 4912.109054 | 24 | 2 | 2 | P | 0 |
| 88 | Polygon | 92 | 26 | 24 | 74.118204 | 388.703651 | 24 | 2 | 1 | P | 549.416382 |
| 18 | Polygon | 19 | 0 | 25 | 6492.417459 | 1028017.71933 | 25 | 2 | 3 | D | 0 |
| 89 | Polygon | 93 | 0 | 25 | 554.807542 | 16660.635002 | 25 | 2 | 1 | D | 20391.567627 |
| 90 | Polygon | 94 | 0 | 25 | 3544.495323 | 54915.498913 | 25 | 2 | 2 | D | 0 |
| 19 | Polygon | 20 | 28 | 26 | 2768.839832 | 124640.481049 | 26 | 2 | 3 | D | 0 |
| 91 | Polygon | 95 | 28 | 26 | 129.145046 | 755.193596 | 26 | 2 | 1 | D | 135.83606 |

**Fig. 6** **a** pre-sliding landslide surface; **b** current landslide surface; **c** altimetric differences between the two surfaces; **d** distribution of landslide volume; **f** landslide inventory with the calculate volume for the headscarp (volume field)

A final revision of the landslide volume calculation will be conducted for a large number of landslides in future work to highlight advantages and limitations of the models.

## Discussion and Conclusion

Model-1 reconstructs the pre-sliding landslide surface through a process of interpolation using altimetric values from headscarps and flanks. Since there are no internal altimetric values of the pre-sliding surface, its reconstruction will not be precise and will vary depending on the types of interpolation used. Nevertheless, Model-1 is the basis of an integral methodology to estimate landslide sediment production in areas of Washington that lack of pre-sliding landslide information. Model-2 is more accurate in the volume calculation since it uses the actual topographic pre-sliding surfaces. Its application will depend on the availability of high-resolution topographic data.

**Acknowledgements** The authors thank authorities from the Washington State Department of Natural Resources (DNR) Washington Geological Survey, and the Department of Physical Geography from the Institute of Geography, UNAM for their approval and help. This sabbatical research was supported by the Programa de Apoyos para la Superación del Personal Académico de la UNAM (PASPA) de la Dirección General de Asuntos del Personal Académico (DGAPA), UNAM.

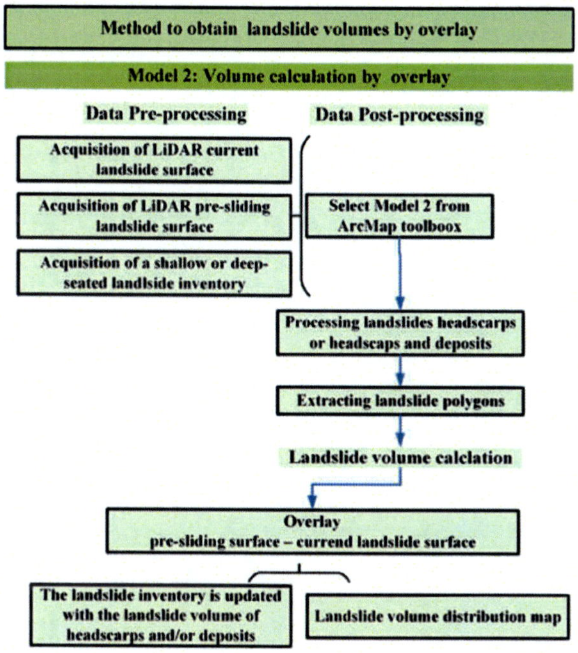


**Fig. 7** General procedure for conducting landslide volumes calculation by overlay

**Fig. 8** Landslide volumes distribution for deep-seated landslides

## References


Blahut J, Van Westen CJ, Sterlacchini S (2010) Analysis of landslide inventories for accurate prediction of debris-flow source areas. Geomorphology 119(1–2):36–51

Boyd TG, Vaugeois LM (2003) On the development of a statewide landslide inventory [abs.]: Geological Society of America Abstracts with Programs 35(6):18

Catani F, Tofani V, Lagomarsino D (2016) Spatial patterns of landslide dimension: a tool for magnitude mapping. Geomorphology 273:361–373

Chen Q, Cheng H, Yang Y, Liu G, Liu L (2014) Quantification of mass wasting volume associated with the giant landside Daguangbao induced by the 2008 Wenchuan earthquake from persistent scatterer InSAR. Remote Sens Environ 152:125–135

Di Traglia F, Fornaciai A, Favalli M, Nolesini T, Casagli N (2020) Catching geomorphological response to volcanic activity on steep slope volcanoes using multi-platform remote sensing. Remote Sens 12(3):438

Guzzetti F, Ardizzone F, Cardinali M, Rossi M (2009) Landslide volumes and landslide mobilization rates in Umbria, central Italy. Earth Planet Sci Lett 279:22–229

Hervás J, Bobrowsky P (2009) Mapping: inventories, susceptibility, hazard and risk. In: Sassa K, Canuti P (Eds) Landslides—Disaster risk reduction. Springer, Berlin, ISBN 978-3-540-69966-8, pp 321–349

Iverson RM, George DL, Allstadt K, Reid ME, Collins BD, Vallance JW, Baum RL (2015) Landslide mobility and hazards: implications of the 2014 Oso disaster. Earth Planet Sci Lett 412:197–208

McKenzie-Johnson AS (2004) Kinematics of the Swift Creek Landslide, Northwest Washington

Montgomery DR, Dietrich WE (1994) A physical based model for the topographic control on shallow landslides. Water Resour Res 30:1153–1171

Pack RT, Tarboton DG, Goodwin CN (2001) Assessing terrain stability in a GIS using SINMAP. In: Proceedings of the 15th annual GIS conference, Vancouver, British Columbia

Perkins JP, Reid ME, Schmidt KM (2017) Control of landslide volume and hazard by glacial stratigraphic architecture, northwest Washington State, USA. Geology 45(12):1139–1142

Sarikhan, Isabelle Y, Stanton KM (2008) Washington geological survey GIS statewide landslide database—From design to implementation. In digital mapping techniques '08—workshop proceedings, Moscow Idaho, pp 2009–1298

Shaw SC, Vaugeois LM (1999) Comparison of GIS-based models of shallow landsliding for application to watershed management. Timber/Fish/Wildlife report TFW-PR, pp 10–99

Šíma J, Seidlová A (2014) The comparison of the calculation of the volume of the foundation pit. Procedia Eng 91:328–333

Slaughter SL, Burns WJ, Mickelson KA, Jacobacci KE, Biel A, Contreras TA (2017) Protocol for landslide inventory mapping from lidar data in Washington State: Washington Geological Survey Bulletin 82, 27 p. text, with 2 accompanying ESRI _le geodatabases and 1 Microsoft Excel _le. https://www.dnr.wa.gov/Publications/ger_b82_landslide_inventory_mapping_protocol.zip

Tang C, Tanyas H, van Westen CJ, Tang C, Fan X, Jetten VG (2018) Analysing post-earthquake mass movement volume dynamics with multi-source DEMs. Eng Geol 248:89–101

Travelletti J, Malet JP (2012) Characterization of the 3D geometry of flow-like landslides: a methodology based on the integration of heterogeneous multi-source data. Eng Geol 128:30–48

UNEP (2013) "A new eye in the sky: Eco-drones". UNEP Global Environmental Alert Service. May 2013. https://na.unep.net/geas/getUNEPPageWithArticleIDScript.php?article_id=100

Wartman J, Montgomery DR, Anderson SA, Keaton JR, Benoît J, dela Chapelle J, Gilbert R (2016) The 22 March 2014 Oso landslide, Washington, USA. Geomorphology 253:275-288

Washington State Department of Natural Resources (DNR), Forest Practices Division (2006) Landslide Hazard Zonation (LHZ) Mapping Protocol, version 2.0. https://www.dnr.wa.gov/BusinessPermits/Topics/LandslideHazardZonation/Pages/fp_lhz_review.aspx

Wenkey D, Jen CH, Böse M, Lin JC (2011) Assessment of sediment delivery from successive erosion on stream-coupled hillslopes via a time series of topographic surveys in the central high mountain range of Taiwan. Quatern Int. https://doi.org/10.1016/j.quaint.2011.02.018

# CRE Dating of Torrential Alluvial Deposits as an Approximation to Holocene Climate-Change Signatures in the Northwestern Andes of Colombia

2020 Kyoto Japan

Santiago Noriega-Londoño, Maria Isabel Marín-Cerón, Julien Carcaillet, Matthias Bernet, and Isandra Angel

**Abstract**

Understanding long-term hillslope and fluvial processes in tropical mountainous settings is key for improving models of local landscape evolution and natural hazard assessment, particularly in relation to mass-wasting and torrential events. Cosmic Ray Exposure (CRE) dating is a powerful tool to constraining timing and rates of geomorphic processes at the Earth's surface. In this study, we calculate the ages of torrential alluvial deposits of the Farallones River located on the eastern slopes of the Western Cordillera of Colombia; a region characterized by high local relief (>3800 m), steep slopes, and relatively high mean annual precipitation (>2000 mm). Ages of alluvial deposits were calculated using a CRE approach based on in-situ produced 10Be concentrations from boulder samples and varying erosion rates between 0 and 1 mm/y. Rock boulder ages range slightly between 6.1 ± 0.5 to 7.9 ± 0.8 ky, despite assuming contrasting erosion rates, and between 6.1 ± 0.5 to 8.5 ± 0.9 ky with erosion rates of 0.01 mm/y. These results indicate that the torrential alluvial deposits of the Farallon River were exposed to cosmic radiation during the middle Holocene, and probably after the after the 8.2 ky event. Moreover, modern catchment morphology and precipitation patterns records are consistent with the long-term geomorphic control on mass movement and torrential behaviour of this mountainous river. The integration of these first CRE ages with quantitative geomorphology analysis allows us to delimit the nature and the chronology of major torrential and landslide events in the tropics, and specifically in the Northern Andes where planning approaches to reduce natural hazards are in strong need of accurate quantitative data.

**Keywords**

Debris flow • CRE • Landscape evolution • Western andean cordillera • Colombia

## Introduction

During the Quaternary period, the Earth's surface evolution depends on isostatic or tectonic as well as external forcing processes, which contribute to the magnitude and frequency of geomorphic processes. Besides, the hydrological balance, the reorganization of the drainage network, and mountain belts changes are controlled by 10 to 100 thousand years glacial and interglacial periods (Allen 2008). Thus, the quantification of the landscape chronology provides robust conceptual and numerical models needed for land management and risk assessment strategies, and moreover we can use them as paleoclimatic proxies (Pánek 2015).

The Cosmic Ray Exposure (CRE) dating is a technique that addresses the surface exposure age of discrete geomorphic features. This is particularly suitable for dating of deposited rock material (Bierman et al. 2002).

When samples are quartz rich, CRE ages could be obtained using the in-situ produced $^{10}$Be terrestrial cosmogenic nuclides (TCN) concentration. Because the $^{10}$Be is a long-lived isotope with a half-life of ~1.39 Ma (Korschinek et al. 2010), it is a useful technique to constraint the

S. Noriega-Londoño (✉) · M. I. Marín-Cerón
Departamento de Ciencias de la Tierra, Universidad EAFIT, 050022 Medellín, Colombia
e-mail: snoriegal@eafit.edu.co

M. I. Marín-Cerón
e-mail: mmarince@eafit.edu.co

J. Carcaillet · M. Bernet
ISTerre, Université Grenoble Alpes, 38000 Grenoble, France
e-mail: julien.carcaillet@univ-grenoble-alpes.fr

M. Bernet
e-mail: matthias.bernet@univ-grenoble-alpes.fr

I. Angel
Departamento de Ciencias de la Tierra, Universidad Simón Bolívar, Caracas, 1086, Venezuela
e-mail: isandrangel@usb.ve


V. Vilímek et al. (eds.), *Understanding and Reducing Landslide Disaster Risk*, ICL Contribution to Landslide Disaster Risk Reduction,
https://doi.org/10.1007/978-3-030-60319-9_42

chronological framework of the Earth's surface changes during quaternary glacial and interglacial periods.

Particularly, the northwestern Andes of Colombia constitute an exceptional case of convergence of tropical climatic setting with a complex oblique convergent margin and Late Cenozoic orogenic processes that predispose hillslopes and alluvial extreme events whose magnitude and frequency are also controlled by climatic factors. In this study, we propose an integration of CRE ages from the Farallones river torrential alluvial deposits with the morphometric, paleoclimatic, stratigraphic and historical features of the Farallones River at the Western Cordillera (WC) of Colombia.

In synthesis, this place becomes a natural laboratory for CRE analyses. This approach seeks to apply CRE dating as a quantitative geomorphology contribution to the dating-based on growing evidence at some reconstructed landslide activity phases that followed major Holocene short-term climatic fluctuations (e.g. Mayewski et al. 2004; Wanner et al. 2011).

## Study Site

The Farallones River is located in the WC, northwestern Andes of Colombia, between ~150 and ~4020 m above sea level (Fig. 1A). Mean annual temperature range from ~6 °C to ~24 °C, from the páramo and the lower canyons respectively. In addition, the mean annual precipitation of ~2300 mm per year indicating the influence of the El Niño Southern Oscillation (ENSO) (Poveda et al. 2001), and the bimodal rainfall seasons with peaks in May and October (e.g. Monsalve 2015) which correlates landslide activity phases that followed major Holocene short-term climatic fluctuations well documented around the world at '8.2' and '4.2' ka BP events (e.g. Mayewski et al. 2004; Wanner et al. 2011).

The sierra-like topography of the Farallones site separates the Cauca, Atrato and San Juan rivers and record the glacial and periglacial events through the Quaternary. On its eastern side, the San Juan River, and the Cauca river tributaries control the topographic base level. Glacial landforms such as U-shaped valleys, moraines, tarns, horns, among others, remains above 3200 m on the surrounded mountain peaks, while coarse fluvio-glacial deposits are documented in the lower river channels and terraces.

The study site is composed of Cretaceous volcano-sedimentary sequences locally intruded by calc-alkaline Miocene bodies, such as the Farallones batholith (Gómez-Tapias et al. 2015). Crystallization U/Pb ages from zircon grain and low temperature thermochronological data from zircon fission track and apatite (U-Th)/he suggest a marked event of magmatism rapidly followed by and exhumation and erosion pulse during Middle Miocene (Restrepo-Moreno et al. 2013).

Regional faults such as the La Mansa Fault cross the study site with a NW to NNW strike and left-lateral component (Fig. 1B), but its associated seismicity is considered as low to middle grade (IGAC and Ingeominas 2001).

In this area, the mass movements and associated debris flow are recurrent due to the strong rainfall season, steep hillslopes, and channel morphology (Naranjo et al. 2019). Historical occurrences of mass movements such as rockfall and debris flow have been documented by Piedrahita and Hermelin (2005), and Pérez-Hincapié (2019), highlighting the debris flow events of April 14, 1878, April 27, 1943, June 4, 1991, April 25, 1993.

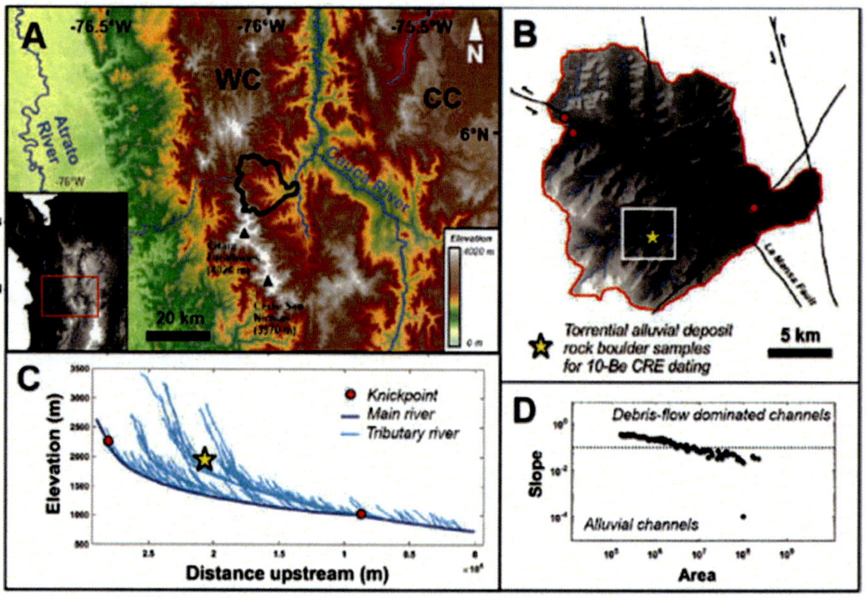


**Fig. 1** Simplified setting of the Farallones river catchment. (A) Relief map of the Farallones site. (B) Detail of the Farallones River catchment and location of rock boulder samples. (C) Longitudinal river profile of the Farallones catchment. (D) Slope-Area plot of the Farallones River

A geomorphological mapping of the alluvial terrace levels and stratigraphic analysis of debris-flow deposits on the Farallones area have been presented by Pérez-Hincapié (2019). In this study, we focus on the torrential deposits that constitute the higher terraces, which exhibit relative elevations of ~42 m above the present river bed, lengths between 50 and 150 m, and slopes between 8° and 25°. The rock boulders constitute the 80% of the deposit and are embedded in a mud matrix (Fig. 2).

Paleoclimatic records of local significance have been reported in the region. In the WC, a pollen-based Late Pleistocene-Holocene paleoclimatic reconstruction from the Páramo de Frontino (PF) indicate millennial scale variations of dry/cold and wet/warm climate since the last 17 ky (Velasquez and Hooghiemstra 2013; Monsalve 2015; Muñoz et al. 2017) while on the Páramo de Belmira (i.e. Central Cordillera, CC), the palynological record from lacustrine deposits document the last 35 ky (Castañeda Riascos 2013).

## Materials and Methods

### Sampling Procedure

To estimate torrential alluvial deposit CRE ages, we measured the $^{10}$Be TCN concentration of four rock samples from the Farallones River debris flows deposits (Fig. 2). We used a chisel and hammer to sampling ~1 kg of quartz-rich tonalite. We focused our sampling on the top of well anchored and high (> 0.5 m) boulders in order to avoid post abandonment displacement and surficial deposits. Boulders are placed on the undulated higher terrace surface of the Farallones River. Three samples (i.e. Far-01, Far-02 and Far-04) were taken from tonalitic boulders with <1 m of diameter while the Far-03 sample, correspond to a bigger and felsic rock boulder (>1 m) located ~50 m upstream from the others samples. As the rock boulder shows rounded shapes and relatively high diameters, we assume that the rock source corresponds to rock walls of the upper granitic massif which feed the Farallones river debris flows after discrete rock fall events.

### Cosmic Ray Exposure Dating

Samples were crushed and sieved at the thermochronology laboratory at the EAFIT University (Colombia). Quartz isolation and preparation were completed in the GTC platform (ISTerre, Grenoble-Alpes University France), following procedures to obtain TCN $^{10}$Be were adapted from Brown et al. (1991) and Merchel and Herpers (1999). The $^{10}$Be/$^{9}$Be ratios were measured at the French National (AMS) Accelerator Mass Spectrometry facility (Arnold et al. 2010). The $^{10}$Be/$^{9}$Be ratios have been corrected from chemical blank of $7.23 \pm 0.13 \times 10^{-15}$ for samples Far-01, Far-02 and $6.36 \pm 0.7 \times 10^{-15}$, to correct samples Far-03, Far-04. The topographic shielding factors were measured for each sample site. We used the topographic shielding calculator version 2 from (Balco et al. 2008). We used the online Cosmic Ray Exposure program (CREP) (Martins et al. 2017) to calculate surface exposure ages from in situ $^{10}$Be TCN concentrations. For this, we employed a local SLHL production rate of $3.74 \pm 0.09$ atm/g/y (Martin et al. 2015), and the Lal (1991)/Stone (2000) scaling scheme, with the ERA-40 (Uppala et al. 2005), and the geomagnetic record of Lifton 2016 VDM (Lifton 2016).

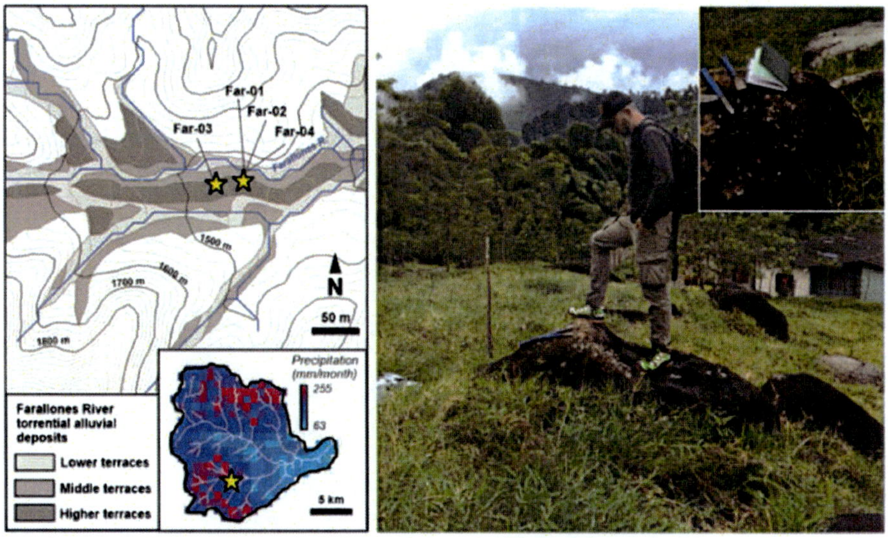


**Fig. 2** Torrential alluvial deposits from the Farallones River

**Table 1** Geographical information and in situ $^{10}Be$ concentration of the Farallones River samples

| Sample ID | Long. W (°) | Lat. N (°) | Altitude (m) | Sample thickness (cm) | Topographic shielding | $^{10}Be/^{9}Be$ ($10^{-14}$) | $^{10}Be/^{9}Be$ Uncert. ($10^{-15}$) | $^{10}Be$ Concentration (at/g) | $^{10}Be$ Uncert. (at/g) |
|---|---|---|---|---|---|---|---|---|---|
| Far-01 | 5.992 | 76.028 | 1440 | 4.5 | 0.993 | 2.818 | 2.222 | 34,291 | 3174 |
| Far-02 | 5.800 | 76.027 | 1438 | 2.5 | 0.993 | 1.433 | 1.640 | 37,875 | 5644 |
| Far-03 | 5.799 | 76.028 | 1443 | 4 | 0.993 | 3.22 | 1.521 | 263,780 | 13,732 |
| Far-04 | 5.799 | 76.027 | 1440 | 4 | 0.993 | 1.072 | 0.801 | 46,128 | 4594 |

As erosion processes can condition the CRE ages, we computed CRE ages under different erosion rates scenarios varying between 0 and 1 mm/y.

## Results

Measured in situ $^{10}Be$ TCN concentrations from rock boulders related to the oldest torrential alluvial deposit varies from 3.4 to 26.4 × $10^4$ atoms $g^{-1}$ (Table 1). Uncertainties were < 10% for all samples except for the Far-02 sample with uncertainties of ~14%. Exposure age data for the same torrential alluvial deposit varies between 6.1 and 7.9 ky (considering 0 mm/y of erosion) and 6.4–8.5 ky (considering 0.01 mm/y of erosion). External uncertainties of calculated CRE ages were less than 10% for all samples except for the older one with uncertainties of ~13%. Exposure ages can be be divided into two groups, a younger age composed by samples Far-01, Far-02 and Far-04, and an older age from Far-03 sample which is significantly different at any erosion rate, with ages varying between 39.9 to 74.5 ky. This older sample exhibit a lighter colour probably related with grain size and/or local mineralogical variation, and corresponds to the bigger boulder with diameter in excess of 1 m. Because Far-03 is significantly different and older, it could indicate an important heritance. Also, Far-03 is located upward the other samples (3 m in elevation and more than 50 m upward the stream flow) and its leucogranite mineralogy differs from the others samples. In this study, Far-03 was considered as an outlier and it was not considered to discussion. Similarly, computed CRE ages with erosion rates of 0.1 mm/y exhibit ages of ~20.53 ± 1.6 ky for the Far-01 sample.

## Discussion and Conclusion

### CRE Ages as Function of Erosion Rates

The two main scenarios of erosion rates considered in this study provide contrasting results. For the low erosion rate scenarios, (<0.1 mm/y) three of the four samples (i.e. Far-01, Far-02, and Far-04) correspond to Middle Holocene ages ranging between 6.1 to 8.5 ky (Fig. 3). In contrast, calculated CRE ages with erosion rates above 0.1 mm/y showed abnormal values, mostly due to the reach of the saturated area, where the low production rates are equal to the relatively high erosion rates plus the radioactive decay. Just the Far-01 sample indicate a Late Pleistocene CRE age of 20.53 ± 1.59 ky.

Although the Holocene erosion rates of the northwestern Andes remain relatively undocumented, some researchers have been documenting long-term erosion rates. For example, Restrepo-Moreno et al. (2009) using low temperature thermochronology (apatite (U-Th)/He), indicates erosion rates for the Central Cordillera of ~0.04 mm/y for the last 25 million years, specifying a maximum of ~0.2–0.4 mm/y during rapid cooling pulses in the early Oligocene and middle Miocene (Restrepo-Moreno et al., 2009). Vertical incision rates from Pleistocene to Pliocene alluvial terraces located on the top of the Central Cordillera suggest values of 0.01–0.08 mm/y (Page and James 1981). Finally, numerical models of Cataño and Vélez (2015) suggest values of 0.0054–0.0020 mm/y for mountain basins in the region. Under this framework, our preferred scenarios are erosion rates ranging from 0.01 to 0.1 mm/y (Table 2).

### Torrential Alluvial Deposits and Holocene Climate Signature

All samples that are treated according to the low erosion rate scenarios correspond with the global interglacial period after

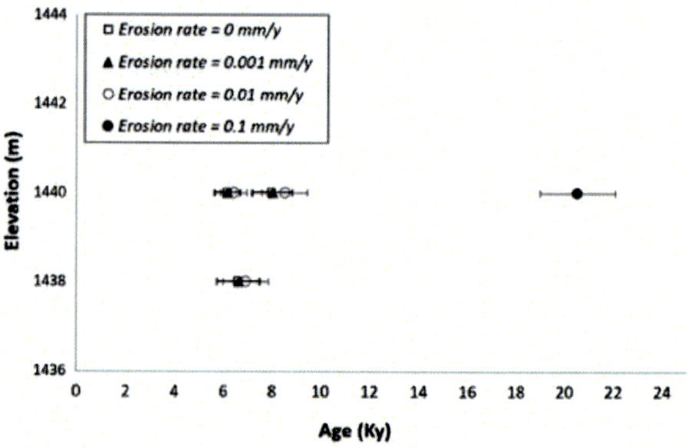


**Fig. 3** Geochronology of the torrential alluvial deposits of the Farallones River from in situ $^{10}Be$ CRE dating

**Table 2** In situ $^{10}$Be CRE ages from the evaluated erosion rates scenarios

| Sample ID | Erosion rate = 0 mm/y | | | Erosion rate = 0.001 mm/y | | | Erosion rate = 0.01 mm/y | | | Erosion rate = 0.1 mm/y | | |
|---|---|---|---|---|---|---|---|---|---|---|---|---|
| | Age (ky) | Internal uncert | External uncert | Age (ky) | Internal uncert | External uncert | Age (ky) | Internal uncert | External uncert | Age (ky) | Internal uncert | External uncert |
| Far-01 | 6.14 | 0.48 | 0.49 | 6.17 | 0.48 | 0.5 | 6.41 | 0.51 | 0.52 | 20.1 | 1.59 | 1.54 |
| Far-02 | 6.59 | 0.84 | 0.85 | 6.62 | 0.85 | 0.86 | 6.92 | 0.9 | 0.92 | – | – | – |
| Far-03 | 39.86 | 1.39 | 1.54 | 40.81 | 1.42 | 1.61 | 74.48 | 4.07 | 4.49 | – | – | – |
| Far-04 | 7.96 | 0.76 | 0.78 | 8.00 | 0.78 | 0.8 | 8.51 | 0.89 | 0.91 | – | – | – |

the LGM. We interpret these ages as debris flow events that transport and expose bedrock boulders during interglacial periods. This is consistent with the Holocene paleoclimatic records of the mountain peaks of the Farallones River and its surroundings. Particularly, this Holocene record is highlighted from the top of the Western and Central Cordilleras, and partially from its inter-Andean basins.

Records from the Páramo de Frontino (PF) at Llano Grande Basin, located 70 km north of the study site, indicate moist to wet climatic conditions between 8000 cal yr BP until 5360 cal yr BP (Velasquez and Hooghiemstra 2013; Muñoz et al. 2017) determined precipitations increased between 7500 cal yr BP and 4500 cal yr BP, with a maximum between 5000 and 4500 cal yr BP. Also, pollen-based analysis from the Páramo de Belmira (PB) in the Central Cordillera indicate a period of wet and warm climate during the Middle Holocene also (Castañeda-Riascos 2013).

From the Cauca River (CR) Canyon, fluvio-lacustrine deposits record a period of sedimentation since 3.5 ky until around 1,0 ky that is related to the Atlantic Multidecadal Oscillation (AMO) (Martínez et al. 2015) and partially correlate with the Late Holocene radiocarbon data from torrential alluvial deposits on the Farallones River of ~2.7 ky (Pérez-Hincapié 2019) (Fig. 3). Also, this Holocene paleoclimatic signature has been reported in some parts of the Eastern Cordillera, where torrential events show ages ranging between 38.9 and 8.7 ky (Hoyos et al. 2015). In addition, paleoclimatic records of the western flank of the Western Cordillera (i.e. Atrato basin) indicate a wet period between 4.0 and 2.7 ky (Urrego et al. 2006).

In synthesis, the presented CRE ages provide an extended window for the torrential alluvial deposits occurrence during Middle Holocene times beyond the radiocarbon ages from lower deposits of the Farallones River and its surroundings which indicate posterior torrential events (Pérez-Hincapié 2019).

Finally, slope-area relationship as well as sedimentological inventory suggests that Farallones catchment exhibit both, torrential and alluvial behavior (Fig. 1D). Note that the analyzed torrential alluvial deposits are located on a set of steeper tributary streams with elevations below 3500 m inside the Farallones river catchment (Fig. 1C). On the other hand, glacial records of Western Cordillera, documented in the Citará Farallones, Tatamá Farallones (T), and in the Páramo de Frontino (Van Der Hammen 2005), show that ice sheet has shaped the topography above ~3200 m with fluvio-glacial deposits reaching an elevation of ~2200 m. In conclusion, highest levels of torrential alluvial deposits of the Farallones River composed by tonalitic rock boulders exhibit interglacial ages, indicating debris flow deposition and exposure during the Holocene Climate Optimum and more precisely after the 8.2 ky event. Thus, this approach seeks to apply CRE dating as a quantitative geomorphology contribution to the dating-based on growing evidence of landslide activity phases that followed major Holocene short-term climatic fluctuations; e.g. well-known '8.2' and '4.2' ka BP events (e.g.Mayewski et al. 2004; Wanner et al. 2011).

**Acknowledgements** We acknowledge support of this study from ECOS-NORD/Colciencias/ICETEX project C12U01 of M. Bernet and M.I. Marín-Cerón, as well as a BQR SUD. The Projects 710-2015 (19628) and 785-2017 (201820001205) provided by the Departamento de Ciencia, Tecnología e Innovación – COLCIENCIAS. Special thanks to Geovany Bedoya, Marco Gamboa and Ana Pérez from the Department of Earth Sciences at Universidad EAFIT. We warmly thank Karim Keddadouche and Georges Aumaitres for the AMS measurements ASTER, Aix in Provence, France.

## References

Allen PA (2008) From landscapes into geological history. Nature. https://doi.org/10.1038/nature06586

Arnold M, Merchel S, Bourlès DL, Braucher R, Benedetti L, Finkel RC, Aumaître G, Gottdang A, Klein M (2010) The French accelerator mass spectrometry facility ASTER: Improved performance and developments. *Nuclear Instruments and Methods in Physics Research, Section B: Beam Interactions with Materials and Atoms*. https://doi.org/10.1016/j.nimb.2010.02.107

Balco G, Stone JO, Lifton NA, Dunai TJ (2008) A complete and easily accessible means of calculating surface exposure ages or erosion rates from 10Be and 26Al measurements. Quat Geochronol. https://doi.org/10.1016/j.quageo.2007.12.001

Bierman PR, Caffee MW, Davis PT, Marsella K, Pavich M, Colgan P, Mickelson D, Larsen J (2002) Rates and timing of earth surface processes from in situ-produced cosmogenic Be-10. Rev Mineral Geochem 50(1):147–205. https://doi.org/10.2138/rmg.2002.50.4

Brown ET, Edmond JM, Raisbeck GM, Yiou F, Kurz MD, Brook EJ (1991) Examination of surface exposure ages of Antarctic moraines using in situ produced 10Be and 26Al. Geochim Cosmochim Acta. https://doi.org/10.1016/0016-7037(91)90103-C

Castañeda Riascos IM (2013) Paleoecología de Alta Resolución del Holoceno (11000 años), en el Páramo de Belmira, Antioquia (Colombia). *Tesis de Grado (Maestría)*

Cataño Álvarez S, Vélez Upegui J (2015) Modelo conceptual agregado de transporte de sedimentos para cuencas de montaña en Antioquia. *Boletín de Ciencias de La Tierra*. https://doi.org/10.15446/rbct.n39.52888

Gómez-Tapias J, Montes NE, Alcárcel FA, Ceballos JA (2015) Catálogo de dataciones radiométricas de Colombia en ArcGIS y Google Earth. Publicaciones Geológicas Especiales 33:63–419

Hoyos N, Monsalve O, Berger GW, Antinao JL, Giraldo H, Silva C, Ojeda G, Bayona G, Escobar J, Montes C (2015) A climatic trigger for catastrophic Pleistocene-Holocene debris flows in the Eastern Andean Cordillera of Colombia. J Quat Sci. https://doi.org/10.1002/jqs.2779

IGAC, INGEOMINAS (2001) *Investigación Integral del Andén Pacífico Colombiano*

Korschinek G, Bergmaier A, Faestermann T, Gerstmann UC, Knie K, Rugel G, Wallner A, Dillmann I, Dollinger G, von Gostomski CL, Kossert K, Maiti M, Poutivtsev M, Remmert A (2010) A new value for the half-life of 10Be by Heavy-Ion Elastic Recoil Detection and liquid scintillation counting. Nuclear Instruments and Methods in Physics Research, Section B: Beam Interactions with Materials and Atoms. https://doi.org/10.1016/j.nimb.2009.09.020

Lal D (1991) Cosmic ray labeling of erosion surfaces: in situ nuclide production rates and erosion models. Earth Planet Sci Lett. https://doi.org/10.1016/0012-821X(91)90220-C

Lifton N (2016) Implications of two Holocene time-dependent geomagnetic models for cosmogenic nuclide production rate scaling. Earth Planet Sci Lett. https://doi.org/10.1016/j.epsl.2015.11.006

Martin LCP, Blard PH, Lavé J, Braucher R, Lupker M, Condom T, Charreau J, Mariotti V, Arnold M, Aumaître G, Bourlés DL, Keddadouche K, Davy E (2015) In situ cosmogenic 10Be production rate in the High Tropical Andes. Quat Geochronol. https://doi.org/10.1016/j.quageo.2015.06.012

Martínez JI, Obrochta S, Yokoyama Y, Battarbee RW (2015) Atlantic Multidecadal Oscillation (AMO) forcing on the late Holocene Cauca paleolake dynamics, northern Andes of Colombia. Clim Past Discuss. https://doi.org/10.5194/cpd-11-2649-2015

Martins AA, Cabral J, Cunha PP, Stokes M, Borges J, Caldeira B, Martins AC (2017) Tectonic and lithological controls on fluvial landscape development in central-eastern Portugal: insights from long profile tributary stream analyses. Geomorphology. https://doi.org/10.1016/j.geomorph.2016.10.012

Mayewski PA, Rohling EE, Stager JC, Karlén W, Maasch KA, Meeker LD, Meyerson EA, Gasse F, van Kreveld S, Holmgren K, Lee-Thorp J, Rosqvist G, Rack F, Staubwasser M, Schneider RR, Steig EJ (2004) Holocene climate variability. Quatern Res 62 (3):243–255. https://doi.org/10.1016/j.yqres.2004.07.001

Merchel S, Herpers U (1999) An update on radiochemical separation techniques for the determination of long-lived radionuclides via accelerator mass spectrometry. Radiochim Acta. https://doi.org/10.1524/ract.1999.84.4.215

Monsalve C (2015) Climatic change during the late glacial in the northwestern part of Colombia, based on palynology and microfluorescence-x. Université de Geneve

Muñoz P, Gorin G, Parra N, Velásquez C, Lemus D, Monsalve MC, Jojoa M (2017) Holocene climatic variations in the Western Cordillera of Colombia: a multiproxy high-resolution record unravels the dual influence of ENSO and ITCZ. Quatern Sci Rev. https://doi.org/10.1016/j.quascirev.2016.11.021

Naranjo Bedoya K, Aristizábal Giraldo EV, Morales Rodelo JA (2019) Influencia del ENSO en la variabilidad espacial y temporal de la ocurrencia de movimientos en masa detonados por lluvias en la región Andina. Ingeniería y Ciencia. https://doi.org/10.17230/ingciencia.15.29.1

Page W, James M (1981) The Antiquity of the erosion surfaces and the Late Cenozoic deposits near Medellin, Colombia: implications to tectonics and erosion rates. Revista CIAF 6(1–3):421–454

Pánek T (2015) Recent progress in landslide dating: a global overview. Prog Phys Geogr 39(2):168–198. https://doi.org/10.1177/0309133314550671

Pérez-Hincapié (2019) Zonificación de amenazas por avenidas torrenciales a partir del análisis geomorfológico de los depósitos asociados y su datación. Caso de estudio : cuencas de la vertiente oriental de la Cordillera Occidental Colombiana, Andes del Norte. Universidad EAFIT

Piedrahita I, Hermelin M (2005) La avenida Torrencial del Río Tapartó Antioquia de 1993. In Desastres de Origen Natural en Colombia 1979–2004, pp 109–120

Poveda G, Jaramillo A, Gil MM, Quiceno N, Mantilla RI (2001) Seasonality in ENSO-related precipitation, river discharges, soil moisture, and vegetation index in Colombia. Water Resour Res 37 (8):2169–2178. https://doi.org/10.1029/2000WR900395

Restrepo-Moreno SA, Min K, Bernet M, Barbosa A, Marín-Cerón MI, Mesa J, Hardwick E, Pardo A, Kamenov G (2013) Thermotectonic history of the Farallones del Citará Batholith (Colombia's Western Cordillera) through multi-system, vertical profile thermochronology/geochronology: Tectonic, geomorphic and climatic implications. XIV Congreso Colombiano de Geología

Restrepo-Moreno S, Foster D, Stockli D, Parra N (2009) Long-term erosion and exhumation of the "Altiplano Antioqueño", Northern Andes (Colombia) from apatite (U–Th)/He thermochronology. Earth Planet Sci Lett 278(1–2):1–12. https://doi.org/10.1016/j.epsl.2008.09.037

Stone JO (2000) Air pressure and cosmogenic isotope production. J Geophys Res: Solid Earth. https://doi.org/10.1029/2000jb900181

Uppala SM, Kållberg PW, Simmons AJ, Andrae U, da Costa Bechtold V, Fiorino M, Gibson JK, Haseler J, Hernandez A, Kelly GA, Li X, Onogi K, Saarinen S, Sokka N, Allan RP, Andersson E, Arpe K, Balmaseda MA, Beljaars ACM, … Woollen J (2005) The ERA-40 re-analysis. Quart J Royal Meteorol Soc. https://doi.org/10.1256/qj.04.176

Urrego LE, Molina LA, Urrego DH, Ramírez LF (2006) Holocene space-time succession of the Middle Atrato wetlands, Chocó biogeographic region, Colombia. Palaeogeography, Palaeoclimatology, Palaeoecology. https://doi.org/10.1016/j.palaeo.2005.10.018

Van Der Hammen T (2005) Zonal ecosystems of the west and east flanks of the Colombian Western Cordillera (Tatamá transect). En: Van Der Hammen T, Orlando Rangel J, Cleef AM (Eds) La Cordillera Occidental Colombiana Transecto Tatamá. Estudio de Ecosistemas Tropandinos 6:935–972

Velasquez C, Hooghiemstra H (2013) Pollen-based 17-kyr forest dynamics and climate change from the Western Cordillera of Colombia; no-analogue associations and temporarily lost biomes. Rev Palaeobot Palynol 194:38–49

Wanner H, Solomina O, Grosjean M, Ritz SP, Jetel M (2011) Structure and origin of Holocene cold events. Quatern Sci Rev 30:3109–3123. https://doi.org/10.1016/j.quascirev.2011.07.010

# Features of Construction in Areas with Deep Block-Type Landslides

2020 Kyoto Japan

Andrey Kazeev and German Postoev

**Abstract**

Understanding of landslide mechanisms is necessary for hazard and risk assessments and designing proper protection measures, including monitoring. It is the key to the sustainable development of urban and mining areas with existing landslide hazards. The mechanisms and main features of deep-seated block landslides of compression-uplifting type are considered in this paper. The scientific methodology for the design of effective reinforcement and planning for sustainable development of landscapes with deep block-slides is provided. The suggested protection strategy is based on earlier received theoretical solutions concerning the reasons of landslide occurrence (the genesis of considered landslide type), on taking into account the mechanism features of deep block-slides, and also on our experience of investigation such landslides in different engineering geological conditions. New protection technologies for areas with deep-seated landslides of compression-uplifting type are discussed in the article.

**Keywords**

Landslide mechanism • Deep block-type landslides • Reinforcement

## Introduction

Landslides annually cause significant economic, social and environmental impacts to all countries, as well as Russia. Over 700 cities in Russia are susceptible to landslide hazard (including deep-seated landslides). Destructive deep landslide activations recently occurred in Russia (in Moscow in 2006, 2009, in Ulyanovsk in 2016; Cheboksary in 2012), in Ukraine (Odessa region in 2017–2018) for example.

Typical deep landslides can be found in the U. K. (Folkestone Warren) (Bromhead and Ibsen 2004; Bromhead 2005; Sassa et.al. 2007), Ukraine (Odessa, Mariupol) (Naumenko 1977), etc. They also occur in mining areas on slopes of open pits and dumps, e.g. in Russia (Kuzbass 2015), Spain (Las Cruces mine 2019), Brazil (Brumadinho tailings dam landslide 2019), etc.

Deep block-type landslides are extremely widespread, especially on coastal slopes and river banks. Particularly serious problems concerning deep landslides appear in urban areas. They often occur suddenly and unexpectedly. On one hand, slow landslide displacements (with the rates on an order of mm/year) provide an illusion of relative stability of such areas, with the possibility of safe development. Typical examples are a construction of an important sports complex on a lucratively picturesque site or an infrastructure (bridges, tunnels) within severe site constraints. From the other hand, periodic destructive landslide activations, affecting significant areas, sometimes make it impossible to protect these structures and nature landscapes.

The design of reinforcement of slopes susceptible to deep block-type landsliding may be very complicated due to uncertainties associated with their characterization, large sizes of landslide amphitheatres, and significant slip-surface depths. The assessment of the stress–strain state of a landslide ground mass, composed of displaced landslide blocks, and it`s interaction with the original mass situated above may present a challenge (Postoev and Kazeev 2014).

A. Kazeev (✉) · G. Postoev
Sergeev Institute of Environmental Geoscience of Russian Academy of Science, Moscow, 101000, Russia
e-mail: kazeev@yandex.ru

G. Postoev
e-mail: opolzen@geoenv.ru



V. Vilímek et al. (eds.), *Understanding and Reducing Landslide Disaster Risk*,
ICL Contribution to Landslide Disaster Risk Reduction,
https://doi.org/10.1007/978-3-030-60319-9_43

Russian experience of urbanization in landslide-prone areas shows that it is practically impossible to reach absolute stabilization of slopes with deep landslides. The most common protective measures are: reinforcement of the river banks to reduce erosion, improving drainage, benching or buttressing a slope, and in more rare cases–pile works. However, rejuvenation and re-initiation of catastrophic movements of deep landslides periodically occurs even on thought-to-be protected slopes, resulting in deformation of existing infrastructure. For example, landslides with deformation of lower Cretaceous clays (with slip surface depths greater than 100 m) periodically occurred in Saratov (Russia) on the reinforced slopes of Sokolova Mountain, causing destructive structural deformations within the city area and the uplifting and bulging of the Volga River bottom. The volume of involved landslide masses was about 10 M cubic meters (Tihvinsky 1988).

Thus, effective protective measurements and land development within the areas potentially prone to landslide hazard should be developed with a clear understanding of the mechanism of the development of deep block-type landslides, which is further addressed in the article.

## The Activation of Deep Landslides in Moscow

In 2006–2009 catastrophic deep-block landslide failures occured at the two sites in Moscow: Khoroshevo-1 (Fig. 1) and Moskvorechye (Fig. 2). At the Khoroshevo-1 site, the width of a destructive landslide along the Moscow River reached 330 m. The cost of the designed protection measures was estimated to exceed 30 M US$.

There are other sites with a known risk of landslide activation. A new large Ski Jump is currently being constructed on the slopes of Vorobyovy Hills, which is a known largest landslide complex in Moscow. The site is located in the area of landslide terraces. Deep block-type landslides are characterized by constant slow deep movements (up to 5 mm/year). Therefore, some protective and/or mitigative measures need to be developed.

In addition, a design of a new escalator to be located on the same slope has recently been completed. The escalator will connect the metro station "Vorobyovy Hills" situated downslope with the Kosygin Street upslope on the plateau. Of note, that the previous escalator at this location had a service life of nearly 25 years (1959–1983) before it was taken out of service as a result of slow slope creep deformations. This is an example of slow structural deformation and damage as opposed to catastrophic active landsliding.

## Landslide Classifications

There are many classifications of landslides based on different characteristics, such as landslide mechanisms; geometric shape, size or volume; age; depth of slip surface; occurrence within a certain geologic formation; triggering mechanism, etc.

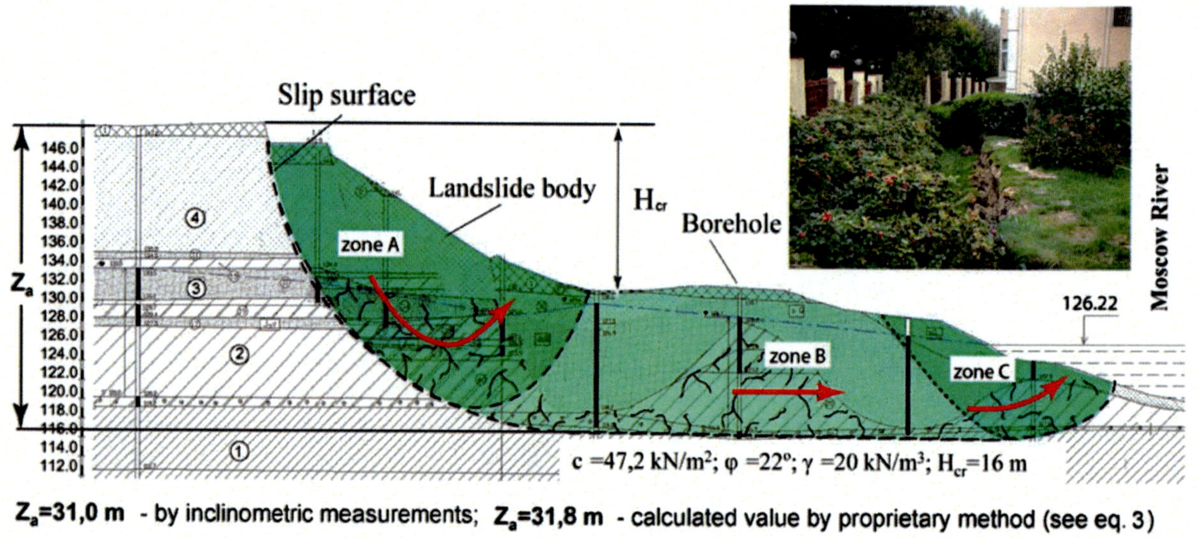


**Fig. 1** A schematic cross-section of central part of a landslide amphiteatre Khoroshevo-1, Moscow (Russia) (Kazeev and Postoev 2017), where 1—Jurassic Oxfordian clays $J_3ox$; 2—Jurassic Tithonian clays $J_3tt$; 3—Cretaceous sands $K_1$; 4—Quaternary sands and loams $Q_{IV}$. $\sigma_{str}$—material (soil or rock) strength of a considered layer (characterized by strength parameters $\varphi$ and $c$); $\gamma$—average unit weight of overlying soils; $Z_a$—slip surface depth in the central part of a landslide amphiteatre; $H_{cr}$—the slope height above the landslide terrace

**Fig. 2** The view of a landslide body and the head scarp formed in 2009 at the site "Moskvorechie" (Moscow, Russia, photo by A. Kazeev, 2018)

Mechanism-based landslide classifications are the most common (Demin 2009; Hungr et al. 2014; Petrov 1987). For example, landslide classification based on a movement type proposed by Cruden and Varnes (1996) is quite popular. The mode of movement of a landslide mass considered in this classification (sliding, rotation, translation, flow, etc.) is coupled with material type (rock, debris or earth). However, the underlying processes happening during the preparatory (pre-activation) stage of landslide formation (the driving forces causing the loss of stability of a soil mass and separation of a landslide body) are not being considered. Therefore, this classification is not particularly usable in recognizing the landslide type, because different landslide types may have a similar mechanism of movement in the intermediate stage of their development, such as movement of a landslide body along the slip surface.

Thus we consider that the mechanism of landslide formation or morphogenesis is more important than the displacement mechanism of already separated part of the groundmass. Moreover, when classifying landslides, it is considered reasonable to take into account the driving and resisting forces and the mode of deformation of the groundmass during preparatory stage of the landslide process, which would generally determine the morphogenesis mechanism of a landslide. Therefore, landslides fall into three main categories (Kozlovsky 1984): (1) deep-seated compression type; (2) shear–sliding type; (3) liquefaction–flow type.

## Landslide Mechanism Features

Deep-seated block landslides of compression type (Kazeev and Postoev 2017) are considered in this research. This landslide type can be correlated in most cases with rotational slides, spreads and complex landslides from Varnes' classification (Hungr et al. 2014).

In the existing landslide amphiteatre, a new landslide block separates from the upper part of the slope and propagates downwards along a steeply curved slip surface. This is the catastrophic phase of the landslide cycle, which is the stage of the main displacement of the landslide. The inclination of the slip surface flattens downslope and may become nearly horizontal. Furthermore, this new sliding block causes the displacement of the landslide terrace composed of previously separated blocks.

Landslides of the compression type can be formed practically in any type of soils or rocks if the following condition in the groundmass is satisfied:

$$\sigma_{str} < \gamma h \qquad (1)$$

where $\sigma_{str}$—vertical compressive strength of a soil or rock layer in consideration; $\gamma$—unit weight of overlying soils; $h$—depth to the top of potentially deformable layer.

Three zones inside a landslide can be identified (refer to Fig. 1). Each one is characterized by a different mode of deformation (Kozlovsky 1984). A formation of new landslide block and its rotation occur in the first zone A. This zone is characterized by open extended cracks, as well as by steep landslide ditches and grabens, formed as a result of wedge-shaped subsidence of back fragments of the separated block (Hungr et al. 2014; Sassa et al.2007; Kazeev and Postoev 2017).

The second zone B, or the transit area, is the largest within the landslide mass and typically extends all the way to the toe. Here, the horizontal displacement vectors significantly prevail over the vertical ones.

The third zone C is the area of compression and uplifting. It occupies a strip along the toe of the slope. This compression zone forms due to resistance of soils in front of the landslide. The formation of an uplifting ridge and a landslide tongue can appear in rivers (refer to Fig. 2 for an example).

## Development of Protection Measures Based on Landslide Features

The planning and constructing on slope areas susceptible to deep-seated landslides of compression-uplifting type should be carried out taking into account the specific features of the considered landslide type.

1. The long-lasting and continuous displacement in a landslide amphitheatre results from the upslope position of the landslide block adjacent to the head scarp (caused by the weight-load of the uppermost block, Fig. 3a). The characteristics of the displacement depend on the size of sliding blocks, soil parameters and other factors. On high slopes composed of rocks this process may continue for many centuries. This is an important feature of deep displacements that should be taken into account when planning the construction in landslide areas, even if deformations on the slope are visually absent.

2. The translational and rotational displacement of landslide blocks is a specific feature of compression-uplifting landslides. It is explained by the compressional mechanism of formation of landslide blocks. The compressed soils are squeezed out towards the slope, with subsequent separation and rotation of the landslide block (Fig. 3b). Based on the results of landslide investigations on the Odessa coastline of the Black Sea, Naumenko (1977) has shown that the rotational movement of blocks occurs along nearly circular surfaces, as they continue to move downslope. The inclination of a rigid marker layer (6–13 m thick limestone layer of Pliocene, Pontik stage) was increasing with progressive landslide displacement. Characteristic extrusion ridges may form in front of the blocks (Fig. 3b). Therefore, it is necessary to take into account the complexity of the movement of the blocks. Due to significant sizes of landslides, the possibility of continuous deformations on slopes should be considered during the design stage, as well as maintenance of buildings and infrastructure to ensure their continuous functionality.

3. A compressional bulge may be formed on a landside terrace (a middle compressional bulge) as a result of the pressure from a new settling block (Fig. 3c). Landslide deformations downslope the bulge are usually insignificant.

This feature may influence the requirements for selecting the type and location of protective structures. The landslide activity and displacements are maximal in the upper part of the slope (i.e. location of the youngest landslide block).

4. The effect of self-developing is a specific feature of deep-seated block-type landslides. Each new landslide block pushes the landslide masses away from the head scarp. This leads to the formation of deep tension side scarps (Fig. 3d) which undercut adjacent landslide-prone ground masses. This is usual, for instance, for long landslide amphiteatres along coastal lines. The phenomenon can be illustrated by catastrophic re-activations in the same landslide complex in Odessa in 1964 and 1965 (Naumenko 1977).

River erosion influences the magnitude of deep landslide displacements. But as a result of "self-propagation" is that once the landslide process has started, it may continue even without the influence of such factors as undercutting or erosion at the toe. Thus, the mechanism of self-propagation should be considered in designing an appropriate mitigation and protection measures.

5. Landslide displacements on the slope stops when a flat landslide terrace is formed in the middle part (see Figs. 1, 3e). This represents the end of a landslide cycle (Kozlovsky 1984). The head scarp reaches its critical height ($H_{cr}$) at the axis of a landslide amphiteatre. The landslide body comes to a stop and the original ground mass (upslope the landslide mass) reaches its limit-state (a new landslide block may form).

The equation for the limit-state assessment of an original ground mass (Fig. 3e) was developed (Kazeev and Postoev 2017) as:

$$Z_a - \frac{\sigma_{str}}{\gamma} = \frac{\pi}{2}(Z_a - H_{cr}) \tag{2}$$

where $\gamma$—average unit weight of overlying soils (above potentially deformable horizon); $Z_a$—slip surface depth in the central part of a landslide amphiteatre: $H_{cr}$—the slope height above the landslide terrace; $\sigma_{str}$—the soil structural strength (Postoev and Kazeev 2014) of the deformable horizon, $\sigma_{str} = 2c \cdot tg(45 + \varphi/2)$.

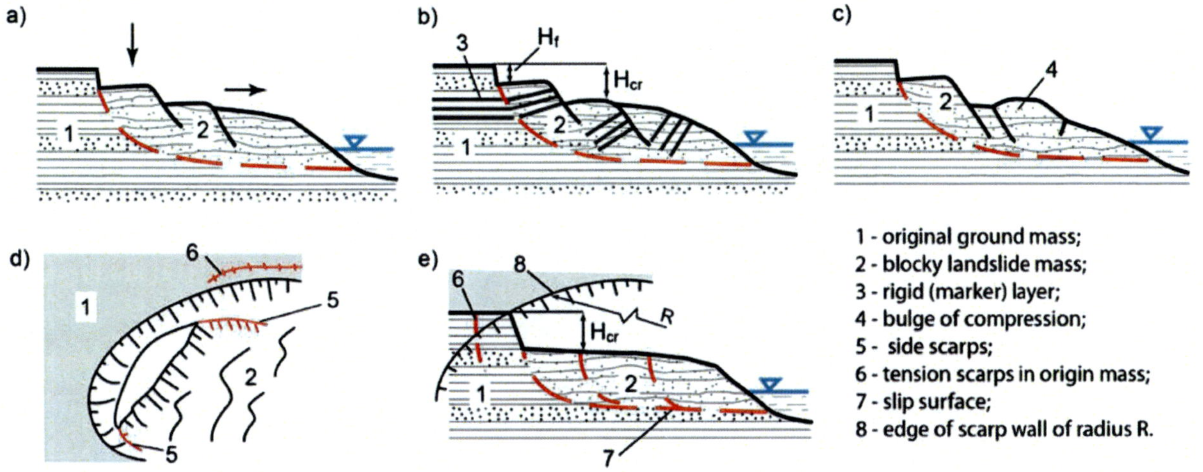


**Fig. 3** Deep-seated block-type landslides (schematic). **a** High position of the upper sliding block; **b** translational and rotational block movements; **c** formation of a compressional bulge within a landslide terrace; **d** formation of side tension cracks and scarps; **e** the limit-state of the native mass above the head scarp. $H_f$ and $H_{cr}$—factual and critical height of the scarp wall

The slip surface depth can be calculated from Eq. (2):

$$Z_a = \frac{\frac{\pi}{2} H_{cr} - \frac{\sigma_{str}}{\gamma}}{\frac{\pi}{2} - 1}. \tag{3}$$

And the Factor of Safety ($K_{st}$) can be defined as:

$$K_{st} = \frac{Z_a - H_f}{Z_a - H_{cr}}. \tag{4}$$

refer to Fig. 3 for definitions.

Thus, protective measures in areas susceptible to deep-seated block-type landsliding should include: (1) timely recognition of landslide-prone original ground masses, which are close to a limit-state condition; (2) the designing of protection measures for slopes with low Factor of Safety, to prevent landslide triggering (separation of a new block) and development of destructive landslide displacements.

## New Protection Measures for Areas with Deep-Seated Landslides

Several new methods were developed to increase slope stability, based on understanding of the mechanisms of deep block-type compression-uplift landslides. The first one represents the construction of artificial cuts similar to naturally occurring erosional downcuts (example- erosional gully or ravine) (Postoev and Kazeev 2011). The purpose of these cuts is to break long continuous slope front above the head scarp, which is considered as one of the necessary conditions of formation of massive compression landslides (Fig. 4). Also the drainage of the groundmass due to the cut becomes significantly improved resulting in the reduction of hydrostatic and hydrodynamic pressures. The groundwater flow changes direction towards the thalweg of the artificial gully. Thus, the stability of the sidewalls, and the overall slope stability is increased. Potentially hazardous slope ground mass with long continuous edge is strengthened by introducing artificial local "gully" cuts and creating local stable zones. At each such modified site the calculated slope stability is increased 1.5 times, as compared to the initial morphology.

Another possible way to stabilize a landslide-prone ground mass (1) effectively is to replace it with a underground structure (3), on the back of the landslide (2) (Fig. 5) (Osipov and Postoev 2007). It should be noticed, that a similar structure can be constructed in an artificial gully cut, shown on Fig. 4. Thus the parking, which causes periodic activation of landslide displacements on the site Moskvorechie (see Fig. 2) could be safely placed underground (Fig. 5).

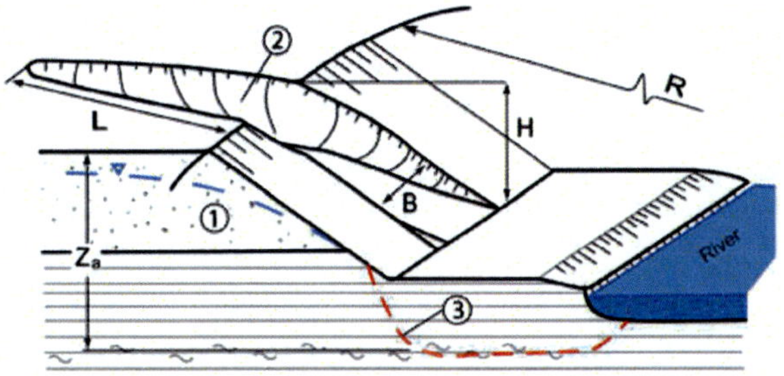


**Fig. 4** Artificial "gully" cuts in a landslide-prone mass (Postoev and Kazeev 2011). 1—Landslide-prone ground mass; 2—Artificial local "gully" cut; 3—Slip surface; H—height of a scarp wall; L & B—length and width of a "gully" cut; R—radius of the scarp wall; $Z_a$–depth to a potentially deformable horizon

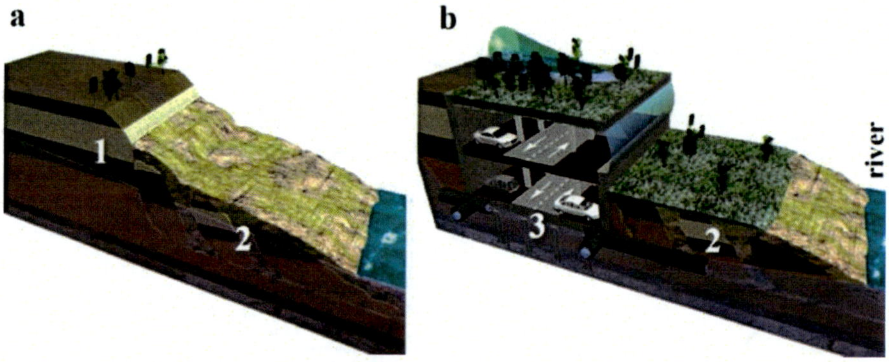


**Fig. 5** The scheme of a natural landslide slope (**a**) and the reinforced slope (**b**). 1—original landslide-prone ground mass; 2—landslide body consisted of landslide blocks; 3—underground structure (construction)

## Conclusions

Deep landslides of compression-uplift mechanism represent a major hazard to urban and mining areas, infrastructure, and structures constructed on the slopes. They have a complex mechanism of development, significant volume and weight, and pose challenges in the design and implementation of protection measures.

The landslide hazard assessment, reliable methodology of evaluating landslide activations, and the implementation of effective protective measures are impossible without identification of the landslide mechanism and assessment of stress–strain state changes in the landslide mass. Construction planning needs to consider specific features of deep landslides and timely recognition of the limit states of ground masses on slopes.

One of the most important things in a stability analysis and designing of protection measures is to define the slip surface depth and therefore the landslide size. This can be made by using Eq. (2).

New methodology for landslide slope stabilization was developed by taking into account the main features of landslide development of the considered landslide type. In most cases, it is very difficult to completely stabilize deep landslides. But constructing on landslide slopes should account the features of deep block type landslides in order to be able to develop slope areas safely and with a minimal risk to structures and public health.

**Acknowledgements** The research was supported by Russian Scientific Foundation (Project № 18-77-00065) and partially by the Ministry of Education and Science of Russian Federation (theme AAAA-A19-119021190077-6).

## References

Bromhead EN, Ibsen ML (2004) Bedding-controlled coastal landslides in Southeast Britain between axmouth and the thames estuary. *Landslides* 1(2), 131–141

Bromhead EN (2005) The stability of slopes, 2nd edn. Taylor and Francis e-Library, p 406

Cruden DM, Varnes DJ (1996) Landslide types and processes. In: Special Report 247: Landslides: Investigation and mitigation, transportation research board, Washington D.C.

Demin AM (2009) Landslides in quarries: analysis and forecast. Moscow GEOS, p 79. (in Russian)

Hungr O, Leroueil S, Picarelli L (2014) The Varnes classification of landslide types, an update. Landslides 11(2):167–194

Kazeev A, Postoev G (2017) Landslide investigations in Russia and in the USSR //. Nat Hazards 88:81–101. https://doi.org/10.1007/s11069-016-2688-z

Kozlovsky EA (1984) Landslides and mudflows. Centre Int. Projects GKNT, vol 1. Moscow, p 352. (in Russian)

Naumenko PN (1977) Conditions of formation and geotechnical characteristics of landslides on the Black sea along the coast of Odessa. Landslides on the black sea coast of Ukraine. Nedra, Moscow, pp 57–100. (in Russian)

Osipov VI, Postoev GP (2007) Method of reinforcement of a landslide slope. Patent of invention RUS 2340729. 28.05.2007.

Petrov NF (1987) Landslide system. Simple landslides (aspects of classification). Shtiintsa, Kishinev,p 164. (in Russian)

Postoev GP, Kazeev AI (2011) Method of strengthening of a landslide-prone slope ground mass. Patent of invention RUS No. 2413056. 27.02.2011. Bul. No. 6. (in Russian)

Postoev GP, Kazeev AI (2011) Method of determining in a landslide prone mass of potentially deformed horizon depth. Patent RF No. 2412305. Bull No. 5. (in Russian)

Postoev G, Kazeev A (2014) The criteria of strength and strain of soils in a groundmass. In: International multidisciplinary scientific geoConference surveying geology and mining ecology management (SGEM) 2(1), pp 901–905

Sassa K, Fukuoka H, Wang F, Wang G (2007) Progress in landslide science. Springer, Berlin, Heidelberg, p 378

Tihvinsky IO (1988) Evaluation and prediction of stability of landslide slopes. Nauka, Moscow, p 144. (in Russian)

# Rock Glaciers and Landslides in the Waste Dump of High-Altitude Kumtor Goldmine (Kyrgyzstan)

2020 Kyoto Japan

Isakbek Torgoev and Salamat Toguzbaev

**Abstract**

The high-altitude Kumtor Goldmine is located in the Tien Shan nival-glacial belt at the altitudes of 3,900–4,400 m above sea level. Harsh climatic conditions of the region, active glaciers and permafrost, complicated geological and tectonic structure of the deposit combined with high seismic and geodynamic activity of the region cause instability of the huge dumps of waste stored on the glaciers and landslides' activity. In recent years, the systematic and safe mining at Kumtor has been seriously complicated by deformations of waste and ice dumps and their sliding down the glacier valleys towards the Kumtor river bed. The first gigantic landslide of more than 200 million m$^3$ in volume at the Kumtor mine dumps with huge economic damage was recorded in spring 2013, in the Davydov glacier valley. In December 2019, collapse and sliding of the rock-ice mass of more than 30 million m$^3$ in the Lysyi glacier valley killed two people. This paper describes the structure and features of the Kumtor mountainous dumps deformation and the main causes of the December 1, 2019 landslide.

**Keywords**

Kumtor gold mine • Glaciers • Permafrost • Waste dumps • Landslide

I. Torgoev (✉)
Insitute of Geomechanics and Mining of the National Academy of Sciences of the Kyrgyz Republic, 98, Mederova str, Bishkek, 720053, Kyrgyz Republic
e-mail: isakbektor@mail.ru

S. Toguzbaev
Kumtor Gold Company, 24, Ibraimov Street, Bishkek, 720031, Kyrgyz Republic
e-mail: Salamat.Toguzbaev@centerragold.com

## Introduction

The high-altitude Kumtor gold mine is located in the Central Tien-Shan Mountains, at the altitudes of 3,900 to 4,400 m (Fig. 1). The project is implemented by the joint Kyrgyz-Canadian venture "Kumtor Gold Company" (KGC) in the zone of active glaciers and continuous permafrost rocks. This mine is located in the Ak-Shyirak mountain range, which is one of the largest areas of the modern glaciation in Kyrgyzstan. This area is characterized by extreme continental climate. The average annual temperature in the mine area is −7.8 °C with the average minimum temperature ranging from −28.8 °C in January up to 1.7 °C in July. The average annual precipitation is 350–400 mm, a quarter of which falls as rain and the rest in the form of snow. Harsh climatic conditions in the area, mountainous glaciers and permafrost, diversity of terrain, rock types, seasonal freezing and melting of the rocks and soil together with global warming and large-scale anthropogenic impact cause active development of cryogenic gravitational processes.

These processes include gravitational displacements of the consolidated masses of waste and ice stored in dumps. Approximately 2 billion tons of overburden and coarse waste, as well as over 170 million tons of crushed glacier ice were excavated from quarries and dumped during the construction and operation of the open pit mine. Throughout the entire period of the deposit development, these materials were accumulated directly on the Davydov and Lysyi glaciers, or were dumped on nearby mountain slopes and in the glacier valleys.

A gigantic landslide (Fig. 1) of over 700 million m$^3$ in volume was formed on the giant waste and ice dumps stored on the Davydov glacier in the Central Valley, the displacement of which started in 2011 along the inclined surface (12°–16°) of the ice-rich permafrost below a layer of the Davydov glacier bottom moraine (Torgoev 2016). Moraine displacement under the load of thick (200 m) dumps was


V. Vilímek et al. (eds.), *Understanding and Reducing Landslide Disaster Risk*,
ICL Contribution to Landslide Disaster Risk Reduction,
https://doi.org/10.1007/978-3-030-60319-9_44

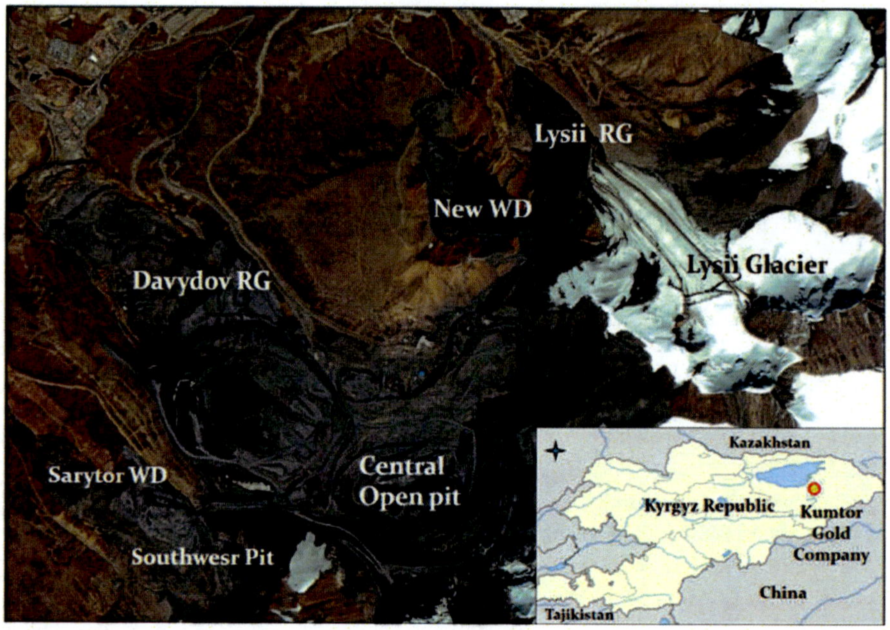


**Fig. 1** Space image of the Kumtor gold mine in 2014. The inset shows the mine location in Kyrgyzstan

due to the presence of groundwater and taliks associated with the Chon-Sarytor channel underflow in the moraine deposits.

In April 2013, accelerated sliding of the landslide of about 200 million m$^3$ occurred due to a combination of increased shear stress on the roof of frozen bedrock due to huge weight of the dumps and the influx of subglacial melt water. This displacement destroyed a number of large facilities of the mine with economic damage exceeding US $100 million (Torgoev et al. 2014). Due to continuous displacement of the landslide down the Chon-Sarytor river valley, the shift lane, electrical substation, meteorological station and other infrastructure facilities of the mine were moved to another place.

On the 1st of December 2019, the dumps in the Lysyi Glacier basin collapsed and slide down at the Kumtor mine following almost the same scenario. As a result of the collapse and displacement of the ice-rock mass over 30 million m$^3$ in volyme, two KGC employees were killed. The main causes of dumps collapsing and displacement of the technogeneous rock glaciers (TRG) in the Lysyi glacier basin are discussed hereafter.

## The Lysyi Glacier Area

The Lysyi valley glacier located between the Petrov and Davydov glaciers (Fig. 2) was the first glacier exposed to direct ore mining (man-made pressure) at the Kumtor mine. First, ice was extracted and removed in the upper part (the western circus) of the glacier on the site within the Central Quarry, and then dumping of waste mixed with glacial ice started on the periphery of this glacier on the western side of the valley (Torgoev 2016).

Based on the survey of the Lysyi glacier in 1984–85 by glaciologists of the Institute of Geography of the Russian Academy of Sciences, ice of this glacier had homogeneous thermal state with a temperature below freezing point−below 0 °C (cold glacier) and was frozen to the bed of the glacier (underlying rocks and bottom moraine) before mining operations had stated. From the point of view of the dynamics and kinematics of the ice and dumps displacement on the glacier, it means that at the time of the survey in 1988 there was no component of glacier sliding along its bed, and the relatively slow plastic flow of ice was the main factor for its displacement downward. The slope of the bed of the Lysyi valley glacier is 4°. According to glaciologists, the glacier gradually receded from the terminal moraine at an average speed of 10.3 m/year from 1932 to 2003 under the influence of climate warming, (Torgoev 2016).

## Waste-Rock Accumulation on the Surface of Lysyi Glacier and in Its Valley

Historically, the very first waste dumps No. 3.4 began to form in September 1996 on the western side of the valley of the Lysyi creek. Overburden was first dumped on the inclined surface of the slope, and then on the surface of the upper part of the glacier (in its circus). By 1999, most of the inner circus of the Lysyi glacier, including the ice accumulation zone, was covered with waste. Based on the

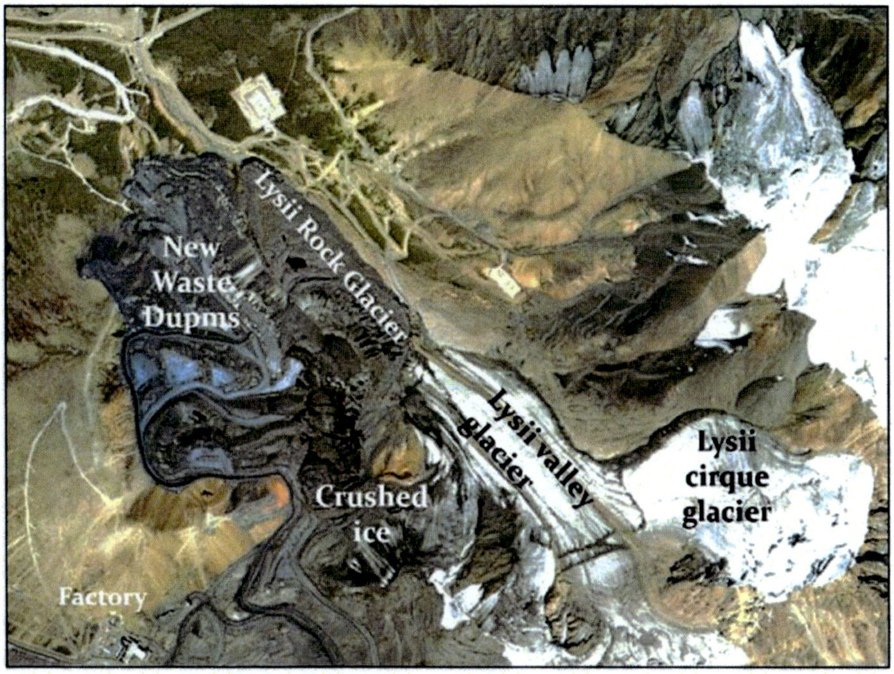


**Fig. 2** Lysii Glacier area: waste and ice dumps, rock glacier as of the summer of 2019

interpretation of the high resolution satellite images Jamieson et al. (2015) performed the retrospective analysis of the accumulation of rocky and icy waste in the Lysyi glacier basin for the period from 1999 to 2014 (Fig. 3a) and identified the style of the glacier and dumps displacements.

In 1999–2002 as the dumps volume increased, these accumulations of rocks and ice transformed into the "Lysyi" Technogenrous Rock Glacier (TRG) and started to move down the slope, towards the thalweg of the glacial valley.

According to the KGC for 1996–2003, 14.7 million $m^3$ of rocks and about 0.5 million $m^3$ of glacier ice were dumped in the Lysyi glacier basin. By 2004, the front part of the Lysyi TRG displaced by 700 m (Jamieson et al. 2015), while in 2003 it blocked the creek valley and the main branch (trunk) of the Lysyi valley glacier (Fig. 3a).

The TRG Lysyi displaced at a relatively constant speed of 230 m/year (about 0.64 m/day). In 2004–2010 the rocks and ice dumping in this area was suspended and, therefore, the speed of the TRG and the Lysyi glacier in this period decreased to very low and relatively constant values. In 2011, the volumes of rock and ice waste from the Central Quarry began to increase steadily, and some of them again were stored on the dumps in the Lysyi glacier basin. The volume of dumped materials particularly increased in 2013, when due to the dumps' sliding, it was prohibited to dump materials in the Davydov glacier valley (Central Valley). For these reasons, in 2012–2013 TRG and the Lysyi glacier began to move down the valley, and the TRG constantly loaded on top (Fig. 3b) moved much faster than the Lysyi valley glacier, ahead of it.

Intensive dumping on the western side of the Lysyi river valley began in 2013 directly on the hillside (Fig. 3b). Moreover, new Northeast dumps No. 3 (NewWD) were dumped using negatively proven top-down method and without preliminary preparation of the foundation for dumps (Torgoev 2016). Due to the fact that the dumps were located on steep slopes composed of fine-grained moraine material with high ice content, they moved from the very beginning along inclined surfaces towards the channel of the Lysyi stream (Fig. 3b). According to the monitoring of the waste dumps displacement by the KGC geotechnical service, the average speed (NewWD) in October 2013 sometimes reached 1.0 m/day. Thus, starting from 2013, a situation in the Lysy glacier basin was characterized by the presence of two different groups of dumps. The first group includes the so-called "historical dumps", which, as the volume of dumped rocks and glacier ice grows, are transformed into the "Lysyi" TRG (Fig. 3b). The average thickness of TRG in October 2017 was 45 m, varying from 41 to 48 m. This TRG displaced mainly along the thalweg of the Lysyi stream valley in the north-western direction towards the Kumtor river bed with a relatively constant speed of 0.22 m/day (Kurylo et al. 2018).

The second group included the abovementioned waste dumps (NewWD), stored on the western side of the Lysyi creek valley, in the near-watershed section of the steep slope

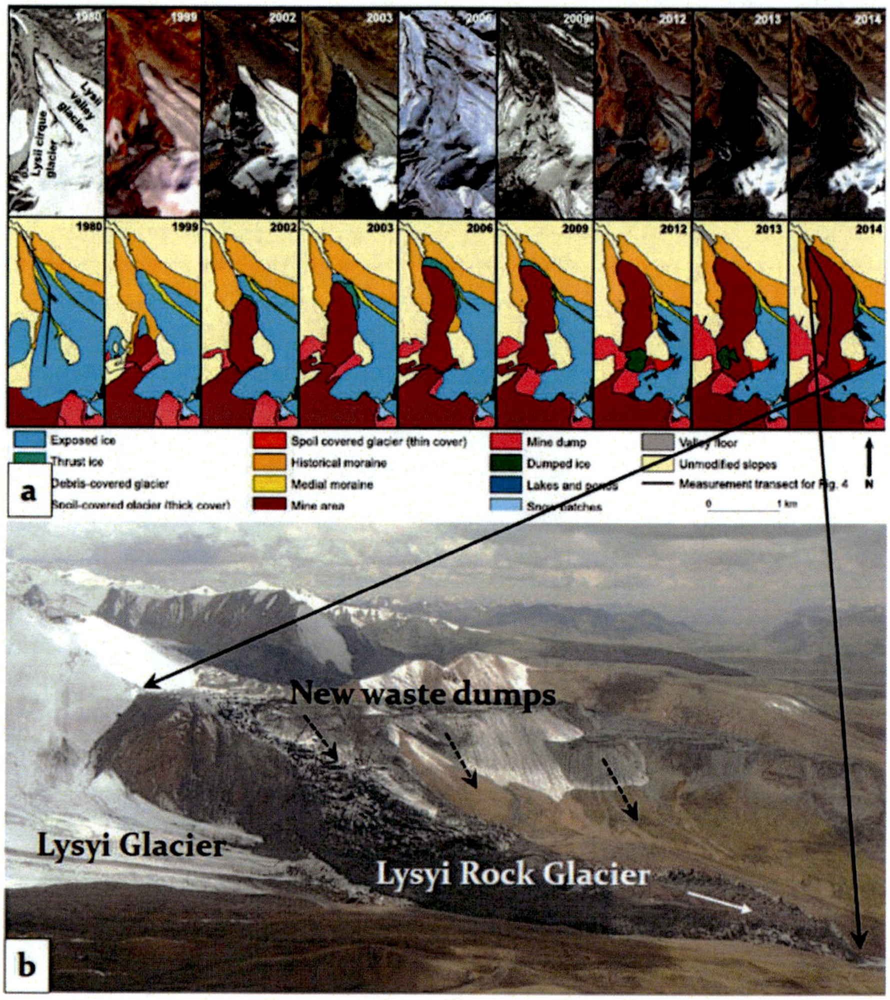


**Fig. 3** Formation of waste dumps in the basin of the Lysyi glacier. **a** Selected time series of glacier and landcover change for the Lysyi cirque Glacier, which hangs above the main Lysyi Glacier between 1999 and 2014. Top: high-resolution satellite imagery from a range of sensors. Bottom: schemes of surficial geology and glacier ice/debris cover (Jamieson et al. 2015). **b** Photo of a Lysyi rock glacier, August 3, 2014. Arrows–the direction of the offset dumps and Lysyi rock glacier

(Fig. 3b). Displacement of both groups of the dumps was measured using the surface monitoring points and automatic laser scanning. As in the case of TRG displacement, the monitoring points on the displacing dumps of NewWD 3 shifted along the slope fall. Displacement rates were predetermined by the steepness of the slope surface, ranging from 0.24 to 0.55 m/day with higher velocity on the steeper upper sections of the slope.

Measurements using inclinometers showed that the dumps (NewWD) displaced towards the Lysyi stream course along the surface of the ice moraine at a speed of 0.76 m/day (Kurylo et al.).

From 1996 to 2019, over 105 million m$^3$ of rocks and almost 50 million m$^3$ of glacier ice were dumped in the Lysyi glacier basin, which ultimately caused the collapse of the dumps (NewWD) and strong displacement of the Lysyi TRG on December 1, 2019.

## Reasons for Instability and Displacement of Dumps

It should be noted that NewWD were accumulated on the slope with steepness of 15–20° composed from the surface to a depth of 1.7–2.5 m by moraine deposits, represented by a layer of weak fine-grained sometimes gravelly soils with ice content from 10–15% to 35%. This layer of weak soils coincides in thickness with a seasonally thawed layer (STL), which lies on permafrost (frozen rocks). In the frozen state, fine-grained soils covering slopes, where waste has been dumped, are stable at low temperatures and are very unstable during melting and under the load, which leads to a loss of their firmness, deformation and subsidence. For this reason, the top-down formation of dumps on steep slopes poses the risk of potential displacement of the dumps along the slope fall inevitably. As evidenced by the dumping practice on the Kumtor slopes, over time they move down the sliding surface, which is confined to the roof of frozen bedrock.

As the dumps' material contains a relatively large number of fragmented rocks and fines that do not allow water to pass through, it formed the specific hydrological conditions: the rock glacier displaced from the western side to the Lysyi stream valley became a kind of a dam accumulating meltwater of the glacier and TRG. It should be borne in mind that meltwater from rock and ice dumps, infiltration of precipitation (most of the precipitation falls from May to September) and groundwater, as well as sumps (small bodies of water on the dumps body and periphery) are the main sources of dumps flooding. These waters leaking to the bed and the periphery of the dumps, especially during the hot season, flood the dump rock mass in its middle and lower parts and acting as a liquid lubricant, increase their displacement rate.

## Sliding in the Waste Dumps

According to the monitoring data on the dumps displacement in spring 2019, gradual increase in the average speed of the dumps displacement (NewWD) was recorded from 0.2 m/day in March to 0.63 m/day in November. Figure 4 shows graphs of the change in the displacement rate of the NewWD-3 dumps at two monitoring points wd3-995 and wd3-947 fixed in the middle and lower parts of the dumps, respectively, during November 2019 preceding the collapse of the dumps on December 1, 2019.

The daily monitoring results presented in graphical form in Fig. 4 indicate of a steady trend of a significant increase in the displacement rate in the middle part (NewWD) (point wd3-995) from 0.51 m/day (31.10.19) to 1.75 m/day (29.11.19). At wd3-947 located in the contact zone of the lower part of the dump (NewWD) with TRG, the change in the displacement rate is hardly noticeable; however, there is a tendency to its gradual increase, especially several days before the collapse, when the displacement rate increased from 0.05 to 0.25 m/day.

Low displacement velocities at wd3-947 are most likely due to the fact that the displacing dump rested against the TRG body. During instrumental observations of natural rock glaciers, it was noticed that the speed in their lateral parts is much less than that of the middle (axial) part. For this reason, on the surface of active (mobile) rock glaciers, including the Kumtor technogenic glaciers, arc-like levees and hollows are formed (Jamieson et al. 2015).

There is good reason to assume that the above dump displacements are largely due to creep deformation of the underlying moraine deposits with a high content of ice and TRG ice material. With this movement mechanism, a significant increase in the displacement rate of rock-ice dumps (Fig. 3) may indicate of the transition of deformations from the stage of established (secondary) to the stage of progressive (tertiary) creep, which inevitably results in the collapse of creeping dumps, when the load (weight of rocks and ice) increases, which happened at NewWD-3 in the early morning of December 1, 2019.

Analysis of the Sentninel high-resolution satellite images showed that at first there was sliding of dumps at 3,970 and 3,940 m a.s,l. (NewWD) down the slope, towards the Lysyi creek. The survey carried out immediately after the dumps' failure confirmed that NewWD sliding occurred as a result of sliding on the roof of permafrost with high ice content below

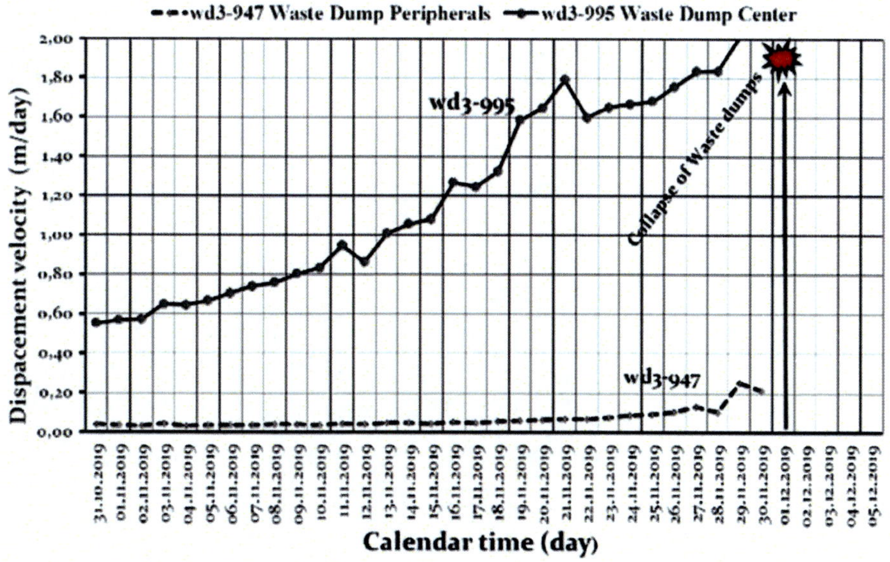


**Fig. 4** Sliding velocity dynamics for different part of waste dumps Lysyi in November 2019—before their collapse

the of moraine deposits layer that was found in the lower part of the section.

Following the NewWD sliding, due to the loss of the counterforse, huge ice-rocky dumps in the upper part of the TRG at the level of 4,116 m collapsed. When this part of the dumps collapsed, a deep headscarp back wall (Fig. 5a) more than 50 m high formed with a cylindrical sliding surface and tilting of the sliding body. Examination of this part of the landslide showed that displacement and sliding area in the upper part passes inside the dump, not at the contact of the dump base with the slope. Such peculiarities of the displacement and sliding area may indicate that the displacement occurred also due to creep deformation predetermined by high ice content inside the upper layered part of the dump, where ice constantly dumped in the so-called "ice dump". The ice was alternated with the layers of rock waste, which were loaded above the ice to prevent slipping of the dump trucks and bulldozers.

Finally, as a result of the strong loading of the TRG body by the sliding NewWD masses, as well as ice-rocky material from its upper part, the TRG (old dumps) displaced, the front part of which (tongue) moved ahead by 500–550 m, sliding onto the retaining prism tray from above (Fig. 5b) without moving them.

In the tongue part of the TRG, arched levees typical of moving rock glaciers were formed. It should be noted that of the tongue of the technogenic landslide advance continued after the collapse for a few more days with gradual deceleration. If on December 1, the displacement speed was 1.1 m/h, on the following days it decreased to 0.6 m/h on December 2, to 0.15 m/h on December 3 and to 0.07 m/h on December 4, 2019. In addition, the ongoing landslide displacement after the dump collapse was due to unloading (excavation) of the dump rocks in the tongue part (Fig. 5b) carried out to search two missing people.

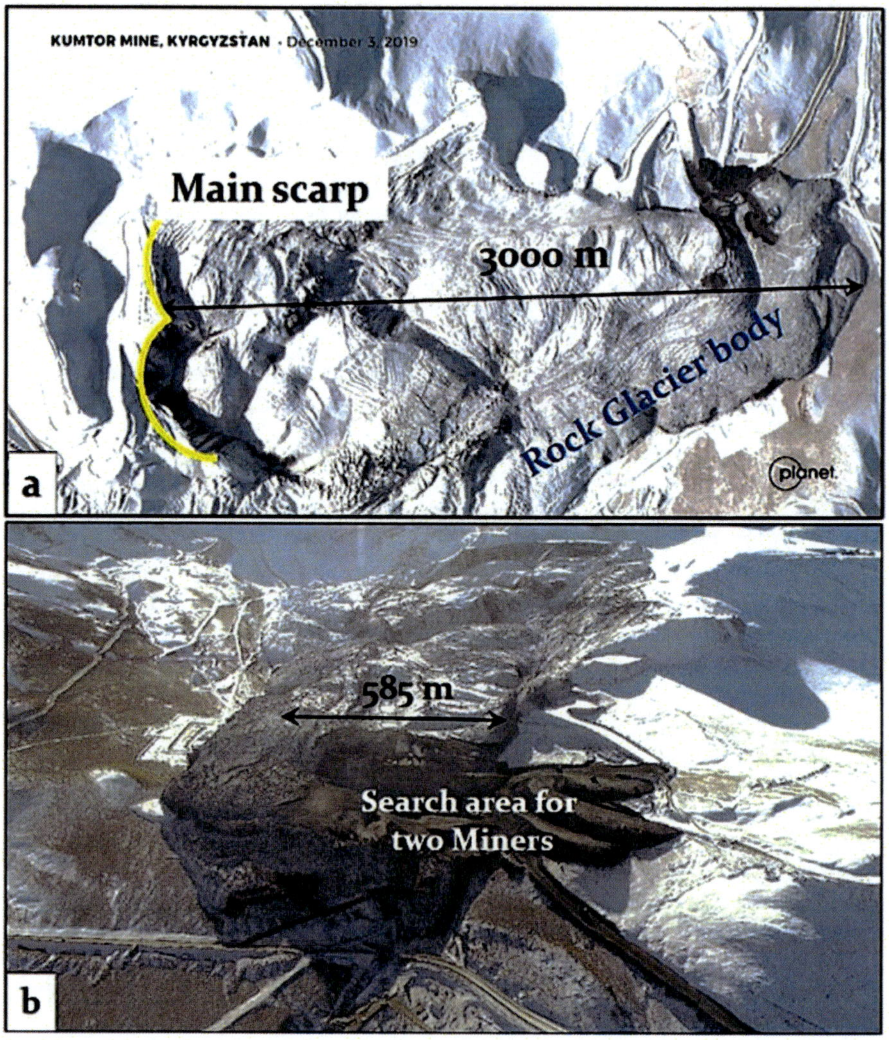


**Fig. 5** General view of the landslide that occurred on December 1, 2019 at the waste dumps in the Lysyi glacier basin: **a** Planet Labs SkySat image of the Kumtor gold mine landslide. **b** Photo of the landslide in the Lysyi Waste Dump as of December 4, 2019

The total length of the landslide from the separation wall in the head to the tongue was 3000 m, the average width – 585 m, the average thickness of the landslide was 35–40 m, and the total volume of the collapsed and sliding dumps exceeded 30 million $m^3$.

Thus, the main cause of the dumps collapse and sliding was the overloading of their upper part. In October–November 2019, waste was dumped on the upper part on the Lysyi glacier mainly, while the retaining prism were formed in its lower part. The bulk of the waste rocks was unloaded in the upper part of the dump with the formation of several steps (benches), and in the lower far part of the dump, the scarps (safety berms) were formed, which probably should have become a kind of the ccounterforce for TRG, which displaces along the valley talweg since 2003.

Based on the geodetic monitoring and satellite imagery data, the first signs of the forthcoming collapse of the dumps appeared a few days before. In particular, two days before the collapse, i.e., on 29 November 2019, there was a slight sliding on the very top shelf of the upper dump. This collapse with the formation of a main scarp of the future landslide, as well as increase in the displacement rate became a harbinger of the dump moving from the stage of the established secondary creep to the dangerous stage of tertiary progressive creep, i.e. the moment of dump collapse is approaching. For the KGC engineering-geological department specialists, who worked at the mine at that time, these harbingers should have been an early warning sign of the forthcoming threat of a dangerous collapse.

## Conclusion

Placement of the gold mining waste on the active glaciers and permafrost soils of Kumtor is a unique, but doubtful experiment in the history of the natural resource mining in nival-glacial regions of the world. At present, large-scale displacements and landslides of the rock-ice dumps are a source of big problems for future activities of the KGC and upcoming restoration of the territory of the high-mountainous open-pit mine. This gravitational displacement of the technogenic rock glaciers on the dumps in the Lysyi, Davydov and Sary Tor glacier valleys may result in potential environmental threats and pollution of surface waters in the Kumtor-Naryn-Syrdarya river basin.

The main causes of geotechnical problems at the dumps of the Kumtor mine is erroneous choice of the dumps storage place, including storage of waste on the surface of active glaciers. Errors also included incomplete recording of glaciological features, characteristics of permafrost and geological environment in the Tien Shan glacial-nival belt.

One of the main reasons for collapsing dumps in the Lysyi glacier basin on December 1, 2019 was excessive overload of the dumps in the upper part of the slope characterized by a steepness of up to 20° and on the surface composed of ice-saturated perennial frozen rocks vulnerable to loading with huge dumps and climate warming.

Based on the different satellite observations and measurements of the NewWD dump displacement it was concluded that the observed displacement was the result of both sliding along the slope surface (sliding on the base of the dumps) and along the localized surface of shear (creep) deformation inside the dump foundation.

## References

Jamieson SR, Ewertowski MW, Evans DJ (2015) Rapid advance of two mountain glaciers in response to mine-related debris loading. J Geophys Res Earth Surf 120:1418–1435

Kurylo J, Rykaart M, Lizcano A (2018) Creep displacements induced from waste rock loading. In: Proceedings of the Tailings and mine Waste 2018. Keystone, Colorado, USA pp 175–186

Petley D (2019) The Kumtor gold mine landslide: high resolution planet labs images. https://blogs.agu.org/landslideblog/2019/12/12/kumtor-gold-mine-landslide.

Torgoev I, Omorov B (2014) Mass movement in the waste dump of high-altitude Kumtor goldmine (Kyrgyzstan). In: Sassa k et al. (eds) Landslide science for a safer geoenvironment, vol. 3. Springer Cham, pp 517–521

Torgoev IA (2016) Glaciers, gold and geoecology of Kumtor. Bishkek, p 197. (in Russian)

# Geosynthetic Reinforced Soil Structures for Slope Stabilization and Landslide Rehabilitation in Asia

Pietro Rimoldi, Matteo Lelli, Pietro Pezzano, and Fabrizia Trovato

**Abstract**

The paper presents successful experiences using geosynthetic reinforced soil structures for slope stabilization and landslide rehabilitation in Myanmar, Indonesia, and Japan. In all the presented cases, reinforced soil structures, combining high strength uniaxial geogrids and metallic gabion units provided with integrated reinforcing tails, were built. The first project highlights the challenges faced in the design and construction of a reinforced soil structure built to rehabilitate the collapsed portion of a national road in Myanmar. The second case study focuses on the application of reinforced soil structures for the reconstruction of road slope failures in Indonesia. The last case study describes the technical solution selected to stabilize a 14 m high slope for the Harima Expressway embankment located in Hyogo prefecture, Japan.

**Keywords**

Reinforced soil structures • Slope failure • Landslide remediation

P. Rimoldi (✉)
Consultant, Milan, 20121, Italy
e-mail: pietro.rimoldi@gmail.com

M. Lelli
PT Maccaferri Indonesia, Jakarta, 12310, Indonesia
e-mail: m.lelli@id.maccaferri.com

P. Pezzano · F. Trovato
Officine Maccaferri Spa, Zola Predosa, 40069, Italy
e-mail: p.pezzano@maccaferri.com

F. Trovato
e-mail: f.trovato@maccaferri.com

## Introduction

Gabion wall have been used widely in the world in numerous erosion control projects, soil reclamation work, stream channels construction and retaining structures (Giani 1992; Kandaris 1999). Metallic gabions have been used for more than 100 years to build gravity type retaining structures in all environments and climates (Toprak et al. 2016). Gabions are rectangular boxes manufactured from hexagonal woven steel wire mesh, laced together and filled with stones at the jobsite. Gabions have been widely used for different infrastructure sectors such as roads, railways and airports construction. Moreover, gabions have been successfully used for landslide mitigation and slope stabilization works.

Since 1980 geosynthetics, and particularly geogrids, have been extensively used for the construction of Reinforced Soil Structures (RSS), and in many projects of landslide stabilisation, even for 60 m high slopes (Cambiaghi and Rimoldi 1991; Cazzuffi et al. 1995; Coluzzi et al. 1996; Dikran and Rimoldi 1994; Manni and Rimoldi 2006; Rimoldi 1996; Rimoldi and Ricciuti 1992; Rimoldi and Scotto 2012). RSS have been designed and built even in high seismic conditions (Rimoldi 2018), and using cohesive/marginal fills (Giroud et al. 2014).

In the last 30 years "hybrid" RSS have been developed: these structures, made up of steel wire meshes and polymeric soil reinforcing geogrids, can be built combining either wrap-around green units (made of double twist wire mesh) or with gabion units, combined with high strength geogrids, as shown in Fig. 1. Geogrids are considered to work as primary reinforcements to provide the overall and internal stability of the structure, preventing potential failure surfaces to develop inside the reinforced soil mass; while the wire mesh units (either wrap-around green or gabion units) behave as a secondary reinforcement providing the local stability of the facing, ensuring that no local mechanism of direct sliding, pull-out or rotational failure can occur.


V. Vilímek et al. (eds.), *Understanding and Reducing Landslide Disaster Risk*,
ICL Contribution to Landslide Disaster Risk Reduction,
https://doi.org/10.1007/978-3-030-60319-9_45

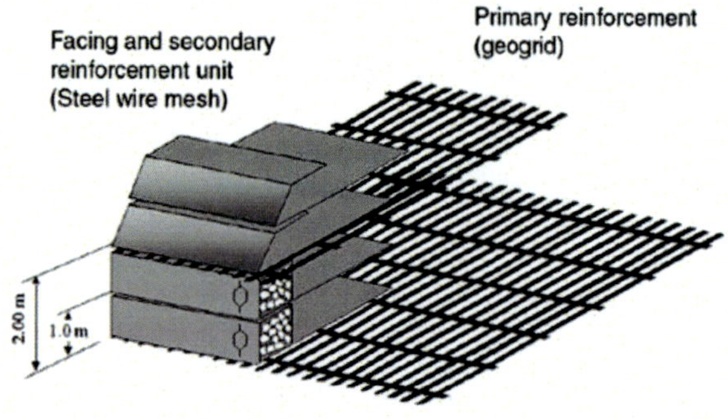


Fig. 1 Typical layout of a hybrid MSE structure

These hybrid RSS are particularly suitable for landslide stabilisation projects (Ferraiolo et al. 1997), thanks to their flexibility and ease of construction.

The connections between gabions and geogrids have been thoroughly investigated through laboratory and field tests (Chen and Tang 2011). The performance of geogrid or hybrid reinforced soil structures for landslide stabilisation projects have been checked by instrumenting several projects (Coluzzi et al. 1997; Montanelli et al. 1999; Rimoldi 1988; Tanyu et al. 2016); moreover, centrifuge testing (Lin et al. 2013; Ling et al. 2016) and shaking table tests (Lugli et al. 2019) provided further confirmation of the high degree of stability of RSS in static and seismic conditions. Finally, numerical analyses of hybrid reinforced soil structures (Gouw et al. 2016; Gu et al. 2017), based on all previous research, allowed to get a detailed view of the reinforcement mechanisms and the stability afforded by these structures.

## Wall as a Landslide Remediation Measure in Myanmar

Myanmar is a geologically active Country; every year several slope failures happens in the territory due to many reasons related to location and geological conditions. In hilly area the main cause is usually prolonged rainfall and soil stratification. On low land the main factors that contribute to the slope failure are slope steeping and scouring, pore water pressure build-up and human activities. Above all, one of the main reasons that cause slope failures in Myanmar is earthquake, caused by the Indian tectonic plate and the enormous Sagaing fault, one of the most active in the world.

Since hybrid RSS offers economic advantages in comparison with conventional retaining walls, the Ministry of Construction of Myanmar proposed the construction of a trial wall along the river Myittha in Kalewa, as a remedial work of an existing collapsed concrete masonry wall.

The 900 $m^2$ surface area wall with 400 $m^2$ of scouring protection have been completed within a year (Fig. 2). The simplicity of the installation, combined with the economic benefits of the structure, gave an enormous confidence to the local authority.

The system consists of a facing made of gabions unit and uniaxial high strength geogrids, where geogrids were considered as a "primary" reinforcement and the double twisted wire mesh tail of the gabion as a secondary reinforcement. A drainage geocomposite, placed along the back of the reinforced soil block, was installed as a replacement for the traditional drainage layer. The construction began with the compaction of the foundation soil; then rock filled woven wire steel gabions, steel mesh extension as secondary reinforcement, were placed at the face of the structure and connected to high strength welded geogrid as primary reinforcement; vertical spacing of reinforcement was 1 m; fill was laid and compacted behind the gabion gacing, and the procedure was repeated until the full wall height of 13 m was reached. Rock filled mattresses were installed at the toe of the wall as a scour protection apron (Fig. 2).

The Kalewa wall represents the successful completion of the first trial hybrid wall build for the ministry of Construction. After that, seven walls were completed in Myanmar in 2019 along the from Naypyidaw-KanPyar-Magwe driveway. The tallest wall is 16 m high and has been built by using gabion units with a height of 1 m in the upper layers and 0.5 in the lower layers to prevent bulging. The structure is reinforced with ParaGrid HF geogrids consisting of a biaxial array of composite geosynthetic strips characterised by a new geometry to optimise the bonding performance. The high strength geogrids are comprised with a core of high tenacity polyester fibres in longitudinal direction, encased in a polyethylene sheath. Geogrids with an Ultimate Tensile Strength (UTS) of 300 kN/m were used at the bottom while geogrids with UTS of 150 kN/m were used for the upper layers. Figure 3 shows a cross section of the designed section.

In this project a drainage geocomposite layer have been designed as a replacement of the traditional drainage layer

Fig. 2 Hybrid RSS after trial intervention

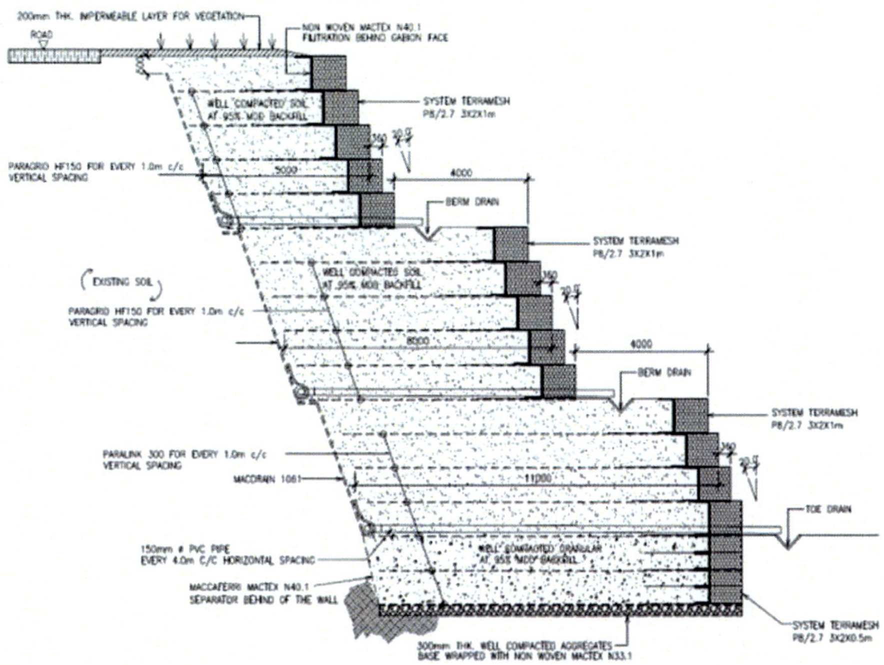


**Fig. 3** Naypyitaw-KanPyar-Magwe wall cross section

behind the back of the reinforced soil block (Fig. 3): a drainage geocomposite with a thickness of 6 mm was used as an equivalent granular layer of 300 mm thickness, based on hydraulic equivalency between geosynthetic and mineral drainage, analysed by using the MacFlow proprietary software.

## Emergency Works and Landslide Remediation in Garut and Batam (Indonesia)

In Indonesia the most of cases of landslides occur during the heavy rainfall. The saturation process caused by rainfall bring to a weakening of the slope stability due to its excess pore water pressure. Furthermore, due to a rise of ground water level, the pore water pressure will increase and finally reduce the shear strength of the soil until the retaining structure or soil slope cannot withstand its own mass. Two similar failures happened in Garut, in Java island, and in Batam island.

### Leuwigoong Dam Access Road—Garut

Leuwigoong Dam was built for irrigation purpose for citizens agricultural lands in Garut regency, West Java Province, Indonesia. As seen from Fig. 4, the stone masonry at the access road along 45 meters collapsed on 24th November 2017. This incident happened after heavy and continuous rain occurred the night before.

This structure allegedly collapsed due to poor quality of backfilling compaction so that the run-off water from the rain can infiltrate inside the embankment, and to the poor drainage system behind the retaining wall which was not enough permeable to allow water flow from the backfill to outside the structure.

The client needed a low-impact, environmentally friendly and time-cost effective integrated solution to prevent any construction delay of the restoration of the gravity wall due to the urgency of the usage of the access road.

The designed solution is an 8 m high reinforced soil structure, using woven geogrids as a primary reinforcement and gabions with the function of front wall fascia unit to maintain stability in the front of the structure and protect the structure from surface erosion. The facing elements are gabions made up of hexagonal double twisted steel wire mesh type 8 mm × 10 mm, which are provided with a tail of wire mesh. The steel wire used for reinforcing units was heavily galvanized and a polymer coating having a nominal thickness of 0.50 mm was applied to provide additional protection against environmental agents (UV rays, rain, etc.). The gabion units (Fig. 5) are manufactured from one continuous panel of mesh; the back panel is factory connected to the base panel. The facing section of the unit is formed by connecting the back panel and diaphragms to the main unit; the gabion facing is divided into uniformly partitioned cells by internal diaphragms. Steel ring fasteners or coated lacing wire are used to connect units and to close and secure them.

Gabions were filled at the site with suitable stones to form a flexible and permeable modular fascia system. Polyester

**Fig. 4** Site condition on 27th November 2017

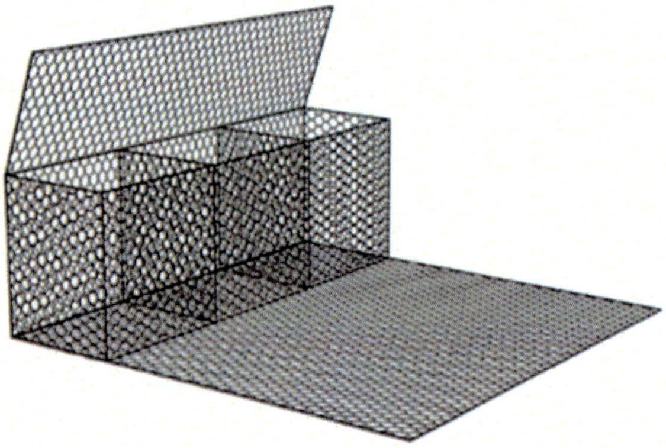

**Fig. 5** Steel gabion units

woven geogrids with an Ultimate Tensile Strength (UTS) of 150 kN/m were used as primary reinforcement. Geogrids were installed with a vertical spacing of 1 m and a length of 6 m. To prevent the backfill soil to pipe out to the facing, non-woven geotextiles which acts as separator and filter were applied behind the gabion facing. Figure 6 shows the typical cross section of the designed solution.

The backfilling soil was transported by trucks and unloaded with the help of heavy equipment. Vibratory rollers of adequate dimensions were used in function of the filling material. A light vibration compactor has been used for the area close to the facia to minimize any damage to the facia units. A drainage system inside behind the structure has been built, by placing a 0.3 m thick gravel layer wrapped with nonwoven geotextile as filter (Fig. 7). The construction process is repeated until it reaches the designed height. This reinforced soil structure was completed in 6 weeks of total construction time. Figure 8 shows a view of the RSS during construction and almost completed.

## Bukit Kemuning Road—Batam

The project site is located in Bukit Kemuning, Batam, and adjacent to the local housing. The area is subjected to high intensity rainfalls and surrounded by hills and mountains. Due to heavy rainfalls and poor drainage system of the road, the rain water runoffs from the mountain and triggered a massive landslide (Fig. 9). The landslide affected the local residence near the area. Some houses were damaged, and residents were evacuated. Moreover, the accident has also disrupted the traffic. The Ministry of Public Works planned to repair the damage road and rehabilitate the existing soil profile in order not to cause greater losses.

The Public Works Department decided to repair the road and improve the drainage system, so that this problem can be avoided in the future. It is important to note that it was fundamental to restore the function of the road as soon as possible to reduce the risk of loss of the local community. Hence a hybrid reinforced soil structure, with the combination of gabion facing and geogrid reinforcement, with 14 m maximum height, was selected by the design engineer. The hybrid RSS (Fig. 10) is reinforced by high strength geogrids with an ultimate tensile strength (UTS) of 300 kN/m. The system was provided with an internal drainage system made up of a 30 cm thick rock layer wrapped with geotextile.

Static and seismic analysis had been performed, checking global stability, internal stability and external stability as retaining wall (sliding, overturning and bearing capacity) using a L.E.M. software. Figure 11 shows the installation phase, while Fig. 12 shows the completed structure.

## Harima Expressway Embankment in Hyogo Prefecture, Japan

The Harima Expressway (11.4 km length) is designed to provide connection between the Chugoku Expressway (E2A) and the Sanyo Expressway (E2). Due to the morphology of the terrain, several tunnels and viaducts were designed. Among those, the Tokishige tunnel and viaduct were one of major infrastructures, developing for 1.2 km along the alignment (Fig. 13).

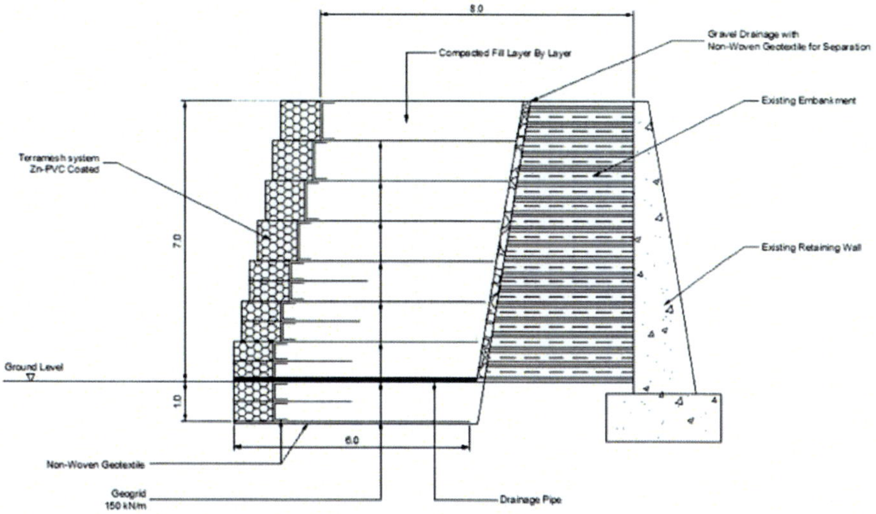


**Fig. 6** Typical cross section

**Fig. 7** Installation of gravel layer wrapped with nonwoven geotextile as drainage material

**Fig. 8** View of the reinforced soil structure during construction (left) and almost completed (right)

**Fig. 9** Rain water runoff situation (left) and condition after collapse (right)

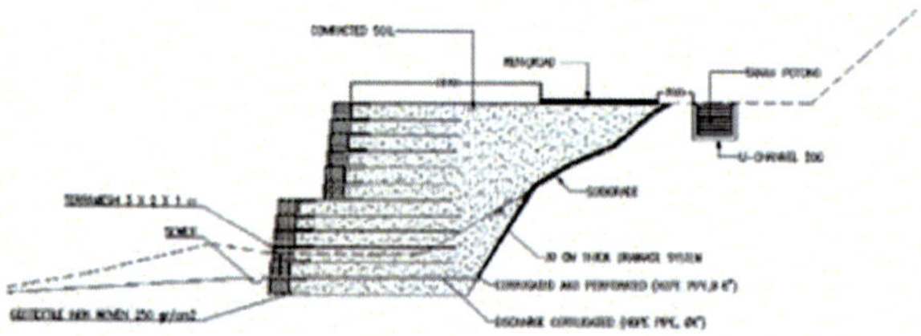

**Fig. 10** Typical Cross section of reinforced soil structure for Bukit Kemuning Road in Batam

During the construction of the infrastructures, a temporary road had to be built to allow the access to the jobsite, which required a retaining wall to be built to support the upstream slope and avoid potential landslides. A hybrid RSS was selected because of the many advantages, including durability, structural stability, economical and logistical feasibility, easy and fast installation. The hybrid structure supports the lower part of the temporary access road. The pictures in Figs. 14 and 15 show the RSS during construction and after completion. The total height of the RSS structure is approximately 8.5 m.

## Conclusion

Remediation works after landslide needs to be performed immediately to minimize the risk or losses that can be caused. Hybrid reinforced soil structures, described in the present paper, represent an attractive solution to restore the slopes after landslide and any other emergency works related. In facts, this method proved to be more effective in terms of construction time and cost compared to other solutions.

All the presented case histories show the successful construction of hybrid reinforced soil structures for slope stabilization and landslide rehabilitation in Myanmar, Indonesia, and Japan.

**Fig. 11** Installation phase

**Fig. 12** View of the completed access road remediation

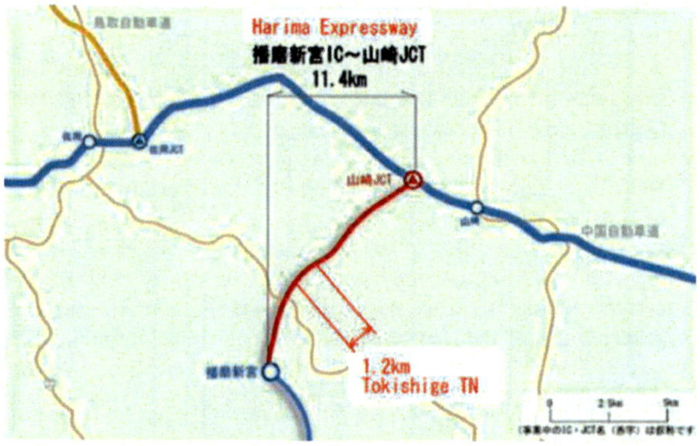


**Fig. 13** Planimetric view of the Tokishige infrastructure location

**Fig. 14** Reinforced soil structure during construction

**Fig. 15** Overview of the completed structure

## References

Cambiaghi A, Rimoldi P (1991) The use of geogrids in landslide control works: a case history from Valtellina (Northern Italy). In: Proceedings of the International conference on slope stability engineering, developments and Applications. Isle of Wight, UK

Cazzuffi D, Giusti G, Rimoldi P (1995) Lessons learned from the loss of performance of a 47 m high geotextile reinforced slope subsequently replaced with a geogrid reinforced slope. In: Giroud JP, Soderman KL (eds) Geosynthetics: lessons learned from Failures. IFAI, Roseville, MN, USA

Chen CW, Tang A (2011) Evaluation of connection strength of geogrid to gabion wall. Geotechnical Special Publication No. 220. ASCE, pp 231–238

Coluzzi E, Montanelli F, Recalcati P, Rimoldi P, Zinesi M (1997) Preliminary results from an instrumented geogrid reinforced slope for the stabilisation of the Montone hill in central Italy. In:

Proceedings of the International conference on mechanically stabilized backfill (MSB). Denver, Colorado, USA

Coluzzi E, Recalcati P, Rimoldi P (1996) Landslide repairs with green faced reinforced walls. In: Proceedings of the International seminar on environmental geotechnology with geosynthetics (INTSE-MEGG'96). New Delhi, India

Dikran S, Rimoldi P(1994) The use of geogrids to increase the stability of slopes. In: Proceedings of the International conference on landslides, slope stability & the safety of infra-structures. Kuala Lumpur, Malaysia

Ferraiolo F, Vicari M, Kulkarni TP, Gharpure AD (1997) Landslide road protection and river training works with gabion structures. In: Proceedings of the Geosynthetics (Asia'97) conference. Bangalore, India

Giani GP(1992) Rock slopes stability analysis. Balkema, Rotterdam, Netherlands, 347 p

Giroud JP, Naughton PJ, Rimoldi P, Scotto M (2014) Design of reinforced slopes and walls with low-permeability fills using draining geogrids. In: Proceedings of the 10th International conference on geosynthetics (ICG'10). Berlin, Germany

Gouw TL, Lelli M, Meinata LE, Laneri R, Cerro M, Rimoldi P (2016) High hybrid reinforced soil slope as runway support Tana Toraja airport case study. In: Proceedings of the GeoAsia 2016 conference. Delhi, India

Gu M, Collin JG, Han J, Zhang Z, Tanyu BF, Leshchinsky D, Ling HI, Rimoldi P (2017) Numerical analysis of instrumented mechanically stabilized gabion walls with large vertical reinforcement spacing. In: Geotextiles and geomembranes. Elsevier, pp 1–13

Kandaris PM (1999) Use of gabions for localized slope stabilization in difficult terrain. In: Proceedings of the 37th U.S. Symposium on rock mechanics (USRMS). Vail, Colarado, pp 1221–1227

Lin H, Ling HI, Li I, Collin JG, Leshchinsky D, Rimoldi P (2013) Centrifuge modeling of gabion walls reinforced with geosynthetics. In: Proceedings of the International symposium on design and practice of geosynthetic-reinforced soil structures. Bologna, Italy

Ling H, Xu L, Leshchinsky D, Collin J, Rimoldi P (2016) Centrifugal modeling of reinforced soil retaining walls considering staged construction. In: Proceedings of the GeoAmericas 2016 conference. Miami, USA

Lugli G, Sordo S, Trovato F, Potì A, Romeo D, Martino L, Rimoldi P (2019) Full scale shaking table tests on a reinforced soil wall with high tenacity polyester geostrips. In: Proceedings of the VII International conference on earthquake geotechnical engineering (VII ICEGE). Roma, Italy

Manni E, Rimoldi P (2006) A complex segmental concrete block retaining wall structure for the reconstruction of a historical bridge. In: Proceedings of the 8th International conference on geosynthetics (ICG 8). Yokohama, Japan

Montanelli F, Recalcati P, Rimoldi P (1999) An Instrumented geogrid reinforced slope in central Italy: field measurements and fem analysis results. In: Proceedings of the International seminar on field instrumentation for soil and rock. Nashville, Tennesse, USA

Rimoldi P (1988) A review of field measurements of the behaviour of geogrid reinforced slopes and walls. In: Proceedings of the International symposium theory and practice of earth reinforcement. Kyushu, Japan

Rimoldi P (1996) Geogrid reinforced walls, slopes and landslides: technical solutions and construction methods. In: Proceedings of the International seminar on environmental geotechnology with geosynthetics (INTSEMEGG'96). New Delhi, India

Rimoldi P (2018) Design and construction of reinforced walls in Italy in complex static and seismic conditions. In: Proceedings of the 11th International conference on geosynthetics (11ICG). Seoul, Korea

Rimoldi P, Ricciuti A (1992) The role of geogrid reinforced embankments in landslide stabilization: theory and practice in Italy. In: Proceedings of the 6th International symposium on landslides. Christchurch, New Zealand

Rimoldi P, Scotto M (2012) Hybrid reinforced soil structures for high walls and slopes. In: Proceeedings of the GeoAmericas 2012 conference. Lima, Peru

Tanyu B, Abbaspour A, Collin J, Leshchinsky D, Han J, Ling H, Rimoldi P (2016) Case study: instrumentation of a hybrid MSE wall system with up to 2 m vertical spacing between reinforcements. In: Proceeedings of the GeoAmericas 2016 conference. Miami, USA

Toprak B, Sevim O, Kalkan I (2016) Gabion walls and their use. Int J Adv Mechan Civil Eng ISSN: 2394-2827 3(4)

# Cutting-Edge Technologies Aiming for Better Outcomes of Landslide Disaster Mitigation

2020 Kyoto Japan

Kazuo Konagai

The International Consortium on Landslides (ICL) and The Global Promotion Committee of the International Programme on Landslides (GPC/IPL) have been responsible for organizing the World Landslide Forums (WLFs) every three years since 2008. Ever since the 1st WLF, the forums have long been the arena for landslide researchers and practitioners to exchange up-to-date information of recent devastations caused by landslides, cutting-edge technologies for landslide disaster mitigations and early warnings etc. to establish synergies among all participants worldwide.

Though the upcoming WLF5 has officially been postponed by one year to 2 to 6 November 2021 due to the global disruption caused by the coronavirus pandemic, the WLF5 will be all the more important with the Kyoto Landslide Commitment 2020 (KLC2020) to be launched as planned in the final online signatory meeting on 5 November 2020; the KLC 2020 is intended to be our action goals as the further advanced successor of the 'Sendai Landslide Partnerships 2015-2025 for Global Promotion of Understanding and Reducing Landslide Disaster Risk' in line with some of 17 Sustainable Development Goals (SDGs), particularly SDG 11, "Make cities and human settlements inclusive, safe, resilient and sustainable," of the United Nations.

For these important goals, the ICL has been inviting sponsorship from industries, businesses, and government agencies; all leading players in landslide science and technologies. They have been supporting a variety of the ICL/IPL activities such as publishing the International full-color journal "Landslides (Journal of the International Consortium on Landslides), full-color books for WLFs, exhibiting their cutting-edge technologies in WLFs, etc. Here follow short introductions of their activities with their names, addresses and contact information:

## Marui & Co. Ltd.

1-9-17 Goryo, Daito City, Osaka 574-0064, Japan
URL: http://marui-group.co.jp/en/index.html
Contact: hp-mail@marui-group.co.jp

Marui & Co. Ltd. celebrates its 100th anniversary in 2020. Marui, as one of the leading manufacturers of testing apparatuses in Japan, has been constantly striving to further improve its service since its foundation in 1920, thus contributing to the sustainable development of our nation and society. Our main products cover a wide variety of destructive and non-destructive testing apparatuses in the fields of geotechnical engineering, concrete engineering (mortar, aggregates, etc.), and ceramic engineering. Of special note is that Marui has been helping manufacture ring-shear apparatuses half-century long based on the leading-edge idea of Dr. Kyoji Sassa, Professor Emeritus at the Kyoto University. Marui has delivered total 7 ring-shear apparatuses to the Disaster Prevention Research Institute, Kyoto University, and 2 to the International Consortium on Landslides. Also the apparatuses were exported to the United States of America, China, Croatia and Vietnam.

Marui & Co. Ltd. takes great pleasure in developing, manufacturing, and providing new products of high value sharing the delight of achievement with our customers, and thus contributing to the social development. The whole staff of Marui & Co. Ltd. are determined to devote ceaseless effort to keep its organization optimized for its speedy and high-quality services, by the motto "Creativity and

K. Konagai (✉)
Organizing Committee of the Fifth World Landslide Forum, International Consortium on Landslides, Kyoto, 606-8226, Japan
e-mail: konagai@iclhq.org


V. Vilímek et al. (eds.), *Understanding and Reducing Landslide Disaster Risk*,
ICL Contribution to Landslide Disaster Risk Reduction,
https://doi.org/10.1007/978-3-030-60319-9_46

Revolution", and strive hard to take a step further, as a leading manufacturer of testing apparatuses, to answer our customer's expectations for the 22nd century to come.

## Nippon Koei Co., Ltd.

5-4 Kojimachi, Chiyoda-ku, Tokyo 102-8539, Japan
URL: https://www.n-koei.co.jp/english/
https://www.n-koei.co.jp/english/contact/input

Nippon Koei Co., Ltd. and its group companies conduct many projects to support the growth of developing countries in Asia, Africa, the Middle and Near East, Latin America and other regions. Examples of their efforts include environmental measures to combat global warming, development of regional transportation infrastructure to support the rapid growth of emerging economies, and reconstruction assistance for regions affected by conflict and/or natural disasters.

## OSASI Technos, Inc.

65-3 Hongu-cho, Kochi City, Kochi 780-0945, Japan
URL: http://www.osasi.co.jp/en/
Contact: cs@osasi.co.jp

OSASI Technos, Inc. has been making its best efforts to develop its cutting-edge technologies for landslide early warning. Its unique compact and lightweight sensors making up the Landslide Early Warning System enable long-term monitoring of unstable landslide mass movements, precipitations, porewater pressure buildups, etc., in a remote mountainous area where commercial power is often unavailable. OSASI Technos, Inc. is also proud of its advanced technology to transfer observed data even in areas with poor telecom environments as proven in the successful implementations in South Asia.

All stuff members of OSASI Technos work together for mitigation of landslide disasters worldwide.

## Godai Corporation

1-35 Kuroda, Kawasaki, Kanagawa 215-0004, Japan
URL: https://soft.godai.co.jp/En/Soft/Product/Products/LS-RAPID/
Contact: pp-sales@godai.co.jp

Ever since its foundation in 1965, Godai Kaihatsu Co., Ltd., a civil engineering consulting firm, has long been providing a variety of software and measures particularly for natural disaster mitigation. With its rich expertise in both civil engineering and information technology (IT), the company has its primary goal to address real world needs of disaster mitigation. All the staff of Godai Kaihatsu Co., Ltd. feel it more than happy that their cutting-edge technologies help mitigate natural disasters.

## Japan Conservation Engineers & Co., Ltd

3-18-5 Toranomon, Minato-ku, Tokyo 1050001, Japan
URL: https://www.jce.co.jp/en/
Contact: go_info@jce.jp

Japan Conservation Engineers & Co., Ltd. (JCE) is a general consulting firm working on landslide prevention research and consulting. JCE provides various disaster prevention technologies for debris flows, landslides, slope failures, rockfalls, etc. In addition, JCE is proud of its expertise having been conducting surveys and consulting works on coastal erosions and tsunami countermeasures for about 20 years. JCE contributes to the world through its activities in the realm of both structural and non-structural measures to build a resilient society.

## OYO Corporation

7 Kanda-Mitoshiro-cho, Chiyoda-ku, Tokyo 101-8486, Japan
URL: https://www.oyo.co.jp/english/
Contact: https://www.oyo.co.jp/english/contacts/

OYO Corporation, the top geological survey company in Japan established in Tokyo in 1957, is well known as one of leading companies providing cutting-edge technologies and measures for natural disasters such as landslides, earthquakes, tsunamis, and floods. Not just developing and selling measuring instruments related to disaster prevention, OYO also delivers a market-leading services in 3D ground/geological modeling and 3D exploration technologies.

## Kokusai Kogyo Co., Ltd.

2 Rokubancho, Chiyoda-ku, Tokyo 102-0085, Japan
URL: https://www.kkc.co.jp/english/index.html
Contact: overseas@kk-grp.jp

Kokusai Kogyo Co., Ltd. as a leading company of geospatial information technologies, has long been providing public services with its comprehensive expertise to address real world needs and cutting-edge measurement technologies. Kokusai Kogyo Co., Ltd. helps rebuild "Green Communities," which has been of our great concern in terms of "environment and energy," "disaster risk reduction" and "asset management". Kokusai Kogyo Co., Ltd. offers the advanced and comprehensive analyses of geospatial information for developing new government policies, maintaining and operating social infrastructures safe and secure, and implementing low-carbon measures in cities.

Influenced by the recent global climate change, extreme rainfall events have become more frequent worldwide and resultant hydro-meteorological hazards are creating more deaths and devastations particularly in many developing countries where effective advanced countermeasures are not readily available. Kokusai Kogyo Co., Ltd. is proud of its achievements in establishing resilient infrastructure systems and implementing effective monitoring/early warning systems in developing countries, which have long been helping reduce the risks from natural hazards.

## Geobrugg AG

Aachstrasse 11, 8590 Romanshorn, Switzerland
URL: www.geobrugg.com
Contact: info@geobrugg.com

Swiss company Geobrugg is the global leader in the supply of high-tensile steel wire safety nets and meshes—with production facilities on four continents, as well as branches and partners in over 50 countries. True to the philosophy "Safety is our nature" the company develops and manufactures protection systems made of high-tensile steel wire. These systems protect against natural hazards such as rockfall, landslides, debris flow and avalanches. They ensure safety in mining and tunneling, as well as on motorsport tracks and stop other impacts from falling or flying objects. More than 65 years of experience and close collaboration with research institutes and universities make Geobrugg a pioneer in these fields.

## Ellegi srl

Via Petrarca, 55 I-22070 Rovello Porro (CO) Italy
URL: http://www.lisalab.com/engl/?seze=1
Contact: info@lisalab.com

Ellegi srl provides worldwide monitoring services and produces Ground Based synthetic aperture radar (GBInsAR) for remote measurement of displacements and deformations on natural hazards and manmade buildings using its own designed and patented LiSALab system.

Its activities started in 2003 as a spin off project to exploit commercially the Ground Based Linear Synthetic Aperture Radars technology developed by European Commission's Ispra Joint Research Centre and based on the results of more than 10 years of research. Since then Ellegi has industrialized and developed the core technology of the LiSALab system and latest LiSAmobile system represents the 5th generation of development.

In 2003 it was the first commercial company in the world to provide GBInSAR measurements of natural hazards and structure.

Ellegi srl offers:

- Displacement fields measurement, control and monitoring of the deformation caused by natural hazards, like landslides, rockslides, sinkhole, volcanic deformation in every operative condition, including emergencies,
- Structural strain fields measurement, control, monitoring and diagnosis of the deformation affecting buildings, bridges, viaducts, dams.
- GBInSAR monitoring systems, installation, management and maintenance in order to provide information about natural hazards or anthropic activity, that can generate or cause slopes failures or buildings instabilities.

In all the above-mentioned activities Ellegi srl uses the GBInSAR LiSALab technology that represents a real "break-through".

## Chuo Kaihatsu Corporation

3-13-5 Nishi-waseda, Shinjuku-ku, Tokyo 169-8612, Japan
URL: https://www.ckcnet.co.jp/global/
Contact: https://www.ckcnet.co.jp/contactus/

Chuo Kaihatsu Corporation (CKC) was founded in 1946, and has been aiming to become the "Only One" consultant for our customers. We engage in the hands-on work that will "Remain with the earth, Remain in people's hearts, and Lead to a prosperous future". We focus on road, river and dam engineering to flesh out industrial infrastructures specifically by means of geophysical/geotechnical/geological investigations, civil engineering surveys and project implementations. In recent years, we make significant efforts on earthquake disaster mitigation, sediment disaster prevention/mitigation and ICT information services. Many achievements of ours have already contributed to mitigation of natural disasters such as landslides, earthquakes and slope failures in Japan, Asia and the Pacific Region.

## IDS GeoRadar s.r.l.

Via Augusto Righi, 6, 6A, 8, Loc. Ospedaletto, Pisa, Italy, 56121
URL: https://idsgeoradar.com/
Contact: info@idsgeoradar.com

IDS GeoRadar, part of Hexagon, provides products and solutions, based on radar technology, for monitoring applications including landslides, rockfalls, complex structures, mining and civil engineering. The company is a leading provider of Ground Penetrating Radar (GPR) and Interferometric Radar solutions worldwide. IDS GeoRadar is committed to delivering best-in-class performance solutions and to the pursuit of product excellence, through the creation of application-specific, innovative and cost-efficient systems for a wide range of applications.

## METER Group, Inc.

2365 NE Hopkins Court, Pullman, WA 99163, USA
URL: https://metergroup.com/wlf5
Contact: https://bryan.wacker@metergroup.com

METER Group provides accurate, rugged, and dependable instrumentation to monitor moisture in all its phases within an unstable slope. METER specializes in instrumentation for near real-time monitoring of incoming moisture in the form of rain and weather. In addition, we provide real-time below-surface monitoring of existing moisture conditions like moisture content and soil suction which show how the soil profile is filling with water to saturation, including the transition to positive pore water pressure.

The ZL6 advanced cloud data logger works together with ZENTRA Cloud data software to simplify and speed up data collection, management, visualization, and alerting. Our well-published instrumentation is used worldwide in universities, research and testing labs, government agencies, and industrial applications.

For almost four decades, scientists and engineers have relied on our instrumentation to understand critical moisture parameters. We've even partnered with NASA to measure soil (regolith) moisture on Mars. Wherever you measure, and whatever you're measuring, rely on METER for accuracy, affordability, and simplicity that will make your job easier.

## Asia Air Survey Co., Ltd.

Shinyuri 21 BLDG 3F, 1-2-2 Manpukuji, Asao-Ku, Kawasaki, Kanagawa 215-0004, Japan
URL: https://www.ajiko.co.jp/en/
Contact: service@ajiko.co.jp

Asia Air Survey (AAS), as one of the leading engineering and consulting companies, has long been providing disaster prevention and mitigation services for over 65 years, particularly in the fields of landslide, debris flow, erosion control, etc. AAS is proud of being the inventor of Red Relief Image Map (RRIM), which is a cutting-edge 3D terrain visualization method allowing great geomorphological details to be visualized in one glance, thus has been used in various facets of disaster prevention and mitigation.

## Kiso-Jiban Consultants Co., Ltd

Kinshicho Prime Tower 12 Floor, 1-5-7 Kameido, Koto-ku, Tokyo 36-8577, Japan
URL: https://www.kisojiban.com/
Contact: kisojiban-contactus@kiso.co.jp

Kiso-Jiban Consultants, established in 1953, is an engineering consulting firm especially well known in the field of geotechnical engineering. The areas of its comprehensive services are listed below:

- Geological and Geotechnical Survey
- Geotechnical Analysis and Design
- Disaster Prevention and Management
- GIS (Geographic Information Systems)
- Soil and Rock Laboratory Tests
- Instrumentation and Monitoring
- Geophysical Exploration and Logging
- Distribution of Geosynthetics Products.

Much-talked-about new service is Kiso-SAR System allowing accurate estimation of both extent and rate of landslide movements based upon a comprehensive interpretation of InSAR results from geotechnical and landslide engineering viewpoint (see the one-page introduction of Kiso-Jiban Consultants Co., Ltd.). With Kiso-SAR system, the following pieces of important geotechnical information can be provided:

(1) Extent of a deforming landslide mass (and the rate of its movement.
(2) Consolidation buildup in soft clay underlying a fill.
(3) Deformation buildups induced by slope cutting.

## Okuyama Boring Co., Ltd.

10-39 Shimei-cho, Yokote City, Akita 013-0046, Japan
URL: https://okuyama.co.jp/en/
Contact: info@okuyama.co.jp

Okuyama Boring Co., Ltd. is proud of its achievements in various projects to help solve many landslide problems. The company has been offering services in geological surveys and analyses, developing rational countermeasures against various geotechnical problems as well as safe workflow diagrams, and providing necessary pieces of advice for ensuring safety during landslide countermeasure works. For this purpose, Okuyama Boring Co., Ltd. works on monitoring, observations, field surveys, numerical analyses, countermeasure works, etc. of landslides.

## Kawasaki Geological Engineering Co. Ltd.

Mita-Kawasaki Bldg, 2-11-15 Mita, Minato-ku, Tokyo108-8337, Japan
URL: http://www.kge.co.jp/
Contact: post-master@kge.co.jp

Kawasaki Geological Engineering Co., Ltd. as one of the leading members of SAAM Research Group, has proactively been involved in developing "Sustainable Asset Anchor Maintenance (SAAM, hereafter) System," enabling easy maintenance of ground anchors. Its unique jack, weighing about half the weight of a conventional jack, together with a newly developed jig, can be applied to any type of anchor even with a short extra length, thus allowing for in situ lift-off tests on these anchors. The SAAM system also has an optional weight meter that can be installed after performing a lift-off test.

## Nissaku Co., Ltd.

4-199-3 Sakuragi-cho, Omiya-ku, Saitama 330-0854, Japan
URL: https://www.nissaku.co.jp/
Contact: survey@nissaku.co.jp

Nissaku Co., Ltd., founded in 1912 as a well drilling company, provides services for far-flung fields of not only groundwater exploitation but also measures for landslides. Having its rich expertise in these fields, Nissaku Co., Ltd. offers general reliable one-stop technical services including designs, investigations, analyses, constructions, and maintenances.

Full-color presentations from the above seventeen exhibitors focusing on their landslide technologies are shown on the following pages. Their cutting-edge technologies have of course been instrumental in the progress that we have made in landslide risk-reduction worldwide, and we want to exert even greater effort to aim high given the KLC 2020 as our new action goals. The International Consortium on Landslides seeks volunteers willing to support our activities introducing their brand-new technologies for landslide disaster mitigation in our international journal "Landslides," full color books for WLFs, exhibitions at WLFs, etc. If you are interested in being engaged in supporting ICL activities, please contact the ICL secretariat <secretariat@iclhq.org> .

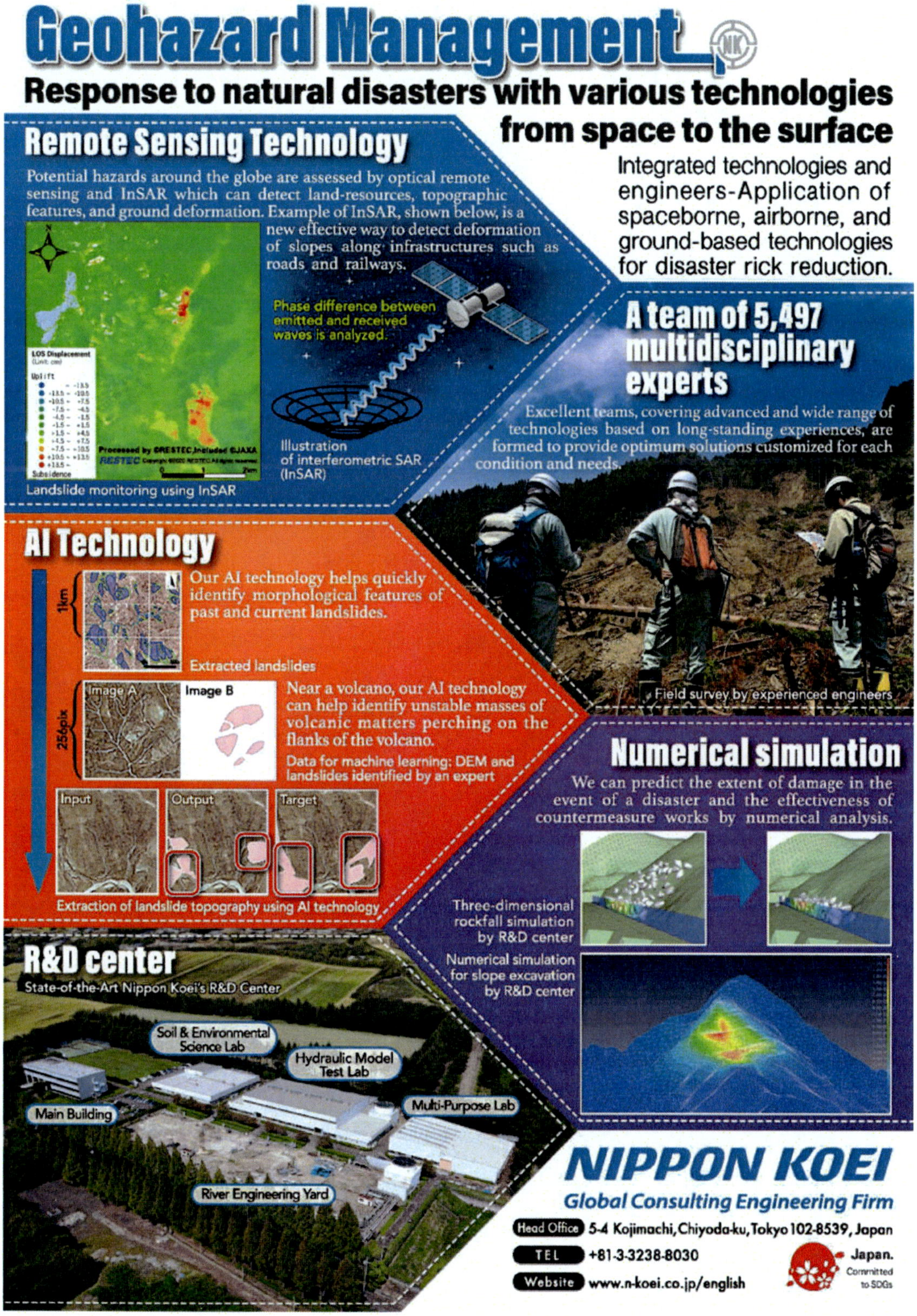
Geohazard Management
Response to natural disasters with various technologies from space to the surface
Integrated technologies and engineers-Application of spaceborne, airborne, and ground-based technologies for disaster rick reduction.
Remote Sensing Technology
Potential hazards around the globe are assessed by optical remote sensing and InSAR which can detect land-resources, topographic features, and ground deformation. Example of InSAR, shown below, is a new effective way to detect deformation of slopes along infrastructures such as roads and railways.
Phase difference between emitted and received waves is analyzed.
Illustration of interferometric SAR (InSAR)
Landslide monitoring using InSAR
A team of 5,497 multidisciplinary experts
Excellent teams, covering advanced and wide range of technologies based on long-standing experiences, are formed to provide optimum solutions customized for each condition and needs.
Field survey by experienced engineers
AI Technology
Our AI technology helps quickly identify morphological features of past and current landslides.
1km
Extracted landslides
Image A
Image B
256pix
Near a volcano, our AI technology can help identify unstable masses of volcanic matters perching on the flanks of the volcano.
Data for machine learning: DEM and landslides identified by an expert
Input
Output
Target
Extraction of landslide topography using AI technology
Numerical simulation
We can predict the extent of damage in the event of a disaster and the effectiveness of countermeasure works by numerical analysis.
Three-dimensional rockfall simulation by R&D center
Numerical simulation for slope excavation by R&D center
R&D center
State-of-the-Art Nippon Koei's R&D Center
Soil & Environmental Science Lab
Hydraulic Model Test Lab
Main Building
Multi-Purpose Lab
River Engineering Yard
NIPPON KOEI
Global Consulting Engineering Firm
Head Office 5-4 Kojimachi, Chiyoda-ku, Tokyo 102-8539, Japan
TEL +81-3-3238-8030
Website www.n-koei.co.jp/english
Japan. Committed to SDGs

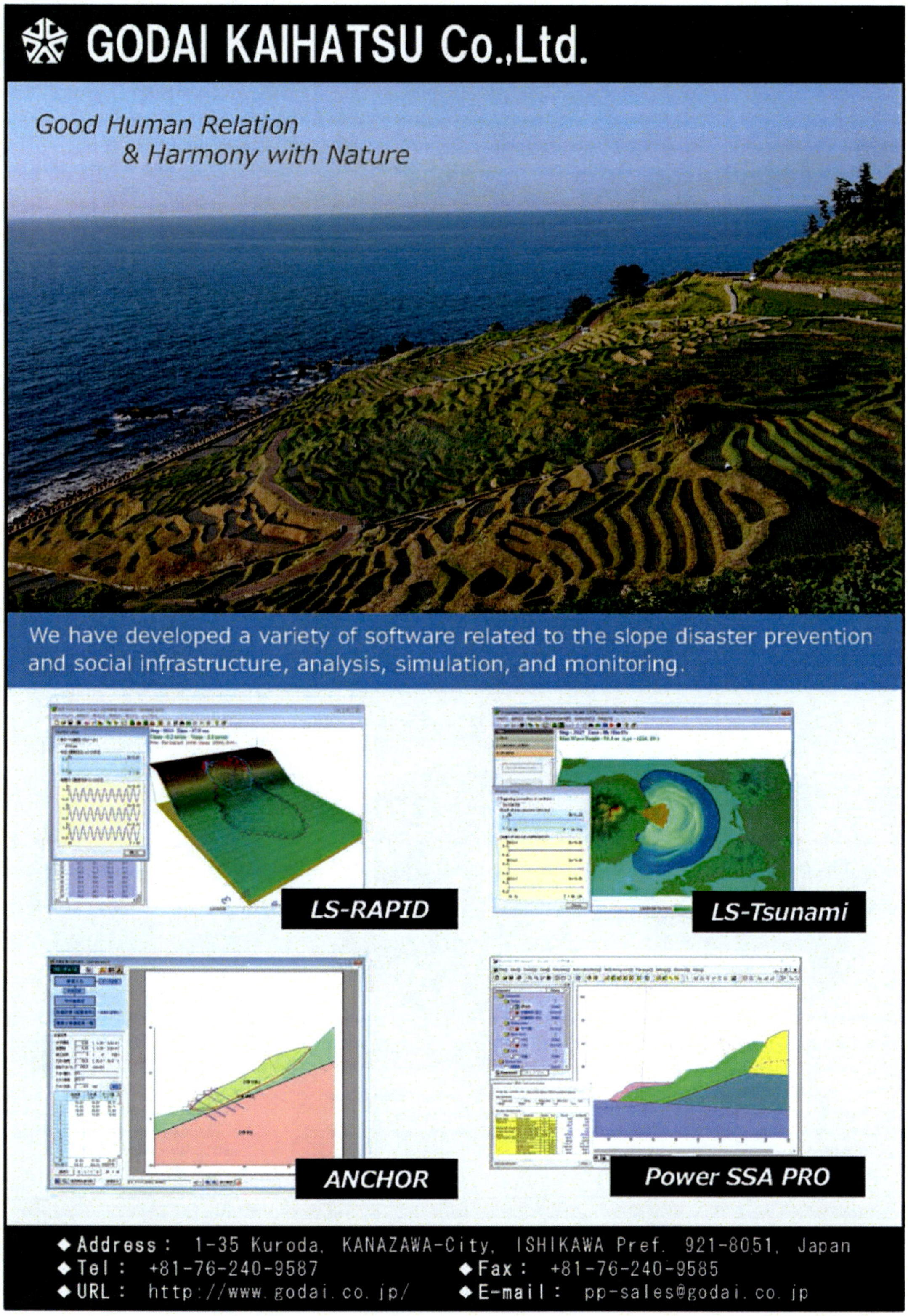
GODAI KAIHATSU Co.,Ltd.
Good Human Relation
& Harmony with Nature
We have developed a variety of software related to the slope disaster prevention and social infrastructure, analysis, simulation, and monitoring.
LS-RAPID
LS-Tsunami
ANCHOR
Power SSA PRO
◆Address : 1-35 Kuroda, KANAZAWA-City, ISHIKAWA Pref. 921-8051, Japan
◆Tel : +81-76-240-9587
◆Fax : +81-76-240-9585
◆URL : http://www.godai.co.jp/
◆E-mail : pp-sales@godai.co.jp

OYO
Find the best answer to the future of people and the earth.
To realize a sustainable society.
3D Geological Modeling
GEO-CRE® /GEO-CRE®PRO
OCTAS® Modeler
OCTAS Modeler is a system to support the utilization of subsurface information. It helps to grasp the positional relationship of the borehole logs and create various 3D models including borehole, soil, bearing layer, and terrain.
Inclinometer
Clino-pole
Tilt sensor for Sediment disaster
3D Geophysical Survey
McSEIS-AT
McSEIS-AT is an epoch-making microtremor exploration system that can measure S-wave velocity structures in three dimensions.
oyo corporation
7 kanda Mitoshiro-cho,Chiyoda-ku,Tokyo
101-8486,JAPAN
https://www.oyo.co.jp/english/
Copyright © 2020 OYO Corporation. All rights reserved

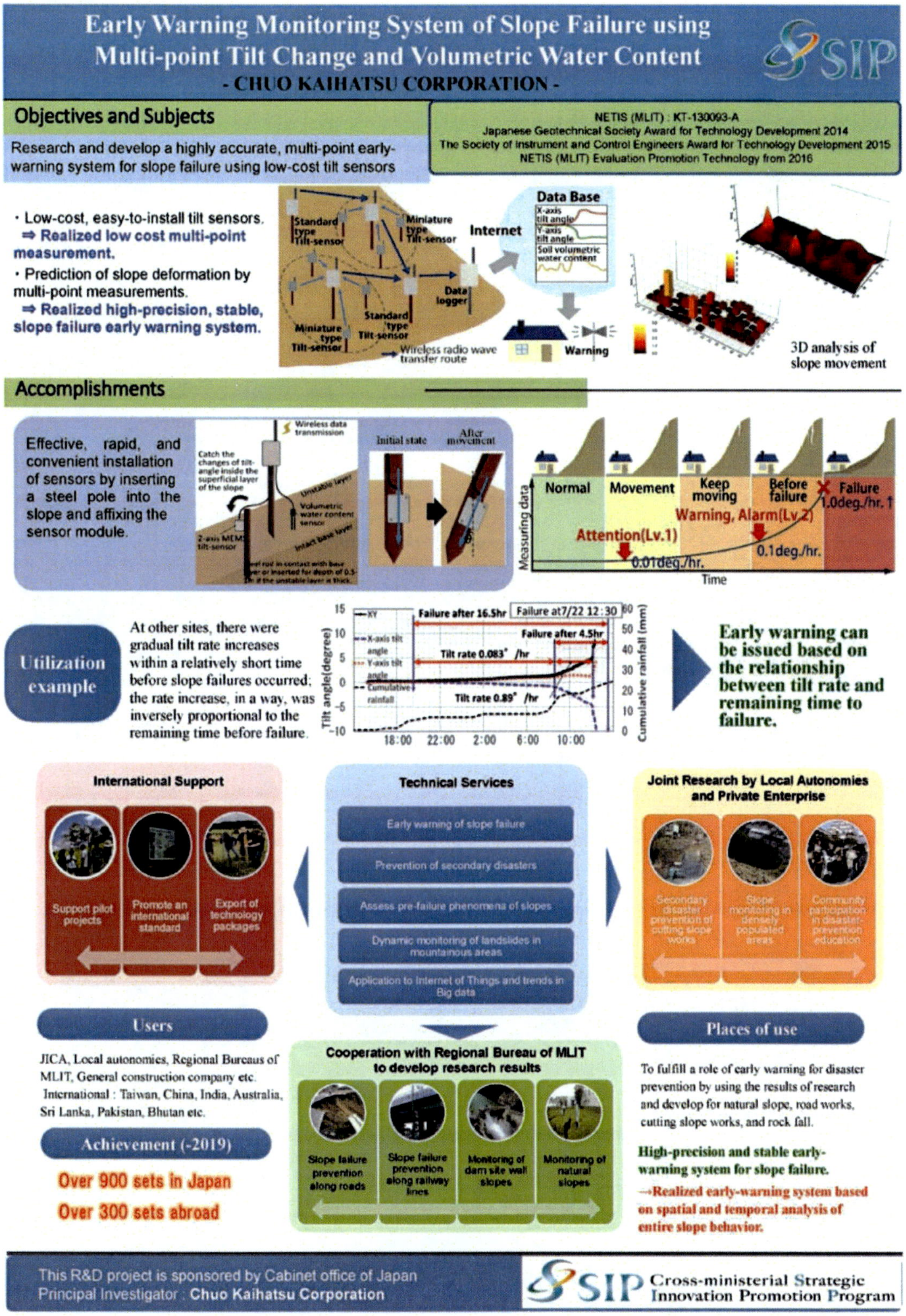

Early Warning Monitoring System of Slope Failure using Multi-point Tilt Change and Volumetric Water Content
- CHUO KAIHATSU CORPORATION -
SIP
Objectives and Subjects
Research and develop a highly accurate, multi-point early-warning system for slope failure using low-cost tilt sensors
NETIS (MLIT) : KT-130093-A
Japanese Geotechnical Society Award for Technology Development 2014
The Society of Instrument and Control Engineers Award for Technology Development 2015
NETIS (MLIT) Evaluation Promotion Technology from 2016
・Low-cost, easy-to-install tilt sensors.
⇒ Realized low cost multi-point measurement.
・Prediction of slope deformation by multi-point measurements.
⇒ Realized high-precision, stable, slope failure early warning system.
Standard type Tilt-sensor
Miniature type Tilt-sensor
Internet
Data Base
X-axis tilt angle
Y-axis tilt angle
Soil volumetric water content
Data logger
Wireless radio wave transfer route
Warning
3D analysis of slope movement
Accomplishments
Effective, rapid, and convenient installation of sensors by inserting a steel pole into the slope and affixing the sensor module.
Wireless data transmission
Catch the changes of tilt-angle inside the superficial layer of the slope
Unstable layer
Volumetric water content sensor
2-axis MEMS tilt-sensor
Intact base layer
Initial state
After movement
Normal
Movement
Keep moving
Before failure
Failure
Attention(Lv.1)
Warning, Alarm(Lv.2)
0.01deg./hr.
0.1deg./hr.
1.0deg./hr.
Measuring data
Time
Utilization example
At other sites, there were gradual tilt rate increases within a relatively short time before slope failures occurred; the rate increase, in a way, was inversely proportional to the remaining time before failure.
Failure after 16.5hr
Failure at7/22 12:30
Failure after 4.5hr
Tilt rate 0.083° /hr
Tilt rate 0.89° /hr
XY
X-axis tilt angle
Y-axis tilt angle
Cumulative rainfall
Tilt angle(degree)
Cumulative rainfall (mm)
18:00 22:00 2:00 6:00 10:00
Early warning can be issued based on the relationship between tilt rate and remaining time to failure.
International Support
Support pilot projects
Promote an international standard
Export of technology packages
Technical Services
Early warning of slope failure
Prevention of secondary disasters
Assess pre-failure phenomena of slopes
Dynamic monitoring of landslides in mountainous areas
Application to Internet of Things and trends in Big data
Joint Research by Local Autonomies and Private Enterprise
Secondary disaster prevention of cutting slope works
Slope monitoring in densely populated areas
Community participation in disaster-prevention education
Users
JICA, Local autonomies, Regional Bureaus of MLIT, General construction company etc.
International : Taiwan, China, India, Australia, Sri Lanka, Pakistan, Bhutan etc.
Achievement (-2019)
Over 900 sets in Japan
Over 300 sets abroad
Cooperation with Regional Bureau of MLIT to develop research results
Slope failure prevention along roads
Slope failure prevention along railway lines
Monitoring of dam site wall slopes
Monitoring of natural slopes
Places of use
To fulfill a role of early warning for disaster prevention by using the results of research and develop for natural slope, road works, cutting slope works, and rock fall.
High-precision and stable early-warning system for slope failure.
→Realized early-warning system based on spatial and temporal analysis of entire slope behavior.
This R&D project is sponsored by Cabinet office of Japan
Principal Investigator : Chuo Kaihatsu Corporation
SIP Cross-ministerial Strategic Innovation Promotion Program

# Monitoring Service for Landslide by Kiso-SAR System

## What's SAR ?

SAR (Synthetic Aperture Radar) is a technique that utilizes interference of radio waves for precise determination of distance. InSAR (interferometric SAR), the phase of the received backscattered signal from two images of the same scene are used to measure path length differences with an accuracy of a few millimeters.

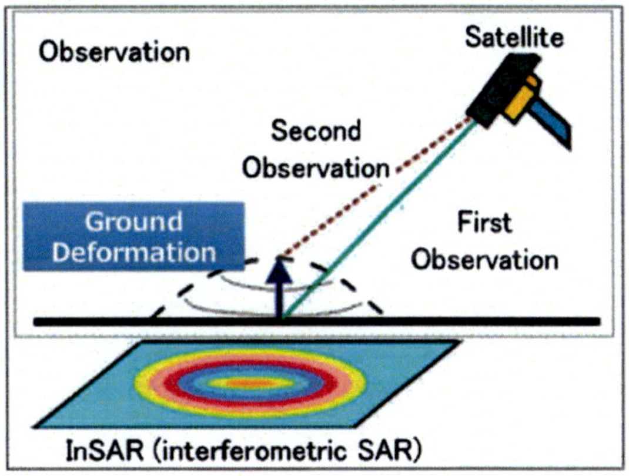


## Ground Deformation Estimation in Landslides

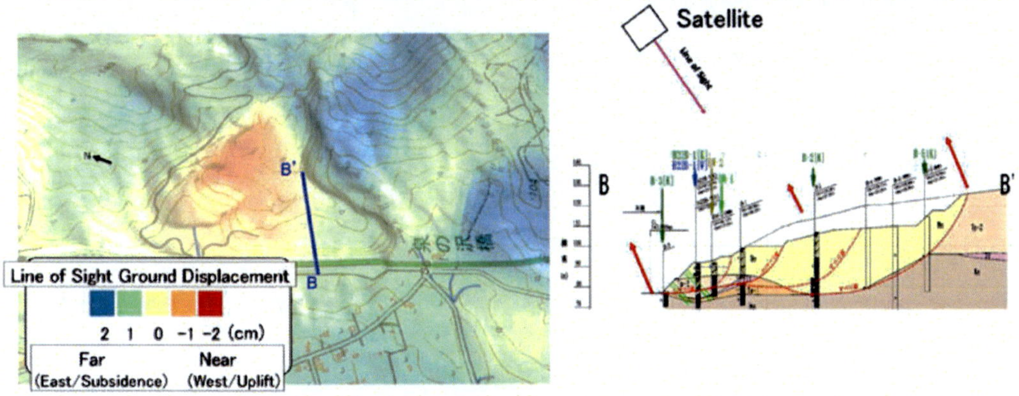


Ground Deformation of Landslide Observed by Kiso-SAR System

# Sustainable Asset Anchor Maintenance System

Earth Doctor

## ■ What's "SAAM System"?

SAAM stands for "Sustainable Asset Anchor Maintenance".

SAAM Jack, only half the weight of a conventional jack, enables lift-off tests in various onsite conditions.

SAAM, thus, allows for easy maintenance of ground anchors.

SAAM Jack

SAAM System

## ■ SAAM-A (PAT No. 5971596)

Lift-off tests on anchors with a short extra length can be performed with a newly developed jig.

Installation of Inner Coupler

Removal of Wedge

## ■ SAAM-L (PAT No. 5440772)

SAAM load cell can be attached to any anchors without loosening them.

Load Meter

## ■ Planar Distribution of Anchor's Tensioning Force

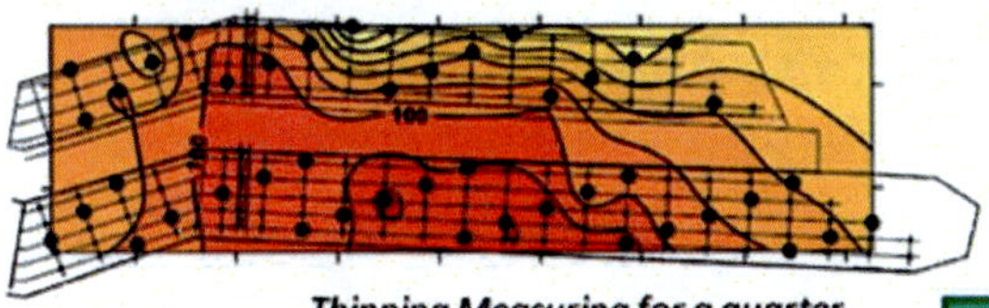


Thinning Measuring for a quarter

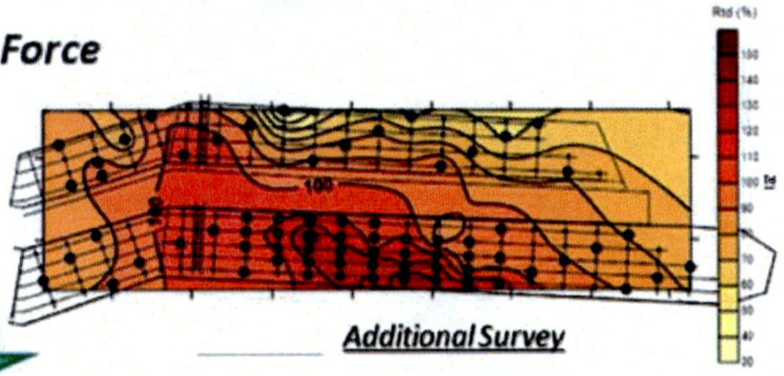


Additional Survey

Evaluation based on Soundness of Anchors

| | Evaluation | Status | Coping example |
|---|---|---|---|
| 0.9 Tys | E+ | Risk of Breakage | Emergency measures |
| 1.1 Ta | D+ | Risk of Damage | Implementation of countermeasure work |
| Allowable anchor force (Ta) | C+ | Exceeds the allowable value | |
| Design anchor force (Td) | B+ | | Follow up |
| Tensile force when fixed (Pt) | A+ | Good | |
| 0.8 Pt | A | | |
| 0.5 Pt | B− | | Follow up |
| 0.1 Pt | C− | Functional deterioration | Implementation of countermeasure work |
| | D− | Dysfunction | |

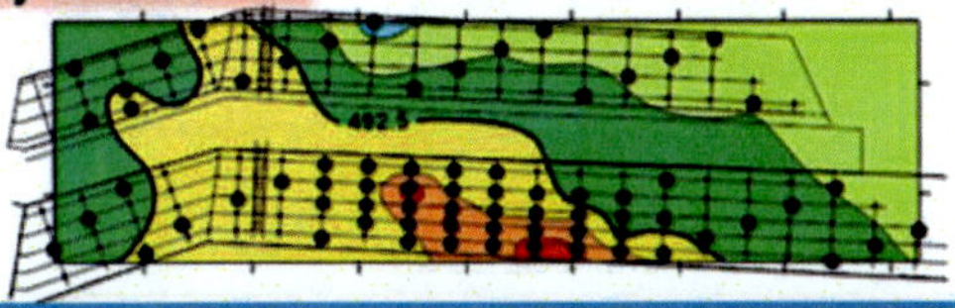


SAAM 合同会社アンカーアセットマネジメント研究会
Society of Anchor Asset Management

川崎地質株式会社
Kawasaki Geological Engineering Co., Ltd.
Marketing & Sales Department
URL : www.kge.co.jp TEL : +81-3-5445-2077

Creation of Happy Society with
our Traditional and Newest Technology

# International Consortium on Landslides

## International Consortium on Landslides

An international non-government and non-profit scientific organization
promoting landslide research and capacity building for the benefit of society and the environment



**ICL Full Members:**

Geotechnical Engineering Office, Hong Kong Special Administrative Region, China
UNESCO Chair for the Prevention and the Sustainable Management of Geo-hydrological Hazards - University of Florence, Italy
Korea Institute of Geoscience and Mineral Resources (KIGAM)
University of Ljubljana, Faculty of Civil and Geodetic Engineering (ULFGG), Slovenia

Albania Geological Survey / The Geotechnical Society of Bosnia and Herzegovina / Center for Scientific Support in Disasters – Federal University of Parana, Brazil/ Geological Survey of Canada / University of Alberta, Canada / Northeast Forestry University, Institute of Cold Regions Science and Engineering, China / China University of Geosciences / Chinese Academy of Sciences, Institute of Mountain Hazards and Environment / Tongji University, College of Surveying and Geo-Informatics, China / The Hong Kong University of Science and Technology, China / Shanghai Jiao Tong University, China / The University of Hong Kong, China / Universidad Nacional de Colombia / Croatian Landslide Group (Faculty of Civil Engineering, University of Rijeka and Faculty of Mining, Geology and Petroleum Engineering, University of Zagreb) / City of Zagreb, Emergency Management Office, Croatia / Charles University, Faculty of Science, Czech Republic / Institute of Rock Structure and Mechanics, Department of Engineering Geology, Czech Republic / Brown Coal Research Institute, Czech Republic / Cairo University, Egypt / Technische Universitat Darmstadt, Institute and Laboratory of Geotechnics, Germany / National Environmental Agency, Department of Geology, Georgia / Universidad Nacional Autonoma de Honduras (UNAH), Honduras / Amrita Vishwa Vidyapeetham, Amrita University / Vellore Institute of Technology, India / National Institute of Disaster Management, India / Agency for Meteorology, Climatology, and Geophysics of the Republic of Indonesia (BMKG Indonesia) / University of Gadjah Mada, Center for Disaster Mitigation and Technological Innovation (GAMA-InaTEK), Indonesia / Parahyangan Catholic University, Indonesia / Building & Housing Research Center, Iran / Italian Institute for Environmental Protection and Research (ISPRA) - Dept. Geological Survey, Italy / University of Calabria, DIMES, CAMILAB, Italy / Istituto de Ricerca per la Protezione Idrogeologica (IRPI), CNR, Italy / DIA–Universita degli Studi di Parma, Italy / University of Torino, Dept of Earth Science , Italy / Centro di Ricerca CERI - Sapienza Università di Roma, Italy / Kyoto University, Disaster Prevention Research Institute, Japan / Japan Landslide Society / Korean Society of Forest Engineering / National Institute of Forest Science, Korea / Korea Infrastructure Safety & Technology Corporation / Korea Institute of Civil Engineering and Building Technology / Slope Engineering Branch, Public Works Department of Malaysia / Institute of Geography, National Autonomous University of Mexico (UNAM) / International Centre for Integrated Mountain Development (ICIMOD), Nepal / University of Nigeria, Department of Geology, Nigeria / Moscow State University, Department of Engineering and Ecological Geology, Russia / JSC "Hydroproject Institute", Russia / University of Belgrade, Faculty of Mining and Geology, Serbia / Comenius University, Faculty of Natural Sciences, Department of Engineering Geology, Slovakia / Geological Survey of Slovenia / University of Ljubljana, Faculty of Natural Sciences and Engineering (ULNTF), Slovenia / Central Engineering Consultancy Bureau (CECB), Sri Lanka / National Building Research Organization, Sri Lanka / Landslide group in National Central University from Graduate Institute of Applied Geology, Department of Civil Engineering, Center for Environmental Studies, Chinese Taipei / National Taiwan University, Department of Civil Engineering, Chinese Taipei / Asian Disaster Preparedness Center, Thailand / Ministry of Agriculture and Cooperative, Land Development Department, Thailand / Institute of Telecommunication and Global Information Space, Ukraine / California State University, Fullerton & Tribhuvan University, Institute of Engineering, USA & Nepal / Institute of Transport Science and Technology, Vietnam / Vietnam Institute of Geosciences and Mineral Resources (VIGMR).

**ICL Associates**

State Key Laboratory of Geohazard Prevention and Geoenvironment Protection (Chengdu University of Technology), China / Czech Geological Survey, Czech Republic / Department of Earth and Environmental Sciences, University Aldo Moro, Bari, Italy / Department of Sciences and Technologies, University of Sannio, Italy / Department of Earth and Environmental Sciences – University of Pavia, Italy / Geotechnical Engineering Group (GEG), University of Salerno, Italy / Niigata University, Research Institute for Natural Hazards and Disaster Recovery, Japan / Ehime University Center for Disaster Management Informatics Research, Japan / Tian-Shan Geological Society, Kyrgyzstan / Institute of Environmental Geoscience RAS (IEG RAS), Russia / Russian State Geological Prospecting University n.a. Sergo Ordzhonikidze (MGRI-RSGPU) / TEMPOS, environmental civil engineering Ltd., Slovenia / Institute of Earth Sciences – Faculty of Geoscience and Environment, University of Lausanne, Switzerland / Middle East Technical University (METU), Turkey / North Dakota State University, USA

**ICL Secretariat:**

Secretary General: Kyoji Sassa
International Consortium on Landslides, 138-1 Tanaka Asukai-cho, Sakyo-ku, Kyoto 606-8226, Japan
Web: http://icl.iplhq.org/, E-mail: secretariat@iclhq.org
Tel: +81-75-723-0640, Fax: +81-75-950-0910


V. Vilímek et al. (eds.), *Understanding and Reducing Landslide Disaster Risk*,
ICL Contribution to Landslide Disaster Risk Reduction,
https://doi.org/10.1007/978-3-030-60319-9